The cultural landscape during 6000 years in southern Sweden

– the Ystad Project

Ecological Bulletins No. 41

The cultural landscape during 6000 years in southern Sweden
– *the Ystad Project*

Edited by
Björn E. Berglund

Editorial Board:
Björn E. Berglund, Lars Larsson,
Nils Lewan, Sten Skansjö

Managing Editor:
Mats Riddersporre

Ecological Bulletins

ECOLOGICAL BULLETINS are published in cooperation with the ecological journals Holarctic Ecology and Oikos. Ecological Bulletins consist of monographs, reports and symposia proceedings on topics of international interest, published on a non-profit-making basis. Orders for volumes should be placed with the publisher. Discounts are available for standing orders.

Editor-in-Chief and Editorial Office:
Pehr H. Enckell
Ecology Building
University of Lund
S-223 62 Lund, Sweden

Editorial Board
Björn E. Berglund, Lund
Tom Fenchel, Helsingør
Erkki Leppäkoski, Turku
Ulrik Lohm, Linköping
Nils Malmer (Chairman), Lund
Hans M. Seip, Oslo

Published and distributed by:
Munksgaard International Booksellers
and Publishers
P. O. Box 2148, DK-1016 Copenhagen K
Denmark

Suggested citation:
Author's name. 1991. Title of chapter. – In: Berglund, B. E. (ed.), The cultural landscape during 6000 years in southern Sweden – the Ystad Project. – Ecol. Bull. (Copenhagen) 41: 000–000.

Maps:
Information from official maps is used in some of the maps in this book. Publication permission has been granted by the National Land Survey (91.0142). Clearance for dissemination 16 July 1991.

ISBN 87-16-11049-8

Preface

This volume presents the results of the project "The Cultural Landscape during 6000 Years", based on interdisciplinary collaboration between six disciplines at Lund University during the period 1982–90. Before this project was initiated, researchers and teachers had long shown an interest in studies of the interaction of society and landscape in a historical perspective, but this interest led to only limited collaboration between the humanities and the natural sciences. One of the enthusiasts behind such collaboration in Lund has been Professor Berta Stjernquist, who also played a major role in designing the present project with its theoretical background and investigation area, the Ystad area.

Interdisciplinary research is time-consuming and expensive. Sponsorship is needed from a financial body with an interest in linking the social and natural sciences together – and we found such a sponsor in the Bank of Sweden Tercentenary Foundation (*Riksbankens Jubileumsfond*). Since 1982 our project has been generously supported during a six-year research period and a two-year publication period. We are particularly grateful to the Foundation's former chairman of the board, Professor Staffan Helmfrid, who launched the project idea with a national perspective, and to the director of the Foundation, Professor Nils-Erik Svensson, who has also shown a vivid interest in this kind of research. Our personal contact with the Foundation office has created an atmosphere of openness, which has also helped us to overcome administrative and financial problems.

Our scientific source material has been sought in the field, in museums, and in archives. The local population in the investigation area around Ystad has been very cooperative in providing information about such matters as archaeological finds and modern land use. We are indebted to many landowners in the Ystad area for their willingness to permit archaeological excavations, corings, and similar fieldwork. We are especially grateful to one of the main landowners, Count Claes C. Piper of Krageholm estate, for his generosity during many excursions to the castle and for his decision to deposit the valuable estate archive of Krageholm at the Provincial Archives in Lund, thereby making the archive accessible to our scholars. In addition, the general interest in our research shown by the authorities, museums, schools, and private societies in the Ystad area has been a stimulus to us. We would also like to mention two specialists who have been of great assistance to us: Dr Göran Hallberg at the Institute of Dialect and Place-Name Research in Lund, and State Geologist Esko Daniel at the Swedish Geological Survey in Lund. They have provided us with unpublished information on place-names and soil geology.

The project has been organized in a series of subprojects under the direction of the scholars mentioned in Table 1.1:2. These subproject leaders also formed a project board with responsibility for coordination, budget, etc. The board was assisted by the project secretary Gösta Regnéll and the administrative secretary Karin Price. It has been pleasant cooperation during almost a decade.

We are indebted to the Editorial Board of Ecological Bulletins who accepted this synthesis as a monograph in their well-reputed series. This is an honour for the project and a recognition of the research field of landscape history, which has difficulties in finding its publication niche. It would not have been possible to publish this volume without our colleague Mats Riddersporre, whose graphical skills and sensitivity for the interaction of text and figures have been invaluable. He has produced numerous drawings, including all the coloured maps. Among those who helped us with illustrations we may particularly mention Christin Andreasson and Anna-Carin Linusson. Our manuscripts have been revised or translated by Alan Crozier. His feeling for language, his word-processing workshop, and his lively interest in our subject have been a great support to us. We are very pleased that we found Alan at the right moment.

We extend our general thanks to all colleagues and technical personnel for generous help and patience during the time-consuming editorial work. We have gained a wealth of experience during these years!

Lund, December 1990

Björn E. Berglund   Lars Larsson   Nils Lewan   Sten Skansjö

Contents

Ecological Bulletins 41: 11–28. Copenhagen 1991

1 Overview of the Ystad project

1.1 The project – background, aims, and organization

Björn E. Berglund

Background

The present cultural landscape resulting from the inter-action between man and environment can be explained only after interdisciplinary studies focused on changes in time and space. This was emphasized at a national symposium on "Man, the Cultural Landscape, and the Future", held in Stockholm in February 1979. The main convenor, Professor Staffan Helmfrid, presented the idea of regional interdisciplinary studies of the development of the cultural landscape in a long-term perspective (Helmfrid 1980). A team of scientists from Lund University met after this symposium for discussions on a project devoted to the cultural landscape in southern Sweden. Similar projects were initiated for other parts of Sweden, like the Lule Älv Project in northern Sweden (Baudou unpubl.) and the Barknåre Project in south-central Sweden (Sporrong unpubl.). We have regarded these as our sister projects and experience and stimulus have been exchanged throughout the years in joint symposia, excursions, guest lectures, and so on.

The southern Swedish project was planned by a team of about twenty scientists representing the following disciplines at Lund University: Quaternary Geology (palaeoecology), Plant Ecology, Prehistoric and Medieval Archaeology, History, and Human Geography. This group met for several "brainstorming" discussions during the years 1979 and 1980. We agreed on the aim, methods and field research area for a project named "The Cultural Landscape during 6000 Years", colloquially known as "The Ystad Project" after the geographical area. In January 1981 a research application was sent to the Bank of Sweden Tercentenary Foundation. Our proposal was approved in spring 1982, which led to a contract for a six-year grant, July 1982 to June 1988, later followed by a one-year extension for publication work. Besides the present synthesis, two archaeological monographs are to be published in 1989–91. A popular version of this book is also in progress. Finally, an international symposium is planned for evaluating the experiences of this project, for comparing results from other regions, and for identifying new problems in our understanding of the interaction between society and environment in a long-term perspective. Table 1.1:1 shows the general time schedule for the project.

General aims

The general aims of the project have been as follows:

1. To describe changes in society and the landscape within a representative area of southern Sweden.

2. To analyse the causes behind these changes and especially to emphasize the relation between land use, vegetation, primary production and consumption on the one hand, and population pressure, social structure, economy, and technology on the other hand.

3. To correlate and compare the investigation area with other areas in Sweden as well as other areas in Europe.

4. To contribute to a scientific exchange between participating disciplines, particularly concerning research approach, methods, and terminology.

5. To contribute to the management of the natural environment, the cultural landscape, and ancient monuments.

Table 1.1:1. Time schedule for the Ystad Project

1.	Project idea	February 1979
2.	Planning – aim, methods, area selection	1979–1980
3.	Grant application	January 1981
4.	Grant contract	May 1982
5.	Project research, phase 1 (initiation, detailed planning)	July 1982–June 1983
6.	Project research, phase 2 (main analytical research)	July 1983–June 1986
7.	Project research, phase 3 (continued research in key areas, correlation and summing up)	July 1986–June 1988
8.	Compilation for publications: Interdisciplinary synthesis Monograph on prehistoric time Monograph on the Middle Ages	July 1988–March 1991
9.	Compilation of project results for a popular book	1991–1992

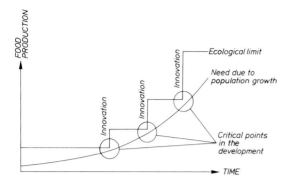

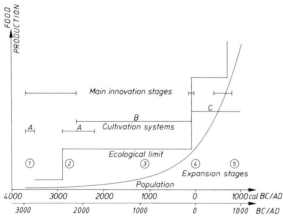

Fig. 1.1:1. Theoretical model for the development of the cultural landscape in southern Sweden based on the relation between population pressure and technological innovations as well as agrarian production and geographical expansion. From Welinder 1975. Time-scale in radiocarbon years as well as in calendar years (cal. BC/AD).

General hypothesis

The development of the agrarian landscape as seen in a long-term perspective is characterized by phases of expansion, consolidation, and regression. This pattern has been identified and described by palaeoecologists (e.g. Berglund 1969, 1986, Aalen 1983, Digerfeldt and Welinder 1988), by archaeologists (e.g. Welinder 1975, 1977, 1979, 1983, Gräslund 1979, Kristiansen 1984, Am-

brosiani 1984b), and by human geographers (Sporrong 1985). In the survey paper on the human influence on the prehistoric landscape in southern Scandinavia Berglund (1969) described four expansion periods. A discussion of the causality was later taken up by Welinder (1975 and later). Inspired by Boserup (1973), he emphasized the importance of technical innovations for increasing the carrying capacity in a society with growing population. Colonization and utilization of marginal and non-central settlement areas were also of great importance in Welinder's model (Fig. 1.1:1). As a consequence of population growth, the technological level was raised to increase agrarian production, and more people were able to survive within a restricted area. This is a model where the social factors play a leading role. Other researchers have emphasized the environmental factors (e.g. Enckell et al. 1979, Emanuelsson

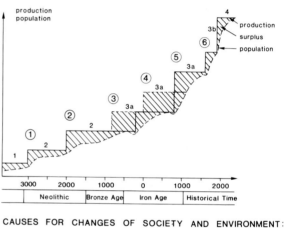

CAUSES FOR CHANGES OF SOCIETY AND ENVIRONMENT:

Population pressure			Climate		
Social organization	}	Social	Hydrology	}	Environmental
Economy		factors	Biota		factors
Technology			Soils		

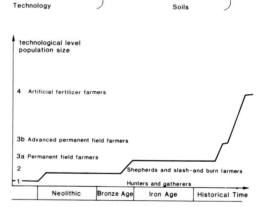

Fig. 1.1:2. Multicausal model for long-term changes of the cultural landscape used as a general hypothesis for the Ystad Project (modified after Berglund 1969, 1986, 1988). Encircled figures 1–6 refer to expansion phases sometimes indicated in pollen diagrams (e.g. Berglund 1988). In the lower figure agrotechnological levels have been transferred into a corresponding scheme (cf. Emanuelsson 1987, 1988). Numbered levels also indicated in the upper figure. Time-scale in radiocarbon years.

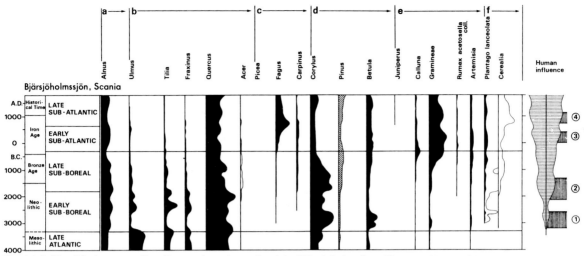

Fig. 1.1:3. Simplified survey pollen diagram from the ancient lake Bjärsjöholmssjön. Changes of human influence indicated by the curve to the right. Phases 1–4 correspond to so-called expansion phases. Phase 3 has been subdivided into two phases in the model of Fig. 1.1:2. Time-scale in radiocarbon years. From Berglund (1969), after T. Nilsson's (1961) original diagram.

1987, 1988). Emanuelsson has further developed Welinder's model by linking the steps in the increased production to different strategies of mobilization of the soil nutrients (Fig. 1.1:2).

In our project on the long-term changes in the cultural landscape in southern Sweden we have proposed a dynamic multicausal theory for the observed changes during prehistoric and historical time. We have illustrated this as a staircase for cultural landscape development (Berglund 1984, 1986), and the causal processes are to be found in social as well as environmental factors (Fig. 1.1:2). The causal factors behind changes in the cultural landscape can be further classified, following Stjernquist (1983), Germundsson (1984), and Berglund (1988). Different phases of settlement development – colonization, expansion, stability – have been identified by Sporrong (1985); see the definition in Ch. 2.9.

From our study area in southernmost Sweden an old

pollen diagram from the ancient lake of Bjärsjöholmssjön was already available (T. Nilsson 1961, Berglund 1969). In the diagram reproduced in Fig. 1.1:3, four expansion periods are indicated. The following main questions related to our project hypothesis were raised against this background:

1. Is the pattern of human influence through time repeated in new pollen diagrams from the larger area?

2. Is it possible to document spatial differences in human influence, for instance, between central settlement areas and marginal areas?

3. If an expansion/regression pattern occurs in the past landscape development, is it then possible to find a correspondence in the archaeological and historical source material?

4. Finally, what are the causes of the changes in the landscape and the society, and especially, what is the link between society and landscape?

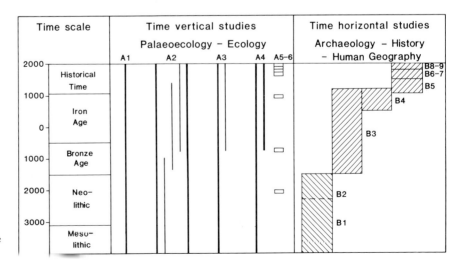

Fig. 1.1:4. Project organization with subprojects plotted against time. Time-scale in radiocarbon years. See Table 1.1:2 for details.

Table 1.1:2. Survey of the Ystad Project with its subprojects

Title of Subproject		Leaders	Discipline	Department	Collaborators
A1 A2	Regional and local changes in vegetation	Berglund	Palaeoecology	Quaternary geology	Bartholin Engelmark Gaillard Göransson Hjelmroos Kolstrup J. Regnéll
A3	Palaeohydrological changes	Digerfeldt/ Gaillard	Palaeoecology	Quaternary geology	Håkansson
A4	Soil erosion	Dearing	Palaeoecology	Quaternary geology	Sandgren
A5	Reconstruction of former vegetation and land use	Olsson/ Malmer	Plant ecology	Plant ecology	Bengtsson-Lindsjö Berlin Ihse Linusson G. Regnéll
A6	Bioproduction available to people during different periods	Olsson	Plant ecology	Plant ecology	–
B1	Changes during the introduction of agriculture	M. Larsson	Prehistoric archaeology	Archaeology	–
B2	Expansion during the Middle Neolithic period	L. Larsson	Prehistoric archaeology	Archaeology	–
B3	Settlement changes during the Bronze and Iron Ages	Stjernquist/ Tesch	Prehistoric archaeology	Archaeology	Olausson Riddersporre
B4	Village consolidation in the Late Iron Age	Callmer	Prehistoric archaeology	Archaeology	–
B5	Settlement development, production, and social organization during the Middle Ages	Cinthio/ Andrén/ Andersson	Medieval archaeology	Medieval archaeology	Marit Anglert Mats Anglert Billberg Lindroth Reisnert Sundnér Wienberg
B6	Estate management and settlement expansion since the mid 16th century	Skansjö	History	History	–
B7	The agrarian landscape from the end of the 17th century until the enclosures during the 19th century	Persson	History	History	–
B8	Agrarian expansion in the 19th century	Lewan	Human geography	Human geography	Möller
B9	From production society to service society during the 20th century	Lewan	Human geography	Human geography	Erlingson Germundsson Ihse Mårtensson Nilsson

Project organization

The organization of the project was originally based on the hypothesis of the assumed expansion phases in the cultural landscape. The focus has therefore been on these phases, but research has also included the intermediate periods. The research projects have been of two kinds: (1) time-vertical studies, dealing with landscape changes in a long-term perspective, and (2) time-horizontal studies, which are multidisciplinary studies dealing with selected periods of special importance for changes in society and landscape.

The time-vertical studies are mainly palaeoecological and ecological in character, and provide the framework for the time-horizontal studies. The arrangement of the subprojects in time is shown in Fig. 1.1:4. The details of the subprojects are shown in Table 1.1:2. Although one discipline has been mainly responsible for each subproject, scientists from different disciplines have collaborated within each. This means that there has been close collaboration between researchers from six departments at Lund University during the period 1982–88, in some cases also including specialists at other universities or institutes. In addition, scientists from outside Sweden have been invited to cooperate. Altogether 25 to 30 scholars have been involved in the project, about ten full-time, others part-time.

The subproject leaders formed a project board with responsibility for scientific approach as well as finances. During the last two years the main project responsibility was delegated to a committee with four members representing the disciplines of ecology-palaeoecology, archaeology, history, and human geography (Björn E. Berglund, Lars Larsson, Sten Skansjö, Nils Lewan). This group has also acted as an editorial committee. The board has been assisted in the administration by a project secretary (Gösta Regnéll) and in the editing by a managing editor (Mats Riddersporre), both active workers on the project. The board and the committee have been chaired by Björn E. Berglund, assisted by the administrative secretary Karin Price.

The general framework and working plan for the project was set up by the board, but the subprojects have acted quite independently. Planning and discussions was organized during board meetings and at open seminars. Several excursions to field-study sites were arranged, sometimes in connection with national or international meetings. Discussions between individual scholars and field visits have often been of an informal character. The project has also included educational activities, such as field courses, examination papers, and research for doctoral theses.

Project publications

During the running of the project preliminary results have been presented continuously by the individual scholars in national as well as international journals, conference volumes, etc., and the board has stimulated the scientists to do so. Mainly for internal communication, a series of reports has been distributed under the heading "En arbetsrapport från Kulturlandskapet under 6000 år" (G. Regnéll ed.). A project symposium in September 1984 resulted in a special volume, also published in this series. All publications with results from the project during the period 1982–1989 are compiled in a list of project publications, (see Papers from the project).

The main final project publications besides this integrated synthesis are two archaeological monographs, one for prehistoric times (Callmer et al. eds) and one for the Middle Ages (Andersson and Anglert eds). A popular book in Swedish is in progress; it is intended to be a well-illustrated book for a wider audience.

During the course of the work the individual researchers have also presented the project and its results in numerous lectures, organized in the project area (in societies, schools, etc.), at Lund University, and at many other places in Sweden and abroad. A project exhibition with posters and other material was compiled in 1986 and displayed at museums in Lund and Ystad during 1986–87. During the university year 1988–89 a lecture series for a wider audience was arranged (convener Hans Andersson).

1.2 The project area

Björn E. Berglund

Choice of project area

The study area was chosen within the southernmost province of Sweden, Scania (Sw. Skåne), to meet the demands of the project's aims as well as the requirements concerning source material for the six main disciplines. This meant that the project area was to include a central settlement area on soils favourable for agriculture and contrasting to a marginal settlement area less favourable for agriculture. If possible the study area was to be representative of the Scanodanian nature region which includes Denmark as well as south-western Scania (cf. Nordiska Ministerrådet 1984). The team of scientists responsible for the choice of the area made a reconnaissance with the help of geological and phys-

icogeographical maps as well as maps illustrating settlement history. This is shown here by the map sequence for Scania which illustrates topography, bedrock, subsoils, phytogeography, and geomorphology in Figs 1.2:1–6 and settlement/population distribution patterns for the Late Neolithic, Late Iron Age, Middle Ages, 17th century, and 1980 in Figs 1.2:7–11. Having considered the various problems of available source material we finally chose the Ystad area – the hundreds of Ljunits and Herrestad. This area seemed to represent a physicogeographical gradient as well as a settlement density gradient for the entire time-span to be studied in the project. However, the choice of area was a compromise and therefore it has not been without drawbacks.

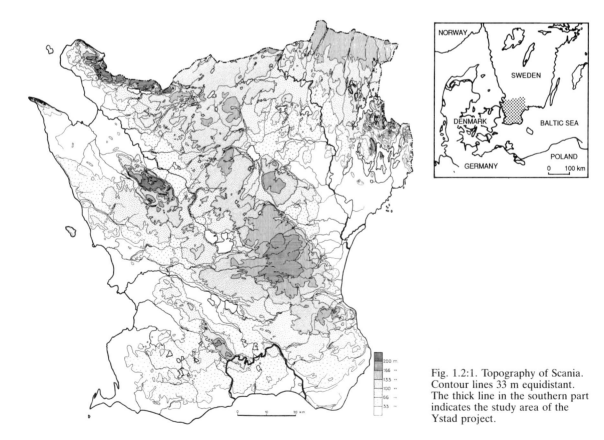

Fig. 1.2:1. Topography of Scania. Contour lines 33 m equidistant. The thick line in the southern part indicates the study area of the Ystad project.

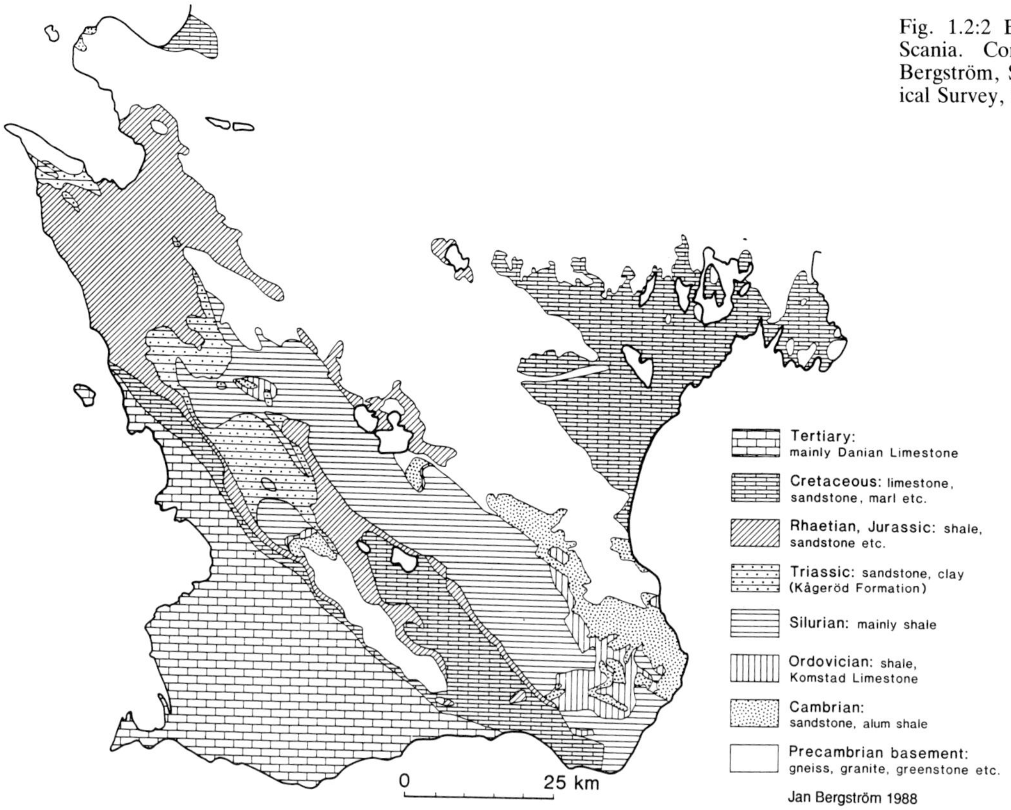

Fig. 1.2:2 Bedrock map of
Scania. Compiled by Jan
Bergström, Swedish Geolog-
ical Survey, Lund.

Tertiary:
mainly Danian Limestone

Cretaceous: limestone,
sandstone, marl etc.

Rhaetian, Jurassic: shale,
sandstone etc.

Triassic: sandstone, clay
(Kågeröd Formation)

Silurian: mainly shale

Ordovician: shale,
Komstad Limestone

Cambrian:
sandstone, alum shale

Precambrian basement:
gneiss, granite, greenstone etc.

Jan Bergström 1988

0   25 km

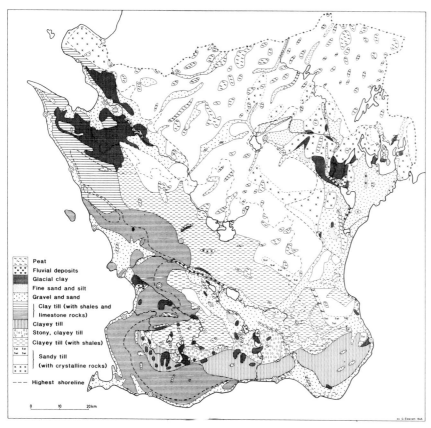

Peat
Fluvial deposits
Glacial clay
Fine sand and silt
Gravel and sand
Clay till (with shales and
limestone rocks)
Clayey till
Stony, clayey till
Clayey till (with shales)
Sandy till
(with crystalline rocks)
Highest shoreline

0   10   20 km

Fig. 1.2:3. Subsoil map of
Scania. From Ekström 1946.

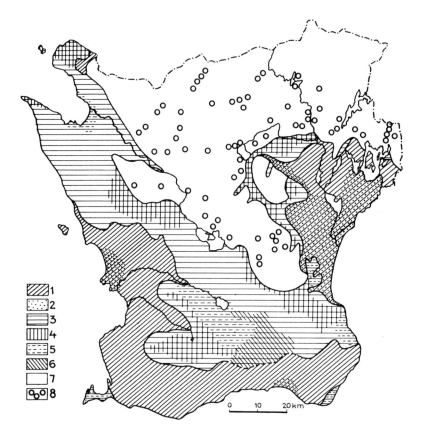

Fig. 1.2:4. Topsoil differentiation in Scania. Legend: 1 – Areas with alkaline soils (pH 7.0–8.0); 2 – acid, leached soils within area 1; 3 – Areas with slightly acid soils; 4 – Areas with frequently occurring acid soils within area 3; 5 – Acid, leached sandy soils within area 3; 6 – Areas with slightly alkaline soils are frequent within area 3; 7 – Areas with acid soils (pH 4.0–5.9); 8 – Scattered occurrence of neutral soils within area 7. From Waldheim 1947.

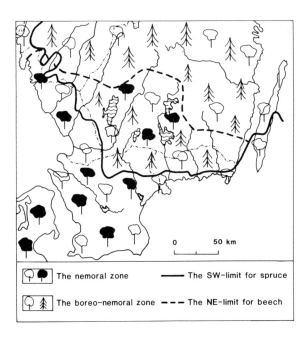

Fig. 1.2:5. Phytogeographical subdivision of southern Sweden according to Sjörs (1963, 1965).

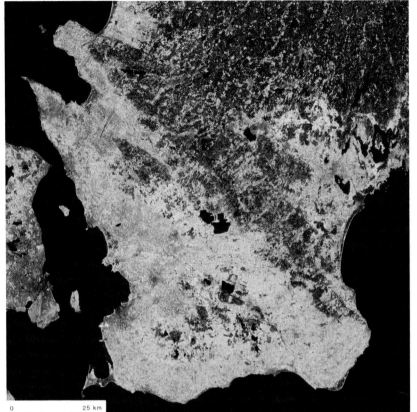
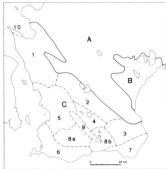

Fig. 1.2:6. Satellite image (TM image from Landsat 5) of Scania with neighbouring provinces. The key map defines the visible borders between main geomorphological regions of Scania, mainly according to Nelson (1936):

A. Precambrian bedrock plateau and horst ridge in north and central Scania.
B. The Kristianstad plain
C. The plain of SW Scania
    1.   The Ängelholm-Helsingborg plain
    2.   The shale plain of central Scania
    3.   The esker/valley area of SE Scania
    4.   The sand plain of Vomb
    5.   The Lund-Landskrona plain
    6.   The plain of SW Scania, "Söderslätt"
    7.   The sandy esker/moraine landscape
    8.   The hummocky moraine landscape: 8 a = SW area, 8 b = SE area
    9.   The horst ridge Romeleåsen
    10.   The horst ridge Kullaberg

Satellite image published by permission of Satellitbild i Kiruna AB.

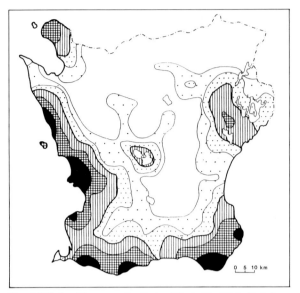

Fig. 1.2:7. Isarithm map for finds of Late Neolithic flint daggers in Scania. Shaded areas defined by the frequencies of fling daggers, indicated by a relative numerical scale. From M. Malmer 1975.

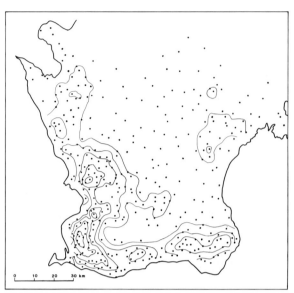

Fig. 1.2:9. Isarithm map for medieval churches in Scania indicating central settlement areas. Each dot represents one church. From Rosborn 1984.

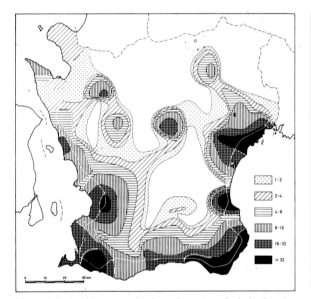

Fig. 1.2:8. Isarithm map for Late Iron Age finds in Scania. Shaded area defined by the frequencies of grave-fields, settlements and stray finds, indicated by a relative numerical scale. From Strömberg 1961.

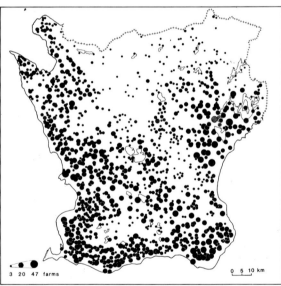

1.2:10. Map for mid 17th century in Scania illustrating the distribution of farm villages. From Dahl 1940.

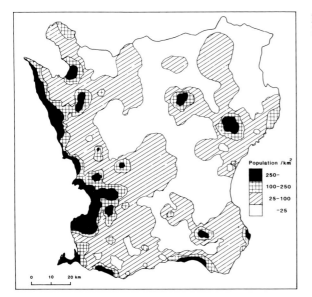

Fig. 1.2:11. Population density in Scania 1975. From Möller 1979.

Population /km^2

■ 250–

100–250

25–100

–25

0   10   20 km

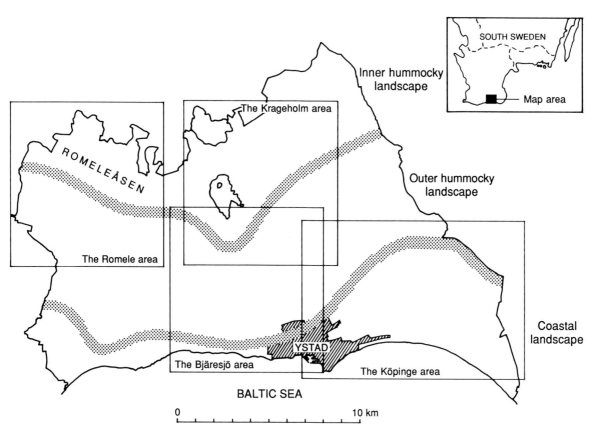

Fig. 1.2:12. Study area around the town of Ystad with the three landscape zones and four main focal areas indicated.

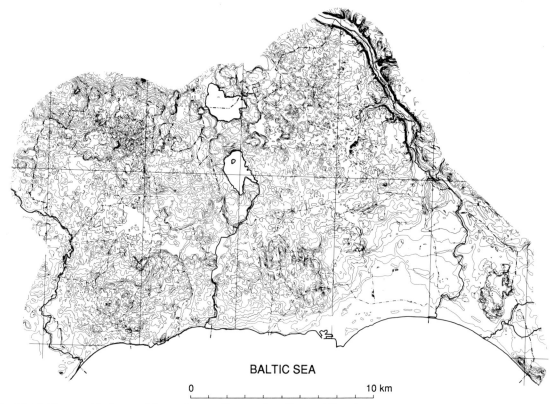

BALTIC SEA

0   10 km

Fig. 1.2:13. Topographical map of the study area. Contour lines 5 m equidistant.

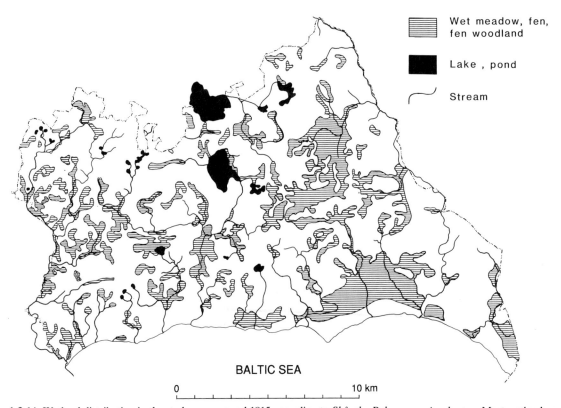

Wet meadow, fen,
fen woodland

Lake , pond

Stream

BALTIC SEA

0   10 km

Fig. 1.2:14. Wetland distribution in the study area around 1815 according to *Skånska Rekognosceringskartan*. Most wetlands are unwooded areas. In general the wetlands occur on peat soils.

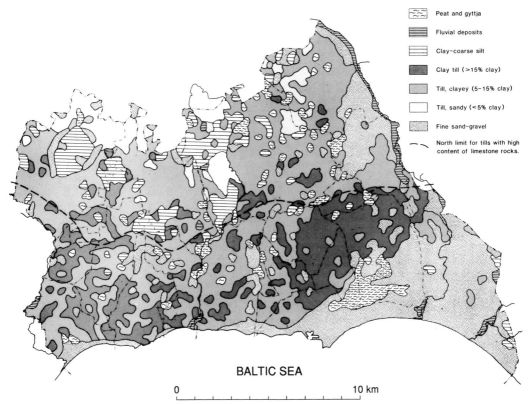

Fig. 1.2:15. Simplified subsoil map of the study area. Compiled by Daniel, 1989, Swedish Geological Survey, Lund (based on detailed maps by Daniel 1986, 1989).

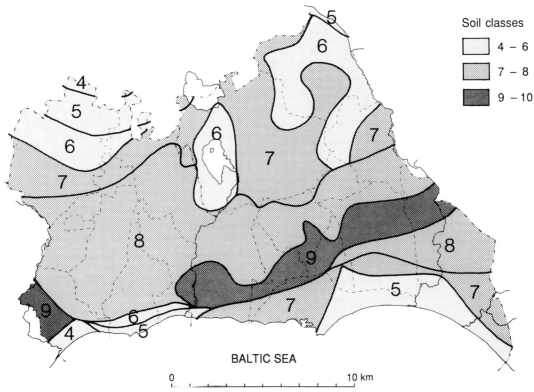

Fig. 1.2:16. Soil classification map of the study area. Agrarian soils classified on a 1–10 grade scale with 10 being the highest production capacity under present use. Extracted from Fysisk riksplanering i Skåne 1977, Länsstyrelsen i Malmöhus län.

Fig. 1.2:17. The historic settlement situation in the study area. Black dots represent the historically known villages. The manors indicated (M) are those that exist in Modern Time. Shaded areas are the built up areas of today. The boundary between Ljunits Hundred in the west and Herrestad hundred in the east follows the river Svartån from the Baltic and continues further north to the west of the lake Krageholmssjön. The remaining boundariew within the area are parish boundaries.

Description of the project area

The Ystad area, with Ljunits hundred in the west and Herrestad hundred in the east, is situated around the town of Ystad on the south coast of Scania. It covers an area of about 28,900 ha; the length of the coast is about 25 km, the west-east distance is also 25 km, and the north-south distance about 20 km. The maximum altitude is around 150 m.

Physicogeographically the area can be divided into three landscape zones from the coast towards the inland (Fig. 1.2:12):

(1) a coastal landscape with sandy soils beside ancient lagoonal basins with peat, low relief (below 25 m elevation), today fully exploited for agriculture and settlement;

(2) an outer hummocky landscape with clayey silty soils, hummocky relief (mainly 25–75 m elevation), today fully exploited for agriculture;

(3) an inner hummocky landscape with a mosaic of clayey silty soils and sandy soils, all with lower carbonate content than in zones 1 and 2, hummocky relief (mainly 75–100 m, but in the north-west reaching 150 m), today partly forested (native beech woods) and less suitable for agriculture. This zone can be divided into

three areas: the Romeleåsen ridge in the north-west, the lake basin depression in the central part, and the Baldringe plateau in the north-east.

The landscape character is shown in the map sequence in Figs 1.2:13–16, which illustrates the topography, the hydrology before artificial drainage, the subsoil geology, and the agricultural soil quality.

The zonation implies a gradient from central to marginal settlement, which would also apply to prehistoric and historical times. The entire area is representative of the present-day cultivated plain extending through southern and western Scania. With archaeological-historical-geographical and palaeoecological studies in each zone, it will be possible to make correlations and comparisons in time and space. Since medieval times, Ystad has been the commercial centre, and in modern times the area has been dominated by big estates; it is therefore possible to study the role of the estates in the development of modern agriculture and the relationship between the town of Ystad and the surrounding countryside during most of historical time.

Fairly ample source material is available for the different disciplines, for research into nature as well as into changes in society. However, there are gaps and errors which are discussed in Ch. 2. One particular drawback

26

is found in the present-day vegetation: because of modern intensive farming activity, very little has been preserved of the pre-industrial land-use pattern.

Inventories have been made in the entire area, but the main research has been focused on four key study areas, also shown on the map in Fig. 1.2:12: the Köpinge area in the coastal landscape, the Bjäresjö area in the outer hummocky landscape, the Krageholm area and the Romele area in the inner hummocky landscape.

These study areas form a diagonal transect from south-east to north-west.

In Fig. 1.2:17 is given an overview of the historic settlement situation within the Ystad area. The map includes various forms of settlement agglomerations and manors. The dispersed settlements that, before the enclosure movements in the 19th century, predominantly existed in the north-western part of the area are not indicated on the map.

1.3 Outline of the synthesis volume

Björn E. Berglund and Mats Riddersporre

This volume, which summarizes the results of the Ystad Project, is edited as an integrated book. The contributors have from an early stage been working according to a plan that emerged from discussions and working groups within the project. This plan has of necessity been flexible in its details, but the various chapters and sections of the book are to be seen as interrelated parts rather than a set of independent articles and papers. A deviation from this is, to some extent, Ch. 5, where more specific topics are dealt with. The sections of this chapter can also be read as more independent contributions. The text follows in general the basic chronological sequence.

In the following, Ch. 2 includes comments on the hypotheses, source material, methods, representativeness, and the like, all described for each participating discipline. The settlement and the landscape mapping presented in Ch. 3 is commented on from a methodological point of view. Also scientific and linguistic terminology is dealt with in Ch. 2.

The project documentation includes Ch. 3, 4, and 5. For Ch. 3 ten time-sections have been chosen to describe the cultural landscape of the entire area through time. The sections are chosen so that important changes occur between each period. The interdisciplinary documentation is concentrated to maps (1:125,000) with brief accompanying texts. The maps are edited so that they should be graphically comparable over time, although the quality of the sources varies from map to map (cf. Ch. 2.7 and 2.8).

In Ch. 4 the four main focal areas are documented, each one described from Mesolithic to modern time. The focal areas are chosen so that they together cover a transect from the coastal plain in south-east, via the outer hummocky zone, to the inner hummocky landscape and the Romeleåsen ridge in the north-west (Fig. 1.2:12). The maps in Ch. 4 (1:50,000) permit a more precise presentation of the settlement and, for prehistoric time, a documentation of the actual source material. On the other hand, a detailed reconstruction of the prehistoric landscape and land use is not possible for these relatively large areas. In some cases hypothetical reconstructions, covering more delimitated subareas, have been carried out (1:25,000).

For Ch. 5 specific problems related to the cultural landscape have been selected and discussed in greater detail, first environmental, then societal problems in prehistoric as well as historical time. The final section, Ch. 5.13, summarizes long-term landscape-ecological changes.

In Ch. 6 conclusions are drawn on the causes behind the landscape changes. Ecological and social factors are discussed and weighed against each other. Finally, in Ch. 7, the project team looks back, discussing experience from this project as well as visions of future research related to society/environment in a long-term perspective. We also discuss the importance of this interdisciplinary project for conservation and management of the natural and cultural environment.

At the end of the book sources of different kinds are presented. Besides a joint list of quoted literature, we have compiled a list of our own project publications.

Pollen diagrams are essential for the palaeoecological interpretation. The chronology is based on a radiocarbon time-scale obtained through radiocarbon datings of investigated sequences as well as on cross-correlation with reference sites outside the Ystad area. The basis for this chronology, in general and specifically for each pollen diagram, is documented in Appendix A (cf. Ch. 2.1).

As Appendix B there is also a fold-out map (1:50,000) showing the landscape of the Ystad area in the 18th century. This map, which is compiled from large-scale land survey maps, represents the earliest historical landscape that can be described from contemporary maps.

Ecological Bulletins 41: 29–62. Copenhagen 1991

2 Source material and methods

2.1 Palaeoecology

Björn E. Berglund

Hypotheses

The main aim of the palaeoecological studies in the Ystad Project is to recover the landscape/vegetation changes on a subregional scale – areas with a radius of 1 to 5 km. The chief method is pollen analysis of lake/mire deposits. On the basis of earlier work on vegetation history and the knowledge of modern forest ecosystems, we work with several hypotheses concerning past landscape changes. The following are of fundamental importance for our interpretations:

1. There is an overall complex interaction of climate-man-environment during the Holocene (the last 10,000 years). It is one of the aims of the project to study this in the Ystad area and the palaeoecological research strategy reflects this. We concentrate on the man/environment relation. There are difficulties in obtaining independent information on palaeoclimate from the area; only the palaeohydrological studies can give us some indirect information on climate (humidity) conditions (Ch. 5.1). We therefore have to rely on the general knowledge of Holocene climatic changes in Europe (cf. Lamb 1977)

2. Large-scale Holocene forest vegetation is assumed to be in equilibrium with climate (cf. von Post 1946, Prentice 1986b, Webb 1986, Huntley 1988). This means that the composition of the tree flora for a large region like southern Scandinavia during any time of the Holocene is conditioned by the climate (cf. Solomon et al. 1981, Huntley and Prentice 1988). However, the time-space scale has to be considered (Delcourt et al. 1983, Birks 1986). A sudden climatic change may cause a remodelling or replacing of a forest ecosystem with a time-lag of 100–500 years (Wright 1984).

3. Forest composition in an area is dependent on the past history of each species because they behave individually (Davis 1981a, Jacobson et al. 1987, Bennett and Lamb 1988). This is the main reason why the composition is different between interglacials. In a similar way, we can expect that the plant composition to some extent differed during mid-Holocene from modern conditions (cf. Ch. 5.13).

4. Holocene long-term vegetation dynamics are also influenced by the soil development, which means a gradual soil deterioration throughout the Holocene (cf. Iversen 1964, 1969, Birks 1986a). Human disturbance may cause accelerated vegetation impoverishment and soil leaching, particularly on light soils, poor in nutri-

ents (Berglund 1966, Aaby 1983, Andersen 1984, Behre 1988).

5. Holocene forested landscapes in southern Scandinavia have stable soil conditions (cf. Bormann and Likens 1979). Human disturbances since the Neolithic will have caused soil erosion, owing to forestry and agricultural technology.

6. Human impact is the dominant factor behind deforestation and other disturbances in the late Holocene (cf. Iversen 1941, Barker 1985, Birks 1986, Delcourt and Delcourt 1987, Behre 1988). Long-term deforestation is a gradual process dependent on societal development, technology, etc. (see Ch. 1.1). This is reflected in pollen diagrams as expansion/regression phases, as defined in Ch. 2.9.

7. Land-use practices in prehistoric and historical times have been applied also in recent time (Emanuelsson et al. 1985, Pott 1988). Therefore, present-day relict cultural landscapes may be used as modern analogues to past ones. This means that the indicator species approach (Birks and Birks 1980, cf. Behre 1981) or the multivariate correlation technique (Berglund et al. 1986) can be applied for deriving information on past human-caused plant communities and land use. But we also have to consider the problem that people in the past had land-use practices unknown to us today.

8. The distribution of forest communities and their dominant tree species during the Holocene is dependent on soil conditions (nutrients, water, texture, etc.) within a restricted area like the Ystad area. Past potential vegetation (Tüxen 1956, Trass and Malmer 1973) can be reconstructed from a combination of pollen stratigraphy and the knowledge of topography, subsoils, hydrology, and other edaphic factors (see Ch. 2.8).

Source material

The source material for our palaeoecological analyses is found in lake/mire deposits. When choosing suitable sites for comparisons with archaeological and historical information we always have to consider the scale problem, because these disciplines operate at different time-space scales – wind-dispersed pollen for pollen analyses, local artefacts, bones, and the like for archaeological sites, written records for single farms, villages, etc. (see Birks and Moe 1986, Groenman-van Waateringe 1988). This also has an impact when we define the following demands for palaeoecological sites:

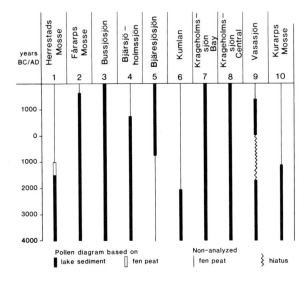

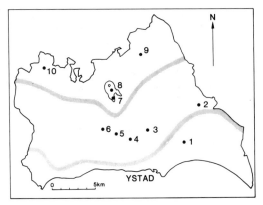

- site with pollen diagram

Fig. 2.1:1. Survey scheme illustrating the time-span for each palaeoecological site with a pollen diagram. Sites nos. 2, 4, 5, 7, and 10 are represented by pollen diagrams in this volume. Site location is indicated on the survey map. Time-scale in radiocarbon years.

1. Lake sediments with the time-span 6000 BP to the present, to ensure comparability between sites and subregions. Such lake basins also act as sediment traps for soil erosion. – Only two sites fulfil these demands, namely Krageholmssjön and Bussjösjön, because of early overgrowing of ancient lake basins. In addition, one site, Bjäresjösjön, represents a lake environment for late Holocene (the last 2700 years). Available peat sequences are not suitable for correct dating and recovery of reliable pollen spectra, because only minerotrophic fens have been found in this area. If good peat sequences have occurred at some sites, these have been destroyed because of fuel exploitation. The time-span for all pollen diagrams is shown in Fig. 2.1:1.

2. Lake sediments with high accumulation and good dating possibilities. – Owing to the situation within a calcareous region the lake sediments in general contain pre-Quaternary carbonate and therefore radiocarbon dates are affected by the hardwater error (giving ages that are too high). Only in the case of Bjärsjöholmssjön was a good series of dates obtained (see Ch. 4.2.2). This lake has a non-calcareous autochthonous gyttja sequence. At other sites the absolute chronology has been obtained through pollen-analytical cross-correlation, using the Bjärsjöholmssjön diagram and the Ageröds Mosse reference chronology (T. Nilsson 1964, see Gaillard 1984, J. Regnéll 1989).

3. Lakes of moderate size, evenly distributed in the three landscape zones. As far as pollen analysis is concerned, there is a certain relation between basin size and source area (see Jacobson and Bradshaw 1981, Prentice 1985, 1988). Our lakes ought to be small enough (basin radius <100–200 m) to reflect the local environment of the key study areas. – All studied sites fulfil this demand except for Krageholmssjön, which is a big lake representative of a large area. Therefore, the pollen diagram from the central part of Krageholmssjön reveals the regional vegetation changes. Most of our palaeoecological sites are distributed along a transect from the coast in the south-east towards the Romeleåsen ridge in the north-west. Unfortunately, it has been impossible to find a basin with full time-coverage on this ridge as well as in the upland area of the north-eastern study area. In this inner hummocky landscape several peat basins occur, but the peat has been exploited. Another problematic area is the coastal plain around Ystad-Köpinge-Kabusa. The only basin here is Herrestads Mosse, with lagoonal sediments deposited until about 3500 BP. However, in the border zone to the hummocky landscape lies the ancient lake of Fårarps Mosse. It is regarded as representative also of the coastal plain, and the lake sediments cover the time until about AD 1600.

4. Small hollows or mor humus of so-called closed-canopy sites (Bradshaw 1988, cf. Andersen 1984) with records of vegetation/landscape changes within a radius of about 30 m. Such sites have been searched for, but only hollows with high biological activity and low pollen preservation have been found, for example, in the wooded areas of the inland zone.

5. Organic deposits in direct contact with occupation layers of excavated sites, containing for instance waste from settlements. Such material has rarely been found during the excavations. Exceptions are charcoal and carbonized seeds at some sites, particularly from the Köpinge area.

A survey of the entire palaeoecological source material is presented in Table 2.1:1. Additional charred plant macroremains and charcoal come from archaeologically excavated sites (Bartholin and Berglund 1991, Engelmark and Hjelmquist 1991).

Table 2.1:1. Information on palaeoecological sites.

Site name	Herrestads Mosse	Fårarps Mosse	Bussjösjön	Bjärsjöholmssjön	Bjäresjösjön	Kumlan	Krageholmssjön bay	Krageholmssjön central	Vasasjön	Kurarps Mosse	Archaeological excavations at various sites
Site no.	1	2	3	4	5	6	7	8	9	10	
Environment											
Elevation, m	5	25	24	50	48	35	43	43	55	95	–
Distance to coast, km	2	6	4	3	3.5	4	8	8	12	12	–
Site identity (age of main change, years BP)	lagoon/3500/fen	lake/300/fen	lake	lake/100/fen	carr/2700/lake	lake/4000/fen	lake	lake	lake/600/fen	lake/3000/fen	–
Basin size, ha	ca. 500	15 → 3	2 → 0,3	18 → 11	2	120	ca. 6 (bay)	220	ca. 20	10	–
Lake radius, m	ca. 700	80	70 → 25	150	60	350	100 (bay)	600	100	80	–
Lake catchment, ha	ca. 3000	115	200 → 25	125 → 115	25 → 15	ca. 1800	940 (lake)	940	ca. 200	160	–
Lake: catchment ratio	1:6	1:8	1:250 → 1:85	1:7	1:6,5	1:5	1:4 (lake)	1:4	1:10	1:16	–
Soil type of catchment	sand + clay till	sand + clay till	clay till	sand + claey till	clayey till	clayey till + clay till	clay + coarse silt	claey till + clay-silt	sandy + clayey till	sandy + clayey till	–
Analyses with investigator's signature											
Sedimentology	JM	MH	JR	HG	MJG	MJG	MJG	JR	MH	EK	–
Detailed lithostratigraphy	–	–	JD, PS	–	MJG	MJG	MJG	JR	–	–	–
Magnetic parameters	–	–	JD	–	JD	–	–	JD	–	–	–
Sediment infflux	–	–	–	–	JD	–	–	JD	–	–	–
Geochemistry	–	–	–	–	ME	–	–	–	–	–	–
¹⁴C dating	SH	SH	–	GS	SH	GS	–	–	SH	–	–
²¹⁰Pb, ¹³⁷Cs dating	–	–	FED	–	FED	–	–	FED	–	–	–
Pollen, %	EK	MH	JR	HG	MJG	MJG	MJG	JR	MH	EK	–
Pollen influx	–	–	JR	HG	MJG	–	MJG	JR	–	EK	–
Plant macrofossils	–	–	–	–	MJG	MJG	MJG	–	–	–	–
Charred seeds	–	–	–	–	–	–	–	–	–	–	RE
Charcoal (macroscopic)	–	–	–	–	–	–	–	–	–	–	TB
Diatoms	HH	–	HH	–	HH	–	HH	HH	–	–	–

TB = Thomas Bartholin, RE = Roger Engelmark, JD = John Dearing, FED = Farid El-Daoushy, ME = Magnus Enell, MJG = Marie-José Gaillard, HG = Hans Göransson, HH = Hannelore Håkansson, SH = Sören Håkansson, MH = Mervi Hjelmroos, EK = Else Kolstrup, JM = Jan Mikaelsson, JR = Joachim Regnéll, PS = Per Sandgren, GS = Göran Skog

Methods and methodological problems

Our research strategy follows the multidisciplinary approach applied in some international programmes dealing with long-term environmental change (see Berglund 1986b). The project area is also a palaeoecological reference area within the project Unesco-IGCP 158. This means that several parallel analyses should be conducted on the same sediment core. However, logistics restricted us to applying this ideal strategy only at a few places, mainly the three lakes existing today: Bussjösjön, Bjäresjösjön, and Krageholmssjön. Otherwise analyses were concentrated on pollen and sedimentological survey analyses. Table 2.1:1 gives an outline of the methods applied at the different sites.

Sediment analysis. Besides conventional methods for sediment classification and measuring the organic minerogenic and carbonate content (physical analyses), magnetic parameters were also measured at the modern lake sites. This information was used for correlating and calculating the sediment influx, a measure of the catchment erosions (Dearing 1986, Ch. 5.2). Sediment descriptions have sometimes been supported by analysis of plant macrofossils. For palaeohydrological interpretation a combination of several stratigraphical methods has been applied to a cross-section with littoral profiles: sediment description, physical analysis, macrofossil analysis, and pollen and diatom analysis (Gaillard 1984, Ch. 5.1). Geochemical analyses were performed only at Bjäresjösjön (Enell in Gaillard et al. in press).

Chronology. The chronology in the various palaeoecological profiles (pollen diagrams etc.) is based on a radiocarbon time-scale. As mentioned above, some content of pre-Quaternary carbonate in the lake sediments causes erroneous dates. However, the Bjärsjöholmssjön sediments are characterized by high-organic autochthonous gyttja, and a series of 12 ^{14}C dates was obtained for the last 7000 years (Göransson 1991). A pollen-analytical cross-correlation with the Ageröd sequence, 50 km north-west of our area, indicates that these dates are quite reliable. The same well-dated sequence from the bog Ageröds Mosse (T. Nilsson 1964) has been used for transferring the Scanian pollen zone system (T. Nilsson 1935, 1961) to our new pollen diagrams (see discussion by Gaillard 1984, J. Regnéll 1989). Pollen-analytical indicator levels have obtained absolute ages so that it has been possible to construct time-depth curves for each pollen-analysed sediment core. For more information on the chronology, see Appendix A.

We normally use an uncorrected radiocarbon time-scale throughout this synthesis (for time-scales see Ch. 2.8). Top cores have been dated by means of ^{210}Pb and ^{137}Cs analyses (El-Daoushy unpubl.). Palaeomagnetic stratigraphy (palaeosecular variation) was a tool used for approximate dating at Bussjösjön (Sandgren and Björck 1988).

Pollen analysis. The analytical technique follows traditional procedure, but a goal has been to optimize the identification of herb pollen taxa to use modern analogues for reconstructing past vegetation as accurately as possible. Methodological problems related to the search for evidence of early agriculture have been discussed by Berglund (1985). Pollen indicator species for different land use have been discussed by J. Regnéll (1989). All diagrams are constructed as percentage diagrams, with taxa ordered and classified into ecological groups (vegetation/land-use groups). The diagram type is named "human impact diagram" following Berglund and Ralska-Jasiewiczowa 1986 (for details see Gaillard unpubl.). The selected diagrams published in Ch. 4 are simplified survey diagrams based on the original diagrams, with percentage values plotted against time. They are subdivided into the following sections:

A. Total diagram, a diagram cumulative for the following groups:
1. Woodland trees, shrubs, and woodland herbs
2. General open land apophytes (native plants favoured by man)
3. Pasture and meadow apophytes (dry to mesic meadows)
4. Wet meadow apophytes
5. Anthropochores (arable and ruderal land).
By apophytes we mean native species favoured by man, while anthropochores are species introduced by man, like cultivated plants and weeds (Linkola 1916). Occasionally AP is used as an abbreviation for arbor pollen and NAP as an abbreviation for non-arbor pollen (mainly groups 2–5 above).

B. Special diagrams, with curves for selected taxa ordered according to groups 1 to 5 above.

C. Pollen sum and number of taxa. The pollen sum is a measure of the statistical confidence. The number of taxa can be used for rarefaction analysis, which is a tool for estimating biotic diversity (Birks et al. 1988, Ch. 5.13).

D. Charcoal diagram, charcoal fragments with size >25 μm.

For some sites with uniform sediment accumulation rate, pollen analyses have been made on volumetric samples. In these cases it has been possible to calculate the pollen accumulation rate (PAR) and to construct *pollen influx or PAR diagrams* (see Berglund and Ralska-Jasiewiczowa 1986). These diagrams are not published here, but they are used in the interpretation because they are a tool for understanding the tree population size (cf. Davis 1969, Delcourt and Delcourt

1987, Bennett and Lamb 1988). In the synthetic correlation of Ch 5.13 influx curves are presented for tree pollen groups and herb pollen at five sites.

Cumulative diagrams covering the apophytic-anthropogenic plant groups (2–5 above) can be used for a schematic correlation between sites (Berglund and Ralska-Jasiewiczowa 1986). We name these *synthetic human impact diagrams* and apply this correlation technique to data from five sites in Ch. 5.13. A similar correlation of human impact intensity can be made with the help of canonical correspondence analysis (Birks et al. 1988). This has been applied to some sites and the results will be published elsewhere (Birks et al. unpubl.).

Numerical technique is available for classification and ordination of pollen data (Birks and Berglund 1979, Birks and Gordon 1985, Birks 1986b). Our proposed *pollen zones* are, however, conventionally defined pollen assemblage zones (PAZ), fully described in our original publications. In the survey diagrams published in Ch. 4 boundaries are indicated between time periods (approximate PAZ) described in the text under separate headings. To some extent it is possible to calibrate pollen spectra for a better understanding of the forest composition (Prentice 1986a). This technique has only been applied here in a regional survey diagram reproduced in Fig. 5.13:1).

Plant macrofossil analysis. This has been applied particularly for tracing displacement of shore vegetation as a consequence of water-level fluctuations (Krageholmssjön, Bjäresjösjön, Kumlan). Particularly in the case of Bjäresjösjön, these analyses have given important information on vegetation changes in the lake as well as in the lake surroundings. A conventional procedure has been applied (cf. Gaillard 1984).

Charred seeds. Organic material from occupation layers at archaeological sites often contains preserved charred seeds which give additional information on cultivated plants, weeds, etc. Particularly excavations in the Köpinge area and in Bjäresjö have revealed rich material.

This is published in the archaeological monograph by Engelmark and Hjelmquist (1991) but the results have been available for this synthesis.

Charcoal. Charcoal from hearths etc. at archaeological sites reveals information on the local tree-shrub vegetation used for fuel. Rich material was obtained particularly from the Köpinge area covering the time-span 1200 BC to AD 1200. This is published in the archaeological monograph (Bartholin and Berglund 1991), but the results have been available for the discussion in Ch. 4.1.2. The methodological problems are also discussed in an earlier case study (Bartholin et al. 1981).

Diatom analysis. Diatom stratigraphy reveals important information on lake trophy and its changes through time, which in turn can be related to land use, deforestation, soil erosion, and so on. Particularly in brackish water lagoons, water-level changes can be traced from the diatom stratigraphy. Diatom analysis has been applied in the main cores in Krageholmssjön, Bjäresjösjön, Bussjösjön, and Herrestads Mosse. The conventional procedure includes also numerical treatment of diatom data (Håkansson 1984, 1989, Håkansson and Kolstrup 1987).

Representativeness

The broad vegetation/landscape development in the Ystad area is regarded as representative of southernmost Sweden, particularly the south-west of Scania. The project area belongs to the palaeoecological type region "Plains and coast of Scania-Blekinge" according to the subdivision of project IGCP 158 (Ralska-Jasiewiczowa 1986, Digerfeldt in press). There are also great similarities with the corresponding type region of the Danish islands (Andersen et al. 1983). However, we may expect local differences in human impact, particularly between southern Scania and the areas in the north, northern Scania-Blekinge-Småland, which may be looked upon as a marginal settlement area.

2.2 Plant ecology

E. Gunilla A. Olsson

The general aim of the plant-ecological efforts in this project is to investigate the role of human utilization of ecological resources in the development and change of the landscape.

Since the investigation period is 6000 years it is easy to realize that such an attempt will suffer from using very different sources with varying reliability and varying accuracy.

The work can be divided into five main parts:

1) Modelling agroecosystems:
   human nutrition (protein)
   plant nutrients
2) Landscape-ecological reconstructions: prehistoric and historic time
3) Vegetation mapping: today's landscape
4) Special study of grassland changes during the 20th century:
   area
   plant species composition
5) Experimental investigation of productivity and nutrient flow in simulated three-course crop rotation

In the following the methods used in *modelling agroecosystems* (1) will be described. *The landscape-ecological reconstructions* (2) and the *mapping of recent vegetation* (3), are dealt with in Ch. 2.8. The *grassland study* (4) is presented in Ch. 5.12 where methods are also described. *The experimental works on crop production* (5) will be published elsewhere.

General hypothesis

The studies of agro-ecosystems in the context of the Ystad project concern two main questions:

1. To what extent can expansion and decline of human settlements and farming activity be related to the use of environmental resources? Specifically, are periods of decline in agricultural intensity or land-use reorganization a result of exceeding the carrying capacity of the area?

2. Are differences in settlement patterns and farming practices between different habitation sites or villages related to differing availability of ecological resources?

Models were constructed for different prehistoric agro-ecosystems – see below. Further, the 18th century agro-ecosystems were modelled in terms of plant nutrients.

The aims of the prehistoric models

For the prehistoric periods the aim is to reconstruct land use and the cultural landscape around habitation sites in the key areas and to study changes between different prehistoric periods. To achieve this it is neccessary to model agricultural production, cropping routines, stock keeping, etc.

Hypotheses for the prehistoric models

1) The prehistoric agro-ecosystems were organized as subsistence economies. All material and resources were derived from the same ecosystem, or withdrawn from neighbouring systems. The human population collected food items, produced crops, and maintained stock primarily for their own nutrition and survival.

2) An alternative hypothesis used is that the human population maintained surplus stock for social, religious, or other reasons.

3) Cultivation was performed on very small areas. It is assumed that in the Neolithic a form of agro-forestry was practised where trees and coppices were kept in combination with the crops (Göransson 1988). Coppicing is associated with a dieback of parts of the root systems of the trees. The crops could benefit from released plant nutrients from the decaying tree roots.

4) Cultivation was practised without manure until the beginning of the Iron Age. There is some evidence of manuring from the Late Bronze Age in the Ystad area (Engelmark and Hjelmquist 1991). This consists of small amounts of seeds from nitrophilous weeds found among storage finds of cereals. This is, however, rather weak evidence for the practice of manuring the arable. These weeds could just as well have grown in the cattle enclosures or settlement yards.

Assumptions and sources for the prehistoric models

1) The *number of humans* living in each settlement site was estimated from the archaeological finds (Lepiksaar 1969, Strömberg 1975, Larsson and Larsson 1986, M.

Larsson 1987, Tesch 1988, Callmer 1990, M. Larsson in press, Olausson et al. unpubl., Tesch in press, Ch. 3.2, 3.3, 3.5, 4.1.2, 4.1.3 and 4.1.4). For the Neolithic and the Bronze Age the mean human population size per settlement site was estimated at 5 adults and 5 children. For the Viking Age the mean human population per settlement site was estimated at approx. 40 individuals, children included – see sources listed above.

2) The *composition of the human diet* was estimated from the archaeological and palaeoecological data (e.g. animal bone remains, fish bones, nuts, cereal remains etc.) (Larsson 1984, Larsson and Larsson 1986, Larsson 1987, Engelmark and Hjelmquist 1991, Olausson et al. unpubl., Engelmark unpubl.). From the composition of finds of animal skeletal material, crop remains, and other palaeoecological evidence a percentage was estimated (subjectively) for the division between animal and plant food.

3) The *protein content of the food intake,* and the *human protein need* was extracted from Anon. (1978, 1983).

4) The *species composition of the domestic stock* and of the animals consumed was adjusted according to the species composition of bone remains at the excavated sites or at analogous sites (Lepiksaar 1969, Strömberg 1975, Larsson and Larsson 1986, Larsson 1987, Callmer 1991, M. Larsson 1991, Olausson 1991, Tesch 1991).

5) The *populations of domestic stock* were calculated to admit the yearly *subsistence consumption* of the human population according to the animal percentage from (2) – and with the necessary yearly stock recruitment taken into consideration. For the Bronze Age and Iron Age alternative calculation for enlarged numbers of stock (according to hypothesis 2 above) were also made.

6) The *size of the different domestic animals* during prehistoric time was interpreted as multiples of animal sizes in the 18th century. Figures for the 18th century were extracted from the following sources: Noring (1836, 1842), Åkerblom (1891), Nannesson (1914), Husdjursskötsel (1919), Seebohm (1927). The mean weight for cows during the early 18th century (south Scandinavia; England, north France, north Germany) was estimated to ca. 250 kg live weight (Slicher van Bath 1963). For the different domestic animals during all the prehistoric periods the following mean weights were used: cattle 167 kg, horse 240 kg, pig 55 kg, and sheep 33 kg. These sizes were estimated to coincide with the interpretations of the skeletal bone finds in various parts of northern Europe from prehistoric time (Barker 1985).

7) The *size of the used fodder areas* was calculated from:
* vegetation types in the area interpreted from pollen evidence (Ch. 4.2.2, Håkansson and Kolstrup 1987, Gaillard and Berglund 1988, Hjelmroos unpubl.).

* geological features extracted from recent geological maps from the area (Daniel 1986, 1989).
 Soil and bedrock qualities give different potential for plant biomass productivity.
* estimates of the biomass production of the plant communities
* the number of stock according to (5) above
* estimates of fodder needs expressed in energy and protein units found in ecological literature for the different types of livestock (Nannesson 1914, Van Dyne and Meyer 1964, Cook et al. 1967, Coupland 1979, Steen 1980, Van Dyne et al. 1980, Ring et al. 1985, Pratt et al. 1986). The estimated sizes of the animals were taken into consideration (cf. (6) above).

8) The *seed:yield* relationship was assessed from knowledge of the soil quality at the different sites, the implements used, and whether or not manure was used, and related to literature on recent experimental cultivation and exhaustion experiments (Garner and Dyke 1969, Cooke 1976, Dam Kofoed and Nemming 1976, Lüning and Meurers-Balke 1980, Olsson unpubl.). Several alternative calculations of seed:yield are presented and discussed for the different sites.

9) The *cropping system and fallow regime* were assumed to include cultivation on abandoned habitation sites and thus benefiting from accumulated organic material and nutrients, especially phosphorus. For the Neolithic and the Bronze Age a cultivation period of 2–5 years on the same site was assumed. Further, it was assumed that the arable sites shifted and each site was fallowed up to 50 years. These assumptions are founded upon the indications that (1) the settlements were moved at ca 30–50 year intervals (Ch. 3.2, 3.3, 4.1.3, 4.1.4) and (2) there was surplus of non-forested land to use for cultivation (Ch. 4.1.3, 4.1.4).

10) The *species composition of cultivated crops* was determined from the occurrence of different palaeobotanical remains, crop remains, and traces of seeds (Hjelmquist 1955, Hjelmroos 1985, Engelmark 1989, Engelmark and Viklund 1991, Engelmark and Hjelmqvist 1991, Hjelmquist unpubl.).

11) The *botanical composition of the environments* surrounding the habitation sites used for the reconstruction of the landscape on the maps in Ch.4.1 and Ch.4.2, was derived from the pollen diagram from the areas (Ch. 4.1.2, Ch. 4.2.2, Hjelmroos 1985, Håkansson and Kolstrup 1987, Gaillard and Berglund 1988, Hjelmroos unpubl.), and from the charcoal finds (Bartholin and Berglund 1991).

Sources for the medieval agro-ecosystems

Data about agricultural production during the Middle Ages in the area are very sparse. The most important written source for this area is the *Landeboken* (1540) reviewing the sizes and the harvest on all the land belonging to the archbishop in Lund, including the investigation area. This survey embraces the years around 1540.

In the absence of other relevant sources on the medieval agro-ecosystems in the Ystad area, analogies are drawn with British medieval agriculture using primarily the review by Titow (1972) on 52 estates in England. Comparisons on yields and livestock numbers are made with the Ystad area (Ch. 5.3).

Aims of the 18th century agro-ecosystem models

The aim of the 18th century agro-ecosystem models is to investigate the degree of exploitation of the ecosystem by focusing on the flow of the major plant nutrients, nitrogen and phosphorus.

Hypotheses for the 18th century agro-ecosystem models

1) By quantifying the flows of the major plant nutrients in the arable ecosystem it is possible to estimate conditions of productivity and the degree of vulnerability of the agro-ecosystem.

2) The productivity of the arable fields is a function of the natural soil fertility, the slope of the field, the actual addition of manure and fertilizers, the cropping regime practised, and the genetic constitution of the seed crops used.

Assumptions and sources for modelling the 18th century agro-ecosystems

1) The land survey maps from the period ca. 1690–1800 were the primary source for information on *land-use categories* and various information on *soil fertility, content of humus*, etc., as well as *botanical information* about the fodder areas.

2) Information on *crop diversity, cropping system*, and *fallow regime* was also obtained from the land survey maps.

3) *Yield of cereals, other crops*, and *hay harvests* was to a great extent also extracted from the same land survey maps. Since this type of information was often generalized by the surveyor, these data were checked by using other sources like tithe and tax tables contemporary with the land survey maps (Ch. 4.1.7.1, 4.2.7.1, 4.4.7.1, Persson 1986a,b 1987, 1988.

4) The data on *number of humans* living at the different sites (villages, hamlets, and single farms) were obtained from written records (Swedish *mantalslängder*). These data sets were investigated by A. Persson (Ch. 4.1.7.1, 4.2.7.1, 4.4.7.1, Persson 1987, 1988).

5) The *number of livestock* was obtained from written records (Swedish *boskapslängder*) (Persson 1987, 1988, Olsson and Persson 1987, Olsson 1988b).

6) The *size of the livestock* appropriate for the 18th century was obtained by combining livestock data from the following sources: Noring (1836, 1842), Åkerblom (1891), Nannesson (1914), Husdjursskötsel (1919), Seebohm (1927), Slicher van Bath (1963). The mean weight of cows (South Scandinavia) during the first half of the 18th century was estimated at 250 kg live weight, amounting to 50% of the values for Swedish cows in the 1950s (Johansson 1953, cf. Olsson 1988a). For the other domestic animals the mean figures for Scandinavian standards during the 1950s (Johannesson 1953) were used with the following weight reductions: horse 50%, pig 50% and sheep 20%.

7) When calculating the *amount of manure* produced by the livestock and used for the arable the following assumptions were used:
* stabling time is reckoned to be five months
* the 18th century livestock produced the same amount of manure per unit time as modern cattle. The smaller size of the 18th century livestock (see (6) above) was taken into account.

8) *Nutrient contents in manure (nitrogen and phosphorus)* were obtained from Bengtsson and Kristiansson (1955), Jansson (1956), Grahn and Hansson (1959). Those sources were chosen in favour of more recent ones with the intention of obtaining conformity with the data on livestock sizes from that period (Johansson 1953 – see above).

9) *Nutrient losses in manure* due to inefficient storage and distribution were assessed at 50% for nitrogen and 20% for phosphorus (cf. Bonsdorff 1939, Schrøder 1985).

10) *Contents of nutrients* (nitrogen and phosphorus) *in the biomass* – seeds, crops, and hays – were extracted from modern plant-ecological literature (Mengel and Kirby 1981, Fitter and Hay 1983).

The nutrient content of seaweed used as fertilizer was taken from Bonsdorff (1939).

Sources for the periods after 1800

For the periods since 1800 there is a variety of agricultural statistics. There is, however, a general lack of data on the level of the individual farm or village. Especially for the latest periods of the 20th century it is impossible to obtain detailed information on the level of single farms or villages. There are several reasons for this; for example, the borders of both the parishes and hundreds have changed during the 19th and 20th centuries. The sources of statistics for this period were: Hushållnings-sällskapet i Malmöhus län: Agricultural statistics; unpublished data from the years 1886, 1892, 1898; Leufvén 1914, Germundsson and Persson 1987, Statistiska Centralbyrån 1917, 1927, 1937, 1951, 1961, 1971, 1986, 1987a,b, 1988, 1989, 1990a,b.

The 19th century land survey maps were used, but they are generally less detailed than the older maps. The military reconnaissance map from 1814 was used for interpretation of landscape features and land use.

Other written sources on practices and yield were Zachrison 1914, 1920, 1922, Nilsson 1920, Nannesson 1920, Weibull 1923.

Methods and methodological problems

The prehistoric agro-ecosystems

By combining the human nutrition need in terms of protein consumption and the assumptions about diet composition listed above, the magnitude of resource areas was calculated. These areas were used for discussing the degree of exploitation of the natural ecosystems. They were also used for a comparison with the interpreted vegetation areas from the pollen diagrams.

For all these early periods it is certainly a problem that the work had to be founded on several untestable assumptions. However, the methods presented and used in this work give a basis for further discussion and for the formulation of new hypotheses.

The medieval period

No model is presented, but the scanty data are used to make comparisons with older periods on a general level.

The 18th century agro-ecosystems

For the agro-ecosystems of the 18th century a method was used to quantify the recycling of plant nutrients within the ecosystems. By quantifying the nutrient input from manure, fertilizers and seed, and the outputs via crop harvest, it is possible to calculate a nutrient quotient for the arable, the Nutrient Use Efficiency, (NUE). This method was first developed for use on individual plant level (Chapin 1980), but is used here on the arable field level. It is thus defined as: *the amount of above-ground crop biomass produced divided by the amount of mineral nutrient supplied in a given period.* For any mineral nutrient (n) NUE can be written:

$$NUE = \frac{OUTPUT-(n)}{INPUT-(n)}$$

Output: nutrient content in crop biomass recorded as harvest. *Input:* nutrient content in supplied manure and fertilizers.

The application to historical production data requires relatively detailed data and is described by Olsson (1988a) and applied in Olsson (1988b, Dodgshon and Olsson 1988, and in Ch. 4.1.7.3, 4.2.7.3, 4.4.7.2, 5.3).

This method allows the quantification of nutrient flows also for areas where no field sampling is possible. However, several difficulties arise; the interpretations of the resulting quotients must involve considerations of the importance of other variables like soil fertility, regional and local climate, exposition of the actual fields, crops grown, and the abundance of weeds.

The agro-ecosystems after ca. 1800

Since it was impossible to obtain detailed data on discrete units for this period, no model calculation of NUE-quotient is presented. Important input sources of nutrients and other crucial variables are presented in time-abundance diagrams. The yields of crops and major land-use categories are given and related to conditions during other periods in this area.

2.3 Archaeology

Lars Larsson

Hypotheses

From the very beginning, the intention in the project was to favour Quaternary biology among the interacting sciences, since its demands and preferences were the most important when choosing the area of study. This has to be borne in mind as much as the adjustment of the archaeological hypotheses to the general aim of the project – the interaction between society and the environment. The precondition of the project and the formulation of hypotheses were influenced by a scientific milieu with a marked emphasis on the interaction of archaeology and natural sciences – the impact of New Archaeology (Stjernquist 1991, Larsson 1991a).

The basis used for the division and direction of the archaeological studies was the model by Berglund (1969) based on the variation of elements proving or indicating human impact studied in the pollen diagrams. In the model, Berglund identified four maxima of human impact during the prehistoric period (Ch. 1).

The main aim of the studies was to find out in what sense these four maxima could be related to the results of archaeological research. Which factors, both natural and human, could be of importance in these changes? Did the variation of human impact really coincide with changes that could be recorded in the archaeological analyses? If so – did they actually prove major changes in society? If not – how could discrepancies between Quaternary biological and archaeological results be interpreted?

In understanding prehistoric change, it was also important to study the intervals of regression or stagnation of human impact according to the pollen diagrams with the same set of questions as above.

Given the period to be studied, the archaeological problem was therefore to examine different processes that were supposed to be more or less contemporaneous with the four stages of human impact on the landscape:

Stage 1. The early part of the Neolithic. There has been intense discussion about the process by which agriculture was introduced and the role of cattle-breeding and agriculture in relation to traditional modes of subsistence like fishing and gathering. This discussion has produced a series of models.

Stage 2. The late part of the Neolithic. The interaction of different cultural groups as an explanation for the establishment of a cultural landscape, dominating the environment, has been of main interest in the discussion.

Stage 3. The Roman Iron Age. The establishment of the farmstead and the formation of infield/outland systems based on more stable agriculture with manuring has been a major topic, as has the interaction between single farms.

Stage 4. The Viking Age. Interest has centred in part on the development of settlement to villages and especially on the establishment of permanent, historically known settlements combined with fields in rotation systems.

It was also of vital importance to analyse contemporaneous societies in different settings, which means a special emphasis on the study of different environments. Particular attention was therefore paid to the study of the distribution, formation, and function of the settlement on the coastal plain in relation to the inner hummocky zone.

Starting from this theoretical base, the prehistory of the area was to be studied in four different subprojects:

B1. Early Neolithic society.
B2. Society during a later part of the Neolithic.
B3. Society during the Bronze Age, Iron Age, and Early Middle Ages in a coastal area.
B4. Late Iron Age and early medieval society in an inland area.

Source material

At the start of the project, the factual base for what we knew about prehistoric settlements was extremely varied. Detailed investigations had already been carried out during the 1970s in one of the areas selected for special study, Stora Köpinge (Tesch 1983). Also available were the findings of a number of studies of the coastal strip in the eastern part of the project area (Lönnberg 1929, Strömberg 1968, Widholm 1974, Lindahl 1976, Wihlborg 1976, Nilsson 1983, Olausson 1986). This clearly identifies the part of the project area which had been investigated, and also the area which had previously attracted the researchers' interest. Archaeological work in the west of the project area had previously been of a highly sporadic nature, and had been restricted exclusively to the narrow, sandy coastal strip (Hansen 1924, Stjernquist 1948, Nagmér 1983). In contrast, the hinterland which dominates the west and north of the project area had attracted very little archaeological interest. This clear difference between the coastal plain and the hinterland has also had a highly noticeable influence on the approach adopted towards, and the conditions affecting archaeological fieldwork.

A contribution in the form of the comprehensive results of a comparative survey has been made through the investigations over many years in the parishes to the east of Stora Köpinge, the findings of which have recently been published (Strömberg 1982). The wealth of published material from the Hagestad project and projects in other parts of south-east Scania (Strömberg 1980 with bibliography, 1985, 1988) has also been of considerable help in processing the material from the project area. The intensified excavation work on Bornholm has also been a source of inspiration, and has provided important comparative material (Nielsen and Nielsen 1985, Kempfer-Jørgensen and Watt 1985, Watt 1985a,b).

Methods and representativeness

As a part of the subsidiary project on the oldest settlement, a considerable amount of effort was devoted to the compilation of an *inventory,* both of the surface area and of the collections of archaeological finds. The project area was too large for a detailed investigation of the whole area, and consequently certain areas were examined intensively in order to reveal traces of prehistoric settlements through detailed inspection of the ground. An examination of existing collections has revealed variations in the interest in collecting prehistoric remains. There are two major private collections, which date from the turn of the century and are now part of museum collections; they provide a distorted view of the inventory of archaeological finds in the project area.

Visits to farms have also revealed that a considerable number of archaeological finds can no longer be traced, having been dispersed by inheritance, by change of ownership, or by being sold. The difference between the hinterland and the sandy coastal plain was less significant in this respect.

Another factor is of critical significance, however – the effect of erosion on the prospect of being able to find prehistoric relics. Cultivation of the soil has caused relatively little damage in the sandy coastal plain. No purpose is served here by ploughing to a greater depth. On the other hand, cultivation of the soil combined with the effect of the surface water has brought about distinct changes in the topography in the more undulating parts of the area. Material from the top of the hill is brought down to the foot of the hill. Bearing in mind that the settlements were very often situated on high ground, this change means that the layers containing finds and features from buildings have been disturbed. Moreover,

material is carried down the slope, where it might cover prehistoric remains with thick erosion layers. The result could be that other locations than settlements at the top of hills cannot be identified during field reconnaissance. This leads to a major source of error, which influences the criteria of representativeness, in the study of the extent of prehistoric remains.

It is also clear that traces of prehistoric settlements from different periods differ with regard to identification. Flint waste, which is practically indestructible, means that remains from Stone Age settlements and, to a certain extent also, from Bronze Age settlements are considerably easier to observe than pottery or other fragile materials (Larsson 1986a). Fire-cracked stones or sooty deposits, which often mark the sites of settlements dating from the Late Bronze Age and the Iron Age, are more difficult to observe, and consequently require that field inspection is carried out under quite specific conditions, such as very humid weather (Olausson 1988b). This can result in a situation in which different inventories must be combined to avoid distortion in the allocation of the localized remains of settlements to their periods.

One practical difficulty which was encountered in conjunction both with the compilation of an inventory in the field and with the excavations is the very rapid crop rotation sequence which is now practised, in which many of the fields are without some form of crop cover only for short periods of the year. The problem of true representativeness is also associated with the fact that large areas within certain sites containing a wealth of prehistoric relics in the form of surface finds were not accessible because of their situation on pasture land or in military training areas with a total grass cover.

Other methods were applied to locate the prehistoric settlements, such as phosphate mapping. Here the results of large-scale testing carried out during the 1930s for the purpose of investigating fields suitable for growing sugar-beet in large parts of Scania were of importance (Arrhenius 1934), but new testing was also conducted. Old maps were consulted in an attempt not only to identify fossil settlement elements, but also to identify place-names which can reflect a former function which has since disappeared. Aerial photography was also used to a limited extent to identify prehistoric relics situated below the surface.

A number of different factors have also influenced the layout of the *excavations.* The clayey surface layers of the hinterland pose problems for attempts to demonstrate the effects of prehistoric man, since the boundaries of pits, for instance, do not appear as clearly as they do in light sand. This meant that various solutions had to be adopted with regard to the preparations for and the execution of the excavations. It thus proved more difficult to undertake the uncovering of large, continuous areas at the surface when excavating the path of the plough in the more undulating clay soils, than it was on the gently undulating sandy plain. Other practical operations, such as sifting, were almost impossible when investigating the clay soils. These differences have clearly marked effects on the interpretation of the settlement pattern.

In view of the comparatively limited areas in which the site density was high, what often happens is that finds from several different periods are made in one and the same area, with regard to both settlements and burial sites. The result is that more recent buildings have influenced older remains to a greater or lesser degree and in so doing made the job of processing the material more difficult. Not least, this entails consid-

erable uncertainty in the interpretation of the shape of house remains or the selection of charred material for further analysis.

Macrofossil and osteological analysis of both animal and human material was of great importance for the study of prehistoric settlement.

To sum up, the study of the prehistoric settlements has revealed considerable differences in the procedures to be adopted in the undulating hinterland and in the sandy coastal plain.

For a more detailed insight into the aspects mentioned above see Callmer et al. 1991.

2.4 Medieval archaeology

Hans Andersson

Research problems

Within the framework of the project's hypotheses and in collaboration with other scholars working on the project, medieval archaeologists have devoted particular attention to three question. The first concerns the *villages* – their chronology and utilization of resources; the second is the *manors*, their chronology, character, and location in the landscape; and the third is the potential use of the *churches* as sources in the study of the development of the cultural landscape.

Work during the first part of the project was primarily concentrated on the villages. The aim was above all to shed light on the medieval expansion of settlement in the inner hummocky landscape through studies of Baldringe parish (Christophersen 1984). The scope was then broadened by the selection of a coastal site, Lilla Tvären, as a contrast to Baldringe. Both places were the scene of archaeological investigations of surviving occupation layers (Billberg 1989).

In the later stages of the project the emphasis was shifted to the question of how medieval archaeology can contribute to a discussion of resource utilization in the landscape. Of great importance here has been the aim of tying findings to the results obtained by natural scientists for the project as a whole. The opportunity of reconstructing the landscape and analysing its production potential provides a broader basis for discussing the rural economy than would previously have been possible (see, for example, Billberg 1989). For medieval archaeologists in Sweden this is a relatively new research trend. It would have been valuable if it had been possible to supplement this with studies of the character of settlement. This, however, has not been possible within the framework of the project. The connection between the shape of settlement and the orientation of production is an interesting and important question. For medieval archaeologists the project has laid the foundation for further progress, as regards both developing methods and testing them on empirical material (Billberg 1989).

The medieval manors have also been studied (Reisnert 1989a, Skansjö et al. 1989, Riddersporre 1989b), although there have been few excavations. Bjäresjö, which was studied as part of subproject B4, represents the Early and High Middle Ages, while Bjersjöholm and Krageholm represent the Late Middle Ages (Skansjö et al. 1989, Callmer 1991b). They have been a valuable supplement to the written sources and the retrogressive analysis of the oldest maps (Skansjö et al. 1989, Riddersporre 1989). In broad terms, it is above all localization and datings that have been clarified in this way. What has not emerged, except at Bjäresjö, is the concrete structure and character of the settlement, which would naturally have been of great significance, not least when one wishes to discuss how it related to other settlement. Important archaeological work remains to be done.

The church investigation came relatively late in the course of the project. It had as its general starting-point the view that a detailed knowledge of the churches and the ecclesiastical landscape would provide a background for answering questions about stability and change in the surrounding cultural landscape, as well as resource utilization and communications.

Source material and methods

The main material used in medieval archaeology is remains of settlement and occupation layers under ground, along with wholly or partially preserved buildings from the period under study. The chronological demarcation of the medieval period is the same here as the traditional definition of the Middle Ages in Nordic historiography, namely from ca. 1050 to the Reformation in the first decades of the 16th century. Besides conventional archaeological methods, the arsenal of the medieval archaeologist also includes the study of surviving buildings, with the aim of illuminating their structural and chronological development. Unlike the archaeology of earlier periods, there is also often the possibility of interpreting the archaeological material in the light of or in combination with written evidence. The latter is rather limited for the Early Middle Ages, but it increases during the course of the period. This naturally broadens the scope of the archaeologist's interpretations, making it possible to incorporate them into a historical framework. At the same time, however, it involves methodological problems.

For the study of villages and manors, traditional archaeological methods have mostly been used. No sites have been totally excavated, but only small, selected parts of medieval settlement sites. The aim has been to obtain answers to questions about chronology and the nature and content of occupation layers. The interest in ecological and functional aspects has meant that special emphasis has been placed on the collection of macro-fossils and the osteological material found in the occupation layers (Lindroth 1987, Billberg 1989). The archaeological results have then been combined in various ways with written sources and the oldest land survey maps, the evidence of the latter being subjected to retrogressive analysis (Riddersporre 1989).

The first part of the study comprised detailed documentation of every surviving church building, which included an analysis of the internal relative chronology of each church (Anglert 1989a, Sundnér 1989a). This material has then formed the basis of thematic studies (Anglert 1989b, Sundnér 1989b, Wienberg 1989, Ch. 5.7).

This type of regional study is unusual. In Sweden Bonnier (1987) has considered the medieval churches of Uppland with these questions in mind. For the area around Ribe in Jutland a project has also been started with the aim of providing a regional overview of the churches (Nyborg 1986). What is special about the church investigation of the Ystad Project is the close connection to the cultural landscape, along with the fact that it covers the entire Middle Ages and in some cases even comes down to modern times (Wienberg 1989). The longer perspective makes it easier to see what the churches represent at different times, although we must naturally be aware that the churches, like all other archaeological material, must be interpreted in their chronological context.

It goes without saying that there are problems of source criticism as regards the use of church material from the Ystad area. Not all the churches have been accessible for the detailed investigations which they deserve. For some churches there is little material; some have been demolished and some have not been archaeologically investigated (Sundnér 1989a, Wienberg 1989).

Moreover, problems of methodology and analysis still confront the researcher who wants to use churches to shed light on the questions of interest here. The churches can be said to express several different factors. They were built for the exercise of the Christian religion. The organization of space and the furnishings are a physical expression of the liturgy. Changes in the latter also affect the former. Yet the church building also marks the position of the church-builder – his economic resources and his claims to power. This is also reflected in the building in various ways. Changes in responsibility for the church, from one powerful family to a parish collective, also have consequences for the shape of the church.

The connection of a church with its surroundings can be seen in microperspective in the archaeological mate-rial, as has happened in Bjäresjö, but it can also be analysed through retrogressive studies of old land survey maps. In a macroperspective, the siting of the church in relation to landscape types and cultivation potential can bear indirect witness to the economy and demography of the area (Wienberg 1986a, 1989, Sundnér 1989a).

A new feature of the church studies is the *systematic* use of dendrochronological datings. Dendrochronologists have worked in parallel with the archaeologists on documentation. This work has given not only methodological experience of great value but also firm landmarks for the chronology of the churches (Bartholin 1989, Sundnér 1989a,b, Wienberg 1989), and the material also gives a chance to assess the type of woodland from which trees were taken.

Representativeness

The investigations carried out by medieval archaeologists show various degrees of representativeness in relation to the area studied by the Ystad Project. Although the villages of Baldringe and Lilla Tvären represent different parts of the project area, it would be too bold to draw conclusions from them regarding the development of settlement in the entire area. There is nevertheless the possibility of using comparisons and analogies with other areas to formulate hypotheses about what happened in the early phase in the Ystad area (Grøngaard Jeppesen 1981, Skansjö 1983).

The same can be said of the investigations of the manors. The archaeological excavations are of primary relevance to the investigated objects themselves, although interest was confined to certain specific questions, primarily concerning the chronology and type of structure. Nevertheless, the studies of the manors have an indirect general relevance since it is possible to link the findings with other evidence and use the resulting material for a more general discussion of the area as a whole.

The churches can be said to be representative, since they cover the entire project area. There are complications, however: although most medieval churches have survived, some churches or chapels have disappeared, while others have been demolished and replaced by new buildings. This deficiency has been compensated for as far as possible through old written sources and earlier archaeological investigations.

Note: The medieval archaeological investigations have been published in greater detail in a separate volume edited by Andersson and Anglert (1989).

2.5 History

Anders Persson and Sten Skansjö

Hypotheses and research problems

The historians working on the Ystad Project have examined the general problems studied by the project for the period from the Middle Ages to about 1800. In practice this has meant finding written sources which provide information about settlement, land use, conditions of ownership, population, and so on. The sources have been analysed and interpreted, and the results have been presented in reports which have served as a basis for further historical study and as additional information for scholars working in neighbouring disciplines.

The interest in *settlement* has been concentrated on problems associated with the historically known settlement pattern in the Ystad area from the High and Late Middle Ages (the date of the oldest written sources) to the time just before the enclosures: to what extent was the area a pure village landscape, to what extent was there dispersed settlement, and how has the settlement pattern changed through time? Other questions concern the variation in the number of farms per village and the cultivation resources of the farms. How is the distribution of the early historically known settlement related to the varying agrogeological conditions – the sandy coastal plain, the outer and inner hummocky landscape, and the Romeleåsen ridge? For early historical times the study of these problems has been closely connected with the work done in medieval archaeology (investigations of villages) and human geography (retrogressive analysis of maps).

It quickly became apparent that the estate-dominated character of the area required a historical investigation of the roots of this phenomenon. This led to an interest in *ownership*, in so far as this could be attested for the earliest period, not least through written sources. Attention was also devoted to the way ownership conditions changed through time, and to the establishment of manors and castles. Attempts to explain development here have adduced certain societal conditions as fundamental factors behind the observed changes in the cultural landscape: demographic, social, political, and economic conditions, and the way these have changed through time.

The analysis and interpretation of written sources have also contributed information to the project's central questions about *changes in the cultural landscape*. This is possible even for the medieval period, despite the fragmentary sources, but the gradually increasing quantity of source material from the post-medieval period allows a more qualified description of production trends, the distribution of arable and meadow, and changes in these aspects of land use. From about 1700 these observations can be supplemented by more detailed demographic information for the Ystad area.

Source material

The written sources from the Middle Ages consist of extant documents such as deeds of sale, wills, and other bills of transaction, as well as terriers and estate registers from various private and institutional landowners in the area. The aim has been to find and use all relevant published and unpublished medieval sources for the Ystad area. Many of these observations have been presented parish by parish (Skansjö 1987, 1988; for the problem of the manors see also Andersson and Anglert 1989). The availability of written sources for the Ystad area in the Middle Ages is rather poor compared with other parts of Scania, however, and the material we do have is unevenly spread over time and space. If one were restricted to the use of written sources there would be little material to answer the central questions posed by the project. On the other hand, the medieval conditions demonstrated by historical sources have often played an important role for further interdisciplinary analysis.

The post-medieval sources are of a different character, since there are many more fiscal records covering the entire area, not least after Sweden took over the administration of the province from Denmark following the Peace of Roskilde in 1658. Land survey documents, with detailed descriptions of the mapped villages, allow the calculation of the area of cultivated land, using a method which will be described below. Population before 1700 can only be estimated on the basis of the pattern and density of settlement, but from 1700 onwards we also have taxation assessment rolls, and after 1750 statistical sources.

Methods

The basis of the historians' method has been the traditional approach of tracing the relevant sources in archives and libraries, analysing how they came about, and critically assessing their value as sources and the information they provide in relation to the problems studied. A special feature of the studies has been to communicate observations to researchers from adjacent disciplines collaborating on the project, and similarly to evaluate and utilize the results obtained by scholars from other disciplines. This applies not least to the archaeological investigation of early manors in the area (Skansjö et al. 1989).

A special method using a database was applied to investigate changes in cultivated areas ca. 1570–1700: the quantities of seed corn used by a large number of vicarages around 1570 were converted into land areas, and the results were compared with the areas for these vicarages around 1700. The investigation reveals that a considerable part of the arable was cultivated at the expense of meadow in the plains villages as far back as the period 1570–1700. This result is also confirmed by palaeoecological findings.

For the investigation of the period ca. 1700–1810 in particular, a database was used in a special method to describe the structure of the cultural landscape and its changes through time, and the resources of the agrarian population in the form of grain supply. In the latter case the objective was to assess the area of productive land in relation to population, and also the productivity of the cultivated land. Population was calculated on the basis of taxation assessment rolls in combination with other demographic evidence and information from statistical sources, while the area of productive land was estimated with the aid of extant land survey maps with descriptions. When the thousands of field strips in the villages and manors were fed into the database, the details were sorted and totalled according to type of cereal, type of meadow, etc. It was thus possible to compile details of field areas, productive capacity, and production of grain and hay, for both entire villages and farms.

Data at farm and village level around 1700 were then generalized for larger regions, using a technique by which the data for three parishes – Stora Köpinge, Villie, and Bjäresjö, which represent three typical regions in the project area – were compared with a contemporary taxation assessment. The comparison revealed the area of land ploughed annually per *mantal* assessment unit for these parishes, and these data were

then generalized for other parishes. The calculations obtained by this model were checked for several other parishes for which detailed land survey documents survive for the same period. In the parishes where the model can be checked by such control data, the general correspondence between the results obtained by the two methods shows that the model is valid.

Population in 1700 was calculated on the basis of taxation assessment rolls, counted on the level of the individual. Allowances were made for omissions from the rolls, and population figures for each parish were obtained. The results were then compared with birth and death statistics for a number of parishes, and the figures were adjusted slightly. This longitudinal comparison indicates that the productive area per individual in the peasant population kept decreasing throughout the 18th century. This conclusion must be correct, as are the conclusions about population growth in the parishes, since population at the first measurement point (ca. 1700) has probably been estimated too high. Population figures from about 1750 onwards are based on more reliable sources. This methodological approach helps to confirm a development throughout the 18th century from a relatively egalitarian to a more socially differentiated society.

Representativeness

In several respects the historical investigations have covered the entire Ystad area. This applies, for example, to the collection and preliminary processing of medieval evidence (Skansjö 1987, 1988), and a description of the oldest known settlement pattern and ownership conditions in the area in the mid 17th century (Skansjö 1985). These studies have shown that the area is fairly representative of the village landscape of southern Scandinavia, and that there are interesting differences in settlement pattern and density which can be linked to variations in natural conditions. One distinctive feature – with roots going far back in history, but with an accentuated development from the 16th and 17th centuries – has proved to be the influence of noble ownership and the dominance of estates in the Ystad area.

2.6 Human geography

Nils Lewan

Hypotheses

Human geographers have looked upon the last two centuries as a period of continuous change, with agricultural modernization and commercialization, with the development of transportation facilities, and with relocation of both settlement and employment opportunities. However, this does not mean that there have not been periods of stagnation and fast expansion.

We also assume an increase of dependence on outside decisions, be they national policies or international influences.

Another hypothesis has been the importance of land ownership, and we have taken a great interest in the decisions of individual estate owners. Thus the organization of work through time inside separate farming units has also been of interest to us. For older periods, with a rural system largely characterized by villages and open fields, the earliest land survey documents attract our attention. This material provides detailed information about the 18th century landscape. It is, however, also a hypothesis that the material, through retrogressive analysis combined with archaeological and historical evidence, should make it possible to identify traces of the medieval and late prehistoric landscape and society.

From a geographer's view, the landscape itself also serves as a framework for putting indications from different sources into their context.

Sources and criticism of sources

The oldest land survey documents from the Ystad region were produced at the end of the 17th century, some forty years after the province of Scania was handed over from Denmark to Sweden. These are large-scale maps with detailed descriptions, making retrogressive analysis possible. In particular, field-names enable the spatial identification of other information, such as the records from the 16th century concerning land belonging to the archbishopric of Lund.

A specific problem is the difference in time between village surveys. While some villages were surveyed around 1700, material concerning others is 50 to 100 years younger. The implications for comparisons are obvious. Limitations are also set for the retrogressive analysis, with later documents generally having less potential.

From the late 18th century, sources concerning both landscape and population abound in Sweden. As to population, statistics are available from 1749 on. Enclosures produced numerous large-scale maps, while from the 1810s onwards topographic maps cover the whole area at intervals. A property map from around 1910 allows us to identify the owner and tenant of each plot of land. Vertical air photographs have been produced at intervals of five to ten years, beginning in 1938.

Information about agricultural practices becomes available from the beginning of the 19th century, in accounts from the county governors, while private archives have been made available concerning certain estates. The last few decades, of course, add another important source, namely the living memory of people.

Problems with sources still have to do with availability as well as interpretation. The first overall map, the reconnaissance map from 1812 to 1820 (*Skånska rekognosceringskartan*) was meant for military purposes, and over the area it varies considerably in quality. Apparently much of it was taken from older local and village maps, while other information was founded upon field investigation. Nevertheless, this map gives the first view of the whole area in one single map from one single period, thus also disclosing the incipient modernization of agriculture through the earliest enclosures (Nelson 1935, Lewan 1982, Emanuelsson and Bergendorff 1983).

Other sources, such as private archives and interviews, have their usual limitations, giving much local data, which are often difficult to put into a combined picture. Still another problem is that in some cases the first and thereby most interesting period of an innovation is not covered by sources still available. This applies to both under-drainage and artificially watered meadows.

Limitations in applicability

Looking upon the project area compared with other parts of the province of Scania, especially the plains, two of the most important differences seem to be the frequency of the three-field system and of large-scale land ownership.

The open field pattern in Scania varied quite a lot, with the three-field system dominating in the southeastern and south-western plains, while systems were more varied in other parts of the province. This applied also to some areas close to the Ystad region (cf. Campbell 1928).

Large landed estates have dominated the Ystad region much longer and to a greater extent than elsewhere, in Scania and in Sweden in general. Thus, the importance of the individual decision-maker should also be different from elsewhere. What has it meant to the landscape that some of the large landowners functioned as innovators, while others lagged behind?

Another factor which cannot be passed over is the location of the project area in relation to urban centres. Changes of landscape and settlement should not be the same during modern times around a relatively small and stagnant place like Ystad compared with regions closer to the much more thriving and expanding urban centres in western Scania, around Malmö, Lund, and Helsingborg. This factor might have played a certain role even earlier, but certainly in our century, and especially since World War II, changes in the countryside have become much more dependent upon the influence of urban centres of different size.

A cultural factor much discussed concerns a kind of time-lag between the western and eastern parts of Scania. This seems to apply not only to such modern phenomena as rural urbanization, but also to the relative frequency of horses used in agriculture, of thatched roofs and timber-framed houses still left, and of other local traditions. However, so far it has not been established whether such a time-lag existed even before our century. During the 19th century, some estate-owners were among the earliest innovators, and today we find some of the most thoroughly modernized large agricultural units close to Ystad.

Summing up, it is obvious that the project area for some periods and in some respects might be representative of much of the Scanian plain, while the region in other respects does not represent much more than itself. Discussing this, one should also be aware of the importance of scale. Looking upon the Ystad region from e.g. the Common Market point of view, it might not differ significantly from western Scania, while contemplated from the city of Malmö the area seems very different from the plain between Malmö and Lund, at least nowadays.

Finally, this also means that it is justified to apply different scales of study, from single farms and villages via parishes and hundreds to larger areas.

2.7 Mapping settlement patterns

Tomas Germundsson, Deborah Olausson and Mats Riddersporre

Mapping settlement patterns in a time perspective of several thousand years is a complicated task. Within the chronological framework of the Ystad project the settlement pattern has gone through several changes: from more or less dispersed prehistoric settlements, via the nucleated medieval villages and the first towns, to the complex pattern of today, a combination of dispersed settlements and various forms of agglomerations. The concept of settlement may, in this context, be confined to the sites where people lived or conducted some kind of long-term activity.

The identification of settlement sites is no major problem for later periods. From around 1700 and onwards contemporary maps give reliable information. For prehistoric times the identification is more vague, and reconstructions must to different extents rely on fragmented material. It is here perhaps more adequate to talk about a delimitation of settled areas. In spite of diverse source materials and differing possibilities of pin-pointing the exact settlement sites for different times, the ambition has been, in the maps for Ch. 3, to present the material in such a way that the information is graphically comparable between maps representing different times. The basic source material is presented on a more detailed scale in Ch. 4.

Mapping prehistoric activity

A number of methods and sources were used to obtain as much information as possible about the now-buried prehistoric cultural landscape (cf. Ch. 2.3). The information thus gained is the starting-point for interpretations leading to the picture of the prehistoric human use of the landscape shown on the maps in Ch. 3 and 4. Information was compiled from archives, museum collections, and fieldwork. In-depth study of land survey documents from the 18th century, as well as parish descriptions from the 18th and 19th centuries and aerial photographs, have also added information of value for the prehistoric periods. The sites of the historical villages were of course important starting-points for research concerning the development before the formation of these villages. A time-consuming but potentially useful source of information about the spatial components of prehistoric activity was field surveys. Interviews with local farmers also provided valuable information about possible areas to investigate further.

The features plotted on the maps in Ch. 4 and the more generalized picture shown in Ch. 3 thus represent information of varying reliability. Only after a stringent source-critical evaluation are we justified in claiming that "empty" areas shown on the maps were in fact devoid of settlement during the period in question. The uneven quality of the sources themselves, and the ways in which the scholars making the maps have interpreted the sources, inevitably means that strict comparability, whether synchronous or diachronous, cannot be assumed, however.

Due to the coarse nature of our archaeological dating, the prehistoric maps do not give snapshot pictures of moments in the past, but rather they represent an accumulation of a number of snapshots, each laid on top of the previous one. For the Neolithic and Late Bronze Age in particular, with their mobile settlement patterns, this may mean that the maps picture settlement as being denser than it in fact was at any one time in the past. To avoid this problem, a lower degree of magnification has been chosen for the Ch. 3 maps, allowing us to delimit settled areas rather than individual settlements. Research from field survey, map study, and excavation have further enabled us to suggest differences in the density of settlement within these areas for the periods in question.

Medieval and 18th century villages

In this period the prerequisites for mapping settlement are different. With few exceptions all settlements are now to be found in villages. Most of these villages are known from medieval written documents, and in the church-villages the medieval church is normally still standing. Owing to the stable settlement pattern in this period the village sites, both medieval and later, can be identified on large-scale cadastral maps from the 18th century. The 18th century villages will also represent the medieval villages on the maps in Ch. 4. There may, however, have occurred changes in the settlement structure within the villages, and the spatial extension of the village cores may have differed slightly.

Churches and manors

From the Early Middle Ages on, churches and manors are important features of the settlement picture. The identification of the churches that were in existence in the Middle Ages relies on research done primarily by medieval archaeologists (Anglert 1989, Bartholin 1989, Sundnér 1989a, cf. Ch. 5.7, Wienberg 1989). On the maps in Ch. 3, the churches that were in active use as parish churches in each period are presented. In the 14th century, as in other periods, there may also have been other churches or chapels in use. The maps in Ch. 4 include all standing churches that were in use during the period in question. These churches are identified on contemporary maps.

The manors presented in the Ch. 3 map of the medieval settlement are those that have produced contemporary documentary and/or archaeological evidence. In addition to these, there may have existed others, today known only from later sources or entirely unknown. The identification of the medieval manors, as well as their location, is based on a study of historical, archaeological, and geographical sources (Skansjö et al. 1989). For later periods the manors are well known through a variety of sources.

Settlement in the 19th and 20th centuries

Information about settlement in this period is based on official maps which cover the entire project area. On these maps, presented below, every settlement unit outside built-up areas is marked separately.

The basis for the settlement picture of the early 19th century is *Skånska rekognosceringskartan*, for which fieldwork was carried out 1812–20. This map, produced on a scale of 1:20,000, was made for military purposes, with information on terrain, settlement, and roads being of great importance (cf. Nelson 1935, Lewan 1982). It forms the basis for the mapping of settlement in both Ch. 3 and Ch. 4.

The settlement picture of the early 20th century is in Ch. 3 based on *Generalstabens topografiska karta* (scale 1:100,000) from 1907, which in turn is a revised edition of a map from 1864. The original map was produced with large-scale cadastral maps as a base, to which information from detailed field observations was added. The measuring up was made on a scale of 1:50,000. Every settlement unit is represented by a symbol on the finished map. The settlement mapping in Ch. 4 is for this period based on the economic map from 1915 (scale 1:20,000).

The picture of the settlement in the late 20th century is in Ch. 3 based on the modern topographic map (scale 1:50,000), most recently revised in 1982. It is produced with modern cartographic methods and gives an exact picture of the location of today's settlement. In the

more detailed presentation in Ch. 4 the late 20th century settlement is mapped with the modern economic map (version 1972/73) at a scale of 1:10,000 as a base.

In Ch. 4 the material has been used to show the categories of manor, farm, and smallholding with little or no arable land. Furthermore, such features as mills and rural industries have been marked. In Ch. 3 the number of settlement units within an imaginary squared net has been counted and represented by closer or thinner hatching. The size of the squares has been chosen to acquire a suitable level of information on a scale of 1:125,000. The number of settlement units within squares having a side-length of 600 m has been counted. More than 10 units per square is considered as relatively dense dispersion, whereas the interval 2–10 units is regarded as less dense. Also agglomerations in the form of villages and built-up areas have been marked. Consequently the maps in Ch. 3 show only a very summary picture of the spread of settlement in the project area at different points in time.

Boundaries and roads

The parish and hundred boundaries used for the maps in Ch. 3 are derived from the modern topographic map that was revised in 1982. Although the exact locations of boundaries certainly have changed over time – and still do – the modern boundaries have been included on all maps in Ch. 3 in order to facilitate comparison between maps representing different times. For the older periods, the boundaries of today have, of course, less or no contemporary significance.

On the maps covering the key study areas in Ch. 4, the boundaries are, from the 18th century onwards, taken from contemporary maps. For the early 19th century the chief source, *Skånska rekognosceringskartan*, gives no complete information about parish boundaries. For this period the parish boundaries have been derived from a topographic map printed in 1864. For the maps showing the prehistoric settlement, boundaries are taken from the economic map surveyed in the 1970s.

The development of roads and transportation has not been studied for more than limited parts of the project area. Hence roads and railways have been omitted on all maps.

Contour lines, when used, are taken from the modern topographic map.

2.8 Landscape-ecological mapping

Björn E. Berglund and E. Gunilla A. Olsson

The reconstructions of vegetation in the maps in Ch. 3 are based on various sources. For historical time environmental information is available in written records including maps. This is valid for the periods 1985, 1915, 1815 and the 18th century, which makes this series of maps quite accurate. For the 1985 map a field inventory of vegetation was also performed – see below. The 14th century map is based on some written documents besides knowledge of ancient settlement and results of archaeological excavations and palaeoecological studies. The vegetation is also interpreted from a retrogressive analysis of the 18th century map. In a similar way the Viking Age landscape is reconstructed on the basis of a combination of archaeological/palaeoecological information and a retrogressive analysis of the 14th century map. However, the vegetation of the five prehistoric periods is mainly based on palaeoecological and archaeological results from this project. This means that the human impact is traced by means of pollen-analytical correlations as well as archaeological inventories. As a whole the vegetation reconstructions for the five prehistoric periods in Ch. 3 should be regarded as approximate, tentative mappings. They rely on information from a few investigated sites which are regarded as representative of areas with similar ecological conditions. The source material and some assumptions related to the interpretations of the landscape in the Ystad area in the maps in Ch. 3 and in the focal areas are maps in Ch. 4 are described in the following.

For the more detailed reconstructions of landscape and vegetation presented in Ch. 4 other methods were used. The local agro-ecosystems are reconstructed for smaller areas (approx. 5 × 5 km). For the Bjäresjö and the Köpinge sites relevant palaeoecological information from nearby pollen diagram sites was used. The landscape reconstruction is based on a protein-based model for human nutrition described in Ch. 2.2. Information on qualitative human dietary consumption was obtained from the archaeological finds and from palaeoecological evidence (remains of charred cereal grains, seeds etc.). These data originate from excavated sites in the area, but sometimes also from corresponding sites outside the Ystad area. The reconstruction of the medieval landscape is in Ch. 4 based on a retrogressive analysis of land survey documents from the 17th and 18th centuries (cf. Ch. 4.1.6.2, 4.2.6.2).

Sources for landscape-ecological mapping

Quaternary soils. The Geological Survey of Sweden (E. Daniel, branch office in Lund) has provided the project with unpubl. field maps (1:10,000) as well as a published map (1:50,000). These are based on the soil quality at 0.5 m depth. There are no topsoil maps available, but some information on topsoils is known from botanical studies of woodlands or pastures.

Hydrology. The present hydrology is available in the standard topographical maps (1:50,000). However, artificial drainage has changed the surface water pattern enormously during the last two centuries (cf. Möller 1984). The reconnaissance map from 1812–1820, *Skånska rekognosceringskartan* (1:20,000), is assumed to show an almost natural drainage pattern. We regard this as reflecting also the distribution of surface waters and wetlands existing in prehistoric time – except for artificial ponds dug in historical time.

Palaeohydrology. Climatically conditioned water-level fluctuations have caused shore displacements at lakes and flooding or drying out in mires. On the scale used for our maps this means only minor displacements of vegetation borders. However, the raised groundwater level after 1000 BC is noticeable in the wetlands – earlier wooded fens were flooded and transformed into open fens. Cf. Ch. 5.1.

Palaeogeography. Palaeogeographical changes on this scale and during this time-span are mainly restricted to shore displacement caused by water-level changes of the Baltic Sea and land uplift (references in Ch. 3). In addition, the number of inland lakes has changed from Mesolithic to Modern Time from about 20 to 3 (taking into account only lakes with a water surface >2 ha). This is caused by infilling and overgrowing of ancient lake basins, and by artificial drainage since the 19th century. Our information is based on reconnaissance corings and datings of the sediment transition gyttja/peat, along with the stratigraphic results from the main palaeoecological sites.

Past vegetation
The reconstruction of vegetation and landscape during

prehistoric time and partly also for medieval time is based mainly on pollen diagrams, but to some extent also on charcoal identification (cf. Ch. 2.1). The interpretations of the spatial distribution of the vegetation types are based on several assumptions mentioned in Ch. 2.1, particularly the fact that all natural vegetation is to some extent a reflection of soil fertility, topography, and hydrology, This can be applied also to the conditions in the past. Another assumption is that the human impact on the natural ecosystems in an agrarian economy leads to deforestation and creation of secondary woodlands and grasslands (cf. Rackham 1986, Pott 1988). When subdividing the ground into two categories according to soil moisture, we recognize two main developmental trends along with increasing human impact:

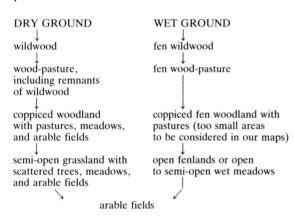

DRY GROUND

wildwood

wood-pasture, including remnants of wildwood

coppiced woodland with pastures, meadows, and arable fields

semi-open grassland with scattered trees, meadows, and arable fields

WET GROUND

fen wildwood

fen wood-pasture

coppiced fen woodland with pastures (too small areas to be considered in our maps)

open fenlands or open to semi-open wet meadows

arable fields

For historical periods series of written documents were used. The first landscape and vegetation map based to some extent on written sources is the one from the 14th century. For the reconstruction of this map (Fig. 3.6:1) the following sources with information on settlement, landuse and vegetation were used:
* village names found in written sources, and place-name chronology for these names (Skansjö 1987, 1988)
* archaeological finds and the dating of existing churches and chapels from the 14th century (Andersson and Anglert 1989)
* 17th and 18th century land survey maps for the whole Ystad area (cf. Linusson and Olsson in Appendix B)
* palaeoecological finds:
   1) remains of charred cereal grains and plants (Engelmark and Hjelmquist 1991)
   2) pollen records for lakes and basins (Gaillard and Berglund 1988, Ch. 4.1.2, 4.2.2, 4.3.2, 4.4.2, 5.1).
The main sources for the vegetation of the 18th century were the land survey maps available on a large scale of 1:4,000. These maps are very detailed and drawn with great accuracy. They contain information on physical soil properties, soil fertility, vegetation, and hydrology.

The interpretation of the 18th century maps and the methods for the reconstruction of the overview map are described in detail by Linusson and Olsson in Appendix B.

For the 19th century landscape *Skånska rekognosceringskartan*, dated to 1812–1820, was the main source. The scale of this map is 1:20,000. Since this map was created to chart possible lines of advance for military troops, the vegetation information is biased. Thus the information on hydrology, wet and moist vegetation is much more detailed than that on the dry and mesic vegetation types. It is not always possible to distinguish between dry meadow, dry pasture and fallow land (cf. Emanuelsson and Bergendorff 1983, Ch. 2.6, 2.7). Moreover, different cartographers used different symbols and different accuracy; the map is thus not uniform all over the area. In spite of these weaknesses this source is a useful tool for an introduction to the 19th century landscape.

The reconstruction of the early 20th century landscape is based on an economic map from 1915 (scale 1:20,000). This map was made with the purpose of mapping land use and land ownership. The map is accompanied by tables of acreages for the different land-use categories and/or vegetation types. The map is made with great accuracy and gives a reliable depiction of the landscape.

Modern vegetation
Present-day vegetation in the whole Ystad area was mapped by field inventories. The field inventory was preceded by interpretation of aerial photographs (IR colour, scale 1:30,000), which allowed preliminary limits for rough vegetation units to be drawn. All areas with natural and semi-natural vegetation were then checked in the field. For the fieldwork modern economic maps on a scale of 1:10,000 were used. This detailed vegetation map was later transformed into a survey map at a scale of 1:20,000 where every subarea was classified, accompanied by a list of the plant species that occur.

Vegetation units used in vegetation mapping were: woodlands (classified according to dominating tree species); shrubland; grassland; fenland; arable.

The map in Ch. 3.10 presents the result of this survey inventory, although heavily simplified.

A detailed vegetation inventory was carried out for the four key study areas presented in Ch. 4. Special effort was devoted to grassland areas and woodlands. Vegetation units used for this mapping were the same as for the survey map, except that the field-layer vegetation was also classified according to moisture and soil fertility. This detailed map and accompanying vegetation lists are unpublished, but available at the County Administration Board in Malmö and at the Department of Ecology, Plant Ecology, Lund University.

Settlement information

The distribution of settlement in prehistoric as well as historical time has been outlined in survey maps (methods and sources discussed in Ch. 2.7). For landscape reconstructions covering prehistoric time this information is fundamental for indications of occupied places, particularly the density and, as a consequence of this, the land use and vegetation differentiation in the landscape. This is a complement to the palaeoecological information which is better suited for general landscape changes through time than precise spatial changes.

2.9 Terms and concepts

Björn E. Berglund, Lars Larsson, Nils Lewan, E. Gunilla A. Olsson, Mats Riddersporre and Sten Skansjö

Some terms and concepts have been of fundamental importance for communication between the scholars in this interdisciplinary project. We have discussed these at an early stage of the project, defined them, and applied them in verbal and written reports. A selection of these is presented here with their definitions under a few main headings.

Time units

Conventional archaeological terminology for southern Sweden has been applied (after Hedeager and Kristiansen 1985). This series of units is shown in Table 2.9:1. In some tables abbreviations have been used. These are also given in Table 2.9:1. When speaking about the rather diffuse time units of prehistoric and historical time the border is placed at AD 1050.

Besides the terms in Table 2.9:1 we use the following subunits:

EN 1,2 (M. Larsson 1985)
MN I-IV (Becker 1954), MN A,B (Nielsen 1977)
LN A-C (Lomborg 1973)
BA I-VI (Montelius 1883)
EI (Early Iron Age) = PRIA, RIA
LI (Late Iron Age) = MP, VP, VA.

The absolute time-scale is based on radiocarbon years – the ages of the unit borders are given in ^{14}C years, named conventional time-scale (conv.). We use this time-scale throughout this synthesis. There is only one exception, in Ch. 5.13, where pollen accumulation rates are discussed. The map sequence in Ch. 3 has calibrated ages (years cal. BC) besides radiocarbon years. Table 2.9:1 also gives calibrated ages for the unit borders (cal.). This calibration follows Stuiver (1986). In some figures we use two time-scales – years BC/AD or years BP. BP age always means "years before 1950".

Sometimes geological time units have been used following the Blytt-Sernander terminology (cf. Mangerud et al. 1974). The Atlantic/Subboreal border corresponds to the Mesolithic/Neolithic border, the Subboreal/Subatlantic border corresponds to the Bronze Age/Iron Age border. Holocene is a time unit referring to the last 10,000 years, sometimes with a tripartite division: Early (10,000–8000 yr BP), Middle (8000–2500 yr BP) and Late (2500–0 yr BP) Holocene.

Landscape concepts

Through time and space we identify different kinds of landscapes. We may look upon these terms in a hierarchic way:

Natural landscape is a landscape more or less untouched by man. Originally wildwoods occupied most of the natural landscape in this region. "Natural" processes caused the changes of such a landscape.

Cultural landscape is a landscape affected by man's activities. Besides natural elements it includes human-caused elements like buildings, arable land, meadows, pastures, ditches, roads, etc. Natural processes are influenced by human activities. Natural vegetation is normally replaced by other vegetation. A special term is used for a geographically delimited cultural area, namely settlement region (Sw. *bygd*, defined on p. 59).

Agrarian landscape (or farming landscape) is a landscape shaped by the farmer who exploits the landscape for arable production and/or stock-raising. When arable production dominates we are dealing with an arable landscape, when grazing dominates we are dealing with a pastoral landscape. In contrast to these more or less woodless landscapes we may identify a woodland landscape exploited for its woodland products: timber, fuel, leaf-fodder, pastures, etc. Normally Swedish agrarian landscapes consist of a mixture of arable, pasture, and woodland landscape.

Vegetation and landscape ecology

Deforestation – succession – regeneration. Deforestation means a destruction of wildwoods via half-open woodlands, wood pastures, coppiced woods towards treeless pastures or arable land. The species composition and the vegetation structure are changed. The biomass production tends to be moved from the tree canopy towards the field layer.

Ecological successions are of two main types. *Primary succession* occurs on virgin areas without a soil layer,

Table 2.9:1. Archaeological time-scale for southern Sweden (slightly revised after Hedeager and Kristiansen 1985). Geological periods indicated (after Mangerud et al. 1974)

Archaeological/historical period	Radiocarbon years	Calendar years	Geological periods
Modern Time (MoT)	AD	AD	
	1500	1500	
Middle Ages (MA)			
	1050	1050	
Viking Age (VA)			
	800	800	
Vendel Period (VP)			
	550	550	Subatlantic
Migration Period (MP)			
	400	400	
Roman Iron Age (RIA)			
	0	0	
Pre-Roman Iron Age (PRIA)	BC	BC	
	500	500	
Late Bronze Age (LBA)			
	800	1000	
Early Bronze Age (EBA)			
	1500	1800	
Late Neolithic (LN)			Subboreal
	1800	2300	
Middle Neolithic (MN)			
	2600	3300	
Early Neolithic (EN)			
	3100	4000	
Mesolithic Time (MT)			Atlantic

soon after exposure to climatic and biological variables, like new land in front of retreating glaciers, exposed alluvial sediments or areas subjected to land upheaval. Primary succession means development of virgin soil and a successive colonization of plants and animals.

Secondary succession occurs on previously vegetated land exposed to some kind of disturbance. Such areas can be clear-cut areas, semi-natural grasslands where the grazing or mowing have ceased, abandoned arable fields etc. Invasion and colonization of plants and animals will occur and the result will – to a certain extent – be a reflection of (1) former vegetation via the buried seed bank in the soil, (2) former land use, (3) surrounding vegetation and land use.

Except for salt-water shorelands, the final outcome of all successions in temperate regions will be different woodland communities – *woodland regeneration*. In landscapes subject to human disturbances biological diversity in terms of the number of species and absence of dominance of certain species is largest in transition stages of succession, and in edge areas, before the final development of a closed canopy. When dealing with ecological succession in a very long perspective of evolutionary time, diversity is largest in the latest stages of succession, as in tropical forests.

Biological diversity refers to the number of species (species richness) and their abundance in a given land area or vegetation type.

Habitat diversity refers to the number of habitat types and their distribution pattern in a landscape sequence.

Landscape diversity refers to the number of landscape types like flatlands, hilly areas, water, agricultural land, woodlands, etc.

Carrying capacity (Odum 1971) signifies the equilibrium level of biological production at which an ecosystem can support a population of animal species or a human population. Concerning human population in this context subsistence economy is assumed. Human-influenced and human-maintained ecosystems, like agro-ecosystems subject to management changes, will achieve different levels of carrying capacity depending on the type of management.

Wildwoods is used in this book as a term for native woodland principally unaltered by humans. Minor openings may occur because of clearings caused by hunter/gatherers.

Woodland signifies all sorts of wooded area, both wildwood as well as wood affected by human influence.

Wood-pasture (Rackham 1986) is a woodland where grazing animals (domestic or wild animals) are kept. This duality in land use involves a conflict between grazing management and the growth of trees. Given the appropriate number of animals, an equilibrium can exist where wood-pasture as a habitat can be stable over long periods of time.

Semi-natural grassland. In a Scandinavian context there are hardly any terrestrial non-wooded natural vegetation types outside the arctic region, except salt marshes. Natural grasslands in the sense of those demanding no management cannot be found in Scandinavia. Semi-natural grassland denotes the whole variety of grasslands managed either by grazing (by domestic stock) or by mowing by humans. As soon as the management ceases the secondary successions will occur and there will be a development into woodland.

Open pasture is defined as semi-natural grassland with a flora characterized by stress-tolerant species – graminids and some herbs, shrubs and dwarf-shrubs. The species composition is depending on grazing pressure, besides nutrient and water content of the soils. With increasing grazing/browsing pressure wood-pastures develop into open pastures. Intense grazing leads to soil erosion.

Meadow is used in this book – as well as in the oral tradition (Sw. *äng*) and in the historical sources – as a term for natural fodder areas. Meadows are managed by mowing or a combination of mowing and grazing after hay harvest. Meadows are composed of grassland areas which typically have very specialized plant and animal communities developed over a long time of continuous management. The natural meadow biotopes are very rich in species and today are among the most threatened biotopes. The threat is generally the cessation of management which leads to forest invasion – or changes in management, e.g. conifer plantations or cultivation to arable.

Meadows are winter fodder areas while the pastures are summer fodder areas. Both meadows and pastures belong to the category of *semi-natural grasslands* (cf. above).

Water-meadow signifies meadows managed by mowing and grazing after hay harvest, *and* that water floods the area at least once a year. This flooding is organized by leading natural brooklets or small rivers in an intricate system of successively shallower channels over the meadow. A similar intricate channel system exists to bring back the water at the distal edge of the meadow. The system of water-meadows was in use in the Ystad area during the latter half of the 19th century. It was invented as a method to raise the biomass production of the meadows in the absence of commercial fertilizers. The moving water brought nutrients and dissolved oxygen to the grassland communities and thereby raised the biological production.

Expansion – stagnation – regression. This concept has been used in Swedish research on Holocene vegetation dynamics during the last two decades, mainly according to the definition by Berglund (1969, 1986; see Ch. 1.1). In a parallel way geographers and archaeologists have applied similar terms for the history of settlement and society (Fig. 2.9:1). The vegetation dynamic terms refer to human utilization of the landscape for agrarian production, the transformation of woodlands into an open landscape. In this process the object is the agrarian landscape with arable, meadows, and pastures.

$$
\begin{array}{ccc}
& \text{forest regeneration} & \\
\text{Expansion} & \longrightarrow & \text{Regression} \\
\text{phase} & \longleftarrow & \text{phase} \\
& \text{deforestation} &
\end{array}
$$

Because of changes in settlement history, land-use pattern, etc., we do not expect a continuous deforestation, but there are periods and places with retarded or decreased utilization leading to stable conditions (stagnation) or reforestation (regression).

This concept is used when interpreting the pollen diagrams. They are defined in the accompanying table (according to Berglund 1985, 1986).

Land use

Mesolithic/Neolithic economy. Mesolithic: an economy based on the local conditions for using the natural flora and fauna, in other words, hunting, fishing, and gathering, corresponding to the economy of *hunter/gatherers.* The only domesticated animal is the dog. There may have been some deliberate changes of the environment, such as the burning of forests to increase the productivity of wild plants and animals. Trade and exchange are of little importance in the economy.

Neolithic: an economy based on animal husbandry and tillage. This can involve the domestication and cultivation of local animals and plants, but for northern Europe this economy is based on the introduction of animals and cultivated crops domesticated elsewhere. Trade and exchange may have some significance for the economy, but they can be of great importance for the establishment and maintenance of social structures. It should be pointed out here that the economy for considerable parts of the Neolithic also includes hunting, fishing, and gathering, the importance of which varies in relation to the farming economy.

Transhumance. Seasonal usage of marginal regions for grazing, for instance summer cattle/sheep grazing in uplands or in inland areas outside the central settlement regions. Such areas have a mobile settlement or at least more primitive houses (shielings) outside the more permanent winter settlements. One form of still existing transhumance is the summer pasturing in Scandinavian mountains.

Woodland utilization has been multiform and several terms are related to this (see Rackham 1986, cf. also Pott 1986). *Girdling* or bark-peeling refers to the custom of ring-barking trees to favour the development of shoots used for leaf-fodder or cattle browsing. *Coppicing* is the practice of cutting trees near ground level in produce low-branching underwood trees (= a coppice). Coppicing favoured access to timber, fuel and leaf-fodder, and the opening of the environment also favoured the development of hay-meadows and pastures. *Pollarding* is the cutting of trees at 2–4 m above the ground to produce successive crops of wood and leaf-fodder. *Shredding* is the custom of cutting side branches of a tree except at the top. Coppicing and pollarding still occur in some places in Scandinavia (Emanuelsson 1987, Birks et al. 1988).

Slash-and-burn cultivation refers to a primitive technique of cultivating woodlands after tree cuttings and burning. Such clearings were used for a few years before the forest regenerated. This technique has been prac-

	Pollen characters	Vegetation dynamics	Land-use pattern
Expansion	Low values of *Ulmus, Tilia, Fraxinus, Quercus* (sometimes also of *Fagus* and *Carpinus*). High values of *Corylus, Betula,* apophytes, and anthropochores	Open, unstable vegetation, early succession stages. High diversity	Clearings, grazing, coppicing, etc. agriculture
Regression	Values opposite	Closed (forested), stable vegetation, late successions. Low diversity	Weak human impact or restricted to grazing and coppicing

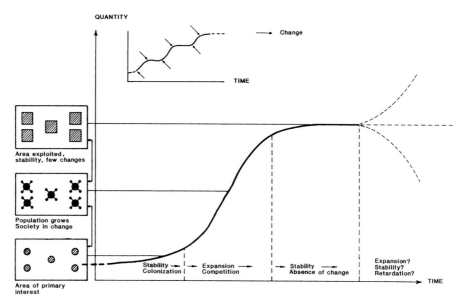

Fig. 2.9:1. Definition of the settlement-historical terms colonization – expansion – stability – retardation according to Sporrong (1985).

tised in Scandinavian coniferous forests until this century (Kardell et al. 1980). It has also been documented in pollen diagrams from Finland (Vuorela 1986).

Intensive/extensive cultivation. Intensive cultivation refers to the production of crops in permanent fields with short rotation – with a maximum fallow period of two years. This means that some kind of fertilizing was needed to maintain production. Extensive cultivation refers to the production of crops in mobile fields with long rotation – normally with a fallow period of 20–50 years. The harvesting and transport of mineral nutrients were balanced by the long fallow periods. Extensive cultivation means low production per unit area and therefore large areas were affected by this kind of cultivation. So-called Celtic fields are one form of extensive cultivation.

Field rotation systems. Before the time of enclosures and agricultural innovations of the 19th century, the predominant field rotation system in the Ystad area was the three-field system. The general feature of this system was that the arable was divided into three separate fields (Sw. *vångar*). Two of these fields were used each year for growing grain, mainly rye and barley, whereas the third was fallowed and could be used for grazing. The three-field system dominated in the southern plains of Scania but in other parts there existed by this time also two- and one-field systems (Campbell 1928). The two-field system could have a crop/fallow cycle of either two or four years (Dahl 1975), whereas the one-field system comprised a field that was sown each year. A mixture of the two latter systems was in use in the north-western parts of the Ystad area, on the Romeleåsen ridge (cf. Ch. 3.7).

Depending on which system was practised, there were different ratios of arable to other vital resources such as hay-meadows and pasture, so as to reach a balance between the needed and obtainable amount of

manure. Within the individual fields of the different systems there was not only arable that followed the yearly rotation of the system; there was usually also arable that was used with longer periods of fallow than the system proper (cf. Ch. 4.1.7.2, 4.2.7.2). This means, for instance, that within the single field of a one-field system, not all arable land was under crop each year (cf. Frandsen 1988). Furthermore, a large part of the hay meadows could be found within the commonly enclosed fields, following the rotation cycle of the system. In addition to that there was in many villages also a separate meadow field where hay could be harvested every year.

Infield/outland (outfield). By the time of the enclosures, more or less all land suitable as arable was used for growing grain. In earlier times the arable was, together with hay meadows, confined to the infields (Sw. *inägor*). The infields were separated by a fence from the surrounding outland (Sw. *utmark*), which was mainly used for pasture, collecting fuel, and other extensive activities. In these times the outland was probably common land. In the north-western parts of the Ystad area there was still in the 18th century a large pasture common in existence, documented on contemporary maps (cf. Ch. 3.7). Although much of the former outland in the Ystad area had been subdivided and transformed into arable land by the 18th century, it was in many cases still used more extensively than the original infields, often with longer periods of fallow and sown with oats. This use of the old outland may in fact correspond to the essence of the English outfield (cf. Ch. 2.10, 4.1.7.2, 4.2.7.2, see also Widgren 1983:73).

Settlement

Site, seasonal site, base camp. A settlement or habitation site refers to a place where human beings have

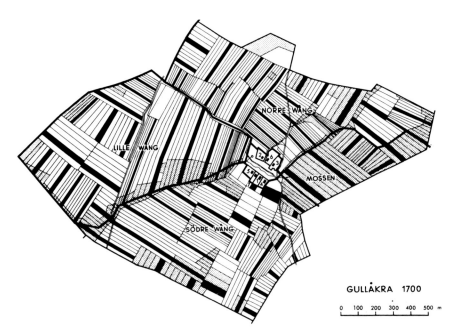

Fig. 2.9:2. Gullåkra, a three-field village on the plains of SW Scania, in the year 1700. The three fields are: *Lille Wång, Södre Wång,* and *Norre Wång* (the Small Field, the South Field, and the North Field). In addition to these fields there was also a fen, *Mossen.* The fields and the fen were subdivided among the farms. The black strips belonged to farm number 1 (Wester 1960).

GULLÅKRA 1700

0 100 200 300 400 500 m

left traces from a shorter or longer period of occupation. *Seasonal site* refers to a settlement site which has been used for shorter, seasonal periods to exploit special resources, such as game or fish. *Base camp* is a settlement site used throughout the year or during longer periods of the year.

Permanent/mobile settlement. "Permanent" denotes settlement at the same site throughout the year, while "mobile" denotes settlement which shifts between different sites throughout the year depending on the natural resources (hunting, fishing, grazing, etc.).

Continuous/discontinuous denotes whether settlement at one site was without any break or was used periodically. Continuity may refer either to site continuity or regional continuity. Historians distinguish between two kinds of deserted settlement: temporary and permanent.

Settlement region (Sw. *bygd*) is a geographically delimited area with its settlements and including intensively as well as extensively utilized areas, often woodless as well as partly wooded areas.

Central/marginal settlement region. Central settlement region refers to an area which has attracted man and society with a pre-industrial agrarian economy so that a continuous settlement of greater or lesser density has developed. A marginal settlement region is less attractive because of its lower potential production. It has no permanent settlement or very sparse settlement. It is often a woodland area which may be utilized for its forest products or its value as extensive pasture. Marginal settlement regions have always been potential colonization areas (cf. Skansjö 1983).

Colonization – expansion – stability – retardation (concentration) are concepts generally used by archaeologists, historians, and geographers about settlement de-

velopment and changes. Definitions have been given by Sporrong (1985), illustrated in Fig. 2.9:1 (Gissel et al. 1981, see also Ch. 1.1). Colonization means that the population invades a new area outside a settlement region. Expansion is a result of increased population pressure, leading to settlement clusters and settlement dispersal (colonization) outside central settlement regions. During expansive periods the society is unstable but it is followed by a period of equilibrium (stability). A new change may lead to expansion or retardation (concentration) of the settlement.

Village (hamlet). In Swedish there is no terminological distinction between village and hamlet. Agrarian settlements comprising more than a single or double farm are normally referred to as *by.* In this book we have adopted the term village as a general translation of the Swedish *by* (cf. Ch. 2.10). In terms of settlement size there may thus be only a minor difference between a small village, as the term is used here, and a single farm of substantial size. In general terms, what is to be regarded as a village consists of an agglomerated agrarian settlement with individual households. A village is, however, not only an agglomeration of farmsteads and cottages – the village core – but also the territory at the disposal of the associated farms – the township. Before the enclosure movements of the 19th century, village life moreover meant a strong interrelationship in agrarian activities. In these times a village was both a social community and an economic entity.

For historical times villages and single farms are identified in contemporary sources, medieval and later, and no further classification has been considered necessary. The pre-enclosure organization of villages is well known through land survey documents, including large-scale maps, from the 17th century onwards. The field-rotation systems of these villages and the organization of infields and outfields/outland are described above. Normally the township was partitioned into the number of fields needed for the rotation system in use (Fig. 2.9:2). There may also have been a separate meadow field. Within these fields the land – both arable and meadow – was subdivided among the farms of the village. This form of division meant that activities like ploughing, sowing, harvesting, and the grazing of fallow fields had to be coordinated among all the farmers of the village.

In many cases neighbouring villages had their field-rotation system adapted so that fallow fields of the same year were adjacent. This meant that no fence was needed between these fields and that the grazing cattle could move over large areas. This system (Sw. *vångalag*) meant an interdependence above the village level and could knit together several villages through the intercommoning of grazed fallow fields (Campbell 1933, cf. Ch. 4.1.7.2, 4.1.7.3). In many villages on the plains this was the only available pasture, since nearly all land had developed into infields by the 17th century. In cases where there was still a pasture common left, it was connected to the village green by a cattle-path (Sw. *fägata*). In many villages there was no woodland left by this time and the peatbog was the only resource for fuel. Several villages were by this time dependent on external resources such as timber, fuel, and summer pasture.

The main features of the village organization known from the 17th century can be traced back to the Middle Ages through documentary evidence (cf. Ch. 4.1.6.1, 4.2.6.1). With the enclosure movements of the 19th century the spatial interdependence between the farms of the villages was abolished. In many cases the settlement was also split up, but most of the villages survived as communities and as settlement agglomerations – often with new functions added (cf. Ch. 5.10). In everyday Swedish the word *by* has thus come to denote the settlements as such, although some of them are no longer agrarian settlements at all.

For prehistoric times the identification of villages is more complicated. In terms of settlement a village, or village-like settlement, should be archaeologically recognizable as a site where several separate dwelling-houses or farmsteads can be identified (cf. Ch. 5.6). A critical point is, however, to distinguish whether the different houses of an excavated settlement are really contemporary or the result of a single farm successively moving around in the area (cf. Ch. 5.5). Concerning territories connected with individual settlements, and various forms of cooperation within them, the source situation is very poor in the Ystad area. There are, for instance, no prehistoric fences or field systems preserved.

A central theme in the research into villages is the question of how and when the historically known villages were established and how they relate to the prehis-

toric settlement picture. Aspects of the foundation of the historical villages are illuminated by several authors in this book (esp. Ch. 4.1.5.1 and 5.6). The establishment of villages not only affected the settlement pattern and the further evolution of farming practices, but also the conditions for social life – as did the dispersion of settlement that followed the enclosures of the 19th century.

Manor (Sw. *huvudgård*, later *herrgård*). The meaning of this term varies from period to period. In the Late Middle Ages it denotes a farm which was occupied or could be occupied by a noble; it could be relatively large, although it did not need to be, but it often had subordinate peasant farms and it was always exempt from taxes to the crown and tithes to the church. A noble could often own several manors, after one of which he titled himself, while the others were administered by bailiffs within an estate complex. The corresponding Latin term for the time from about 1350 is *curia principalis*. Before this time the sources use the designation *mansio* or *curia*. These magnates' farms in the Early Middle Ages were not infrequently larger cultivation units than their successors in the Late Middle Ages. The medieval manors differed from the peasant farms not only through their privileges but also in many cases through fortifications such as mottes, moats, and the like. Medieval ecclesiastic landowners could also own "manors", in the sense of large farms functioning as administration centres within an estate complex (e.g. *skudgårdar*, the late medieval estates owned by the archbishopric). In the 16th and 17th centuries the extent of tillage practised by the manors often developed into what was for the time large-scale management, at the same time as the manor houses took on the dimensions of castles. The exemption from tax which had previously been granted only to manors could be extended in the Middle Ages to apply also to some of their tenant farms, a development which accelerated in the course of the 16th century.

Several of the manors known from early historical times still exist in modern times. In recent times, however, they have not had the special feature of tax exemption and are not necessarily under noble ownership. They are still, however, remarkably large agricultural units and can still have tenant farms. In the 18th century and particularly in the 19th century there were substantial investments in land and buildings. This period also saw the creation of large new farming units, often parcelled off from earlier manors, but these are not reckoned in the category of manor.

Estate (Sw. *gods*) denotes the total property belonging to a major landowner, in early historical times an individual noble or an ecclesiastic landowner. An estate could consist of one or more manors and a varying number of their tenant farms, with varying degrees of independence (cf. Ch. 5.9).

2.10 The rendering of Swedish terms in English

Alan Crozier

Rather than retaining specifically Swedish terms in their original form, the practice in this volume is to use the closest corresponding English term. This should not be taken to imply complete equivalence. The various *skifte* reforms, for example, are referred to as "enclosures", although in Sweden the idea was not so much to enclose common land as to consolidate scattered land parcels.

The English terms "parish", "hundred", and "county" are used to render the administrative units of *socken, härad,* and *län.* Note that the latter originally denoted a "fief", a feudal benefice which could be of any size (cf. Ch. 4.1.6.1).

In cases where the closest English equivalent could lead to misunderstanding, Swedish terms have been translated literally. For example, the system of *inmark/utmark* is not the same as the English infield/outfield system. Although "infield" corresponds closely enough to the cultivated *inmark* (also known as *inägor*), "outfield" is not appropriate for the Swedish *utmark,* since English village outfields could be cultivated for occasional periods, while the *utmark* was uncultivated outlying land, used as pasture and for collecting fodder and fuel. Nor is it always suitable to translate the concept by the alternative Swedish word *allmänning* "common", which might suggest too great a similarity with the English village common. When necessary, then, we refer to the Swedish *inmark/utmark* as an "infield/outland system" (cf. Ch. 2.9).

Some terms are better left in the original Swedish. These include the unit of square measure, the *tunnland* (0.4936 ha), and the fiscal unit of land assessment, the *mantal,* originally the land needed to support one family.

Place-name elements

Readers who do not understand Swedish may benefit from translations of some of the common elements in place-names. The qualifications *Norra, Södra, Östra,* and *Västra* denote the compass-points: north, south, east, and west. The equivalents of English "Great" and "Little" are seen in *Stora* and *Lilla Köpinge* respectively. Common place-name elements denoting natural features are *sjö* "lake", as in Krageholmssjön, *å* "river", as in Svartån, *mosse* "bog", as in Kurarps Mosse, *hult* "wood, copse", as in Lindhult ("lime wood"), and *ås* "ridge", as in Romeleåsen.

The names of settlements can include generic elements which have either disappeared from the language or else acquired new meanings since the names were coined. The word *stad,* for example, which now denotes a town, historically with borough privileges, is a medieval loan from German. In older place-names the native word *stad* appears to mean simply "place" (like its English equivalent *stead*); Ystad probably means a homestead or farmstead where yew-trees grew.

In modern times *köping* means a small market town; in names like Stora Köpinge the original sense is a market-place of some sort. The word is cognate with English *Chipping* in place-names.

Names in which the element *by* "village" is found appear to have been agglomerated settlements as far back as the evidence goes. It should be noted, however, that the Norwegian and Icelandic cognates of the word are used of isolated settlements, so it cannot be categorically assumed that the Swedish (and Danish) word has always referred exclusively to agglomerated settlements. Note also that *by* is always rendered as "village" in this book, even when the actual settlement might more accurately be described as a "hamlet" (cf. Ch. 2.9); a village in a somewhat stricter sense can be distinguished as *kyrkby* "church-village".

The element *-löv* had ceased to be productive in place-names before the Viking Age. It is related to English *leave* and refers to the inheritance of property. Gundralöv, for example, was either bequeathed or inherited by a man named Gundrath.

The modern word for "farm" is *gård.* It is common in place-names, some of which – like Rydsgård – now refer to larger settlements.

The word *torp* in modern times denotes a smallholding occupied by a person with labour obligations towards the landowner; it is rendered as "croft" here. In medieval place-names, however, where *torp* is the most common settlement term, it is assumed to mean an offshoot of a parent settlement, a colonization of outlying land. Examples of the varying forms of this element in the Ystad area include Svenstorp, Fårarp, Sjörup.

Not all settlement names ending in *-sjö* refer to lakes. Bjäresjö, for example, is a corruption of an earlier Berghusa, which consists of a form of the word *berg* "hill" and *husa* "house(s)". According to one hypothesis (see Ch. 4.1.5.1), place-names in *-husa* represent royal administrative centres, like the *husaby* locations in Sweden; this interpretation is contested by Hallberg (1989).

The element *-lösa* (as in Katslösa and Varmlösa) is

not recorded except in place-names, so its meaning has to be conjectured on etymological grounds. It may be related to the obsolete English *lease* "pasturage, meadowland". Alternatively, it may be related to Swedish *ljus* "light", perhaps denoting an open area in woodland. A place-name element which certainly denotes a clearing is *-ryd* or *-röd* (as in Långaröd), which is related to the obsolete English *ridding* "clearing, assart".

The settlement names of one part of the Ystad area (Ljunits Hundred) have been studied in detail by Hallberg (1975).

Translation of terms

Important Swedish concepts are rendered by the following English terms throughout the book. The list has been compiled in consultation with B. E. Berglund, E. G. A. Olsson, and M. Riddersporre. Some of these terms are explained in greater detail in Ch. 2.9.

English	Swedish
agricultural labourer	statare
alder carr	alkärr
arable	åker
bark-peeling	ringbarkning
barrow	gravhög
beach ridge	strandvall
Black Earthenware	svartgods
bog	mosse
bog iron ore	sjömalm, myrmalm
broad-leaved trees	ädla lövträd
cadastral register	jordebok
camp site	lägerplats
cattle-path	fägata
cemetery	gravfält
church-village	kyrkby
Civil Survey	Jordrevningsprotokoll
clear-cut woodland	hygge
clergyman's report	prästrelation
common	allmänning
coppice-pasture	betad skottskog
coppiced woodland	skottskog
corvée	hoveri
cottage	gatehus
cotter	gatehusman
croft	torp
crofter	torpare
day-labour	dagsverk
day-labourer	daglönare
day-labourer's croft	dagsverkartorp
dispersed/isolated farm	ensamgård
dolmen	dös
dry meadow	hårdvall
estate	gods
estate tenant	frälsebonde
fallow	träda
fen	kärr(mark)
fen wood-pasture	betad sumpskog
fen woodland	sumpskog
field	vång, åker
Funnel Beaker culture	trattbägarkulturen (TRB)
girdling	ringbarkning
grassland	gräsmark
habitation area	boplatsyta
habitation site	boplats
half-open woodland	gles skog
hay-meadow	slåtteräng
hide	bol (obsolete unit of land measurement)
hummocky landscape	backlandskap
isolated farmstead	ensamgård
land survey documents	lantmäteriakter, -handlingar
large tenant farm	plattgård
ley	vall
Lund diocesan terrier	Lunds stifts landebok
manor	huvudgård, säteri, herrgård
marsh meadow	strandäng
meadow	äng, slåtteräng, slåttermark
meadow field	ängsvång
mesic meadow	frisk äng
occupation layer	kulturlager
outland	utmark
owner-occupier home	egnahem
pasture	betesmark
Pitted Ware	gropkeramik
pollarding	hamling (lövtäkt)
post-built house	stolpburet hus
prefect of a fief	länsman
royal naval organization	ledung
sedge fen	starrkärr
settlement area	bosättningsområde
settlement region	bygd
shieling	fäbod
shredding	hamling
single farm	ensamgård
slash-and-burn	svedjning
stately home	herrgård
summer pasturing	fäboddrift
sunken-floor hut	grophus
taxation assessment roll	mantalslängd
tenant	landbo
turbary	torvtäkt
underwood	lågskog
vicarage	prästgård
village core	bykärna
village site	bytomt
wandering arable fields	vandrande åkrar
wet bottom	fuktstråk
wet meadow	fuktäng, sidvall
wetland	våtmark
wildwood	naturskog
wood-pasture	betesskog, hage (hagmark)
woodland	skog
woodland with standard trees	högskog

Ecological Bulletins 41: 63–105. Copenhagen 1991

3 Society and environment through time

3.1 The Late Mesolithic landscape

3.1.1 Landscape, land use, and vegetation

Björn E. Berglund

Physical geography

The regional climate during this period (ca. 4000–3200 BC) was milder than today, which means approx. 2°C higher summer temperature, along with milder winters, longer growing periods, and less humid conditions (Lamb 1977, Huntley and Prentice 1988, Ch. 5.1). The Holocene thermal optimum in central Scandinavia is often placed at ca. 4000 BC (Kullman 1988). Minor fluctuations occurred before 3500 BC.

The relative sea level of the Baltic was about 3–4 m higher than today, with a transgression dated to the very end of the Atlantic Period (Digerfeldt 1975, Christensen 1982, Gaillard et al. 1988). From a palaeogeographical point of view, this meant that minor bays were formed along the coast, and a larger, shallow lagoon in the present Öja-Herrestads Mosse area (Håkansson and Kolstrup 1987, Mikaelsson unpubl.). This was sheltered towards the exposed sea in the south by a sandy beach ridge peninsula, which may sometimes during transgression periods have been dissected into sandy bars or islands. The trophic conditions in these shallow, brackish environments favoured a rich fauna (Gaillard et al. 1988, Jonsson 1988). In the inland area there were several lakes (about 20 with a size >2 ha indicated on the map), which made the landscape more diverse than today.

Vegetation and land use

Wildwoods with broad-leaved trees dominated the landscape. We distinguish four different kinds of woodlands according to the soil conditions. This ecological pattern persisted until historical times, although long-term land use changed the wood structure, even causing replacement of wood ecosystems.

1. Woods with broad-leaved trees – oak, lime, elm, ash, hazel, etc. – in the canopy and a species-rich understorey and shrub layer (*Euonymus* (spindle-tree), *Frangula* (alder buckthorn), *Viburnum* (guelder rose), etc.) growing on rather damp, fertile clayey to silty soils. This kind of woods dominated the landscape, particularly in the outer hummocky zone and the northern lake area.

2. Woods with a mixture of broad-leaved trees and birch and pine on slightly drier, less fertile, sandy till soils, These woods were slightly more open than the previous type. They occurred mainly in the hilly areas of the inner hummocky zone.

3. Mixed woods dominated by oak, birch, and pine on dry, less fertile, sandy plains at the coast and the sandy hummocks in the inland zones.

4. Fen woods, mainly alder carrs, on peaty ground in depressions in the landscape, often enclosing small lakes, rivulets, brooks, etc. In the border zones of these wetlands, gentle slopes with gley soils were covered by ash woods. Open woodless fens occupied only very small areas, as sloping mires surrounding these alder carrs. The wooded wetlands with alder and ash covered large areas, particularly in the outer hummocky zone.

The evidence of settlement is concentrated in the coastal zone, particularly the sandy islands in the Öja-Herrestad lagoon, or the sandy peninsula nearby, and at some of the lakes, indicating that fishing and hunting were most important for the Mesolithic economy (Ch. 3.1.2). People also utilized the woodlands for gathering plant products. There may have been some garden cultivation during the final phase of the Mesolithic (Ch. 4.1.2, 4.2.4, 4.4.2). If such cultivation occurred we can assume that this mainly affected the mixed wood on sandy soils (type 3). Altogether this means that the wildwoods were affected by man only in the vicinity of the settlements. If woods were coppiced during this time, this meant a transformation of wildwoods only within restricted areas (cf. Göransson 1986, 1988).

3.1.2 Settlement and society

Lars Larsson

Background

The human activity which took place in the time between the melting of the Glacial ice sheet and the Atlantic period is represented in the Ystad area by very little evidence. The Late Palaeolithic Bromme culture during the Allerød interstadial period is substantiated only by a couple of characteristic tanged points, which cannot be associated with any particular place (Larsson 1984c).

The Maglemose culture during the Preboreal and Boreal periods is represented by small numbers of widely distributed remains of settlements within essentially the whole of the project area. It is also possible to include amongst the evidence a number of bog finds of bone and antler items, like slotted bone points (Larsson 1973), which belong to this cultural period. The Ystad area does not include any really large lake which was continuously being filled with vegetation remains during the Boreal period – this was an environment which was otherwise particularly attractive to the people of the Maglemose culture in southern Scandinavia (Welinder

1971, Larsson 1978, Andersen et al. 1982). A similar pattern of distribution of small remains of settlements was found in south-east Scania (Strömberg 1986).

The sea level during the Early Mesolithic time was very much lower than it is today, for which reason the settlements along the coast are now at a submarine level (Larsson 1983).

Settlements

Lagoons and deltas throughout large areas of south Scandinavia were formed and enlarged by repeated transgressions during the Atlantic period. The settlements were concentrated in these environments in particular. Settlements on lagoons or at the mouths of rivers are well documented along the coasts of south Scandinavia (Larsson 1980, 1982b).

Settlement within the investigated area was focused on the area which today is covered by the bog Öja-Herrestads Mosse, immediately to the east of Ystad. The lagoon which was created here was an extremely attractive site for settlement. A number of factors contributed to this. The proximity of fresh water, brackish water and salt water provided a varied and highly productive fish fauna (Gaillard et al. 1988, Jonsson 1988). The headlands and sandbanks which were formed at the mouth of the lagoon were important resting places for seals. The sparse woodland on the sandy soils around the lagoon contained lush undergrowth and was thus an important habitat for large woodland animals.

Molluscs provide the most concrete traces of coastal settlement in southern Scandinavia (Andersen and Johansen 1986, Larsson 1989e). No kitchen middens are found along the south coast of southernmost Sweden, however. Late Mesolithic coastal settlements dating from the Ertebølle culture, from that part of the project area with which we are concerned, have been documented in two areas of the lagoon which today is Öja-Herrestads Mosse: (1) headlands or islands out in the lagoon; and (2) on sandbanks at the mouth of the lagoon. There are examples of detailed study of the settlement in each of these environments (Nilsson 1983, M. Larsson 1986b), although they are based on rather a small number of excavated sites (Lindahl 1976). The Ystad area provides some evidence of habitation in the interior close to lakes, some of which were at the stage of being filled up by vegetation, such as the boggy area known as Kumlan, adjacent to the river Svartån, or next to open lakes, as in the case of the remains of settlements along the shores of lake Ellestadssjön and on the islands on the same lake (Larsson 1990b).

These settlements represent both an old and an intermediate phase of the Ertebølle culture (Vang Petersen 1983). The most recent part of the Ertebølle culture, on the other hand, is not at all well documented.

The settlements vary in size from a couple of hundred square metres to about a thousand square metres. The Bredasten settlement, which is situated on a headland in the lagoon, contains the remains of at least one structure which has been interpreted as a hut (M. Larsson 1986b).

Use of resources

The position of the large settlements, the archaeological finds and the ^{13}C measurements indicate that fishing was the main source of food (Larsson 1982b, 1988d). Nevertheless, the utilization of the available resources extended to a very much larger geographical area than could be reached in one day's travel from the base settlement. The settlements on the sandbanks are regarded as seasonal camps which were used in conjunction with specific hunting activities such as seal-hunting and fishing. As far as the settlements which have been found in the area behind the coast are concerned, the gathering of vegetables during the autumn may have been an important activity (Larsson 1983a). The watercourses and the coast were of great importance for transport within the resource area.

Social structure

Settlement around the ancient lagoon in the Ystad area can be dated in close chronological association with the studies of the Skateholm project. This project covers the investigation of a number of Late Mesolithic settlements close to an ancient lagoon approximately 20 km west of the bog Öja-Herrestads Mosse. A combination of settlement and cemetery, both of considerable size, was documented at two sites, both of which are situated on islands (Larsson 1984a,e, 1988b,c). These are regarded as base camps.

The presence of cemeteries close to the settlement is interpreted as a way for a community to mark its right to utilize a resource area, providing an indication of the existence of some kind of confrontation between the communities (Larsson 1984e). This may have been of a social nature rather than governed by a situation involving a lack of food. Evidence of large cemeteries, which were in use for several centuries, indicates that the settlement was of a fairly permanent nature.

In conjunction with any movement of the settlement, due to transgression, a new settlement was selected in the vicinity of the old one. The sites appear to have been used by a population which included a number of family-sized groups.

The wide distribution of this form of settlement can be appreciated from the fact that similar sites of finds with a combination of settlement and cemetery are encountered along the whole Atlantic coast of western Europe (Pequart et al. 1937, Pequart and Pequart 1954, Albrethsen and Brinch Petersen 1977).

The presence of cemeteries and their function of demarcating territories indicate the existence during the Late Mesolithic period of certain conflicts between communities. This may have been caused by a slight,

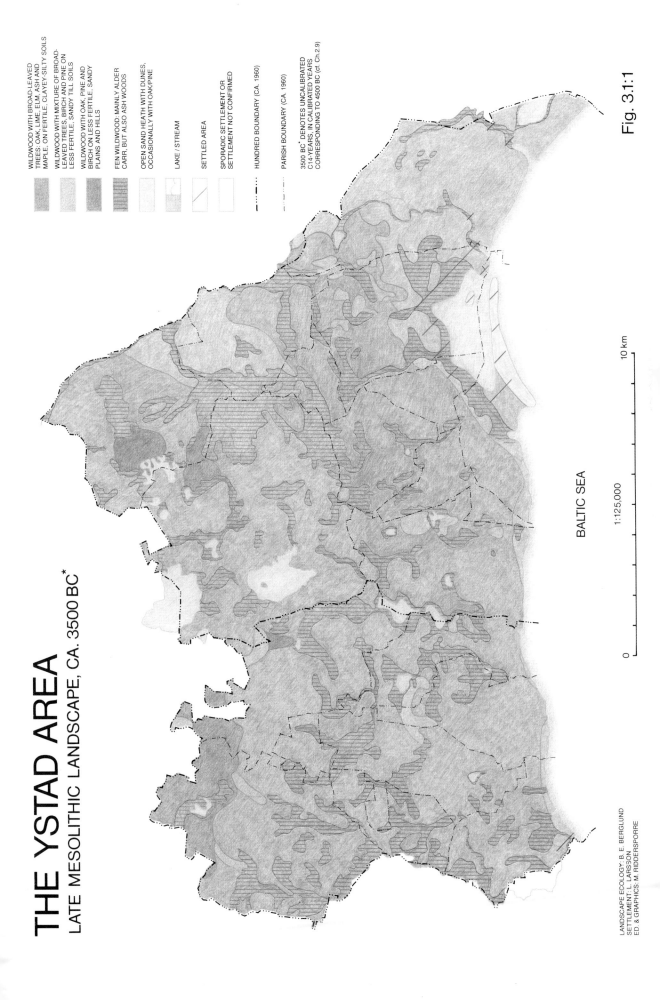

THE YSTAD AREA
LATE MESOLITHIC LANDSCAPE, CA. 3500 BC*

WILDWOOD WITH BROAD-LEAVED
TREES: OAK, LIME, ELM, ASH AND
MAPLE, ON FERTILE, CLAYEY-SILTY SOILS

WILDWOOD WITH MIXTURE OF BROAD-
LEAVED TREES, BIRCH AND PINE ON
LESS FERTILE, SANDY TILL SOILS

WILDWOOD WITH OAK, PINE AND
BIRCH ON LESS FERTILE, SANDY
PLAINS AND HILLS

FEN WILDWOOD: MAINLY ALDER
CARR, BUT ALSO ASH WOODS

OPEN SAND HEATH WITH DUNES,
OCCASIONALLY WITH OAK/PINE

LAKE / STREAM

SETTLED AREA

SPORADIC SETTLEMENT OR
SETTLEMENT NOT CONFIRMED

--·--·-- HUNDRED BOUNDARY (CA. 1960)

--···--··· PARISH BOUNDARY (CA. 1960)

3500 BC* DENOTES UNCALIBRATED
C14-YEARS, IN CALIBRATED YEARS
CORRESPONDING TO 4500 BC (cf. Ch.2.9)

BALTIC SEA

1:125,000

0   10 km

LANDSCAPE ECOLOGY: B. E. BERGLUND
SETTLEMENT: L. LARSSON
ED. & GRAPHICS: M. RIDDERSPORRE

Fig. 3:1:1

but in the longer term noticeable increase in the population (Welinder 1982), combined with changes in the environment due to transgressions in the vicinity of the groups of settlements. The estuaries and lagoon either lost their protected beach ridges or were almost totally closed by new and larger beach ridge systems which drastically changed the productivity of the important fishing when the ration of fresh to salt water changed considerably (Larsson 1987e). The utilization of previously marginal resources, such as the hunting of large game in the inland areas, may have increased during the latter part of the Mesolithic period. Certain minor direct changes to the environment can also be identified (Göransson 1988). This may have coincided with the introduction of arable farming (Jennbert 1984).

Skills and raw materials may have been introduced from the west, and also from the south. Communities based on both arable farming and animal husbandry already existed in the northern parts of Continental Europe during the middle period of the Ertebølle culture (Whittle 1985). Concrete finds dating from this time point to contacts between these agricultural communities and south Scandinavia (Fischer 1982, Jennbert 1984, Larsson 1988a).

3.2 The Early Neolithic landscape

3.2.1 Landscape, land use, and vegetation

Björn E. Berglund

Physical geography

The regional climate changed around 3000 BC towards more continental conditions, at least temporarily, for the Early Neolithic period. This meant colder winters, possibly warm summers, and low humidity (Iversen 1944, Lamb 1977, Ch. 5.1). In the Scandinavian mountains the timber-line was distinctly lowered around 3000 BC (Kullman 1988) and glaciers advanced around 3100–2500 BC (Karlén 1988). Evidence from lake levels in southern Sweden indicates low humidity for most of the period 4000–2500 BC (Ch. 5.1). Possibly there were unstable thermal and hydrological conditions with sudden climatic changes around 3000 BC (Ralska-Jasiewiczowa and Starkel 1988). However, after 2500 BC the climate continued to change slowly towards more maritime conditions – cool/humid alternating with warm/dry conditions.

The relative sea level of the Baltic was about the same as during the preceding period, although with a distinct regression about 3100 BC followed by a transgression peak about 2700 BC. The coastline reached approximately 4 m a.s.l. (Christensen 1982), about the same as before with a lagoon in the Öja-Herrestad basin. During the transgressive period about 2700 BC, however, the sandy peninsula was probably dissected. In the inland area some lakes disappeared because of dry conditions which favoured overgrowing. The number of lakes with a size >2 ha decreased from 20 to 13. A temporary rise of groundwater level is recorded in Scania around 3100 BC (Digerfeldt 1988).

Vegetation and land use

Around 3150 BC the woodlands on dry soils were suddenly deforested, the event named the "elm decline", which we regard as a catastrophic event caused by a complex of interacting factors, among them sudden climatic change, pathogen outbreak (elm disease), soil leaching – followed by increased human impact (Troels-Smith 1960, Watts 1961, Iversen 1973, Rackham 1980, Groenman-van Waateringe 1983, Huntley and Birks 1983, Göransson 1983, Berglund 1985, Birks 1986a). For a period of about 600 years (3100–2500 BP) the landscape was characterized by half-open, patchy woodlands where birch and hazel on dry ground and alder and willows on wet ground colonized the wood glades instead of elm, lime, ash, and oak. Open areas with grass- and herb-rich communities were also more common than before.

The ratio of dry/wet wood communities was about the same as before, but probably the sudden and short-lasting water-level rise at 3100 BC caused a brief expansion of fen woods with alder, around earlier carrs as well as in former woodland depressions with stands of broad-leaved trees (e.g. Fårarps Mosse, Ch. 4.1, Bjärsjö-holmssjön, Ch. 4.2). However, at the time of the Early Neolithic map (Fig. 3.2:1) the wet communities had a reduced distribution compared to the time of the Meso-lithic map (Fig. 3.1:1).

The human impact was caused by small-scale farming – stock-raising and farming on small fields. This activity was concentrated to the coastal zone and the inner hummocky landscape (Ch. 3.2.2). Possibly the inland areas with wood-pastures and sparse settlement were connected with similar areas on the Vomb plain north of the Romeleåsen ridge (Digerfeldt unpubl.). More occasionally also areas with light soils in the outer hummocky landscape were utilized, like the Bjersjöholm area (indicated on the map as one example of an area with human disturbances in this landscape zone). In these areas the landscape was characterized by wood-pastures: these are woodlands with half-open areas caused by grazing and browsing, around settlement places with more open conditions – arable fields, pastures, etc. Probably the woods within and around settled areas were coppiced in order to obtain leaf fodder, building material, etc. Regenerating woodland – underwood – was particularly well suited for coppicing. The mobile settlement character, for instance within the Köpinge area, meant that quite large areas (in relation to the population size) were affected by some kind of exploitation – dynamic wood-pastures with coppice (Ch. 4.1.3). Because of the light soils with less productive vegetation on the sandy Köpinge plain, we assume that this landscape was the first area to change to a half-open pasture landscape with coppiced trees. The coastal area east of the river Skivarpsån also had a similar character. There is some indication from the pollen diagrams of expanding agrarian land use throughout the period 3100–2700 BC.

Landscape changes throughout the Neolithic. Pollen diagrams as well as archaeological evidence speak in favour of a concentration of settlement and agrarian land use to the coastal area during the period 2700–2300 BC, corresponding to the closing phase of the Early Neolithic and the start of the Middle Neolithic. This interpretation means that a phase with woodland regeneration after 2700 BC was caused by (1) the natural recovery after the earlier deforestation crisis, and (2) decreased human impact in inland woodlands (Berglund 1969, 1985). The concentration of settlement to the coast meant, however, a locally expanded human impact also during this time. After 2300 BC there was an expansion, particularly of pastures, and coppiced woodlands became more common also in the outer hummocky landscape. But still areas with wet or heavy clay soils were untouched in this landscape zone. This gradual deforestation corresponds to an agrarian expansion covering the late Middle Neolithic, Late Neolithic, and Early Bronze Age. The coastal area around Öja-Herrestad-Köpinge became the most utilized landscape with an open character of pastures, arable fields, small coppiced woods or single trees already from the Late Neolithic onwards. It is probable that the inland zones were mainly used for coppicing and grazing.

3.2.2 Settlement and society

Mats Larsson

The research situation

Becker's "Mosefundne lerkar" (1947) was for many years the standard work for the chronological discussion of the Early Neolithic in this area. More recently the discussion has to a certain extent been continued from other points of view. The increasing number of ^{14}C datings has led to a more diversified picture, and greater emphasis has been placed on local developments observed in various areas of southern Scandinavia (Ebbesen and Mahler 1980, Madsen and Petersen 1984, M. Larsson 1984, Nielsen 1985). Local groups have been identified, and it is now difficult to discuss common Scandinavian development.

The last decade has seen the publication of a number of works which deal with various aspects of settlement, graves, and social structure in the Early Neolithic Funnel Beaker culture. One important aspect is the view of the megalithic graves as social manifestations (Randsborg 1975, Jensen 1979, Kristiansen 1982, Strömberg 1982, Tilley 1984, Hårdh 1986, M. Larsson 1988b).

The view of Early Neolithic agriculture as based entirely on slash-and-burn has been discussed in great depth in recent years. The basis for the discussion is represented by Iversen's work (1941). Later works have placed more emphasis on changes in economy, climate, and social factors (Rowley-Conwy 1981, Göransson 1982, 1986).

Distribution of settlement

Compared with the Mesolithic, our knowledge of settlement during the Early Neolithic is based on much larger material. Field surveys have been made of most of the Ystad area, revealing the existence of a large number of sites. Some of these have been investigated (Tesch 1983, Larsson and Larsson 1984, 1986, Larsson 1985, M. Larsson 1987). The area is characterized by small settlement sites. A combination of sandy soil, gentle heights, and access to water and wetlands is also typical. Similar patterns of settlement have been observed in other parts of southern Scandinavia (Bekmose 1977, M. Larsson 1984, Skaarup 1985). Of particular interest is the change in the pattern of settlement which can be observed between the older (Period 1) and the younger (Period 2) parts of the Early Neolithic. From a situation of small, widely distributed sites both along the coastal strip and in the hummocky landscape during Period 1, it is possible to distinguish a noticeable concentration of settlement in certain parts of the Ystad area from about 2700 BC. This change in the pattern of settlement can be studied in three areas in particular: the Mossby area, Herrestads Mosse, and the Köpinge/Kabusa area. Several investigations have shown that the wetlands, river mouths, lagoons, and archipelagos were attractive for settlement during Period 2 of the Early Neolithic (Bekmose 1977, Jensen 1979, Madsen 1982, M. Larsson 1984, 1988b). The first dolmens were also built in areas like these (Bekmose 1977).

Sites and finds

Distribution and structure of sites. The sites are small, rarely larger than 800 m². Houses have been found at some of them (Larsson and Larsson 1984, 1986). Never more than a single contemporaneous Early Neolithic house or hut has been found at any site in the project area. The best-preserved house was found at Mossby (Västra Nöbbelöv parish). The house had a single row of roof-bearing posts and was about 12 m long and about 6 m wide, with a slightly curved wall line. The house has been dated to Period 1 of the Early Neolithic, and has been allocated to the Mossby group (Larsson and Larsson 1986). Radiocarbon dating puts the age of the site at between 3280 (Ua-429) and 2965 (Ua-753) BC).

Graves. No flat-earth graves dating from the Early Neolithic have been found in the Ystad area. A funnel beaker from Fredriksberg (Stora Herrestad parish) may possibly come from a grave. One megalithic grave has been preserved in the area, the dolmen Trollasten in Stora Köpinge parish (Strömberg 1968). Studies of old land survey documents have revealed that there were at least three more in the Stora Herrestad/Stora Köpinge area alone (Wallin 1951, Riddersporre 1987). A further dolmen was identified in Stora Köpinge parish (Skogsdala). Its presence was substantiated only through coloration of the sand (Jakobsson 1986). A similar situation relating to sources is encountered in other parts of the project area. There is information about a dolmen in Snårestad parish, although its location has never been established (Bruzelius 1866). A severely disturbed

THE YSTAD AREA
EARLY NEOLITHIC LANDSCAPE, CA. 2700 BC*

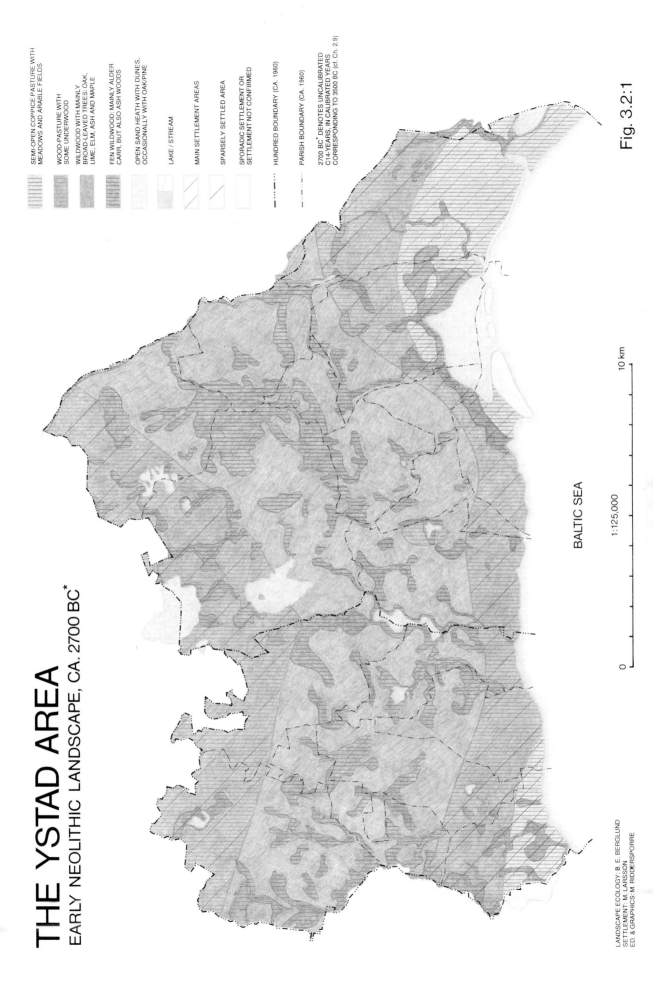

SEMI-OPEN COPPICE PASTURE WITH MEADOWS AND ARABLE FIELDS

WOOD-PASTURE WITH SOME UNDERWOOD

WILDWOOD WITH MAINLY BROAD-LEAVED TREES: OAK, LIME, ELM, ASH AND MAPLE

FEN WILDWOOD: MAINLY ALDER CARR, BUT ALSO ASH WOODS

OPEN SAND HEATH WITH DUNES, OCCASIONALLY WITH OAK/PINE

LAKE / STREAM

MAIN SETTLEMENT AREAS

SPARSELY SETTLED AREA

SPORADIC SETTLEMENT OR SETTLEMENT NOT CONFIRMED

HUNDRED BOUNDARY (CA. 1960)

PARISH BOUNDARY (CA. 1960)

2700 BC* DENOTES UNCALIBRATED C14-YEARS, IN CALIBRATED YEARS CORRESPONDING TO 3500 BC (cf. Ch. 2.9)

BALTIC SEA

1:125,000

0   10 km

LANDSCAPE ECOLOGY: B. E. BERGLUND
SETTLEMENT: M. LARSSON
ED. & GRAPHICS: M. RIDDERSPORRE

Fig. 3.2:1

megalithic grave lies to the south of Charlottenlund (Bruzelius 1866).

Hoards. A number of hoards of thin-butted axes have been recorded in the Ystad area (Hellerström 1988). These are distributed over a relatively large area, although a concentration can be observed in the Herrestad area. The significance of these finds has been the subject of much debate in recent years (Rech 1979, Olausson 1983, Nielsen 1984, M. Larsson 1985).

Economy

Arable farming. Arable farming in the area is relatively well documented through studies of macrofossils and impressions found in sherds. The main crops were emmer and einkorn, although some naked barley was also grown. This accords well with the observations made in other areas (Jørgensen 1982, Hjelmqvist 1979).

Animal husbandry. Our knowledge of animal husbandry is severely restricted because of the poor conditions for the preservation of osteological material in large parts of the Ystad area. There is evidence of pigs, domestic cattle, and sheep/goats from the Mossby site. It has been suggested that the importance of the pig was at its greatest during the Early Neolithic (Madsen 1982). This has recently been questioned (Nyegaard 1985). There is a tendency for the importance of animal husbandry to increase during Period 2 of the Early Neolithic and the early part of the Middle Neolithic (Madsen 1982, Nyegaard 1985).

Trade and external contacts

Trade in its true sense did not occur, with the exception of certain objects; long, thin-butted axes, copper axes, battleaxes, and amber were in circulation as symbols of prestige within a small segment of society (Kristiansen 1982, M. Larsson 1988a).

Social organization

Individual sites. Sites were small, of family size. Solitary houses/huts, hearths, pits, and occupation layers have been found on them. Different activity zones have been identified on the sites in certain cases, where work such as hide-processing may have been done.

Settlement area. Within the project area, the coastal strip was clearly utilized most intensively during Period 2 of the Early Neolithic. Five areas have been identified around Herrestads Mosse/Stora Köpinge/Kabusa, each of which exhibits continuity of settlement from the earliest part of the Early Neolithic up to the Middle Neolithic (M. Larsson 1988b). Each individual territory contained during each period one site, and in certain cases two, sites inhabited by a family-sized group of people. As an estimate, the Herrestad/Köpinge area may have been occupied by 25–35 persons at one and the same time. Other territories of a similar kind have been identified in the Mossby area, around Skivarpsån. However, the area between this western and the aforementioned eastern part of the Ystad area was much less widely settled. The area around Svarte may also have been a central settlement area.

Concluding remarks

The main problem is to relate the archaeological material to the stages of expansion observed in the pollen diagrams: there is a clear correlation between these and the increase in the intensity of settlement discussed above, which includes a late part of the Early Neolithic. This does not necessarily mean that the population increased dramatically, but simply that one particular area was utilized more intensively. The reorganization of arable farming associated with the introduction of the ard plough may be one such contributory factor. Settlement was also concentrated in a different way into small, ecologically favourable areas, where settlement may be regarded as having remained largely constant from a late part of the Early Neolithic and throughout the rest of the Neolithic.

3.3 The Late Bronze Age landscape

3.3.1 Landscape, land use, and vegetation

Björn E. Berglund

Physical geography

There was a major climatic change around 1000 BC towards more maritime conditions with lower summer temperatures and increased maritime humidity (Lamb 1977, Ch. 5.1). Paludification is frequent at this time in southern Scandinavia (Aaby 1976). In the Scandinavian mountains the timber-line was distinctly lowered and the glaciers were advancing (Kullman 1988, Karlén 1988). In general terms, the warm Subboreal climate changed towards the cool Subatlantic climate around the start of the first millennium BC. In earlier archaeological and palaeobotanical discussions the change was dated to about 500 BC and named "Fimbulvintern" (see Bergeron et al. 1956).

The relative sea level of the Baltic was distinctly lower than during the preceding periods. This is the time following the complex Littorina transgression. At the south coast of Scania the sea level was about 2 m higher than today, and the coast reached about the same configuration as in historical time. This meant that the former lagoonal basins became infilled and overgrown. Around 1500 BC the Öja-Herrestad basin became a marsh fen surrounded by alder carrs. The former sandy peninsula became overgrown by coastal oak-pine woods. In the inland area further lakes disappeared because of infilling. At this time the number of lakes with a size >2 ha had decreased to 8. The big Kumlan lake in the Svartån river valley became an alder carr around 2000 BC. The rise of the groundwater level around 800 BC caused flooding of some basins, for instance the Kumlan basin. At this time also a new lake was formed, namely Bjäresjösjön. It is probable that flooding caused a general expansion of wetlands.

Vegetation and land use

Around 800 BC, just at the transition Early/Late Bronze Age, a new deforestation phase affected the landscape, with the greatest changes in the outer hummocky zone. The coastal area was still the most utilized area, particularly the Köpinge plain which more or less lacked woods – coppiced stands or single trees occurred particularly on wet ground (cf. Ch. 4.1.4.3). However, alder carrs were partly transformed to open sedge fens or herb-rich meadows. We also assume that the western part of the coastal zone was more wooded than the Köpinge plain, possibly with the exception of a narrow coastal zone between Svarte and Ystad.

In the outer hummocky landscape the dry ground of hilly areas was settled and used for agriculture in the same way as the coastal area. Pasture-coppice with a mosaic of open pastures, meadows, and arable fields surrounded by wooded wetlands (carrs) characterize this landscape zone, which also included the eastern part of the Krageholm area (cf. Ch. 4.2.4.2), However, the alder carrs were partly replaced by open sedge fens or herb-rich meadows because of the growing need for fodder. Some parts of deep basins became flooded and for this hydrological reason alder stands were replaced by fen vegetation (reeds, etc.). This has been shown for the Kumlan basin by Gaillard (Ch. 5.1) but assumed for other large fen basins.

Most of the inner hummocky landscape was a marginal woodland area with occasional settlement used as wood-pasture – a common resource for the settled areas in the south. At least in some areas it still had the character of wildwood with untouched, tall trees, but in other areas grazing and some coppicing had led to at least temporary openings of the woods. However, we are not able to pinpoint the location of such places. A possible example is the area east of lake Krageholmssjön. It is reasonable to regard this inland zone as an area for transhumance wood-pasture (cf. Chambers et al. 1988).

The woods on dry ground had until now been dominated by the traditional broad-leaved tree species – oak, elm, lime, ash, hazel, and some maple – but from this time onwards also *Fagus* (beech) and *Carpinus* (hornbeam) belonged to the tree flora. They started to colonize at least the wood-pastures of the inland area, but had difficulties in coming into the more exploited landscape at the coast, which is evident from the Bronze Age map. In fact, our pollen diagrams from the south-eastern coastal area indicate that beech and hornbeam have always occurred sparsely here.

3.3.2 Settlement and society

Sten Tesch and Deborah Olausson

The research situation

Our view of society and the cultural landscape in southern Scandinavia in the Bronze Age has long been heavily influenced by the material found in graves. It was not until the 1950s that the first long-houses dated to the Late Bronze Age were found in Denmark, and it took several years before it was generally accepted that this was the normal Late Bronze Age form of settlement –

stable, year-round dwellings instead of tents or huts. The latter fitted the notion that Bronze Age society was semi-nomadic, with a population consisting of wandering herdsmen, whose burial places were a fixed point in their settlement territory. Since then the evidence from habitation sites has been rapidly increasing, and in the last ten years long-houses from the Early Bronze Age and even the Late Neolithic have also become common (Draiby 1985, Ethelberg 1987, Näsman 1987, Jensen 1988, Kristiansen 1988).

In Sweden too, it was long assumed that the remains of little huts revealed by earlier excavations represented the normal dwellings of the Bronze Age (Hyenstrand 1976). However, since the end of the 1960s, large-scale rescue excavations in several places in the Mälaren region have revealed traces of stable long-houses (Tesch 1980, 1983d, Jaanusson 1981, A.-L. Nielsen 1987, Wigren 1987, Ullén 1988a, b). Research into settlement in Scania had to depend for a long time on the evidence of graves (Stjernquist 1961, 1983, Welinder 1977), although the fragmentary material from habitation sites had also been studied (Strömberg 1954, Stjernquist 1969). It was only towards the end of the 1970s that the first sites with long-houses began to be discovered, both around Malmö and in the Ystad area (Tesch 1979a,b, Tesch et al. 1980, Björhem and Säfvestad 1983). This new evidence has led to intensified research into Bronze Age settlement patterns in Scania too (Tesch 1980, 1983b, Widholm 1980, Strömberg 1982, Björhem 1986, Olausson 1988).

The habitation sites discovered in southern Scandinavia in recent decades have thus meant that a new picture of Bronze Age society has emerged. Briefly, settlements appear to have been relatively stable, and tillage was more important than was previously assumed. The decisive changes in production and social structure that led up to Early Iron Age society seem to have begun in the Late Bronze Age. There is no evidence of any significant increase in population in the Bronze Age, so the causes of the change are assumed to have been more of an ecological and social kind (Kristiansen 1978, 1988, Jensen 1979, 1987a, 1988, Wigren 1987).

Settlement pattern and land use

At the start of the project there was fairly ample Late Bronze Age archaeological material from the coastal region, from habitation sites with long-houses and from cemeteries. The settlement pattern was assumed to have been relatively stable, with single farms moving short distances within naturally delimited settlement areas of about one square kilometre. In the hummocky landscape, on the other hand, the archaeological material consisted only of scattered hearths, pits, caches, and bog finds; this was taken to indicate that this was an outlying buffer zone, utilized for occasional settlement and summer pasturage, as well as for specialized activities such as harvesting foliage, hunting, lake fishing, and the exercise of cults (Tesch 1983b).

The western area
Field survey and limited excavation have produced evidence, in the form of concentrations of hearths/cooking pits or larger refuse pits containing up to 10 kg of pottery, for Late Bronze Age settlement at scattered spots in the area of barrow concentration, which stretches from the coast to about 4.5 km inland in the western part of the area (Fig. 3.3:1). Flotation samples from one such pit at Bjäresjö yielded evidence for the cultivation of barley, emmer wheat, millet, and gold of pleasure (Ch. 4.2.4.1). Efforts in the western area have however not uncovered any direct evidence for structures comparable to those on the Ystad plain to the east. Whether this difference is due to the necessarily limited nature of excavation in the west or to an actual difference in the type of settlement between the two regions is impossible to say on the basis of our present knowledge.

In the face of this dilemma, we can turn to other evidence, namely burials, in our attempts to reconstruct Late Bronze Age patterns of settlement. Here the barrows constitute the most obvious landmarks. Few in the western area have been excavated; nevertheless, from parallels with excavated barrows from other regions we assume that they represent continuity from the Early Bronze Age (when most such barrows were erected) in that they continued to be used for burial in the Late Bronze Age. We assume further that the barrows represent an investment in time and effort, and their location on prominent sites in the landscape indicates they were meant to be seen. For these reasons we presume that effort was made to maintain their visibility in the landscape during the Late Bronze Age: barrows represented continuity and probably territorial markings. As there is much evidence in support of the postulate that there was close spatial agreement between barrows and Bronze Age settlement, we conclude that Late Bronze Age settlement coincides with the area of barrow concentration. The numerous Late Bronze Age flat-earth burials at the coast might then represent the fixed point in a subsistence system in which family groups supported themselves by a mixed farming strategy in this zone. We note further that the zone follows the coast: perhaps marine resources or access to transportation routes were of decisive economic importance as well, and played a part in the choice of settlement location.

Further inland, beyond this primary zone, the evidence for settlement diminishes considerably. Barrows are extremely infrequent, and no flat-earth burials are known here. Indications of settlement consist of scattered concentrations of hearths/cooking pits found during surface survey and one clearance cairn east of lake Krageholmssjön. The pollen evidence attests to clearance occurring at this time (Ch. 4.3.2), but it is difficult to reconcile this with the scanty archaeological evi-

THE YSTAD AREA
LATE BRONZE AGE LANDSCAPE, CA. 800 BC–500 BC*

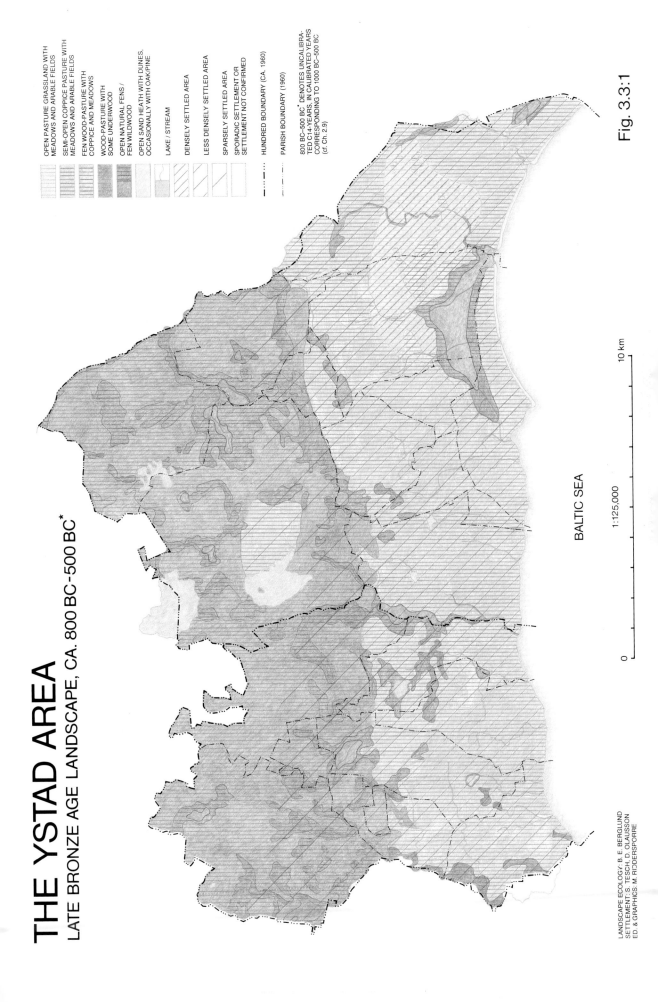

OPEN PASTURE GRASSLAND WITH MEADOWS AND ARABLE FIELDS

SEMI-OPEN COPPICE PASTURE WITH MEADOWS AND ARABLE FIELDS

FEN WOOD-PASTURE WITH COPPICE AND MEADOWS

WOOD-PASTURE WITH SOME UNDERWOOD

OPEN NATURAL FENS / FEN WILDWOOD

OPEN SAND HEATH WITH DUNES, OCCASIONALLY WITH OAK/PINE

LAKE / STREAM

DENSELY SETTLED AREA

LESS DENSELY SETTLED AREA

SPARSELY SETTLED AREA

SPORADIC SETTLEMENT OR SETTLEMENT NOT CONFIRMED

HUNDRED BOUNDARY (CA. 1960)

PARISH BOUNDARY (1960)

800 BC–500 BC* DENOTES UNCALIBRATED C-14-YEARS. IN CALIBRATED YEARS CORRESPONDING TO 1000 BC–500 BC (cf. Ch. 2.9)

BALTIC SEA

1:125,000

0   10 km

Fig. 3.3:1

LANDSCAPE ECOLOGY: B. E. BERGLUND
SETTLEMENT: S. TESCH, D. OLAUSSON
ED. & GRAPHICS: M. RIDDERSPORRE

dence. One possible interpretation is that settlement activity in this area was more sporadic and much more mobile than nearer the coast. The region could have been used for grazing to a greater extent than during the preceding period. It is however not possible to determine whether settlement was located in this zone, or whether what we see here is a forerunner to the system of transhumance practised in Sweden during the Middle Ages. The presence of Late Bronze Age hoards at scattered spots in this zone is, however, another possible indication of a territorial claim showing a sense of spatial continuity here.

The eastern area
In the Late Bronze Age there were very clear topographical preferences for the localization of settlement in the Köpinge area, the 4 × 6 km large Ystad plain (Ch. 4.1.4.1). The pattern included well-drained sites on low rises, light sandy soils, and proximity to wetlands, rivers, and the sea. It was more important, however, to be near the river Nybroån than to be near Öja-Herrestads Mosse or the sea.

Excavations have been carried out at eight of these habitation sites, or rather habitation areas. At six of these a varying number of long-houses have been discovered. Only slightly more habitation areas are known from the Late Bronze Age than from the Early Bronze Age or the Late Neolithic. Most of the habitation areas also show continuity since the Late Neolithic. Assuming that there was a roughly similar pattern of settlement, this does not imply any great increase in population since the Late Neolithic. On the other hand, a far greater number of camp sites on the periphery of the central area, towards the hummocky landscape or the sea, have been dated to the Late Bronze Age. This suggests that people were forced to look further afield for certain essential resources.

A few key sites are located just over a kilometre from the coast, on either side of the river Nybroån. At nearby habitation areas west of the river, excavations have uncovered a total of nine long-houses, measuring 12–16 × 7–8 m, and some smaller buildings. Three long-houses, measuring 13–16 × 7–8 m, have been excavated east of the river, as well as some smaller buildings. The long-houses are in general slightly smaller than those from the Early Bronze Age. At a glance these two sites look like villages with the houses arranged in rows. In the western settlement area, however, there are in fact three cases where houses are superposed on older houses, and a large number of ^{14}C datings show that the houses can be dated to the entire period. A more likely interpretation would thus be that settlement consisted of a single farm on either side of the river. Each farm moved a short distance, a few hundred metres at most, within a habitation area that was centrally situated in a settlement area which was about one square kilometre and relatively well defined topographically (Ch. 4.1.4.1). Since the sea and the groundwater were at

higher levels than today, every settlement area was more or less surrounded by wetlands. It has been estimated that the number of settlement areas – and hence the number of farms – on the Ystad plain in the Late Bronze Age cannot have been much more than about ten.

It has been difficult to use cemeteries for this study, since only one has been investigated in its entirety. Most of them are known only in the form of indications – ploughed-up cremation graves. We can nevertheless see a close spatial connection between graves and dwellings. The excavated cemetery is assumed to have served one farm (Olausson 1987, Ch. 4.1.4.2), which is further evidence for the single farm as the agrarian management unit.

Several interacting factors, not least a cooler climate, suggest that people began to stall livestock and manure the arable fields in the Late Bronze Age. However, byre partitions in long-houses have only occasionally been observed in Denmark. A change in the internal structure of the house, with pairs of posts placed closer together at the eastern end, is usually seen as an indication that animals were stalled at this end of the house (Draiby 1985, Näsman 1987). This functional division of the long-houses in the Köpinge area coincides with the start of the climatic change. The byre sections here are relatively small, no more than 4–7 m long, and they cannot have held more than 10–18 animals. Small byre sections have also been documented in some regions of Denmark, where they have been assumed to reflect a composite economy with tillage, cattle-raising, sheep farming, and fishing, rather than an economy based mainly on cattle-raising.

Late Bronze Age farmers grew more types of cereal than farmers in any other prehistoric period: naked barley, hulled barley, spelt, rye brome, oats, millet, etc. Often most species are found at one and the same habitation site, which suggests varied and well-planned agriculture. *Camelina sativa* (gold of pleasure) was also cultivated, possibly for its oil (Engelmark and Hjelmquist 1991).

No traces of ridge-and-furrow created by long-term ploughing have as yet been found in southern Scandinavia, which indicates that Bronze Age farmers had a mobile cultivation system, tilling fields for only a few years at a time (Jensen 1988).

In the course of the Late Bronze Age hulled barley became the main cereal in the Köpinge area. Since it requires more nutrient-rich soils than the cereals that were previously grown, it is assumed that some form of manuring was introduced during the period (Engelmark and Hjelmquist 1991). Manure is most easily obtained in an animal fold or by stalling cattle. People may also have used the defiled site of a former dwelling for new cultivation with an increased nutrient yield. Cultivation plots and houses were thus moved together over short distances within a single habitation area.

The old varieties of cereals, however, were high in

protein and very suitable for the light sandy soils, so they cannot have been abandoned without good reason. It is possible that the continuous deterioration in the climate had a detrimental effect on the yield of the old cereals and their resistance to disease (Engelmark and Hjelmqvist 1991). It is very hard to determine whether it was the colder climate or the need for manure that was the main reason for assuming that cattle were stalled, at least during the winter.

The end of the Late Bronze Age and the start of the Early Iron Age is usually viewed as a period of crisis in the development of the cultural landscape in southern Scandinavia. The factors adduced as the cause of this crisis, whether individually or in combination, are population pressure, overexploitation, and a worsened climate (Kristiansen 1978, 1988, Jensen 1979a, Burenhult 1983).

The gradually moister and cooler climate from around 1000 BC, with its climax around 600 BC, ought not to have led to any short-term worsening of grazing conditions on light soils, but in the long term the increased precipitation could have caused leaching of cal-cium and other mineral nutrients. The presumed growth in the herds of sheep and goats would also have reduced the sward and the broad-leaved woodland. As the woodland disappeared, people were forced to look further afield to collect foliage to feed the animals in their winter stalls. What is read in the pollen diagrams as an expansion of the cultural landscape could equally well be interpreted, from the archaeological evidence, as a society under stress. Yet farmers did not abandon the light soils. People in the Köpinge area adapted their farming methods to suit the changed conditions. Although the changes did not lead to a fully developed system of infield and outland with haymaking, as in the Early Iron Age, it can still be said that the foundation was laid for a more differentiated agrarian society. The transition to the Iron Age was far from being a break in the development of the cultural landscape. Both the structure of the habitation sites and the appearance of the houses indicate continuity.

Note: The section on The western area was written by Deborah Olausson.

3.4 The Roman Iron Age landscape

3.4.1 Landscape, land use, and vegetation

Björn E. Berglund

Physical geography

The Roman Iron Age falls within a period of milder climate than before and afterwards (Lamb 1977, 1982). In the Scandinavian mountains the glaciers were retreating, but began to expand again around AD 400 (Karlén 1988). The timber-line was also lowered at about the same time.

A major low-water period in south Swedish lakes also indicates low humidity for the period AD 0–750 (Ch. 5.1).

The sea level of the Baltic was about 1 m higher than today, which means that the coastline was similar to the present day. In the inland area the same lakes existed as during the Bronze Age. Possibly some swampy areas dried out because of the low humidity, indicated on the map (Fig. 3.4:1) by slightly reduced areas of wetlands.

Vegetation and land use

The Late Pre-Roman Iron Age and the Early Roman Iron Age are characterized by expanding woodlands at the expense of open land (pastures, arable land). Possibly this was caused by a concentration of the settlement to permanent farms/villages. These were situated in the coastal area and in favourable parts of the outer hummocky landscape. Our interpretation means that some areas of the outer hummocky landscape were deserted and colonized by secondary wood/shrub land, possibly forming two wood lobes west of the Svartån river valley and east of the Krageholm-Bjäresjö area. It is also possible that the utilization of the wood-pastures in the marginal inland zone decreased, described above as extensive transhumance pasture land. Wetlands with alder woods were continuously being transformed into open meadows to meet the increasing demand for winter fodder. This means that the area of alder carrs was reduced, while untouched wooded swamps became more fragmented.

All this means that the Roman Iron Age landscape, broadly speaking, was similar to the Late Bronze Age landscape. On a smaller scale, however, it differed because of the organization/land-use pattern related to the permanent villages: infields with arable fields and meadows around the farms, outland with pastures and coppiced woods. Although this system was already realized in part during the preceding period, it was fully applied during the Roman Iron Age. This was a more

intense and more productive agrarian system which made the landscape less dynamic than before.

Landscape changes around the Migration Period

During the period AD 300–600 there seems to be a continued woodland regeneration – expansion of birch, oak, beech, and hornbeam – probably as a consequence of settlement concentration as in the Roman Iron Age. It is likely that this affected the openness of the woodland landscape mainly in the inner part of the hummocky landscape. Possibly this change is related to concentration of the main agrarian production to farm/village infields, allowing woodland regeneration in the surrounding outland (Ch. 4.3.2).

3.4.2 Settlement and society

Sten Tesch

The research situation

Earlier archaeologists, who mostly worked with grave-finds, viewed the Pre-Roman Iron Age as a regression period. Thanks to a different kind of source material – habitation sites – modern archaeological research has been able to paint a completely different picture of Early Iron Age society in southern Scandinavia. Settlement development from the Bronze Age to the Early Roman Iron Age now appears to have been continuous. The centuries around the birth of Christ, when the warm, dry climate was favourable for cereal cultivation, was a very expansive period with a noticeable increase in population as a result (Burenhult 1984, Näsman 1987, Widgren 1987, Hedeager 1988). However, the continuous development in settlement came to an end at the transition to the Late Roman Iron Age. A time of change, known as the crisis of the Migration Period, began in the 3rd century and lasted right until the 7th century (Näsman 1988).

For the Bronze Age we have no clear evidence of villages. On the other hand, the large-scale excavations of recent years in Jutland have shown that the normal agrarian unit in the Early Iron Age was the "migrating village" (Becker 1971, 1983, Näsman 1987, Hedeager 1988, Hvass 1988). These villages normally consist of 10–15 farms, but a few are much larger (Lund 1988). Up until the Early Roman Iron Age the farms were assembled within a common enclosure. In the Late Roman Iron Age a noticeable break took place, with the

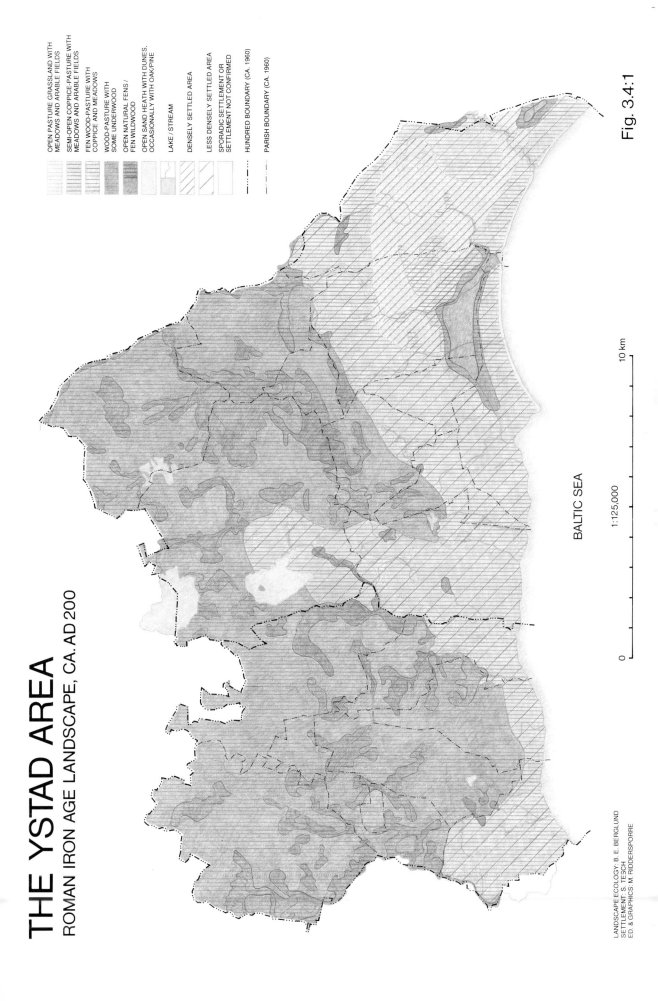

THE YSTAD AREA
ROMAN IRON AGE LANDSCAPE, CA. AD 200

OPEN PASTURE GRASSLAND WITH MEADOWS AND ARABLE FIELDS

SEMI-OPEN COPPICE-PASTURE WITH MEADOWS AND ARABLE FIELDS

FEN WOOD-PASTURE WITH COPPICE AND MEADOWS

WOOD-PASTURE WITH SOME UNDERWOOD

OPEN NATURAL FENS / FEN WILDWOOD

OPEN SAND HEATH WITH DUNES, OCCASIONALLY WITH OAK/PINE

LAKE / STREAM

DENSELY SETTLED AREA

LESS DENSELY SETTLED AREA

SPORADIC SETTLEMENT OR SETTLEMENT NOT CONFIRMED

HUNDRED BOUNDARY (CA. 1960)

PARISH BOUNDARY (CA. 1960)

BALTIC SEA

1:125,000

0   10 km

LANDSCAPE ECOLOGY: B. E. BERGLUND
SETTLEMENT: S. TESCH
ED. & GRAPHICS: M. RIDDERSPORRE

Fig. 3.4:1

farms becoming more differentiated and each one being individually enclosed – "regulated" – at the same time as land became private (Hvass 1988).

In Scania too, the last decade has seen a great many excavations in present-day arable land, with large habitation areas being stripped (Björhem and Säfvestad 1983, Wihlborg et al. 1985), and not least the study of the Ystad plain as part of this project (Ch. 4.1.4.4).

These excavations have shown that there is a noticeable difference in habitation site structure between, on the one hand, western Denmark (Jutland) and, on the other hand, eastern Denmark (including Scania). In eastern Denmark there are also villages, but they are often much smaller. In addition, the isolated farm is still a common form of settlement (Rønne 1986, Hedeager 1988).

Descriptions of land use in the Early Iron Age have long been based on areas where a fossil cultivation landscape has survived or can be detected by aerial photography, as in Öland, Gotland, Östergötland, and Jutland (Carlsson 1979, Widgren 1983, Stumman Hansen 1984). It is normally thought that most regions in southern Scandinavia went through the same development at roughly the same time, as regards not just land use (Berglund 1969) but also society and politics. In recent years it has become increasingly clear that we must reckon with large regional and local variations and that this is the reason why fossil cultivation landscapes can be found in some places but not in others. Variations are found not only between areas with and without a fossil cultivation landscape, but also between areas within these two categories (Widgren 1988). The differences between western and eastern Denmark concern not only settlement structure but also the agrarian organization that goes with it. In the west there are large, relatively extensively cultivated field systems of the Celtic field type. Although occasional large field systems do occur in the east, this region is characterized by smaller, perhaps more intensively exploited field systems (V. Nielsen 1984).

The restructuring of the agrarian system that affected the whole of southern Scandinavia as a result of the crisis of the Migration Period was far from being synchronous. There are widely varying opinions about the causes and effects of the process, and indeed about whether there actually was a crisis (Näsman 1988). There was a general intensification of cultivation, which laid the foundation for increased social stratification and the emergence of centres of wealth, such as Stevns in Zealand and Gudme in Funen. In Gotland the large, extensively utilized field systems were abandoned in favour of small fields with intensive single-course cultivation, which require heavy manuring if the land is not to be impoverished. The agglomerated settlements were split up into isolated farmsteads, and houses were now built with stone foundations (Carlsson 1979).

Settlement pattern and land use

The western area

Great effort has been expended in trying to find remains of Early Iron Age settlement in the narrow coastal strip in the west of the project area, and in the hummocky landscape as a whole (Olausson 1988). Indications have been found in both the Bjäresjö and Krageholm areas (Ch. 4.2.4, 4.3.4), but the material is not only small but also difficult to interpret. It is hard to say whether this reflects the actual situation or is due to the effects of modern agriculture and the difficulty in documenting remains in the soil types of the hummocky landscape. The localization of cemeteries in the coastal region should be a good guideline as to where the permanent farm settlements in particular must have been. In the area studied by the Hagestad Project, further to the east, settlement was mainly located on light soils close to the cemeteries (Strömberg 1982).

The settlements that have been found are concentrated in the Svartån river basin and to the east of lake Krageholmssjön in the outer hummocky zone. The hilly topography meant that settlement was on a small scale, either isolated farmsteads or clusters of a couple of farms. The small and fragmentary remains of houses that have been found hitherto give no clear indication of the prevailing habitation site structure or settlement pattern. Perhaps the settlement pattern was more mobile here than in the coastal region, rather like what Olausson has assumed for the Late Bronze Age (Ch. 3.3.2). Perhaps there was some sort of seasonal settlement, such as shielings for summer pasturage. The occurrence of large amounts of slag around these sites shows that iron production was one of the activities carried out here.

The eastern area (the Ystad plain)

A total of 10 habitation sites or areas have been dated to the Pre-Roman Iron Age and 12 to the Roman Iron Age. This is an increase on the number of Late Bronze Age sites investigated. As regards topography, settlements are localized according to roughly the same criteria as in the Late Bronze Age: they moved around within topographically delimited areas. One difference is that the settlement areas nearest the coast were less attractive in the Pre-Roman and Early Roman Iron Age. In general it can be said that there is distinct continuity in settlement development from the Late Bronze Age to the Roman Iron Age, whereas the Late Roman Iron Age, being a transition to the Late Iron Age, saw a break in this continuity.

As has been said above (Ch. 3.3.2), the transition to the Pre-Roman Iron Age did not bring about any sudden restructuring of agriculture in this area. The farms were still isolated, with long-houses of the same size as during the Late Bronze Age, with room for one extended family. This contrasts with what happened in

Jutland, where long-houses became much smaller in the Pre-Roman Iron Age, with room for only one family household, but with many houses clustered together like villages.

Although the deterioration in climate in the Late Bronze Age led to an overexploitation of the cultural landscape in the Ystad area, with a consequent deterioration in the soil, Bronze Age society nevertheless had the strength to confront the crisis with a gradual adaptation and reorganization. Although the changes – stalling animals, manuring, choice of cereals, and agricultural techniques – did not lead to a fully developed system of infield/outland with haymaking, it can still be said that the foundation was laid for the more differentiated and dynamic agrarian system of the Iron Age.

The Iron Age archaeological material suggests a considerable increase in population since the Bronze Age. Around the birth of Christ there is evidence in several places of agglomerated settlements – villages – comprising two or more farms. For several centuries these settlements appear to have been permanent in one place. A key site is located on the tofts of Lilla Köpinge village. Each farm here appears to have had its own fixed site, on which several layers of long-houses can be observed. The long-houses are relatively large: 17–26 × 5.5–6 m. Their overall area is not much larger than that of long-houses in the Late Bronze Age, but the greater length of the buildings made it possible to house larger numbers of livestock. The farms also have some smaller buildings, including sunken-floor huts, which were used primarily for weaving. In Denmark the first sunken-floor huts do not appear before the Late Roman Iron Age (Näsman 1987). In the Köpinge area, by contrast, there is sure evidence of them from the Pre-Roman Iron Age (Tesch 1988, Ch. 4.1.4.4).

In the Ystad area there are no surviving traces of fencing systems and arable parcels. Nevertheless, the changed pattern of settlement that can be interpreted on the basis of the archaeological material from the Köpinge area must be due to an intensified cultivation system. Like the rise in population, this is linked to an increase in agrarian productivity, especially grain cropping. Analyses of carbonized plants show that the cultivation of hulled barley predominated totally, although there is occasional evidence of rye cultivation (Engelmark and Hjelmqvist 1991). A question that remains to be answered is how the agrarian development that has been assumed for the Köpinge area is to be viewed in comparison with the "large field systems", for example in Gotland, which meant relatively extensive cultivation, with the only fertilization coming from household waste and animals grazing on fallow land (Carlsson 1979). The long continuity of the habitation sites in the Köpinge area could equally well indicate a fully developed infield/outland system with one-course tillage from as far back as the Early Roman Iron Age or even earlier, and that the basis of this system had been laid in the Late Bronze Age (Ch. 3.3.2).

In both Denmark and the Köpinge area there are few habitation sites showing continuity between the Early and Late Roman Iron Age, in contrast to the very common continuity in habitation sites between, on the one hand, the Pre-Roman and the Early Roman Iron Age and, on the other hand, the Late Roman Iron Age and the Migration Period (Näsman 1987). As in the Köpinge area, the number of documented habitation sites is much lower in the Late Roman Iron Age than in the Early Roman Iron Age. In the Köpinge area the agglomerated settlements are dissolved, and the isolated farm becomes the most common settlement form in the Late Roman Iron Age. There is a certain contraction of settlement, with a concentration to the settlement areas near the coast. This tendency becomes even clearer in the Migration Period, when the known settlement in the Köpinge area is largely concentrated in a single settlement area (Ch. 4.1.4.4, 4.1.5.1).

As we have seen, more intensive forms of cultivation can have emerged in the Early Roman Iron Age or even earlier, which illustrates the difficulty in drawing parallels between developments in different areas.

By contrast, changes in society and land use in the Late Roman Iron Age show the following two diametrically opposite features:

1) The well-controlled ecological system which had functioned for several centuries, and which had allowed the greatest population density hitherto in prehistoric times, came adrift and collapsed. The mutual dependence of tillage and animal husbandry (the source of manure) made the system vulnerable. The dispersion of settlement in single farms and the noticeably reduced number of known habitation sites bear witness to this, suggesting that there was a drop in population and a regression in the development of the cultural landscape. The decline is also evident in a pollen diagram from the fringe of the Köpinge area (Fårarp). The period is also characterized by a drop in groundwater level, infilled lakes, and the spread of the beech. The fall in the groundwater level that has been observed during the period AD 0–500 implied a long-term deterioration in the conditions for haymaking, which could in turn have affected the nutrient balance.

2) The increased social stratification suggests that the restructuring allowed a margin for the production of a surplus, which must at the same time have meant increased productivity in grain cultivation. In the Köpinge area a few very long houses have been measured, and a relatively lavish grave has been found.

The development of the cultural landscape in the Ystad area in the Roman Iron Age corresponds, broadly speaking, with the general development in southern Scandinavia. Sometimes the meagre or one-sided source material can make it difficult to synchronize the course of events here with that in other areas. At the same time, the evidence also suggests that there were both regional and local variations.

3.5 The Viking Age landscape

3.5.1 Landscape, land use, and vegetation

Björn E. Berglund

Physical geography

After a cool and humid period AD 400–900 there were in general warm conditions during the time AD 900–1200, that is the Viking Age and the Early Middle Ages (Lamb 1977, 1982). The favourable climatic conditions for agriculture in northern latitudes like Iceland and Greenland are well known (cf. Arbman 1962). This is also confirmed by recent studies of glacier changes (Karlén 1988) and tree growth at the timber-line (Schweingruber et al. 1988). In southern Swedish lakes the water level rose after about AD 750, indicating continuous humid conditions (Ch. 5.1).

The sea level and the coastline were about the same as today. The inland lakes were the same as before, and as far as we know, artificial damming of ponds had not yet started.

Vegetation and land use

During the period AD 700–900 there was a large expansion of the agrarian landscape. The more or less open village landscape expanded northwards so that the coastal zone as well as most of the outer hummocky zone became a woodless area with arable fields and meadows. Remnants of the coppiced woodlands were left in the northern part of the outer hummocky landscape, particularly in two lobes west and east of the Svartån river valley. In the marginal inner hummocky zone a wide wood-pasture was still left, probably used as common land for cattle and sheep belonging to the villages in the south. Possibly minor areas of wildwoods still occurred in this inland area, for example in the north-eastern corner.

For the entire area there was a noticeable reduction of broad-leaved trees on dry ground and alder in wet swamps. From this time onwards most alder carrs in the coastal area and in the outer hummocky landscape were transformed into open meadows used for mowing or grazing. Beech and hornbeam continued to expand during the period AD 500–800, mainly in wood-pastures belonging to the inner hummocky zone and outlying areas of the outer hummocky zone. Some pastures, particularly on light soils, became impoverished and developed – at least in the Early Middle Ages – into heaths with heather, juniper, and other heath elements (Ch. 4.2.2).

The land use and the pattern of the agrarian village landscape was about the same as during the High Middle Ages (AD 1300). This is described in greater detail in Ch. 3.6.2.

The village areas formed islands of a more open landscape, like the wide-open coastal plain around Köpinge – indicated on the map (Fig. 3.5:1) as "pasture grassland". The village settlement was surrounded by infields with arable fields, covering larger areas than before, and meadows for haymaking. The mowed meadows covered dry ground as well as wetlands in depressions. Surrounding the infields were the outland pastures, which were open or covered with scattered trees. They continued gradually into more remote coppice-pastures forming half-open woodland fringes between the village areas. Only in a few cases were the infield areas of two adjacent villages not separated by outland pasture.

3.5.2 Settlement and society

Johan Callmer

The research situation

The changes in the cultural landscape and settlement in the Late Iron Age and the Early Middle Ages are a problem that has been discussed by representatives of various disciplines, especially among historians, place-name scholars, archaeologists, and historical geographers. In recent years Quaternary biologists have also joined the debate. The discussion of these problems in southern Scandinavia is part of a research field that encompasses the whole of northern and north-west Europe. For southern Scandinavia historians and place-name scholars were the earliest to present elaborate theories about settlement in the Late Iron Age. A theory based on toponymic evidence in particular claimed that Late Iron Age settlement consisted of villages and that their locations were identical to those of the historical villages (e.g. Arup 1926:64ff.). This view, which scarcely considered archaeological findings, was heavily criticized in the 1970s and 1980s. The criticism was justified in that the older view had a poor empirical foundation. It was now maintained instead that there was discontinuity between the settlement of the Late Iron Age and that of the Early Middle Ages. One scholar who has emphatically argued the case for discontinuity, on the basis of a series of excavations of historical village cores in Funen, is Grøngaard Jeppesen (1981). Many archaeologists and historians then seconded his view that there was a noticeable change of structure between the Late Iron Age and the Early Middle Ages (e.g. Skansjö 1983, Stjernquist 1987). It

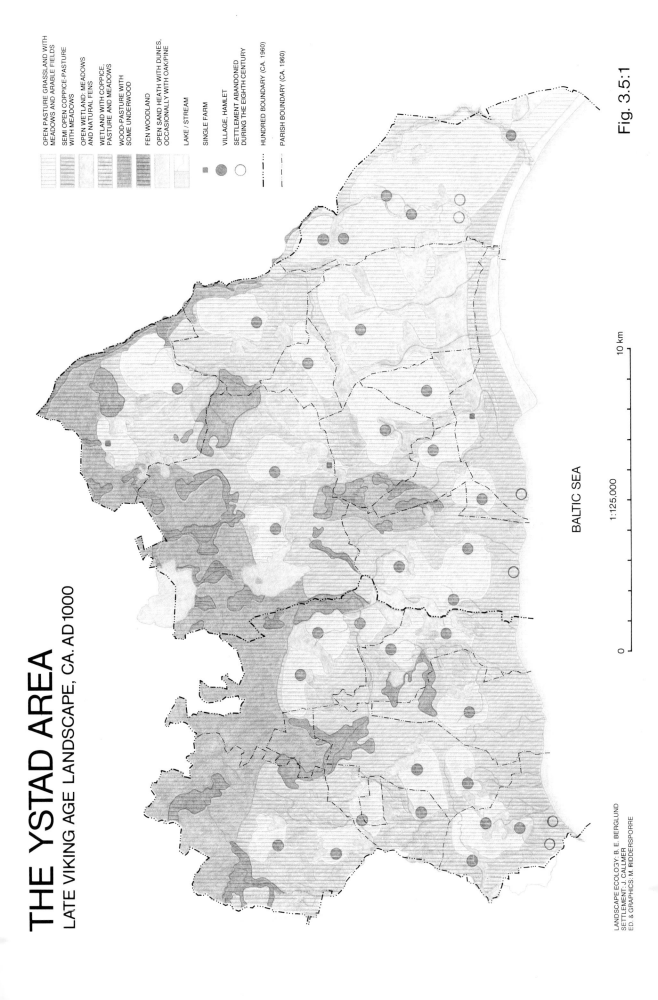

THE YSTAD AREA
LATE VIKING AGE LANDSCAPE, CA. AD 1000

OPEN PASTURE GRASSLAND WITH
MEADOWS AND ARABLE FIELDS

SEMI OPEN COPPICE-PASTURE
WITH MEADOWS

OPEN WETLAND: MEADOWS
AND NATURAL FENS

WETLAND WITH COPPICE,
PASTURE AND MEADOWS

WOOD-PASTURE WITH
SOME UNDERWOOD

FEN WOODLAND

OPEN SAND HEATH WITH DUNES,
OCCASIONALLY WITH OAK/PINE

LAKE / STREAM

SINGLE FARM

VILLAGE, HAMLET

SETTLEMENT ABANDONED
DURING THE EIGHTH CENTURY

HUNDRED BOUNDARY (CA. 1960)

PARISH BOUNDARY (CA. 1960)

BALTIC SEA

1:125,000

0   10 km

LANDSCAPE ECOLOGY: B. E. BERGLUND
SETTLEMENT: J. CALLMER
ED. & GRAPHICS: M. RIDDERSPORRE

Fig. 3.5:1

was also considered that cultivation history supported this interpretation. A dramatic expansive phase for cultivation was claimed to be synchronous with the structural change. In recent years a new discussion has developed against the background of comprehensive excavations and charting of indications of habitation sites and graves from the Late Iron Age (Callmer 1986b). In parts of the central settlement regions there is often a close connection between Late Iron Age sites and graves and historical village cores. The *pattern* of Viking Age and early medieval settlement developed mostly during the Late Iron Age. On the other hand, there appears to have been a greater difference between the Early and the Late Iron Age. There is now a well-founded and widely accepted view that Viking Age habitation sites in central plains settlements in southern Scandinavia consisted of village-like agglomerations.

Source criticism and methods

Settlements from the Late Iron Age are very difficult to identify by most surveying methods (Grøngaard Jeppesen 1981:22, 117–119). There are often no finds on the surface, and the limited use of hearths means that very little fire-cracked rock can be seen by surface survey. In undulating terrain the tops of the hills have lost all traces of occupation layers, which have been removed by cultivation, whereas the slopes are often covered by a thick layer of soil brought down by recent erosion. For these reasons the phosphate evidence can also be lacking. It appears, however, that phosphates are generally preserved at such a depth that mapping and calculations of intensity often give satisfactory results when the erosion cover is not too thick to make investigation impossible (Callmer 1980, Tesch 1982). For the compilation of the settlement map, the primary value has been ascribed to positive indications in the form of artefacts (from potsherds to runic stones). After that the phosphate evidence has been studied on the basis of the map by Arrhenius (1934), and the relative terrain locations of the historical villages have been analysed. Place-name evidence has also been taken into consideration (Hallberg 1975 for Ljunits Hundred and Hallberg pers. comm. for Herrestad Hundred).

Distribution of settlement in time and space

Settlement indications are spread fairly evenly over the central and coastal parts of the Ystad area. However, some localization preferences can be detected: lakes, watercourses, and the border zone between the coastal plain and the hummocky landscape. The picture of settlement also shows a fairly clear number of concentrations separated by areas lacking indications.

To the west there is a certain concentration by the river Skivarpsån. This must have been separated from settlement to the east by a lobe of woodland extending from the north and reaching almost as far as the sea. This western settlement area covered a rather wide space: as far as Norra Villie, possibly even Trunnerup, to the north, and to the east into the hummocky landscape as far as Norra Vallösa and Sjörup. In the 10th century two or more old settlements by the coast were abandoned (Larsson and Olausson 1986). One of these probably moved only one kilometre inland, while the other may have led to the establishment of one or more new settlements, such as at Västra Nöbbelöv – the name Nöbbelöv is a corruption of *nyböle* "new settlement" (Strömberg 1961:43).

In the central part of the Ystad area settlement is located on both sides of the river Svartån, and it extends as far north as Gussnava. Sites on the actual river-bank were avoided. Settlement does not extend very far either east or west. East of the river the northernmost site in the continuous settlement region is at Bjäresjö. Between this area along the river Svartån and that at Skivarpsån lies Snårestad, relatively near the coast, but with no clear-cut connection to either of these two settlement areas. There may have been one or more Iron Age coastal settlements by the mouth of the river Svartån, abandoned in the 10th century, although we have no certain evidence for this. Further east we find two historical villages with locations very similar to conditions at the mouth of Skivarpsån, and here we have evidence of two coastal sites that were deserted in the 10th century (Strömberg 1961:25–26).

In the east of the Ystad area there is a row of settlements along the eastern edge of the hummocky landscape, bordering on the plain. Three indications are also found a little further into the hummocky zone, but with links to the outer chain of settlements.

The settlements we find on the eastern edge of the Ystad area are localized according to quite different principles. By Nybroån the settlements are very near the river. This contrasts with the west of the area, but it agrees well with what is found beside a number of lakes and rivers elsewhere in southern Scandinavia. A good example of this is the lower course of Nybroån, with Lilla Köpinge and Stora Köpinge. The location of the two villages follows the pattern of pairs of villages, known also from bigger river valleys. At the end of the Viking Age both Stora and Lilla Köpinge were settlements of village type composed of units which had developed locally in the immediate neighbourhood of the village site and which had thus a long continuity, and units which had completely or partly constituted separate settlements closer to the mouth of the river (Callmer 1986:45–48). These contractive patterns appear to continue with varying intensity in southern Scandinavia in the Late Iron Age and into the Early Middle Ages.

In the interior of the Ystad area there are two small isolated settlements: Baldringe is an isolated outpost to the north, while Årsjö and Sövestad together constitute a small, continuous settlement area on the eastern shore of lake Krageholmssjön. Two observations from the parishes of Sövestad and Baldringe indicate that there were smaller settlements consisting of isolated farms or double farms in the interior (Strömberg 1961:25, 27). In an early phase of the Late Iron Age, settlement near Årsjö consisted of a small group of farms or a single farmstead (Callmer 1986a:48–49). The two settlements that have been investigated at Baldringe and Sövestad appear to have ceased to exist in the Late Viking Age or the Early Middle Ages. It is uncertain how common isolated farms like this were in the colonization phase of the interior, but it is not improbable that there were more than we know of at present and that our examples do not paint the complete picture.

The physical extent and appearance of the habitation sites

On account of the poor preservation of habitation sites in the hummocky landscape, and because of the somewhat unsatisfactory conditions for investigating other places on the coastal plain, no site has been excavated in its entirety. Data from the area nevertheless show that settlement consisted largely of village-like agglomerations, with some isolated farms in the interior. As regards population and surface area, the habitation sites on the plain and in the outer hummocky zone were larger than those further inland. We lack the evidence necessary to estimate the number of farms in each settlement.

Dwelling houses in the Late Viking Age continued to be long-houses – rectangular or with convex long walls. These were supplemented by rectangular wooden buildings and sunken-floor houses for various purposes. A farmstead covered a large area, since its buildings were loosely grouped. In the Middle Ages the big long-houses were replaced by smaller rectangular houses in closer groupings. It is probable that in the Late Viking Age – and perhaps long before – there were manor-like complexes which distinguished themselves by their size and the quality of their buildings. In both the coastal plain and the hummocky landscape it was very common that the local élite marked their social status and their claim to land in runic texts inscribed on the stones they

erected at boundaries, roadsides, and perhaps also at cemeteries in the immediate vicinity of the settlement (Christophersen 1982).

The pagan cemeteries, with their stones set in geometrical shapes and their often exposed locations, were a conspicuous element in the landscape. They have almost entirely been destroyed as a result of recent cultivation, and only a few fragments survive today.

Cultural landscape and subsistence economy

The landscape around the habitation sites, closest to the farmsteads, consisted of arable land, as well as cemeteries and perhaps sacrificial wells. The predominant cereal was rye, with barley also being common. By all appearances, the grain was no longer cultivated in single courses in the same fields; some form of rotation system was in use. Unlike in the Early Iron Age, arable fields took up considerable areas of land. The fields were well fenced, and a cattle-path led from the pastures to the farm. The extensive areas where fodder was collected were located in the best terrain for the purpose. Grazing lands were still further away, and it is likely that stands of woodland at quite a distance from the farms were also used for pasturage.

The villages' arable and grazing land thus constituted large, continuous stretches of almost totally open landscape. There is virtually no evidence of any trees growing in this cultural landscape, but there was woodland further away from the farms. As we have seen, however, there was a wedge of woodland projecting between the settlements along the rivers Skivarpsån and Svartån, and the areas in the north and north-east were dominated by woods.

The population of the area had excellent chances to develop a system of regional exchange. The hummocky landscape and the plain provide natural bases for different forms of specialization. There were also contacts with regions beyond the Ystad area. The archaeological material reveals socially integrated exchange contacts at high levels (Hårdh 1976:33, 70). Links with long-distance trading contacts can also be documented in the finds.

LATE MESOLITHIC, CA. 3500 BC

ROMAN IRON AGE, CA. AD 200

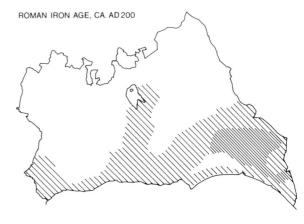

EARLY NEOLITHIC, CA. 2700 BC

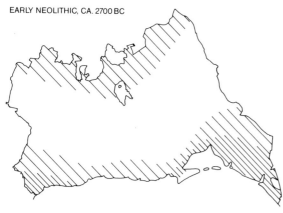

LATE VIKING AGE, CA. AD 1000

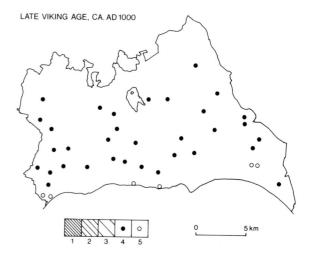

LATE BRONZE AGE, CA. 800 BC-500.BC

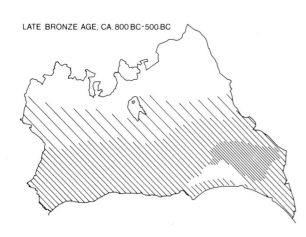

Fig. 3.5:2. Settlement changes from the Mesolithic to the Early Middle Ages. These maps are intended to give a picture of the most important changes in settlement.

Late Mesolithic settlement is concentrated in base camps by a prehistoric lagoon in the south-east of the area. The use of other parts of the area is seasonal.

Early Neolithic settlement is at its densest on the flat coastal plains. Settlement is also found in the inner hummocky landscape with its varying soil types, while the clayey outer hummocky landscape is sparsely populated.

Concentration of settlement on the coastal plain is accentuated in the Late Bronze Age. In addition, considerable parts of the outer hummocky landscape are used, as is clear from the Quaternary biological invistigations but less noticeable in the archaeological evidence.

A tendency towards concentration of settlement can be detected in the Roman Iron Age. Growth in this period mainly takes the form of internal expansion.

In the Viking Age there is a marked expansion of settlement in the outer hummocky landscape. The medieval villages are established. Because of recurrent hostile raids from the sea, the coastal sites are abandoned. 1. Densely settled areas. 2. Less densely settled areas. 3. Sparsely settled areas. 4. Village, hamlet. 5. Settlement abandoned during the eighth century. The dates 3500 BC. 2700 BC, and 800 BC–500 BC are in uncalibrated radiocarbon years (cf. Ch. 2.9).

3.6 The 14th century landscape

3.6.1 Settlement and society

Hans Andersson

The period around 1300 is a distinct watershed in the history of settlement in the area. The map shows how medieval settlement was established mostly from the 11th century until about 1300. The few archaeological studies of medieval villages in the area, like other investigations in the region, indicate that the earliest village sites can be dated to the end of the 10th century and the beginning of the 11th (Skansjö 1983, esp. pp. 145ff., Billberg 1989, Ch. 4.1.5.1, Ch. 4.2.5.1). At the beginning of the 14th century, village settlement covered the central and south-western parts of the area in particular. By contrast, the northern part – the Romeleåsen ridge and the northern parts of the Baldringe and Sövestad area – was largely uninhabited and probably used only (at least partly) as pasture (Ch. 4.4.5, 4.4.6). Conditions were similar in the wedge of land projecting down north of lake Krageholmssjön, which seems to have been the basis of the boundary between the hundreds of Ljunits and Herrestad, although the exact course of the boundary at this time is unclear (Hallberg 1975:20, 58, Skansjö et al. 1989).

At the time represented in the map the division of the area into parishes was already established. This is shown by the churches in the area. Behind the picture of the distribution of the churches lies a process which probably began with main churches or minsters in Skårby and Stora Herrestad in the 11th century, under the patronage of the archbishop and perhaps the Crown. Most of the Romanesque stone churches which arose after this, their west towers having such features as private galleries, can be associated with the patronage of local magnates from the mid 12th century. In most villages where there are churches the presence of manors within the village settlement can be demonstrated. Occasionally it is possible to show a direct link between manor and church, a particularly clear example being found in Bjäresjö, where the cellar of the manor is built of the same material as the church (Skansjö 1986, Skansjö et al. 1989, Mats Anglert 1989b, Callmer 1991b, Riddersporre 1989a, Sundnér 1989a, Wienberg 1989, Ch. 4.2.5, Ch. 5.7).

In view of the picture provided by the map, the economic base for settlement must have varied within the area in different combinations of arable farming and animal husbandry. The placing of the churches of Stora Herrestad and Skårby in areas with good conditions for both suggests an interest in a composite use of the land.

But it is also essential to point out that the new land brought into use up to the Early Middle Ages shows a predominance of grain crops, not least in an area like Bjäresjö-Högestad (Ch. 3.6.2). It is within this area that a three-field system can be observed (at Lilla Tvären) as early as the late Viking Age (Billberg 1989, Engelmark 1989, Engelmark and Hjelmqvist 1991).

We might also expect to find watermills in the area. The oldest written evidence of these is from the end of the 14th century (Ch. 4.1.6.1), but finds of ponds which were in use at Lilla Tvären from the 11th century to the 13th show that there were watermills in the area as far back as the Early Middle Ages (Billberg 1989).

The map also shows the town of Ystad. This is a new feature of the landscape, which probably came into existence around 1250. At the time shown on the map it was well established, probably with two parish churches and a Franciscan friary. Right from the beginning the role of the town was as a channel for the distribution of surplus produce from agriculture and the herring fishery (Kraft 1956, Tesch 1983c,d, Ch. 4.1.5.2, 5.8).

The emergence of the town of Ystad should be seen both as an expression of changed economic circumstances and as a motive force in the changes that can be read. With the town the central government also sought to control trade. The Ystad area was linked with the international commerce of the Baltic region, which was expanding and changing character in the 13th century (Ch. 5.8).

But the birth of the town is also a sign of the change that was beginning to take place in the area, and which became especially clear after the period represented by the map. Another indication of change is that the west towers of the churches ceased to display the signs of the magnates; this part of the church now belonged to the common people more than before. The churches were rebuilt; the one in Köpinge was given a new, very large chancel. Many churches around Ystad were given brick vaults, which should perhaps be connected with the building of brick churches in Ystad from the middle of the 13th century (Sundnér 1989a, Ch. 5.7). Gradually, beginning in the mid 14th century, several manors moved out of the villages (for example, Bjäresjö: Skansjö 1986, Skansjö et al. 1989, Ch. 4.2.6). Ystad and the surrounding district were also bound closely together; in the Late Middle Ages we note contacts between Ystad and the magnates in the surrounding countryside (Reisnert 1989b).

The transformation of the manor structure and the hints of increased grazing pressure which can be seen in the pollen diagram from Bjäresjö (Gaillard and Berglund 1988) may mean a greater emphasis on stock-raising in the Late Middle Ages. The surplus from this could have been distributed via Ystad. Yet the picture is not uniform throughout the area. In Baldringe parish in the north-east there is indirect evidence that arable land increased at the expense of meadow in the 14th and 15th centuries, and that pigs became more important than cattle (Billberg 1989). A shift in agriculture from intensive grain cultivation to extensive animal husbandry could also be seen as a sign of regression, but there are only vague indications of desertion in the project area, except in the Romele area in the north (Ch. 4.2.6.1, 4.4.6). It must be underlined, however, that the sources are too meagre to allow any definite conclusions. There is no written evidence of any trade in cattle in the area before the 17th century (Ch. 5.8), but from other parts of Denmark we know of cattle-trading from as far back as the 15th century (Poulsen 1987). For the Ystad area we must content ourselves with the observation that there *may* have been a trade in cattle in the Late Middle Ages (Ch. 5.8).

The restructuring that is seen in the relocation of manors from villages, and in some cases the equalization of farms in the villages as well (see Ch. 3.6.1), reflects a reinforcement of the political and economic status of the élite. The area had already borne a feudal stamp in the Early Middle Ages – the manors in the villages and the displays of aristocratic status in the churches – but the big new manors marked more clearly than before the social segregation in the cultural landscape. The estate landscape of later times began to take shape in the 14th century (Skansjö 1986, Ch. 5.9).

3.6.2 Landscape, land use, and vegetation

E. Gunilla A. Olsson

Introduction

The task of reconstructing vegetation and land use for the Middle Ages to be presented on a map is indeed a challenge. This time is almost 400 years before available maps and a reconstruction must rely on a variety of sources. The reconstructed landscape of the early 14th century (Fig. 3.6:1) is based on archaeological (Billberg 1989, Sundnér 1989, Callmer 1991, Ch. 4.2.4.2) and palaeoecological findings (Gaillard and Berglund 1988, Engelmark 1989, Ch. 4.2.2, 4.3.2, 5.1, Hjelmroos unpubl.), sparse written records (Skansjö 1987, 1988), and retrogressive analysis of the 18th century map (Appendix B). A detailed description of the methods for this reconstruction is given in Ch. 2.8.

Description of subregions

Below is presented a description of the medieval landscape shown on the map in Fig. 3.6:1. The division into separate subregions is shown in Fig. 3.6:2. The parish

borders on the map are the modern ones, but there are several indications that these were roughly the same in the 14th century (Sundnér 1988, Anglert 1988).

The Ystad area at this time can be divided into six subregions, each with its own landscape characteristics.

1. The Romele area – the large common. The area comprises the northern part of the present parishes of Villie, Skårby, and Sjörup parishes. The esker of Romeleåsen was part of a large common extending from Lund to Ystad. Stock from all the surrounding villages were grazed here. At the same time this area also functioned as the main course of communication where all land transports of goods and information occurred between the towns (cf. Sjöbeck 1964). The landscape was a wood-pasture common (cf. Rackham 1984) containing scattered trees and groves of predominantly beech and oak (cf. Ch. 4.3.2). Coppicing was not practised owing to the difficulty in protecting the regrowth from grazing stock, but many oaks were probably pollarded (Rackham 1986). Patches of genuine heathland vegetation occurred frequently, and junipers and thorny shrubs like wild roses and hawthorn were abundant. The area contained no settlements and it was indeed a huge "no man's land".

2. The Baldringe-Sövestad area – the woodlands. This area was distinguished by the large deciduous woodlands of beech, elm, and oak, both east of lake Krageholmssjön (cf. Ch. 4.3.2) and in Baldringe parish. The existence in Baldringe of a woodland for 2000 pigs, equalling approx. 800 ha (calculated according to Rack-

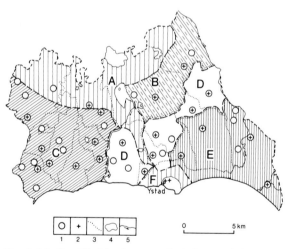

Fig. 3.6:2. Simplified landscape-ecological map of the Ystad area in the early 14th century. This map is a generalization of Fig. 3.6:1 where land use and vegetation form the basis for the classification of the different subareas. A. The Romele area - the large common. B. The Baldringe-Sövestad area - the woodlands. C. The south-western corner - the coppice area. D. The Bjäresjö-Högestad area - the grain-cropping area. E. The south-eastern corner - the sand and fen plain. F. Ystad common. 1. Village. 2. Church, chapel. 3. Parish boundary. 4. Lake. 5. Stream. After Olsson 1989.

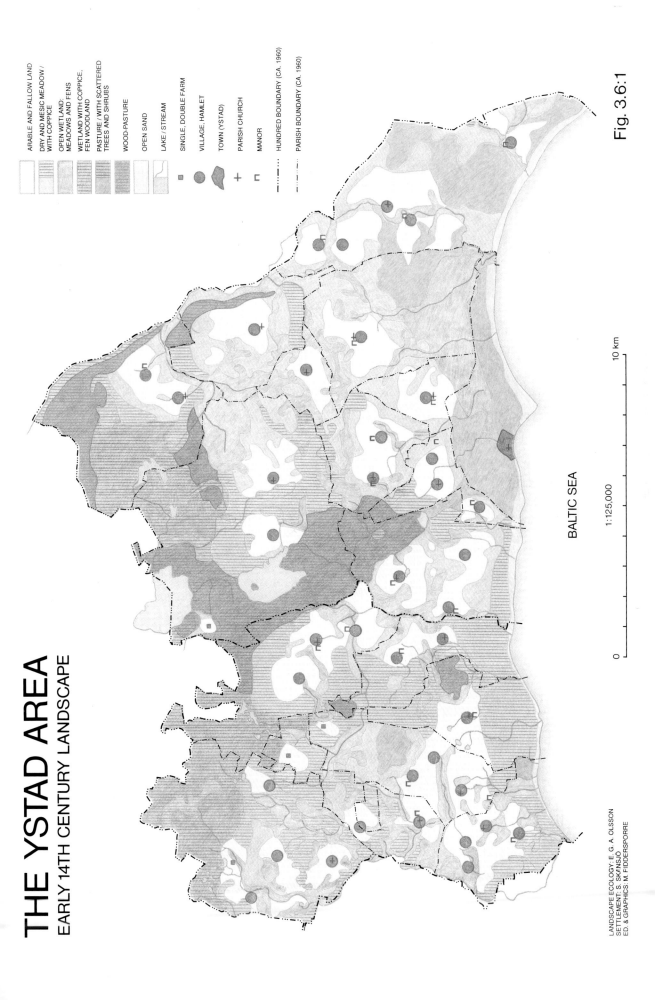

THE YSTAD AREA
EARLY 14TH CENTURY LANDSCAPE

ARABLE AND FALLOW LAND

DRY AND MESIC MEADOW /
WITH COPPICE

OPEN WETLAND:
MEADOWS AND FENS

WETLAND WITH COPPICE,
FEN WOODLAND

PASTURE / WITH SCATTERED
TREES AND SHRUBS

WOOD-PASTURE

OPEN SAND

LAKE / STREAM

SINGLE, DOUBLE FARM

VILLAGE, HAMLET

TOWN (YSTAD)

PARISH CHURCH

MANOR

HUNDRED BOUNDARY (CA. 1960)

PARISH BOUNDARY (CA. 1960)

BALTIC SEA

1:125,000

0   10 km

Fig. 3.6:1

LANDSCAPE ECOLOGY: E. G. A. OLSSON
SETTLEMENT: S. SKANSJÖ
ED. & GRAPHICS: M. FIDDERSPORRE

ham 1980) was noted from 1611 (Skansjö 1988). A large common constituting almost half of the area of the present parish of Baldringe and extending towards lake Ellestadssjön was presumably connected to the Romele common. Sövestad and Baldringe parishes were remarkably large, but contained together only four villages and one large estate (Billberg 1989). This situation was reflected in the landscape by the existence of relatively small areas of arable and larger meadows and woodlands than in the more densely populated parishes of the south-west.

3. The south-western coppice area. This area comprises Västra Nöbbelöv, Katslösa, Sjörup, Snårestad, and Balkåkra parishes. We encounter here a concentration of neighbouring villages, sometimes four villages in one parish as in Västra Nöbbelöv. Nowhere else in the Ystad region was the coppice area so large as here. Particularly large coppices were found in Snårestad, Balkåkra, and Villie.

In the southern part of this area there existed a wooded common (cf. Rackham 1984) which was included in the common called Kroppsmarken. It is reasonable to interpret this as similar to the large wooded common west of the Ystad area called Simremarken (Sjöbeck 1964). Commons like these were characterized by coppices, but also grazing land and occasional small arable fields predominantly used for growing oats (Sjöbeck 1964, 1973). According to Sjöbeck (1964) the coppices on such areas were cleared some time during the period 1550–1750.

The rolling topography here meant that small wetlands were abundant and scattered all over the area. The villages in this region were well provided with winter fodder and wood. However, the arable was increasing and already by this time arable lands belonging to different villages were approaching each other and the former outer circle of meadows had been ploughed to arable.

4. Bjäresjö-Högestad – the grain-cropping area. This region includes the parishes of Bjäresjö, Hedeskoga, Bromma, Borrie, and Högestad. This was a fertile, open landscape dominated by cereal-cropping in three-course rotation on the fertile calcareous clay soil (Olsson 1989). Arable fields dominated the landscape and took up larger areas than the meadows. Woodlands were already very scarce by this time but small areas containing standard trees were found along the gully of Fyleån in Högestad parish. A portion of the large Romele common extended into the north-eastern part of the present Bjäresjö parish, transversing the area of lake Bjärsjöholmssjön and even reaching Ystad (Fig. 3.6:1). This was indicated by the relatively large presence of *Juniperus* pollen for the Middle Ages in the stratigraphy of Bjäresjö (Gaillard and Berglund 1988). Field-names including the element *fälad* ("common pasture") within the present village of Bjäresjö also indicate former common land here. Coppices were rare

but occurred along the coast in Bjäresjö and north-west of Hedeskoga village. Wet meadows yielding winter fodder were abundant in Bjäresjö, Högestad, and Borrie, but scarce in Hedeskoga and Bromma.

5. The sand and fen plain. Öja, Herrestad, and Stora Köpinge parishes are included in this region. A large sand plain and peat deposits determine the features of the medieval landscape here. By this time the area was more or less totally deforested. Two large fenlands, Öja-Herrestads Mosse in the south and the Högestad fenland in the north, framed the terrestrial area. These fenlands were essential fodder areas, used both for grazing and for hay harvests. Wells with nutrient- and lime-rich water fertilized the meadows in the northern part of Öja-Herrestads Mosse and provided high hay yields. The fenlands also provided peat for building and burning as well as reeds for thatching and building purposes, such as wattle-and-daub walls (Sw. *klineväggar*).

Stora Köpinge parish was dominated by the great sand plain with open drift-sand areas both along the shore and inland. The arable was situated close to the villages. It is reasonable to believe that three-course rotation was already used here at this time (cf. Fox 1984, Ch. 4.1.6.2). The meadows were probably smaller than the arable (Fox 1984). The infields were encircled by extensive sandy commons, dominated by grass and heather heathlands (cf. Sjöbeck 1964). All three parishes used the sandy areas along the shore for grazing. The rivers of Kabusaån and Nybroån transversing the parish were of crucial importance, contributing to high-yielding wet meadows, supplying power to watermills, and providing fishing waters. Four out of the five villages in this parish were located along these rivers.

6. Ystad town territory. The borders of the territory of the town of Ystad shown on the early 18th century maps were probably the same back in the early 14th century when Ystad had the utilization rights to a large tract of land outside the town. This area functioned mainly as common grazing land for stock belonging to residents in Ystad. The area was mainly treeless; as well as being used for grazing it was also the site of several windmills.

Conclusions

As the section above and the map in Fig. 3.6:1 have shown, the Ystad area was strictly differentiated in terms of land use and vegetation in early medieval times. To a large extent this was a result of the differences in available natural resources but also of different distances to main communication and marketing sites, like the town of Ystad. These differences between the subregions obvious in the Middle Ages were maintained over the centuries and it was not until the last century that they were levelled out to some extent. An extended version of this text is presented in Olsson (1989a). For a review of agricultural practices in this area during the Middle Ages see Olsson (1989b).

90

3.7 The early 18th century landscape

3.7.1 Settlement and society

Anders Persson and Sten Skansjö

Settlement

Not until around 1700 can we obtain a detailed picture of the basic structures of the cultural landscape in the Ystad area as a whole, such as the pattern and distribution of settlement, cultivation systems, production trends, and property distribution. Some of these basic structures certainly had their roots in the Middle Ages, but it is only around the end of the 17th century that the sources provide evidence for the entire project area.

It is well known that the pattern of agrarian settlement in the plains areas of southern Scandinavia before 1800 differs from the rest of Scandinavia in that it consisted of nucleated, relatively large villages surrounded by cultivated lands lacking settlement. This Danish-Scanian pattern is closer to what we find in north-west Europe than in, for example, central Sweden (Sporrong 1970:18ff., Helmfrid 1985:66ff.).

It is also known that Scania in the mid 17th century was dominated by village settlement: of the roughly 14,000 farms in the province, 86% were in villages (i.e., nucleated settlements of three or more farms), while only 14% were dispersed settlements. There were, however, important differences within the landscape. On the one hand, we have areas with exclusively nucleated settlement, such as the south-western moraine area, where almost 100% of the farms were in villages. By contrast, dispersed settlement clearly dominated in some parts of the Pre-Cambrian moraine (Dahl 1942a:12ff.).

Around 1700 the Ystad area was dominated by village settlement: no less than 93% of the total 587 farms were in villages, while only 7% of the farms were dispersed settlements (Skansjö 1985:17ff., Fig. 3.7:2). These figures show that the proportion of villages in the Ystad area was higher than the average for Scania, but still not as high as in the area of exclusive village settlement. The Ystad area was nevertheless fully representative of the southern Scandinavian village landscape.

In the project area there are significant differences in the size of villages. According to Dahl's (1942) calculations, the average number of farms per village in Scania was 10. However, there could be large variations between different types of area: on the Pre-Cambrian moraine of northern Scania there were only 3.5 farms per village on average, whereas on the sandy Kristian-stad plain there were 20.7 farms per village (Dahl 1942a:12ff.). The Ystad area, with its 13.1 farms per village, was well above the average for Scania. Villages here varied in size between 3 and 45 farms, but about half of the village farms in the area were in villages of 8–15 farms.

If we examine the distribution of nucleated and dispersed settlement within the project area, we find that dispersed settlement was clearly restricted to a belt running from west to east across the north of the area (Fig. 3.7:2). To be more specific, the 38 dispersed settlements are found only in the Romele area in the north (Skansjö 1985:19ff.). Whereas most of the villages in the area are documented in medieval sources, the isolated farms in the north are more recent than 1300. Most of them can be assumed to have arisen in the period from the mid 16th century to the first few decades of the 17th century, and can be interpreted as a wave of settlement colonizing marginal areas. This phenomenon is well known in the provinces of Götaland in Sweden, and elsewhere in marginal areas of Scania and Denmark. This colonization can be seen as a reflection of increased population pressure, which is known to have been general in Europe at this time (cf. Skansjö 1987a:106, 111).

Apart from the villages and isolated farms, one feature that dominates the picture of settlement in many ways is the presence of ten noble manors. Some of these can be traced back to medieval origins, as the seats of knights and squires: Marsvinsholm (previously Borrsjö), Krageholm, Herrestad, Bjersjöholm, Gundralöv, and Hunnestad. Before the Reformation (1530s) three of the manorial estates were among the property owned by the archbishopric of Lund: Högestad, Snårestad, and Baldringe. The last manor to be created, Rydsgård, was not founded until the 1680s. Whereas in the Late Middle Ages there were many small manors scattered through the area, these were now replaced by a smaller number of manors, several of which practised farming that was on a large scale for the day. Although the smallest manors – Hunnestad and Gundralöv – were equivalent to about 4.5 normal farms in the area, the biggest manors – Marsvinsholm and Krageholm – were equivalent to over 20 farms. Marsvinsholm, Krageholm, and Bjersjöholm were separate farms with their own lands, whereas the others were still situated in or near the villages with the same name as the manors; in these cases the manors' arable and meadow land was intermingled with that of the village farms (Skansjö 1985:14f.).

In the time from the Reformation to the early years of the Swedish period (ca. 1680) the Ystad area came heavily under the dominance of the nobility. Most of

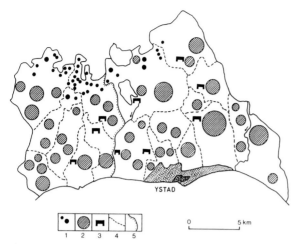

Fig. 3.7:2. Settlement structure in the study area around 1670 (from Skansjö 1987a). 1. Isolated farms or double farm. 2. Village with 3–45 farms. 3. Manor. 4. Parish boundary. 5. Hundred boundary.

the Renaissance-style castles were built around this time, in conjunction with increases in the scale of estate management. It was also from this time onwards that the nobility, through purchases and exchanges, came into possession of over 80% of the farms in the area. This figure can be compared with that of 54% for Scania as a whole. The development can be explained above all by the demand for agricultural produce in Europe, in combination with social and political changes in the Danish society of the time (Skansjö 1985:38ff., 1987a:106ff., cf. Ch. 5.9).

Yet another characteristic feature of the settlement picture is the cottages or *gatehus*. These were adjacent to the farmsteads in the villages, owning little or no land. We know nothing of their existence or number in the Middle Ages. Around 1700 the number of cottages was highest in parishes with manors; this must reflect the labour needs of the estates (Persson unpubl.).

Another type of settlement with special functions consisted of the 12 mills in the area. Six of the watermills are known from medieval sources, and a few medieval mills had disappeared before 1700. On the periphery of the agrarian society we also find five little fishermen's settlements: Brunstorp, Krutboden, Sannhus, Skarviken, and Svartebo. There is no medieval evidence for these; they can be assumed to date from the time after the Reformation (cf. Bunte 1977).

In comparison with the picture of settlement around 1300, minor changes in parish division have occurred in the area. Two churches which may have been medieval parish centres were closed, namely, Borrsjö and Lilla Tvären. The closure of the latter may be connected with the redistribution of church property by the Crown after the Reformation. Chapels in Gussnava and Kabusa were also closed (Ch. 5.7).

Cultivated land

The area of village settlement in the project area was predominantly geared to cereal cultivation. In the dispersed settlement in the Romele area animal husbandry was more important, and grain cropping here was extensive rather than intensive (Fig. 3.7:3). Productivity, however, was largely the same as in the villages, since the isolated farms had more plentiful supplies of manure, having access to outlying land that could be used for pasture. In the villages to the south, the last remains of common land and other outlands had largely disappeared since the Middle Ages.

Cultivation in the villages still normally followed the three-field system, and rye and barley were grown on lands that were regularly manured. On the villages' sandier lands, often on the periphery, oats were cropped. Rye and barley fields were normally cropped for two years out of three. Longer fallow periods for rye and barley fields were practised not only by the dispersed farms but also by some villages in the south-east of Herrestad Hundred (the Köpinge area), where buckwheat was also cultivated. Rye and barley crops usually brought a seed return between 4 and 6. Oats were cultivated in more extensive forms, giving a much lower seed return of 2 to 4. In the villages only 20%–35% of the land was tilled each year. The corresponding figure for the region of dispersed farms was 10%–15%. Differences in these landscape zones are clearly seen in Fig. 3.7:3, which shows the figures for each parish. The south of the project area had an open landscape around 1700, whereas the north was not as intensively exploited. The difference can be explained by the fact that the proportion of fallow land was greater on the dispersed farms in the north, and especially by the fact that there were larger areas of woodland and other outlands in the northern parishes.

Arable as a proportion of village land was usually

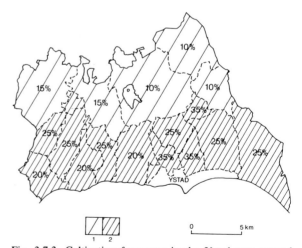

Fig. 3.7:3. Cultivation frequency in the Ystad area around 1700. 1. 10–15% of the area ploughed each year. 2. 20–35% of the area ploughed each year. Parish boundaries indicated.

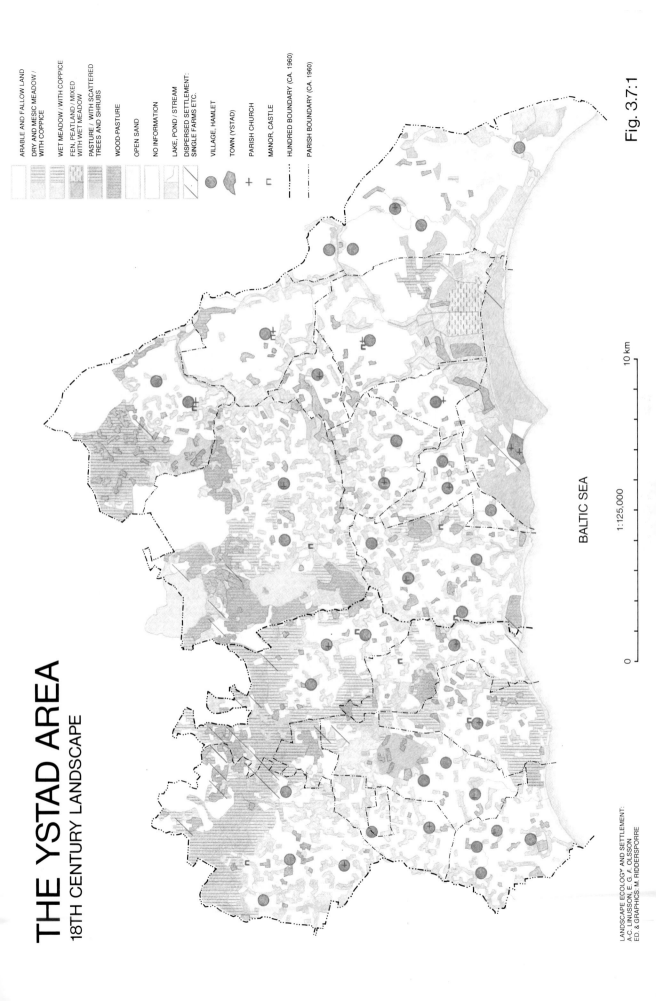

THE YSTAD AREA
18TH CENTURY LANDSCAPE

ARABLE AND FALLOW LAND

DRY AND MESIC MEADOW /
WITH COPPICE

WET MEADOW / WITH COPPICE

FEN, PEATLAND / MIXED
WITH WET MEADOW

PASTURE / WITH SCATTERED
TREES AND SHRUBS

WOOD-PASTURE

OPEN SAND

NO INFORMATION

LAKE, POND / STREAM

DISPERSED SETTLEMENT:
SINGLE FARMS ETC.

VILLAGE, HAMLET

TOWN (YSTAD)

PARISH CHURCH

MANOR, CASTLE

HUNDRED BOUNDARY (CA. 1960)

PARISH BOUNDARY (CA. 1960)

BALTIC SEA

1:125,000

0   10 km

Fig. 3.7:1

LANDSCAPE ECOLOGY AND SETTLEMENT:
A-C. LINUSSON, E. G. A. OLSSON
ED. & GRAPHICS: M. RIDDERSPORRE

about 50%–75%, leaving 25%–50% as meadow. In the Köpinge area in the south-west there were even larger proportions of arable in relation to meadow land. Since the mid 16th century the extent of arable had increased by an estimated 50%. This figure is based on an intensive study of a number of village farms at two points in time – 1570 and ca. 1700. The change appears to be connected with changes in population, which found expression in the colonization of the north-west of the project area by dispersed farms (Persson and Skansjö unpubl.).

The size of farms in the project area was normally between 25 and 50 ha. In the Romele area (Villie parish) the farms were smaller, normally 10–30 ha. Villages owned entirely by manors had often been equalized, giving all the farms the same amount of dues and labour duties to the lord of the manor, and the same area, normally 30–40 ha.

Immediately before the end of the 18th century, there was a fall in the amount of grain cultivated in the Ystad area. This was partly due to a decline in the agrarian economy in the period 1658–1700 (apart from the 1680s), and partly to the ravages of the Scanian war, which afflicted the area directly from the summer of 1676 to the summer of 1678. In the first decades of the 18th century, however, there was an improvement in the agrarian economy.

Communications

According to the so-called Buhrmann map of Scania from 1684, there were seven main highways running through the project area towards Ystad, which was already the market centre of the area. A finer mesh of local roads linked the villages. Land survey documents also provide evidence of local roads leading from the villages to the infields.

3.7.2 Landscape, land use, and vegetation

E. Gunilla A. Olsson

The land survey maps from the 18th century give us a unique vision of the agrarian landscape around the year 1700. These documents contain not only land-use data but also information on vegetation. A review of the land use and vegetation units is given by Linusson and Olsson in Appendix B and the detailed interpretation can be seen in the map there. A generalized version of this map is presented in Fig. 3.7:1. The division into subareas in Fig. 3.7:3 is related to the analogous division for the 14th century presented in Fig. 3.6:2. The early 18th landscape displayed a fine-grained, patchy pattern with grasslands (fodder areas like commons, pastures, and meadows) of different sizes and shapes intermin-

gled with coppice areas, wood lots, and arable fields surrounding the nucleated village settlements. The marked differences in land use and vegetation between the inner hummocky area and other parts, which were so obvious during the Middle Ages (Fig. 3.6:2), persisted. This is seen in the occurrence of a relatively large common pasture on the Romeleåsen ridge and in the Sövestad-Baldringe area, and the greater frequency of coppice meadows in the northern and the south-western part. Away from the Romele-Baldringe area only one common pasture existed: a small common between the villages of Norra and Södra Vallösa.

Wood-pastures and parks with timber trees of oak and beech were kept mainly as hunting grounds and now occurred only around the large estates of Krageholm, Baldringe, Marsvinsholm, Bjersjöholm, and Gundralöv. The rest of the Ystad area was an open landscape dominated by the arable fields, treeless meadows, and fenlands. Ash and elm were often planted around settlements, and rows of pollarded willows along roads and in field borders were common all over the Ystad area.

Characteristic of this time was the abundance of water in the landscape – numerous small streams and brooklets, ponds, lakes, and wetlands. Along the rivers of Skivarpsån, Svartån, Nybroån, and Kabusaån there were large wet meadows with good grass production (the mean yield from wetlands at this time was approx. 1000 kg ha^{-1} yr^{-1} – Ch. 5.3). However, the highest yield recorded in this area, 2200 kg ha^{-1} yr^{-1}, was obtained from meadows in the large mire-fenland of Öja-Herrestads Mosse in the south-east (Ch. 4.1.7.3). This fertile wetland dominated the landscape here and formed an important hay-fodder area. However, by 1704 drainage of these fenlands had started; this is visible on land survey maps as a system of open, widely spaced channels.

The differences in vegetation at this time were largely determined by three factors: environmental conditions, cropping systems, and land tenure conditions. The inner hummocky zone, with its marked relief and sandy till soils with relatively low lime content, was less favourable for intensive cropping than the fertile calcareous clay soils in the south-western part of the area, where a three-course rotation system was practised. The southeastern coastal plain composed of calcareous sandy soils was sensitive to drought but quite fertile due to the lime content, and three-course rotation was dominant also in this region. This intensive cropping system, concentrating on cereal production, kept most of the land as open arable fields. Fodder-producing areas, coppices, and woodlands were by this time reduced in size, and in the three-course villages they were generally much smaller than the arable. No commons or pastures were available here, and cattle had to be grazed solely on the fallows (Linnaeus 1751, Olsson 1988). On the eastern sandy plain the fallow periods were extended to up to 4–12 years. Hence the relation between open arable and

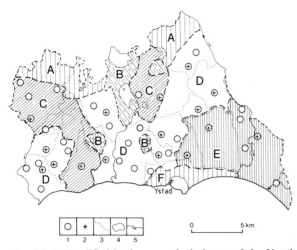

Fig. 3.7:4. Simplified landscape-ecological map of the Ystad area in the early 18th century. This map is a simplification of Fig. 3.7:1 and the Appendix B map. It should be compared with Fig. 3.6:2. A. The Romele area – common grazing land. B. The woodlands. C. The coppice area. D. The grain-cropping area. E. The sand and fen plain. F. Ystad common. 1. Village. 2. Church. 3. Parish boundary. 4. Lake. 5. Stream. After Olsson 1989.

fallow grassland was different from the situation on the clay soil areas.

The decrease of woodlands and coppices led to a pronounced shortage of timber and fuel. Existing peat deposits were continuously exploited for fuel. Fences were built from thorny shrubs, turf, soil, or seaweed (Linnaeus 1751). So severe was the shortage of fencing material that thorny shrubs like *Prunus spinosa* (blackthorn) were almost extinct in the areas around Bjäresjö and Marsvinsholm during the mid 18th century (Sjöbeck 1973). In order to reduce the need for fencing material, several three-course villages synchronized their cropping cycles and practised communal grazing on their fallows (Sw. *vångalag*). This landscape was prevalent in the villages in the south and south-east: Kabusa, Stora Köpinge, Stora Herrestad, Öja, Bjäresjö, and Hedeskoga (Fig 3.7:4).

Some three-course villages in the outer hummocky zone – Snårestad, Skårby, Gussnava, and Årsjö – still maintained large coppice meadows with predominantly hazel and oak and sometimes lime and willows (Fig. 3.7:4). Ash and hornbeam were rare coppice species in the Ystad area.

A conspicuous feature at this time was the large driftsand field east of Ystad. The citizens of Ystad and in the villages of Kabusa, Stora Köpinge, and Lilla Köpinge used it for grazing stock which triggered soil erosion and enhanced the sand movement. Similar areas, but smaller in size, also existed along the shores west of Ystad, although here they were framed inland by windswept coppices of oak and hazel belonging to the villages of Mossby, Nöbbelöv, and Snårestad.

On the Romeleåsen ridge, the cropping systems used (Dahl 1989) were a mixture of two-course rotation and a one-field system (field-grass system, cf. Frandsen 1983). Such systems depended on the availability of manure to maintain the productivity of the arable, and cattle-rearing was thus a prerequisite. The relatively large wooded meadows in Norra Villie, Egarp, Rynge, and at the single farmsteads on the ridge maintained coppices of hazel and oak. The commons on Romeleåsen were dominated by heathland vegetation with *Calluna vulgaris* (heather) and scattered shrubs of juniper, roses and hawthorn. Occasionally some wide-crowned beeches shaped by the grazing also existed here.

The large variation in land-use types and hence in vegetation types resulted in large landscape diversity (Pickett and White 1985, Emanuelsson et al. 1987) with the availability of many different succession stages and habitats (cf. Ch. 5.11). Also species diversity was greater in the 18th century landscape. The patchy landscape facilitated the dispersal, establishment, and survival of plant and animal species, many of which are endangered or extinct in the modern agrarian landscape. Another factor which favoured and maintained species diversity in this landscape was the continuity in management, not specifically on the same site, but in the same area. This is valid for species belonging to semi-natural vegetation types like the manifold grasslands in this region which were first developed during prehistoric time (cf Ch. 3.2, 4.2). Another group of species prevalent in this pre-industrial landscape was the arable weeds which were favoured by the cultivation methods, and by the variability in soil conditions within each field. Such species included *Agrostemma githago* (corn cockle) and *Melampyrum arvense* (field cow-wheat).

The coppiced or pollarded trees along roads and in the meadows reached high ages, which favoured hole-dwelling birds, bats, and small mammals like the hazel dormouse (Ahlén and Tjernberg 1988), as well as mosses and lichens associated with this environment (Andersson and Appelqvist 1987). The abundance of surface water and wetlands constituted suitable habitats for waders, of which the stork, now extinct in Sweden, was very common and the most prominent.

3.8 The early 19th century landscape

3.8.1. Settlement and society

Nils Lewan

During the last two centuries the pace of change has rapidly increased, affecting both landscape and settlement. The causes of these changes differ, however. Some are general in kind and are to be seen in most of Western Europe, others are national in kind, while still others may be looked upon as local or specific. In the sections of Ch. 3 covering the last two hundred years, the findings concerning the general development and its causes are reviewed against the background of three generalized settlement and vegetation maps, which are based upon more detailed maps and separate investigations. As public maps cover most of the period – in contrast to earlier periods – specific investigations of settlement changes have not been as thorough as for earlier periods; much more is already known.

In order to simplify comparison a series of settlement maps on smaller scales have been included as Fig. 3.10:2 in Ch. 3.10.

Settlement and landscape

The first years of the 19th century give us the earliest published landscape survey for the whole area at one point in time, including villages and scattered settlements, roads and physical details of the area. The source is the 1:20,000 military reconnaissance map (Lewan 1982, Emanuelsson and Bergendorff 1983, Ch. 2.8). Although technically heterogeneous and based on various sources, this is a unique landscape source. Information from that map has been included in Fig. 3.8:1, although in heavily generalized form.

Most settlements are nucleated villages or hamlets, as they have been for hundreds of years, while the town of Ystad covers only a very small piece of land and still has no real harbour. The growth period of towns and cities has not yet begun. There are, however, exceptions to the common pattern. In the very north-west the partly medieval pattern of scattered farms and cottages lies as before, shown in lighter shading on the map. In the centre, and in the south-west, the same hatching has other reasons. A royal decree concerning *enskifte*, i.e. land amalgamation and enclosure, was promulgated for the province of Scania in 1803; four years later it was made general for Sweden (Helmfrid 1961). As a result, estate-owners as well as freehold farmers started asking the help of surveyors to get a more effective organization of both fields and settlement. According to the given rules, the land of each farm was to be concentrated into a single plot. As the farmsteads were to be sited inside their alloted piece of land, many had to move from their old site inside the village (Helmfrid 1961).

Up to the 1810s only a few villages in this area had been enclosed, while enclosure by the same time had gathered much greater momentum in the south-western plains of the province, dominated by freehold farmers (Dahl 1941). The very first enclosures in the area were initiated by the nobility, owning land in the parishes of Bromma (Bussjö village) and Öja.

Organized before 1803, they actually fell under the rules given in an earlier and revised decree concerning *storskifte* in 1783. In effect, however, they were *enskifte* enclosures. In the Ystad area most of the decisions stayed with the still dominating huge estates and with the nobility, and it was only in the south-westernmost parish that freehold farmers could and did take the initiative to implement *enskifte* themselves (1803).

Enclosure meant important changes per se, and opened up possibilities for many more to come, concerning both land use and other agricultural practices, affecting both the town of Ystad and its surrounding area. So far, however, changes mainly lay within the traditional land-use system. The open field system still was almost intact, the courses of roads remained the same as before, where enclosure had not yet occurred, and new practices of individual farmers had not yet made themselves felt to any extent. As may be seen from Ch. 5.9 and Fig. 5.9:6, the area in 1825 was still totally dominated by a small number of large estates owned by the nobility. The absence of change, and of innovations in land use, also meant that population figures remained small, even though the century before had seen a slow increase.

Until this period the appearance and the development of the landscape in the region may be looked upon as following many of the general tendencies common for continental Europe, if only the open plains are considered. This also means that the causes of change were much the same as elsewhere. Generally speaking, in 1800 an agricultural subsistence economy still dominated Swedish rural areas, as also in this region. However, despite the fact that settlement, road networks, and the like, remained much the same during the 18th century, important changes took place in land use.

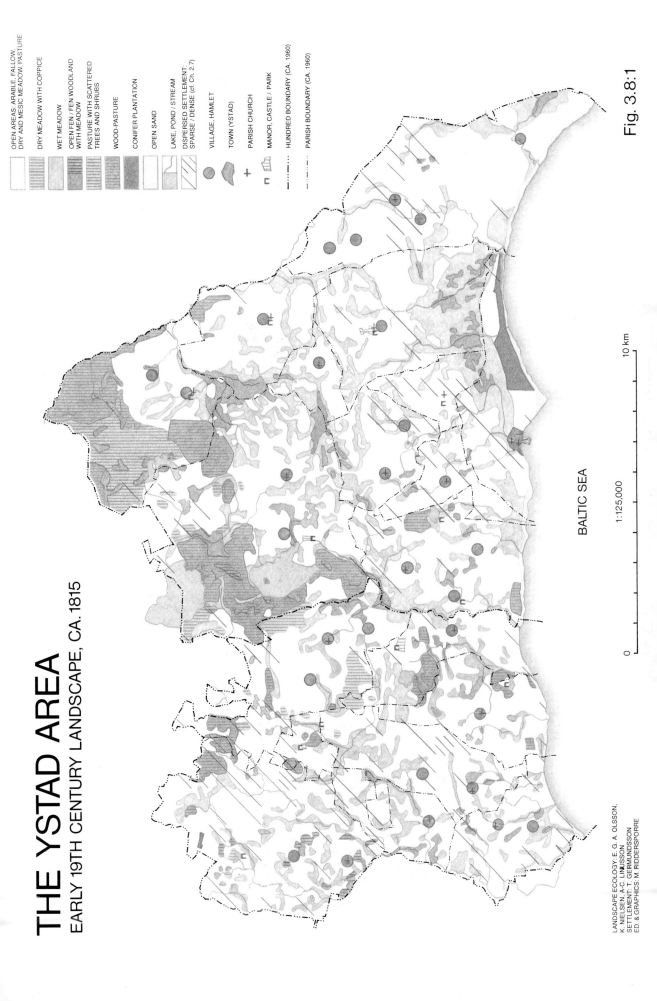

THE YSTAD AREA
EARLY 19TH CENTURY LANDSCAPE, CA. 1815

OPEN AREAS: ARABLE, FALLOW,
DRY AND MESIC MEADOW PASTURE

DRY MEADOW WITH COPPICE

WET MEADOW

OPEN FEN / FEN WOODLAND
WITH MEADOW

PASTURE WITH SCATTERED
TREES AND SHRUBS

WOOD-PASTURE

CONIFER PLANTATION

OPEN SAND

LAKE, POND / STREAM

DISPERSED SETTLEMENT:
SPARSE / DENSE (cf. Ch. 2.7)

VILLAGE, HAMLET

TOWN (YSTAD)

PARISH CHURCH

MANOR, CASTLE / PARK

HUNDRED BOUNDARY (CA. 1960)

PARISH BOUNDARY (CA. 1960)

BALTIC SEA

1:125,000

0   10 km

Fig. 3.8:1

LANDSCAPE ECOLOGY: E. G. A. OLSSON,
K. NIELSEN, A-C. LINJUSSON,
SETTLEMENT: T. GERMUNDSSON
ED. & GRAPHICS: M. RIDDERSPORRE

3.8.2 Landscape, land use, and vegetation

E. Gunilla A. Olsson

In 1820 the Ystad region had many similarities with the landscape here 100 years earlier. The enclosure activity had started, with its resultant changes in land use, and the majority of the villages had undergone the *enskifte* enclosures before 1820. However, the effects of this extensive land-use reform on the landscape were not yet fully visible.

The *shortage of woods and trees* was amplified by this time and deciduous plantations had been created in the north-east at Baldringe and Ebbarp. Wooded meadows and coppice areas had generally decreased, remaining only in Gussnava village, in villages around Krage-holmssjön, and in the Romele area. The relatively large coastal coppices at Snårestad and Mossby had also been reduced to one small enclosed wood lot inside an arable field. On the other hand, the establishment of three new estates in the late 18th century – Rynge, Kadesjö, and Charlottenlund – meant a change of infield wooded meadows in the original villages to wood-pastures and parks with standard trees surrounding the new estates. Parks and wood-pastures at other estates remained unchanged. In Årsjö village in 1770 several arable fields were excluded from cultivation and left for spontaneous forest succession, possibly with the aim of encouraging forest development.

Most of the peat deposits, especially in villages practising three-course rotation, had been exhausted and turned into a variety of water-filled pits and ponds. Sward turf was cut from meadows and pastures in the coastal villages of Stora Köpinge and Nöbbelöv. Sward-cutting stimulated the growth of heather instead of herbs and grasses on the meadows, which impaired fodder production and enhanced soil erosion and sand drift (cf. Linnaeus 1751). Consequently coastal sand drift areas had been enlarged since the early 18th century. The extreme shortage of fuel in the coastal villages was reflected in the use of seaweed, mixtures of manure and straw, and even bulky weeds for fuel, as described by Linnaeus (1751).

A general *reduction of the grassland areas,* including both pastures and meadows, had occurred during the 18th century. The large common land on the Romeleå-sen ridge, which had been used for grazing and communication between the towns of Malmö and Ystad since the Early Middle Ages (Sjöbeck 1964, Ch. 3.6.2), was continuously being reduced. By 1759 it was divided into separate parts belonging to different estates, villages, and isolated farmsteads. The smaller area of common land at Vallösa had been transformed into arable by the early 19th century. A desperate need for grazing land was expressed in 1788 at Nöbbelöv, where meadows were opened for cattle grazing. Generally there was an increase of arable at the expense of meadow. The patchy landscape was beginning to take on a simpler

pattern, a process which is visible, for example, at Högestad, Stora Köpinge, and Bjäresjö. Back in 1749 Linnaeus had complained of how small the meadow areas were compared to the large arable fields in this region. In villages where *enskifte* enclosures had been implemented the fodder areas seldom reached more than 40% of the arable area. Most of the mesic and dry meadows had been ploughed by this time, although villages with a relatively large share of wet meadows still kept considerable fodder areas. This can be seen from a comparison of the neighbouring villages of Stora Herrestad and Stora Köpinge. The former used the wetlands of Öja-Herrestads Mosse for fodder collection and the fodder areas amounted to 65% of the arable. Stora Köpinge on the other hand, largely lacked wet meadows and its fodder areas were only 24% of the arable.

Draining of wetlands. Öja-Herrestads Mosse was still an important hay-fodder area, but was gradually drained with the aim of making cultivation possible. By 1820 an intricate system of channels was spread all over the fenland.

The activity of tapping and draining lakes had also started by the early 19th century. The Sjörup lake in the south-west was totally drained and transformed into wet meadow.

Cropping systems and productivity

Decreasing hay yields during the 18th century (cf. Ret-zius 1799, Osvald 1962) were met by small-scale grass-crop rotation. Hay-meadows were occasionally ploughed and sown with oats and rye for some years (Linnaeus 1751). The management of water-meadows was another way of improving hay yield. By means of extensive systems of shallow, artificial channels the meadows were flooded at certain periods, providing not only irrigation, but also nutrients (Armstrong 1975, Rackham 1986). Such water meadows existed by this time at Ebbarp and Herrestad. The reduction of fodder areas automatically led to a decrease of the stock and thus reduced amounts of manure for the arable. At very low soil nutrient levels, competition between crops and weeds increases (Oldfield 1987, Olsson 1988). Weeds were also very common and could sometimes even dominate over the crop plants in the arable fields (Linnaeus 1751). Reduced crop yields in the late 18th century were related to this fact. Cereal yields of only twice or three times the seed input were common even on fertile soils in this area (Persson 1986, Olsson and Persson 1987). Cultivation of leguminous crops started in the 18th century, although to a small extent, within the traditional cropping system. Already around 1740 pulses and vetches were grown either among the oats or separately on the fallows (Linnaeus 1751). By 1810 the proportion of pulses was about 5% of all crops in Malmöhus County. Potatoes had the same proportion. Among cereals barley dominated, while wheat, a new crop here, amounted to less than 2% (Osvald 1962).

3.9 The early 20th century landscape

3.9.1 Settlement and society

Nils Lewan

Compared with Fig. 3.8:1, Fig. 3.9:1 sums up one hundred years of dramatic change, the 19th century being the most revolutionary period up to then, perhaps even more dramatic than the latest century as regards landscape and settlement. The 19th century took this and other important agricultural regions from an economy dominated by subsistence to one with almost every farmer included in a capitalistic system, this being the century of the agricultural revolution in Sweden (Möller 1989).

The most obvious features are the scattered but now also denser pattern of settlement, a number of "new" agglomerations, and the growth of Ystad; however the density of settlement in the area as a whole was still rather varied. The *skifte* or enclosure has now taken place in every village and hamlet. As a result, the pattern of settlement, if studied in detail, is much more "national" than a hundred years before, although there are still of course also common European features.

The enclosures took place in this area before ca. 1850, and by 1900 they had made their imprint upon every farming region in the whole of Sweden, although later in the more remote areas, less dominated by agriculture. Thus, the pattern of scattered rural settlement had become nationwide, while the villages and hamlets of the plains were reduced in size.

Behind the new, renewed, and extended agglomerations there are common causes, national as well as international, like the advent of the railway, the spread of commercial and other services, as well as of rural industries, such as dairies, machine shops, and brickyards. As another consequence the road network was expanded, although in this case the important changes still lay ahead.

It should be noted that the first railways in Sweden came into being in 1856, while in this region four different railways were opened between 1865 and 1901. In at least one case, the 1874 line between Malmö and Ystad, the estates and the nobility played a leading role as initiators and financiers (Möller 1987, Ch. 5.9).

Another important prerequisite to change was the abolition of restrictions on industrial production and commerce in rural areas between 1846 and 1864; before that such functions were allowed only in cities, towns and, partly, at periodic markets.

As the 19th century was the period of the agricultural revolution in Sweden, it gave birth to a number of important farming and other innovations, affecting both settlement and landscape (Möller 1989).

Among the first to be included was the introduction of *plattgårdar* or big tenant farms on some of the estates, replacing a number of ordinary-sized tenant farms, and at times whole villages. The main reason for this change seems to have been the interest in more efficient use of farm labour (Möller 1986; cf. also Table 5.9:3 and Fig. 5.10:6).

The introduction of *egnahem*, or small freehold cottages, is considerably later. Supported by public money, the own-your-home movement started in 1904, and it led to two major concentrations of these homes, one in the eastern part, another in the central one, where at first a few and later another thirty were established. Again, there were different underlying reasons, including the aspiration to reduce emigration, to further the availability of rural workers, and to help the landless to a small piece of land of their own (cf. Germundsson 1985 and Ch. 5.10).

In addition to the establishment of *plattgårdar* and *egnahem*, quite a number of tenant farms, and of *plattgårdar* were sold off by the estates. Thus a deconcentration of land ownership in the area also became an important tendency during the century, being a reason for some changes in the landscape, and a result of others (cf. Möller 1989, Ch.4.3, Ch. 5.9). This, of course, meant new and different possibilities for innovations and other kinds of change, now furthered by many more landowners.

As the modernization of agriculture gathered momentum during the 19th century, with the estates very much in the lead, a number of other important changes also occurred, with greater or lesser effect on the landscape (Ch. 4.3) Among these changes may be mentioned crop rotation instead of three-field tillage, the introduction of under-drainage and marling, of agricultural machines and of milk farming (Möller 1989 and Ch. 4.3).

Some of the changes affecting agriculture had national causes, while others had to do with the growing international relationship. Both the railway to the provincial capital of Malmö, and the proximity to the new harbour of Ystad became important for the increasing agricultural export from and import to the area (cf. Möller 1987 and 1989).

In the landscape agricultural modernization during the second half of the century meant both reductions in and additions to the area of wetland. While under-drainage greatly reduced such areas, there were additions through marling, the flooding of meadows, and the digging of peat for fuel and other uses.

Still another conspicuous change during the century was that of the road system (Ch. 5.10 and Fig. 5.10.6). In many cases the enclosures also brought about a total restructuring of the local road network, with the new roads following the equally straight farm boundaries, and with dirt roads leading to each single relocated farmstead. In other cases the railway meant the relocation of roads, towards the stations. The main roads, however, remained the same as before.

The 19th century as a whole was a period of rapid population growth, a population peak being reached around 1870 in most parishes. The differences in development between these are great, however. Highly agricultural and estate-dominated parishes were losing population earlier than parishes with rural industries and service facilities (Ch. 5.9, Figs 5.9.9 and 5.9.10). At the same time, Ystad proper continued to grow, as did a couple of other nucleated settlements. Migration to other countries, such a significant feature in much of Sweden from the 1860s onwards, seems to have been of minor importance here, while seasonal in-migration became of interest at least to major farms.

The causes behind changes during the 19th century were both national and international. The resulting landscape bears signs of this, the highly scattered settlement (by comparison with the rest of Europe) being the most obvious feature. The changes have continued in the new century. Some are ebbing away, others increase in importance, while still others have yet to emerge.

3.9.2 Landscape, land use, and vegetation

E. Gunilla A. Olsson

The main landscape characteristic in the early 20th century can be summarized in the term *simplification*. Enclosures had now been implemented in almost every village and accompanied by changes in land use. The introduction of new crops, crop rotation, and the increasing mechanization of agriculture (Germundsson and Möller 1987) led to a pronounced trend towards cereal-cropping in most of the Ystad area (Ch. 5.3). As animal husbandry decreased in importance, fodder areas were strongly reduced and hay meadows remained only as small, fragmented islands in a matrix of arable fields. The cultivation of leys was introduced during the last decades of the 19th century and successively took over the role of hay meadows. Permanent pastures of some acreage remained around 1915 at Baldringe, at Högestad, and as smaller plots in the Romele area.

The 19th century had been the heyday of the creation of well-managed parks surrounding the large manors. Parallel to this development ran the progressive decline of the deciduous woodlands. In fact, woodlands existed now principally only as wood-pastures and parks at the

estates, around Krageholmssjön and at Baldringe. Also wooded meadows had mostly been ploughed to become arable, especially where villages had been transformed into the large *plattgårdar* (Möller 1986, Ch. 5.9). Wooded meadows persisted at only one place, the village of Bellinga near Ellestadssjön. The opposite development – where wooded meadows were abandoned in favour of woodland development – can be seen at Sövestad and Villie. The former role of coppice wood had been taken over by planted conifers, which are now found at several places in the north of the area and on the sand drifts on the coast in the south-east, where some oak had also been planted.

The change of the landscape hydrology was the most dramatic alteration since the early 19th century. The extensive drainage of lakes, ponds, and wetlands in the past hundred years not only changed the landscape hydrology but also the land use. Virtually all lakes and ponds were tapped, except Krageholmssjön, Ellestadssjön, and Bjäresjösjön. The wetland area was reduced by about 90%. The south-western corner was now almost devoid of wetlands. The formerly large area of fenlands at Högestad-Borrie, extending almost 5 km, had now been reduced to one single locality (20% of its former distribution). The great fenland of Öja-Herrestads Mosse, which had been such an important fodder area ever since prehistoric times, was finally drained in the 1880s. During the years 1886–88, 571 ha or 95% of wetlands were completely drained and thereafter used for arable farming (Zachrison 1922). The draining process started early in the Ystad area compared with the rest of Sweden. By 1858 almost 60% of the arable was drained by subsoil clay pipes (Möller 1984). The large estates were precursors in the practice of under-draining by clay pipes. Three estates in the Ystad area – Marsvinsholm, Charlottenlund, and Krageholm – each maintained 200–700 ha of under-drained arable land in the 1860s, thus leading the way both in the province of Scania and in the whole of Sweden (Zachrison 1922).

An interesting contrast to the extensive draining process in this area is the frequent use of water-meadows here. Before the general breakthrough of ley cultivation, the management of water-meadows increased during the second half of the 19th century. A peak was reached in the Ystad area in 1889, when an annual addition of 115 ha was reported. Bjäresjö and Villie parishes had the largest areas (Olsson and Berlin 1990). After 1900 new water-meadows were not reported and many of the existing areas were gradually abandoned. The estate of Marsvinsholm had large areas of water meadows and they also hired staff who specialized in the management of them.

Land use and agricultural production

The reorganization of the villages following the enclosure process was accompanied by a large-scale extension of the arable land. During the period 1805–1914 the

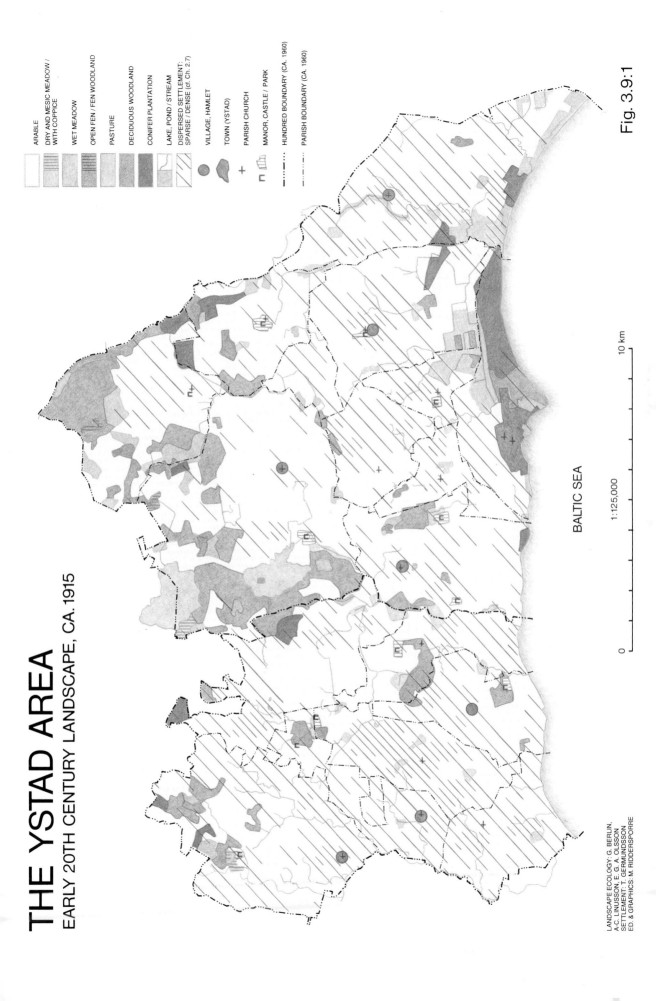

THE YSTAD AREA
EARLY 20TH CENTURY LANDSCAPE, CA. 1915

ARABLE

DRY AND MESIC MEADOW /
WITH COPPICE

WET MEADOW

OPEN FEN / FEN WOODLAND

PASTURE

DECIDUOUS WOODLAND

CONIFER PLANTATION

LAKE, POND / STREAM

DISPERSED SETTLEMENT:
SPARSE / DENSE (cf. Ch. 2.7)

VILLAGE, HAMLET

TOWN (YSTAD)

PARISH CHURCH

MANOR, CASTLE / PARK

HUNDRED BOUNDARY (CA. 1960)

PARISH BOUNDARY (CA. 1960)

BALTIC SEA

0   1:125,000   10 km

Fig. 3.9:1

LANDSCAPE ECOLOGY: G. BERLIN,
A.C. LINUSSON, E. G. A. OLSSON
SETTLEMENT: T. GERMUNDSSON
ED. & GRAPHICS: M. RIDDERSPORRE

arable increased by 340% in Malmöhus County (Zachrison 1922). In this respect the county of which the Ystad area is a part led the way in Sweden. Parallel to these changes there was also a change in cropping systems and the crops cultivated. The subsistence economy and the internal supply of nutrients of the agro-ecosystems had already been changed (Ch. 5.3). To cope with decreasing yields, more efficient manure handling was practised and a variety of organic fertilizers were made available. These included all sorts of organic material such as guano, ground animal bones, agro-industrial wastes, etc. Commercial fertilizers were still rare in 1909 and used in small amounts at only a few of the largest estates. Marling was widely used to improve soil fertility and this practice increased dramatically in the 20th century. During the period 1858–1885 marled arable land increased from 7% to 115% of the total arable in the Ystad area (Zachrison 1914; the figure of 115% includes some arable which was marled several times). The landscape of 1915 (visible in detail on maps in Ch. 4.2.8) displayed a fine-grained pattern of marl-pits as traces of this activity.

Another way of improving crop yield was rotation with nitrogen-fixing species. Cultivation of pulses became more common and reached a peak in this area around 1890 (Olsson and Berlin 1990). Cultivation of leys was introduced during the last decades of the 19th century and successively took over the role of hay-meadows. Among root crops the proportion of potatoes increased generally until 1909, but was thereafter gradually replaced by sugar-beet. Crop yields increased from the middle of the 19th century. The increment was most conspicuous for cereals, where a mean yield for barley around 1910 was 2500 kg ha^{-1} in the Ystad area (Olsson and Berlin 1990). This is a fourfold increase compared with the 18th century.

3.10 The late 20th century landscape

3.10.1 Settlement and society

Nils Lewan

The 20th century has meant a number of important changes to the rural landscape, but the most conspicuous is the growth of Ystad proper, including the harbour area. Although not seen on the map, the thorough modernization of the roads is another very important feature (cf. Ch. 5.12).

Behind the changes lie the general concentration of work to fewer places, the importance of the private car and the lorry, and the often scattered living also of urban people, in other words factors well known over the western world. This means a somewhat sparser density of settlement in most of the area, together with bigger agglomerations than before. It also means that changes are of the same kind but not at all of the same extent as in the more expansive and prosperous western Scania around cities like Malmö, Lund, and Helsingborg. However, at the same time commuting from the Ystad region to the cities in the west has increased and made the region more dependent. This has also helped to keep population figures on a higher level, and caused more houses to remain than would otherwise have been the case (cf. Ch. 5.12.3).

As may be understood from Fig. 3.10:1, the modernization of farming has meant that almost all agricultural land is by now arable. While the number of farm units, big and small, remained stable until the 1940s, it has since decreased fast (Table 5.12:1).

One reason is the national agricultural policy since World War II, aiming at bigger units with incomes for farmers comparable to those of other groups (Lewan 1988). Another reason combined with this is the increasing specialization in farming. By now there are few mixed farms in the plains, which are dominated by large units cropping wheat and barley, rape seed and sugarbeet. On the edge of the area, especially on the undulating slopes of the Romeleåsen in the north-west, smaller units still dominate the picture, often with animal husbandry important to the single farm unit. The future of these units is, however, bleak, which also may mean profound changes to a landscape of great value from a natural, historical, and scenic point of view (cf. Ch. 5.12).

Characteristic of most of the area are the vast fields, rarely interrupted by any obstacles. In this respect there are few differences between this area and western Scania. However, as much of the current over-production of grain comes from areas like these, the future land use and thus the appearance of the landscape might be in the balance, if current political discussions are taken into consideration.

Taken as a whole, today's landscape very much resembles that of regions like East Anglia or the Paris Basin. Current reasons for changes and patterns seem to be much the same, with more decisions than before taken from outside and from a distance, meaning also that the future might be both much the same as elsewhere and much in the balance. To take just one example, a contemplated road-bridge between Malmö and Copenhagen could, if built, mean quite a change in the future of this region, only some 50 km from Malmö.

3.10.2 Landscape, land use, and vegetation

E. Gunilla A. Olsson

The present agrarian landscape was established in the Ystad area in the early 20th century. However, in the landscape of today these tendencies are underlined and manifested as an *agro-industrial landscape*. We encounter here a landscape in which non-cultivated areas are rare exceptions and which is almost totally devoid of natural and semi-natural vegetation. So effective has been the increase of arable that even road verges and the edges of ditches and streams are extinct. A "final" drainage of surface water and wetlands has occurred. The numerous marl-pits, open ditches, and wet depressions which were still so characteristic here around 1915 are now drained and cultivated (Ch. 5.11). The streams are canalized over considerable distances. One single non-agrarian vegetation type has increased – conifer plantations; these take the form of relatively small-scale spruce plantations on former grasslands inside the arable, serving as game refuges. Woodlands were rare here by 1915 and today they exist only around Krageholmssjön and surrounding some estate buildings. However, at several sites beech and oak have been replaced by spruce.

Pastures used today in this area are of two origins: old leys and natural grasslands. Both types are underdrained and heavily fertilized, however, leading to a convergence of the plant species communities. The resulting vegetation is characterized by species-poor communities dominated by a few nitrophilous species and with considerable elements of ruderals and weeds (Ch. 5.11). Natural grasslands of significant size exist only as nature reserves at Baldringe and at Högestads Mosse.

Other natural grasslands remain as minute fragments, often grazed by racehorses, or are merely left without management for spontaneous forest succession. Grasslands also exist at Köpinge and Kabusa, but these areas are military exercise grounds with reseeded and highly fertilized swards.

The exhaustion of the number of sites for natural vegetation and the decrease in landscape diversity is clearly illustrated by the map in Fig. 3.10:1. The non-cultivated spots appear as islands in a matrix of arable fields. There are very rarely any dispersal corridors between these islands, and distances are large. This pattern has fatal consequences for the dispersal and survival of many species of plants and animals inhabiting the remaining sites of natural and semi-natural vegetation. At the same time, all these organisms are subjected to a continuous threat due to the spraying of pesticides and the high load of mineral fertilizers on the adjoining arable fields. This profound depletion of natural vegetation sites and thus of landscape and species diversity has developed over the relatively short period of 100 years. From the ecological point of view this is the most thorough landscape change that has occurred here in the entire time span of 6000 years which the Ystad Project covers.

A prerequisite for the development of the agro-industrial landscape of today is the change from medium-scale mechanization to the industrial scale of management which has occurred. A progressive concentration of land ownership in the hands of fewer persons and companies is a further reason for this development. Generally the multidimensional land use, with both meat and dairy production along with cultivation of fodder for the stock on the fields, has been abandoned. A pure orientation towards commercial cropping has been more pronounced here than in most other parts of Sweden. The highly fertile soils in combination with favourable climate and the increasing availability of fertilizers and pesticides have made this one of the leading regions in Scandinavia as regards crop yield. Wheat and rape now lead in terms of acreage on the clays. On the sandy soils in this area root crops (sugar-beet and industrial potatoes) predominate. The large sugar-processing plant in Köpingebro, which is the centre of this industry, was established at the beginning of the century (Ch. 4.1.8). During the 20th century the yields of winter wheat and barley rose by 230% and 190% respectively (Ch. 5.3). However, this high production requires the input of more than 150 kg of nitrogen per hectare in the form of commercial fertilizer (valid for winter wheat in this area, SCB Report, 1990) along with a high load of pesticides. The present specialized agricultural production demonstrates a high degree of dependence on the import of resources from distant ecosystems and hence a vulnerability to disturbances in this supply.

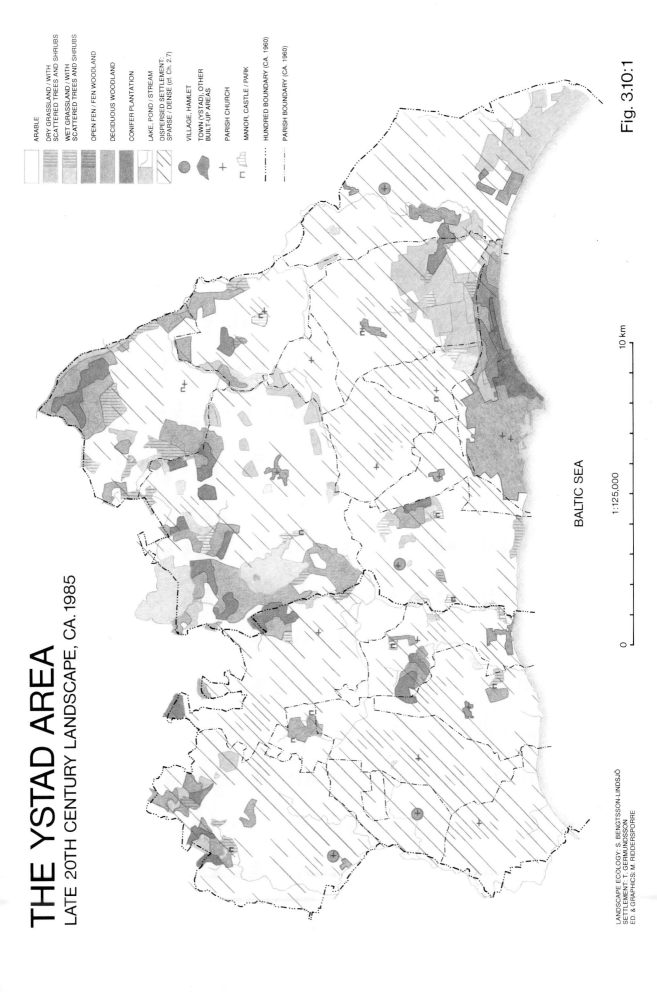

THE YSTAD AREA
LATE 20TH CENTURY LANDSCAPE, CA. 1985

ARABLE

DRY GRASSLAND / WITH
SCATTERED TREES AND SHRUBS

WET GRASSLAND / WITH
SCATTERED TREES AND SHRUBS

OPEN FEN / FEN WOODLAND

DECIDUOUS WOODLAND

CONIFER PLANTATION

LAKE, POND / STREAM

DISPERSED SETTLEMENT:
SPARSE / DENSE (cf. Ch. 2.7)

VILLAGE, HAMLET

TOWN (YSTAD), OTHER
BUILT-UP AREAS

PARISH CHURCH

MANOR, CASTLE / PARK

HUNDRED BOUNDARY (CA. 1960)

PARISH BOUNDARY (CA. 1960)

BALTIC SEA

0   1:125,000   10 km

LANDSCAPE ECOLOGY: S. BENGTSSON-LINDSJÖ
SETTLEMENT: T. GERMUNDSSON
ED. & GRAPHICS: M. RIDDERSPORRE

Fig. 3.10:1

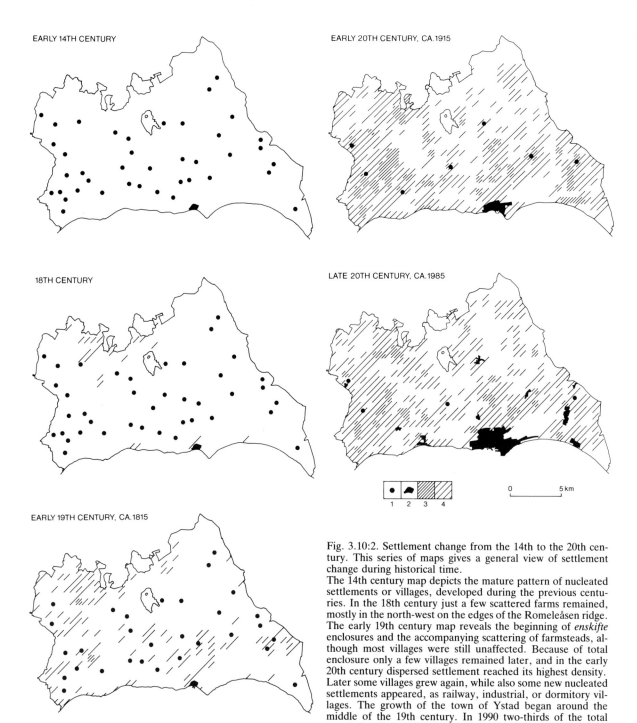

EARLY 14TH CENTURY

18TH CENTURY

EARLY 19TH CENTURY, CA.1815

EARLY 20TH CENTURY, CA.1915

LATE 20TH CENTURY, CA.1985

0   5 km

1   2   3   4

Fig. 3.10:2. Settlement change from the 14th to the 20th century. This series of maps gives a general view of settlement change during historical time.

The 14th century map depicts the mature pattern of nucleated settlements or villages, developed during the previous centuries. In the 18th century just a few scattered farms remained, mostly in the north-west on the edges of the Romeleåsen ridge. The early 19th century map reveals the beginning of *enskifte* enclosures and the accompanying scattering of farmsteads, although most villages were still unaffected. Because of total enclosure only a few villages remained later, and in the early 20th century dispersed settlement reached its highest density. Later some villages grew again, while also some new nucleated settlements appeared, as railway, industrial, or dormitory villages. The growth of the town of Ystad began around the middle of the 19th century. In 1990 two-thirds of the total population live in Ystad proper, compared with one-fourth in 1815, around one-third in 1860, and a little less than one-half in 1915. This depopulation of rural areas can also be followed in the maps from 1915 and 1985. 1. Village, hamlet. 2. Town (Ystad) and other built-up areas. 3 and 4. Dispersed settlement: dense/sparse.

Ecological Bulletins 41: 107–271. Copenhagen 1991

4 Environment and society in selected areas

The arable plain of the Köpinge area. View from Stora Herrestad towards the southeast. Photo: Ingmar Holmåsen/N, Sept. 1988.

4.1 The Köpinge area

4.1.1 Introduction

Björn E. Berglund

Area and sites studied. The Köpinge area, or the Ystad plain, comprises the present parishes of Öja, Stora Herrestad, Stora Köpinge, and parts of Ystad. Our studies deal with the entire area, but the Herrestad-Köpinge-Kabusa area has been examined in greater detail (Figs 4.1:2, 4, 7–9).

The main field study sites are the following:

Palaeoecology: The lagoon basin Herrestads Mosse and the lake basin Fårarps Mosse.

Ecology: Herrestad, Köpinge, and Kabusa villages; the Borrie-Högestad area.

Archaeology: Numerous settlement sites within the entire area. Rich material is available as a result of excavations by RAÄ (Central Board of National Antiquities), particularly in Stora Köpinge and Ystad.

Historical and human geographical studies deal with source material from Stora Köpinge parish in particular.

Geomorphology and hydrology. The greater part of the area is situated within the coastal landscape, which is a smooth sand plain rising about 25 m a.s.l. A beach ridge (5 m a.s.l.) formed during mid-Holocene runs from Ystad eastwards and delimits a former lagoonal basin in the north, today called Öja-Herrestads Mosse. This was a Baltic Sea bay or lagoon, about 5000–2000 BC (Håkansson and Kolstrup 1987). South-east of Köpinge a hilly area with glaciofluvial sand and gravel reaches 25 to 40 m a.s.l.

From a hydrological point of view the area is dominated by small brooks running southwards towards Öja-Herrestads Mosse and by the rivers Nybroån and Kabusaån, which both dissect the sandy plain. With the exception of the large peatland occupying the former lagoon, the area is well drained, Most brooks have today been replaced by pipe drains (Möller 1984). Fens occur mainly in the brook and river valleys. The only ancient lake basin found so far is Fårarps Mosse. It is situated in the border zone between the coastal landscape and the outer hummocky landscape.

Subsoil distribution. The Quaternary soils are dominated by sand and gravel, but along the border towards the hummocky landscape there is a mosaic of sandy soils and clayey tills. Fen peat has a vast distribution in the Öja-Herrestads Mosse basin. Silty and sandy fluvial deposits occur in the river valleys. See Fig. 1.2:15 (Daniel 1986, 1989).

Conditions for cultivation. The sandy soils were favourable for primitive cultivation, but also sensitive to leaching. This applies to the sandy plain around Köpinge, where the agricultural soil quality is rather poor (values 5–8 on the 10-grade scale, Fig. 1.2:16).

The transitional zone between the hummocky landscape and the coastal area is characterized by diverse soil and water conditions, and can therefore be expected to have been more favourable for long-term cultivation. The soil quality here is comparable to the conditions in the outer hummocky zone (value 9, Fig. 1.2:16). The river meadows and the deforested peatlands just inland from the sandy beach have been a particularly important resource for hay production and grazing.

4.1.2 Vegetation and landscape through time

Björn E. Berglund, Mervi Hjelmroos and Else Kolstrup

The vegetation history of the Köpinge area is mainly based on the pollen diagrams from the ancient lake basin of Fårarps Mosse (Hjelmroos 1985, unpubl.), covering the last 6000 years until about AD 1600, and from the lagoon basin of Herrestads Mosse (Håkansson and Kolstrup 1987), covering the time-span 4000 to 1400 BC. The problem of finding a complete lake sequence from the Köpinge plain has been discussed in Ch. 2.1. In addition, a charcoal diagram based on samples from archaeological excavations in the Köpinge area has been useful for the time 1200 BC to AD 1200 (Bartholin and Berglund 1991. The following description is based on an integrated interpretation of these sources. Six main periods of vegetation history have been distinguished. The pollen diagram from Fårarps Mosse is shown in Fig. 4.1:1.

1) 4000–3150 BC. The coastal forest landscape and early human impact

During the Mesolithic the Köpinge area was forested. On dry ground there grew *Quercus* (oak), *Tilia* (lime), *Corylus* (hazel), probably also with scattered *Pinus* (pine) trees, particularly along the sandy coast. On damper ground there were *Ulmus* (elm) and *Fraxinus* (ash), and on wet, peaty ground *Alnus* (alder), forming alder carrs in depressions along streams and along the brackish lagoons at the coast. In general, the diagrams show low herb pollen values without any conspicuous changes. This, together with the constant percentages of

FÅRARPS MOSSE
POLLEN DIAGRAM

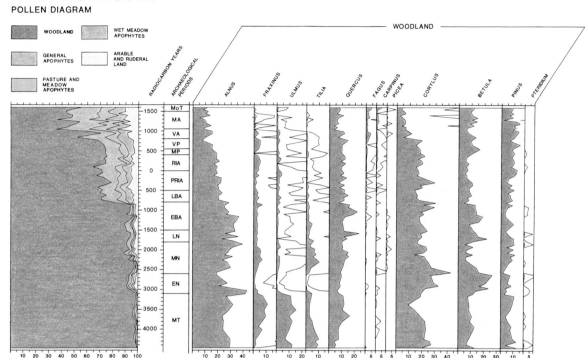

Fig. 4.1:1. Pollen diagram from Fårarps mosse.

the tree pollen curves, does not indicate any obvious disturbances of the forest cover. Kolstrup (in Håkansson and Kolstrup 1987) comments on a charcoal layer followed by occurrence of *Urtica* (nettle), *Galium* (bedstraw), etc. in the Herrestads Mosse diagram a little above the 3500 BC level. Hjelmroos (unpubl.) mentions that in the Fårarps Mosse diagram, at a level correlated to 4300 BC, indicators of open habitats occur – single grains of a cereal pollen type and of *Plantago lanceolata* (ribwort plantain) together with slight maxima for Poaceae and *Pteridium* (bracken). This may be interpreted as minor human influence related to husbandry during the Late Mesolithic. However, we must also assume openings caused by human hand around the Mesolithic hunting-and-gathering settlements along the coast (Ch. 4.2.3).

Owing to the Littorina Sea transgression, the Baltic shoreline along the Scanian south coast was about 3 m above the present shoreline during this period (Nilsson 1935, Gaillard et al. 1988). A brackish sea bay with fluctuating water level occupied the present marsh area south of Öja-Herrestad-Köpinge – the Herrestad lagoon – which was partly sheltered by a sandy beach bar pointing eastwards from the present town area of Ystad. The river Nybroån and minor streams had their outlets along the northern and eastern shores of this lagoon. The shallow bay where the brackish and fresh water became mixed was a favourable environment for a rich

fauna with fish, birds, and seals (cf. Jonsson 1988). This favoured the Mesolithic settlement at the coast (cf. Ch. 5.4).

2) 3150–2500 BC. Small-scale farming in a coastal woodland landscape

This was a very dynamic period which started with an opening of the forest as a result of the "elm decline" catastrophe: elm, lime, and ash decreased heavily, which seems to have favoured hazel and birch on dry ground and alder and willow on wet soils. Finds of several cereal pollen grains *(Triticum)* indicate cultivation during the whole period, and there is a distinct increase of *Plantago lanceolata*, Poaceae, and others, which may point to an expansion of pastures and possibly also of arable land (cf. Groenman-van Waateringe 1986) about 200 years later. The new composition of forest trees indicates a rather light-open underwood with much birch and hazel throughout the period. Pastures and arable fields probably occurred near settlements.

Also during this period the Herrestad lagoon existed with shallow water sheltered by a coastal bar. However, the lagoon gradually became more shallow because of sediment infilling. It is possible that decreased water depth and lower salinity in the Baltic Sea reduced ma-

110

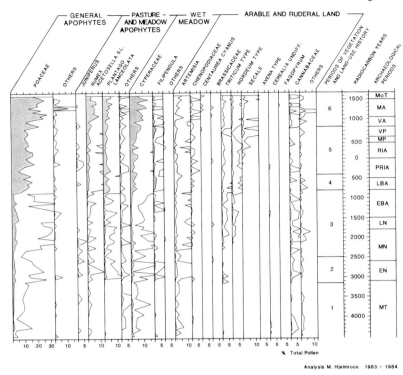

Analysis M. Hjelmroos 1983 – 1984

rine organic production during the beginning of this period and later.

3) 2500–800 BC. Expanding grazing and cultivation in an opening woodland landscape

This period starts with distinctively higher pollen percentages of the broad-leaved trees elm, lime, and ash, at the expense of hazel and birch. During the period ca. 2500–2300 BC there are low values of pasture indicators and no cereal pollen grains. This may be interpreted as weak human impact, mainly based on stock-raising, in a landscape with coppiced woods (cf. Berglund 1969, 1985). An alternative interpretation has been proposed by Göransson (1982, 1986, 1988): wandering arable fields in a coppiced underwood with human impact of the same extent as before, but not detectable through pollen analysis because of pollen filtration (cf. Tauber 1965).

The continued high and/or varying values of elm, lime and ash, together with peaks of hazel and birch, indicate unstable forest conditions, typical of regeneration forest or woodland utilized by people as coppiced woods. There is again a rise of pasture indicators as well as arable field indicators (*Triticum, Hordeum,* weeds) from about 2300 BC, possibly with a slight decrease about 1200 BC. The most probable interpretation is that

for most of this time, i.e., Middle and Late Neolithic and Early Bronze Age, the areas with dry soils were covered by a half-open coppiced woodland with pastures and arable fields. Between 1800 and 800 BC lower values for the broad-leaved trees together with higher values for hazel and birch indicate an expansion of a half-open woodland landscape.

During the Neolithic and the Early Bronze Age, wetland areas were forested, mainly by alder. Because of the Baltic Sea regression after 1500 BC, large areas at the Herrestads Mosse lagoon became overgrown by fen vegetation, possibly a large reed fen (*Phragmites, Carex*) surrounded by alder.

During the early half of the period there was still brackish water in the lagoon, and a mosaic of trees surrounded it. In the sandy areas pine had become more frequent, and alder was frequent in the swampy areas. During the middle part of this period – some time around 1500 BC – the combined effect of gradual infilling, overgrowth, and regression of the sea level put a stop to the marine influence and the marshes expanded. Inevitably this changed the conditions for fishing and hunting in and around the lagoonal area.

4) 800–400 BC. Expansion of pastures and arable fields in a more open landscape

This and the following periods are not represented by reliable pollen data from Herrestads Mosse, and the following outline is therefore primarily based on the diagram from Fårarps Mosse. In that diagram there is a distinct increase of Poaceae, *Plantago lanceolata, Rumex acetosella, Artemisia* and Chenopodiaceae together with *Triticum* type and *Hordeum* type. At the same time there is an expansion of sedges (Cyperaceae) and *Filipendula* (probably *F. ulmaria* – meadowsweet). Among the tree pollen the decrease of *Alnus* is particularly notable. The co-occurrence of all these elements points to intensified land use with expansion of arable fields as well as pastures on dry and wet ground, suggesting a greater demand for fodder, probably including winter storage.

On the whole, during this period the landscape continued its change from woodland towards an increasingly open pasture landscape in which there were meadows with coppiced trees together with fields presumably in a long-term circulation pattern. Undisturbed alder forests may have remained locally only in the wettest areas. Beech and hornbeam had difficulties in colonizing the area. However, the charcoal diagram indicates that firewood was collected from rather closed woodlands still existing until ca. 300 BC, possibly along the coast or in some stream valley.

5) 400 BC-AD 800. Continued expansion of pastures and crop cultivation

This period is a continuation of the expansive period initiated during the Late Bronze Age. However, it starts with a decrease of open land indicators, which may be interpreted in two ways: either a real regression of human impact and production, or a concentration of agriculture to restricted areas around the settlements. The latter alternative could mean a change of organization to permanent settlements with adjacent permanent, enclosed fields for yearly harvesting, surrounded by common land for grazing and utilization of forest products. The increase of hazel pollen supports both alternatives. This period lasts for about 200 years.

From about 200 BC onwards there is a continuous, strong human impact indicated by an increase of elements representing stock-raising and crop cultivation, possibly with a slightly larger proportion of cereal cultivation than during the Late Bronze Age. There may have been a slight decrease of human impact about AD 300.

Pastures with coppiced trees characterized the landscape, on dry as well as on wet soils. Most wetlands were half-open meadows with alder on peatland and ash on surrounding slopes. There seems to be a new decrease of alder around AD 500–800. On dry ground pastures had coppiced trees, like maple and lime, some

hazel and other shrubs of broad-leaved forest. Stands of oak also occurred here. Pine was still available in dry, sandy places (evidence from charcoal analysis). Beech and hornbeam never became common in this area, in contrast to their expansion in the inland area. This certainly reflects the high human pressure on the coastal landscape.

The arable fields, assumed to have been manured areas with permanent cultivation (one-field system), had crops of cereals (*Triticum* and *Hordeum*) and from the Roman Iron Age onwards also of *Secale* (rye).

6) AD 800–1600. Expanding crop cultivation and declining pastures in an open landscape

This period was initiated during the Viking Age by a new increase in the human impact. This time it was mainly expressed as an expansion of arable fields, but the whole landscape was affected. Stands of trees as well as scattered trees became less common from the Viking Age onwards – all the broad-leaved trees as well as hazel became scarcer according to the pollen diagram as well as the charcoal diagram. Also during this time beech and hornbeam seem to have been rare in the area. The curves of the pollen diagram reflect the situation in the inland. Charcoal findings, however, seem to indicate that at least scattered trees occurred around Stora Köpinge. The trees that were present during this period were oak, mainly on dry soils, and alder, along streams and in some fens. In early successions of periodically devastated land (fallows etc.) birch, hazel, and aspen occurred. Some increase of *Calluna* (heather) in the Middle Ages indicates impoverished soil conditions, at least locally. Sandy areas along the coast probably developed into open heath during the Viking Age and Middle Ages.

The area of arable land expands on to earlier pastures. Possibly a two-field rotation system is connected with the expansion in the Viking Age or the Early Middle Ages. The earlier dominant cereals (*Triticum, Hordeum*) are now cultivated together with *Secale. Cannabis* (hemp) is common, at least during the Viking Age and the Middle Ages.

The palaeoecological material from the Köpinge area does not point to any striking changes in cultivation during historical times.

4.1.3 The Neolithic
4.1.3.1 The establishment of agriculture in the Köpinge area

Mats Larsson

Stray finds and the cultural landscape

The chorological distribution of the stray finds in the three parishes of Stora Herrestad, Stora Köpinge, and Öja is discussed in this section. The study also includes the town of Ystad and its surrounding area. The main chronological emphasis here is on the Early Neolithic (EN) and Middle Neolithic Funnel Beaker (MN FB) culture. In terms of ^{14}C dating, this section relates to the period from approximately 3000 to 2100 BC.

Compared with other focal areas, the number of recorded stray finds is considerable in the Köpinge area. A total of 610 artefacts have been dated here. This material covers the whole of the Neolithic period, from the Early to the Late Neolithic. The following totals emerge, when distributed over the periods with which we are concerned in this section:

	EN	MN FB (=TRB)
Number	107	194
%	36	64

The vast majority of the stray finds come from Stora Herrestad/Ystad (Hellerström 1988). A total of 78% of all the artefacts were found in this area. The finds in museum collections were supplemented in the years between 1982 and 1986 by a comprehensive field survey (Larsson and Larsson 1984, 1986, M. Larsson 1987, Hellerström 1988). The aim was to cover all the accessible ground in the area. Most of the surveying was done in the early spring, and only in exceptional cases during the late summer/autumn. A considerable number of sites were recorded in this way. The stray finds will now be discussed as an introduction.

As far as the Early Neolithic is concerned, it is mainly the pointed-butted and thin-butted axes that are able to tell us something about the extent of settlement in the period.

The number of pointed-butted axes is relatively small, amounting to 26 items in total. These come from two main areas: Öja-Herrestads Mosse and, to a lesser extent, also from the Kabusa area. The stray finds point to settlement along the coast (M. Larsson 1987a, Hellerström 1988). The close connection between the pointed-butted axe and the Early Neolithic Oxie group has been discussed elsewhere (Nielsen 1977, 1985, Hernek 1985). It has been claimed that there is a tendency for this type of axe to be associated primarily with the inner hummocky zone. Thin-butted axes are considerably more numerous than pointed-butted axes. A total of 81 axes of this type have been dated to the Early

Neolithic (Hellerström 1988). These include Nielsen's types I-IV (Nielsen 1977). The area around Herrestad also has a particular concentration of this type of axe.

There is a distinct increase in the number of recorded stray finds during the Middle Neolithic Funnel Beaker culture. A similar increase in the number of stray finds has been observed in various other areas, such as north-west Zealand, Langeland, and south-east Scania (Mathiassen 1959, Skaarup 1985, Strömberg 1985). Particularly noticeable is the fact that the items concerned are the later types of axe from the Middle Neolithic IV-V (Lindø and Store Valby). This is a pattern which tends to repeat itself in all the focal areas.

The Neolithic settlement region

What is important in this context is the distribution of the stray finds compared with the evidence presented by the settlements. The total number of sites is 79, of which 24 have been excavated. The number of sites which can be dated to the Funnel Beaker culture is 20 (Fig. 4.1:12). The sites were excavated partly within the framework of the Ystad Project and partly under the auspices of the Central Board of National Antiquities (Tesch 1983).

Early Neolithic. For the purposes of this survey, the period has been divided into two subperiods: Early Neolithic 1 refers to the oldest part of the period (3000 – ca. 2700 BC); Early Neolithic 2 denotes the youngest part of the period (2700–2600 BC).

We were able to document only five reliably dated sites from the Early Neolithic 1 in the area – Robertsdal, Karlshem, Karlsfält, Kabusa IV, and Kabusa 7:2 (Larsson 1985, Larsson and Larsson 1986, M. Larsson 1987a). Two of these can be classified as inland sites, Robertsdal and Karlsfält, whereas the others are near the coast. Fragments of pointed-butted axes have been found at a couple of the sites. All the investigated sites can be classified in the distinct Mossby Group, which is characterized above all by cord ornamentation (Larsson and Larsson 1986, M. Larsson unpubl.). Radiocarbon datings performed at the site which gives the group its name, Mossby in Västra Nöbbelöv parish, put its age at between 3060 (Ua-430) and 2975 (Ua-755) BC. The datings from this site are amongst the oldest Neolithic datings in southern Scandinavia, and are comparable in every respect with the oldest Danish datings (Madsen and Petersen 1984).

The settlement sites from the Early Neolithic 1 are few in number and scattered. They are characterized by their localization on sandy soils, preferably at a certain elevation and in the vicinity of water. The sites are small; the surface area of those which were investigated does not exceed 600 m^2. The pattern is characterized by widely distributed settlements in the form of small units, to all intents and purposes of family size.

In terms of the available material, the Early Neolithic

2 represents a clear expansion compared with the Early Neolithic 1. The increase in the number of sites is clearly related to the aforementioned increase in the number of thin-butted axes. A similar relationship has also been observed in many other parts of southern Scandinavia, for example on north-west Zealand and Langeland (Mathiassen 1959, Skaarup 1985).

Twenty-five sites are known from the Early Neolithic 2, although some of these may belong to the Middle Neolithic I, since the fragments of the polished axes are frequently too small to permit accurate dating. Of these sites, six have been investigated (Tesch 1983, Larsson and Larsson 1984, 1986, Jakobsson 1986, M. Larsson 1987a, Hellerström 1988).

A distinct increase in the density of occupation can be observed in the area. This is true in particular of the area around Herrestads Mosse, the area north-east of this bog, parts of Tingshög near Köpingebro, and the Kabusa area. The tendency towards larger sites which was observed by Skaarup (1985) on Langeland cannot be found in this area, however. The sites are still small, with relatively small quantities of finds. A small house, which is partly below ground level, has been investigated at one of the sites, Piledal, at Stora Herrestad (Larsson and Larsson 1984).

Is interesting that a tendency can be seen for the growth of local territories in the area. Most likely the first megalithic graves, the dolmens, should be understood in this context. One theory which has been frequently advanced is that the megalithic graves constituted a form of territorial boundary for a particular group of people (Madsen and Jensen 1982, Strömberg 1982, Renfrew 1984, M. Larsson 1985, 1988b). In addition to having functioned as burial places, the dolmens also served as a symbol for the right of a specific group to occupy a particular area. Only one megalithic grave is preserved today in the area with which we are concerned, Trollasten in Stora Köpinge parish (Strömberg 1968). It has proved possible to document a further three with the help of old land survey maps (Riddersporre 1987). Yet another was discovered, at Skogsdala (Jakobsson 1986), in conjunction with the excavation of a ploughed-out Bronze Age barrow.

In summary, it can be stated that a noticeable difference exists between the older and the younger parts of the Early Neolithic in terms of settlement structure, social structure, and economy (M. Larsson 1985). From a system made up of widely distributed, small sites with access to large areas of territory, in which both the coastal and inland areas were utilized, a noticeable change takes place during the latter part of the period. An increase in the density of settlements can be observed, accompanied by a structural change in which the growth of local territories plays a major role. In this respect, the megalithic graves had a significant part to play in a manifestation of social harmony. The coastal strip in particular was used during this part of the Early Neolithic, whereas the inland area appears to have been used only sporadically. Settlement during the Early Neolithic was associated with light soils in the vicinity of wetlands. In addition, the outer hummocky zone with its heavy clay soils had small areas of lighter soil suitable for settlement. This is a feature which was also observed by Skaarup (1985) on the islands to the south of Funen.

Middle Neolithic. As mentioned above, there is a noticeable increase in the number of stray finds from the Middle Neolithic. Nevertheless, a clear discrepancy exists between the number of stray finds and the number of dated sites. Only 20 sites have been dated to the period, of which 11 have been investigated. The majority can be dated to the Middle Neolithic I. In a couple of cases it is possible to study the continuity from the previous period on the same site. It is clear that the Middle Neolithic I represents continued expansion with regard to settlement. The largest sites within the area date from this period. The Kabusa IV site covers approximately 1800 m^2, for example (M. Larsson 1987a). A number of ^{14}C datings put its age at between 2590 BC (Lu-2773) and 2550 BC (Lu-2801).

The pattern of settlement during the period gives the impression of having been centred around larger sites, known as base sites. Furthermore there is evidence on the coast and in the area around Herrestads Mosse of what may be described as hunting stations, or satellite sites (Skaarup 1973, Madsen and Jensen 1982, M. Larsson 1984). One such small site has been investigated within the boundary of the present-day town of Ystad. This site was situated directly next to the then existing shore (Tesch 1985c). Sites of this type appear to have been characteristic of the later part of the Early Neolithic and the early part of the Middle Neolithic (Skaarup 1973).

The major problem is to establish what happened in the later part of the Middle Neolithic. The number of stray finds increases, not only within the area with which we are concerned here, but also in neighbouring areas (Strömberg 1985). The number of dated sites from Periods II to V of the Middle Neolithic is small, however. This is particularly true of Periods II and III, which are represented by only a couple of investigated sites (Piledal and Kabusa II). However, to judge from the Kabusa II site, which has been thoroughly investigated, the settlements during the period are relatively large (M. Larsson 1987a).

The actual increase in the number of stray finds, in particular thick-butted axes, can be dated to the Middle Neolithic IV/V. It is interesting that the number of sites investigated has also increased. A few sites dating from the end of the Middle Neolithic Funnel Beaker culture have been investigated – Piledal, Karlsfält, and Tingshög (Larsson and Larsson 1984, 1986, Larsson 1985). A large house structure has also been excavated at the Piledal site.

The settlement pattern does not give the impression of having undergone any distinct change from the previ-

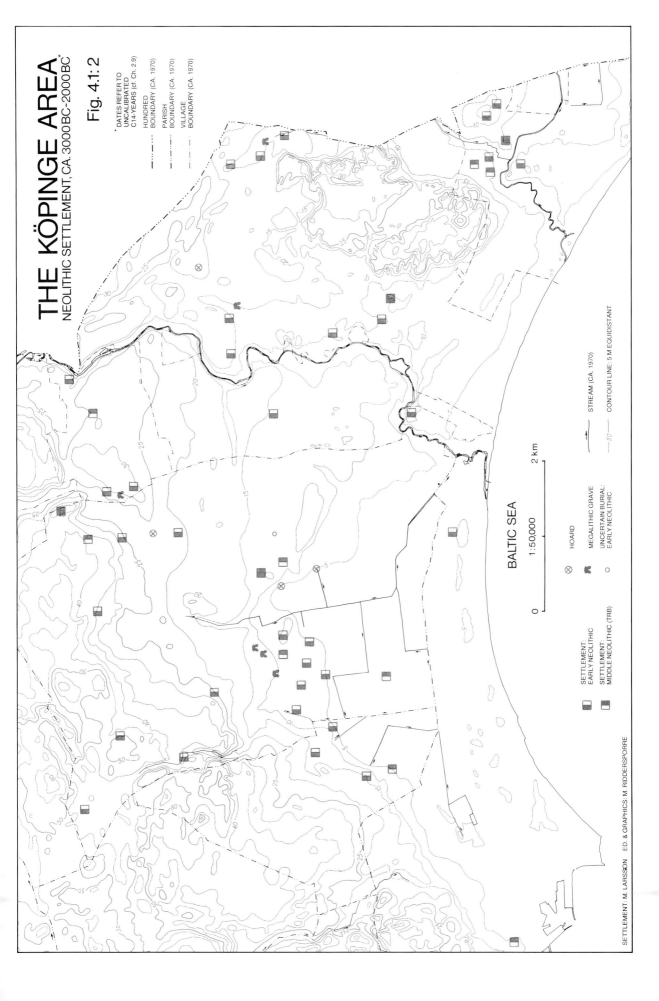

THE KÖPINGE AREA

NEOLITHIC SETTLEMENT, CA. 3000 BC–2000 BC*

Fig. 4.1:2

* DATES REFER TO UNCALIBRATED C14-YEARS (cf. Ch. 2.9)

—··—··— HUNDRED BOUNDARY (CA. 1970)

—···—···— PARISH BOUNDARY (CA. 1970)

—·—·—·— VILLAGE BOUNDARY (CA. 1970)

BALTIC SEA

1:50,000

0   2 km

⊗ HOARD

MEGALITHIC GRAVE

UNCERTAIN BURIAL: EARLY NEOLITHIC

○ EARLY NEOLITHIC

SETTLEMENT: EARLY NEOLITHIC

SETTLEMENT: MIDDLE NEOLITHIC (TRB)

STREAM (CA. 1970)

—50— CONTOUR LINE: 5 M EQUIDISTANT

SETTLEMENT: M. LARSSON   ED. & GRAPHICS: M. RIDDERSPORRE

ous pattern. The sites appear to have been more widely distributed during the period, with settlement even extending as far as the outer hummocky zone. This area of clayey soils appears to have been utilized to a greater extent, a situation which has also been observed on Langeland (Skaarup 1985). The stray finds clearly indicate an increase in the utilization of the more peripheral areas. It is difficult to explain these changes in the sites on the basis of the relatively small amount of available material. However, similar indications to the effect that the sites may have been larger, of the kind observed by, amongst others, Davidsen (1978) and Skaarup (1985), cannot be found in the Ystad area. Settlement may instead have been concentrated in certain areas, as has been observed in the Malmö area (Svensson 1986). Some of the older territories were still occupied during the late Funnel Beaker culture, whereas others, such as the Kabusa area, had no settlement during the period.

The pattern of settlement which emerges during the late Funnel Beaker culture is one of relatively small, widely distributed sites, with use also being made of the more peripheral areas. There is also evidence of hunting stations. A small site of this kind has been found in the middle of Herrestads Mosse. What is interesting is the discovery here of a type A flake arrowhead and a cylindrical flake block. Items such as these are usually associated with the Pitted Ware culture. There is a good deal of evidence, however, to suggest that they were an integral part of the late Funnel Beaker culture (Skaarup 1985).

A pattern of settlement consisting of widely distributed small sites can tell us a great deal about the structure of society and the factors which had a restrictive influence on it (Renfrew 1983). Whittle (1985), for example, maintains that a settlement pattern with widely distributed units may point to changed social circumstances and the onset of more distinct social differentiation. Factors such as the need for larger open spaces for animal husbandry may have played a part in this (M. Larsson 1987a). The economy of the late Funnel Beaker culture also gives the impression of having been structured entirely around animal husbandry, in particular for the production of meat (Nyegaard 1985).

4.1.3.2 Late Middle Neolithic and Late Neolithic in the Köpinge area

Lars Larsson

Stray finds and the cultural landscape

The Late Middle Neolithic, also referred to as MN B, in this case corresponds to the Battleaxe culture, which is one of a number of cultural manifestations within the northern European Battleaxe culture (Larsson 1989b). The period covered by this section is from 2200 to 1800

BC (2800–2300 cal. BC). The Late Neolithic phase (LN) constitutes a uniform cultural complex for the period 1800–1500 BC (2300–1800 cal. BC).

With regard to the period with which we are concerned, types of implement have been found which, as in earlier periods, provide relatively narrow dating, although there are also other types for which the dating parameters are broader than those indicated above. This is particularly true of those types of artefact which are referred to as being Late Neolithic, but which in many cases were in use for a period extending over shorter or longer sections of the Bronze Age.

The artefacts are distributed as follows in the three parishes discussed here:

	MN B		LN	
	No.	%	No.	%
Öja	5	24	9	43
Stora Herrestad (inc. Ystad)	58	13	142	32
Stora Köpinge	19	13	66	43
Total	82	13	217	36

The vast majority of the artefacts come from Stora Herrestad, as is the case for earlier periods, although we have also included the finds from a fairly large private collection which are listed only as being from the Ystad region; some of these items may possibly belong to the other two parishes.

When related to the subdivision of the categories of finds by Neolithic epochs, the distribution of the artefacts by year is as follows (here calendar years are used). EN (700 years): 0.15; MN (600 years): 0.32; MN B (500 years): 0.15; and LN (500 years): 0.43.

On the basis of the above reservation as to the age of the Late Neolithic artefacts, the figure of 0.43 for LN is rather high. With the exception of MN B, there is a tendency towards an increased level of production of objects, which may possibly be on a par with a small increase in the population.

As far as the distribution of settlement is concerned, the individual finds should indicate a distribution pattern which is different for MN B. Since large parts of the parish of Öja consist of hummocks, the increased proportion of finds from this period may point to the spread of settlement into this type of landscape (Ch. 5.4). Areas of hummocky landscape are also to be found in the two other parishes. However, it has been established that the stray finds in both parishes are distributed within the area of the slightly undulating sandy plain.

During the Late Neolithic period, with its more abundant finds, the distribution of the various types of dagger indicates that the more recent part of the period is better represented than the older part.

The Neolithic settlement region

Sites dating from MN B are very few in number, as in the rest of southern Scandinavia (Hvass 1977, 1986, Ch. 5.4). Objects which are characteristic of this phase have been found at only four sites, all on sandy soil. One of these, Stora Köpinge 25:1, has been excavated (Larsson and Larsson 1986, Larsson 1989b, 1991b). The sites of a number of buildings have been found here, including a shallow depression with post-holes, which presumably marks the position of a house. The number of finds was very limited; this is also a typical feature of sites dating from this period. The site can be dated to the end of MN B, period 5, in accordance with Malmer's classification (Malmer 1962). Two grave sites have been dated to MN B. One of these, a flat-earth grave at Kabusa (Oldeberg 1952), was situated on the site under investigation, whereas the other was discovered in conjunction with the investigation of the Trollakistan (Trollasten) dolmen (Strömberg 1968). Both graves are from a late part of MN B.

The number of sites containing finds which are characteristic of the Late Neolithic is large in relation to MN B. Late Neolithic artefacts have been found on a total of eleven sites; six of these sites have additionally been found to contain items dating from other periods. Several of the sites have been investigated, and the remains of houses have been found at five of them (Herrestad 68:87, Lilla Köpinge 7:3, Lilla Köpinge 19:1, Lilla Köpinge 64:1, and Stora Köpinge 21:25) (Tesch 1983, Larsson and Larsson 1984, 1986; M. Larsson 1987a). Two different types of house can be distinguished. The first is in the form of the site of a house with a sunken floor area. Discoloration left by posts has been found here, both along the edge and in the central area of the pit. The other type of house has a row of posts supporting the roof and wall posts. Occasionally the wall posts next to the central post-holes are inset slightly. This type of house is encountered in two different sizes, one of up to about 20 m long, and another which may be up to 40 m long. Houses of these types have also been documented at other locations in western Scania (Björhem and Säfvestad 1987) and on the island of Bornholm (Nielsen and Nielsen 1985). This suggests different forms of social units within the community. Both types of house were in existence during an early part of the Early Bronze Age. The absence of occupation layers at the sites which have been investigated means that we have only an uncertain idea of the size of the sites. Nevertheless, the distribution of the stray finds does not appear to indicate that they were any larger than during the earlier parts of the Neolithic.

A total of six graves are known from the Köpinge area (Larsson 1989b). These were discovered as primary graves in barrows and as flat-earth graves, one of which was situated directly adjacent to a habitation site. A number of Late Neolithic finds which have been made at the site of a destroyed megalithic grave point to secondary interments in this type of grave.

4.1.3.3 The agro-ecosystem of Neolithic farmers at Kabusa

E. Gunilla A. Olsson

Introduction

The Early and Middle Neolithic settlements at Kabusa cover a period of 900 years during the time 3000–2100 BC (M. Larsson 1987, Ch. 4.1.3.1). The various excavations here have yielded large amounts of artefacts and remains from human activity. It cannot be doubted that this activity had an impact on the landscape, but the extent can be discussed. To understand the conditions for life of Neolithic humans, we must consider their environment with its range of ecological resources and limitations. Below is presented a reconstruction of the landscape at Kabusa during the Early and Middle Neolithic.

Site description

Sandy soils and a flat topography characterize the area (Fig. 4.1:3). The Kabusa settlements are encircled to the west and north by the river Kabusaån. Peatbog and fenland, the present Ingelstorps Mosse, meet in the east, and the area is bordered to the south by the sea. In the Early Neolithic, when the sea level was 5 m higher than today, the settlements were thus situated very close to the actual seashore on points of land protruding into a narrow and shallow sea bay (Figs 3.2:1, 4.1:1). The present Ingelstorps Mosse presumably formed a mosaic of fertile brackish marsh and alder carr (M. Larsson 1987, Daniel unpubl.).

Methods and sources

Assumptions for the reconstruction. Interpretations of compiled field data from archaeological excavations at Kabusa and from nearby settlements at Mossby (Larsson and Larsson 1986), Herrestad (M. Larsson unpubl.) and Östra Vemmenhög (L. Larsson unpubl.) are combined with palaeoecological data from this and adjacent areas (Håkansson and Kolstrup 1988, Engelmark and Hjelmqvist in press). This collection of data constitutes the basis for a protein-based model reconstruction of the agro-ecosystem at Kabusa (see also methods in Ch. 2.2).

It is assumed that the human population living at Kabusa consisted of a relatively small group of about 5 adults and 5 children. This number is in accordance with data from archaeological excavations on similar sites (cf. Larsson 1987). Their dietary composition (Table 4.1:1) is interpreted from the sources specified above. In total this makes a dietary composition based on 60% animal protein and 40% vegetarian protein. Mean human adult weight is calculated at 65 kg. Daily dietary protein intake is estimated as 0.8 g kg^{-1} body weight,

MIDDLE NEOLITHIC LANDSCAPE IN KABUSA, CA. 2400 BC*

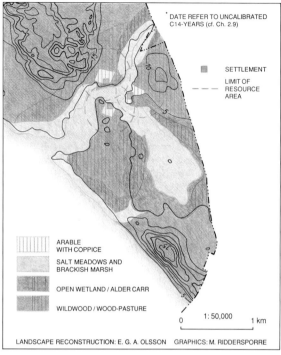

* DATE REFER TO UNCALIBRATED C14-YEARS (cf. Ch. 2.9)

■ SETTLEMENT

— — — LIMIT OF RESOURCE AREA

▦ ARABLE WITH COPPICE

▦ SALT MEADOWS AND BRACKISH MARSH

▦ OPEN WETLAND / ALDER CARR

▦ WILDWOOD / WOOD-PASTURE

0   1:50,000   1 km

LANDSCAPE RECONSTRUCTION: E. G. A. OLSSON   GRAPHICS: M. RIDDERSPORRE

Fig. 4.1:3. Landscape-ecological map of the Kabusa area during the Middle Neolithic, ca. 2400 BC. Only rough vegetation types can be marked owing to the lack of a pollen core in the centre of the area. The circles surrounding the settlements denote the estimated resource area (arable and grazing land) when subsistence level for 5 adults and 5 children is assumed. Source: see text and Ch. 2.2.

which corresponds to the modern Swedish recommended intake (Näringslära 1983). This level was chosen to avoid underestimates. The protein content of the different products was taken from modern literature (Livsmedelstabeller 1978).

Einkorn and emmer wheat were the cultivated cereals (M. Larsson 1987, Engelmark and Hjelmquist in press).

For the seed yield two alternative calculations were made: with estimated seed return 4 and with seed return 6.

Results and discussion

To produce the necessary subsistence amount of cereals for the people inhabiting a Kabusa settlement (Table 4.1:1) a field area of 0.3–0.4 ha should have been harvested annually (Table 4.1:2). The smaller area relates to a seed return of 6. This is highly probable considering the type of agriculture practised here and experimental results from the cultivation of cereals without manure (Garner and Dyke 1969, Lüning and Meurers-Balke 1980, Olsson unpubl.). Manuring in the sense used later during the Bronze Age and Iron Age was not practised, but it is reasonable to imagine that abandoned settlement sites were used for cultivation. Thirteen such adjacent sites from the Early and Middle Neolithic have been discovered within distances of less than one kilometre, and several of these have been excavated (M. Larsson 1987, Ch. 4.1.3.1). The settlement are of equal size, which leads to the assumption that the human population was of roughly equal size during the whole period. Cultivation on such sites could have benefited from higher levels of soil nutrients, especially phosphorus and to a lesser extent nitrogen. If the cultivation was also combined with coppice cycles with deciduous trees and hazel (cf. Göransson 1988), it was fully possible to maintain a proper plant nutrient circulation on the arable sites. The estimated area of one settlement site was at most 1500 m^2 (M. Larsson 1987) and the calculated field area, based on the assumption that cereals contributed to 30% of the human dietary composition (cf. above and Table 4.1:1), was thus about twice this area (Table 4.1:2). The varieties of ancient wheat – einkorn and emmer – were exceptionally rich in protein content, around 20% of dry matter weight (Engelmark and Hjelmquist in press). In fact, these cereals had as high a protein content as meat and fish. Obviously, this fact influences the size of the calculated areas for cereal cultivation.

The calculated amount of annually consumed meat,

Table 4.1:1. Inferred dietary composition during the Middle Neolithic in Kabusa, as suggested by the archaeological findings. The protein content of the different food items is extracted from modern data (Livsmedelstabeller 1978). The data below are calculated for subsistence level for a settlement site inhabited by 7.5 adult units.

Item	% protein contribution	kg protein adult^{-1} yr^{-1}	kg food adult^{-1} yr^{-1}	Total no. animals consumed yr^{-1}	No. of stock
einkorn	30	5.7	28.5	–	–
nuts etc.	10	1.9	15.1	–	–
fish	9	1.7	8.2	–	–
cattle	15	2.9	13.6	1.2	5
sheep	15	2.9	14.2	6.4	20.5
pigs	15	2.9	14.2	3.8	6
game, eggs	6	1.1	7.6	56.6	–
Total	100				

Table 4.1:2. The relationship between seed return of einkorn and emmer and arable area. Calculation valid for subsistence consumption (according to Table 4.1:1) of 7.5 adults.

Crop	Seed return	Arable area (ha)
einkorn	4	0.44
einkorn, emmer	6	0.28

transformed into the number of slaughtered animals, is listed in Table 4.1:1. These figures imply the maintenance of a stock consisting of (given as adults units) 5 cows, 6 pigs, and about 20 sheep. The fodder for cattle and sheep was obtained from grazing and browsing an area of approximately 40 ha (combined data from Coupland (1979) have been used for this calculation). If leaf-fodder from coppices was also harvested and used as extra support during the winter grazing, the foraging area needed would have been smaller. The present figures relate to all-year grazing, since the climate did not require winter stalling. In fact, the period after 3000 BC had a slightly drier climate than before and after, as indicated by lower lake-water levels in this area (Ch. 5.1).

Hazel-nuts were consumed (Larsson 1984) and young hazel rods were used for the construction of dwellings (M. Larsson 1987). It is probable that elm and hazel were coppiced also to give building material and more proliferate fruit crops of hazel (Bartholin 1978, Göransson 1988).

Unfortunately, there is no pollen diagram for Kabusa. The diagram from Herrestads Mosse, with similar ecological conditions (Håkansson and Kolstrup 1988), shows increased values for hazel and birch pollen accompanying the elm decline around 3000 BC. This is coincident with higher values for cereal pollen (wheat varieties) as well as pollen from open grasslands (Poaceae and Cyperaceae). The same tendency is found in Fårarps Mosse, situated 8 km north of Kabusa (Hjelmroos 1985). This development can be interpreted as an opening of the deciduous forest by factors such as the coppicing of elm and hazel and other deciduous trees accompanying small-scale cultivation of wheat (Göransson 1988).

The woodlands on this sandy site were a mixture of elm, lime, oak, and hazel (Håkansson and Kolstrup 1988). The vegetation surrounding a settlement can be interpreted as a sort of combined woodland, pasture, and coppice areas. A closed canopy probably did not exist, but a mosaic of coppice groves and trees affected by browsing. In this environment there was enough light to allow patches of grassland vegetation, which was maintained by the grazing stock. The small cultivated areas, making up less than 1% of the resource area, were probably fenced for protection from the free-grazing stock. When the cultivated areas were left for spontaneous succession they were overgrown by birch. This could explain the higher birch pollen curve for the pe-

riod and is a particularly appropriate path of secondary succession on this sandy site. The large alder carr in the east, indicated by preserved peat layers and high levels of *Alnus* pollen (Håkansson and Kolstrup 1988), provided proliferate grazing possibilities. The brackish marshes and the salt-water meadows along the seashore had high biological production and supplied nutritious fodder for the stock, especially appreciated by the cattle. Finally, the shallow sea coves harboured fish and seafood.

The settlement site chosen by Early and Middle Neolithic farmers at Kabusa contained an ecologically varied environment and provided rich possibilities for sustainable subsistence farming. These conditions are indirectly supported by the archaeological data: (1) people lived here and used this settlement site for 900 years; (2) the size of the human population and possibly also the extent of the farming activities remained unchanged during this period, as interpreted from the size of the settlement areas.

4.1.4 The Bronze Age and Early Iron Age
4.1.4.1 A central settlement region on the coastal plain in the Early and Late Bronze Age

Sten Tesch

Introduction

The investigation of the Köpinge area as part of subproject B3 spans almost half the temporal framework of the Ystad Project: the Bronze Age, Iron Age, and Early Middle Ages, up to the emergence of the town of Ystad at the beginning of the 13th century (Tesch 1988).

When the project began there was already habitation site material from both the Bronze and Iron Ages (Tesch 1979a,b, 1982a, Tesch et al. 1980); this varied and, for Sweden, relatively ample material was the result of excavations and phosphate mapping by the Central Board of National Antiquities in the Köpinge area (Stora Köpinge parish). We not only had a good general picture of the topographical location of the sites and the nature of settlement in various periods, but also a hypothesis about a mobile pattern of habitation; every settlement, at least up to the Early Iron Age, had regularly shifted a short distance to a new site within a limited area (Tesch 1980, 1983c). It was possible to discern distinct geographical preferences for the choice of habitation site, in relation to terrain, soil, and wetlands. Low heights, often glacial beach ridges, on the gently undulating sandy plain were preferred to clayey soils or noticeable heights like the Köpinge hills. The localization of settlement was also determined by the proximity of wetlands and above all the river Nybroån.

In relation to the hummocky landscape, the Köpinge area appears to have been a central area with continuous settlement. There does not seem to have been any definitive colonization of the inner hummocky zone before the Viking Age. On the other hand, this landscape had been a buffer for the plains to varying extents under different periods. It was used for temporary settlements, such as shielings, and as outland not only for sheep farming, gathering, lake fishing, and hunting, but also for special activities such as iron extraction and cult (Tesch 1980, 1983c, Tesch et al. 1980, Tesch and Wihlborg 1981).

For the Medieval Town Project (Tesch 1983a) the then existing archaeological material was used as a basis for a separate hinterland study, an analysis of settlement development in the entire Ystad area up to the emergence of the town (Tesch 1983b). Settlement colonization of the hummocky landscape, the establishment of villages, and the urbanization process in the Viking Age and Early Middle Ages have been described elsewhere (Tesch 1985b). In the course of the project the representativeness of the archaeological material has also been discussed (Tesch 1985a).

The years of the Ystad Project have seen a considerable increase in the amount of archaeological material, partly through the investigations of subproject B3 (Tesch 1988), but above all through the Central Board of National Antiquities' continued rescue excavations in conjunction with the exploitation of arable land. Further archaeological material from the Bronze and Iron Ages has also come as a by-product of studies of the Köpinge area as part of subprojects B1 and B2 (Larsson and Larsson 1984, 1986, Larsson 1987a,b).

There is no room here to consider in detail the work of subproject B3 in the Köpinge area. It consisted not only of excavations (Tesch 1988) but also of aerial archaeology and a retrogressive analysis of the cultural landscape on the basis of the oldest land survey documents (Riddersporre 1983, 1985, 1986, 1987, 1988, 1989, Ch. 4.1.6.2, 4.1.7.2). There were also analyses of various kinds: macrofossils (Engelmark and Hjelmqvist 1991), wood anatomy (Bartholin and Berglund 1991), [14]C (Stockholm, Lund, and Uppsala), osteology (Szalay 1987, Jonsson 1989). The results of these analyses, and the studies of the entire subproject, will be presented by the respective authors in the archaeological monograph (Callmer et al. 1991). The map in Fig. 4.1:4 shows the distribution of Bronze Age settlement.

The source material

The investigations of habitation sites in southern Scandinavia in the latest decades have greatly changed our ideas about prehistoric times. The sites as source material give a much richer picture of the living society than we obtain from grave-finds (Näsman 1987:69f.).

The analysis of settlement development in the Bronze and Iron Ages in the Köpinge area comprises 58 sites with a total stripped area of over 100 m². It is important to point out that it is rarely a question of single dwell-

ings, but rather of habitation areas which have been occupied recurrently, either several times during one period or in some or all of the periods. There are thus far more than 58 chronologically delimited dwellings. The places occupied in several different periods are centrally located in the Köpinge area. There are remains at these sites of occupation from on average three of the periods (from the Late Neolithic to the Early Iron Age). In three of the areas that were stripped traces were found of habitation only from the time before the Late Neolithic. Five smaller sites were not datable.

A total of some 15 ha was stripped, of which 87% was in Stora Köpinge parish. This figure includes over 25 km of trial trenches almost 2 m wide. If we include the area between these trenches, which mostly ran parallel (two or more at intervals of 10 m), then a total of some 45 ha has been investigated. In all 13,000 structures were documented: hearths, pits, wells, and not least post-holes. In the maze of post-holes it has been possible to distinguish almost a hundred certain long-houses. In addition, 54 sunken-floor huts have been investigated. These figures can be compared with, for example, the 47 ha of stripped surface within an area of 60 ha of the Fosie IV excavation in Malmö (Björhem 1986:90) or the 20 ha of stripped surface within an area of 1 km² of the Vorbasse excavation in Jutland (Hvass 1987:49). About 4400 structures were studied at Fosie, distributed among six more or less clearly delimited habitation areas. Around a hundred long-houses and over twenty sunken-floor huts from the Late Neolithic to the Viking Age were discernible. At Vorbasse the settlement comprised eight villages from the 1st century BC to AD 1100, or what was actually one village moving around within the area almost every century.

There are thus many more habitation sites in the Köpinge area. They are also spread over a much larger area, although most of the most important investigations are concentrated to an area of about 3 km² along the river Nybroån. This spatial dispersion need not be a disadvantage for the analysis of land use in a long-term perspective.

Graves were found at a quarter of the habitation sites in the Köpinge area, but they constitute only 0.3% (40 in number) of the total structures. We should expect this proportion to be higher if it were random; this raises the question of how large a share of the population was buried at different periods. Furthermore, the grave-finds as a whole in the Köpinge area show a highly uneven chronological distribution. Only in the Bronze Age is the material sufficiently representative to provide evidence for settlement archaeology (Ch. 4.1.4.2).

Although many archaeological investigations in the Köpinge area were undertaken as part of the project, it is nevertheless the extensive investigations conducted by the Central Board of National Antiquities which enable any discussion of settlement development and the use of the cultural landscape in the Köpinge area during the Bronze and Iron Ages. These rescue excava-

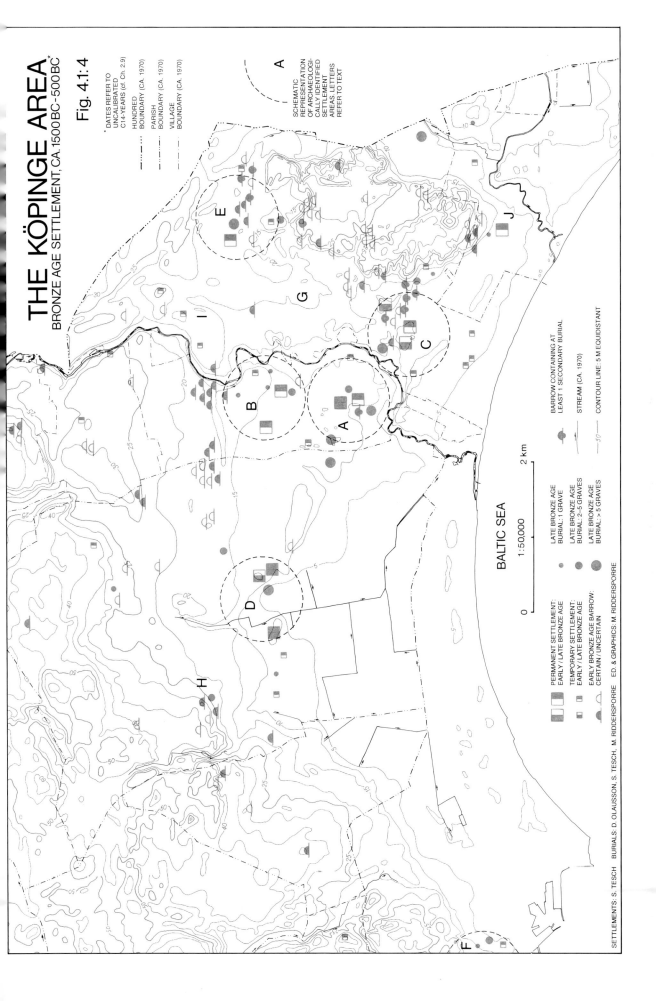

THE KÖPINGE AREA
BRONZE AGE SETTLEMENT, CA. 1500BC–500BC*

Fig. 4.1:4

* DATES REFER TO
  UNCALIBRATED
  C14-YEARS (cf. Ch. 2.9)

– · · · – HUNDRED
  BOUNDARY (CA. 1970)

– · · – PARISH
  BOUNDARY (CA. 1970)

– · – VILLAGE
  BOUNDARY (CA. 1970)

SCHEMATIC
REPRESENTATION
OF ARCHAEOLOGI-
CALLY IDENTIFIED
SETTLEMENT
AREAS. LETTERS
REFER TO TEXT

PERMANENT SETTLEMENT:
EARLY / LATE BRONZE AGE

TEMPORARY SETTLEMENT:
EARLY / LATE BRONZE AGE

EARLY BRONZE AGE BARROW:
CERTAIN / UNCERTAIN

LATE BRONZE AGE
BURIAL: 1 GRAVE

LATE BRONZE AGE
BURIAL: 2–5 GRAVES

LATE BRONZE AGE
BURIAL: > 5 GRAVES

BARROW CONTAINING AT
LEAST 1 SECONDARY BURIAL

STREAM (CA. 1970)

–50– CONTOUR LINE: 5 M EQUIDISTANT

BALTIC SEA

1:50,000

0   2 km

SETTLEMENTS: S. TESCH   BURIALS: D. OLAUSSON, S. TESCH, M. RIDDERSPORRE   ED. & GRAPHICS: M. RIDDERSPORRE

tions, unlike those of the Ystad Project, are dictated by the pressure of exploitation and are therefore a random selection, which need not be a disadvantage in this connection.

From the point of view of source criticism it is important to distinguish between different types of habitation sites (Björhem et al. 1982, Tesch 1983b:58ff.). In my view it is possible to distinguish three types: (1) year-round habitation site; (2) temporary camp site; (3) seasonal site. The year-round sites, henceforth referred to as habitation sites, are the same as the agrarian production unit, the farm, comprising one or more long-houses and smaller buildings. This type of site occurs in central locations in the resource area, at the shortest possible distance from suitable lands for tillage, meadows, and pasturage. Camp sites, on the other hand, are usually on the periphery of the resource area; they were used for grazing or quarrying clay in outlands, or salt-water fishing. Traces in the ground consist of hearths and isolated pits. Stray finds and remains of hearths turned up by ploughing may indicate a habitation site, but could equally well denote a camp site – to judge with certainty we would need to excavate. Finally, seasonal sites go hand in hand with some form of specialized activity, such as trading, summer pasturing, or iron extraction; they can be far away from the "permanent" habitation. Concrete traces of settlement often consist of smaller long-houses, huts, or sunken-floor huts and hearths and pits.

For an understanding of land use in a long-term perspective it is important to assess with the aid of the archaeological evidence what patterns of settlement prevailed in different periods. One must also bear in mind that a mobile pattern of settlement leaves many more habitation sites in the landscape than a more stationary pattern. We have three conceivable models to choose from: (1) that settlement moved very short distances within the same habitation area; (2) that settlement moved short distances within a resource area or a settlement area often limited by topographical features; (3) that settlement moved long distances within the entire central area, the settlement region.

By combining the archaeological material with soil and phosphate maps (Arrhenius 1934, Tesch 1983b) it is possible to distinguish, at any rate for the Bronze Age and the Early Iron Age, at least six centrally located and topographically delimited settlement areas. The pattern of settlement probably varied between the first two of the above models, depending on the organization of the agrarian economy. The assumed size of the settlement areas, about 1 km², agrees well with what has been shown to be reasonable in comparable contexts (Welinder 1977, Strömberg 1982). As regards the camp sites, these are usually peripheral to or outside the assumed settlement areas, in the outlands in or near the hummocky landscape, or near the seashore.

Settlement areas (Fig. 4.1:4)	Settlement periods					
A. Southern Köpingebro	LN	EBA	LBA	PRIA	RIA	LIA
B. Lilla Köpinge village	LN	EBA	LBA	PRIA	RIA	LIA
C. Tingshög	LN	EBA	LBA	PRIA	RIA	LIA
D. Piledal in Stora Herrestad	LN	EBA	LBA	PRIA	RIA	–
E. Piledal in Stora Köpinge (graves)	LN	EBA	LBA	–	RIA	LIA?
F. Ystad	–	–	LBA	PRIA	RIA	LIA

By combining the evidence of phosphate and soil maps one can tentatively distinguish other possible settlement areas; the existing archaeological material, however, is too scanty for any certainty about these areas:

G. Between areas C and E
H. Stora Herrestad village (perhaps more than one)
I. Stora Köpinge village
J. Kabusa (partly outside the project area)

The curve for the number of habitation sites shows a steady rise from the Late Neolithic up to and including the Late Iron Age. There are 2.5 times as many at the end of this time-span, which may be a rough measure of population development, even though there are many uncertain factors. The number of camp sites, on the other hand, shows a noticeable increase in the Late Bronze Age and the number is also high during parts of the Early Iron Age, which testifies above all to mobility in the use of peripheral areas.

	LN	EBA	LBA	PRIA	RIA	LIA	Undated
Habitation sites	6	6	8	11	12	15	–
Camp sites	1	1	19	8	13	3	5

The houses and farms are among the best sources we have for analysing economic and social conditions in prehistoric times. The development of the long-house from the Neolithic to the Viking Age, indeed, almost up to our own day, shows the conservatism of peasant society. The construction through the ages undergoes only a few major changes or interruptions in development, but many minor alterations, showing that people were capable of adapting to new conditions within the traditional framework (Näsman 1987:69f., 78).

In the Neolithic and up to the Early Bronze Age period I, most buildings are two-aisled long-houses. In the transition to period II, however, there is a decisive break in development, which occurred at the same time in various parts of northern Europe. The inset posts or supporting posts on either side of the central posts are moved into the room, meaning that the central posts can be set on the crossbeam, thereby giving a structure with a main central aisle and two side aisles. This house type subsequently reigned virtually supreme until the end of the Viking Age.

However, the posts supporting the roof were gradually moved further into the room, at the same time as the distance between them was reduced. In the Köpinge area the distance between the posts was about 4 m during the Early Bronze Age periods II–III, about 3 m during the Late Bronze Age, and about 2 m during the Roman Iron Age. In the Late Iron Age the main change in the structure is the more curved wall, and the house can also have a more varied appearance. Such changes, whether large or small, in the construction and plan of the long-house, if they are common to large areas, are an excellent dating tool.

Arable fields, meadows, and pasture on the one hand and settlement on the other are different aspects of the same economic situation. Since people must have preferred to have the essential means of production as close to their habitation as possible, it is crucial for us to obtain knowledge of the localization of settlement. Insight into prehistoric economy and society requires details of the size of the houses, the presence and size of a byre section (perhaps with stalls), outhouses and other buildings suggesting differentiated spheres of work, and so on. We must also know whether the production units consisted of dispersed farms, agglomerated settlements, or villages. To assess the forms of land use in various periods, whether settlement was "permanent" or shifting, consecutive strata and exact datings are important.

Henrik Thrane (1987:7) has described the method of settlement archaeology as an attempt to establish synchronic situational overviews which can be combined into a diachronic course of development, like the individual pictures in a film. These situational descriptions are not as static as they may appear, since archaeological dating techniques are still too rough. We cannot possibly determine which settlements existed contemporaneously in a certain year, such as AD 125; instead we have to content ourselves with broad dating frameworks, such as 0–AD 200 or AD 200–400, or even wider spans.

What may appear to be a densely settled area may in reality not even represent continuous occupation. It is important to bear this in mind when reading the following description by periods of the development of settlement and land use in the Köpinge area over almost three thousand years.

Early Bronze Age 1500–1000 BC

Source material. When the project started, at least 150 barrows were known in the Ystad area, of which three-quarters were in the Köpinge area (Tesch 1983b:25), but not a single habitation site dated to the Early Bronze Age. However, three habitation sites with house re-

mains from the Late Neolithic had been investigated. These sites were centrally located in the area. The areas investigated were big enough to show clearly that these were dispersed farms with a mobile settlement pattern. A two-aisle long-house measuring 11.6 × 6 m and a few smaller buildings with sunken floors (Tesch 1983b: Fig. 30A a-b) were ^{14}C-dated to around 1700 BC. Since these dwellings were in the same settlement areas (A, C) and in a few cases on the same habitation sites as those from the Late Bronze Age, there were good reasons for assuming a similar pattern of localization and settlement in the Early Bronze Age.

Through a combination of ^{14}C datings and analyses of pottery and house typology, it has been possible to identify habitation sites with remains of houses from the Early Bronze Age, not only in the material that has come to light in the course of the project but also in older material. Similar results have been obtained by the analysis of older habitation site material from the Early Bronze Age in Denmark (Ethelberg 1987:164).

In the Köpinge area we now have six habitation sites and a camp site dated to the Early Bronze Age, the same number as we have from the Late Neolithic. The sites are located in the central settlement areas nearest the coast (A, C, D, compared with the A, B, C, D of the Late Neolithic). There is no immediate spatial link with the great lines of barrows. A somewhat uncertain habitation site is located near four barrows, two habitation sites have only one known barrow in the vicinity, and at three sites there are no barrows at all in the presumed settlement areas.

Settlement. Excavations have been carried out at the known habitation sites in the Köpinge area of about ten long-houses of greater or lesser certainty. Only one house is attested from the Late Neolithic/Early Bronze Age period I(?), in area D. It was 30 m long and 6 m wide (Larsson and Larsson 1986:30ff.). For periods II–III the houses investigated are of varying size: some are large, 22–26 m long; others are smaller, 10–16 m long. In one settlement area (C) the levelling of land for a new road revealed no less than four long-houses in an area about 300 m wide. Since there were no overlying layers or material for ^{14}C dating of all the houses, it was impossible to ascertain whether any of the houses could be contemporary.

The change in construction undergone by the long-houses between periods I and II is assumed to be functionally determined: a three-aisle house is better suited than a two-aisle house for the stalling of animals, since it allows partitions along both the long sides of the house. There is no certain division of the houses into dwelling and byre sections, still less evidence of stall partitions. Traces of the pairs of posts that supported the roof are also evenly spaced, not closer as we might expect in the presumed eastern byre section. We thus have no answer to the question of whether the whole house was used for human habitation or whether animals were brought indoors during the winter, but without stalls. At any rate, Early Bronze Age houses are often relatively large, which might suggest the latter alternative, and cooking pits are often found in the western end (Ethelberg 1987:165f., Näsman 1987:78f.). The climate was warm enough, however, to allow the livestock to remain outside the whole year round (cf. Ch. 5.1). The existence of very long houses even in the Köpinge area thus raises the question of what functions can have been combined in them. Late Neolithic houses up to 44 m long and 7.5–8.5 m wide (= approx. 350 m²) have been studied on Bornholm (Nielsen and Nielsen 1986:36ff.). As dwelling houses they are too big for one family. One idea that has been suggested is that they functioned as a farm for several related nuclear families in a communal household (Larsson and Larsson 1986:32, Ethelberg 1987:166), which might suggest a more or less egalitarian society. However, the existence of both large and small houses in the Late Neolithic and Early Bronze Age complicates the picture of the social structure. It could as easily be interpreted as evidence of a stratified society. A third possible interpretation is that the big houses represent habitation bases and the little houses some sort of satellite or seasonal dwellings.

Settlement patterns and land use. The cultural landscape of the Early Bronze Age in southern Scandinavia has long been associated with animal husbandry and the pastoral economy. The proximity of the habitation sites to natural wetlands, as in the Köpinge area, is generally seen as evidence of an economy based mainly on stock-raising (Jensen 1979a:47). Our knowledge of the expansion of animal husbandry in the Late Neolithic and Early Bronze Age is essentially based on pollen analyses (in the Köpinge area, see Ch. 4.1.2). The building of barrows presupposes an open landscape, not only because the barrows were built of sods, but also because they were meant to be seen. Much larger herds of cattle than were needed for consumption must have been kept to bring about this opening of the landscape. The cattle probably had great value as a status symbol and may also have been used for the exchange of wedding partners. The large herds of cattle roved freely and in the course of the Bronze Age transformed the half-open wood-pastures into large, open, heath-like grazing lands. This shifted the ecological balance, which meant that new farming methods had to be brought into use (Kristiansen 1988:81, 107).

The role played by cereal cultivation in the Early Bronze Age is unclear. It was probably subordinate to stock-raising, and perhaps originally had a ritual role. Traces of ard ploughs have been found under Bronze Age barrows in many places in Denmark; the areas exposed have been 400–1000 m² (Thrane 1984:10). Investigations of field boundaries in eastern Denmark show the development of cultivation as far back as 1200 BC (V. Nielsen 1984:161). In the Köpinge area there is a slight decline in indicators of grazing and cultivation at

the same time (Ch. 4.1.2). This, however, could be interpreted as an expression of an intensification of tillage, with certain fields being manured, which would have led to a somewhat more stable pattern of settlement. This also involved a slight reduction in the number of livestock. The time coincides also with the major change we have seen in the construction of the longhouses, probably with the aim of making it easier to stall animals.

Arable farming is also attested in the Köpinge area in the form of ard traces and analyses of carbonized plant material. Ard traces have been found twice, one under a Bronze Age barrow (area E; see Jacobsson 1986:106f.) and one at a habitation site (area C; see M. Larsson 1987b:10) above a Late Neolithic house and closely associated with a long-house from period II. The analysis of carbonized plant material from the habitation sites in the area show that, as in the Late Neolithic, emmer and naked barley, but also einkorn and *Bromus secalinus* (rye brome) have been found. However, not all kinds of cereals thrive on poor soil without manure (Engelmark and Hjelmqvist 1991). Macrofossil analyses do not therefore provide any direct support for the hypothesis that there was an intensification of tillage in the course of the Early Bronze Age. It has also been suggested that tillage may have been part of a coppicing system (Ch. 4.1.2). In the osteological material the introduction of the horse is attested as early as the Late Neolithic (Jonsson 1989).

Both cereal cultivation and animal husbandry combined to give a mobile settlement pattern, perhaps with a certain stabilization around 1200 BC. The archaeological material can be interpreted as showing that people moved either short distances within one habitation area (area C) or within a delimited resource area (area D).

The Late Bronze Age 1000–500 BC

Source material. Unlike the previous period, there was already extensive archaeological material in the Köpinge area before the start of the Ystad Project, as regards both habitation sites and graves from the Late Bronze Age. The material was interpreted to show that there were up to five isolated farms in the area with a mobile settlement pattern based on animal husbandry as the dominant source of livelihood. Whenever the land around a habitation site had been exhausted, the farm moved a short distance to another site, circulating like this within a topographically delimited settlement area (Tesch 1980:86ff., 1983b:46f.). The cemetery, by contrast, appears to have been a central and relatively fixed point in the settlement area (Tesch 1983b:51).

The Late Bronze Age is the period which is most likely to be found when a habitation area is stripped (it is closely followed by the Roman Iron Age). Remains of dwellings from the Late Bronze Age have been found at nearly half (27 = 46.6%) of the habitation areas investigated. This figure is misleading, however, since it is largely a question of camp sites (19 = 32.7%), whereas only a small fraction (8 = 13.7%) can be described as habitation sites, in other words, not more than during the Late Neolithic and Early Bronze Age.

Habitation sites are attested in four of the settlement areas (A, B, C, D). In a fifth area (E) a cemetery with 62 burials has been investigated in its entirety (Tesch 1983b:45, Olausson 1987, Ch. 4.1.4.2).

Settlement. During the Late Bronze Age there were several changes in the long-houses and in farm structure, pointing the way forward to Iron Age settlement. However, there is no certain evidence of village-like groupings on the habitation sites. Instead, the isolated farmstead appears to be the normal unit (Näsman 1987:82), although the converse has also been claimed (Draiby 1985:168ff.). In some cases it has been possible to document undoubted traces of byre stalls in one end of the long-house. Reasonable conclusions can also be drawn from the variation in the positioning of pairs of posts in the long-house: the eastern end has more frequent and regular posts than the west, suggesting a division into a byre and a dwelling. In the Early Bronze Age the roof-bearing posts had been placed at relatively frequent and regular intervals. Near the long-house there were often smaller buildings of "Dutch barn" type, consisting of four posts holding up a roof; these are interpreted as stores or workshops, and especially as barns for the storage of the harvest and winter fodder. In rare cases there were sunken-floor huts. There are also traces of fences (Näsman 1987:79, 82).

The structure of the long-house in Denmark and Scania can be divided into two types. Type 1 has sturdy wall posts positioned at wide intervals and a very curved gable; the walls, probably of wood, resemble certain house types in the Early Bronze Age. Type 2 has less sturdy and more closely positioned wall posts and a straight or only slightly rounded gable with rounded or cut corners; the walls were probably wattle and daub, anticipating Early Iron Age house types.

In the Köpinge area about ten long-houses of type 2, some smaller houses, including four-post barns, and perhaps a sunken-floor hut have been documented. Although the houses are in groups or in a row, the stratification and datings (Tesch 1983b: Figs 28–29) suggest that this picture is rather the result of the movement of a single farm within a large habitation area or between habitation areas within one settlement area. The long-houses are 12–16 × 7–8 m (Tesch 1979a: Fig. 1, 1979b: Fig. 2, 1983b: Fig. 30A c, Wallin 1986:31). House sizes thus correspond to those of long-houses from the Late Neolithic and the Early Bronze Age, although these earlier periods also had houses which were much bigger. The wall posts are relatively closely spaced, at intervals of 0.7–1.2 m. The following are the important changes in the internal construction of the long-house: the very long span (4.5–5 m) between the pairs of posts supporting the roof at the western end of the house, contrasting

with the closely spaced pairs at the eastern end. The latter change appears to suggest a functional division of the house into a dwelling and a byre. There is often a hearth in the centre of the dwelling section, but no sure traces of stalls have been found in the presumed byre.

The main settlement areas where house remains have been documented lie in a row near the coast. They are all built on the remains of one glacial beach ridge, running WNW-ESE parallel to the shore, and the settlement areas are demarcated by Nybroån and other watercourses. This is also a direct continuation of the line of settlement in the adjacent area of the Hagestad Project (Strömberg 1982: Fig. 107). The level of the habitation areas roughly follows the 10 m contour. During the investigation of the habitation area in C it was found that groundwater seeped up when the topsoil was being removed; this was a full 75 m from the top of the ridge on the 250 m long southward slope down towards a stream separating this from the next ridge. On the lower remains of the beach ridge closer to the sea, the high groundwater level meant that only camp sites could be found, in the form of collections of hearths, dated by ^{14}C to different periods in the Late Bronze Age.

Settlement patterns and land use. Habitation sites appear to have been localized in virtually the same places as in the Early Bronze Age. We have more archaeological material from the Late Bronze Age, but it mostly shows increased mobility in the periphery of the settlement area and in the outlands. The evidence does not allow us to detect any significant increase in population. From the habitation area in the southern part of the present-day village of Köpingebro (A) it is clear that settlement did not consist of village-like clusters but of isolated farms with households of 6–10 people, who either migrated a short distance at a time within a habitation area or between different habitation areas within a settlement area. The distance between each habitation area need not have been more than a few hundred metres (Tesch 1983b: Fig. 28a). The settlement pattern observed here has been used to show that Bronze Age farmers were semi-nomads exploiting the fertilizing effect of woodland clearance, which is obtained by shifting tillage in a coppiced environment (Emanuelsson et al. 1985:36).

The settlement areas on either side of the river Nybroån (A and C) had long-houses with different types of wattle-and-daub wall. This speaks against any movement between these presumed settlement areas. However, as no habitation has been totally investigated, the possibility that two or more farms could have existed at the same time in one settlement area cannot be entirely ruled out. The cemetery investigated in area E, however, has been interpreted as a burial place for a farm (Olausson 1987:140f., Ch. 4.1.4.2).

The superposition of the strata does not appear to be random; it suggests that the builders knew the location of the previous house. It can scarcely be the case, however, that they built a new house directly on the site of the old one. The old site must have been so defiled and unwholesome an environment as to make people avoid such a procedure. Moreover, the old site offered the possibility of reclaiming the land for cultivation with increased nutrients. There are documented examples of the way cultivation plots were also moved short distances within a habitation area, and the way old houses were ploughed over and cultivation plots were right up against the houses (Draiby 1985:168f.). In the Köpinge area there are also examples of the digging of graves on the old habitation site, and the construction of houses over graves; this shows the close association of all the facets of a settlement.

The archaeological evidence nevertheless indicates a relatively stable settlement pattern, which might mean that people had started manuring the fields. Manure is obtained by stalling cattle. It is hard to say if this was a consequence of the increasingly damp and cool climate of the Late Bronze Age or if the need for manure was the primary factor; it was probably a combination of the two.

The analyses of carbonized plants show that hulled barley dominated more and more in the course of the Late Bronze Age. Since hulled barley requires more soluble nutrition than naked barley, spelt, and emmer, it is reasonable to assume that some form of manuring was introduced towards the end of the period (Engelmark 1988, Hjelmqvist 1987, Engelmark and Hjelmqvist 1991). The old cereals, however, were high in protein, as well as being well suited to the poor sandy soils of the Köpinge area, so there was no good reason to abandon them. The changes have been seen as a result of the continual deterioration of climate, which had a detrimental effect on the yield of the old cereals and their resistance to disease.

The presumed byre sections in the Köpinge area are relatively small, however, and can hardly have held more than 10–18 head of cattle. The changed farm structure and the small byres suggest a more composite economy in which tillage has assumed greater practical significance. We also know from other areas that deep-sea fishing for cod and seal-hunting were other sources of food (Kristiansen 1988:85, Ullén 1988). Identification of bone material from a number of Danish and Scanian habitation sites shows that sheep and goats became more common in the course of the Late Bronze Age, with a corresponding decline in the number of cattle (Kristiansen 1988:85f.). The small byre sections in the Köpinge area might also be evidence of this. People did not house more animals than they had winter fodder for. The forest – the staple source of fodder – was becoming smaller and increasingly remote. People thus adapted to the lack of winter fodder in a dramatically opened landscape, as is evident from the pollen diagrams (Ch. 4.1.2). Cattle became smaller during the period, which may be due to the reduced supply of fodder and an adjustment to the more cramped stalls.

This may have contributed to the eventual change to the keeping of cattle primarily for milk (Kristiansen 1988:88f.) – strainers for cheese-making are common finds from the Late Bronze Age.

There has been discussion of the extent to which the Late Bronze Age was a society in crisis, and whether this was caused by factors like population pressure, overexploitation, climate deterioration, or a combination of these (Kristiansen 1988:103ff.).

We have already observed that the archaeological material does not indicate any noticeable increase in population in the Köpinge area. This agrees with calculations made on the basis of the Danish material (Jensen 1979a, Kristiansen 1988:106). The gradual change to a damp, cool climate, culminating around 600 BC, did not reduce the grazing potential in the short term, but the increased precipitation brought about a long-term leaching of calcium and minerals. Intensified grazing made the landscape more open. The growing herds of sheep and goats had a serious effect on the sward. All this together may have led to an overexploitation of the thin, sensitive cover of sandy mould around the settlement areas, possibly resulting in wind erosion and the development of heath. This can have caused an agricultural crisis in the Köpinge area, as evidenced in the heavier exploitation of the outlands. Here people gathered the winter fodder that was no longer available in the settlement area. What in the pollen diagrams is interpreted as an expansion of the cultural landscape might thus be a sign of a society under pressure. The worsened climate was the last straw which tipped the sorely strained ecological balance. However, alterations in the construction of the long-house and the testimony of the macrofossils suggest that people were already actively trying to come to terms with the changed conditions in the Late Bronze Age by changing their farming methods. Although this did not lead to a fully developed system of infield/outland with haymaking, as in the Early Iron Age, it can nevertheless be said that the foundation was laid for a more highly differentiated agrarian society. In the Köpinge area the transition to the Iron Age was least of all a break in the development of the cultural landscape. Both the structure of the habitation sites and the appearance of the houses (including their size) suggest continuity. How this very slow process took place is hard to say. Was it based on the farmers' proverbial common sense, or was it influenced by ideas from without? The answer probably is that it was an interaction of these factors.

4.1.4.2 Bronze Age burial

Deborah Olausson

The concentration of Early Bronze Age barrows in the Köpinge area is not markedly higher than in the Bjäresjö area: 1.8 barrows vs. 1.6 km^{-2} in Bjäresjö (Ch. 4.2.4). For Stora Köpinge parish alone, the figure is higher: 4.1 barrows km^{-2}. A minimum of 123 graves, in 85 tumuli, dating to the Early Bronze Age can be identified in this region. A preference for higher ground and/or glaciofluvial deposits is one noticeable characteristic of barrow distribution. Another is the linear patterning of some barrows: these probably lie along Bronze Age transportation routes (Mathiassen 1948:11, M. Jørgensen 1982:142). Further, barrow placements shows a distinct preference for the sandy soils; those which are located in the clay till area inland lie on "islands" of sandy soils or glaciofluvial deposits.

The pattern shown by Late Bronze Age burials shows a shift towards the coast and a tendency towards burial in cemeteries, although there are occasional solitary burials as well. The distribution pattern evident during the Early Bronze Age is further accentuated: only one cemetery (containing three burials) lies outside the sandy plain of Ystad. To some extent this pattern may reflect the greater intensity of excavation on the plain than inland. However, a similar pattern was evident in the western areas, where surface survey was specifically aimed at augmenting our knowledge of the inland areas (Ch. 4.2.4.1, 4.3.4, 4.4.4).

Barrows in the Stora Köpinge area appear to have filled a bounding function, while Late Bronze Age flat-earth cemeteries served as centralizing elements (a pattern which can be observed in other Bronze Age contexts as well: Petré 1984:57; cf. Hansen 1938:14, Bradley 1981:102). An apparent change towards more direct association between graves and settlement can be seen in the Late Bronze Age here. Barrows seem to have been placed where they were easily visible along the edges of, for instance, the northern limit of the sandy coastal zone or on the glaciofluvial soils of the hills known as "Köpinge backar" (a pattern which is also seen in neighbouring Valleberga: Strömberg 1975a:78). In addition to their function as markers, burial mounds also involved destruction of the topsoil and permanent deprivation of at least the area underneath (Mathiassen 1948:110, Glob 1971:107, Thrane 1980:169, Furingsten 1985:45). The shift from peripheral to central burial towards the Late Bronze Age may well reflect an increased desire to affirm land rights (Chapman 1981:80, Hodder 1982:196) in connection with a period of economic stress (cf. Kristiansen 1975:90).

A minimum of 184 individuals is represented in the Late Bronze Age burials in this area, compared with 123 known for the Early Bronze Age. This evidence therefore does not indicate any substantial population increase here during the Bronze Age.

While many agglomerations of Late Bronze Age burials can be identified in the area, only two cemeteries have been investigated in some detail. Assuming that cemeteries reflect settlement intensity and population better than individual graves do (Hyenstrand 1979:142), these two cemeteries may represent somewhat different levels of settlement in the Stora Köpinge/Stora Herrestad area. Larsson and Larsson (1986:33) claim that the cemetery at Stora Herrestad was completely excavated, although the possibility that it was originally much larger and that many graves have been lost during modern ploughing cannot be ruled out. The seven burials recovered included four females, one male, and one child, and datable burials were from periods V and VI (Larsson and Larsson 1984:36ff., 1986:33). The cemetery at Piledal contained ten Early Bronze Age burials and 69 from the whole of the Late Bronze Age, as well as burials from the Stone and Iron Ages (Tesch 1983a:51, Olausson 1987). The first cemetery, located in a peripheral area, used for a short time only, and containing few graves, would seem to represent burials connected with a short-term settlement (cf. Dehn 1985:200), while the Piledal cemetery served more long-term settlement. Goldstein (1981:61) has suggested that the practice of burying the dead in a cemetery indicates a corporate group structure affirming its rights to limited resources. Using such a model, Piledal can be interpreted as part of a centralized burial complex used by an extended family over the course of many generations. Its primary resource area should have been located within a kilometre or less of the cemetery (Kristiansen 1978:33, Bradley 1981:100, Strömberg 1982:61). This area includes both sandy soils for cultivation and clay till for grazing: ideal conditions for Bronze Age settlement. The population data for the Piledal cemetery are in agreement with other Late Bronze Age burial concentrations in Scania (Stjernquist 1961:125, Strömberg 1975a:211, 262f.). These can be used to verify an interpretation of a household of about ten people as the basic economic unit in the Late Bronze Age (e.g. Strömberg 1976:25, 1982:11, Ellison 1981:432, Jensen 1981:54).

A census of the burial data available for the settlement areas identified by Tesch, although the data are incomplete, yields support to the idea that these areas were occupied by one extended family. Most of the evidence points to the areas being used for burial throughout most of the Bronze Age, indicating site or area continuity during the period.

Area	EBA	LBA	Dating (LBA)
	(no. of known burials)		
A	1	30	IV-VI
B	1	4	V-VI
C	4	32	IV-VI
D	1	9	V-VI
E	46	83	IV-VI

Using Kristiansen's formula for calculating contemporary population from cemetery data (1985a:125), we can estimate a population of about six adults for area E, the most completely investigated, at any one time during the Late Bronze Age.

Acknowledgement. I am grateful to M. Riddersporre for placing the unpublished data on barrows at my disposal.

4.1.4.3 The agrarian landscape of the Köpinge area in the Late Bronze Age

E. Gunilla A. Olsson

Introduction

The ecological content of the agrarian landscape can be expressed in terms of the area of land-use categories and vegetation types. Such information should help us to interpret the conditions for life for prehistoric inhabitants of the Köpinge area in the Late Bronze Age. However, a reconstructed landscape with such information cannot be directly inferred from the pollen diagrams or from other data sets available for prehistoric periods.

The following study is based on the whole range of Late Bronze Age data available for this area. It will present a landscape reconstruction for the Köpinge area based on the archaeological findings related to the number of humans and their diets, number of stock, etc. These findings are combined with the palaeoecological data to reconstruct the composition of the human diet and the cultural landscape.

Site description

The area used for the reconstruction (7 × 6 km) comprises the large wetland of Herrestads Mosse and the adjoining sandy plain between the seashore in the south and the hummocky boulder clay terrain in the north. Sand-hills in the south-east reach 35 m a.s.l. The river Nybroån transverses the area from north to south, and the river Kabusaån forms the eastern borderline.

Methods and sources

The model is based on archaeological findings (Lepiksaar 1969, Tesch 1988, Ch. 4.1.4.1, 4.1.4.2, 4.2.4) and palaeoecological data: pollen analysis from Herrestads Mosse (Håkansson and Kolstrup 1988), from Fårarps Mosse (Hjelmroos unpubl.), charcoal findings (Bartholin and Berglund 1991), botanical macrofossils from Piledal and Köpinge (Engelmark and Hjelmqvist 1991) and interpretations of the palaeoecological data sets by Berglund (Ch. 4.1.2). See also the methods described in Ch. 2.2

Assumptions for the reconstruction. Using the interpretation by Tesch (Ch. 4.1.4) only sites which show evidence of post-holes indicating houses erected to withstand more than one or two seasons are used here. Five such settlement sites are distinguished. The number of humans living at each site is estimated to be a small "family" group of about 5 adults and 5 children (Ch. 4.1.4). This size conforms to the situation in other Scandinavian Late Bronze Age settlements (Ch. 4.1.4.2).

Protein-based calculations are used for the amounts of the different food substances consumed and from these figures is inferred the area required for stock-raising and tillage. The same method is used for Kabusa during the Middle Neolithic (Ch. 4.1.3.3) and for Bjäresjö in the Late Bronze Age and Iron Age (Ch. 4.2.4.2, 4.2.5.2). The dietary contribution of different foodstuffs is interpreted from the archaeological find-

Table 4.1:3. Inferred dietary composition during the Late Bronze Age in Köpinge as suggested by the archaeological findings. The protein content of the different food items is extracted from modern tables (Livsmedelstabeller 1978). The data below are calculated for subsistence level for a settlement site inhabited by 7.5 adult units.

Item	% protein contribution	kg protein adult^{-1} yr^{-1}	kg food adult^{-1} yr^{-1}	Total no. consumed animals yr^{-1}	No. of stock
barley	30	5.7	40.7	–	–
emmer	15	2.9	14.2	–	–
Camelina	1	0.2	1.1	–	–
nuts etc.	1.5	0.3	2.3	–	–
fish	5	1.0	5.4	–	–
cattle	28	5.3	25.3	2.3	8
sheep	8.5	1.6	8.1	3.6	12
pig	4.5	0.9	4.3	1.2	2.5
horse	4.5	0.9	4.1	0.3	2
game, eggs	2	0.4	2.0	–	–
Total	100.0				

Table 4.1:4. The relationship between seed return and arable area. Calculation valid for subsistence consumption (according to Table 4.1:3) of adults

Crop	Seed return	Arable area (ha)
barley	4	0.64
emmer, spelt	4	0.22
Camelina	4	0.02
Total arable		0.88
barley	2	1.53
emmer, spelt	2	0.53
Camelina	4	0.02
Total arable		2.08

ings and from the palaeoecological remains (Engelmark and Hjelmqvist 1991).

A protein-based model of dietary composition was preferred to energy-based models (cf. Poulsen 1980). The intake of food components in terms of essential energy is much more difficult to relate to environmental impact. Many available edible items do not imply any great effect on the environment (e.g. apples and nuts) as a result of human consumption.

The dietary composition for the Late Bronze Age in Köpinge is listed in Table 4.1:3. In total the dietary composition is based on animal and vegetarian protein proportions of 48% and 52% respectively. The protein content of the different products was taken from modern literature (Livsmedelstabeller 1978).

Mean human adult weight was set to 65 kg. Daily dietary protein intake was estimated as 0.8 g kg^{-1} body weight, which corresponds to modern Swedish recommended intake (Näringslära 1983). This level was chosen to avoid underestimation.

area (ha)

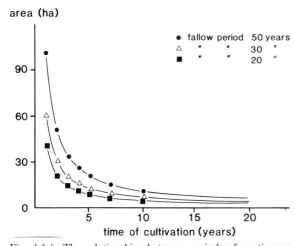

Fig. 4.1:6. The relationships between periods of continuous cultivation of a field and total area needed for the rotating arable. Three levels of fallow duration are shown. If a group of 5 adults and 5 children is assumed then 2 ha of arable had to be cropped annually (cf. Tables 4.1:3 and 4.1:4) Equation for the graph: $\log Y = 1.6 - \log X$.

When estimating the plant biomass for forage intake by stock it was assumed that open grasslands of different moisture levels and half-open oak and hazel grasslands were available. Year-round grazing was assumed. The size of the required foraging areas for two levels of stock numbers was calculated with information compiled from the following sources: Van Dyne and Meyer (1964), Cook et al. (1967), Coupland (1979), Steen (1980), Van Dyne et al. (1980), Ring et al. (1985), Pratt et al. (1986).

The area of arable land was deduced from the human cereal consumption. Cultivated cereals were *Triticum dicoccum/spelta* (spelt/emmer wheat) and both *Hordeum vulgare* v. *xx* (hulled barley) and *Hordeum vulgare* v. *nudum* (naked barley). The macrofossils of cereal grains from this area do not permit conclusions about the proportions of cultivated naked and hulled barley during the Late Bronze Age. Findings from the period at other sites in Scania (both clayey and sandy sites) reveal traces of *Avena sativa* (oats), *Bromus secalinus* (rye brome), *Panicum miliaceum* (millet) and *Camelina sativa* (gold of pleasure). These findings indicate small scale and perhaps infrequent cultivation of such crops (Hjelmqvist 1955, Engelmark and Hjelmqvist 1991) and are probably valid also for Köpinge.

Results and discussion

The reconstruction of vegetation and land use is shown in Fig. 4.1:5. Mapping vegetation reveals the problem of simultaneously visualizing a particular time period and a longer time span corresponding to the archaeological and the palaeoecological data sets.

Cultivation and the extent of arable. This reconstruction assumes shifting cultivation, where the arable sites were used for a short time and then temporarily abandoned for nearby sites. The question of whether manure was used at this time is controversial. Only very few seeds of *Chenopodium album* (fat hen) exist from Köpinge (Engelmark and Hjelmqvist 1991). This weed is favoured by high levels of soil nitrogen, but it is risky to use those few seeds as proof of the existence of manured arable fields.

Two alternative calculations were made for the crop yield (Table 4.1:4). For constructing the map (Fig.

Table 4.1:5. Relationship between the number of stock and their foraging area. Foraging area is calculated for human subsistence consumption of stock, and for stock number on an enlarged level: 5 times subsistence consumption of cattle, horses, and pigs, 10 times the human subsistence consumption of sheep.

Number of stock	Foraging area (ha)
Subsistence level from Table 4.1:3	54.5
Enlarged level: 5 × (cattle, horse, pig) 10 × sheep	326.5

4.1:5) the low seed return of 2 for the main crops was assumed, implying an arable area of 2 ha. This is defensible for a situation with sandy soils without manure. However, assuming twice the seed return would imply a rather small decrease of the arable (0.9 ha), which is insignificant compared with the total resource area (Tables 4.1:3, 4). The model used here and visualized on the map assumes a cropping period of 5 years and 50 years of fallow. Despite such a long fallow period, not more than 20 ha would have been necessary for the total cultivation area, if 2 ha were cropped annually (Fig. 4.1:6). Assuming a shorter cropping period, such as two years at the same site, would require 100 ha as the total area for the shifting arable; a shorter fallow period would mean a smaller total area need (Fig. 4.1:6). An area for shifting cultivation of 20 ha is marked adjacent to the arable (Fig. 4.1:5). Within this area also the houses were moved from time to time (Ch. 4.1.4.1).

There are no finds of agricultural implements from Köpinge. However, ard marks have been found at two sites (Ch. 4.1.4.1) and ard-working of the arable would also have required weeding. Thus, ard-working, weeding, and rotating cultivation – whether with or without coppicing – all give the prerequisites for a reasonably high seed return. A seed return of at least 4 would be probable (cf Lüning and Meurers-Balke 1980, Olsson unpubl.). The figure of 2 ha for the arable area annually used would thus be a maximum.

Number of stock and foraging areas. The number of stock was calculated at two levels. The subsistence level as inferred from the dietary composition resulted in a stock number of 8 cattle, 12 sheep, 2 horses, and a few pigs for each settlement unit (Table 4.1:3). The necessary foraging area to maintain this stock level was calculated to 55 ha (Table 4.1:5) and is expressed in Fig. 4.1:5 as solid circles. Year-round grazing was assumed for this model since no data exist that can firmly demonstrate winter stalling of the stock. Continuous grazing implies a larger foraging area than winter stalling, so the present figures will display maximum foraging areas.

However, subsistence level of the stock implies that only 10% of the Köpinge area would have been used for intensive grazing and cultivation (Fig. 4.1:5). The erection of the barrows during the Early Bronze Age and their widespread distribution meant that considerable areas of open grasslands existed here, at least when the barrows were created. Furthermore, the pollen data indicate a significant decline of forest areas during the Bronze Age (Hjelmroos unpubl.). Obviously there are clues to the presence of grassland areas in Köpinge to an extent above what the subsistence level of stock could bring about and maintain.

Anthropological data from present-day subsistence cultures allow hypothetical analogies with Late Bronze Age settlements. These studies suggest a considerably larger stock size, partly for social and religious reasons like marriage and cult (Callmer 1989, Ch. 4.1.4.1). Assuming a larger number of stock (5–10 times subsistence level) generates a required foraging area of 327 ha (Table 4.1:5; Fig. 4.1:5, broken circles). This means that almost all of the sandy plain and the swampy and fen areas would have been used for grazing, coppicing, collection of firewood, and tillage. This model is well in accordance with the interpretation of the pollen diagrams (Ch. 4.1.2).

The Bronze Age landscape. The pattern of excavated settlements both with and without remains of buildings (Ch. 4.1.4.1) and the barrows and cemeteries (Ch. 4.1.4.2) quite clearly shows that all of the sandy plain and the wetlands were utilized by Bronze Age humans. Evidence of secondary burial in several barrows during the Late Bronze Age (Ch. 4.1.4.2) indicates that other reasons existed than pure forage need to manage the barrows and their surroundings. It is thus conceivable that open grasslands surrounded the barrows – also during the Late Bronze Age. The pollen diagram from Fårarps Mosse depicts the adjacent landscape as consisting of a mosaic of grasslands, oak-hazel woodlands, and wetlands. Parts of the wetlands were overgrown with *Alnus glutinosa* (alder). The Fårarp core was collected at a site just on the border between the sandy plain and the clayey hummocky area. Thus the information from the pollen data for the sandy plain landscape still remains elusive.

The charcoal samples (Bartholin and Berglund 1991) collected from the excavated settlements on the sandy plain are dominated by oak and hazel, but contain also alder, some lime, ash, and small amounts of *Prunus* species and wild apples. The species composition of these samples suggests a similar vegetation to the vicinity of Fårarp. From this it can be deduced that the same woody species existed also here on the plain, although probably in smaller numbers than in the hummocky landscape. The vegetation dominating the sandy plain is thus interpreted as a mosaic of dry and mesic grasslands and of oak-hazel wood-pastures. The wetlands were mostly deforested, although groves of alder existed.

The arable areas used were of modest size (1–2 ha). Although situated on sandy soils, the risk of depletion of nutrients within these systems seems to have been rather small. The area offered space enough and possibilities to practise fallows of 50 years' duration (see above). This would have been quite enough to guarantee the recovery of the major soil elements nitrogen,

Fig. 4.1:5. Landscape-ecological map of the Köpinge area during the Late Bronze Age, 800–500 BC. Only rough vegetation types can be marked owing to the lack of a pollen core from the centre of the area. The five settlement sites with remains of buildings are used for reconstruction of cultivation and foraging areas. All other excavated settlement sites are also marked, as well as existing barrows, vanished barrows, and other graves. The solid circles denote resource areas if subsistence level is assumed. Hatched circles denote resource areas when enlarged numbers of stock are assumed (5 × subsistence level) – see text. Source: see text and Ch. 2.2.

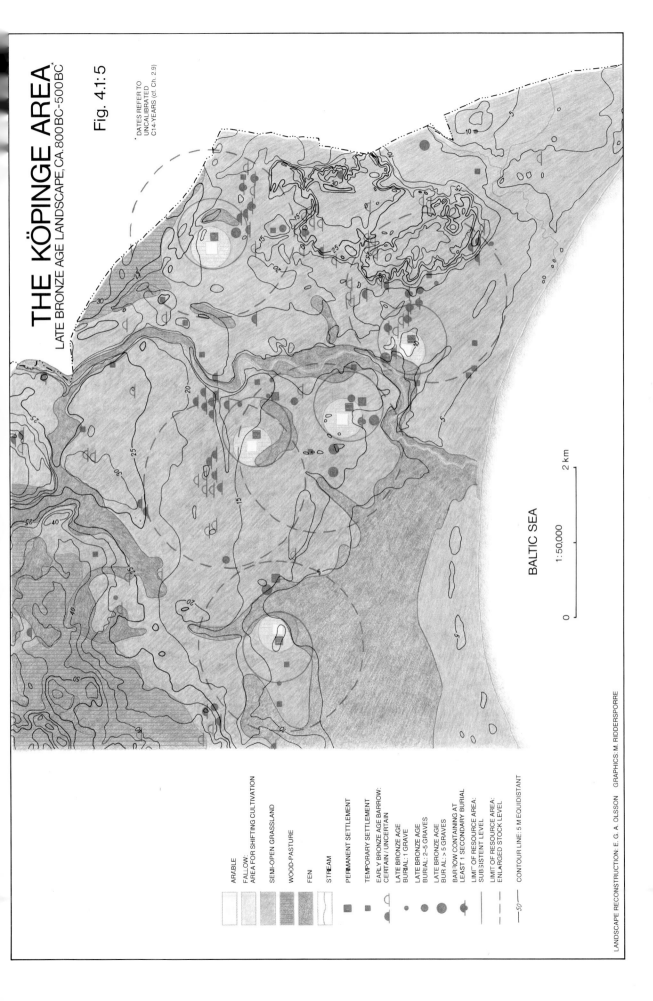

THE KÖPINGE AREA
LATE BRONZE AGE LANDSCAPE, CA.800BC–500BC*

Fig. 4.1:5

*DATES REFER TO
UNCALIBRATED
C14-YEARS (cf. Ch. 2.9)

BALTIC SEA

1:50,000

0   2 km

ARABLE

FALLOW:
AREA FOR SHIFTING CULTIVATION

SEMI-OPEN GRASSLAND

WOOD-PASTURE

FEN

STREAM

PERMANENT SETTLEMENT

TEMPORARY SETTLEMENT

EARLY BRONZE AGE BARROW:
CERTAIN / UNCERTAIN

LATE BRONZE AGE
BURIAL: 1 GRAVE

LATE BRONZE AGE
BURIAL: 2–5 GRAVES

LATE BRONZE AGE
BUR AL: > 5 GRAVES

BARROW CONTAINING AT
LEAST 1 SECONDARY BURIAL

LIMIT OF RESOURCE AREA:
SUBSISTENT LEVEL

LIMIT OF RESOURCE AREA:
ENLARGED STOCK LEVEL

—50— CONTOUR LINE: 5 M EQUIDISTANT

LANDSCAPE RECONSTRUCTION: E. G. A. OLSSON   GRAPHICS: M. RIDDERSPORRE

phosphorus, and potassium. Long-term experiments from Denmark on sandy soils show that even after 80 years of continuous cereal cultivation on the same site, the decrease in yield was only 13% compared with the first years (Dam Kofoed and Nemming 1976). Twenty years of continuous barley cultivation on sandy soils without manure in England yielded an annual harvest of 1250 kg ha^{-1}, equivalent to a seed return of 6 (Cooke 1976). Severe nutrient depletion as a result of long-term cropping and overgrazing would have resulted in the development of *Calluna* heathlands which were extensive in the Köpinge area during the 18th century (Ch. 4.1.7.3). However, the pollen evidence for the presence of such vegetation here during the Bronze Age is lacking (Hjelmroos unpubl.).

The assumed number of settlements, level of cultivation, and number of stock led to the creation of a pastoral landscape with relatively large areas of open grasslands and small-scale cultivations adjacent to the settlements. According to the long-term Danish experiments referred to above, this level of exploitation could well admit a balanced ecosystem – even over long periods like the Late Bronze Age.

However, the situation would be quite different if we interpret *all* the discovered settlements in this region (Fig. 4.1:5) as more or less permanent sites and not as occasional camp sites. In this case there would be reason to consider severe soil nutrient depletion associated with the development of heathland ecosystems.

4.1.4.4 A central settlement region on the coastal plain in the Early Iron Age

Sten Tesch

Pre-Roman Iron Age 500 BC – 0

Source material. At the start of the project our knowledge of the cultural landscape in the Pre-Roman Iron Age was very patchy. A few long-houses had been studied, but there was little other archaeological material. However, it was not thought that there had been any break in settlement continuity, although a temporary reduction in population could not be ruled out (Tesch et al. 1980:20, 1983b:46, 50).

Today, however, the situation is quite different. At a large number of the habitation areas investigated (18 = 31%) there have been found remains dated to the Pre-Roman Iron Age. House remains were documented at ten of these places. The material shows a distinct increase towards the end of the period, and many habitation sites show continuity into the Early Roman Iron Age. Grave-finds, however, are virtually non-existent. See the map in Fig. 4.1:7.

Settlement. In comparison with the Danish material, the long-houses in the Köpinge area are unusually big: 20–26 (30,40)× 6–7 m. By contrast, most of the long-houses from the early part of the Pre-Roman Iron Age which have been unearthed in conjunction with the Danish natural gas project are no bigger than 8–12 × 5 m, while in the later part of the period there is a trend towards longer houses, sometimes up to 22 m in length. The long-houses in the Köpinge area are not very different in size from those of the Late Bronze Age. The break in the development of house construction that is evident in the Danish material cannot be detected here. Instead we see small, gradual changes. Continuity in settlement development in the Köpinge area – as in Denmark – from the Pre-Roman Iron Age to the Early Roman Iron Age is seen in the way many sites are dated to both periods (Näsman 1987:79f.).

A difference from the Danish material is that no stall partitions have been documented. Nor does the relatively regular placing of the pairs of roof-bearing posts give us any guidance in this connection. On the other hand, the length of the houses and the location of the hearths seem to indicate that one end was used as a byre, and that more animals were housed than in the Late Bronze Age.

No less than three sunken-floor huts have been safely dated – by finds and ^{14}C dating – to the Pre-Roman Iron Age, and one more to around the birth of Christ. Hitherto the general view has been that sunken-floor huts do not appear until the Late Roman Iron Age in both Denmark and Scania (Hvass 1979:38, Näsman 1987:82). Yet the idea as such should not have been foreign, since houses with sunken floor-levels occur in both the Late Neolithic and the Late Bronze Age (Tesch 1983b:57). The sunken-floor huts were rounded and their dimensions were 4.0 × 3.2–2.0 × 2.0 m. In three of the sunken-floor huts there are spindle whorls and in one of them loom weights, clearly showing the function the houses had. In addition, smaller post-built houses and four-post barns have been documented. The latter, however, are hard to distinguish in the cluster of postholes and difficult to date.

As regards the structure of settlement, the archaeological material clearly demonstrates that we are dealing with isolated farmsteads. Only towards the end of the period do we find evidence of agglomerated settlements of two or more farms. This is very different from conditions in Jutland, where the large migrating villages are characteristic of the period (Becker 1971, Hvass 1985, Hedeager 1988). These villages usually consist of 10–15 farms, but a few are much larger (Lund 1988:158). In the east of Denmark, where investigations have been nothing like as intensive, it has only recently become clear that the shifting isolated farm was a more common form of settlement that was previously believed (Rønne 1986:11ff., Hedeager 1988:118, 128ff.). If there were clustered settlements, they were – with the exception of a few central sites – not much bigger than those in the Köpinge area. There is thus a clear differ-

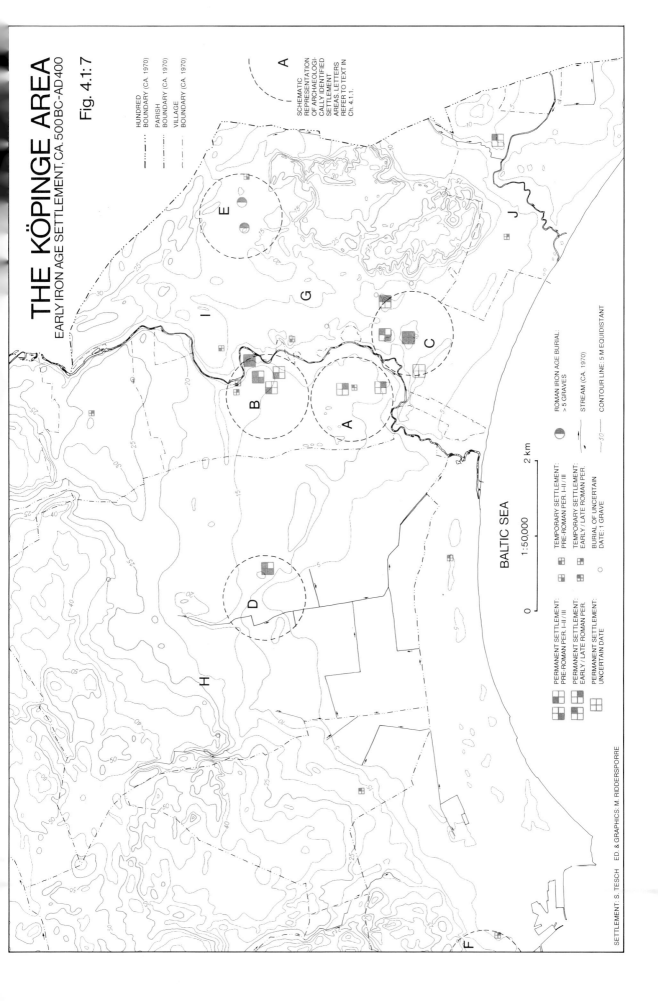

THE KÖPINGE AREA

EARLY IRON AGE SETTLEMENT, CA. 500 BC–AD 400

Fig. 4.1:7

——···—— HUNDRED BOUNDARY (CA. 1970)

——·—— PARISH BOUNDARY (CA. 1970)

——— VILLAGE BOUNDARY (CA. 1970)

SCHEMATIC REPRESENTATION OF ARCHAEOLOGICALLY IDENTIFIED SETTLEMENT AREAS. LETTERS REFER TO TEXT IN Ch. 4.1.1.

PERMANENT SETTLEMENT: PRE-ROMAN PER. I–II / III

PERMANENT SETTLEMENT: EARLY / LATE ROMAN PER.

PERMANENT SETTLEMENT: UNCERTAIN DATE

TEMPORARY SETTLEMENT: PRE-ROMAN PER. I–II / III

TEMPORARY SETTLEMENT: EARLY / LATE ROMAN PER.

BURIAL OF UNCERTAIN DATE: 1 GRAVE

ROMAN IRON AGE BURIAL: > 5 GRAVES

STREAM (CA. 1970)

—50— CONTOUR LINE: 5 M EQUIDISTANT

BALTIC SEA

1:50,000

0   2 km

SETTLEMENT: S. TESCH   ED. & GRAPHICS: M. RIDDERSPORRE

ence in the structure of Early Iron Age habitation sites between, on the one hand, western Denmark and, on the other, eastern Denmark, including Scania.

The central settlement areas were probably occupied continuously. The material from the first half of the period is limited, however, which might suggest that the escalating ecological crisis of the Late Bronze Age had not yet been overcome. Towards the end of the Pre-Roman Iron Age there is a dramatic increase in material. This expansion from around 200 BC is also evident from the pollen diagrams (Ch. 4.1.2).

Settlement patterns and land use. Settlement patterns are seen to take on increasingly fixed forms towards the end of the period. Evidence for this comes, above all, from area B, adjacent to the site of Lilla Köpinge village. At the transition to the Early Roman Iron Age, the archaeological material indicates the existence of a clustered settlement comprising two or more farms side by side. Unfortunately, it is impossible to say how many, since the modern sugar mill at Köpingebro has destroyed our chances of further investigation. Whether or not this settlement could be called a village is uncertain. At any rate, there is no form of shared fence, as is usual in Jutland. On the other hand, the regular placing of the farms indicates that there was some form of cooperation. The length of the long-houses might prove that a greater number of cattle were stalled than during the Late Bronze Age, suggesting that a system of hay-meadows had been introduced. The harvesting of hay in the centuries around the birth of Christ was made more efficient with the introduction of the scythe and the rake (Myrdal 1984:89).

Manuring of more permanent arable fields influenced the pattern of settlement, leading to less mobility. The greater number of settlement units and their ever more stationary nature can only be interpreted as showing an increase in the population of the Köpinge area during the period.

The period 200 BC-AD 200 is characterized by a warm, dry climate favourable for cereal cultivation (Hedeager 1988:165f.).

Descriptions of the cultural landscape and land use of the Early Iron Age have long borne the stamp of the archaeological material from Jutland and areas preserving a fossilized cultivation landscape, such as Gotland and Östergötland (Carlsson 1979, Widgren 1983, Stumman Hansen 1984). It is usually thought that most regions in southern Scandinavia underwent the same development at roughly the same time, not just as regards the cultural landscape (Berglund 1969) but also in social and political terms; it is only as a result of differences in the form of the natural landscape in different regions that this development can be studied today in only certain regions. In recent years it has become more evident that we must reckon with large regional and local variation, not just differences between areas with

and without a fossilized cultivation landscape, but also between areas within these categories (Widgren 1988).

As we can see a decisive difference between the settlement structure of eastern and western Denmark, there is also a marked difference in the field systems. In the west there were large field systems, relatively extensively cultivated. Although this type occurs occasionally in the east, the characteristic type here is small, perhaps more intensively exploited field systems (V. Nielsen 1984:160).

Since no fossil fields, fences, or other kinds of field divisions have been documented from the Köpinge area (and are rare in Scania as a whole), it is impossible to know whether the farms in the area cooperated in any form of joint fencing, or what type of cultivation system was used.

Analyses of carbonized plant material from the habitation sites show, however, that hulled barley was now the main crop and that weeds like *Chenopodium* became more common, indicating manured fields. The presence of meadow plants in the material can be interpreted as hay waste and the stalling of animals (Engelmark and Hjelmqvist 1991).

Traces of the production and working of iron have been documented on several occasions in the area. It is not clear, however, whether the demand for bog ore could be satisfied from local sources. Iron ore may have been mined at Fyledalen (Strömberg 1981:7f., 29, Tesch 1983b:51). Iron extraction may also have led to the establishment of special habitation sites, as in the Krageholm area (Ch. 4.3.4).

The Roman Iron Age 0-AD 400

Source material. Also as regards the Roman Iron Age, the source material has grown so much during the years of the project as to transform our picture of settlement patterns and land use. Admittedly, habitation sites with long-houses from both the Early Roman Iron Age and the Late Roman Iron Age were known before (Tesch 1979b: Fig. 2, 1983b: Fig. 30A e), but the material was not extensive enough to allow us to assume any great increase in population.

The archaeological material now available shows that the noticeably expansive phase which began during the latter part of the Pre-Roman Iron Age (period III) continued in the Early Roman Iron Age. Remains of habitation sites from the Roman Iron Age as a whole have been found at 25 (43%) of the habitation sites (second only to the Late Bronze Age). Unlike the Late Bronze Age, where most of these were camp sites, almost half (12) of those from the Roman Iron Age were habitation sites, most of them (10) with sure house remains. In view of the increasingly stable settlement pattern in the Early Iron Age, these figures show that there must have been a considerable rise in population after the Late Bronze Age. Towards the end of the period, however, there is a marked reduction in material.

Settlement. House remains, most of which are dated to the Early Roman Iron Age, are still relatively large: 17–26 × 5.6–6 m. The occurrence of some very long houses, 30–38 m (area B) and richer graves (Stjernquist 1955:77) are also signs of social stratification in the Köpinge area in the beginning of the Late Roman Iron Age.

In a few areas (B, D, possibly C) evidence was found of agglomerated settlement, that is, two or more farms in the same habitation area, cooperating in some way in agricultural production. These could have arisen through the division or amalgamation of farms. In the Late Roman Iron Age we appear to see a breakup of the agglomerated settlements. In the most expansive area, adjacent to Lilla Köpinge village (B), it is even hard to prove the existence of any settlement in the latter part of the Late Roman Iron Age. The same tendency is also noticed in another couple of central settlement areas (D and E). Already here we can detect a contraction of settlement to the areas nearest the coast (A and especially C), a feature characteristic of the following period (Ch. 4.1.5.1). This picture of the Late Roman Iron Age as a phase of change and transition to the Migration Period also emerges from the Danish archaeological evidence (Näsman 1987:80, Hedeager 1988:145).

Settlement patterns and land use. Unfortunately, the archaeological evidence from the Köpinge area does not allow us to make a detailed analysis of land use, since only the houses have left any traces on the ground.

It is uncertain at what point in the Early Iron Age there arose a more fixed pattern of settlement combined with a system of infields/outland with permanent, intensively cultivated fields. Indirect evidence for such a system may be adduced from the lasting agglomerated settlements which existed on the same site for a large part of the period 200 BC–AD 200 (areas B and D). Long-term use of the poor soils of the Köpinge area must have required highly intensive cultivation, in other words, a large amount of manure. Increased fertilization also presupposes larger herds, which in turn necessitates a greater availability of hay and manpower. This heavy exploitation of the land must have been untenable in the long run. On the light soils of the Köpinge area it can have led to leaching and podzolization.

Analyses of carbonized plant material from the habitation sites show only minor changes since the Pre-Roman Iron Age. Hulled barley continues to dominate, a full 97% of the material from the Early Iron Age. Occasional seeds of rye, oats, wheat, and *Camelina sativa* (gold of pleasure) testify to the cultivation of these crops, but on an insignificant scale. The few plants in the material associated with meadows and grazing suggest that wet meadows and coastal meadows were used (Engelmark and Hjelmqvist 1991). A remarkable feature in the osteological material is the presence of seal bones (Jonsson 1989).

A restructuring of the agricultural system in the Köpinge area in the Late Roman Iron Age is evident from the decline and the dissolution into smaller settlement units, probably dispersed farms. The pollen diagrams show possible hints of a slight decline in human impact on the landscape around AD 300 (Ch. 4.1.2), which coincides roughly with the archaeological dating of the restructuring.

The restructuring of the agricultural system was far from synchronous throughout southern Scandinavia. What has been called the crisis of the Migration Period lasted from the 3rd to the 7th century. The causes and effects of this process are a subject of discussion, as regards, for example, whether the phase of change was also a period of crisis (Näsman 1988).

4.1.5 The Late Iron Age and Early Middle Ages

4.1.5.1 From habitation site to village site in the Köpinge area

Sten Tesch

Source material

At the start of the project there was habitation site material with definite long-houses only from the Tingshög area (C). The remains on the low rise here were so extensive that it was estimated that not one but several farms had existed at the same time over a large part of the Migration Period and Vendel Period. This concentration of long-term settlement in one place was associated with a restructuring of farming into an intensive cultivation system with heavy manuring, possibly accompanied by some form of crop rotation. The restructuring is interpreted in part as a way of reacting to an unfavourable ecological change in the Migration Period (Tesch 1983b:53ff.).

There were seven medieval villages in the Köpinge area: five in Stora Köpinge parish – Stora Köpinge, Lilla Köpinge, Kabusa, Svenstorp, and Fårarp (formerly in Nedraby Parish) – and Stora Herrestad and Öja in their eponymous parishes.

Before the project began, our knowledge of when and how the villages arose and what the district was like in the Viking Age was highly diffuse, since archaeological material was largely lacking.

In view of the location in an old prehistoric central settlement area, it was thought likely that most of the villages in the Köpinge area belonged to the oldest stratum in the Ystad area. For example, Herrestad is mentioned in Saint Knut's deed of gift (1085) to the church in Lund, which received no less than 8 hides *(mansi)* in the village (Skansjö 1988:43). On the other hand, Svenstorp and Fårarp – two names in *torp*, originally meaning a secondary settlement or outlying farm – have been thought to belong to a later phase (Tesch 1983b:90).

The churches here are also among the oldest in the Ystad area – before the mid 12th century. The great church in Stora Herrestad is even earlier, from around 1100. It is built of an easily worked sandstone found only in Stora Köpinge parish, whereas many other churches in the Ystad area are of calcareous tufa, a material found only at Fyledalen a little to the north. The presence of this material in churches in Lund and Dalby suggests royal ownership of the quarry at Fyledalen (Ch. 5.7). The quarry lay near what might have been a royal common adjacent to the king's administrative centre *(kongelev)* at Arvalund in Nedraby parish, four kilometres up the river Nybroån from Stora Köpinge (Kraft 1956:34, Sjöbäck 1965:208ff., Tesch 1983b:87ff.).

The trial excavation programme of the subproject aimed mostly at the localization of habitation sites from the Late Iron Age. The existing archaeological material was not only scanty; for some periods it was non-existent. We deliberately avoided conducting any archaeological investigations of the village sites in the Köpinge area. It would have required excavations on a scale beyond the scope of the subproject to enable us to say anything more than that some form of settlement was established during the 11th and 12th centuries.

Today we have material from the Late Iron Age from almost a third (18 = 29.3%) of the habitation areas. Fifteen of these can be described as habitation sites, which is the highest number found for any period.

As for village formation, our working hypothesis was that the Viking Age settlement which had preceded the emergence of the regulated medieval village must be sought in the village's *tofts*; this term denotes the plot (presumably originally fenced in) adjacent to every farmstead in the village. The tofts were arable fields in the 18th century, and the adjacent site where the farmhouse stood was called *tomt*. The medieval provincial laws show that the size of the toft indicated the share the farm had of the common resources of the village (Riddersporre 1988).

The hypothesis was based on habitation site material found during an excavation of the historical village site of Lilla Köpinge (Arkeologi i Sverige 1984:290f., 1985:269). The analysis of land survey documents from the 18th century showed that large parts of the excavation area were judged to be the original toft belonging to one of the farms in the village (Riddersporre 1988). From the latter part of the Pre-Roman Iron Age and the Early Roman Iron Age there was settlement on a large scale here, which, at any rate during the later phase, consisted of at least two farms situated side by side (Ch. 4.1.4.4). There was also evidence, however, for agglomerated settlement from the end of the 10th century and the 11th century, consisting of both long-houses and

sunken-floor huts, which preceded the emergence of the regulated village. There was no evidence to show whether there was continuity in the settlements from the Early to the Late Iron Age.

According to this hypothesis, settlement in the tofts is an important link between the prehistoric migrating farm or agglomerated settlement and the stationary medieval village. Trial excavations in tofts by the villages of Stora Köpinge, Kabusa, and Svenstorp have revealed remains of Viking Age and early medieval habitation sites, which at least do not refute the hypothesis. To prove it, however, we would need more comprehensive investigations (Riddersporre 1988, Tesch 1988).

Settlement

In the Köpinge area the farms in the earlier part of the Late Iron Age (the Migration Period and Vendel Period) often consist of several small long-houses, 12–19 metres in length, but probably with different functions, along with one or two sunken-floor huts. On one of the low heights in the Tingshög area, however, a 35-metre-long house was found, dated by ^{14}C to AD 580 ± 90 (Tesch 1988:219ff.). The plan of the house, its long walls heavily curved, has close parallels in recently discovered contemporary house remains from Zealand (Mahler 1985, Rønne 1987). Three entrances and at least one dividing wall suggest a division of functions in the interior. The farm also has a sunken-floor hut. Because of the limited excavation area, it cannot be stated with certainty if the farm comprised more houses or if there were more farms on the site. In a field just north of the excavation site some forty hearths have been ploughed up, but stray finds such as brooches and an ornamental guard from a sword (Strömberg 1961:27) also indicate a cemetery. The total archaeological material suggests the presence of a magnate's farm in the area for a large part of the Late Iron Age. The way the settlement faces the river Nybroån and the many sunken-floor huts investigated in the vicinity might also indicate that this was a place for the exchange of commodities. At any rate, here two kilometres upstream there was a well sheltered landing-place for boats. Tingshög can, as later, have functioned as an intersection for land communications, one route being over the river Nybroån at Barevad (*vad* means "ford"). The place could be seen as a predecessor of the *köpinge* or "market-place" that was established during the second half of the 10th century a little further upstream.

The settlement from the Late Viking Age/Early Middle Ages which has been documented in the Köpinge area consists, as before, of long-houses and sunken-floor huts. The long-houses can be hard to identify in the cluster of post-holes, however, since building practices were highly varied during this period. For example, the posts supporting the roof stand in either one row or two, or they can be absent altogether. A house at least 12 m long in the tofts adjacent to Lilla Köpinge

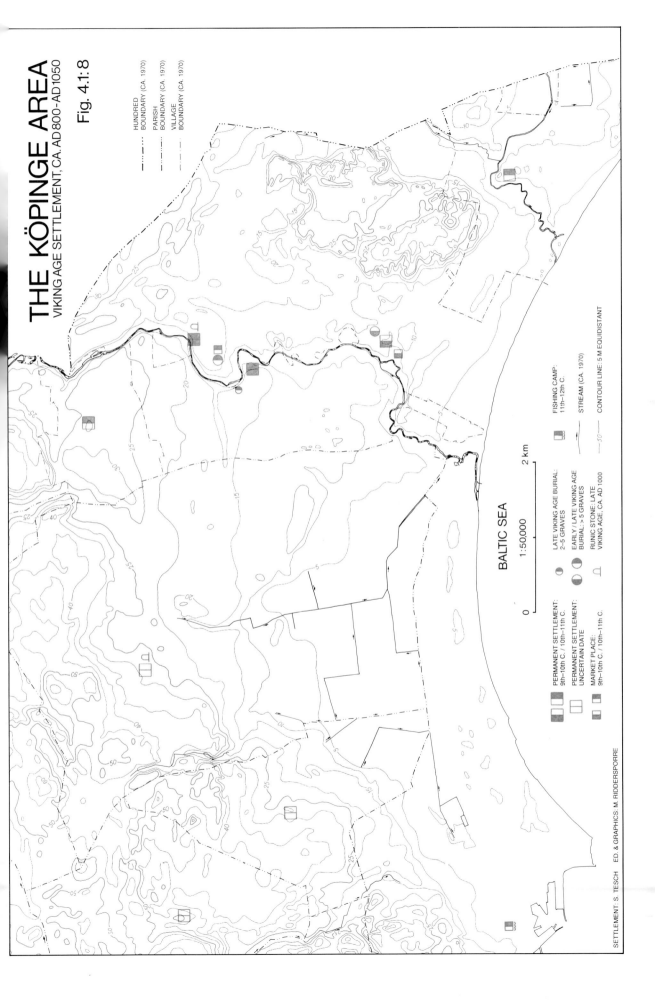

THE KÖPINGE AREA
VIKING AGE SETTLEMENT, CA. AD 800–AD 1050

Fig. 4.1: 8

HUNDRED
BOUNDARY (CA. 1970)

PARISH
BOUNDARY (CA. 1970)

VILLAGE
BOUNDARY (CA. 1970)

BALTIC SEA

1:50,000

0   2 km

PERMANENT SETTLEMENT:
9th–10th C. / 10th–11th C.

PERMANENT SETTLEMENT:
UNCERTAIN DATE

MARKET PLACE:
9th–10th C. / 10th–11th C.

LATE VIKING AGE BURIAL:
2–5 GRAVES

EARLY / LATE VIKING AGE
BURIAL: > 5 GRAVES

RUNIC STONE: LATE
VIKING AGE, CA. AD 1000

FISHING CAMP:
11th–12th C.

STREAM (CA. 1970)

—50— CONTOUR LINE: 5 M EQUIDISTANT

SETTLEMENT: S. TESCH   ED. & GRAPHICS: M. RIDDERSPORRE

village shows great similarity to house remains from the 11th and 12th centuries in the village of Vorbasse in Jutland (Kieffer-Olsen 1987, house 1). In the same toft a house more than 26 m long has also been measured. The occurrence at this time of houses of impressive size in southern Scania is shown by a 50 × 37 m courtyard farmstead with houses on all four sides over 30 m in length. This magnate's farm, situated in Ilstorp village just north of the Ystad area, also had a number of sunken-floor huts (Arkeologi i Sverige 1982–1983:318, 364, Tesch 1985b:84ff.). The existence of early medieval structures like sunken-floor huts in the tofts at Svenstorp village means that the settlement unit of Svenstorp must have been established as early as the 11th century. By analogy with Ilstorp (Igul's *torp*), the original settlement may have consisted of a magnate's farm (Sven's *torp*) (Tesch 1988:225). In Kabusa village a hypothetical early medieval royal administrative centre has been reconstructed through retrogressive analysis of the oldest land survey maps on the basis of the place-name element *husa* (Riddersporre 1983, 1986:54ff.). There were five further units of this kind in the north-western part of the Ystad area: in Bjäresjö, Kadesjö, Mattusa, Munkesjö, and Ugglesjö, all originally *husa*-names (Tesch 1983b:64ff., 1985b:89ff., see also Ch. 4.2.5.1). The Kabusa hypothesis also includes the possibility that there was a harbour for the royal naval organization (*ledung*) and a shipyard (Riddersporre 1986:59). The hypothesis has been tested by a number of trial digs within the scope of the subproject, but these neither confirmed nor disproved the hypothesis (Tesch 1988:203ff.).

Through retrogressive analysis of the oldest land survey maps it has been possible to reconstruct an original, regulated toft structure (in certain cases perhaps planned but never completed), associated with *bolskifte*, the way land was divided in of all the villages in the Köpinge area. There are great similarities between the toft structure investigated in Lilla Köpinge and what has emerged from Danish village studies (Riddersporre 1988: Figs 9–11).

Settlement patterns and land use

The existing archaeological material from the Migration Period suggests a restructuring and concentration of settlement to the Tingshög area (C), relatively near the coast. However, this change was initiated during the Late Roman Iron Age, which is in many ways an introductory phase of the Late Iron Age. We may hypothesize that a similar concentration of settlement took place in Stora Herrestad parish, but there have been no excavations to test this possibility.

This restructuring of the settlement pattern is part of a phase of change throughout southern Scandinavia, which is seen in the temporary decline in human pressure on the landscape, the so-called crisis of the Migration Period. This phase was not synchronous, however, but was spread over a long period stretching from the

Late Roman Iron Age to the Early Vendel Period (Näsman 1988:240ff.). In the Köpinge area a slight decrease in human impact on the landscape is observed around AD 300 (Ch. 4.1.2), which coincides with the changes that are observed in the archaeological material (Ch. 4.1.4.4). There is little agreement about the causes of these changes. The changes were not synchronous, so the crisis cannot have been due to a deterioration in climate. Rather than being an acute crisis, it was probably more of an adjustment to cultural and natural circumstances that had undergone general changes through time (Näsman 1988:249ff.).

In the 8th century we can detect an incipient expansion of settlement in the Köpinge area and a reclaiming of former cultivated lands, a process also occurring in other parts of southern Scandinavia. The Tingshög area (C) still appears to be the centre of the settlement region into the 10th century, although the concentration of settlement in this area may have been dispersed to some extent. New settlement units are found, some adjacent to the village tofts (Stora and Lilla Köpinge), and other smaller units not directly connected to the old settlement areas. The paucity of archaeological material makes it impossible to describe the development in any detail; it might be interpreted as showing a transition to a more mobile pattern of settlement with increased emphasis on animal husbandry. Pollen analysis shows, however, that this coincides with an accelerating opening and cultivation of the landscape (Ch. 4.1.2), a feature common to the whole of southern Scandinavia.

When we come to the medieval village landscape, only Stora Herrestad and Stora and Lilla Köpinge are situated in the old settlement district, more precisely in the transition zone between plain and hummocky landscape, in other words, between light and heavy soils. There was thus a gradual shift of settlement in the old core area. The former settlement areas nearest the coast were abandoned, and they are later found in the oldest land survey documents as outland pastures or unmanured fields lying fallow for long periods.

Our knowledge of the development of Iron Age settlement and its relation to the medieval villages has undergone a dramatic evolution as a result of the research of the last fifteen years. Yet no total excavation of any village site has been carried out, so we are unable today to assess which factors were most important in making settlement cease to migrate. One suggested cause is changes in the farming system. Other factors to consider are the introduction of a wheeled plough with a mould-board, which made it possible to cultivate clayey soil and to plough manure into the soil, and crop rotation using two or three fenced fields to grow springsown barley and autumn-sown rye as base crops. The structure of the village site as the basis of a fixed taxation system may also have been of decisive importance (Porsmose 1988:228f.).

Carbonized plant material from the Migration Period and the Vendel Period is too scarce in the Köpinge region to enable us to interpret cultivation. However, if this material is combined with the more representative material from the Viking Age, it is seen that a distinct change in the composition of seeds has occurred since the Roman Iron Age. Autumn-sown rye is now cultivated as much as barley. Oats also acquired some importance. Wheat, however, is found in only one instance. The composition of weeds, like that of the cereals, also indicates regular crop rotation. To determine whether this was in two or three courses would require finding a store which could show us the exact sequence of rotation. During the Late Iron Age we see an ever greater division of the functions of the farm among separate buildings. The almost total lack of seeds from meadow plants may suggest that byres and human dwellings were separate buildings (Engelmark and Hjelmqvist 1991).

There is great variation in the osteological material from the infill in the sunken-floor huts. Apart from the usual cattle, sheep, goats, and pigs, there are also bones of tame fowl (hens and geese) and fish (cod, herring, etc.) (Jonsson 1989).

Excavations in fifteen villages on Funen, Denmark, showed that the villages first appeared in their present locations in the transition between the Viking Age and the Early Middle Ages (Grøngaard Jeppesen 1981). Later investigations in Denmark and Scania have qualified the picture of village formation, as a slow process with both regional and local differences (Skansjö 1983, Madsen 1985, Callmer 1986b, Porsmose 1988). Several scholars have also hinted at the possibility of gradual changes in settlement structure on or near the site of the regulated village (Callmer 1986b, Porsmose 1988, Riddersporre 1988).

For the Ystad area I have suggested that village formation in the hummocky landscape was preceded by colonizing settlement in the latter half of the Viking Age. This could have been due to external factors of a political and economic nature, resulting from the desire of the growing Danish state power to control Scania. The first settlements are assumed to have consisted of large isolated farms – cattle ranches – of roughly the same character as Ilstorp.

It may have been in the same phase as the building of the Romanesque rural churches and the fixed parish division that a system arose of regulated villages and tenant farms and with the main emphasis on cereal production. The coming of trade in food commodities and the emergence of the town of Ystad in the first half of the 13th century must have exerted increased pressure on surplus agricultural production in the hinterland, a precondition for improvements in farm management (Tesch 1983b, 1985b).

4.1.5.2 Köpinge and Ystad. Market place in the Viking Age and Early Middle Ages and town in the High Middle Ages

Sten Tesch

In the second half of the 10th century a number of trading sites near the coast ceased to exist. Coming at about the same time over a large part of the Baltic area, this may have been partly due to changing cultural contacts and channels of exchange, and the formation of states, including the expansive ambitions of the Danish king and state.

Some of these trading sites were of considerable size, being not only central localities but also to some extent of an urban character (like Birka). Other large places, some of which were surrounded by ramparts, give the impression of seasonal use, such as Paviken and other sites on Gotland (Carlsson 1988), and Scanian sites at Vikhög in Löddeköpinge (Ohlsson 1976) and at Åhus (Callmer 1984).

There is also a line of sunken-floor hut habitation sites, both big and small, along the Scanian coast, almost at the water's edge. Examples include Tankbåten west of Ystad (Strömberg 1978, 1980, 1981), by the estuary of the river Skivarpsån (Larsson and Olausson 1986), at Östra Torp (Stjernquist 1988), and in the town of Trelleborg (Jacobsson 1987). These settlements are usually interpreted as normal agrarian units later relocated to other settlements (village formation) further from the coast (Callmer 1984). From the agricultural point of view these locations are extreme, and the nature of the settlements (with only sunken-floor huts) and to some extent also of the finds speaks rather for some kind of seasonal sites for special activities, such as exchange. That these places ceased to exist was probably due to the tendencies towards central government which are noticeable in the second half of the 10th century, also bringing about the foundation of towns (like Lund and Sigtuna) as points of support for political interests. Early medieval trading sites with central functions under royal control preceded most of the places which achieved urban status in the 13th century in the Mälaren region, Östergötland, and Småland. Many of these places reveal their commercial function through names ending in -köping (Hasselmo et al. 1987). In the Scanian landscape there are a number of place-names in köpinge which are viewed as trading sites from the Viking Age or the Early Middle Ages. They have been variously interpreted as magnates' trading sites from the Viking Age (Cinthio 1972, 1975, Ersgård 1988:184f.), or trading sites controlled by the Crown in the late Viking Age/Early Middle Ages (Tesch 1983b:73ff.), or as archiepiscopal trading sites from the end of the 12th century, competing with the king's towns at a time when the royal power was weak (Callmer 1984:69ff.). Against the latter theory it must be said that there is no clear evidence before the 16th century that these places belonged to the archbishop. What the *köpinge* localities have in common is not only their name but also their situation: near the sea but a few kilometres up a navigable river, close to hundred boundaries and near royal administrative centres *(kongelev)* and later market towns (Tesch 1983b:75ff.). These shared features bear witness to an overarching central control. Since the archaeological material that has been assumed to belong to the *köpinge* localities is mostly older than the 12th century, this control must have been exerted by the king, not the archbishop.

In the Köpinge area the site of the actual *köpinge* or market-place has been localized to a bend in the river about four kilometres upstream. The archaeological remains consist of sunken-floor huts of Slav type with a smoke-oven in one corner, waste pits, and so on. These have been found on both sides of the river half-way between the villages of Stora and Lilla Köpinge. The finds allow the settlement to be dated to a time from the end of the 10th century to some way into the 12th century. In addition, a cemetery with skeletal graves (probably Christian) has been found near the settlement remains; this was probably a graveyard by an early medieval wooden chapel (Tesch 1979b, 1983b, 1985b, Riddersporre 1985). The osteological material from the sunken-floor huts contains the oldest safely datable evidence for the presence of house mice in Sweden. The house mouse possibly immigrated as a stowaway on the trading ships of the day (Jonsson 1989).

A large but suddenly interrupted extension of Stora Köpinge church, dated to around 1200, has been interpreted as a sign that the vigorous development of the *köpinge* was curbed by the foundation of Ystad (Lundberg 1940:124, Gustafsson and Redin 1969:33). However, a revised dendrochronological dating of 1285 for the new building work on the church (Bartholin 1989, Sundnér 1989a) makes it impossible to speak of the succession passing from Stora Köpinge to Ystad. The church's unusual magnitude for a rural location shows that Stora Köpinge continued to have special functions into the 13th century (Wienberg 1988). These circumstances may suggest (at least planned) parallel development between Ystad and Stora Köpinge during the early part of the 13th century. The main interested parties in this competition must have been the king and the local magnates, or, most likely, the archbishop. The key to the interpretation of the development of the *köpinge* in Stora Köpinge parish must be that we are dealing with two totally distinct temporal horizons. One horizon, from the end of the 10th century to the start of the 12th, represents a general development in all the *köpinge* localities controlled by the king. Another later horizon in the 13th century represents a special development, an incipient urbanization of Stora Köpinge, where the archbishopric controls trade in competition with the royal foundation of the town of Ystad. In view of the types of ships used at this time, this must have

been a rather uneven struggle, from which Ystad quickly emerged the victor. In a clergyman's report from 1729, the vicar records a local tradition that Köpinge, as the name suggests, "in former times was like a little market town in which markets and trading were carried on. This is also related by many old men still living in the village and the parish, who have heard it from their fathers. Otherwise there are here in the village some small rudera which testify to this, such as some remains of cobbled streets, and a wide area in the middle of the village, which is supposed to have been the square" (Skansjö 1988:63). It is hardly likely that such a tradition could have survived from the Early Middle Ages. If we are to pay any heed to this at all, it seems more reasonable to interpret it in terms of the other temporal horizon. Cobbled streets and squares could not be associated with early medieval seasonal trading sites, but with towns of the High Middle Ages. It may also be of interest in this connection that the tradition is communicated by the church. Only future archaeological investigations in Stora Köpinge village can determine whether there is a second temporal horizon in the development of the *köpinge* locality. Nothing has as yet been found to suggest this.

Although the town of Ystad is partly a product of its hinterland, there is no direct link as regards commercial functions between the *köpinge* and the town. The "missing link" might be the 12th century fishing camps. At this time herring came in huge quantities to this part of the Baltic. The population in the agrarian districts gathered in the autumn at appropriate landing sites to obtain their share of the catch. At these places, which were originally unorganized, a chapel was often built, and it is conceivable that market activities were carried on here. One such place was the estuary by the inlet known as Ystad. St Nicholas' Church may originally have functioned as a chapel for the duration of the fishing camp. Archaeological evidence for herring fishing comes in the form of clay-lined pits dug in the sand *(lerbottnar)*, and used in some way for storing and preparing the herring (Stenholm 1981, Tesch 1982b, 1983b, 1985b). These pits have been found in a row about a kilometre long along where the shore once ran at Ystad. Another probable such fishing camp with its chapel existed at Kabusa. In addition, we may mention that a fragmentary clay-lined pit with remains of herring has been found by the early medieval settlement at Lilla Köpinge village (Jonsson 1989).

Since the foreshore was royal property, the Crown had the right to everything that passed the shore. To facilitate control and taxation of the activities associated with herring-fishing, it was in the interests of the Crown to concentrate this in as few places as possible. The best places were favourably located as regards topography and population, like Ystad. At the same time as the selected places became towns – built-up areas with a permanent population of merchants and craftsmen – large town churches were also erected, like St Mary's in

Ystad. The construction in the 13th century of a market town on the shore was one way to adapt to the new situation that had arisen in the Baltic region with the newly won commercial hegemony of the German Hanseatic merchants (Tesch 1982b, 1983b:98ff., 1985b).

With the emergence of a town in the area, the situation of the agrarian hinterland changed. It was to affect not least the appearance of the cultural landscape as a result of the increasingly important role of grain production, as well as the general opportunities of producing foodstuffs for sale. The control of grain cultivation was an instrument of power in the Middle Ages: a telling expression of this was the finding of a mill and a central bakery during excavations in the Pernilla district of Ystad (Arkeologi i Sverige 1984:209ff., Tesch 1985b:94f.).

4.1.6 The Middle Ages and Modern Time until ca. 1670
4.1.6.1 Settlement, cultivation systems, and property distribution in the Köpinge area

Sten Skansjö

The oldest known historical settlement

Although Herrestad parish is the old judicial centre of Herrestad Hundred, it was not until 1891 that the whole of Stora Köpinge parish belonged to this hundred. Before this, the greater part of the parish, the church-village, and Kabusa village belonged to Ingelstad Hundred. The old boundary between the hundreds was the river Nybroån. The village of Fårarp was also divided between two parishes: a few farms paid tithes to Nedraby (previously Arvalunda) parish. Between 1635, when Nedraby parish was dissolved, and 1891, these farms belonged to Övraby parish. In the following section "the Köpinge area" denotes the parish of Stora Köpinge after 1891, along with Herrestad parish.

There is written evidence of the village of Herrestad in King Knut's deed of gift to the church in Lund from 1085. The 14th century sees the first mention in written sources of the following known historical settlements in the Köpinge area: Stora Köpinge (1322), Lilla Köpinge (1348), Kabusa (1366), Fårarp (1385), and Svenstorp (1387). A number of watermills operated along the river Nybroån during the Middle Ages: Barevad mill is mentioned in 1397; Lawerewatsmölla ("low-water mill") is named only once, in 1398, after which it vanishes from the sources; Svenstorp mill is known since 1503; and Köpinge mill is recorded for the first time in 1546. These villages and mills constitute all the historical settlements known before 1670 in the Köpinge area.

The settlement pattern in the Köpinge area, both in the Middle Ages and the mid 17th century, consisted of villages with a varying number of farms. From around

1670 we have details of all the villages in the entire Ystad area; at that time Fårarp had five farms, Svenstorp eight, and Lilla Köpinge and Kabusa ten each. These four villages are below the average size for villages in the Ystad area, which is thirteen, while the church-village of Stora Köpinge, with its 28 farms, was the second largest village in the area, surpassed only by Stora Herrestad (45 farms).

A comparative calculation of settlement density and its variations in the Ystad area (Skansjö 1985) suggests that the parish of Stora Köpinge is near the average for the entire Ystad area: 2.8 "normal" farms km^{-2}. However, with its 2.7 farms km^{-2}, Stora Köpinge has a lower settlement density than other parishes on the coastal plain and the outer hummocky zone, but noticeably higher than the parishes in the inner hummocky zone, such as Högestad, Baldringe, Sövestad, Skårby, and Villie. By contrast, the corresponding figure for Stora Herrestad indicates a settlement density well above the average – 3.4 normal farms km^{-2}, which is on a par with the following parishes in the coastal region and the outer hummocky zone: Snårestad, Katslösa, Hedeskoga, Öja, and Balkåkra.

There are no details in contemporary written sources to suggest any regression in settlement or cultivation in the Late Middle Ages. The late local tradition (1729) of Stora Köping as a former market town with traces of cobbled streets and a square (Ch. 4.1.5.2) may well have been suggested by the etymology of the village name (*Köpinge* means "market town" and is transparently related to *köpa* "buy"). From the same time we are informed that there had once been a chapel in Kabusa.

Apart from late medieval information to the effect that Stora Köpinge formerly had one main farm with a large number of tenant farms (see below), there is no empirical evidence in our sources of any significant changes in settlement.

In Stora Köpinge parish (not the whole Köpinge area) there were some 65 farms around 1570 and still in the mid 17th century. The area thus appears not to have been affected by the expansion of settlement which can be observed in the Romele area at this time. In Herrestad parish we can detect a slight decline in the number of farms between around 1570 and 1670; this drop from 50 farms to 45 is probably due to the expansion of the manor (see below) (Skansjö 1988).

Cultivation according to the oldest written sources

From the early 16th century we know of a three-field system in the church-villages of Stora Herrestad and Stora Köpinge, as well as in Kabusa. According to the Lund diocesan terrier from about 1570 there was in (or near) Stora Köpinge's south field a sandy area (*Saanden*) which was planted with rye every six years.

A comparison of tithe data in the terrier for the various parishes in the Ystad area shows that grain production per tithe-paying unit in Stora Köpinge parish was more than 20% below the average for the entire project area. In this respect Herrestad parish also shows clear signs of being a more central cultivation district – some 33% above average. The differences between the two neighbouring parishes are thus distinct around 1570.

The detailed description in the same terrier of the cultivated land belonging to the vicarages can, for some parishes, be compared with the corresponding descriptions in the oldest land survey documents from the time around 1700 (or slightly later). To the extent that the vicarages can be considered representative of the villages to which they belonged, we obtain clear indications of any changes in their production trends and in the proportion of arable and meadow. We see that there was a noticeable tendency, especially in the villages on clayey soil in the outer hummocky zone, for the area of tilled land to increase at the expense of meadow between 1570 and the early 18th century. The expansion of tilled land affected about 25% of the total area of the vicarages in the villages of Bjäresjö, Hedeskoga, and Öja. Calculated on the basis of the arable land alone, the expansion was around 50%. For Villie parish the expansion was negligible, but here the people had the possibility of expanding beyond the established villages into the outlands, where a number of new settlements arose in the period 1583–1651 (Ch. 4.4.6). Stora Köpinge reveals a similar trend to the villages in the outer hummocky zone, although the development here was not so pronounced; the expansion of arable land between 1570 and 1732 was only 12% as a percentage of the total land and about 25% of the arable alone. One explanation for this might be sought in the fact that there was less room for expanding the area of tillage in the poorer sandy soils by the coast. It is probable that people had to be more restrictive about encroaching on the area of meadowland if they were to retain the productive capacity of the soil. Moreover, Stora Köpinge already had large areas of arable in proportion to meadow in 1570, but, as pointed out above, large parts of the arable were farmed extensively, with long periods of fallow. Lacunae in the source material have made it impossible to carry out a corresponding enquiry into the village of Stora Herrestad (Persson 1986b, Persson and Skansjö unpubl.).

Property distribution

To the extent that we know anything about the medieval distribution of property in Stora Köpinge parish, it underwent changes both during the Late Middle Ages and after the Reformation. The largest known landowner in the Middle Ages was the archbishopric of Lund. The estate held by the archbishop *ex officio* included *Köpinge län* (Köpinge fief), which was registered early in the 16th century in two surviving cadastres. One of these lists the following farms in the fief: in Stora Köp-

inge there was an administrative manor *(curia principalis)* and 31 tenant farms, together with one smaller unit *(colonia)*; in Lilla Köpinge the fief owned three tenant farms and three *coloniae*; and in Kabusa the estate complex owned five tenant farms. The extent of land owned by the manor in Stora Köpinge's three fields totalled about 35 ha, on top of which there were meadows yielding 40 loads of hay annually.

At the Reformation these tenant farms belonging to the archbishopric in the Köpinge area passed into royal ownership. Part of the formerly undivided fief of Köpinge came into private hands in 1587 through an exchange of property between the Crown and Frederik Ulfeldt. He obtained eight farms in Stora Köpinge and one in Lilla Köpinge. The rest of the fief went the same way in 1611 as part of an even larger exchange of property between the Crown and Jochim Byllow, who obtained 27 units in Stora Köpinge, Köpinge mill, four farms in Lilla Köpinge, and Barevad mill, which was then deserted. The greater part of the parish thus fell into the hands of the private nobility, although no noble manor was established in the area after the Reformation.

From the Middle Ages, however, we know of several nobleman's manors in the parish of Stora Köpinge. They are attested – usually in isolated references – in Lilla Köpinge, Kabusa, and Fårarp in the 14th century and in Svenstorp in the late 15th century. In the Late Middle Ages they ceased to function as manors, coming into the possession either of ecclesiastical institutions (Lilla Köpinge and Fårarp) or of large noble landowners (Kabusa). This development, which is typical of the Late Middle Ages, was probably due to the economic problems of the gentry and their difficulty in meeting the demands of their status (on the "crisis of the gentry" see Ch. 5.9).

For Herrestad parish we can observe a development from the differentiated ownership of the Middle Ages to the almost total domination of the parish by one privately owned noble manor.

Medieval ecclesiastical landowners include Lund Cathedral, which in 1085 was awarded eight hides (*bol*) in the village. These lands had been part of the estate which one Öpe Thorbjörnsson of Lund paid to the king for his release from outlawry (*pro pace sua*). The earliest known owner was thus a magnate. In the Early Middle Ages these eight hides belonged to the cathedral chapter in Lund.

The first evidence of noble ownership of land in Herrestad parish comes from the 1340s, but this may be a question of scattered property belonging to estates elsewhere. From 1386, however, there is clear testimony to the existence of a nobleman's manor in Herrestad village. Despite the presence of the manor, there were various noble landowners with tenant farms in the village in the Late Middle Ages.

Post-medieval sources clearly show that the Crown also had farms in Herrestad parish at its disposal; from the 16th and 17th centuries we know of several transactions between the Crown and noblemen involving individual farms in the village. In 1630 the owner of the Herrestad manor received 16 farm units from the Crown in an exchange of property, an indication of the lord of the manor's aspiration to consolidate his estate. We see a change in the distribution of property in the parish, a turning-point which was to be significant for the future: between 1630 and about 1670 almost the whole of Herrestad parish came into the possession of Herrestad manor.

Sources from around 1680 describe the Herrestad manor as of great age. Its fields and meadows were mixed amongst those of the peasants. The manor buildings at that time do not appear to have been in the best condition. The village had 39 farms and 10 cottages, along with a windmill near the village and two fisherman's houses by the shore. There was no woodland or common. It is stated that five or six farms had been incorporated into the manor farm.

In the late 17th century the church retained possession of only five farm units; the rest of the parish consisted of Herrestad manor and about forty subject farms with tenants liable to do corvée (labour duty on the demesne farm) (Skansjö 1988).

4.1.6.2 Landscape and land use in the Köpinge area around 1300. A hypothetical reconstruction based on land survey documents from around 1700

Mats Riddersporre

As shown in Ch. 4.1.5.1, the shift – more or less general in southern Scandinavia – from mobile settlements to permanent villages took place also in the Köpinge area during the Late Viking Age and Early Middle Ages. Most of the villages still exist, and are also known through standing churches, medieval written sources, and land survey documents from around 1700. There is still, however, too little knowledge concerning settlement and land use to allow a detailed reconstruction of the early medieval landscape. Through a retrogressive analysis of the land survey documents – maps and descriptions – it is nevertheless possible to reconstruct a landscape that might reflect the High Middle Ages. The reconstruction in Fig. 4.1:9 is the result of such an analysis and is based on a combination of indications such as field-names, field layout, crop rotations, soil quality, and the physical landscape itself (cf. Ch. 4.1.7.2, Figs 4.1:13 and 4.1:14).

Change and stability – some considerations

To provide a framework for the reconstruction, some basic assumptions are necessary. A general assumption is that there has been an increase over time of tilled land, at the expense of pasture and meadowland. The core area of the arable, physical conditions permitting, should be found near the settlements, and more extensively used land at a more distant location (cf. Hansen 1976).

Around 1700 the crop rotation in all the villages was based on the three-field system. This system is assumed to have been in existence also around 1300. The idea of a general medieval transformation from a two-field to a three-field system has recently been discussed concerning Denmark and England. The conclusions are that there is very limited evidence of this transformation (Fox 1986, 1988, Frandsen 1988).

Further, it is assumed that there are field-names that may go back to the 14th century at least. There has been some criticism of the idea that field-names in general should be of Viking Age or early medieval date (Jørgensen 1984).

Some of the features in the reconstruction show in reality the situation around 1700. The village cores, the watermills, and the boundaries are drawn from the oldest maps. Concerning the settlements there is no evidence to suggest drastic changes between 1300 and 1700. The mill at Barevad is known to have existed in 1397 (Ch. 4.1.6.1). The boundaries, on the other hand, may not have been totally defined in the 14th century.

Infields and common pasture

The reconstruction shows a division of lands into infields (*inägor*) and commons, or outlands (*allmänning, utmark*). In the infields are found arable and hay meadows. The commons were probably mainly used for pasture, but on wet ground there may also have been hay meadows. There are no indications of woodlands other than in the north-eastern part of Stora Köpinge, where one field-name contains the word for wood. The name, *Skogsåkrarna*, probably refers to a wood which still exists on the other side of the parish boundary, and which of course may previously also have covered the neighbouring parts of Stora Köpinge.

The reconstruction of arable is based on the distribution of lands under intensive use around 1700. That is, basically, the parts of the tilled land that were used in two years out of every three-year cycle, and that were sown with rye and barley. The soils here are normally also reported as containing high proportions of mould, i.e. a high content of organic material in the topsoil, a consequence of a long process of manuring and ploughing, etc. (cf. Steensberg 1976:113). The distribution of these areas is, as expected, related to the distance from the village cores (cf. Arrhenius 1934, Dahl 1942a:100, Helmfrid 1962: 221, Hansen 1976, Steensberg 1976:96–97).

The reconstruction of meadows is based on what was meadow in the 18th century, but encompasses also parts of the areas then used for grain, mainly oats, with long fallows. According to pollen analysis from Bjäresjö, oats was not a major crop in that area until the 17th century (Ch. 4.2.2). The existence of oats in the Viking Age is, however, indicated through seed finds from the Köpinge area (Ch. 4.1.5.1). In *Lunds stifts landebok*, a terrier registering all farms belonging to the church around 1570, the growing of oats is reported in almost all the furlongs in Stora Köpinge that can be identified by name and that were sown with oats around 1700. Parts of the meadows – on dry ground in the infields – may have previously, at times, been ploughed and sown with oats.

If the one field that, according to the three-field system, lay fallow each year was used for grazing, a hay harvest was only possible two years out of three on the infield meadows. The meadows on the commons, on the other hand, could be harvested every year.

The two large pasture areas are indicated by field-names referring to cattle, fencing against cattle, common land, and to burning – probably of heather. They are situated on sandy soils with low contents of mould, which in the 18th century were sown with grain, mainly rye, and lay fallow for very long periods: 6–20 years. In Stora Köpinge this area around 1700 is referred to as *Sanden* (the Sand) or *Utmarken* (the Outlying Lands), while the infields are referred to as *Hemmavångarna* (the Homefields). Parts of the grazing areas were still used only as pasture around 1700. They were then covered with heather and small thorny bushes. In several cases it is possible to identify cattle-paths that connect the pasture areas with the village cores.

Kabusa – two three-field systems in one village

The situation in Kabusa is complicated and needs further explanation. To the east of the river Kabusaån is found what has been interpreted as the three original fields of the village (A, B, C in Fig. 4.1:9). There is a fourth field west of the stream, which has been interpreted as the former infield of a vanished unit: a large farm, which might have been part of the early medieval royal estate (cf. Ch. 4.1.5.1). When this unit disappeared is difficult to say, but it must have occurred before 1500. Presumably it had vanished already around 1300. The field was probably divided into a separate three-field system (D1, D2, D3 in Fig. 4.1:9), and around 1700 it was mainly cropped by three farmers in Kabusa having all their ploughland here. The small fields (D4) belonged around 1700 to two smallholdings (*fästor*) and were probably taken in from common land (Riddersporre 1983, 1986).

Boundaries

As regards the village boundaries, it can be seen that the later known boundaries run through what has been interpreted as common pasture. It is likely that the surrounding villages shared common rights to these areas. There would have been no reason for defining boundaries here until the pasture was divided and used for growing grain. Judging by field-names appearing in the diocesan terrier, this process seems to have started before 1570 in Stora Köpinge but not in Stora Herrestad.

A more pronounced linear feature of the medieval landscape was probably the field boundaries. The reconstruction shows an example of infields from two neighbouring villages – Svenstorp and Lilla Köpinge – growing into contact. Providing they were fallowed the same year, there was no need for a fence between these two fields. The main purpose of the fences was to keep grazing cattle away from the growing crop (Campbell 1933, Dahl 1968). This form of cooperation in fencing, *vångalag*, is fully developed in the reconstruction of the landscape around 1700 (Fig. 4.1:13).

1300–1700

The reconstruction shows much of the same organization as the situation around 1700, a condition that may well be in accordance with historical reality. The Danish historian Erland Porsmose (1988) claims that the period up to the 14th century was a time of growth, when the infrastructure of the cultural landscape was established. This structure then remained much the same until the 18th century. In fact, the reconstruction of the 14th century landscape gives an impression of the optimum use of the Köpinge area, assuming a more or less subsistent agrarian economy.

4.1.7 The 18th century

4.1.7.1 Cultivation, population, and social structure in the Köpinge area

Anders Persson

Introduction

The form and use of the cultural landscape in Stora Köpinge Parish can be described on the basis of historical sources from the late 17th and early 18th centuries. We have particularly good information about the villages of Stora Köpinge, Lilla Köpinge, and Kabusa in the south, whereas the information for Svenstorp and Fårarp in the north is not so detailed. A series of assessments and surveyors' descriptions with maps provide details of the cultural landscape during the period ca. 1670–1730. These can be compared with cartographic information from the 1760s and the early 19th century. Settlement and population data have been compiled for the years 1700, 1750, and 1805.

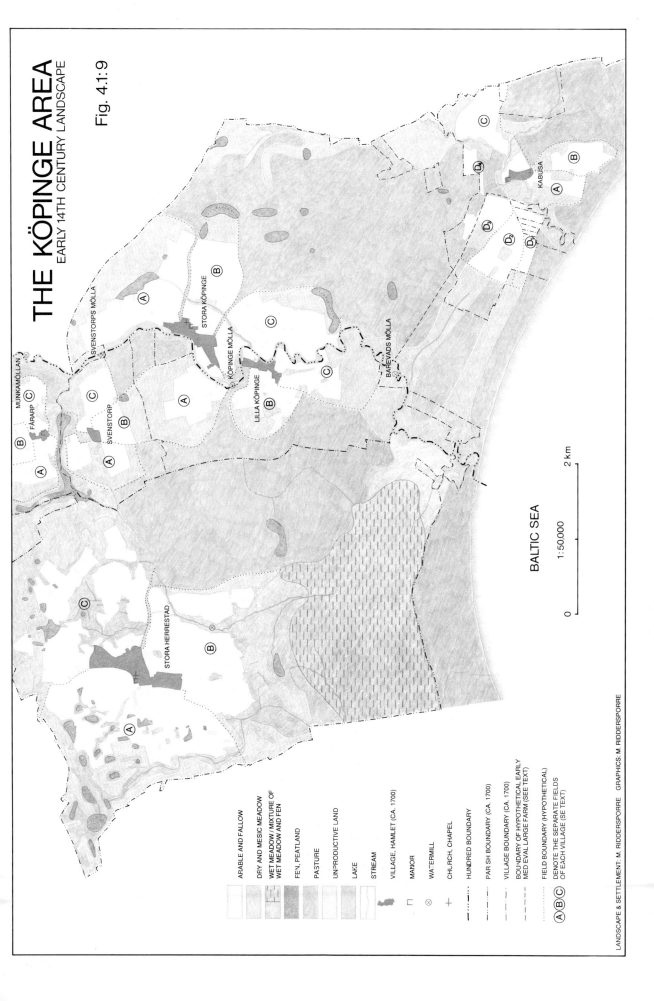

THE KÖPINGE AREA
EARLY 14TH CENTURY LANDSCAPE

Fig. 4.1:9

BALTIC SEA

1:50,000

0   2 km

SVENSTORPS MÖLLA

MUNKAMÖLLAN

FÄRARP Ⓒ

Ⓑ

Ⓐ

Ⓒ

Ⓑ

SVENSTORP

Ⓐ

Ⓐ

STORA KÖPINGE

Ⓑ

Ⓒ

KÖPINGE MÖLLA

LILLA KÖPINGE

Ⓑ

Ⓒ

BAREVADS MÖLLA

STORA HERRESTAD

Ⓒ

Ⓑ

Ⓐ

Ⓒ

Ⓓ₄

Ⓓ₃

Ⓓ₂

Ⓓ₁

KABUSA

Ⓑ

Ⓐ

APABLE AND FALLOW

DRY AND MESIC MEADOW

**WET MEADOW / MIXTURE OF
WET MEADOW AND FEN**

FEN, PEATLAND

PASTURE

UNPRODUCTIVE LAND

LAKE

STREAM

VILLAGE, HAMLET (CA. 1700)

MANOR

WATERMILL

CHURCH, CHAPEL

HUNDRED BOUNDARY

PARISH BOUNDARY (CA. 1700)

VILLAGE BOUNDARY (CA. 1700)

**BOUNDARY OF HYPOTHETICAL EARLY
MEDIEVAL LARGE FARM (SEE TEXT)**

FIELD BOUNDARY (HYPOTHETICAL)

Ⓐ Ⓑ Ⓒ **DENOTE THE SEPARATE FIELDS
OF EACH VILLAGE (SE TEXT)**

LANDSCAPE & SETTLEMENT: M. RIDDERSPORRE   GRAPHICS: M. RIDDERSPORRE

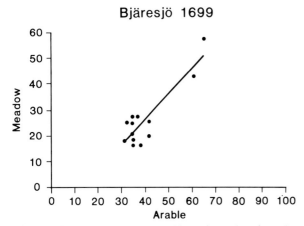

Bjäresjö 1699

Fig. 4.1:10. Relation between arable and meadow (area in tunnland) in Bjäresjö village, 1699. Regression line plotted (correlation +0.89; no. of farms 13).

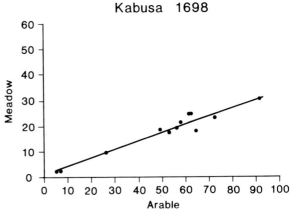

Kabusa 1698

Fig. 4.1:11. Relation between arable and meadow (area in tunnland) in Kabusa village, 1698. Regression line plotted (correlation +0.97; no. of farms 12).

Settlement

For the five villages in Stora Köpinge Parish a total of 84 settlement units are recorded in taxation assessment rolls. Of these units, 67 were farms and 12 cottages. The cottages were thus 14% of the total, which is only about half of the proportion found in parishes with manors (see Ch. 4.2.7.1). This settlement structure changed radically during the 18th century. A slight rise in the number of farms can be noticed, but what we see in particular is a sharp increase during this century in the number of cottages, whose share of the total settlement rose to over 50% by the mid 18th century. In real numbers the changes were as follows: farms increased from 67 in 1700 to 73 in 1805, while cottages increased from 12 in 1700 to 76 in 1805. The near doubling of the total number of settlement units is a reflection of population changes in the 18th century.

Population

The population of the parish can be estimated to about 440 (index 100) in 1700 (a density of 14 km^{-2}). In the mid 18th century the number was about 635 (21 km^{-2}), rising to around 880 (28 km^{-2}) at the start of the 19th century (index: 100–145–200). The figures reflect a doubling of the population in an area with very extensively farmed soils of low productivity (Ch. 4.1.7.3).

The population density of Stora Köpinge parish in 1700, 14 km^{-2}, can be compared with that of other parishes in the same year (not counting the people living on manors and the area of manor lands). For Sövestad the approximate figure is 15 km^{-2}, for Villie 15 km^{-2}, for Bjäresjö 23 km^{-2}, and for Västra Nöbbelöv 24 km^{-2}. The low values for the parishes of Sövestad and Villie can be explained by the greater amount of outlands here at the start of the 18th century. This was not the case in Stora Köpinge, which should rather be compared with parishes like Bjäresjö and Västra Nöbbelöv. In comparison

with these parishes, which were also dominated by villages, Stora Köpinge was a sparsely populated parish at the start of the 18th century.

The agrarian structure

The district as a whole on the basis of taxation records. The farms and cottages in Stora Köpinge parish were assessed in 1671 and 1682; the details of seed-corn and hay supply can be used for comparative purposes, both within the parish and in relation to other parishes in the project area. In the four villages in the parish in 1671 and 1682, rye and barley made up about two-thirds and oats one-third of the total volume of seed-corn. In the village of Kabusa, to the south, the proportion of rye and barley was 6/7 in 1671 (4/5 in 1682) and oats about 1/7 in 1671 (1/5 in 1682). Oat cultivation was thus not as dominant in the coastal village of Kabusa, where more rye and barley were grown. This picture is confirmed by analysis of the land survey documents from 10–15 years

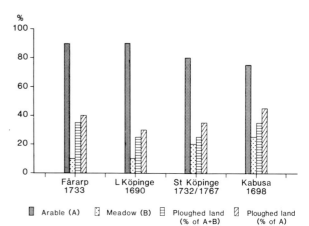

Fig. 4.1:12. Land use in villages in Stora Köpinge parish in the late 17th and the early 18th centuries.

later, which show the same regional difference between the inland villages and Kabusa as regards the distribution of these crops on arable land (see below).

The data in the assessment records from 1671 can also be drawn on for information about the supply of hay in relation to the manured land, that is, the rye and barley fields. We find that the farms in the big villages in the south – Lilla Köpinge, Stora Köpinge, and Kabusa – had at their disposal an average of between 1.1 and 1.3 loads of hay for every barrel of rye and barley seed. For the villages of Fårarp and Svenstorp the volume of hay was less than half, only about 0.5 loads. These figures should be compared with those provided by the same source for the Romele area. Below the south-western slope of the Romeleåsen ridge, five villages had a hay supply of between 1.5 and 2.5 loads per barrel of rye and barley seed; further up the ridge, where settlement consisted of single and double farms, the volume of hay was between 2.5 and 8.5 loads per barrel of seed-corn (Persson 1986a). From this comparison we see that the hay supply in Stora Köpinge parish at the end of the 17th century was small in proportion to the rye and barley fields. This was doubtless a major reason for the less pronounced expansion of arable land in the period 1570–1700 in the Köpinge area compared with other parishes in the project area (Persson and Skansjö unpubl., Ch. 4.1.6.1). The land surveyor also noted that the meadows in Stora Köpinge village were "for the most part dry and very poor" and that the pasturage was "as poor as it could possibly be". It is significant that the Civil Survey for the village in 1671 describes one cotter as a "meadow-ward" (Persson 1987b). The natural conditions in Stora Köpinge parish at this time meant that the amount of livestock in proportion to the annually ploughed and manured land could not possibly be as high as in the better favoured village landscape or the regions with single and double farms to the west (for these see Olsson and Persson 1987). It is also significant that we find in the Köpinge area observations such as that seed-corn "cannot be sown more than three bushels" per *tunnland* (0.5 barrels per 0.5 ha), which is a low density for this time (cf. Dahl 1942, Persson 1982, 1986a,b). The crop yields in Stora Köpinge parish in the late 17th century (Persson 1986b) and the 18th century are not very different from those in the other parishes in the Ystad area. To sum up the situation in this period, we can say that the cereal yields of land in Stora Köpinge were only about half those of the villages studied in the west of the project area. The demographic picture of Stora Köpinge as a relatively sparsely populated parish in 1700 must be seen in this perspective.

The villages in focus. The information in the surveyors' descriptions from around 1690–1730 has been computer-processed to provide details of farm size, the proportion of arable and meadow, areas of ploughed land, and so forth. The details are also presented here in a comparative perspective, chiefly as a contrast to those for

Bjäresjö village in the parish of the same name. In this connection the field areas are sometimes stated in *tunnland* since this unit has a more natural link with historical reality. Further details of the functions of the cultivated land and its spatial distribution are given below (Ch. 4.1.7.2), along with a description and analysis of the plant ecology of the cultural landscape (Ch. 4.1.7.3).

To begin with, it can be said that the farms in Stora Köpinge parish had much more land than those in, for example, Bjäresjö parish west of Ystad. The most normal farm size was between 30 and 50 ha. This can be contrasted with the farms in the villages of Bjäresjö (Fig. 4.1:10) and Kabusa (Fig. 4.1:11). In Bjäresjö most of the farms, with the exception of the vicarage and one other large farm, were about 30 ha. In Kabusa the farms were bigger, most of them falling in the 30 to 50 ha range. The diagrams also show different proportions of arable and meadow: in Bjäresjö the total area of farms, including both arable and meadow, was roughly the same as the total area of arable alone in Kabusa. The villages of Stora and Lilla Köpinge have similar proportions of arable and meadow. Fig. 4.1:12 presents the figures for the size of farms in the villages in question; it shows that arable land made up between 75% and 90% of the mapped area of the villages. Some 25–35% of these fields were ploughed annually, which means that 65–75% of the village lands consisted of fallow or meadowland. An estimated 30–50% of the arable was sown each year. The share of the arable that was sown annually was roughly the same in proportion to the total area of the village in the villages situated on the more clayey soils in the Ystad area. In Stora Köpinge parish, however, the oat lands were smaller than rye and barley lands: the proportion in Stora and Lilla Köpinge was 33:67, compared with 50:50 in, for example, the villages of Bjäresjö and Lilla Tvären west of Ystad town (Persson 1986a,b, 1987b, 1988). The oat lands in the Köpinge area were also alternately planted with buckwheat. Moreover, grain cultivation in the Köpinge area was pursued over larger areas on account of the extensive cultivation methods, which explains the large areas of arable, which comprised buckwheat land to a not insignificant extent (33% according to the map description of Stora Köpinge in 1767). In Lilla Köpinge the meadows were also periodically sown with buckwheat.

The cultivated land through time. Meadowland became arable in the 18th century, so there must have been an expansion of the acreage used for growing cereals. The striking changes in settlement and population can only be explained if there was an increase in production. However, it is difficult to judge the extent of this, since sources from the start of the 19th century give us no information about the proportion of arable that was ploughed each year. It must suffice here to point out that the proportions of meadow and arable changed to some extent. In Kabusa the arable was about 75% in

1698; in 1808 it was an estimated 90%. In Stora Köpinge village the arable was about 80% in 1732 and 1767, rising to ab. 90% in 1810. In Lilla Köpinge village there seems to have been only a marginal growth in arable land in the 18th century; its share was about 90% in 1690 and in 1803. In view of the system of extensive cultivation – the best arable land was already cultivated at the start of the 18th century – these changes cannot have involved a doubling of production. There must therefore have been a social and economic differentiation in the parish during the 18th century as a consequence of the doubling of the population. It is significant that of the 76 cottages in the parish in 1805, only one cultivated grain.

Conclusions

The Köpinge area can be seen as a marginal agrarian area during the period under study here. The peasant population here farmed land of low productivity, and the conditions for a tolerable life were worse than in other parts of the Ystad area (see Ch. 5.9). The area was more sparsely populated than other parishes in the Ystad area. The 18th century saw a heavy increase in population. Since a large share of the people could not earn their livelihood on farms, whether as farmers or servants, they were forced to try to survive in cottages as day-labourers, craftsmen, or beggars. A new lower class had emerged. The scanty resources of the area did not permit any division of the farms or the establishment of new crofts cultivating their own plots. It was not until the beginning of the 19th century that the population density (28 km^{-2}) passed the figure that Västra Nöbbelöv had reached a century earlier (24 km^{-2}), but by then the population density in Västra Nöbbelöv had risen to 42 km^{-2}. These comparisons reinforce the picture of a district where it was primarily the natural conditions that dictated development in the cultural landscape and population in the 18th century and before (Ch. 4.1.6.1). Of special interest here are the findings which show that farming in this marginal area was carried on in a way that ensured a well-balanced circulation of nutrients (Ch. 5.3).

4.1.7.2 Landscape and land use in the Köpinge area according to land survey documents from around 1700

Mats Riddersporre

Our first detailed information about the landscape and its use comes from around 1700. From this time onwards there exist large-scale maps with descriptions of most of the villages in the Köpinge area. On the basis of this material it is possible to make a thorough reconstruction of the early 18th century landscape. Fig. 4.1:13

shows the parishes of Stora Herrestad and Stora Köpinge around 1700. It shall be noticed, however, that the oldest map of the village of Svenstorp was made in 1809–10. Fig. 4.1:14 shows the central part of Stora Köpinge on a more detailed scale.

At the beginning of the 18th century the eastern boundary of the hundred of Herrestad followed the river Nybroån (cf. Ch. 4.1.6.1). This boundary was also the boundary between the counties of Malmöhus and Kristianstad. Thus two villages in the parish of Stora Köpinge – Stora Köpinge and Kabusa – belonged to the hundred of Ingelstad and the county of Kristianstad.

Ownership and settlement structure

The parish of Stora Herrestad consisted of one large village with about 65 units, farms and cottages, and a manor. The parish of Stora Köpinge, on the other hand, included five villages with a total of about 75 units. In Stora Herrestad a greater part of the farms and cottages belonged to the manor. In Stora Köpinge there was no manor at this time, but a majority of the farms were owned by the nobility and belonged to manors in the surrounding parishes (cf. Ch. 4.1.6.1).

Apart from the village farms there existed in Stora Köpinge parish, as separate units, also a couple of millers' farms. The watermill at Barevad was deserted, but its fields were cultivated by the miller in Stora Köpinge village. The mill at Barevad, with its fields, was part of neither Lilla Köpinge nor Stora Köpinge village. The mill in Stora Köpinge, on the other hand, was a farm in the village and had its fields intermixed with the other farms. Documentary evidence is lacking for the mill in Svenstorp, but it seems to have had some grassland along the stream. The Munkamöllan mill belonged to the neighbouring parish of Tosterup. In Stora Herrestad there was no watermill. A windmill here belonged to a farm owned by the manor.

Fields and fences

The arrangement of fields and fences was adapted to the three-field system. All the villages, except Kabusa (see below), had their grain fields within three separated fields, *vångar*. For the villages west of the river Nybroån it can be seen that there was cooperation in the construction of permanent fences. The purpose of the fences was mainly to keep animals grazing in the fallow field away from the growing crop in the other fields. If neighbouring villages managed to adapt their field boundaries and the year of fallow, there would be no need for fencing of adjacent fields from different villages. Instead the cattle could graze over larger areas, and the farmers could save the troublesome and expensive work of building fences. This form of inter-community cooperation was called *vångalag*. There would be an extra incitement for such cooperation in an area like Köpinge. It is pointed out in the descriptions that the

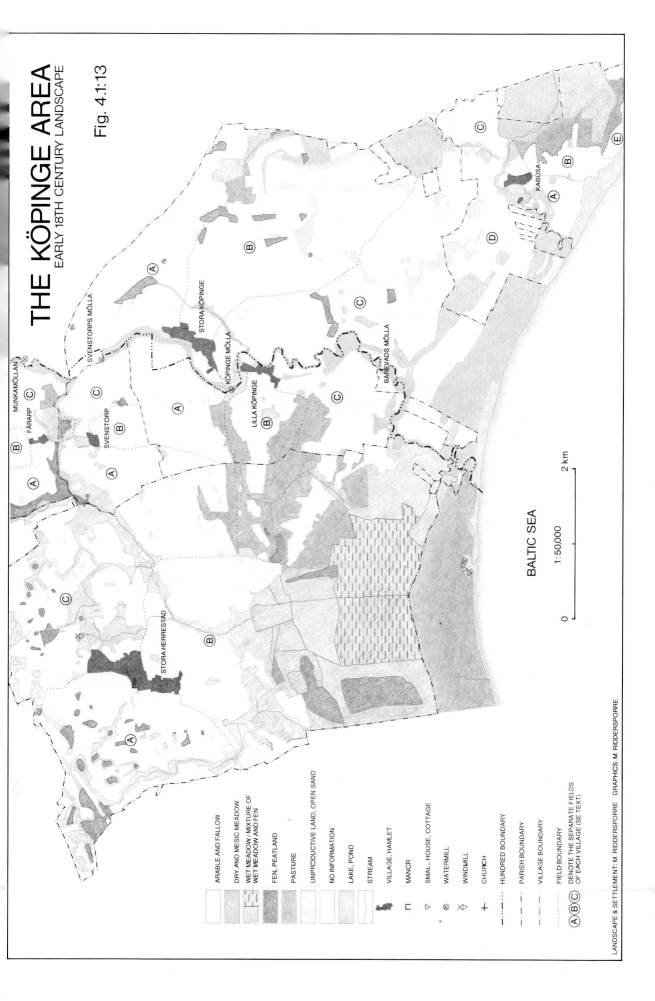

THE KÖPINGE AREA
EARLY 18TH CENTURY LANDSCAPE

Fig. 4:1:13

BALTIC SEA

1:50,000

0   2 km

ARABLE AND FALLOW

DRY AND MESIC MEADOW

WET MEADOW / MIXTURE OF
WET MEADOW AND FEN

FEN, PEATLAND

PASTURE

UNPRODUCTIVE LAND, OPEN SAND

NO INFORMATION

LAKE, POND

STREAM

VILLAGE, HAMLET

MANOR

SMALL HOUSE, COTTAGE

WATERMILL

WINDMILL

CHURCH

HUNDRED BOUNDARY

PARISH BOUNDARY

VILLAGE BOUNDARY

FIELD BOUNDARY

(A)(B)(C)   DENOTE THE SEPARATE FIELDS
OF EACH VILLAGE (SE TEXT)

LANDSCAPE & SETTLEMENT: M. RIDDERSPORRE   GRAPHICS: M. RIDDERSPORRE

MUNKAMÖLLAN

FÅRARP

SVENSTORPS MÖLLA

SVENSTORP

STORA KÖPINGE

KÖPINGE MÖLLA

LILLA KÖPINGE

BAREVADS MÖLLA

STORA HERRESTAD

KABUSA

villages lacked all kinds of wood and timber, for fences as well as for building and fuel. The area also almost totally lacks stones for stone walls.

Apart from pasture areas, all lands were subdivided into strip parcels. An exception was the manor in Stora Herrestad, which also had some large block-shaped parcels. In all the villages it is possible to identify *bolskifte*, a south Scandinavian system of subdividing fields, which is of medieval origin (Hannerberg 1958, 1960, Andersson 1959).

Intensive and extensive land use

Merely by looking at the surveyors' maps one can get the impression of very intensive land use, almost all land being arable. This is in a way a false picture. Within the arable there were great differences in the quality of soil, the intensity of use, and the crop that was grown. In general the fields near the village cores were most intensively used. Here we find the proper three-field rotation with two years of grain – rye and barley – followed by one year of fallow. Here we also find the good soils with a high content of mould, a result of frequent manuring over a long time (Riddersporre 1986). At a more distant location we find fields that were used with longer periods of fallow. These fields were sown with oats, rye, or buckwheat and were seldom or never manured (cf. Hansen 1976, Steensberg 1976:96–97). For parts of Stora Köpinge parish it is possible to reconstruct these relations at a very detailed level (Fig. 4.1:14).

The situation in Stora Köpinge village is most illustrative. The three-field soils are found near the village core in the originally fenced-in fields, referred to as *Hemmavångarna* (the Homefields). In the periphery of the Homefields, and on the rest of the arable, the crop rotation was longer than three years. In the area outside the Homefields the period of fallow could be up to 20 years. What can be seen is an infield-outfield system – that is: an intensively cropped core area, the infield, and an outer area also under grain but in a more extensive form, the outfield – where the outfields spatially correspond to the reconstructed pasture commons of the Middle Ages (cf. Ch. 4.1.6.2, Fig. 4.1:9). Due to sandy soil conditions the outfield area was rather large. The situation is in principle the same in the other villages, allthough in Lilla Köpinge all land, including pasture, had been fenced in. It is explicitly stated that there was no pasture here, other than that in the fallow field.

The situation in Kabusa is more complex. Three of the farms had all their arable in the field west of the river Kabusaån. Furthermore, there were some strips here that belonged to a farm in Stora Köpinge. This field (D in Fig. 4.1:13) has been interpreted as the former infield of a single farm that may have existed in the Early Middle Ages (Ch. 4.1.6.2). The rest of the farms in Kabusa had their arable in the four fields east of the stream. Three of these were used with a rotation

of three years. The fourth field (E in Fig. 4.1:13), with longer fallows, was probably taken in from common land. It was situated on Hammars Backar, a longish glaciofluvial plateau more or less covered with aeolian sand.

Areas for harvesting hay in Stora Köpinge parish were very small. The impression is that meadows only existed where it was not possible to produce grain. Moreover, hay-meadows situated within fields in three-year rotation could only be harvested two out of every three years. There are no indications that hay was produced on the arable that was fallowed for long periods. When fallowed, these areas were used for grazing.

Environmental conditions and the ecological balance of land use in this area are crucial aspects; these are set out in Ch. 4.1.7.3.

4.1.7.3 Environmental resources and subsistence agriculture in Köpinge in Early Modern Time

E. Gunilla A. Olsson

Landscape and vegetation

The landscape characteristics and the differences in Köpinge compared with other parts of the Ystad area, which were already visible in the Middle Ages, were amplified by the early 18th century. The geomorphological conditions clearly set limits for land use. The flat landscape was wide open, treeless, and dominated by three elements: grain fields, grasslands, and sand. Grain fields, predominantly for rye, took up about a third of the infields. Most of the arable was fallow, grazed by cattle, geese, and pigs. The fallows in this region were generally of long duration, up to 30 years on the harshest sandy soils (Riddersporre 1985). Short-term fallow vegetation was composed of a variety of weeds, but was replaced on the long-term fallows by heath vegetation dominated by grasses and *Calluna vulgaris*. Along the shore, areas of sand drift and grass-dominated heathlands prevailed.

Öja-Herrestads Mosse was the most high-yielding and valuable winter fodder area. To make use of the wettest, central parts, small-scale, cautious drainage had started and a sparse grid of open ditches crisscrossed the fenland. The vegetation here was a mosaic of fen species like rushes and sedges, mire vegetation with mosses intermingled with meadow plant communities with herbs and grasses. The river valleys transversing the area were bordered by narrow alluvial terraces maintaining valuable hay-meadows. The outlets of the rivers Nybroån and Kabusaån were surrounded by wet meadows, regularly flooded and with probably botanically rich plant communities. A relatively large hay-meadow on clay soil dominated the landscape northeast of Kabusa village. The information on landscape

LAND USE IN KÖPINGE
CA.1700
Fig. 4.1:14

STORA KÖPINGE

LILLA KÖPINGE

KABUSA

- - - - - FIELD BOUNDARY

--#--#-- FIELD BOUNDARY, FENCE (ACC. TO MAP)

-#--#- OTHER FENCE (ACC. TO MAP)

DRY AND MESIC MEADOW

WET MEADOW

FEN, PEATLAND

PASTURE

UNPRODUCTIVE LAND,
OPEN SAND

ARABLE

THREE-YEAR ROTATION:
BARLEY – RYE – FALLOW

THREE-YEAR ROTATION:
OATS – OATS / FALLOW – FALLOW

3–8 YEARS FALLOW

9–20 YEARS FALLOW

0   1:25,000   500 m

LANDSCAPE & SETTLEMENT: M. RIDDERSPORRE   GRAPHICS: M. RIDDERSPORRE

and vegetation above comes from 18th century land survey maps with descriptions, Stora Köpinge and Stora Herrestad, see Ch. 2.2.

Ecological aspects of land use

All the villages in this region practised three-course rotation, although with modifications. Rarely more than 40% of the arable was used for annual cultivation (Ch. 4.1.7.1) implying that most of the fields were fallows of different ages. Fallow periods of more than about five years were accompanied by the development of plant communities, which, when exposed to continuous grazing pressure, developed into regular pasture vegetation. The plant communities were composed of low nutrient demanding species like *Festuca ovina* (sheep's fescue), *Nardus stricta* (mat-grass), and *Calluna vulgaris,* all giving a low-nutrient quality fodder. Intensive grazing, overgrazing, probably contributed to the formation of the large drift-sand areas in the shore areas of Kabusa, as well as in several areas in Stora Köpinge. Turf-cutting for fencing purposes also exaggerated the formation of open sand areas (Linnaeus 1751).

Rye and oats were the predominant crops which is to be expected considering the sandy soils here. Barley was grown to a small extent and restricted to the small areas of humus-rich soil near the village core. In Stora Köpinge buckwheat was cultivated on one-third of the arable (Riddersporre 1985); this crop is very modest in nutrient demand and responds with reasonable yields even on poor soils. Shortage of manure owing to the low number of cattle permitted the fertilizing of arable land to only a limited extent. In Kabusa village only 39% of the total arable was manured each year (land survey map, Kabusa 1697). In this village seaweed was also used as fertilizer, and generally all the villages recirculated used thatch and different kinds of building material as manure (cf. Dodgshon and Olsson 1989). Buckwheat and oats never received manure, and barley and rye were manured according to the fallow periodicity. For rye this varied from regular three-course rotation, implying the addition of manure every third year. However, rye could also have been cropped here just 2 out of 20 years, with manure addition only once in 20 years. Grain yields were significantly lower here than in other parts of the Ystad area, reaching only half of the amounts for villages on lime-rich soils (Ch. 4.1.7.1).

The area of meadow was very small in all the villages in Stora Köpinge parish, amounting to at most 25% of the infield area (Fig. 4.1:12). This contrasts with the conditions in Stora Herrestad, which was well provided with fodder areas. The description accompanying the land survey map of Stora Herrestad from 1704 notes that this parish has "fully enough grazing lands and also peat deposits". The highest meadow production recorded in the Ystad area is found in the large meadows of Öja-Herrestads Mosse, reaching up to 2200 kg ha^{-1} yr^{-1}. The wet and flooded meadows at the river outlets belong to the same yield class (land survey maps, Stora Herrestad 1704). For Kabusa, on the other hand, more than 55% of the meadows belong to the lowest fertility class, yielding annually about 270 kg hay ha^{-1} (land survey map, Kabusa 1699).

The small fodder areas in Köpinge did not permit a large livestock. Data from the late 17th century (1671) show that the villages kept on average only one or two cows per farm, and oxen were used for most of the agricultural work (Persson unpubl.). The land survey maps from the 18th century also explicitly show that the poor vegetation on many of the fallows did not permit other foraging animals than geese and pigs. The number of livestock in Stora Köpinge parish was roughly equal in all the villages counted per infield area (Ch. 4.1.7.1). The supply of manure, based on the data from 1671 (Persson unpubl.), calculated as available nitrogen for the arable soils would have been around 10 kg ha^{-1} yr^{-1} (cf. Olsson 1989). This is a very low value, but it is reflected in the low crop yields. Nutrient balances for the arable systems in Kabusa and Stora Köpinge show that, in spite of the low levels of plant nutrients added and the leaching disposition of the sandy soils, the recirculation of nutrients appears to have been acceptable and the systems balanced, although at very low plant nutrient levels (Ch. 5.3). Obviously, the use of long-term fallows, crops with modest demands for nutrients, and a relatively low density of human population was able to balance the harsh environmental conditions for agriculture in Stora Köpinge parish.

For Stora Herrestad the environmental conditions were quite different. Besides the big, high-yielding meadows of the large fenland, the village possessed the fertile meadows along the outlet of the river Nybroån. Pastures of different kinds also existed, both grasslands with scattered shrubs and low-productive areas with rough grasses and heather on the sand-hills close to the shore (Fig. 4.1:13). The favourable fodder situation in combination with arable on boulder clay soils provided Stora Herrestad with rich prerequisites for agriculture with a different production potential from the neighbouring Stora Köpinge.

4.1.8 The 19th and 20th centuries
Landscape and settlement changes in the Köpinge area since 1800

Tomas Germundsson and E. Gunilla A. Olsson

Landscape and vegetation

The landscape of the Köpinge area around 1815 (Fig. 4.1:15) had not changed much compared with the early 18th century (Fig. 4.1:13). At this time the enclosure process had started but was not yet visible in the landscape. An intensification of agriculture compared with the preceding century can be traced in the transformation of meadow areas into arable near the villages cores of Fårarp, Svenstorp, and Lilla Köpinge. However, the reverse occurred south-east of Lilla Köpinge, where some arable and meadow areas on sandy soil became grazing land. A shortage of winter fodder related to an increasing number of cattle is reflected in the tapping of the small lake Fårarpssjön and the increased drainage of Öja-Herrestads Mosse. By 1820 this wetland contained an elaborate network of drainage channels. The general pronounced shortage of fuel and the increasing exploitation of turf deposits is visible as several water-filled ponds in Öja-Herrestads Mosse and in the large wet meadows north-east of Kabusa.

Landholding and enclosure

At the beginning of the 19th century most of the Köpinge area (Fig. 4.1:15) was possessed by estate-owners, while only small parts were owned by freeholders. As in Scania in general, most of the villages were still nucleated at this time. The scattered farms that did exist were the result of early enclosures in villages situated on estate-owned land. This can be seen on the domain of Öja estate (Fig. 4.1:15.), and in the eastern part of the mapped area, on the domain of Bollerup estate situated just outside the study area. In the case of Herrestad estate, the traditional nucleated shape of the village still persisted, with only some exceptions just south of the village. The four smaller villages of Fårarp, Svenstorp, Lilla Köpinge, and Kabusa were not under estate domination in the early 19th century. They lay in a north-south oriented corridor of peasant-owned land between the estate demesnes on both sides.

A typical feature of the progress of the enclosure movement in Scania was that, generally, villages dominated by freeholders were enclosed quite early on, preceded only by a relatively small number of villages on estate land. Here the enclosure was autocratically decided by the estate-owner (Fridlizius 1979) – as in the examples mentioned above. After these pioneering achievements, and a law concerning the *enskifte* enclosure in 1803 (Ch. 3.8.1), a movement of land redistribution within the villages began, primarily initiated by the peasants themselves. The *enskifte* enclosures of the four peasant-owned villages mentioned in the Köpinge area were carried out in 1819 (Fårarp), 1825 (Svenstorp), 1812 (Lilla Köpinge), and 1838 (Kabusa). The result of these changes was a dispersed settlement pattern which superseded the nucleated villages (Fig. 4.1:16).

Ecological aspects of land use

By 1850 enclosures had been carried out in all villages in the area. Land-use changes like the cessation of the three-course rotation in the villages and the change to

individual crop sequences were also followed by changes in cultivated crops. In 1850 rye was still the predominant grain crop in Stora Köpinge (33% of the arable), but pulses were grown on 9% of the arable and the newly introduced potato accounted for 8% (Germundsson and Persson 1987). Sugar-beet was introduced by the end of the century, Köpinge being the core area for this crop. The total arable remained stable from the late 19th until the 1960s (Fig.4.1:17). The situation was slightly different in Stora Herrestad. By 1850 this was the main area for potato cultivation within the Ystad region, with 13% of the arable being used for this crop. A peak for arable in Herrestad was reached around 1925, and the present area is approximately the same.

Crop yield rose almost threefold during the period 1700–1850 (cf. Germundsson and Persson 1987, Ch. 3.9.2). This was a consequence of higher nutrient additions made possible by larger stock size, and by increased pH values for the soil as a result of marling and liming. Soil nutrient leaching was diminished by the shorter fallow periods related to the new crop sequences and by the introduction of ley cropping. Available soil nitrogen levels also rose as a result of the increased cultivation of nitrogen-fixing crops like vetches, peas, and beans.

Changes in stock composition were most dramatic for cattle, which increased as a proportion of the stock in Stora Köpinge from 9% around 1700 to 40% at the end of the 19th century (Fig. 4.1:19). Fodder for the increasing number of cattle was obtained from leys with clover and vetches. Oxen were replaced by horses as working animals; by 1855 they amounted to 5.7% of the total stock (Germundsson and Persson 1987) and were extinct by 1886 (Anonymous 1886–1909). When grazing possibilities on the former fallows disappeared sheep decreased in number. Stock composition was different in Stora Herrestad where cows dominated in 1850 (28% of total stock; Germundsson and Persson 1987). Oxen persisted longer in this parish, which contains considerable areas of heavy clay soil.

Population changes

Population development during the 19th century was dynamic. The typical pattern for the Swedish countryside was a considerable population growth from the middle of the 18th century up to around 1900. In estate-dominated areas, however, it has been found that the peak was reached earlier (Jeansson 1966), and in the Ystad region as a whole the population maximum occurred in the 1880s (Germundsson 1987). Population growth was often especially strong in areas where land was sold off from the estates to peasant owners (Fridlizius 1975b), and consequently many new settlements were established here.

Population growth brought about extensive partitioning of farms which was also made possible by more efficient farming techniques. During the whole period

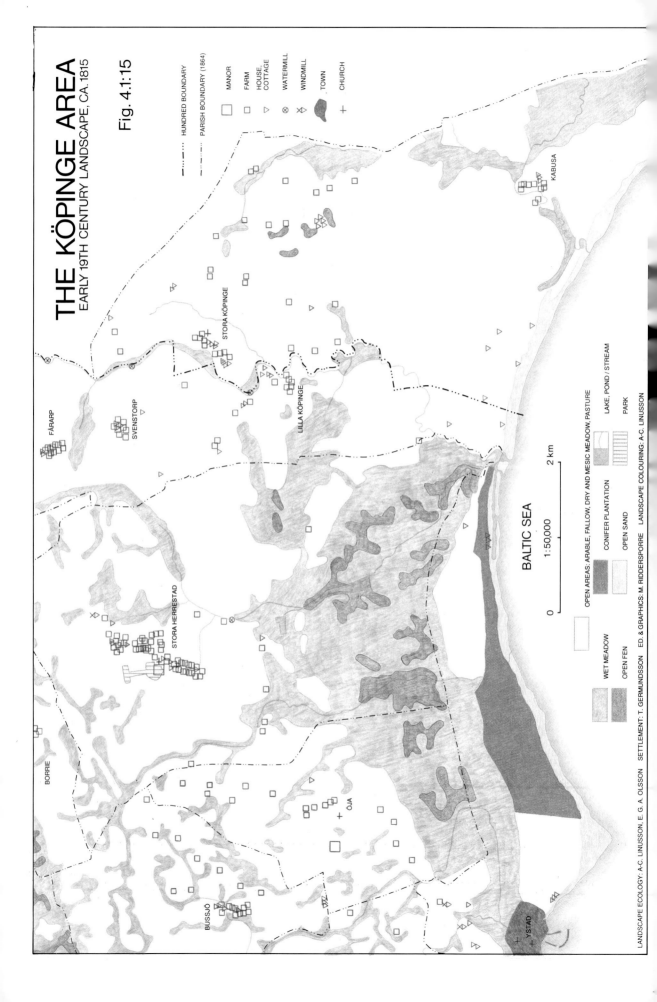

THE KÖPINGE AREA
EARLY 19TH CENTURY LANDSCAPE, CA. 1815

Fig. 4.1:15

———··· HUNDRED BOUNDARY
———··——·· PARISH BOUNDARY (1864)

◻ MANOR
▫ FARM
▷ HOUSE, COTTAGE
⊗ WATERMILL
⋈ WINDMILL
⬤ TOWN
✝ CHURCH

KABUSA

FÅRARP

SVENSTORP

STORA KÖPINGE

LILLA KÖPINGE

STORA HERRESTAD

BORRIE

OJA

BUSSJÖ

YSTAD

BALTIC SEA

0   1:50,000   2 km

OPEN AREAS: ARABLE, FALLOW, DRY AND MESIC MEADOW, PASTURE

WET MEADOW

OPEN FEN

CONIFER PLANTATION

OPEN SAND

LAKE, POND / STREAM

PARK

LANDSCAPE COLOURING: A-C. LINUSSON

LANDSCAPE ECOLOGY: A-C. LINUSSON, E. G. A. OLSSON   SETTLEMENT: T. GERMUNDSSON   ED. & GRAPHICS: M. RIDDERSPORRE   LANDSCAPE COLOURING: A-C. LINUSSON

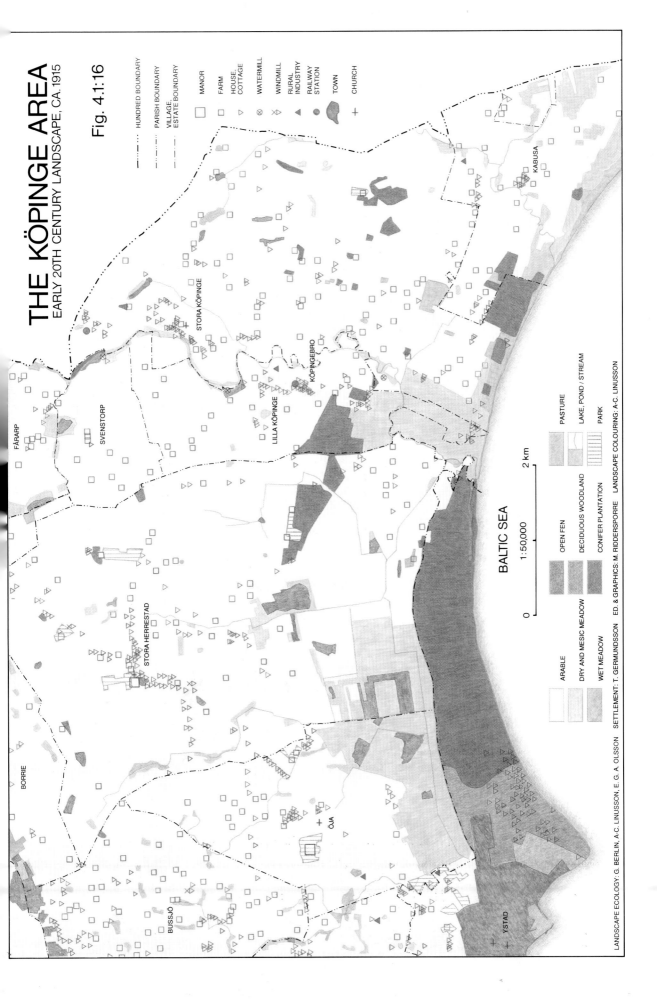

THE KÖPINGE AREA
EARLY 20TH CENTURY LANDSCAPE, CA. 1915

Fig. 4.1:16

- ---··· HUNDRED BOUNDARY
- --··· PARISH BOUNDARY
- --·· VILLAGE,
  ESTATE BOUNDARY

☐	MANOR
☐	FARM
▷	HOUSE, COTTAGE
⊗	WATERMILL
✕	WINDMILL
▲	RURAL INDUSTRY
●	RAILWAY STATION
⬛	TOWN
+	CHURCH

BALTIC SEA

1:50,000

0   1 km   2 km

ARABLE	
DRY AND MESIC MEADOW	
WET MEADOW	
OPEN FEN	
DECIDUOUS WOODLAND	
CONIFER PLANTATION	
PASTURE	
LAKE, POND / STREAM	
PARK	

FÅRARP

SVENSTORP

STORA KÖPINGE

LILLA KÖPINGE

KÖPINGEBRO

BORRIE

STORA HERRESTAD

BUSSJÖ

ÖJA

YSTAD

KABUSA

LANDSCAPE ECOLOGY: G. BERLIN, A-C. LINUSSON, E. G. A. OLSSON   SETTLEMENT: T. GERMUNDSSON   ED. & GRAPHICS: M. RIDDERSPORRE   LANDSCAPE COLOURING: A-C. LINUSSON

KÖPINGE

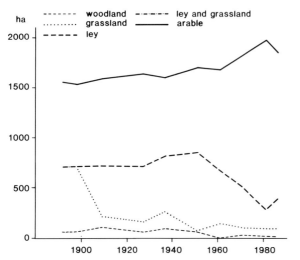

Fig. 4.1:17. Land use in Stora Köpinge parish during the period 1886–1986: arable, grasslands, and woodlands. Source: Agricultural statistics, unpubl. (1886-1909), and publ. (1909-1986) – see Ch. 2.2.

from 1815 to 1915 land was sold off from the estates. Within Stora Herrestad parish, however, a more extensive selling of estate land did not begin until around 1875. When this process started it attracted numerous new settlements, and between 1890 and 1920 the parish experienced a population growth, due to in-migration. At the latter point of time areas with low density of settlement coincide well with estate-owned areas (north-west of Herrestad estate and the eastern part of Herrestad parish). Naturally, settlements were also sparse within the large peatbog area (Öja-Herrestad Mosse) (Fig. 4.1:16).

The railway and agricultural industry

Another aspect of landscape development in the south-eastern part of the region was the introduction of the railway between Ystad and Eslöv in 1865. This line connected Ystad to the main line between Stockholm and Malmö. The railway ran through the western parts of Stora Köpinge parish, with stations in Köpingebro and Svenstorp, and at these places built-up areas with functions connected to the railway developed.

Such a process was very typical of the Swedish countryside in the early railway era, and it was at many places the first cautious step towards urbanized life in this country. The effects of the Ystad-Eslöv railway, within the project area, can partly be seen as the dense settlement in the places mentioned above. The land-holding situation played a great role in the development of the railway network, and the settlements connected to it. It is therefore significant that built-up areas with manifold functions came into being only around the

railway stations mentioned above, as this was land not possessed by estate-owners.

On estate-controlled demesnes the development was different. In 1874 a railway between Malmö and Ystad was opened, and it was to a great extent financed by manorial estates adjacent to the line. Stations were erected close to the estates, but here – in sharp contrast to what happened on non-manorial land – only a few buildings, necessary to handle passengers and goods, were built. Owing to the estate-owners' control of the land, no spin-off effects occurred in settlement or activities (Möller 1987).

Another trait characterizing the south-eastern part of the project area is the development of agricultural in-

KÖPINGE

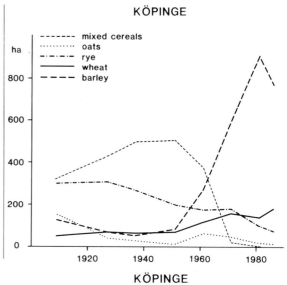

KÖPINGE

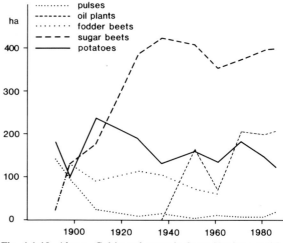

Fig. 4.1:18. Above. Cultivated crops in Stora Köpinge parish during the period 1909-1986: grain crops. Source: Agricultural statistics, unpubl. (1886-1909), and publ. (1909-1986) – see Ch. 2.2. – Below. Cultivated crops in Stora Köpinge parish during the period 1886-1986: oil plants, pulses, and root crops. Source: Agricultural statistics, unpubl. (1886–1909), and publ. (1909–1986) – see Ch. 2.2.

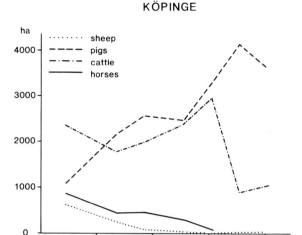

KÖPINGE

Fig. 4.1:19. Livestock in Stora Köpinge parish during the period 1909–1986. Source: Agricultural statistics, unpubl. (1886–1909), and publ. (1909–1986) – see Ch. 2.2.

dustry around the turn of the century. During the 1890s a sugar mill was constructed in Köpingebro and together with some other industrial activities, including the construction of a new railway branch, this made up the base for the development of a small agro-industrial town in Köpingebro. This in turn meant that the population in the parish was stable from around 1900 up to the 1950s. In most other parishes in the area this was a period of steady out-migration (see Fig. 4.1:20).

To sum up, the settlement pattern shown in Fig. 4.1:15 represents a farming structure where much land had been sold off from the estates and where the number of cultivation units had reached a peak. As a result of modernization in agriculture the land was intensively used and the production was oriented towards the market. In conjunction with the railway, agricultural industries and other settlements were built. In areas still cultivated by the estates, settlements were very sparse, and agriculture there was managed directly from the estates.

Landscape and land use in the early 20th century

The map from 1915 (Fig. 4.1:16) shows the physical result of the land enclosures. Specialization in agricultural management is seen in the increasing cultivation of root crops followed by potatoes, sugar-beet, and mangel. Among grain crops, rye loses its dominance to mixed grain (Fig. 4.1:18). Oats and pulses have decreased since the 1880s. Ley cultivation increased just slightly during the same period. The changes in cultivated crops must be related to changes in stock composition, where cattle increased eightfold since 1850, but were outnumbered by pigs already in 1909 (Fig. 4.1:19).

The decrease of arable area during the 1970s is related to the transformation of arable land on sandy soil

south and west of Barevadsmölla into a golf course and also enclosures for military purposes. Several small meadows have also been ploughed, but the great grassland reductions in this area had already occurred in the 18th century. Landscape changes have been least obvious in Kabusa, where considerable grassland areas still prevail.

A great change since the early 19th century is the efficient drainage of Öja-Herrestads Mosse. Only fragments remain of the once extensive fens and wet meadows. On an economic map from 1915 most of the wetlands are classified as "mesic meadows" used for mowing, and some areas are even ploughed and used for cereal cropping. The other spectacular landscape change is the large coniferous plantations covering pastures and drift-sand areas bordering the sea and on former meadows and pastures west of Köpingebro. Mixed deciduous woodland with oak, birch, and willows is found in the sandy areas east of Ystad. Large gardens with groves of deciduous trees have also been created at the recently developed estates of Köpingsberg, Fredriksberg, and Karlsfält. Small coniferous plantations are scattered all over the Köpinge area. Along Nybroån riverside woodlands with *Alnus glutinosa* (black alder) and *Fraxinus excelsior* (ash) have developed on former wet meadows. Woodland, including the coniferous plantations, increased around the turn of the century from almost zero to 10% of the land area in Stora Köpinge parish. We must return to the Middle Ages to find an equal amount of woodland in this area.

Creation of smallholdings

One feature, quite usual in the Swedish countryside in the early 20th century, when the rural population was still growing, was the creation of smallholdings with the help of special loans from the state (Ch. 5.12). These small owner-occupied farms – *egnahem* – were mostly insufficient to support a family the year round, and often some of the family members had to work at other places, such as rural industries, estates, or other big farms. The phenomenon was not unique to Sweden,

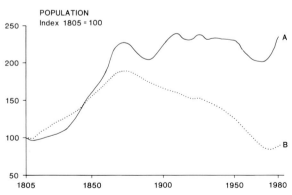

Fig. 4.1:20. Population in Stora Köpinge parish (A) and the Ystad area (B) 1805-1980. Sources: parish registers, official statistics.

having parallels in other European countries. The location of these smallholdings was influenced by the supply of land suitable to subdivide into new plots. It is therefore not unusual to find colonies of *egnahem* on previous estate-owned land, and one illustrative example is found in the parish of Stora Köpinge.

Fig. 4.1:21a shows the large tenant farm (*plattgård*) of Köpingsberg as it looked in 1915. Ten years later the owner of this big farm split up the unit into 28 parts – one large section around the main buildings and 27 smaller plots to be sold as *egnahem*. Some of the new plots were amalgamated, and in total 22 *egnahem* were established here 1924–25. These are still visible on the modern topographic map (Fig. 4.1:21b). The plots were quickly bought by smallholders with the help of loans from the state, and the sale was probably a good stroke of business for the landowner. Even if these tiny holdings (3–5 ha) functioned as small family farms (made possible primarily by the cultivation of sugar-beet) in the 1920s, it soon became obvious that it was impossible to make a living on them and also pay back the loans. When the economic crisis set in during the 1930s the authorities were forced to offer new loans to many *egnahem*-owners, and also a lot of these places were sold. Today only a few of them function as agricultural units, and only with greatly enlarged arable land. Most of the houses are inhabited by elderly people or people working elsewhere (Germundsson 1985).

The formation of today's agro-industrial landscape

The specialization in root crops where sugar-beet has dominated since the 1920s has continued (Fig. 4.1:18). Cultivation of mixed grain has been successively replaced by barley since the 1960s. Oil plants have been greater in extent than potatoes since 1960. Today large-scale cultivation of vegetables like onions is important and well adapted to the light and sandy soils here. Ley cultivation occupies the third largest share of the arable. It has remained relatively stable, accounting for about 30% of the arable which is the same magnitude as in the Romele area (Ch. 4.4.8). Following the general Swedish trend, leys decreased in Köpinge in the 1960s, but increased again and have now about the same extent as sugar-beet (Figs 4.1:17 and 4.1:18). Cattle-raising increased until 1960, but has stabilized at a lower level since the 1970s (Fig. 4.1:19). It is still important, however, and directly coupled to the extensive ley cropping. Sheep and horses disappeared around 1960 and pigs have been dominating in number since the 1920s (Fig.4.1:19).

To generalize, the existing grasslands in Stora Köpinge parish today belong to two rough categories: military grounds – along the shore in Kabusa and southeast of Köpingebro – or golf courses. The military lands are, however, used as pastures for beef cattle most of the time, and subjected to high loads of fertilizers. Öja-Herrestads Mosse has finally met its end, and today

almost nothing remains of fen and wet meadow vegetation. Most of the area contains fertilized leys of varying ages and is used for extensive cattle and racehorse grazing. Parts of the former fenland have also been incorporated into the expanding golf course in the southeast. Natural grassland vegetation is more or less extinct in all this area. Remaining fragments of unploughed grassland are heavily affected by nitrogen fertilizers and vegetation is dominated by nitrophiles, weeds, and ley species (Ch. 5.11). Former hay-meadows on alluvial terraces along the river Nybroån are not kept up today. This vegetation has changed from species-rich meadow communities to monocultures of tall clonal herbs like *Petasites hybridus* (butterbur) and *Urtica dioica* (nettle). Before the introduction of today's agro-industrial landscape, exclusive plant communities of high conservation value containing dry meadow and grass-heath vegetation were characteristic of the large sandy areas in this region. Representative plant species in these communities are *Corynephorus canescens* (grey hair-grass), *Helichrysum arenarium* , *Artemisia campestris* (field wormwood), *Erigeron acer* (blue fleabane), and *Sedum acre* (biting stonecrop). Very minute plots of this vegetation still exist south-west of Köpingebro.

During the 20th century the woodland area again decreased. Many of the plantations from the turn of the century have been cleared without new planting. This is partly due to the creation of planned housing areas in the former shore plantations, as at Nybrostrand. Deciduous woodlands are extremely rare, but in two areas close to Nybroån, at Barevadsmölla and east of Lilla Köpinge, birches and willows have invaded unmanaged grasslands (Fig. 4.1:22). Several stretches along Nybroån are bordered by narrow strips of riverside communities with black alder, some ash and nitrophilic herbs like *Heracleum sphondylium* (hogweed), *Urtica dioica*, *Anthriscus sylvestris* (cow parsley), and *Impatiens parviflora* (small balsam) (Bengtsson-Lindsjö unpubl.).

Landholding and settlement today

The estates in the Köpinge area hold about the same area of land as they did in 1915. Today the estates are highly rationalized production units, as reflected in the huge fields characterizing their domains. Of course, family-based farming has also undergone profound changes, in connection with rationalization and amalgamation. But even though changes in agriculture have been radical since 1915, the settlement pattern has not changed to a corresponding degree (Fig. 4.1:22). Most of the buildings in the area constructed during the period of growth in the 19th and early 20th century still exist, although with new functions and not necessarily inhabited by people working in agriculture.

Comparing the map from 1915 (Fig. 4.1:16) with the one showing the settlement today (Fig. 4.1:22), the biggest change is the creation of built-up areas. This is

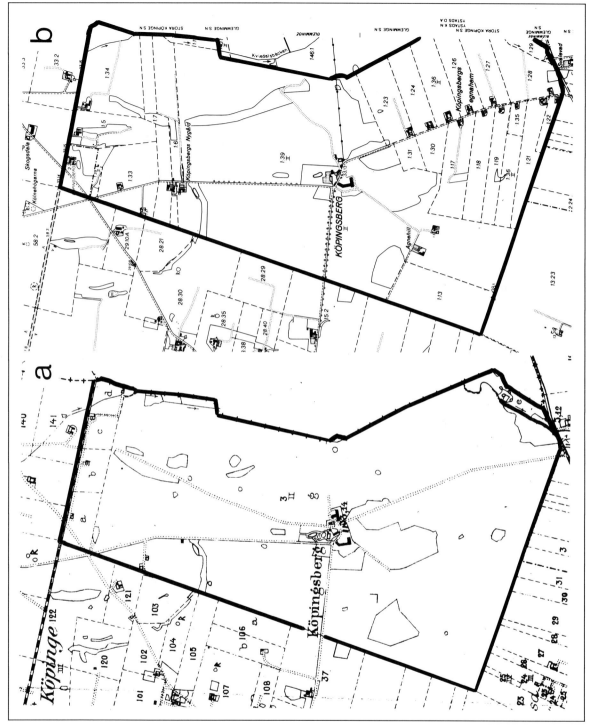

Fig. 4.1:21. Settlement in Köpingsberg 1915 (a) and 1973 (b). Sources: economic map 1915 and economic map 1973. Scale 1:20,000. From Germundsson 1985.

particularly obvious in Nybrostrand on the coast, which is a modern commuter village established in the 1970s. Also Svenstorp and Köpingebro have seen the creation of many chiefly one-family houses. In the village of Stora Herrestad, on the other hand, the changes in settlements have been very modest since 1915.

4.1.9 Conclusions

Sten Tesch and Björn E. Berglund

Man's impact on the cultural landscape in the Köpinge area has been determined by the very special geographical conditions. Unlike the other focal areas, which lie in the outer or inner hummocky zones, the Köpinge area is to a large extent synonymous with the coastal region.

In prehistoric times the Köpinge area was a central settlement region in relation to the other parts of the project area. The easily worked sandy soils of the Ystad plain, the sea, Öja-Herrestads Mosse, the Nybroån and Kabusaån rivers and the wetlands by them, combined to provide a rich and varied environment, creating the basis for a continuous and expansive prehistoric farming society. This was preceded by a Mesolithic phase with numerous hunting and catching sites by the coast. It is still uncertain whether there was any primitive cultivation before 3100 BC.

During the long period covered by the Stone Age it is, of course, possible to detect great differences in human impact on the landscape. The area around the lagoon of Öja-Herrestads Mosse and the Kabusa area were most attractive to Stone Age farmers during the Early and Middle Neolithic. Here they were able to combine cultivation with hunting and fishing.

A significant difference can be observed between the earlier and later phases of the Early Neolithic. The earlier phase shows a system of scattered small habitation sites utilizing both the coast and the inland. In the later phase the coastal sites become denser at the same time as they increase in size, indicating expansion. The structure of society also changed with the emergence of local territorial units around the five known dolmens in the area.

For the Middle Neolithic we have evidence of continued expansion at the beginning and end of the period. In the earliest phase (MN I) the pattern of settlement gives the impression of having been structured around large habitation bases. By the coast and the wetland there were also smaller hunting stations or satellite dwellings. Towards the end of the period (MN IV/V) the pattern of settlement appears to have undergone a distinct change. Not only were the habitation sites more widely scattered, but people also began to exploit the clay soils of the outer hummocky zone. This settlement pattern can be a reflection of the way the economy of

the late Funnel Beaker people was based on animal husbandry, particularly meat production.

Tillage during the Early and Middle Neolithic took the form of temporary cultivation plots broken in coppices. The cereals cultivated were einkorn and emmer. The supply of nutrient to the crops could also have been ensured by cultivating abandoned habitation sites. Forest clearance and tillage were made easier by the lightness of the soil.

Habitation sites from the Battleaxe culture (MNB) are few in number here, as in the whole of southern Scandinavia. East of the river Nybroån, however, something as unusual as a fragmentary house structure has been discovered.

In the Late Neolithic and the first phase of the Early Bronze Age, habitation sites with remains of houses become a relatively common feature of the settlement picture. Two different forms of house can be distinguished. The first consists of a house plan with a sunken floor section. The other consists of a true long-house with a row of posts down the middle supporting the roof. The size of the latter houses can be both normal and very large; the two house types may suggest a stratified society. With the filling up of the lagoon that was Öja-Herrestads Mosse, the centre of the settlement region is shifted to the plains on either side of the river Nybroån, today's Stora Köpinge parish. This shift is also seen in the high concentration of Bronze Age barrows in this parish.

In this phase we can see the probable emergence of a cyclic settlement pattern, which was to survive in broad outline for a few millennia. This meant the recurrent migration of farms and cultivation plots within a relatively naturally delimited settlement area of about one square kilometre. The settlement pattern became increasingly sedentary through time.

The time from the later phase of the Middle Neolithic to the Early Bronze Age was the golden age of stock-raising. The climate was so favourable that the large herds, dominated by cattle, were able to remain outside the whole year round. The herds wandered freely, and before the end of the Bronze Age they had transformed the half-open grazing forests into large heath-like pastures. People had to look further into the hummocky landscape to find winter fodder. Grain cultivation was probably of less importance than stock-raising. Pollen diagrams, macrofossils, and archaeological evidence suggest an intensification of tillage around the middle of the Early Bronze Age, with a corresponding slight decline in the size of the livestock herds. Tillage is attested in the area through both traces of ard ploughs and macrofossils. The Late Neolithic and the Early Bronze Age were dominated by emmer and naked barley, cereals which thrive on poor soil without needing manuring. In the phase of change in the middle of the Early Bronze Age, however, we see signs that people may have been begun to manure certain cultivation plots.

In the Late Bronze Age there are several changes in

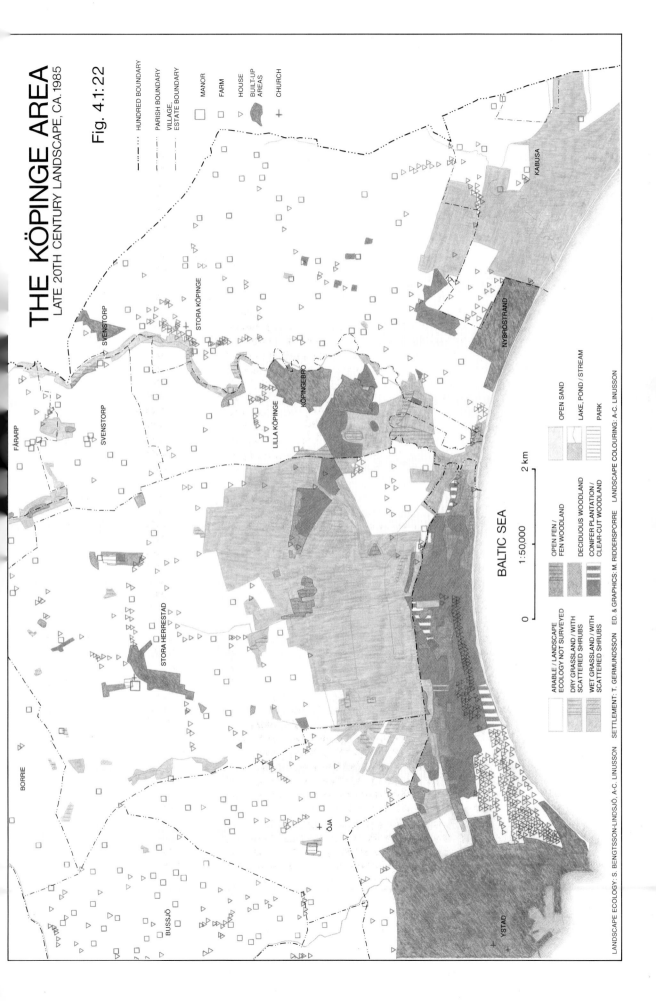

THE KÖPINGE AREA
LATE 20TH CENTURY LANDSCAPE, CA. 1985

Fig. 4.1:22

- –··–··– HUNDRED BOUNDARY
- –·–·–· PARISH BOUNDARY
- –·–·– VILLAGE, ESTATE BOUNDARY

- ☐ MANOR
- ☐ FARM
- ▷ HOUSE
- ⬛ BUILT-UP AREAS
- ✛ CHURCH

FÅRARP

SVENSTORP

SVENSTORP

STORA KÖPINGE

SVENSTORP

LILLA KÖPINGE

KÖPINGEBRO

NYBROSTRAND

KABUSA

STORA HERRESTAD

BORRIE

BUSSJÖ

ÖJA

YSTAD

BALTIC SEA

0   1:50,000   2 km

	ARABLE / LANDSCAPE ECOLOGY NOT SURVEYED	OPEN FEN / FEN WOODLAND	OPEN SAND
	DRY GRASSLAND / WITH SCATTERED SHRUBS	DECIDUOUS WOODLAND	LAKE, POND / STREAM
	WET GRASSLAND / WITH SCATTERED SHRUBS	CONIFER PLANTATION / CLEAR-CUT WOODLAND	PARK

LANDSCAPE ECOLOGY: S. BENGTSSON-LINDSJÖ, A.-C. LINUSSON   SETTLEMENT: T. GERMUNDSSON   ED. & GRAPHICS: M. RIDDERSPORRE   LANDSCAPE COLOURING: A.-C. LINUSSON

the long-houses and farm structure which point the way forwards to the settlement of the Early Iron Age. However, the isolated farmstead still appears to have been the normal form of settlement. The household probably consisted of at least ten people, including children. The locations chosen for habitation sites seem to have been the same as during the Late Neolithic and Early Bronze Age. Settlement was probably somewhat more stationary than before. Less than ten settlement units have been found in the area.

In the course of the Early Bronze Age (EB II) the long-house develops into a three-aisled hall, possibly with the aim of facilitating the stalling of cattle to provide manure for the fields. In the Late Bronze Age the design is altered so that at least the dwelling area is clearly discernible. The presumed byre section is still relatively small and could hardly have housed more than 10–18 cattle. The changed farm structure and the small byre sections suggest a more composite economy, in which tillage has taken on greater practical importance than before. The Late Bronze Age is increasingly dominated by hulled barley, which requires heavy manuring. Complementary sources of livelihood were seal-hunting, catching sea birds, fishing in the sea, lakes, and rivers. Sheep and goats, which could be kept outside the year round, even in the growing cool and damp of the climate, increased in number at the expense of cattle. People did not stall more cattle than they could find winter fodder for. The cattle also became smaller during the period and were now perhaps primarily kept for their milk, as well as the important duty of producing manure. There may have been some surplus production of cattle for social or religious reasons.

People adjusted to the lack of winter fodder in a landscape that was opening dramatically. Wetlands were now reclaimed and the alder carrs began to recede. The winter fodder which was no longer available in the settlement areas was now gathered in the outlands of the hummocky zone. What is interpreted in the pollen diagrams as an expansion of the cultural landscape reveals a society under stress. More severe winters than before meant a growing demand for winter fodder for the stock which had to be kept indoors.

The landscape had been overexploited for a long time. The deteriorating climate may have been the factor which upset the ecological balance. In this crisis, however, the population factor was an independent variable, since there is no evidence of any obvious increase in population during the period from the Late Neolithic to the Late Bronze Age. Yet the slow restructuring of the farming system during the Late Bronze Age shows that the farmers in the Köpinge region were fully capable of adapting to the changed conditions for agrarian production. This also laid the foundation for the more differentiated farming society of the Iron Age, with a fixed system of arable and meadowland. The transition to the Iron Age in the Köpinge area was thus far from being a break in the development of land use. The

archaeological evidence suggests continuity and a relatively smooth development of the cultural landscape.

It is only towards the end of the Pre-Roman Iron Age that we see a change in the settlement structure from that of the Late Bronze Age. It is now possible to detect in a few settlement areas agglomerations of two or more farms. However, this settlement structure is different from that in Jutland, where large migrating villages became common during the early phase of the Pre-Roman Iron Age. In the eastern parts of Denmark, including Scania, isolated farmsteads and small clusters appear to be the characteristic form of settlement. Sunken-floor huts for purposes such as textile production are a new feature of farms in the latter part of the Pre-Roman Iron Age.

In the Early Iron Age the settlement pattern in the Köpinge area assumes an increasingly fixed form with a growing number of settlement units. Against this background it is reasonable to suspect an increase in population during the period. This applies in particular to the time 150 BC–AD 200, which coincides with a period of warm, dry climate favourable to cereal cultivation. Judging by the archaeological evidence, a maximum prehistoric population was reached during this period. The settlement concentrations that existed for much of this time brought about a long-term utilization of the poor sandy soils, which must have needed an ample supply of manure. This hints at the existence of a fixed system of arable and meadow. However, the archaeological material shows only indirect evidence of an infield/outland system with permanent fields and hay meadows. In the Köpinge area there are only house remains, but no surviving traces of the cultivation landscape apart from macrofossils. Hulled barley is the dominant crop. Occasional seeds of rye, oats, wheat, and *Camelina sativa* show that these plants were cultivated but that they were of minor importance. Weeds like goosefoot become more common, indicating manuring. The presence of meadow plants in the material can be interpreted as hay waste, which suggests the stalling of animals and haymaking on wet meadows and marsh meadows. The relatively long houses made it possible to stall more animals than during the Late Bronze Age. By letting cattle graze in the outlands, nutrients were brought every evening via a cattle-path through the fenced fields and meadows to the infields. In the long run, however, this intensive exploitation of the light soils led to leaching and podzolization (cf. Ch. 4.2.2).

A split into smaller settlement units, probably dispersed farms, testifies to a restructuring of the farming system in the Late Roman Iron Age. Occasional very large houses and rich graves suggest social stratification. This process coincides with a temporary decline in human pressure on the landscape. The pollen diagrams indicate that certain regions were afforested at this time. Development in the Köpinge area can be seen in the wider context of what has been called the crisis of the Migration Period in the 3rd to 7th centuries. This crisis, however, was as much an adjustment to new times – socially, economically, and politically.

The restructuring of settlement that began in the Late Roman Iron Age continued during the Migration Period and the Vendel Period. Settlement during these centuries seems to have been concentrated in one area – around Tingshög, nearly two kilometres up the river Nybroån. The settlement concentration lies near what appears to have been a magnate's farm and a place for exchange. There may have been a similar concentration of settlement in the west of the Köpinge area. The reason for this concentration is hard to evaluate. One possible explanation could be that the available archaeological evidence does not reflect the true picture of settlement. It could also be an expression of the introduction of crop rotation in a system of two or three fenced fields cropped with spring-sown barley and autumn-sown rye. The macrofossil material from the Migration Period and the Vendel Period for the Köpinge area is too restricted to allow any interpretation of cultivation. Together with the more representative material from the Viking Age it shows that rye and barley are equally frequent. Oats were also grown, but wheat only to a very small extent. The composition of both the weeds and the cereals indicates regular rotation, but we are less certain about when this was introduced and whether it was in two or three courses.

From the middle of the Vendel Period, both the archaeological material and the pollen analyses show an expansion in the population and the cultivated land. This is not just a matter of settlement colonizing the hummocky landscape, but also a restructuring of settlement in the Köpinge area. Tingshög remains a central locality in the area a little into the Viking Age, but alongside the magnate's farm there were also smaller, relatively short-lived settlement units scattered over the landscape. The first stage of village formation is seen in the gradual establishment of settlement on several of the sites of the later known village cores. A regulation of settlement on these sites probably did not take place until the 12th century.

The localization of the villages in or on the edge of the hummocky landscape testifies to new preferences as regards soil, and should thus be seen against the background of new agricultural techniques. It is also relevant that the light soils were probably leached after four thousand years of continuous use. The landscape was now completely open, with the exception of occasional trees. Beech and hornbeam had no chance of spreading here by the coast.

From having been a central settlement region in prehistoric times, the Köpinge area became almost marginal in historical times. Up until the enclosures, large parts of the old prehistoric settlement area near the coast remained outland pasture and unmanured oatfields with long fallow periods.

In the Early Middle Ages, however, the Köpinge area retained its central importance, although in a different sense. Here, by the river Nybroån, were a few places with a central function. There was not only a marketplace for the exchange of goods under controlled forms – the *köpinge* – but also an administrative headquarters for the hundreds of Herrestad and Ingelstad – the *kongelev* or royal farm of Arvalund. Perhaps Kabusa also played a role in this connection, not least as a harbour for the *ledung,* the royal naval organization.

The foundation of the town of Ystad in the High Middle Ages, however, was a confirmation that even these central functions had been shifted to the periphery of the Köpinge area.

Settlement in the 14th century gives the impression that the cultural landscape was exploited to the full. Agrarian organization appears in large measure to have reached the level revealed in the oldest land survey maps from around 1700. There was cooperation between the villages as regards fencing (the organization of *vångalag*). Fencing was a problem in an area where wood and stone were in short supply and turf a valuable resource. There is evidence of several medieval watermills along the river Nybroån.

There is no information in the written sources to suggest any regression of settlement or cultivation during the Late Middle Ages. On the other hand, the area was not affected by the expansion of settlement that took place in the inner hummocky zone between 1570 and the mid 17th century. Around 1570 Stora Köpinge parish already had excessively large areas of arable in proportion to meadow, so people here were not affected to the same extent by the expansion of arable land that took place in the outer hummocky zone up to the 18th century.

Around 1670 the village of Stora Herrestad had 45 farms while Stora Köpinge had 28; these were the biggest villages in the Ystad area, the others being relatively small. Settlement density in Stora Köpinge parish was low, with 2.7 normal farms per square kilometre, somewhat below the average for parishes on the coastal plain and in the outer hummocky zone, whereas Stora Herrestad, with its 3.4 farms per square kilometre, was well above average density. The difference between these neighbouring parishes is also clearly seen in their grain production per tithe-paying unit. In Stora Köpinge it was 20% below the average for the whole project area in 1570, whereas Stora Herrestad was 33% above the average.

As for land ownership in early historical times, this underwent changes both in the Late Middle Ages and after the Reformation. The largest known landowner in the Middle Ages was the archbishopric of Lund. The archbishop received *ex officio* property including the fief of Köpinge, the manor of Stora Köpinge along with 39 tenant farms and four smallholdings in the parish. In the Köpinge area there is also evidence for a number of nobleman's manors: in Lilla Köpinge, Kabusa, Fårarp, Svenstorp, Stora Herrestad, and Öja. As a consequence of the crisis of the gentry in the Late Middle Ages, these passed into church ownership. After the Reformation the major part of Stora Köpinge parish came into the hands of the private nobility. Most of the farms came to belong to noble estates in the surrounding parishes. In Stora Herrestad virtually all farms and cottages were subordinated to Herrestad manor in the course of the 17th century.

During the time from the 17th century to the 19th, the agrarian economy of the Köpinge area was mostly based on grain production, the predominant crops being rye and oats. Settlement in Stora Köpinge nearly doubled during the period. Although the number of farms increased only slightly, there was a large rise in the number of cottages. In the 18th century the population doubled from 440 to 880, which is remarkable for an area with soil of low productivity used extensively. Because of the low productivity of the meadow and pasture, there were no more than one or two cows per farm. How then is such a pronounced population growth to be explained? The scant space for the expansion of arable land could not have allowed the doubled production necessary to retain the same living standard. Behind the change there is above all an economic and social differentiation, as is clear from the large increase in cottages. The cottages were occupied almost exclusively by landless day-labourers. Stora Köpinge parish at the end of the 18th century was clearly a marginal agrarian area. The conditions for a tolerable life were poorer here than in other parts of the Ystad area. Conditions in Stora Herrestad parish were different, thanks to the abundance of highly productive meadow and pasture in Öja-Herrestads Mosse.

By the mid 19th century the enclosures had been implemented in all the villages in the Köpinge area. The restructuring of the cultural landscape involved more than the moving of farms away from the village cores. When the estates sold off land, new farming units were also created. Harvests nearly tripled their yields in the period 1700–1850. This was made possible by the introduction of new cultivation cycles, ley farming, and marling and liming. Another contributory cause was that manuring could be intensified, since leys allowed farms to keep much larger herds of cattle. Horses replaced oxen as draught animals. Peas, beans, potatoes, mangel, and sugar-beet were now included along with grain in the sequence of crops. Rye was the dominant cereal in Stora Köpinge and barley in Stora Herrestad.

The favourable development of agriculture brought about a considerable growth in population, which reached its maximum at the end of the 19th century. Unlike other parishes in the Ystad area, Stora Köpinge retained this high population until the 1950s. Contributory factors here were the Ystad-Eslöv railway, built in 1865, and the sugar mill that came in the 1890s. Beside the railway there grew up the non-agrarian villages of Köpingebro and Svenstorp. Köpingebro, where the sugar mill was sited, was gradually to become the central locality in the area.

Sugar-beet has naturally been an important feature of the cultural landscape in the 20th century. Oil-producing plants are a new feature. Recent decades have also seen the coming of large-scale vegetable growing, on sandy soils which are well suited to the purpose. As everywhere else, the cultural landscape in modern times has been transformed by the rationalization of farming units. Here, however, amalgamations have primarily affected the fields, leaving much of the settlement picture from the end of the 19th century unchanged to this day. The growing of labour-intensive crops like sugar-beet and vegetables has also meant that small farming units are still relatively numerous. The eastern edge of the area has the special feature of its owner-occupier homes. Stately homes with large parks are another new element in the landscape picture.

Perhaps the greatest changes undergone by the cultural landscape in the 20th century have been of a non-agrarian kind. Large coniferous forests have been planted in the western part of Köpingebro and on the sand-dunes by the sea (Ystad Sandskog and Nybrostrand). Elsewhere there are smaller plantations providing shelter to fields. Large areas of the old meadow and pasture lands on the coast have become golf-links or military exercise grounds, although huge herds of beef cattle still graze on the exercise grounds. Holiday-makers and anglers are a noticeable but seasonal feature of the landscape. In recent years, however, the area of summer houses at Nybrostrand has been transformed into a residential dormitory town for Ystad.

Last but not least, Köpingebro, in the heart of the prehistoric settlement district, is still one of the central localities in the Ystad area, and the municipality has great plans for its future. Where there were once longhouses, people are now building detached houses on the old cultivation lands.

Bjäresjö village with lake Bjäresjösjön. View towards the northwest. Photo: Ingmar Holmåsen/N, September 1988.

4.2 The Bjäresjö area

4.2.1 Introduction

Björn E. Berglund

Area and sites studied. The Bjäresjö area comprises the present parishes of Bjäresjö and Hedeskoga. Our studies are concentrated on Bjäresjö parish, particularly Bjäresjö village and the Bjersjöholm estate (Figs 4.2:3–4, 6–7, 9).

The main field study sites are the following:

Palaeoecology: The lake basins of Bjärsjöholmssjön, Bjäresjösjön, Kumlan, and Bussjösjön (Bromma parish).

Ecology: Bjäresjö village; Lilla Tvären village.

Archaeology: Bjäresjö village; Bjersjöholm castle; the ancient village of Lilla Tvären.

Historical and human geographical studies deal with source material particularly from Bjäresjö parish, including the Bjersjöholm estate.

Geomorphology and hydrology. The greater part of the area is situated within the outer hummocky landscape, which is built up of moraine hills and kame terraces. The elevation varies from about 25 m near the coast to about 70 m in the hilly area in the north-east. The gently undulating countryside is typical of this landscape zone. The highest relief is to be found around Bussjösjön – the lake basin at 25 m situated close to a hill at 70 m.

Most of the area is drained towards the river Svartån or towards the ancient lake basin of Bjärsjöholmssjön, which has an outlet towards the Baltic. Today the only lakes are Bjäresjösjön and Bussjösjön. In former times (see the maps in Ch. 3 and in this chapter) several small lakes or peatlands occupied the basins. Until about 2000 BC a large lake existed in the Svartån valley west of Bjäresjö village, the present basin named Kumlan (see Ch. 5.1). This and other ancient lakes are now infilled and overgrown, and the peat exploited or cultivated. Former brooks have to a great extent been replaced by pipe drains. Dammed water reservoirs have been common in historical time, for example, along the Svartån valley where the Skönadal pond still exists (Möller 1984).

Subsoil distribution. The Quaternary soils are dominated by clayey till, and in depressions also clay till, which is a very heavy soil type. The content of limestone is high, which also means a high calcium carbonate content in these soils (in general >20–30%). Glaciofluvial sand and gravel occur, particularly in the ridges north and south of Bjersjöholm, but also in the fills around Bjäresjö. Fen peat occurs mainly in the Kumlan wetland along the Svartån river, but also in minor depressions which have been occupied by lakes or fens. See Fig. 1.2:15. (Daniel 1986, 1989).

Conditions for cultivation. The soil and water situation means that this landscape originally had a high diversity with favourable conditions for arable land on certain "islands", particularly around Bjäresjö, Bjersjöholm, Hedeskoga, etc. More advanced ploughing techniques and drainage have been needed to cultivate the clayey till soils. The agricultural soil quality indicates very productive soils (values 8–9 on a 10-grade scale, Fig. 1.2:16). The frequent damp soils have been favourable for pastures and mowed meadows. The flooded meadows along Svartån river seem to have been of particular potential. The only area which can possibly be regarded as of poor agricultural worth are the sandy hills (today forested) around Bjersjöholm.

4.2.2 Vegetation and landscape through time

Marie-José Gaillard and Hans Göransson

Results from two of the four sites chosen for pollen-analytical investigations in the Bjäresjö area are presented here: the lakes Bjärsjöholmssjön and Bjäresjösjön (Ch. 2.1 and Figs 4.2:1,2). The pollen diagram from Bussjösjön is published elsewhere (Regnéll 1989), and the study of Kumlan is in progress (Ch. 5.1).

The former lake of Bjärsjöholmssjön was lowered at the end of the 19th century and is replaced today by pasture lands. The pollen diagram published by Nilsson (1961) shows that the lake's sediments date from the Younger Dryas to the Subatlantic period until ca. AD 1400. The new pollen diagram (Göransson 1991) covers the time-span ca. 4000 BC to ca. AD 700 (12 ^{14}C-dated levels, Appendix A).

At Bjäresjösjön, carr peat deposited from Preboreal (7700 BC) to Late Subboreal time, and the creation of the lake has been ^{14}C-dated to 700 BC. Pollen analysis was performed on the lake sediments only (Gaillard and Berglund 1988).

The two pollen diagrams (Figs 4.2:1,2) overlap for the time-span 700 BC to ca. AD 700 and together offer a picture of the landscape in the Bjäresjö area for the last 6000 years. However, Bjäresjösjön and Bjärsjöholmssjön have been small lakes with rather limited catchment areas. Therefore, in both cases, pollen analysis provides mainly a local vegetation history for a radius of about one kilometre around each site. We are, therefore, not justified in generalizing from the Bjärsjöholmssjön pollen diagram to the surroundings of Bjäresjö village for the period 4000 BC to 700 BC. The figure of 1 km suggested for the radius of the pollen-source area is of course a very rough estimation. The pollen-source area varies considerably according to the plant species and to the type of vegetation prevailing around the lake (Bradshaw and Webb III 1985). However, it has been demonstrated by several authors that small basins collect their pollen from a smaller area of the surrounding vegetation than do large basins (Tauber 1967, 1977, Bradshaw and Webb III 1985).

The pollen zones according to Nilsson (1964) and the archaeological periods for southern Sweden (Ch. 2.9) are indicated in the diagrams for comparison. Nine main periods of land-use history may be distinguished, periods 1 to 4 in Bjärsjöholmssjön, period 5 in both Bjärsjöholmssjön and Bjäresjösjön, and periods 6 to 9 in Bjäresjösjön.

1) 4000–3100 BC. The broad-leaved forest and its utilization by Mesolithic man

High values of microscopic charcoal particles and the occurrence of *Pteridium* (bracken) disclose burning of the forest ground. Light-demanding species such as *Populus*, *Sorbus*, *Juniperus*, *Plantago media*, *Rumex acetosella* coll., *Artemisia*, *Viola arvensis* type, Cerealia type (one single grain) and, above all, the high pollen values of *Corylus* indicate that the Late Atlantic forests were not as dense as generally assumed, and that they may have been utilized by Mesolithic man (see also Göransson 1987b, 1988a). Cultivation of cereals on a very small scale may have occurred in the forest.

2) 3100–2500 BC. Scrub of broad-leaved trees and wandering arable fields during the Early Neolithic

The *Ulmus* decline is synchronous over the whole of southern Scandinavia and is dated to ca. 3200 BC (Nilsson 1964, Huntley and Birks 1983, Göransson 1987b). The simultaneous fall of the *Ulmus* and *Fraxinus* curves in the pollen diagram of Bjärsjöholmssjön discloses that the elm was not selectively affected (see also Ch. 4.3.2). Elm disease as the only cause of the elm decline may thus be called in question, as discussed earlier for southern Sweden by Nilsson (1961) and for Europe by Heitz-Weniger (1976), Moore (1984), and Huntley and Birks (1983) among others. It seems probable that many interacting physical and biological factors were involved in the vegetational changes during the start of the Neolithic, man and his livestock being one of those factors, climate and probable disease being others (Nilsson 1961, Göransson 1987b). These vegetational changes suggest that the forest ecosystem was weakened during the incipient Neolithic. During that time, the main resource of prehistoric man, the broad-leaved forest, was damaged (the "destruction phase" of Göransson 1987b;

see also Birks 1986a). For that reason, the Early Neolithic forest farmer had to use larger areas than before the elm decline and the forests of elm, ash, lime, and hazel may have been transformed into scrub. This change in the landscape was thus partly caused by the activity of man and his livestock, and partly by factors lying beyond human control.

The changes in the pollen spectra after the elm decline are of the same type as those found in other diagrams from southern Sweden and Denmark: a rise of the curves of *Betula* and charcoal particles, a rise in *Corylus*, and low values of *Ulmus*, *Fraxinus*, and *Tilia* (see also Ch. 4.3.2). These pollen-analytical features probably reflect clearance fires – and, in more inaccessible areas, natural fires – of the damaged forest (Göransson 1987b). The slash-and-burn technique was certainly not necessary for the cultivation of cereals during the Early Neolithic. The nutrient-rich soils of the south Swedish broad-leaved forests of the Late Atlantic were, without doubt, of sufficient quality for early agriculture (Göransson 1987b, Engelmark and Hjelmqvist 1991). Therefore, the hypothesis of "landnam" (Iversen 1941, 1949), implying slash-and-burn, is not justified for southern Scandinavia (Göransson 1988a, 1989).

The continuous curve of *Plantago lanceolata* indicates that grazing obviously was of importance in the Bjärsjöholmssjön area, as far back as an early part of the Early Neolithic. There are only a few pollen grains of *Triticum* type recorded in Bjärsjöholmssjön. Probably only limited areas around the former lake were suitable for the cultivation of cereals during the Stone Age (Malmer and Regnéll 1986). The study of plant macrofossil remains (carbonized seeds and cereal imprints in pottery) from archaeological excavations in Scania show that, during the Early Neolithic, the main crops were naked barley and wheat (Hjelmqvist 1955, 1979, Engelmark and Hjelmqvist 1991). The spelt wheats *Triticum monococcum* and *T. dicoccum* were also cultivated.

At the very beginning of the Early Neolithic, after the introduction of domestic animals, cultivation of cereals may have taken place on a very small scale, on soils manured by tethered sheep and cattle, and ploughed by pigs, as suggested by Rowley-Conwy (1981), Troels-Smith (1984), and Rasmussen (1988). The archaeological investigations of seven settlement areas in southern Scania support this theory (M. Larsson 1988a). After this initial phase, cereals may have been cultivated in a system of small "wandering arable fields" in hazel coppice woods (Göransson 1987b). "Wandering arable fields" is a term introduced by Göransson (1987b) for the Swedish *vandrande åkrar*. They differ from shifting arable fields in that they do not imply burn-beating.

In our terminology, "coppice woods" are not equivalent to the strict coppice woods such as those described by Rackham (1980) for the Middle Ages in Great Britain, which were characterized by a well-organized management concentrated on wood production. The term "stump-sprout forests" would be more convenient for

168

the type of forests occurring during Neolithic time, in a system where the production of hay was as important as that of wood, and where trees were probably cut in different ways, including shredding, without following a strict management. Coppice woods as they have existed in central Europe and Britain have never existed in Scania, but similar management on a smaller scale is documented from the Middle Ages (Bergendorff and Emanuelsson 1982, Emanuelsson et al. 1985).

Unfortunately, we have no pollen-analytical information from Bjäresjösjön for the period discussed above. But we know from the archaeological investigations that an area north of Bjäresjö was used by man during the Early Neolithic, and that several settlements may have occurred around Bjäresjösjön and the ancient lake of Kumlan (Ch. 4.2.3). There are, however, no remains of settlements close to Bjärsjöholmssjön (Ch. 4.2.3).

3) 2500–1800 BC. Leaf-fodder collection, grazing, and cereal cultivation in the Middle Neolithic

There is a distinct regeneration (or increase in the pollen frequencies) of elm, ash, and lime ca. 700 ^{14}C years after the elm decline (ca. 2500 BC), during the early part of the Middle Neolithic. This event is synchronous over the whole of southern Scandinavia (Göransson 1987b). The pollen diagram from Bjärsjöholmssjön provides no evidence for a regression in human impact on the vegetation. The *Plantago lanceolata* curve remains unbroken, which indicates continuous grazing in the neighbourhood of the site. The forests seem to have been healed after the Early Neolithic destruction phase (see above), but they were most probably still utilized by man. Leaf-fodder collection may have been more common in the Middle Neolithic (see also Ch. 4.3.2). Cereals were possibly cultivated in shifting fields on light soils in a system of coppice groves of different ages (Göransson 1984b, 1986, 1987b). This type of land use, together with interval clearings undertaken by the forest farmer for different purposes, may explain the successive declines and regenerations of elm, ash and lime pollen curves in the Middle Neolithic. The low representation of cereal pollen – a single grain of *Hordeum* type was recorded – may be due to the filtration effect of the "regenerated" forest (Göransson 1986, 1987a,b). Moreover, we know from the analysis of botanical macroremains in archaeological excavations that naked barley was probably the most cultivated cereal during the Middle Neolithic in Scania (Engelmark and Hjelmqvist 1991).

The change in the vegetation (regeneration of forests) and in land use (large areas with coppiced trees and probably intensive fodder collection) corresponds to a change in the settlement pattern during the latest part of the Early Neolithic and in the Middle Neolithic (Ch. 4.2.3). The few settlements situated in the north of the Bjäresjö area in the Early Neolithic seem to have been abandoned during the Middle Neolithic, indicating that settlements were more concentrated in the Middle Neolithic (Ch. 4.2.3).

4) 1800–800 BC. Increased cereal cultivation and grazing in the Late Neolithic and Early Bronze Age

From ca. 1800 BC, a distinct increase in *Plantago lanceolata* and Poaceae indicates that grazing activities became still more important at the expense of the broadleaved forests. From that time onwards, the pollen curve of cereals remains almost unbroken, which speaks in favour of their cultivation in the vicinity of Bjärsjöholmssjön, probably in a system of bush-fallows. The latter is shown by a distinct and long-lasting (1000 years) rise in the hazel pollen frequencies after 1600 BC. Both *Hordeum* and *Triticum* were cultivated. Some artefacts dated to the Late Neolithic period were found around Bjärsjöholmssjön, indicating possible settlements (Ch. 4.2.3) and confirming that cereals were probably cultivated close to the lake after that time. According to Hjelmqvist (1955, 1979) and Engelmark and Hjelmqvist (1991) naked barley and spelt wheat *(Triticum dicoccum)* seem to have been the main crops during the late Neolithic in Scania. Naked barley was again dominant in the Early Bronze Age, and hulled barley also became more common.

During the Early Bronze Age, the settlements seem to have been dispersed in the hummocky landscape zone, and mobile (Ch. 4.2.4.1). The latter would support the hypothesis of the occurrence of "wandering arable fields" in the area. The large number of barrows shows that the Early Bronze Age landscape was already rather open, which corroborates the interpretation of the pollen spectra in terms of coppice woods.

5) 800 BC–AD 600. The transition Early/Late Bronze Age and the creation of large areas for grazing

The two pollen diagrams from Bjärsjöholmssjön and Bjäresjösjön provide a very similar picture of the landscape for the period 700 BC–AD 600. The main differences are the following: (1) trees such as *Ulmus, Tilia, Quercus, Carpinus, Corylus, Betula,* and *Alnus* have much higher pollen values at Bjärsjöholmssjön; (2) herbs such as Poaceae, *Plantago lanceolata, Rumex acetosella* coll., Cyperaceae (sedges), and Cerealia (especially *Hordeum* type) are more frequent at Bjäresjösjön. These differences indicate that the landscape of the Late Bronze Age and Early Iron Age was still rather wooded around Bjärsjöholmssjön and that there were larger areas of woodland pasture in the vicinity of the former lake than in the surroundings of Bjäresjö village, where the landscape was already intensively deforested and exploited, consisting mainly of pasture lands, meadows, and fields of cereals (barley dominant). This difference between the two sites was probably maintained throughout the Iron Age, Middle Ages, and Early Modern Time (Ch. 4.2.6.1, 4.2.6.2, 4.2.7.2, 4.2.7.3, 4.2.8).

BJÄRSJÖHOLMSSJÖN POLLEN DIAGRAM

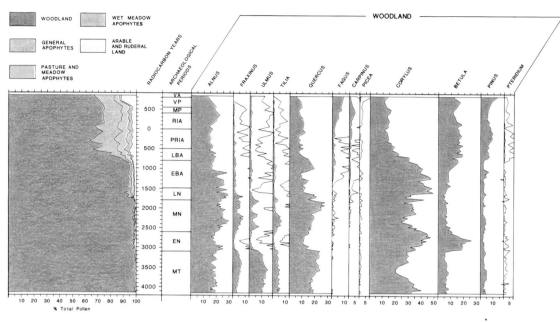

Fig. 4.2:1. Pollen diagram from Bjäresjöholmsjön.

A great part of Bjärsjöholmssjön's surroundings was still wooded around AD 1700. The wooded meadows lost their trees during the 18th century.

The lower part of the pollen diagram from Bjäresjösjön is ^{14}C-dated to the very beginning of the Late Bronze Age. Therefore, the transition Early/Late Bronze Age is best described from the Bjärsjöholmssjön pollen diagram. This period is characterized by the creation of a new type of landscape. Deforestation by burning (increase in charcoal particles) favoured the expansion of birch. The stands of hazel (possibly coppiced) diminished (this is shown in the diagram from Bjäresjösjön as well). The grazed areas became larger than before (increase in Poaceae, *Plantago lanceolata*, Cyperaceae), as did the arable land (higher pollen values of Cerealia and weeds such as *Rumex acetosella* coll.

and *Artemisia*). The first *Secale* pollen grain is found at a level dated to the Early Bronze Age, but it is registered regularly from the transition Early/Late Bronze Age. There is, however, no indication of the cultivation of rye during this period in the record of seeds and pottery imprints from Scania (Hjelmqvist 1955, 1979, Engelmark and Hjelmqvist 1991). Rye may therefore have grown as a weed in the fields of barley (Behre 1981, Körber-Grohne 1987). In both pollen diagrams, *Hordeum* is the dominant cereal pollen type during the Late Bronze Age. These results agree with the archaeobotanical data (Engelmark and Hjelmqvist 1991). The Late Bronze Age is clearly characterized by the increase of hulled barley at the expense of spelt wheat. Hulled and naked barley apparently became of about equal importance. Moreover, new crops were introduced:

BJÄRESJÖSJÖN
POLLEN DIAGRAM

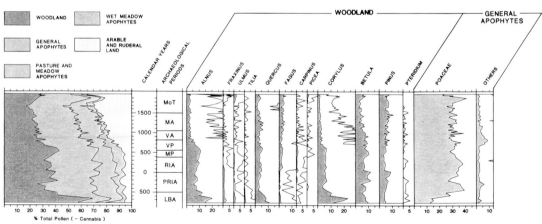

Fig. 4.2:2. Pollen diagram from Bjäresjösjön.

ECOLOGICAL BULLETINS, 41, 1991

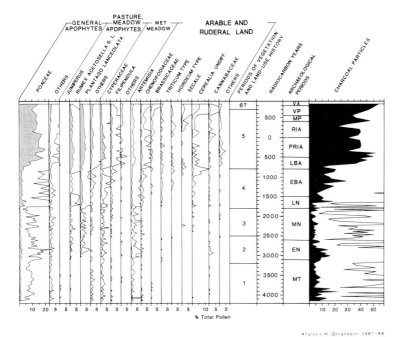

Analysis H. Göransson 1987-88

Avena, Panicum miliaceum (millet, one find from Bjä-resjö) and *Camelina sativa* (Engelmark and Hjelmqvist 1991).

It is tempting to draw a parallel between this main vegetational change and the shift in the use of the landscape demonstrated by the archaeological data, from area-extensive during the Early Bronze Age to more consolidated and concentrated during the Late Bronze Age and the Pre-Roman Iron Age (Ch. 4.2.4.1). The two pollen diagrams do not indicate any significant change in the agricultural and grazing activities throughout the Late Bronze Age, Pre-Roman Iron Age, Roman Iron Age, and Migration Period; in this respect they differ from the development described on the basis of the Krageholmssjön pollen diagram (Ch. 4.3.2). There is also archaeological evidence for settlement continuity north of Bjäresjösjön during the same period (Ch. 4.2.4.3). However, in the pollen diagram from Bjäresjösjön, there is indication of recolonization of open areas by *Populus* followed by birch and hazel,

during the later part of Pre-Roman Iron Age, and mainly by birch during the Migration Period. Whether these pollen-analytical features reflect a general regression in land use must remain an open question. They may also represent strictly local patterns, such as the occurrence of bush-fallows during certain periods.

According to Engelmark and Hjelmqvist (1991), the change towards one-course rotation was completed during the Pre-Roman Iron Age. The earlier, more extensive cultivation methods were replaced by an intensive land use that needed much manuring. Only hulled barley was cultivated at that time, and the recorded weed species suggest one-course rotation. This cultivation system probably led to the impoverishment of soils, the spread of parasites and diseases, and invasion by weeds. Therefore, the fields had probably to be abandoned after some decades and new areas had to be exploited for agriculture (Engelmark and Hjelmqvist 1991). Such phenomena could explain the increase in aspen, birch,

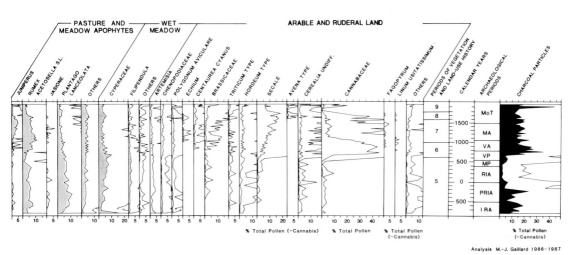

Analysis M.-J. Gaillard 1986-1987

and hazel pollen during the later part of the Pre-Roman Iron Age in the pollen diagram from Bjäresjösjön.

In Scania, the Roman Iron Age is characterized by the development of the one-course rotation system (with hulled barley as the dominant crop) and by the introduction of new crops: *Secale* and *Linum usitatissimum* (Engelmark and Hjelmqvist 1991). The pollen diagrams from Bjärsjöholmssjön and Bjäresjösjön show the regular occurrence of *Secale* pollen since the middle of the Roman Iron Age (increase in the pollen curve around AD 250 in Bjäresjösjön). Rye may have been cultivated on a small scale after that time (Engelmark and Hjelmqvist, 1991). There are no finds of *Linum* pollen grains from this period, but this does not preclude its cultivation at that time. As a matter of fact, *Linum* is normally underrepresented in pollen diagrams (Willerding 1986).

No distinct change in land use is shown for the Migration Period. However, in the pollen diagram from Bjäresjösjön, an increase in percentages of birch and a parallel decrease of several meadow plants and weeds (*Rumex acetosella* coll., *Anthemis* (e.g. corn chamomile), *Chenopodium* (Good King Henry, goosefoot), *Polygonum aviculare* (knotgrass)) are noted. These changes may be due to different land-use practices. There is, nevertheless, no indication of a decrease in cereal cultivation. On the contrary, the percentages of *Hordeum* pollen type increase steadily. Unfortunately, the plant macrofossil data from archaeological sites are very poor for this period, but the results obtained suggest a cultivation system similar to that known for the Roman Iron Age (Engelmark and Hjelmqvist 1991).

It is not possible with the aid of the pollen diagrams to establish the time of the first manured arable fields. The increase in *Fagus* pollen during the middle part of the Roman Iron Age in both diagrams may reflect the spread of *Fagus* in grazed outfield areas, which may in turn imply the occurrence of enclosed fields. This hypothesis is supported by the archaeological evidence for greater stability in settlements since the Pre-Roman Iron Age, and particularly during Roman Iron Age (Ch. 4.2.4.3). However, the increase in *Fagus* pollen may also result from a long-distance transport from north of the Bjäresjö area, like the Krageholm area, where *Fagus* was established in the Late Pre-Roman Iron Age and became, together with *Quercus,* the dominant tree after the Migration Period (Ch. 4.3.2). According to Engelmark and Hjelmqvist (1991), the importance of hulled barley and *Chenopodium* during the Late Bronze Age suggests the introduction of some kind of manuring. The distinct increase in pollen percentages of Chenopodiaceae at the beginning of Late Bronze Age in all pollen diagrams from the Ystad area may well be an indication of the start of manuring in Scania already at that time.

At the beginning of the Late Bronze Age, forest clearing and/or increased humidity produced a rise in the groundwater level, resulting in the creation of the lake at Bjäresjö (Ch. 5.1). The fen existing there since Preboreal time (ca. 7700 BC) was replaced by a small body of water with floating-leaved and submerged macrophytes such as *Potamogeton* (pondweed) and *Myriophyllum* (water-milfoil). A lowering of the water level, probably due to drier climatic conditions, started at the end of the Pre-Roman Iron Age and culminated during the Roman Iron Age and the Migration Period (Ch. 5.1). The small lake was transformed into a very shallow pond overgrown by *Equisetum* (water horsetail), *Nuphar lutea* (yellow water-lily) and *Sparganium* (bur-reed). This regional climatic event does not seem to have influenced land use at Bjäresjö.

6) AD 600–1000. Creation of arable fields and introduction of new crops during the Late Iron Age

A drastic change in the cultural landscape occurred around AD 600, in the middle of the Vendel Period. Trees became very rare in the area of Bjäresjö village. The fall in the alder and hazel pollen values suggests that alder carrs (perhaps used earlier as grazed coppiced woods) and areas with hazel (possibly coppiced) decreased in size and were replaced by open meadows and arable fields respectively. The increase in charcoal particles may indicate the burning of these woods. Moreover, the area of arable land increased at the expense of pasture lands. This development may be compared to the increased stability of settlements known for southern Scandinavia during the Vendel Period (Callmer 1986b), which is also suggested by archaeological observations at Bjäresjö (Ch. 4.2.4.3). This change in the landscape is also contemporaneous with a general rise in the water-table, which is ascribed to more humid climatic conditions (Ch. 5.1). The temporary shallow pond at Bjäresjö (see above) was thus transformed into a deeper water body. From that time onwards, a lake existed at Bjäresjö.

According to the pollen diagram from Bjäresjösjön, rye has been used as a major crop since the later part of the Vendel Period. These results are confirmed by a dated find from an archaeological excavation in Scania showing that the cultivation of rye was as important as that of barley during the Vendel Period (Engelmark and Hjelmqvist 1991). The very high percentages of *Cannabis* pollen indicate that Bjäresjösjön was used as a retting pit for extraction of hemp fibres, which was a common procedure at that time in southern Scania (Gaillard and Berglund 1988, Digerfeldt 1988, J. Regnéll 1989) and in the whole of southern Scandinavia (Påhlsson 1981, Gaillard and Berglund 1988). Therefore, hemp was most probably cultivated in the neighbourhood of Bjäresjö village (Bradshaw et al. 1981). The cultivation of hemp needs deep, calcareous, humus- and nitrogen-rich soils, and good irrigation. The most favourable areas are nutrient-rich river basins and well-drained mires (Körber-Grohne 1987). Such bio-

172

topes were available along the outlet of the lake and in the Svartån valley .

Rye, barley, and hemp were the three main crops during the Viking Age. Moreover, *Linum usitatissimum* and *Fagopyrum esculentum* were probably introduced as cultivated plants. There are no finds of buckwheat remains from archaeological sites older than the Late Middle Ages (Hjelmqvist 1955, 1979, Engelmark and Hjelmqvist 1991). However, the regular occurrence of *Fagopyrum* pollen grains between ca. AD 900 and 1450 in the diagram from Bjäresjösjön is a good indication of the cultivation of buckwheat close to the lake as far back as the Late Viking Age and Early Middle Ages (Gaillard and Berglund 1988).

A distinct change in the weed flora – an increase in *Rumex acetosella* coll., *Polygonum aviculare*, Brassicaceae (e.g. wild radish), *Hornungia* type (e.g. shepherd's purse), *Anthemis* type (e.g. corn chamomile), *Spergula arvensis* (corn spurrey) and the first occurrence of *Centaurea cyanus* (cornflower) – suggests that, since the beginning of the Viking Age (if not earlier, Late Vendel Period), fields were permanent and that cultivation followed a certain rotation, possibly already in three courses. The same conclusions were drawn from the archaeo-botanical data that indicate the establishment of three-course rotation (fallow-rye-barley) during the Viking Age (Engelmark and Hjelmqvist 1991).

Water-retting of hemp in the lake was common during the period ca. AD 700–1000 and decreased between ca. AD 1000 and 1200. During the period of intensive retting, the lake became rapidly very eutrophic (Gaillard et al. 1991), which could explain why the procedure ceased or was replaced by another retting technique, after three centuries of use.

7) AD 1000–1600. Intensive rye cultivation and expansion of juniper in the grazed outfields during the Middle Ages

Rye became more common after AD 1000 and may have occupied larger areas than earlier. Higher pollen values of *Secale* and weeds from winter crops *(Centaurea cyanus, Polygonum aviculare)* may be interpreted in terms of an expansion of the three-course-rotation system (see also Engelmark and Hjelmqvist 1991). It was also around AD 950 that a very marked change took place in settlement distribution in the area, the coastal zone being abandoned (Ch. 4.2.4.3). At Bjäresjö there is evidence for the existence of a manor and a settlement of village type at least since the Late Viking Age.

Since AD 1200 (or even earlier, perhaps since AD 1000?) the retting process for hemp may have moved to another pit or changed in its technique, or the cultivation of hemp may have been partly replaced by other crops. Around AD 1100, another change in the landscape is shown by the regular occurrence of *Juniperus*, *Calluna* and species of dry meadows on poor soils, such as *Jasione* (sheep's-bit). After more than 2000 years of

grazing, leaching of the pastureland soils could have led to a change in the flora, *Juniperus, Calluna,* and *Jasione* being favoured on the poorer sandy soils. An alternative explanation for the expansion of juniper and heather would be that the areas with poor soils were used for grazing after AD 1000 because of the development of agriculture on the rich soils at the expense of pasture lands (Regnéll 1989).

The period AD 1000–1600 was not characterized by any major changes in the cultivation of crops. However, rye became gradually more common, particularly after AD 1250. According to Engelmark and Hjelmqvist (1991), the 13th and 14th centuries were characterized by the exploitation of most of the good soils for cereal cultivation in a strict three-course rotation system.

A slight decrease in *Secale* pollen values between ca. AD 1350 and 1450 may possibly be ascribed to a stagnation in the expansion of cultivation during the Late Middle Ages, although this cannot be compared to a real "agrarian crisis" (Dearing et al. 1987, Ch. 4.2.6.1). However, a slight stagnation in the agricultural development may be related to the destruction of the manor at Bjäresjö and the moving of its residential part to Bjersjöholm (Ch. 4.2.6.1), which resulted in the reorganization of the village of Bjäresjö (Ch. 4.2.5.1). This event was interpreted as being an effect of the period of agitation known in Europe during the 14th century (Ch. 4.2.6.1).

8) AD 1600–1800. The transition to modern agriculture and the replacement of grazed outfield areas by cultivated fields

Between about AD 1600 and 1800, an increase in the cultivated area at the expense of pastures and meadows is shown by very high pollen values of *Secale* and its weeds (*Centaurea cyanus* in particular) and falling percentages of meadow and pasture pollen indicators such as *Plantago lanceolata, Jasione, Sagina* (pearlwort), *Centaurea nigra* (knapweed), Rubiaceae (bedstraw), *Cirsium* (thistle) and *Ononis* type (rest-harrow). According to written sources, the increase in arable land was some 25% of the total Bjäresjö village area during the period AD 1570–1699 (Persson 1986b, Ch. 4.2.6.1). The cultivation system at Bjäresjö is known from 1570, the time of the construction of the Renaissance castle at Bjersjöholm. It was a three-course rotation with a fourth field used as meadow (Ch. 4.2.6.1).

Juniperus pollen values decrease gradually between ca. AD 1550 and 1650 to become very low at the end of the 17th century, when *Avena* pollen appears for the first time. The last outland areas were reclaimed by this time and used for the cultivation of oats (Ch. 4.2.6.1, 4.2.7.3). However, the absence of *Avena* pollen type before the end of the 17th century does not imply that cultivation of oats did not occur earlier at Bjäresjö, on a small scale. Charred seeds of *Avena sativa* were found in the plant macrofossil remains recovered from a Late

Bronze Age layer at Bjäresjö (Engelmark 1988). Cultivation of flax and buckwheat probably decreased considerably at the beginning of the 18th century, since no pollen of these species is registered from that time onwards. We know from historical sources that many farmers abandoned the cultivation of flax and buckwheat after 1738 because of the difficulty of buying good seed in the towns, and because cultivation of rye and barley was more profitable (Anders Persson, pers. comm.).

The period AD 1700–1800 is characterized in the Bjäresjö area by a further marked increase in the area of cultivated fields (Ch. 4.2.7.1), which is shown in the pollen diagram by very high values of cereals and weeds.

9) AD 1800–1983. Plantations of trees, introduction of new crops and decrease in weeds

The last main change in the landscape at Bjäresjö is seen around AD 1800, when trees *(Ulmus, Fraxinus)* and shrubs *(Salix* (willow), *Sambucus nigra* (elder)) increased as a result of plantations around farms and along roads. A planted avenue of elms along the entrance road south of the Ruuthsbo estate (south of Bjäresjö) existed at the beginning of the 19th century, and this practice of planting continued until the beginning of the 20th century (Ch. 4.2.8). The increase in *Picea* (spruce) pollen indicates the planting of this tree in the Ystad area at the end of the 19th century. The pollen diagram also shows that the main cultivated crops were still barley and rye, but wheat and oats were also common. Moreover, rye was not as dominant as before, and the weed flora became very poor. These changes may well correspond to the major agrarian reorganization and innovations that took place during the 19th century. The strict three-course rotation was abandoned and succeeded by crop rotation systems including leguminous crops (Ch. 4.2.8). This would explain the decrease or disappearance of most of the winter crop weeds and the increase in Fabaceae (e.g. red clover and white clover) seen in the pollen diagram. The increase in pollen of Poaceae, dated to ca. AD 1900, is possibly a result of the cultivation of leys in the direct vicinity of Bjäresjösjön. This practice was common in the area during the period 1900–1945 (Ch. 4.2.8, 4.3.2.).

After AD 1945, the high pollen values of Brassicaceae correspond without doubt to the introduction of rape as an important crop after the Second World War (Ch. 4.2.8). Moreover, the pollen percentages of *Secale* are decreasing at the expense of *Hordeum* type and other cereal types (probably mainly *Triticum*), which agrees with the great changes in agricultural production known after the last world war, the four main crops cultivated in Bjäresjö being wheat, barley, rape, and sugar-beet (Ch. 4.2.8).

4.2.3 The Neolithic
The cultural landscape in the Bjäresjö area during the Neolithic

Mats Larsson

Stray finds and the cultural landscape

A major problem associated with the evaluation of the distribution of the stray finds in the Bjäresjö area is the fact that the majority of the material comes from a very large private collection, gathered from the land making up the Ruuthsbo estate. Known as the Jacobæus collection, this contains 223 of the total of 237 recorded items in collections and museums.

The survey of museum collections was accordingly supplemented between 1984 and 1985 by a field survey with the aim of covering the accessible open agricultural land. This work, which was carried out in conjunction with subprojects B3 and B4, produced a more varied picture of the localization of settlement in the area.

The Bjäresjö area includes two of the three landscape zones into which the Ystad area can be divided, namely, the coastal strip and the outer hummocky landscape. In the area with which we are concerned here, the width of the coastal strip, which may be defined as an area of sandy soil, is only about 150–200 m. The outer hummocky zone in this area consists essentially of heavier clay soils. The majority of the stray finds come from the coastal strip, with certain observable concentrations around the beach ridge, which is under cultivation at the present time.

The chorological and chronological distribution of the stray finds is discussed below on the basis of the Early Neolithic material. The types of object with which we are concerned in this context are the pointed-butted and thin-butted axes. The pointed-butted axes, which constitute an older type, represent a kind of implement which is relatively uncommon in the area. Their distribution is interesting, however. Two concentrations can be observed: at Rudolfsfält and Gustavsfält on the coast, and at Kärragården in the north of Bjäresjö parish. No finds of pointed-butted axes have been made anywhere between these two concentrations. The other type of axe which can be dated to the Early Neolithic, the thin-butted axe, presents an entirely different picture. By far the majority of these axes originate from the coastal zone, while only a very small number were found in the interior of the area. However, the thin-butted axe was used over a long period. The dating given to it lies between the Early Neolithic and the first phase of the Middle Neolithic (I–II). What is interesting is the fact that all the datable thin-butted axes fall within the limits of the Early Neolithic, with a certain predominance of types that have been dated to a late part of the period. None of these axes can be attributed to the Middle Neolithic.

The thick-butted axe is predominant during most of

THE BJÄRESJÖ AREA

NEOLITHIC SETTLEMENT, CA. 3000 BC - 2000 BC[*]

Fig. 4.2:3

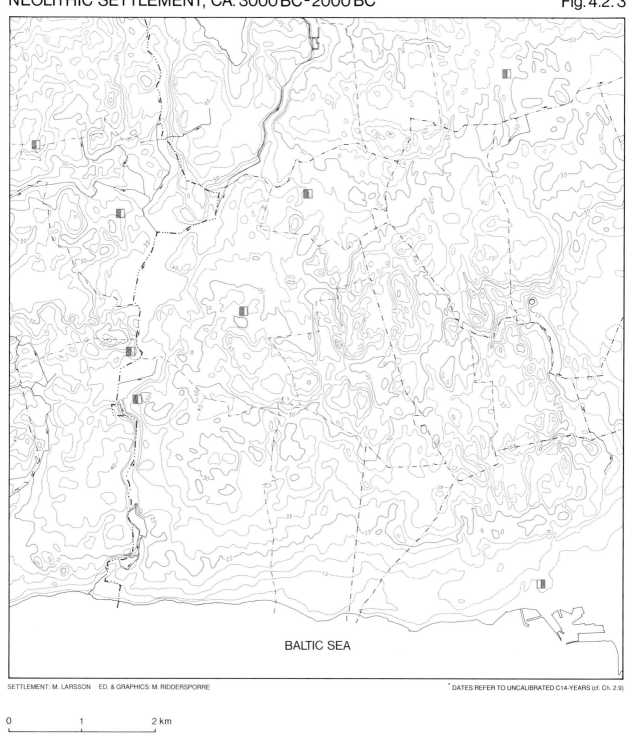

SETTLEMENT: M. LARSSON   ED. & GRAPHICS: M. RIDDERSPORRE

[*] DATES REFER TO UNCALIBRATED C14-YEARS (cf. Ch. 2.9)

BALTIC SEA

0   1   2 km

1:50,000

▣ SETTLEMENT: EARLY NEOLITHIC	⬡ LAKE / STREAM (CA. 1970)	—··—··· HUNDRED BOUNDARY (CA. 1970)
▣ SETTLEMENT: MIDDLE NEOLITHIC (TRB)	—50— CONTOUR LINE: 5 M EQUIDISTANT	—·—·— PARISH BOUNDARY (CA. 1970)
		— — — VILLAGE, ESTATE BOUNDARY (CA. 1970)

the Middle Neolithic. The period is divided into the Middle Neolithic A and the Middle Neolithic B for the purposes of the following discussion (Nielsen 1977). The former relates to the Funnel Beaker culture, and the latter to the Battleaxe culture. The thick-butted axes that can be dated to the Middle Neolithic A are clearly localized to the coastal zone, whereas only a very small number have been found inland. The later Lindø and Store Valby types are totally predominant chronologically. These types are datable to the Middle Neolithic IV and the Middle Neolithic V. Certain thick-butted axes from the Middle Neolithic III (Bundsø) have not been identified in the material.

The finds from the Middle Neolithic B, the Battleaxe culture, are few in number and originate with one exception from the area nearest the coast.

A dramatic increase in the number of recorded stray finds in the Bjäresjö area is associated with the Late Neolithic. Of the datable stray finds, the proportion from the Late Neolithic is 52%. This represents a considerable increase in the quantity of material in relation to the length of the period. Those objects which are easiest to date more accurately within the period, the daggers, are concentrated in the middle of the Late Neolithic B. A small proportion of the daggers belongs to the Early Bronze Age, period 1. The area closest to the coast still gives the impression of having also been of central importance during the Late Neolithic. A certain amount of dispersion over the area can be observed from the fact that a small number of daggers has also been found in the interior part of the area.

In summary, it can be stated that it is possible to distinguish an area of settlement during the early part of the Early Neolithic, which does not give the impression of having been utilized subsequently during the Neolithic period; this is the north of the Bjäresjö area, where a number of pointed-butted axes have been found. The narrow coastal strip and the areas behind it appear to have been the predominant settlement area during the later part of the Early Neolithic and the transition to the Middle Neolithic. This settlement pattern seems to have been characteristic of the whole of the Neolithic. A certain amount of spreading in the direction of the north of the area can be observed, however, during the Middle Neolithic B and the Late Neolithic.

The Neolithic settlement area

The findings discussed here come from field survey and excavations. Neighbouring areas in the parishes of Balkåkra, Hedeskoga, and Skårby have been included to produce a more detailed picture of the extent of settlement.

Twenty-one sites have been localized in the area. Of these, 17 were identified during the field survey, and three were already known to exist. In addition, one site was examined in conjunction with the investigations of

subproject B4 immediately to the north of lake Bjäresjösjön. Two areas emerged as being attractive for settlement during the Neolithic period: the coastal strip and the area around the prehistoric lake known as Kumlan.

Early Neolithic. The variation in distribution between the pointed-butted and thin-butted axes was discussed in the introduction. The pointed-butted axes tend to be more common in the northern, interior parts of the area, whereas the thin-butted axes are concentrated in the coastal strip. This is a tendency which has been observed in other parts of southern Scandinavia (Larsson 1984, Skaarup 1985). No settlement sites containing finds of pointed-butted axes have been encountered in the area, however. The discovery of four pointed-butted axes close to a wetland area immediately north of Kärragården indicates with a high degree of probability the presence of a settlement dating from the oldest part of the Early Neolithic. Three small concentrations of worked flint were found in this area, which may be indicative of sites dating from the period.

By far the majority of the sites, with the exception of those mentioned above, are associated with the extensive boggy areas close to the prehistoric lake. The sites, whose presence is indicated by worked flints, and in certain cases also by fire-cracked stones, were localized in positions close to the shore. There is a clear preference for narrow headlands or areas around bays on the prehistoric lake. Small streams, which previously discharged into the lake, now run through these areas. Very little material has been found on most of these sites, which makes accurate dating difficult. One site is situated in the inner part of the prehistoric lake (Hunnestad). A number of thin-butted axes were found there, all of which are of a type belonging to an older part of the Early Neolithic. A small number of other sites have been found to contain fragments of polished axes which can be included in the group of thin-butted axes.

The only site to be excavated was situated on the land sloping down to Bjäresjösjön. It was discovered in conjunction with stripping of the topsoil on the Bjäresjö estate. The few finds included flints and sherds. Among the flints was one fragment from a thin-butted axe. The sherds in particular serve as the basis for more accurate dating. The small number of ornamented sherds all bear cord ornamentation below the rim. The cord ornamentation is present in the form of horizontal bands, or as a combination of horizontal and vertical bands. The fragmentary appearance of the sherds makes it difficult to establish the shapes of the vessels, although it is obvious that the majority of them are from Funnel Beakers with a short neck and a weakly pronounced transition between the neck and the belly. The nearest parallels to the sherds from this site are to be found in the material from sites such as Mossby and Kabusa IVb (Larsson and Larsson 1986, M. Larsson 1987). The pottery from the

Mossby site in particular is very similar to what was found on the Bjäresjö estate site. Unfortunately, it is not possible to draw any conclusions about the size of the Bjäresjö site, since the area has been subjected to very intensive cultivation, and most traces of the prehistoric settlement have been destroyed by cultivation. Nevertheless, on the basis of the knowledge which we now have relating to the size of the sites in the Ystad area, it is possible to assume that we are here concerned with a small site of perhaps 800 to 1000 m².

In summary, settlement during the Early Neolithic, and essentially during the oldest part of the period, was localized to slightly elevated areas in the vicinity of watercourses. In many cases the sites were positioned on headlands or bays in the vicinity of the prehistoric lake Kumlan.

Since the Bjäresjö area generally lacks sandy areas, apart from the narrow coastal strip, settlement was localized to clay soils or clayey sand. The possibility that settlement during the Early Neolithic may also have been concentrated in areas of clayey soil has been seen elsewhere. Investigations in Jutland have revealed that, in certain cases, the soil type was not of great significance (Skamby Madsen 1984, Vedsted 1986).

Middle Neolithic. No sites that can be reliably dated to the Middle Neolithic A or B were found in the course of our investigations. A large number of thin-butted axes of types which can be attributed to the transition between the Early and Middle Neolithic have nevertheless been found in the area. All were discovered in the coastal strip. By contrast, the number of thin-butted and thick-butted axes dating from the first half of the Middle Neolithic is very small in this area. The fluctuation in the amount of material dating from this period has also been encountered, for example, at Langeland in Denmark (Skaarup 1985). The predominant finds in Denmark are of the later thick-butted axes of the Lindø and Store Valby types. These types can be dated to the late Funnel Beaker culture, Periods IV and V of the Middle Neolithic (Nielsen 1979).

During the time of the Battleaxe culture, the Middle Neolithic B, the distribution of the finds gives a clear impression of continuity from the late Funnel Beaker culture. The central area is still the coastal strip and the area immediately to the north. Only isolated finds have been made in the area around Bjäresjö, for example.

The material is too small to permit a detailed analysis of the settlement pattern, although a tendency towards the concentration of settlement in the late Middle Neolithic A and B can be observed in the Bjäresjö area. The coastal strip gives a distinct impression of having been the most attractive area during this period.

Late Neolithic. A noticeable increase can be observed in the number of finds from the Late Neolithic period compared with earlier periods. The datable daggers are

attributed in particular to the middle of the period, whereas it is most unusual for the early types to be encountered in the material. The field survey discovered no sites which could be dated to this period. Three reliable sites from the land making up the Ruuthsbo estate are known to us, however, from the Jacobæus collection. Unfortunately, the precise location is not known, with a single exception: it has proved possible to localize one site to the land making up the Nymölla estate at the mouth of the river Svartån. The material is small, and includes a spearhead and a shaft-hole axe. The other two sites were probably in the area between Ruuthsbo and the sea. Graves have been documented here, for the first time during the Neolithic period. Further details are given by L. Larsson (1991). These are three presumably flat-earth graves.

There thus appears to have been considerable Late Neolithic settlement in this area. All the evidence indicates a preference for the coastal strip in particular and the area immediately inland from it. No accurate assessment can be made of the size of the sites. We now know of a large number of sites with long-houses dating from this period, in several places in southern Scandinavia (Björhem and Säfvestad 1983, Larsson and Larsson 1984, 1986, Nielsen and Nielsen 1985). This suggests that small, village-like communities existed during the Late Neolithic. It is possible to discern a settlement pattern, with both isolated farms and small clusters of farms.

The attractiveness of the coastal strips during the period is also supported by evidence from Denmark, where the majority of sites in certain areas are situated within a distance of 300 m from the then existing coastline (Skaarup 1985).

In summary, it can be stated that both the Early and the Late Neolithic are well represented in the material from the Bjäresjö area, whereas the Middle Neolithic is less well represented. Although there are gaps in the material, it does permit a discussion of land use and settlement patterns.

The basis for any discussion of changes in the landscape in the Bjäresjö area must be the pollen diagram for Bjärsjöholmssjön (Ch. 4.2.2). The period following the disappearance of the elm around 3150 BC is characterized by an unstable forest ecosystem. It is possible that this may have forced man to utilize larger areas of land than before. Every individual settlement needed a greater area for arable farming and animal husbandry. This is indicated by the wide distribution of the pointed-butted axes, for example. The forests during the Early Neolithic give the impression of having been transformed into coppiced woodland or shrub and bush vegetation. This also means that the old theory of slash-and-burn agriculture may now be regarded as being out of date. One should instead envisage small, well-managed cultivated plots close to the sites. There are no direct indications of sites having existed around lake Bjärsjöholmssjön during the Early Neolithic. This is also supported by the palaeoecological conclusions which have been reached. One reason for this may be the absence of suitable easily cultivated soils around the lake basin.

A noticeable re-establishment of the common broad-leaved trees took place during the Middle Neolithic, from about 2500 BC. This was a synchronous occurrence over the whole of southern Scandinavia. A remarkable feature is that the curve for *Plantago lanceolata* (ribwort plantain) remains unbroken for the whole of the period. This can be explained by the transformation of the forest into coppices. The early part of the Middle Neolithic is associated with a decline in the number of recorded stray finds and sites in the interior of the area. There is a distinct displacement towards the coastal strip. In many areas of southern Scandinavia it is possible to observe a clear increase in the density of settlement within certain ecologically favourable areas (M. Larsson 1988b). A change to a more intensive form of agriculture took place from about 2600/2500 BC. The introduction of a simple form of cultivation based on the ard can be dated to this period (Thrane 1982). There is also a great deal of evidence to suggest that the importance of livestock increased during this period (Madsen and Jensen 1982). It is also possible to observe a move towards a society in which greater emphasis was placed on rituals in the form of monumental graves and the offering of sacrifices (Ch. 5.4). The emergence of locally characteristic groups in a number of areas in southern Scandinavia points to the increased importance of local territories (M. Larsson 1988b).

A noticeable opening-up of the landscape took place from about 2100 BC – the later part of the Middle Neolithic. This can be linked to the increase in the number of finds and sites in the interior parts of the area (Ch. 4.3.2). The growing importance of animal husbandry meant that larger areas of land were taken into use. There is also a good deal of evidence to suggest that entirely new social structures developed during this period, which can be associated archaeologically with the Battleaxe culture. There is an obvious pattern of development towards smaller, family-sized sites.

The cereal pollen curve remains unbroken from the next period onwards, the Late Neolithic. Bushes and shrubby vegetation develop. There is a particularly noticeable increase in hazel pollen. It is possible to imagine a system of coppiced woodland dominated by hazel bushes. The hazel is cut down, to be replaced by the cultivation of cereal grain crops in these areas. The archaeological material does not readily point to any expansion of settlement around the actual lake of Bjärsjöholmssjön. On the other hand, expansion is clearly visible in the area to the south of Ruuthsbo, i.e., the coastal strip. There is clear evidence in this area of increased settlement. The number of graves also increases during the period. At the same time as this settlement in the coastal area, there are also traces of expansion in the inner hummocky zone, often in the form of continuous settlement of those sites which had already been in use during the later part of the Middle Neolithic.

4.2.4 The Bronze Age and Early Iron Age
4.2.4.1 The cultural landscape in the Bjäresjö area during the Bronze Age

Deborah Olausson

This focal area shows marked activity during the Bronze Age, clearly greater than in the areas of Krageholm and Romele. Evidence from burial and settlement shows a pattern somewhat at odds with tendencies for consolidation at the coast seen in the preceding period (Ch. 4.2.3).

The Early Bronze Age

Barrows and burials. It is evident from Fig. 4.1.4, 4.2:4 that barrows are evenly distributed over the landscape within the zone up to about five kilometres from the coast. There are approximately 50 barrows registered in the Bjäresjö region (the parishes of Balkåkra, Bjäresjö, Hedeskoga, Bromma, and Ystad) in the Swedish Register of Ancient Monuments. In addition, 24 names in *hög* (a Swedish word for "mound", the occurrence of which on early maps is a good indication of a barrow, E. Porsmose, pers. comm.) can be found from a study of cadastral maps, earlier written descriptions, etc. (Riddersporre 1987). To this should be added stray finds which can be attributed to Early Bronze Age burials. Of the burials which can be plotted (excluding *hög* names), 12 are located in the sandy coastal zone and the rest in the outer hummocky landscape.

Single finds, hoards. This category is also well represented and it includes at least one outstanding hoard-find. In addition to Late Neolithic daggers there are 14 bronze objects whose find circumstances or type permit an interpretation of deposition in a non-burial context. The most outstanding and well-known Early Bronze Age hoard-find from the area is the "Balkåkra drum" (SHM 1461), found in 1847 in a bog in the village of Balkåkra and considered to be an import from Hungary (Montelius 1917:34, Freij 1977, Lund 1979). Another remarkable find bearing witness to connections with the south is the remains of the "Hedeskoga wagon" (SHM 2791). The bronze vessel has disappeared, but the wheels and undercarriage were found in 1855 in the former lake Bjärsjöholmssjön. Parallels to the type can be found as far south as Milavče, Czechoslovakia (Montelius 1874, L. Larsson 1984a).

Settlement. A total of about 86 ha, or 1.5% of the area, has been surveyed, some fields on repeated occasions

THE BJÄRESJÖ AREA

BRONZE AGE SETTLEMENT, CA. 1500 BC–500 BC[*]

Fig. 4.2:4

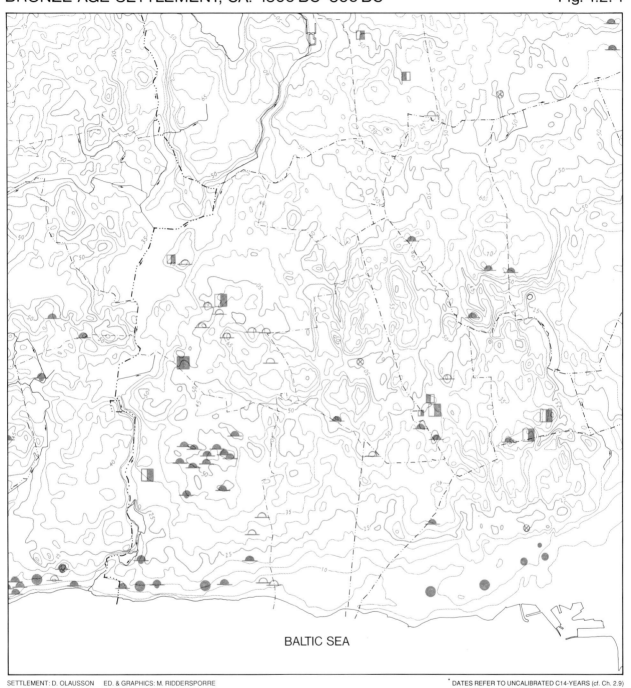

BALTIC SEA

SETTLEMENT: D. OLAUSSON   ED. & GRAPHICS: M. RIDDERSPORRE

[*] DATES REFER TO UNCALIBRATED C14-YEARS (cf. Ch. 2.9)

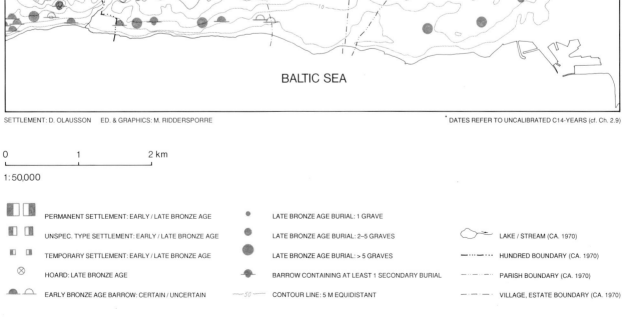

0   1   2 km

1:50,000

	PERMANENT SETTLEMENT: EARLY / LATE BRONZE AGE		LATE BRONZE AGE BURIAL: 1 GRAVE		
	UNSPEC. TYPE SETTLEMENT: EARLY / LATE BRONZE AGE		LATE BRONZE AGE BURIAL: 2–5 GRAVES		LAKE / STREAM (CA. 1970)
	TEMPORARY SETTLEMENT: EARLY / LATE BRONZE AGE		LATE BRONZE AGE BURIAL: > 5 GRAVES		HUNDRED BOUNDARY (CA. 1970)
	HOARD: LATE BRONZE AGE		BARROW CONTAINING AT LEAST 1 SECONDARY BURIAL		PARISH BOUNDARY (CA. 1970)
	EARLY BRONZE AGE BARROW: CERTAIN / UNCERTAIN		CONTOUR LINE: 5 M EQUIDISTANT		VILLAGE, ESTATE BOUNDARY (CA. 1970)

(Larsson and Larsson 1986:11f., Olausson 1988). In addition, the Central Board of National Antiquities (RAÄ, UV-Syd) has been active in the area since the 1970s, overseeing construction in connection with the expansion of the modern village of Svarte which has occurred since then. In spite of this, our knowledge of Early Bronze Age settlement in the area is embarrassingly slight. While reports of ploughed-up hearths from especially the sandy coastal zone are numerous, we have only four instances of concrete evidence for possible Early Bronze Age settlement in the form of surface finds of pits, radiocarbon-dated to this period, found at Bjäresjö and Hedeskoga (Olausson 1988: Table 1).

The Late Bronze Age

Graves. The number of Late Bronze Age burials in the Bjäresjö area is in remarkable contrast to the situation in the Krageholm and Romele areas. A minimum of 211 graves, 77% of which come from the area around the modern village of Svarte (Olausson 1987), can be counted here. Furthermore, there is a tendency towards clustering, in contrast to the dispersed pattern of the Early Bronze Age burial tradition. Another remarkable feature is the fact that of the burials whose location can be established (202), only the seven burials in the barrow at Skönadalssjön plus two other secondary burials in barrows lie outside the narrow sandy coastal zone.

A critical evaluation of the sources is necessary in order to establish if this pattern is real or apparent. Two factors speak in favour of the former. One is the clear preference of Late Bronze Age burial for sandy soils in the Köpinge area (Fig. 4.1:4). Secondly, the large number of graves registered in the parish of Bjäresjö is in part due to the efforts of Gustav Jacobæus, former owner of the Ruuthsbo estate and an avid amateur archaeologist (Olausson 1987:130). While his efforts mean a much higher find density on the Ruuthsbo grounds than in the rest of the Bjäresjö area, the spatial pattern of finds *within* the boundaries of this estate can be considered to reflect a real situation, since the chances of observation during ploughing can be assumed to have been equal on all the Ruuthsbo fields (i.e., both those at the coast and those located inland). Thus the pattern of Late Bronze Age burial being confined to the coast seems to be confirmed by this evidence.

Single finds, hoards. There are ten bronze objects, eight of which are socketed axes, found as stray finds in the area. True to a pattern seen elsewhere (e.g. Romele, Ch. 4.4.4) the axes are found widely scattered. Two hoard finds deserve special mention. The first is the hammered bronze vessel (SHM 7993) found in 1886 in the former lake Bjärsjöholmssjön. Montelius suggests that the Bjersjöholm vessel comes from Italy (Montelius 1889, cf. Thrane 1965). Another remarkable find,

which provides evidence about both settlement and economy, is the bronzesmith's hoard (YM 1388–1415) from Ystad (Oldeberg 1927).

Settlement. Whereas Late Bronze Age burials seem to be concentrated at the coast, evidence for settlement appears both at the coast and in the outer hummocky landscape. References to scattered hearths in the whole of the sandy coastal zone are frequent (e.g. Hansen 1924, Bodén 1977, Tesch 1981, Nagmér 1983, 1987). These are, however, undated and cannot be taken as positive proof of Late Bronze Age settlement.

Looking inland, we see indications of Late Bronze Age settlement at several locations. Settlement remains from excavated sites consist of pits containing large quantities of pottery and settlement debris. Engelmark interprets the contents of a pit at Bjäresjö 19:17 as settlement refuse, and suggests a permanent habitation site in the vicinity (R. Engelmark, pers. comm.). During excavations in Bjäresjö village, Callmer uncovered a number of similar pits filled with household rubbish. These could be dated by their contents to the Late Bronze Age (J. Callmer, pers. comm.).

The Bronze Age cultural landscape

In the virtual absence of direct evidence for Early Bronze Age settlement here, we must be content to rely on the barrows when reconstructing Early Bronze Age land-use patterns. The barrows show a dispersed pattern in the inland zone, with a slight tendency towards agglomerations at Ruuthsbo and Bjäresjö (Fig. 4.2:4). We can estimate that at least 90 Early Bronze Age barrows may have originally existed: a density of 1.6 barrows km^{-2} of land area. For the parish of Bjäresjö alone, the density is 2.2 barrows km^{-2}.

The large number of barrows, commonly dated to the Early Bronze Age, would indicate that the landscape was probably open by the Early Bronze Age. This is because open grassland is a precondition for building barrows (e.g. Thrane 1984a:116), which here in Scania are constructed of turf. Also, a certain degree of open landscape would have been necessary for the barrows to be visible, which we assume to have been one of their functions in the contemporary society. The building of the mounds contributed at the same time to the impoverishment of the land: about 0.8–1.1 ha of topsoil had to be stripped for each mound (Glob 1971:107, Kristiansen 1978:342). Between 1% and 2% of the Bjäresjö area (or 1.8–2.7% of Bjäresjö parish alone) would have been stripped of its topsoil during the course of the Early Bronze Age.

For the Early Bronze Age, the large number of barrows and bronze single finds, and especially the Balkåkra drum, suggests a generous subsistence base in this area, capable of supporting a number of families as well as providing a surplus. A mixed economy, combining traditional elements of Bronze Age subsistence (e.g.

Jensen 1981, 1982) with supplementary sources can be suggested for Bjäresjö. Settlement would have been dispersed in the outer hummocky zone, and mobile. Livestock grazed freely all year round in the open forests and grasslands growing on the clay till of this zone, which may have functioned as common pasture (a pattern seen on the island of Gotland: Lundmark 1986:40). The area even further inland could also provide leaf-fodder and/or grazing (see Ch. 4.3.4), as well as resources obtainable through hunting and gathering. In addition the sea, with its marine resources and possibilities for trade and transportation, must have been an important element. This pattern is comparable to that suggested by Kristiansen for Early Bronze Age Zealand (Kristiansen 1978:334).

The subsistence pattern stabilized in the Late Bronze Age. By now the pollen evidence shows that the landscape around lake Bjäresjösjön was nearly completely open and covered by grass or under cultivation (Ch. 4.2.2, 4.2.4.2). The area continued to be able to support several extended families and provided for surplus production which in turn could be used for importing objects from great distances, such as the Bjäresjö vessel. The subsistence base would have included growing crops such as barley, emmer, millet and gold of pleasure, and keeping livestock. Manuring may have been practised, and the lack of weeds suggests that each field was used only a short time (R. Engelmark, pers. comm.). The phenomenon of burial concentration at the coast may reflect a Late Bronze Age tendency towards a higher degree of settlement consolidation. However, as most studies of Bronze Age settlement have shown that burial rarely occurs more than one kilometre from settlement (e.g. Kristiansen 1978:331, Strömberg 1982:158), the fact that burials are lacking inland, where we appear to have evidence for settlement, seems contradictory. If the explanation is not source-critical factors (see above), I would suggest that the cemeteries at the coast represent a fixed point in a semi-permanent settlement pattern. The existence of a cemetery can be a group's means of affirming control over a territory or over certain resources (Goldstein 1981:61). Kristiansen (1978:331) notes that the younger Bronze Age settlement concentrations in north-west Zealand are 5–10 km apart. The Svarte area lies between two such concentrations: about 15 km west are the Bronze Age settlements on the Ystad plain (Ch. 4.1.4), while the complex at Abbekås/Øremölla lies 8 km to the east.

A mixed economy combining livestock, farming, and fishing complemented by bronze-working (as at Hallunda: Jaanusson 1981:30) can be suggested for the Bjäresjö area in the Late Bronze Age. The quality and quantity of the grave-goods in the burials at Svarte show that this strategic location, where the inhabitants had access to the forests further inland, control of the mouth of the river Svartån, and access to the sea with its marine resources and means of transportation, meant that this area of clay soils was capable of supporting a number of families through the whole of the period (Olausson 1987:141).

4.2.4.2 The agrarian landscape of Late Bronze Age farmers in Bjäresjö

E. Gunilla A. Olsson

Introduction

To illustrate the environmental conditions and the basis of Bronze Age agriculture in the Bjäresjö area, a landscape reconstruction for settlements at lake Bjäresjösjön will be presented below. The representation is based on archaeological artefacts, palaeoecological evidence, geological surveys of bedrock and soil conditions, and recent vegetation maps of the area. A combination of these different data sets and general knowledge of vegetation dynamics is used to generate hypotheses about the ecological content of the agrarian Bronze Age landscape.

Site description

The area of the landscape reconstruction embraces the eastern side of the river Svartån and the surroundings of the small lake Bjäresjösjön within an area of around 2 × 3.5 km (Fig. 4.2:5). The hilly landscape is covered by clayey till soil with high lime content (up to 20%), but coarse sand and loam cover the hilltops as a result of post-glacial water movements. Height differences within the area are about 15 m. In the Late Bronze Age, around 800 BC, the lake was created by continuous embogging of alder carrs in a depression between the hills (Ch. 4.2.2). The western part of the area descends into the wide valley of the present river Svartån. Extensive areas of wetlands including alder carrs were characteristic features here. Existing woodlands were composed of oak, elm, lime, and hazel (Ch. 4.2.2).

Methods and sources

Assumptions for the reconstruction. Compiled field data from excavations in the Bjäresjö area (Callmer et al. 1991, Olausson in press), in the Ystad area (Strömberg 1975, Tesch 1988), and published results from similar areas in southern Sweden and in Denmark (Lepiksaar 1969) are used for assumptions about the size of human population and the composition of their diet. From these sources the mean human population of each settlement has been estimated to about 5 adults and 5 children (cf. Olausson in press).

For this study a protein-based model of composition of dietary intake was used. For details of methods of calculation, published sources, basic assumptions about

LATE BRONZE AGE LANDSCAPE IN BJÄRESJÖ, CA. 800-500 BC*

* DATES REFER TO UNCALIBRATED C14-YEARS (cf. Ch. 2.9)

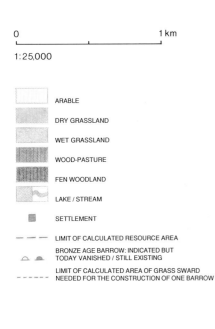

0   1 km

1:25,000

ARABLE

DRY GRASSLAND

WET GRASSLAND

WOOD-PASTURE

FEN WOODLAND

LAKE / STREAM

SETTLEMENT

– – – – LIMIT OF CALCULATED RESOURCE AREA

BRONZE AGE BARROW: INDICATED BUT
TODAY VANISHED / STILL EXISTING

LIMIT OF CALCULATED AREA OF GRASS SWARD
– – – – – NEEDED FOR THE CONSTRUCTION OF ONE BARROW

LANDSCAPE RECONSTRUCTION: E. G. A. OLSSON
GRAPHICS: M. RIDDERSPORRE

Fig. 4.2:5. Bjäresjö. Late Bronze Age landscape in Bjäresjö.

Table 4.2:1. Inferred dietary composition during the Late Bronze Age in Bjäresjö as suggested by the archaeological and palaeoecological findings. The protein contents of the different food items are extracted from modern tables (Livsmedelstabeller 1978). The data below are calculated for subsistence level for a settlement site inhabited by 7.5 human adult units.

Item	% dietary contribution	kg protein adult^{-1} yr^{-1}	kg food adult^{-1} yr^{-1}	total no. consumed	no. of stock animals yr^{-1}
barley	30	5.7	40.7	–	–
emmer	15	2.9	14.2	–	–
Camelina	1	0.2	1.1	–	–
nuts etc.	1.5	0.3	2.3	–	–
fish (pike)	0.5	0.1	0.5	–	–
cattle	28	5.3	25.3	2.3	8
sheep	5	1.0	4.8	2.1	7.5
pigs	5	1.0	4.8	1.3	3
horse	12	2.3	10.9	0.7	3
eggs etc.*	2	0.4	2.0	–	–

* This heading includes not only eggs from wildfowl etc. but also meat from game, although this is a very small item (2%).

Table 4.2:2. The relationship between seed return and arable area in Bjäresjö during the Late Bronze Age. Calculation valid for subsistence consumption according to Table 4.2:1 of 7.5 adults – see text.

Crop	Seed return	Arable area (ha)
barley	6	0.40
emmer, spelt	6	0.14
Camelina	6	0.01
Total arable		0.55
barley	4	0.64
emmer, spelt	4	0.22
Camelina	4	0.02
Total arable		0.88

the body weight of prehistoric humans, etc. see Ch. 2.2, 4.1.3.3, 4.1.4.3. The field finds from the excavations form the basis for the assumption that the human diet consisted to approx. 50% of vegetarian food items, among which cereals dominated (cf. Olausson in press, Tesch in press, Engelmark unpubl.). The animalian contribution was dominated by meat from domesticated animals (Olausson in press).

The size of the resource area and of arable land was calculated from the number of livestock and cereal consumption deduced from the human diet need. The dominating cereal was *Hordeum vulgare* (hulled barley), followed by *Triticum dicoccum/spelta* (emmer/spelt wheat) (Engelmark unpubl.). The large number of seeds from *Camelina sativa* (gold of pleasure) found in Bjäresjö indicates cultivation of this plant as a crop, mainly for its oil (Engelmark unpubl.). Two alternative calculations were made: with estimated seed return 4 and seed return 6 respectively.

Results and discussion

Dietary composition in terms of protein contribution is listed in Table 4.2:1. These estimates of cereal consumption would imply a yearly harvested field area of 0.6–0.9 ha (Table 4.2:2). Most likely the real arable area approached the lower limit obtained with a seed return of 6. Even this could be an underestimation considering the high soil fertility in this area, and especially assuming that manure was used (cf. Garner and Dyke 1969, Lüning and Meurers-Balke 1980, Olsson unpubl.). Seeds of *Chenopodium album,* a nitrophilous weed dispersed with manure, are found in Bronze Age remains in Bjäresjö (Engelmark unpubl.). These findings, together with the change in dominating crop from emmer to hulled barley, are indications of the presence of nitrogen available for the crop plants. The nitrogen need could have been covered by the addition of manure and household wastes, or a combination of manuring with short-term coppice rotation among the arable (cf. Göransson 1988) in almost the same way as postulated for the Middle and Late Mesolithic in Kabusa (Ch.

4.1.3.3). Since we have indications of short-distance movements of settlements sites – only a few hundred metres (Ch. 4.2.4.2, 4.2.4.3, 4.1.4.1) as in the Neolithic (Larsson 1987), it is tempting to assume that the temporarily deserted settlement sites were used for cultivation, hence utilizing the deposited phosphorus here.

The archaeological findings from Bjäresjö (Olausson et al. unpubl., in press, Callmer in press) give almost no indication of fish and seafood in the human diet (Table 4.2:1). The vast majority of the animal protein was obtained from domestic animals. The diet contribution from different stock animals differs from that of the Neolithic, at least judging from the bone remains (cf. Lepiksaar 1969, Larsson 1987, Olausson in press, Ch. 4.1.3.3). During the Late Bronze Age the horse was introduced, and the stock composition changed to favour cattle and horses (Lepiksaar 1969). For a human population of the same size as in the Neolithic (5 adults and 5 children per settlement) the calculated amount of annually consumed meat, transformed into the number of slaughtered animals, implies a stock size of (adults units) 8 cows, 3 horses, 3–4 pigs, and about 8 sheep (Table 4.2:1). The total resource area needed, including arable and fodder collection areas, would then amount to at most 40 ha (combined data from Coupland (1979) have been used for this calculation). This is about the same size as the resource area calculated for Neolithic family groups, although they lived in an environment with sandy soils and ample marine ecological resources (Ch. 4.1.3.3). If leaf-fodder from coppices was collected in large amounts the total fodder area would have been still smaller. There is some archaeological evidence of winter fodder collection, such as indications of dwellings in the Köpinge area suggesting a byre function (Ch. 4.1.4.1), and findings of leaf sickles from the Bjäresjö area (D. Olausson, pers. comm.). Assuming that coppicing of the dominating trees and hazel occurred (cf. Göransson 1988), it is probable that the products obtained – twigs and leaves – were used only partly for stock fodder and partly for building and fuel purposes. Maybe the main purpose was to obtain decent cultivation conditions (in terms of nutrients) founded upon practical experience.

Pollen diagrams from the Late Bronze Age in this

Table 4.2:3. The relationship between the number of stock and their foraging area during the Late Bronze Age in Bjäresjö. Foraging area is calculated for human subsistence consumption of stock, and for stock number on an enlarged level: 3 times subsistence consumption of cattle, horses, and pigs, and 5 times the human subsistence consumption of sheep – see text.

Number of stock	Foraging area (ha)
Subsistence level from Table 4.2:1	38.5
Enlarged level: 3 × (cattle, horse, pig) 5 × sheep	131.0

region and from Bjäresjösjön indicate a landscape containing considerable areas of grassland compared with preceding periods (Ch. 4.2.2). It is possible that quite extensive burning was applied with the intention of creating and maintaining suitable areas for cultivation and for grazing. It is also possible that an open landscape surrounding the barrows may have been maintained for religious reasons. Traces of charcoal in the lake sediment core may indicate fires, although their extent and frequency cannot be determined (cf. Ch. 4.2.2). The activities suggested by the pollen diagrams indicate a more intensive use of the environment – and also an increased human population compared with the Middle and Late Neolithic. Assuming that family groups were of the same size as in the Late Neolithic, and accepting the size of resource area presented here (Table 4.2:1, Fig. 4.2:5), the number of settlements must have been larger than before to produce this heavy environmental impact.

The pollen diagram (Gaillard and Berglund 1988) indicates that existing woodlands in this area were dominated by oak and hazel, with only minor contributions from elm, lime, and ash. Large amounts of black alder pollen testify to extensive alder carr along the river Svartån valley and probably also in the wet bottoms of the numerous hollows between the hills. The environment surrounding the settlement sites contained a mosaic of small cultivated plots (less than 0.5 ha) of which some were in succession with shrubs and coppice of hazel and elm(?), and wood-pasture formed by browsing and grazing stock. Open grassland areas existed in this mosaic, but they were probably only on a large scale in areas where a concentration of barrows occurs, west and south of the lake and along the coastal hummocks (cf. Ch. 4.2.4.1, Fig. 3.3:1). Outside this open grassland regions the Late Bronze Age landscape in the Bjäresjö area was patchy owing to the varied soil conditions and the undulating topography, compared with the uniform sandy grounds in the Köpinge area.

The reconstructed resource areas for the four settlements are small in relation to the cultural influences suggested by the pollen diagrams. Historical documents and maps indicate the occurrence in this region of more barrows than those remaining today. In Fig. 4.2:5 both existing and vanished barrows are marked. The surrounding circles suggest the area of fresh grass sward needed for the construction of a single barrow (2–4 ha, Thrane 1984). The maintenance of the indicated barrow areas is well above the estimated need for grazing land calculated for the four settlements. My model differs from the interpretation of the pollen diagram by Gaillard and Göransson (Ch. 4.2.2) in that it does not account for the extent of deforestation; to arrive at the same interpretation it would be necessary to introduce explanatory factors such as extensive use of fire or larger numbers of stock. Calculations of stock foraging areas related to a number of livestock on levels exceeding human subsistence level have been performed – on

the same principles as the interpretation for Stora Köpinge (Ch. 4.1.4.3). Considering the vegetation and environmental conditions in Bjäresjö in relation to the estimated fodder need for the livestock, the model predicts that livestock at a level three times the human subsistence consumption would have been allowed if the three settlement sites were to be used at the same time (Fig. 4.2:5, Table 4.2:3).

4.2.4.3 Settlement and landscape in the Bjäresjö area in the Early Iron Age

Johan Callmer

The Pre-Roman Iron Age

In the Pre-Roman Iron Age the main features of settlement scarcely changed in any decisive way. Settlement probably became somewhat more stable; movements to new house-sites occurred at longer intervals than before. It is possible that the population grew slightly in this period, as is suggested by an increase in the number of settlements and a consequent expansion of the cultivated landscape. On the other hand, most of the settlements were still relatively small, consisting of a small number of farms. The conditions do not allow a more detailed study, and the finds from the Bjäresjö area are few. The above description is mainly based on results from comparable areas in western Scania (the Malmö region).

From the intensive study area we can observe that settlement by lake Bjäresjösjön did not change to any great extent. Probable evidence of settlement in the period comes from potsherds on the ridge just north of the lake. The area utilized was, as before, the immediate vicinity of the lake. In the Pre-Roman Iron Age (if not earlier) people went over to the cultivation of barley in systematically manured fields (Engelmark 1989). Manuring in itself was also a factor in the stabilization of settlement on account of the investment of nutrients, and the mould-forming process thereby stimulated. Arable land was, as already in the Late Bronze Age, arranged in large systems of block-shaped fields, of which only a few were under cultivation simultaneously (Odgaard 1984, Hedeager and Kristiansen 1988:151ff., Engelmark 1989, Ch. 5.2). It is uncertain whether these field systems in the hummocky landscape were the same as the classical Celtic fields of western Europe. Nielsen's results (1984) from eastern Denmark show instead irregularly arranged rows of stones. On heavier soils as at Bjäresjö, drainage must have been a problem (cf. Ch. 5.2).

The forms of social organization are unlikely to have been either more or less complex than before. Habitation site material from the eastern part of south Scandinavia is not illuminating enough to allow us to draw

THE BJÄRESJÖ AREA

ROMAN IRON AGE SETTLEMENT, CA. 0-AD 400

Fig. 4.2:6

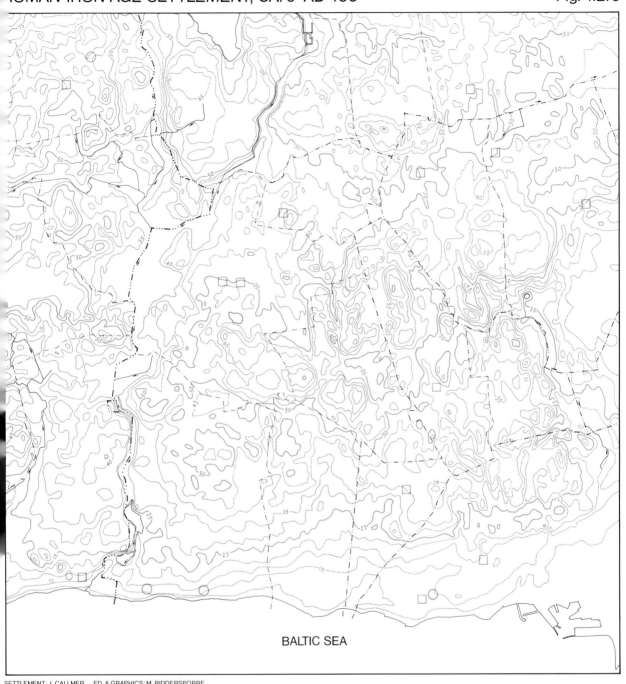

BALTIC SEA

SETTLEMENT: J. CALLMER   ED. & GRAPHICS: M. RIDDERSPORRE

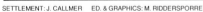

0   1   2 km

1:50,000

○ BURIAL: 1 GRAVE	□ SETTLEMENT	········· HUNDRED BOUNDARY (CA. 1970)
◯ BURIAL: 2–5 GRAVES	⬭ LAKE / STREAM (CA. 1970)	─··─··─ PARISH BOUNDARY (CA. 1970)
◯ BURIAL: > 5 GRAVES	─50─ CONTOUR LINE: 5 M EQUIDISTANT	─ ─ ─ VILLAGE, ESTATE BOUNDARY (CA. 1970)

extensive conclusions. Finds from graves and hoards are also very few and indistinct. Nevertheless, the gradual change towards agglomerated settlements, which we must assign mainly to this period, undoubtedly brought about significant changes when the kin group as a social identification was complemented by the community – and also, in at least some cases, the hierarchy – of the agglomerated settlement (cf. Callmer, Ch. 5.6).

The Roman Iron Age

In the Roman Iron Age we have a clearer picture of settlement than that provided by the material from the preceding period. In southern Scandinavia we see a general trend towards larger buildings, bigger farms, and probably also in many cases somewhat larger settlements (Hedeager and Kristiansen 1988:116ff.). It is unclear whether larger settlements of this kind existed in the area under study here. The topography of the landscape must have favoured small-scale settlement in the Bjäresjö area, and in the hummocky landscape as a whole. Settlement probably consisted in large measure of a few farms grouped more or less closely together. As in the Late Bronze Age, the population of the hummocky landscape was well integrated in a larger social system. In the Roman Iron Age this can have encompassed the greater part of the Ystad area. Observations of boundary defences and the distribution of settlement from Denmark and northern Germany suggest that there were now large tribal territories functioning as strongly integrating regional units in neighbouring areas (cf. Jankuhn 1964). Of course, we do not know which larger community and ethnic identity the Ystad area or its component parts belonged to (see Fig. 4.2:6).

Judging by the number of habitation sites, settlement in the Bjäresjö area can hardly have increased significantly in the Roman Iron Age; on the contrary, it is probable that there was a slight concentration of settlement. Larger agglomerated settlements with more inhabitants appeared. There was a gradual increase in cereal cultivation throughout the Roman Iron Age, but no such increase in the areas used for pasturage and fodder. We may suppose that the systematical manuring introduced during the Pre-Roman Iron Age was mainly practised in a system of Celtic fields. Since the gradual expansion of the area cultivated for grain was not accompanied by any noticeable increase in the size of herds, the supply of manure must have become an ever greater problem. This field system was abandoned throughout north-west Europe in the Roman Iron Age (Kossack et al. 1984). It is difficult to determine what succeeded the Celtic fields in the Bjäresjö area, but it is probable that the more extensive older system was replaced by some form of one-course cultivation.

The population of the hummocky landscape, like the population of the coastal zone and the more cohesive settlement on the plain, benefitted from the socially integrated system of exchange which the age brought.

Roman glass, for example, has been found in a pit near the church in Bjäresjö. Imported beads of Roman manufacture often occur on habitation sites. It is noteworthy that such finds are made despite the limited volume of material from the period. The system of exchange must have played a major role. Ultimately these imports are linked to the production of trading commodities in the Roman Empire. A well-developed system of exchange in the Ystad area could have included some food production. The conditions existed for regional exchange based on some specialization in production in different types of landscape. Important for the society of the day were the changing circumstances affecting iron production. The area was probably supplied with iron mainly by the plentiful iron ore in the southern part of the sandy plain of Vomb. There may also have been local iron production in the hummocky landscape. Iron technology took on a much greater significance in the Roman Iron Age. Axes, knives, and sickles encouraged more efficient production, not least of fodder.

The picture of settlement in the intensively studied area by Bjäresjösjön shows that settlement from the Roman Iron Age (probably the oldest phase) was still exclusively located on the low ridge immediately north of the lake. Unfortunately, the excavations, which were of necessity highly limited, have not enabled us to paint any detailed picture of settlement here. The relative frequency of potsherds shows that these cannot have been spread in connection with manuring, but must be associated with a habitation site in this sector.

Finds from later parts of the Roman Iron Age – a number of pits and possible remains of post-built houses oriented east-west – were also made on the hill north of this settlement location. It is difficult to date the potsherds more exactly. Much clearer datings are provided by the finding of a fibula (the middle of the Roman Iron Age) and a glass bead (later in the period). In the area around the lake itself we may cautiously conclude that settlement moved from the southerly location near the lake to the somewhat higher position on the hill to the north. Whether this is a change of a profound structural nature cannot be deduced from the Bjäresjö material. As we have seen, a number of investigations in comparable areas of southern Scandinavia indicate that there was a significant change in the direction of a more stable cultivation system, which also led to longer occupation of habitation sites. To all appearances, there was a trend towards greater stability at Bjäresjö also. The pottery finds show that the new location of the habitation site was still occupied in the early part of the Migration Period.

This habitation site at Bjäresjö is by no means the only such site in the vicinity. About 1 km NNE, an area of about a hectare with remains of a habitation site from the Roman Iron Age has been documented. The location is very similar to that at Bjäresjö. Pottery and hearths have been found on the top and on the southern slope of a very similar large hill. This site is separated from the Bjäresjö site by only a wet bottom through which flows the stream that drains lake Bjäresjösjön. The pottery finds from the site are probably exclusively from the Roman Iron Age. It is not improbable that this settlement ceased to exist and was incorporated into the settlement in Bjäresjö village, but this cannot be proved with the existing material.

Apart from the material described here, the only finds in the Bjäresjö area from the Roman Iron Age are grave-goods. At Balkåkra 11 near the village of Balkåkra an inhumation grave was uncovered during agricultural work (Stjernquist 1955:169). The grave belongs to the later part of the period. The grave-goods included three fibulae, beads, and a bronze box, so the burial must be considered a rich one. Three cemeteries have been found along the coast, on the littoral terrace (Strömberg 1960:59). The westernmost cemetery is that at Svarte, and about 1 km away lies the one by the road up to Ruuthsbo, and 1.5 km further east is the one at Gustavsfält. These are all cremation graves, some of which may be from the end of the Pre-Roman Iron Age, but most of which are from the Early Roman Iron Age. Remains of contemporary habitation sites have been found close to the cemetery at Svarte. As for the other two cemeteries, it is also very likely that they too were situated near habitation sites. This concrete evidence for human occupation of the coastal zone thus indicates dense settlement with contemporary habitation sites at fairly regular intervals in the landscape.

4.2.5 The Late Iron Age and Early Middle Ages
4.2.5.1 Late Iron Age and medieval settlement development at Bjäresjö

Johan Callmer

The Migration Period and Vendel Period

As in other parts of the hummocky landscape with its heavy soils, it is very difficult to find evidence in the Bjäresjö area of settlement from the early phase of the Late Iron Age. The chance of finding habitation sites from this period by field-walking is practically nil (cf. Grøngaard Jeppesen 1981:87ff.). The actual number of remains of habitation sites is also much smaller than that for the Early Iron Age. Judging by recent results of field investigations in connection with excavations for new buildings and gas pipelines in southern Scandinavia, it is not improbable that the number of surviving habitation sites from the Early Iron Age is at least four times greater than that from the Late Iron Age (excavation statistics UV-Syd, Malmö Museum, Näsman 1987). As has been pointed out, this change appears to have begun in the Roman Iron Age, probably at the transition Early/Late Roman Iron Age. The transitional

phase may very well have been a lengthy process, and the change probably had very different effects in different parts of the country. The material from the Bjäresjö area is, as expected, very modest.

In the part of the area nearest the coast there is at present no evidence from Svarte allowing us to follow the cemetery from the Bronze Age and the Early Iron Age. From the western cremation cemetery at Ruuthsbo, mostly dated to the Early Iron Age, a grave from the Migration Period has been studied (Strömberg 1961:2, 26); this suggests that settlement on nearby habitation sites continued into this period.

According to information from Bruzelius (1869:153) and old maps, there were several stone ships adjacent to the eastern cemetery at Ruuthsbo (Gustavsfält). The form of the graves suggests that this cemetery was used from the Roman Iron Age far into the Late Iron Age. Of the graves and finds exposed at random, a bird-shaped brooch is concrete evidence of the use of the cemetery in this phase. The cemetery was still in use during the Viking Age. It is probable that there was a habitation site belonging to the cemetery right on the coast below. A number of similar settlements are known along the south coast of Scania.

There is otherwise no knowledge of graves from this period in the Bjäresjö area, and remains of habitation sites are highly limited. In conjunction with the investigation of the medieval chapel of Borrsjö (Marsvinsholm) and parts of the adjacent graveyard, finds were made of pottery from the Migration Period, probably from destroyed occupation layers. This site is very interesting in that it lies very near the site of the historical village of Borrsjö. Although there is little other material here from the pre-Viking Late Iron Age, it is very likely, in view of the important localization factor of lake Borrsjösjön, that there was continuity in settlement on the site of the village and in the immediate vicinity.

In the intensive study area comprising the lake basin of Bjäresjösjön, there is only slight evidence of settlement. However, it is found in the same areas which we have been able to show were preferred habitation sites in earlier periods. Excavations on the large hill north of the lake uncovered sherds of an earthenware vessel with a design typical of the early part of the Migration Period. The find was in the eastern end of the excavated area, near the churchyard wall. It was impossible to establish whether the vessel was associated with a grave, a pit, or an occupation layer. In the same area, only 10 m to the west, was found a pit with stamped, hand-made pottery coming from not very late in the Vendel Period, probably the 7th or 8th century (Strömberg 1961:1, 132).

These admittedly modest finds from different intervals indicate continued settlement on the hill up to the Early Vendel Period. By contrast, material from the Late Vendel Period and the Early Viking Age seems not to be represented in the area investigated on the hill. It is impossible to be sure if settlement after the Early Vendel Period remained in the same place without leaving any traces in the archaeological evidence, or if it was shifted to the ridge just north of the lake. Practical reasons made it impossible to undertake anything but partial trial excavations on the top and the southern slope of the ridge. However, finds from the top of the ridge suggest that there are remains of settlement from the Late Iron Age in the area. Unfortunately, the major part of the best conceivable habitation site is occupied by a modern farm and has thus been too badly destroyed to allow investigation.

In the Late Iron Age we have stronger indications of stability in settlement and continuity in small focal areas (Callmer 1986). In the Migration Period and Vendel Period we see the definitive changes leading to the cultivation system typical of the medieval landscape. Particularly noticeable is the upsurge in tillage late in the Vendel Period. The introduction at roughly the same time of autumn-sown rye shows that an important innovation in cultivation arrived at this time (Engelmark 1989, Ch. 4.2.2). It probably had the character of organized rotation of some kind, in two or three courses.

The sharp decline in cultivation in the Migration Period reported from various parts of Scandinavia cannot be traced in the pollen diagrams from the lakes of Bjäresjösjön and Bjärsjöholmssjön. Nor can any such decline be deduced from the purely archaeological evidence.

It is unclear whether this period saw an increase in the number of households within the agglomerated settlement which we can presume to have existed at Bjäresjö since the Late Roman Iron Age. The heightened impact of cultivation that can be discerned may indicate only that cultivation increased. It is probable, however, that there was a real growth in settlement towards the middle of the Viking Age.

The Viking Age and the Early Middle Ages

As we have seen, material from the Early Viking Age is scarcely known from the Bjäresjö area strictly defined. For the Late Viking Age, however, there is more concrete archaeological source material (Fig. 4.2:7).

From the actual coastal zone two or three inhumation graves in the eastern cemetery at Ruuthsbo (Gustavsfält) can be dated to the 10th century (Strömberg 1961:2, 25–26). In the course of the 10th century, however, a drastic change occurred: the coastal district was totally abandoned (Callmer 1986:200). The coast as a habitation location was given up in favour of settlement a kilometre or two inland. This is part of a general abandonment of the coasts throughout southern Scandinavia. The explanation can hardly be economic, but must have something to do with insecure conditions for the population. This rather remarkable change must

also have meant changes for the people already living inland, causing pressure on habitation sites there. We cannot be sure whether the homeless people from the coasts were assimilated by the inland population or were forced to find new land on the periphery of the settlement region.

In the area we have primarily indications of fairly extensive settlement in the intensive study area by Bjäresjösjön. Traces of settlement can also be noted from a number of places in the area, all close to the historical villages. Early pottery of the Black Earthenware type known from the Late Viking Age and Early Middle Ages has been found during excavations at Skårby, Balkåkra, Bjäresjö village, Lilla Tvären (Källesjö), and during field surveys at Gundralöv (Ruuthsbo), Borrsjö (Marsvinsholm), Hunnestad, and Hedeskoga (LUHM and UV-Syd). Runic stones are known in their original sites from the following villages or their immediate vicinity: Skårby, Hunnestad, Gundralöv, Bjäresjö. There is thus every reason to assume that settlement, probably of village type, existed from this time in the immediate vicinity of the historical villages or on the actual village site.

In the intensive study area of the Bjäresjösjön basin, settlement from this period has been documented on the ridge north of the lake, in the westernmost part of the historical village site, and especially on the hill where the church stands today. It thus appears well documented that the greater part of the historical village site, except the parts to the far north and east, was occupied. At the end of the Viking Age a settlement feature was established on the top of the hill; it distinguishes itself not only by the richer finds but also by the continuity with which it later develops into a distinct early medieval manor complex. The settlement appears to consist of a number of big post-built houses, oriented east-west. Among the finds we can draw special attention to early coins (Hardaknut and Sven Estridsen, and German coins from about AD 1000) and belt-fittings in Scandinavian and oriental style (Jansson 1978). These artefacts are closely associated with the warrior aristocracy of the time. Rich finds of pottery were also uncovered, especially during the stripping of the north-eastern part of the area and in sondages east of this.

By all accounts the Late Viking Age settlement lay along the southern slope of the hill and its western spur. A further 150 m to the east, structures and occupation layers probably datable to the Late Iron Age and the Early Middle Ages have been observed. It is thus probable that the settlement at the end of the Iron Age and the start of the Middle Ages stretched in an east-west direction for about 200 m; the topography suggests about 250 m. There was also contemporary settlement on the southern side of the stream which drained the little lake or fen (the modern fire pond). Occupation layers and structures from this period have been documented on the crest of the ridge north of the lake. On account of the difficulties already mentioned (the mod-

THE BJÄRESJÖ AREA

LATE VIKING AGE SETTLEMENT, CA. AD 1000

Fig. 4.2:7

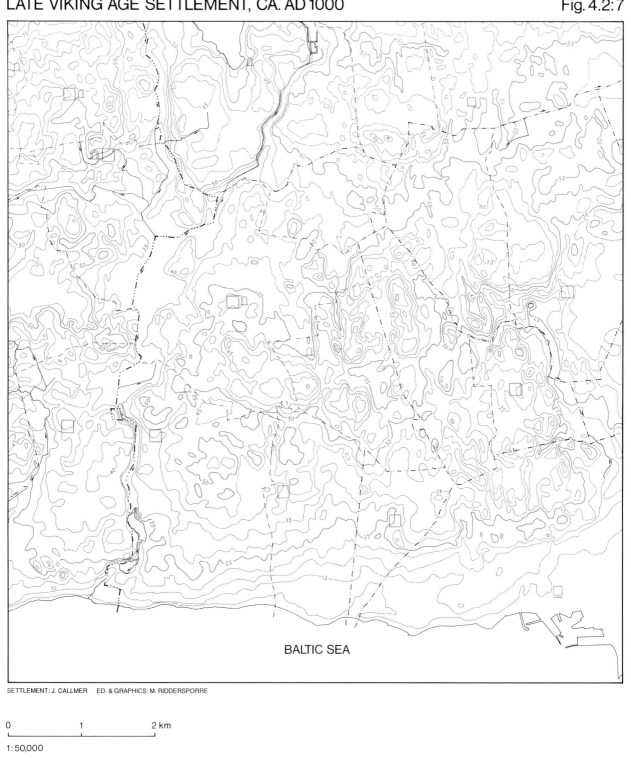

BALTIC SEA

SETTLEMENT: J. CALLMER   ED. & GRAPHICS: M. RIDDERSPORRE

0   1   2 km

1:50,000

☐ PERMANENT SETTLEMENT	
☐ UNSPEC. TYPE SETTLEMENT	
☐ FISHING CAMP	

⌂ RUNIC STONE

⌂ MONUMENT INCLUDING RUNIC STONES AND PICTURE STONES

—50— CONTOUR LINE: 5 M EQUIDISTANT

◠ LAKE / STREAM (CA. 1970)

—··—··· HUNDRED BOUNDARY (CA. 1970)

—·—·— PARISH BOUNDARY (CA. 1970)

— — — VILLAGE, ESTATE BOUNDARY (CA. 1970)

ern farm buildings), it was possible to investigate only small areas. It is probable, however, that settlement stretched for about a hundred metres from the southwest to the north-east.

At the end of prehistoric times, and probably much earlier, settlement covered a considerable area. A village-like cluster lay on either side of the little stream flowing into Bjäresjösjön. The greater part of the settlement was on the northern side, with only a small part on the south. The remaining parts of the historical village site were not yet settled. The manor complex on the hill had a high location in relation to the rest of the settlement. We do not know if there was a heathen cult site at the settlement. What we do know is that a stave church was erected early in the Middle Ages, probably on the site of the present-day church, a short distance southeast of the manor complex (Ch. 5.7). From the close spatial connection between the manor and the church it is clear that the stave church was probably from the beginning attached to the manor and thus not a parish church in the ordinary sense.

In the Viking Age and Early Middle Ages there was a slight increase in the scope of cultivation. This may be associated with a certain extension of settlement beyond hitherto cultivated areas. The settlement may have grown in conjunction with the abandonment of the settlements by the coast. Farmers had to move and either join nearby villages, form new villages, or establish isolated farmsteads on the periphery of the settlement district.

In the 12th century the manor complex and the church underwent a radical change. Although settlement within the manor complex had already distinguished itself by its high location and perhaps also by the nature and size of its buildings, a restructuring of settlement here gave the manor an even more distinctive character. In the middle of the 12th century the old stave church was replaced by a Romanesque stone church. The church appears to have had a western tower, a typical feature of patronized churches, and was decorated with magnificent Romanesque murals. At roughly the same time, a new manor house was built, partly of the same material as the church. The manor and church now constituted a uniform architectural complex, intended to impress (cf. Stiesdal 1980). There were probably no extensive defences. The steep slopes of the hill, especially to the east, made it easy to defend – a simple palisade on the slope would have been a sufficiently difficult obstacle to deter an assailant.

The character of the manor is also evident from the rich artefacts found here, including imported pottery. Perhaps the most distinctive feature is the copious presence of coins. No less than forty coins have been found in the manor area west of the church. No coins have been found outside this area, and finds of coins are otherwise very uncommon in contemporary agrarian environments. All the coins except one come from the period up to the mid 14th century.

In the 13th century the old manor house was demolished and on the site was erected a large house, either timber-framed or built according to a similar technique (cf. Stiesdal 1979). This was not the new manor house. This building was later demolished and the plot replanned at the middle of the 14th century. Among the building debris brickwork is prominent. Some of the bricks were glazed. These were doubtless used for one or more large buildings of which we know nothing today. It is probable that the new manor house was of brick. We have little knowledge of the location of the building, but it is not unlikely that it was closer to the church. About 15 m south-west of the present church tower a probable foundation of large stones and remains of a brick floor were recently observed during the digging of a grave.

It thus appears as if the first half or the middle of the 14th century saw the last of the manor complex and the end of a manor house in the village. By all appearances the residential part of the manor was moved from Bjäresjö village to Bjersjöholm ca. 1 km to the south-east.

In the Late Middle Ages the village was also transformed when the areas on the northern and eastern edges of the historical village site were occupied. We cannot tell whether this was because the number of farms in the village rose or there was a replanning of the site in conjunction with a regulation of the village; the latter alternative is more plausible. By the 15th century, then, Bjäresjö village had probably taken on the appearance it retained until our oldest maps.

In the Bjäresjö area written sources and archaeological observations suggest that a number of villages underwent identical or similar development to Bjäresjö, with manors going back at least to the Late Viking Age. Stable social stratification thus typified a considerable part of the settlement that existed in the area in late prehistoric times and in the Early Middle Ages.

4.2.5.2 The agrarian landscape of Viking Age farmers at Bjäresjö

E. Gunilla A. Olsson

Introduction

The Viking Age landscape of Bjäresjö has been reconstructed by the method used for Kabusa during the Neolithic (Ch. 4.1.3.3), and Bjäresjö during the Late Bronze Age (Ch 4.2.4.2). A protein-based model for human dietary need was applied for calculation of acreages (for details see Ch 2.2, 4.2.4.2). Excavations from Bjäresjö revealed settlements from the Late Viking Age, AD 900, on the hill north of lake Bjäresjösjön, close to the present church (Callmer 1991). There were at least 4–6 dwellings here. Using an estimate of "family groups" of about 7–8 individuals, based on field finds,

VIKING AGE LANDSCAPE IN BJÄRESJÖ, CA. AD 900

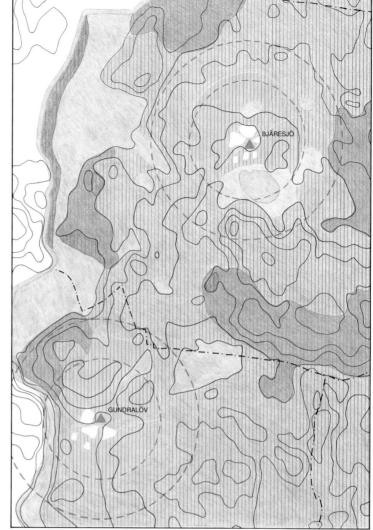

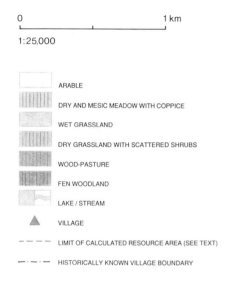

0   1 km

1:25,000

	ARABLE
	DRY AND MESIC MEADOW WITH COPPICE
	WET GRASSLAND
	DRY GRASSLAND WITH SCATTERED SHRUBS
	WOOD-PASTURE
	FEN WOODLAND
	LAKE / STREAM
▲	VILLAGE
– – – –	LIMIT OF CALCULATED RESOURCE AREA (SEE TEXT)
–·–·–	HISTORICALLY KNOWN VILLAGE BOUNDARY

LANDSCAPE RECONSTRUCTION: E. G. A. OLSSON
GRAPHICS: M. RIDDERSPORRE

Fig. 4.2:8. Bjäresjö. Viking Age landscape.

the total human population would have been about 40 individuals, including children, at a minimum (Callmer 1991). The dietary composition is based on archaeological findings from four excavated settlements with generally similar environments to Bjäresjö (Callmer 1991).

The pollen diagram from lakes Bjäresjösjön and Bjärsjöholmssjön (Gaillard and Berglund 1988, Ch. 4.2.2), palaeobotanical investigations by Engelmark (1989, unpubl.) and a recent geological survey (Daniel, unpubl.) provided the further prerequisites for constructing the landscape map.

Results and discussion

Compared with the previous landscape scenery in the vicinity of Bjäresjösjön (Ch. 4.2.4.2), dramatic changes had occurred (Fig. 4.2.8). The scattered settlements in

this area during the Late Bronze Age contracted to one joint site north of the lake, at the location of the historical village. The next settlement was to be found 2 km southwards near the river Svartån, at what later become the village of Gundralöv (Ch. 4.2.4.3). The two settlements are marked on the map (Fig. 4.2:8) and they are assumed to be of roughly equal size and population. The figures below are valid for each of the settlements/villages.

Farming activity was well established and cultivated food crops were rye, barley, and spelt/emmer wheat (Engelmark unpubl., Ch 4.2.2). Assuming that 50% of the human protein need was covered by cereals, arable land of 2–4 ha was harvested each year (Tables 4.2:4, 5; this is estimated on the basis of a mean seed return of 4–6; see also Ch. 4.2.4.2). According to the pollen diagram (Gaillard and Berglund 1988), hemp was

Table 4.2:4. Inferred dietary composition during the Late Viking Age in Bjäresjö as suggested by the archaeological and palaeoecological findings. The protein contents of the different food items are extracted from modern tables (Livsmedelstabeller 1978). The data below are calculated for subsistence level for a settlement site inhabited by a total of 30 human adult units.

Item	% dietary contribution	kg protein adult^{-1} yr^{-1}	kg food adult^{-1} yr^{-1}	total no. consumed animals yr^{-1}	no. of stock
barley	12.5	2.4	17	–	–
emmer	12.5	2.4	11.9	–	–
rye	25	4.8	36.5	–	–
nuts	1.5	0.3	2.3	–	–
fish	0.5	0.1	0.5	–	–
cattle	19.2	3.6	17.4	6.3	20
sheep	11	2.1	10.4	18.8	57.5
pigs	11	2.1	10.4	11.3	14
horses	5.8	1.1	5.2	1.3	5
eggs	1	0.2	1.0	–	–

grown in Bjäresjö at this time. Although it is impossible to give an exact figure for the area of arable land, approx. 0.5 ha could be expected for this cultivation. The hemp fields were probably situated in the wet areas near the lake, since this crop demands relatively high soil moisture. Composition of weed flora and cultivated crops as shown in the palaeobotanical remains (Engelmark unpubl.) and in the pollen diagram (Gaillard and Berglund 1988) indicate that: (1) fields were manured – and increase of *Chenopodium album* (fat hen); (2) rye was the dominating crop; (3) autumn-sown crops were used, indicated by the occurrence of autumn-germinating weeds like *Centaurea cyanus* (cornflower). This in turn indicates (4) some sort of crop rotation. The latter could have been in two or three courses, with fallow periods of unknown duration.

In the Late Viking Age livestock was kept in byres over the winter and consequently there was a need to collect winter fodder. This is indicated by the following evidence: (1) the climate was slightly more humid and cooler than before (Ch. 5.1), making winter grazing unprofitable; (2) there were compartments in the longhouses indicating byre stalls for livestock (Ch. 4.1.4.4); (3) regular manuring was practised, demanding accessible supplies of manure, indicated by increased frequency of seeds from *Chenopodium album*, which are

dispersed by manure; (4) iron scythes and sickles for harvesting hay and leaf-fodder have been found.

The establishment of enclosed infields containing meadows and manured arable probably took place in the Early Iron Age, judging by analogies with similar regions in Sweden (Widgren 1983). Although field boundaries and cultivation marks from the Iron Age have not been found in Bjäresjö, an organization like the one shown in Fig. 4.2:8 is plausible. The meadows in the infields certainly contained open grassland patches, but also wooded areas and coppices. Remains of this organization still existed in the 17th century, as indicated by the land survey map from 1699; south of the lake there are field-names like *Risåker* "coppice or shrubby field".

If domestic animals contributed to the human protein nutrition to the tune of 50%, then a stock of approx. 20 cattle, 5 horses, 58 sheep, 14 pigs (Table 4.2:4) must have been maintained. Calculated winter fodder needs for cattle and horses over 6.5 months (October–April) would require a meadow area of approx. 50 ha. (This calculation is based on a hay production of 750 kg ha^{-1}. Mean hay production in the late 17th century was 500 kg ha^{-1}, but if wooded meadows and coppices were harvested in the Iron Age, the total biomass production would have been larger. The meadow area used here is thus a maximum value.) It can, however, be doubted whether it was possible for a human population of 30 adults to manage and harvest 50 ha of meadows with rather inefficient tools. It is more likely that a considerably smaller meadow area was harvested, because the stock were (1) stalled less than 6–7 months; (2) the harvested meadows yielded more than 750 kg grass ha^{-1} yr^{-1}; (3) the animals were starved during the winter. For the construction of the map (Fig. 4.2:8) a meadow area of 50 ha was used. For summer grazing a foraging area of further 47 ha would have been needed (Table 4.2:4). The total resource area would thus be estimated to approx. 100 ha.

The pollen diagrams suggest an open landscape where woodlands and tree vegetation were declining. Alder, elm, and hazel were heavily affected. Alder carrs

Table 4.2:5. The relationship between seed return and arable area during the Viking Age. Calculation valid for subsistence consumption of 30 adults, according to Table 4.2:4 – see text.

Crop	Seed return	Arable area (ha)
barley	6	0.66
emmer, spelt	5	0.57
rye	8	1.03
Total arable		2.25
barley	4	1.06
emmer, spelt	4	0.74
rye	4	2.28
Total arable		4.08

had now been significantly reduced and replaced by treeless wetlands created by abundant grazing and perhaps also by hay harvesting (Ch. 4.2.2). These productive wetlands and fens in the river Svartån valley may have provided attractive fodder areas for other settlements in the neighbourhood (cf. Callmer 1991). The decline of hazel and elm conforms with the larger number of livestock and hence with a harder grazing pressure. The most palatable species like elm, hazel, and ash would then be expected to show the largest decrease. By the Viking Age this area had been used for extensive grazing and cultivation for more than 3000 years (since the Neolithic). Considerable leaching of the sandy soils on the top of the hummocks might have promoted some colonization of beeches on the former open, dry grasslands here (Ch. 4.2.2). Oaks were abundant, however, and the grazing territories contained both open grassland, grassland with scattered trees and groves of beech, and wood-pasture where oaks were frequent. However, the overall tree pollen contribution (Gaillard and Berglund 1988) for this period is low and the region certainly was more devoid of tree vegetation than ever before during prehistoric time.

It should also be remembered that the stock size estimated here on the basis of pure food consumption requirements could have been larger for quite different reasons. Cattle and horses may have been important, for example, for religious practices or as markers of social status. Such factors would have affected the size of the resource area and thus had other consequences for the environment. As regards crop production too, other reasons than pure food requirements, like production for export, could have been involved and led to larger production areas.

4.2.6 The Middle Ages and Modern Time until ca. 1670
4.2.6.1 Settlement, cultivation systems and property distribution in the Bjäresjö area

Sten Skansjö

The oldest known historical settlement

Written evidence exists from the 14th century of the following settlements in the parish: the church-village of Bjäresjö, the villages of Gundralöv and Stora Tvären, the manors of Berghusagård/Bjersjöholm and Gundralövsgård, as well as the Hersaremölla watermill. Archaeological observations around the historical village sites allow us to say with some probability that the villages of Bjäresjö and Gundralöv are much older, having existed since the Viking Age or the Early Middle Ages (Ch. 4.2.5.1).

The settlement known from the Middle Ages also includes the Gundralöv watermill, attested around

1500. The first securely localizable evidence for the little village of Kärragården comes from 1632, but a place recorded as "Keregardh" in 1470 may refer to this settlement. However, alternative localizations in Scania cannot be ruled out, so Kärragården could also be interpreted as a post-medieval settlement like those known from the Krageholm area and the Romele area (see Ch. 4.3.7 and 4.4.6–7 below).

The oldest historically known settlement pattern, both during the Middle Ages and in the mid 17th century, consisted on the one hand of the two noble manors of Berghusagård/Bjersjöholm and Gundralövsgård, and on the other hand of the villages of Bjäresjö, Gundralöv, Stora Tvären, and Kärragården. In the mid 17th century the church-village of Bjäresjö, with its 13 farms, well represented the average size of villages in the Ystad area. Gundralöv, which had 10 at this time, and Stora Tvären, which had 8, were smaller than average, but since these villages were controlled by a single owner (see below), the farms had been equalized and reduced in number by the 17th century.

A comparative study of settlement density in the Ystad area around 1670 shows that Bjäresjö parish, with 2.7 farms km^{-2}, was very near the average, which has been estimated to 2.8 normal farms km^{-2} for the entire project area (Skansjö 1985).

Changes in settlement in the Late Middle Ages

Palaeoecological material from Bjäresjösjön indicates a stagnation in the intensity of cultivation in the Late Middle Ages, which leads one to think of the late medieval agrarian crisis and its possible repercussions on this area (Ch. 4.2.2). However, there are only few and diffuse signs of crisis in the shape of deserted farms and villages in the written sources from the Ystad area as a whole, although for the Romele area there are indications of a regression in the late Middle Ages (Ch. 4.4.6). For the Bjäresjö area we can observe only that one deserted farm is mentioned, in 1470, when the distribution of the property of the deceased owner of Krageholm, Bonde Jepsen (Thott), included a deserted farm in the village of Stora Tvären. The desertion here may have been temporary, and this isolated reference cannot be interpreted as evidence of any major period of regression in cultivation and settlement. To this can be added a possibly vanished settlement named just once: from 1419 we have a mention of a settlement called "Gunnærsløffsthorp" which belonged to the Krageholm estate; there is a clear toponymic connection with Gundralöv in Bjäresjö parish. It has not been possible to ascertain whether this isolated reference actually denoted a settlement that was later abandoned, or where it may have been located (Skansjö 1988).

An interpretation based on several concordant sources shows that the manor in Bjäresjö village was moved to Bjersjöholm, the islet (*holme*) in Bjärsjöholmsjön in the mid 14th century. Despite the synchrony, this move can scarcely have had any causal connection with the Black Death or the agrarian crisis, but was more probably occasioned by the necessity of defending the manor (Skansjö et al. 1986, Skansjö 1991, Ch. 4.2.5.1).

Changes in settlement in the 16th and 17th centuries

Unlike the inner hummocky zone, which experienced an expansion in settlement as it was colonized with dispersed farmsteads, it appears as if the villages in the Bjäresjö area declined in size, at least as regards the number of farms. In Bjäresjö parish, for example, there were 50 tithe-paying units around 1570, a number which had fallen to 43 in the mid 17th century, while there were only 35 farms in the parish by 1670. The reduction can largely be explained by the closure of farms, equalization, and the like, caused by the development of the estates of the two manors in the parish, Bjersjöholm and Gundralöv (Skansjö 1988, see also below).

Cultivation according to the oldest written sources

From 1570 we know the cultivation system in the church-village of Bjäresjö in the form of a three-field system together with an area of annual meadowland. In other words, there was a special meadow field along with the three arable fields.

Before this, around 1500, we have mention of two areas of forest, Stubbeora and Senholt, which lay outside the village's cultivation system but which the peasants in the village were allowed to use. These forests can be localized in the border zone between Skårby and Krageholm, ownership of which was disputed between the archbishop of Lund and the owner of Krageholm at the start of the 16th century. The areas had been donated to the church in Bjäresjö, but the villagers were entitled to cut whatever wood they needed (Skansjö 1986).

From a comparison of the diocesan terrier of 1570 with the oldest land survey descriptions from the early 18th century (Ch. 4.1.6.1), we can see that there was an expansion of about 25% in the total area of land tilled by the vicarage in Bjäresjö. Calculated on the basis of the tilled land alone, the expansion was about 50%. Unlike Stora Köpinge, Bjäresjö had the chance to expand cultivation on to the productive clayey soil of the outer hummocky zone.

Property distribution and estate development in the Middle Ages

What we know of ownership in the Middle Ages suggests that the area was dominated by the nobility from an early stage. Gundralöv manor is known from the 14th century; it appears to have encompassed the entire village of Gundralöv. Our information on the manor of Bjäresjö/Bjersjöholm is more extensive, based as it is

on the results of interdisciplinary studies as part of the Ystad Project.

As Callmer has shown (Ch. 4.2.5.1), we know the existence of a Viking Age manor complex, whose exact character is hard to assess at present, but where the evidence testifies to the presence of an aristocracy as far back as the Late Viking Age. In the 12th century a distinctively manorial setting emerged here, with contemporary architectural design of the manor house and the church. This constellation no doubt dominated the other settlement, whose existence has also been demonstrated at this time. The manor on the hill, after rebuilding in the 12th century, survived until the mid 14th century, when it by all accounts ceased to exist on this site.

The unanimous evidence of written sources and archaeological observations, both in the village nucleus and on the castle hill (Skansjö et al. 1989, Skansjö 1991), suggests that the manor had moved out of the village by the mid 14th century, to be relocated on the islet near the lake-shore. We know the name of the first owner of Bjersjöholm, Ingemar Karlsen (of the Thott family). He first titled himself as of Berghusagård (1344), which must have been the name of the manor in Bjäresjö village, and then of Berghusaholm (1350). A more famous owner of the oldest Bjersjöholm was the knight Åge Ingvarsen Båd (died 1376), who was for a few years the king's steward in Scania and a close ally of King Valdemar Atterdag. The known parts of Åge Ingvarsen's estate included 13 tenant units (farms or smallholdings) in Herrestad Hundred.

To explain the course of events that has been interpreted as a move from the church-village of Bjäresjö to establish a private castle at a distance from the village, it is worth comparing certain general contemporary developments. We can see that the move was made at a time when private castles were growing up in Denmark to an extent that was unique for the Middle Ages. This general process can be understood partly as a sign of the growing power of the nobility in relation to a weak central government, and partly as a measure taken in an age that demanded fortifications. To ensure up-to-date defensive sites it was sometimes appropriate to move a manor out of its village to an isolated site. Such a move would also have furthered the reorganization of estate structures which we see in this period. In several parts of Denmark this development meant that high medieval village estates consisting of one large farm with its small tenant farms were transformed into villages with tenant farms of roughly equal size. The empirical material from Bjäresjö, however, does not display the complete structure typical of the High Middle Ages; although the manor farm is clearly seen in the physical remains, the nature of the tenant farms is largely unknown before the latter half of the 14th century. It is possible, however, that the oldest land survey sources for Bjäresjö village from 1699 reflect the reduction of an earlier large estate with equalized farms of late medieval type. In addition, the interpretation of the archaeological material suggests that the village may have been replanned and regulated in the Late Middle Ages (Ch. 4.2.5.1).

Details of the farm and estate of Bjersjöholm in contemporary written sources are extremely scanty and tell us little about the period from about 1400 until the time just after the middle of the 16th century. What we find is mostly only a number of personal names along with the information that they titled themselves as lords of Bjersjöholm. Throughout the Late Middle Ages the known owners belonged to the Rotfeld family, who were descended from an ancient noble Jutland family with its seat in Bratskov.

What seems clear is that Bjersjöholm in the 15th century did not function as the headquarters of a nobleman's estate. It was only occasionally visited and used by its owner. It is well known that the Rotfeld family were far from being among the largest noble landowners of their time. Nevertheless, it was typical of the times that the property amassed by a nobleman could have manors widely spread throughout Denmark. The manors in such cases were usually managed by a bailiff as administration centres for the surrounding tenant farms.

Since the family seat at Bratskov was by all appearances the place where the owners resided, there is reason to suspect that the buildings at Bjersjöholm at the end of the Middle Age were relatively plain. This is confirmed by a few traditions recorded in the late 17th century. In view of this, and from what we know of the management of the Bjersjöholm estate during the Renaissance (see below), it is reasonable to assume that the late medieval manor had less land at its disposal than did its successors (Skansjö 1991).

Bjersjöholm in the 16th and 17th centuries

One of the most significant turning-points in the long history of the estates of Bjersjöholm came in the late 16th century, when Björn Kaas, knight, councillor of the realm, and prefect of a royal fief, built a splendid castle at Bjersjöholm, which was largely finished by 1576. There are clear indications that this building work was associated with an extension of the lands farmed by the estate, at the expense of the nearby villages. In addition, the owner of Bjersjöholm acquired a number of tenant farms with corvée obligations in the immediate vicinity of the manor.

From 1632, when the formation of the estate must have been completed and the castle finished, we have our first overall picture of the number of subject peasants obliged to do corvée: there were 25, but only 16 of these were from Bjäresjö parish. In other words, Bjersjöholm controlled less than half of the parish in which the manor lay. This can be compared with Krageholm, which at the same time had 45 subject peasants in Sövestad and thus totally dominated that parish, which had

about 50 tithe-payers; on top of this, Krageholm had 37 subject peasants from outside the parish. Bjersjöholm was thus somewhat below the average of the manors in the area as regards the availability of peasants obliged to do corvée.

We may also note that the area farmed by the Bjersjöholm estate was only of medium size in comparison with other estates in Scania. According to calculations on the basis of the detailed data provided by land survey documents and estate taxation assessments from the end of the 17th century, the demesne lands of the estate were equivalent to only about ten normal tenant farms in the hundreds of Ljunits and Herrestad.

Although we have several such signs that Bjersjöholm really did grow in conjunction with the building of the castle, as regards both the area farmed and the greater number of tenants obliged to do corvée, the estate-building efforts did not achieve more than a mediocre result. It is therefore obvious that the management of the estate was far from being on a par with the grandiose image projected by the manor house. One finding of this study, then, is that the stately home was on a scale that did not compare with the resources at its disposal. A phenomenon like the Renaissance castle of Bjersjöholm cannot therefore be explained without due regard to contemporary changes on several levels of society. This applies to the development of the European agrarian economy, as well as to important political processes in Europe as a whole and in Denmark in particular, where the period 1536–1660 has justly been called "the time of noble dominance" (Skansjö 1987a, 1991, see Ch. 5.8).

4.2.6.2 Landscape, land use, and the moving of a manor. A hypothetical reconstruction based on land survey documents from around 1700

Mats Riddersporre

As demonstrated above, several independent observations indicate that the 16th-century castle at Bjersjöholm replaced an older building that was probably the result of the manor having been moved out of Bjäresjö village in the 14th century. The moving out of the manor raises questions about the organization of land use and what happened to the village. If the move concerned not only the manorial building, but also included the breaking-out of demesne land from the village community, this would have affected the entire village and caused a rearrangement of field divisions. Another question concerns the status, physical and formal, of the area the manor moved to.

To illuminate these questions, the oldest land survey documents for Bjäresjö and Bjersjöholm have been used for a reconstruction of the high medieval landscape

(Fig. 4.2:9). The reconstruction is based on a retrogressive analysis of maps and descriptions from 1698 and 1699 (cf. Ch. 4.2.7.2, Fig. 4.2:13). Key features of the analysis are *field systems* and *field subdivisions*. In the analysis these features are related to other characteristics of the village, such as topography, soil quality, length of fallows, as well as the layout of the village nucleus and its tofts (cf. also Ch. 4.1.6.2). Fig. 4.2:9 shows the situation around 1300, that is, before the moving out of the manor.

The manor in the village

The sometime existence of a manor or large farm in the village core is supported by the toft structure. By the time of the land survey map the tofts were part of the arable but in the Middle Ages they served as building grounds and as a documentation of the relative size of the individual farms in the village. Farm no. 1, which in 1699 was situated immediately to the west of the church, had a toft that was notably larger than the tofts of the other farms (Riddersporre 1989a). This large toft encompasses the site where archaeologists have excavated the remains of a large farm from the Late Viking Age and Early Middle Ages. Furthermore, it looks as if the church was built on ground that originally belonged to this toft – an indication of the close connection between the large farm and the church in the Early Middle Ages (cf. Ch 4.2.5.1).

Next to the eastern side of the church in 1699 was the vicarage, farm no. 2. This farm had no toft here, which might indicate that the vicarage, as a separate farm, is a later addition to the village. The toft structure of Bjäresjö, including the toftless vicar's farm, is almost exactly the same as the village of Stora Herrestad. Here the manor is still in the village core, on its large toft (Riddersporre 1989a,b, 1991).

Changes in the fields

The village territory of Bjäresjö in 1699 was divided into five fields. Three of these fields constituted the traditional three-field system: North, South, and West Fields (A, B, C in Fig. 4.2:13). A fourth field, *Gallfäladsvången*, (D in Fig. 4.2:13) was used only for growing oats. The name contains the word for common pasture, *fälad*. The fifth field was a separately fenced meadow field, *Ängsvången*, (E in Fig. 4.2:13) along the river Svartån. The territory of the Bjersjöholm estate in 1698 was also arranged in a three-field system. There was, furthermore, a separately fenced wood (D1 in Fig. 4.2:13).

Among the fields of Bjäresjö there are structural differences in the way land was subdivided in 1699. In the North Field – except the north-eastern part – the lands of farms no. 1 and 2 were concentrated in a few large block-shaped parcels. The rest of the farms appear to have practised *bolskifte* with rather small strip par-

cels. The characteristic of *bolskifte* is that farms are grouped together into equal-sized component fiscal holdings, *bol*. Larger farms could constitute one *bol* or more by themselves. Every *bol* had its share of strip parcels spread over the township. *Bol* parcels with more than one shareholder were divided into narrower strips (Hannerberg 1958, 1960, Andersson 1959).

In the South and East Fields all land was *bol*-divided; including farms no. 1 and 2, which seem to have constituted at least one *bol* each. The parcels were larger and straighter than the strips of the North Field. Within the Meadow Field it is possible to trace both *bol*-division and large irregular block-shaped parcels. Not only the farms in the village had shares in the block-shaped parcels, but also Bjersjöholm and some farms in Gundralöv, the neighbouring village to the south of Bjäresjö.

Within the *Gallfäladsvången*, the north-eastern parts of the North Field, and the Meadow Field there was a division with a special characteristic. Here farms which in other fields were parts of *bols*, appear as independent units. Here we also find field-names referring to common land.

The structural differences can hypothetically be interpreted as indications of a relative chronology in the subdivision of the fields. The large block-parcels in the North Field could be the remnants of land that belonged to the early medieval manor. At some occasion this land had been split into two halves, one of them becoming a vicar's farm. Assuming that the three-field system already existed, the same pattern was probably to be found also in the South and East Fields. At a probably later stage it was for some reason necessary to rearrange the subdivision of these two fields. They became *bol*-divided, and farms no. 1 and 2 now constituted one *bol* each. It was not necessary, however, to rearrange the North Field.

Later again, when the former common pasture of the *Gallfäladen*, the north-eastern part of the North Field and the Meadow Field was divided, farms which had previously been parts of *bols* had developed into independent units. According to field-names appearing in the diocesan terrier, this seems not yet to have taken place in 1570 (cf. also Persson 1986b, Ch. 4.2.2). At what time the rest of the Meadow Field was divided is hard to say. It seems partly to have been at the disposal not only of the village but also of the parish.

The question that arises is then: why and when were the South and East Fields rearranged? One possibility is that it was the effect of the estate breaking out its lands, after moving out of the village around 1350. The distribution of the good soils, that is, the intensively manured soils sown with rye and barley in three-year rotation, suggests that there had been some changes mainly in the East Field. This field may previously have encompassed areas which around 1700 belonged to the Bjersjöholm estate (cf. Fig. 4.2:13).

If the manor farm claimed parts of the good soils in

EARLY 14TH CENTURY
LANDSCAPE IN BJÄRESJÖ

Fig. 4.2:9

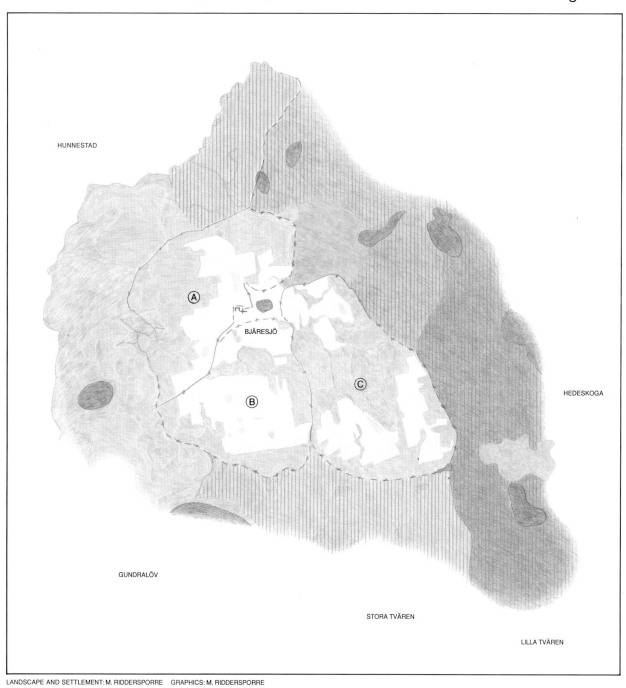

HUNNESTAD

Ⓐ

BJÄRESJÖ

Ⓑ

Ⓒ

HEDESKOGA

GUNDRALÖV

STORA TVÄREN

LILLA TVÄREN

LANDSCAPE AND SETTLEMENT: M. RIDDERSPORRE   GRAPHICS: M. RIDDERSPORRE

0   1 km

1:25,000

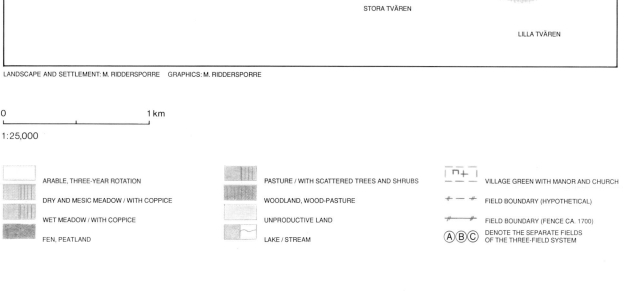

ARABLE, THREE-YEAR ROTATION	PASTURE / WITH SCATTERED TREES AND SHRUBS	VILLAGE GREEN WITH MANOR AND CHURCH
DRY AND MESIC MEADOW / WITH COPPICE	WOODLAND, WOOD-PASTURE	FIELD BOUNDARY (HYPOTHETICAL)
WET MEADOW / WITH COPPICE	UNPRODUCTIVE LAND	FIELD BOUNDARY (FENCE CA. 1700)
FEN, PEATLAND	LAKE / STREAM	ⒶⒷⒸ DENOTE THE SEPARATE FIELDS OF THE THREE-FIELD SYSTEM

the original East Field, that would have created an imbalance between the size of the three fields of the village. There were two ways to handle this situation: either to redivide the entire village, or to expand the arable in the East Field. The latter could have been done at the expense of meadows in the East Field and/or by transferring neighbouring parts of the South Field into parts of the East Field. If a reasonable balance between the fields was reached that way, it could explain why only land in the South and East Fields was redivided.

In this way the manor may have got access to good arable soils within the new demesne, instead of having to break new land. Possibly the estate also, at this time or later, incorporated land from the villages of Stora Tvären, Lilla Tvären, and Hedeskoga. Parts of an original East Field in Bjäresjö may thus have served as the starting-point for an expansion of arable on the outskirts of the surrounding villages, but with the Bjersjöholm manor in a central location.

From common to private

Good soils were not, however, the only important thing for the new location of the manor. The lake and its surroundings were an important factor, with good defensive potential (Reisnert 1989, cf. Ch. 4.2.6.1). Moreover, the area where the manor moved seems to have been part of a larger wood and pasture common. A great part of the estate was still wooded around 1700 and used for grazing horses. The possibilities for hunting in the wood were probably an attractive factor. The status of the area as common may also have facilitated the enclosure – and enlargement – of the demesne land.

This type of location, in peripheral areas, with access to wood, water, and wetland, seems to have been preferred also by other manors at this time (Dahl 1942b). In all parts of Denmark, manors were moved out of village cores – often to locations with physical conditions like these (Stiesdal 1981, Porsmose 1987). From the Ystad area there are indications of a similar process in the villages of Sjörup, Hedeskoga, and Bussjö (Skansjö et al. 1989). The sites indicating the location of these manors have many features in common with the area where the manor of Bjäresjö moved around 1350. Unlike Bjersjöholm they did not develop into estates surviving into modern days.

The wood and pasture common in Bjäresjö probably continued to the north, where indications come from names referring to common land and to pasture. As mentioned above, these parts of Bjäresjö were probably not reclaimed until after 1570. It is here possible to reconstruct a cattle-path connecting village green and pasture. The presence of large common areas presumably also made possible the supposed post-medieval establishment of the village of Kärragården (cf. Ch. 4.2.6.1). The area was probably a common for the surrounding parishes and was connected to the larger wooded areas to the north-west. Here the farmers in Bjäresjö seem to have had some rights to cut wood around 1500 (Ch. 4.2.6.1).

Lakes and land use

The reconstruction of the high medieval landscape of Bjäresjö and Bjersjöholm shows a great difference in land use and vegetation in the surroundings of the two lakes in the study area. Lake Bjäresjön was situated in the central part of the infields of the village. As shown by pollen analysis (Ch. 4.2.2), the surroundings of the lake had long since been used for intensive agriculture. The area around lake Bjärsjöholmssjön, on the other hand, was probably characterized by extensive use and wooded areas before the establishment of the Bjersjöholm estate.

4.2.7 The 18th century
4.2.7.1 Cultivation, population, and social structure in the Bjäresjö area

Anders Persson

Introduction

The form and use of the cultural landscape in Bjäresjö parish can be described in detail as far back as the late 17th and early 18th centuries. Early maps survive from 1700–1720 of the villages of Bjäresjö, Gundralöv, and Stora Tvären, whereas the earliest cartographic information for the little village of Kärragården in the north of the parish is from the 1780s. Maps of the manor of Bjersjöholm exist for about 1700, 1790, and 1815, and of the manor of Gundralöv for about 1720, 1760, and 1810. A series of assessment records for scattered points in the period 1670–1800 supplement the surveyors' descriptions with information about the cultural landscape in the parish. Despite all this, the description of the form of the village landscape during the 18th century is difficult to capture because the early 18th century land survey information for the villages of Bjäresjö and Stora Tvären is not directly comparable with the cartographic data from the time of the *enskifte* enclosures around 1815. Details of settlement and population have been compiled for the years 1700, 1750, and 1805.

The great changes in the parish in the period under study here mostly concern the management of the manors in the second half of the 18th century and the enclosure of the parish villages in the 1810s, which totally changed the settlement structure (see also Ch. 4.2.6.1) and created the conditions for more efficient utilization of the village lands. Unlike Sövestad parish, where the older agrarian system survived in the villages right up to the middle of the 19th century, the progress of the *enskifte* enclosures in Bjäresjö followed the general chronological trend.

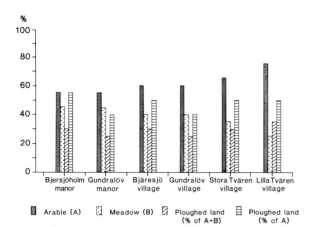

Fig. 4.2:10. Land use on manors and in villages in the Bjäresjö area in the late 17th and the early 18th centuries.

Legend: Arable (A) · Meadow (B) · Ploughed land (% of A+B) · Ploughed land (% of A)

Settlement

The period 1700–1805 saw a growth in the number of settlements. Slightly over 60 settlement units can be registered for 1700 and about 80 for 1801. The number of farms declined somewhat in the 18th century, from 41 in 1700 to 38 in 1801. On the other hand, the number of cottages more than doubled, from around 20 in 1700 to 45 in 1801. This change in settlement structure meant that the proportion of cottages rose from 35% in 1700 to 50% in 1801. Even for 1700, the proportion of cottages was relatively high in the manor parishes of the Ystad area, although Bjäresjö parish in this respect came after the neighbouring parish of Balkåkra to the west (with Marsvinsholm manor) and Stora Herrestad (with Herrestad manor) in the east of the project area. In these two parishes the proportion of cottages in 1700 was as much as 40% of settlement that year. For parishes without manors the proportion of cottages was much smaller, 5–15% of the total settlement, as against 20–40% in the manor parishes (Persson 1989).

Population

The population of the parish can be estimated to around 330 (index 100) in 1700 (23 km^{-2}). By the mid 18th century this had risen to 475 (33 km^{-2}), reaching about 670 (47 km^{-2}) in 1805 (index 100–145-200). These figures reflect the development outside the manors, whose population and lands are not included in the calculation of population density. (Calculations of this kind have in general been made this way when studying parishes with manors.) What we see is a doubling of population in an area which, for its times, had highly productive land that was exploited relatively intensively (Olsson and Persson 1987, Persson 1988a, Ch. 4.2.7.3, Fig. 4.2:10).

The closest equivalent to the population density of Bjäresjö in 1700 is the parish of Västra Nöbbelöv, with a figure of 24 km^{-2}. These figures can be contrasted with

the population density of Stora Köpinge parish, which was 14 km^{-2} in 1700. For both Sövestad and Villie parishes the figure was also low, 15 km^{-2} (Persson 1988b), which could be explained in part as due to the greater amount of outlands here at the start of the 18th century. This was not the case in Bjäresjö, nor in Stora Köpinge, which was also dominated by villages. In this perspective, Bjäresjö must be considered a densely populated parish in the early 18th century.

The agrarian structure

Land use on the manors and in the villages. At the start of the 18th century, grain was cultivated in a traditional three-course system on the two manors and in the four villages in the parish. Work on the manors was done as corvée by peasants whose farms had been equalized, partly to simplify the calculation of labour obligations on the manor, partly to facilitate the calculation and collection of rents.

The cultivation system on the manors of Bjersjöholm and Gundralöv is shown in Fig. 4.2:10. The area of arable and meadow land on each manor was 150–160 ha. Some 25–30% of this was ploughed each year. The grain-producing area on each manor was thus no larger than about 40–45 ha, equivalent to what 4–6 farms in the area ploughed and sowed at the same time (Persson 1986b, 1988a, Ch. 5.9). The farms in the parish were usually about 30 ha. This was much smaller than contemporary farms in Stora Köpinge parish (see Ch. 4.1.7.1, Figs 4.1:10,11).

Around 1700 the proportion of arable and meadow, along with the intensity of cultivation and productivity, were largely the same on the manors as in the villages in Bjäresjö parish. (This does not count the village of Kärragården in the north, for which we lack early land survey documents.) The villages, however, had a greater proportion of arable land than the manors (60–65% as against 55%). Gundralöv village, whose lands at this time were still being farmed along with those of the

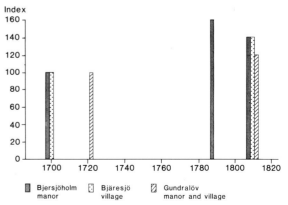

Fig. 4.2:11. The expansion of arable land in Bjäresjö parish 1700–1810. Index: 1700= 100.

manor, was also cultivated less than the villages of Bjä-resjö and Stora Tvären, where the area of land culti-vated annually was greater (Fig. 4.2:10). Almost half the actual demesne land on Bjersjöholm manor, 120 ha, was neither arable nor meadow, consisting of woodland and pasture, much of which was to be transformed into arable land through a massive cultivation drive in the 18th century. The corresponding land on Gundralöv manor, some 20 ha, was mostly bog and other wetland. To sum up, it can be said that 25–30% of the arable and meadow in the parish was ploughed and sown annually around 1700, while 70–75% of this kind of land lay fallow or was used as meadow (cf. Persson 1987b). Conditions were similar in Stora Köpinge parish, but there the amount of arable land farmed each year was much smaller as a proportion of the total arable.

Development from 1700 to 1810. In this period there was a dramatic expansion of arable land. This development has been analysed for the two manors and for two villages, Bjäresjö and Gundralöv. To provide diachron-ically comparable areas, the village and manor of Gun-dralöv are treated together as a unit. In the 18th century the area of arable land was expanded by about a third in these manors and villages. This expansion was greatest on the manor of Bjersjöholm and in Bjäresjö village. The increase here over the century was about 40%, and the manor's arable land was even larger in 1790 than it would be twenty years later.

Fig. 4.2:11 presents a picture of the development of arable area. Before the *enskifte* enclosures of 1815, however, no more than half the total arable land in the villages was tilled each year, while an intensification resulting in increased productivity of grain cultivation on the manor appears to have been carried through in the period 1760–1800. In 1760 the manor of Gundralöv withdrew its lands from the village's communally farmed three-field system and organized them into its own four-field system. In 1799 an assessment of the manor stated that its lands were divided into six fields. Tillage was similarly intensified on Bjersjöholm manor. Although a three-course system was still in use in 1765, seven fields are recorded in 1789 and 1799. Here, how-ever, they were mostly oat fields broken on poorer soil that had once been outland. At Gundralöv the agrarian expansion was manifested in the 18th century in the establishment of a mill of an almost industrial scale on the lower course of the river Svartån. Count Erik Ruuth, owner of Gundralöv manor, had the Hersare-mölla mill closed down and the milling rights trans-ferred to a newly built mill called Skönadal. This was a two-storey brick edifice with a tiled roof and a ground area as big as 240 m^2. The new mill with its large pond was assessed for tax in 1801 and was deemed to be built with "indescribable skill".

Conclusion

Bjäresjö gives the impression of being a parish with relatively productive lands in this period. In comparison with other parishes in the Ystad area, it appears to have been densely populated. In the 18th century the pop-ulation doubled. However, the number of farms did not increase accordingly. On the contrary, the number fell somewhat on account of equalization. When Gundralöv village was equalized in 1760, the number of farms was reduced from ten to eight. The village lands were also reduced from about 340 to 270 ha. Of the land formerly belonging to the village, about 45 ha were allotted to 12 cottages. On Bjersjöholm manor's own lands, as at Gundralöv, a few cottages arose with functions associ-ated with the manor: wood-ward, field-ward, and so on. With information from Bjäresjö parish we can paint a picture of a society in change during the 18th century. The population appears to have grown much more than the extent of arable land. A cautious deduction from this would be that there was a general fall in the capacity of the villages to support their inhabitants, and that this agrarian society underwent greater economic and social stratification. The increases in productivity on the ma-nors no doubt reinforced this picture. A certain concen-tration of settlement on the farms appears to have been possible, even with a decline in the number of farms in the parish. Nevertheless, a great deal of the growing population would have been directed to cottages, the number of which more than doubled during the century. Just before the enclosures in the 1810s, population den-sity was nearly 50 people km^{-2} (not counting the people and lands of the manors). A change in production con-ditions with a consequent increase in production must have been felt to be inevitable around 1800.

4.2.7.2 Landscape and land use in Bjäresjö according to 18th century land survey documents

Mats Riddersporre

In Ch. 4.2.6.2, land survey documents from the end of the 17th century were used as a basis for a reconstruct-ion of the high medieval landscape in Bjäresjö/Bjer-sjöholm. By the end of the 18th century all villages in Bjäresjö parish had been mapped and described by land surveyors. On the basis of this material it is possible to make a thorough reconstruction of the landscape in the 18th century (Fig. 4.2:12). For Bjäresjö village and the Bjersjöholm estate a more detailed reconstruction of land use has also been made (Fig. 4.2:13). This figure is comparable with the reconstruction of the medieval landscape in Bjäresjö/Bjersjöholm (Fig. 4.2:9). It is also comparable with the reconstruction of the Köpinge area in Fig. 4.1:14.

THE BJÄRESJÖ AREA

18TH CENTURY LANDSCAPE

Fig. 4.2:12

LANDSCAPE ECOLOGY & SETTLEMENT: A-C. LINUSSON, M. RIDDERSPORRE   ED. & GRAPHICS: M. RIDDERSPORRE   LANDSCAPE COLOURING: A-C. LINUSSON

0   1   2 km

1:50,000

ARABLE AND FALLOW LAND	WOOD-PASTURE	VILLAGE, HAMLET	TOWN	
DRY AND MESIC MEADOW / WITH COPPICE	UNPRODUCTIVE LAND, OPEN SAND	MANOR	CHURCH	
WET MEADOW / WITH COPPICE	NO INFORMATION	SMALL HOUSE, COTTAGE	HUNDRED BOUNDARY	
FEN, PEATLAND	LAKE, POND	WATERMILL	PARISH BOUNDARY	
PASTURE	STREAM	WINDMILL	VILLAGE, ESTATE BOUNDARY	
			FIELD BOUNDARY	

Land ownership and land-use patterns

The villages in Bjäresjö parish were now, as before, dominated by large estates (cf. Ch. 4.2.6.1). All farms in Gundralöv belonged to the manor in the village. Bjäresjö and Stora Tvären were almost totally dominated by Bjersjöholm, and Kärragården belonged to the large estate of Krageholm which also dominated the parish of Sövestad to the north of Bjäresjö. The owner of Bjersjöholm, by this time the Crown, was also a major landowner in Hedeskoga parish to the east of Bjäresjö.

All the villages, as well as the Bjersjöholm estate, were arranged according to the three-field system. Apart from the three essential grain fields, they could also have additional fields. In several villages there was a separate meadow field. In Bjäresjö there was also the small field, *Gallfäladen* (D in Fig. 4.2.13), only used for growing oats. At Bjersjöholm there was a separately fenced wood, *Djurhagen* (D1), used for grazing horses and harvesting some hay. To the north of that was a small grain field (D2) that had recently been reclaimed in 1698.

Land use in Bjäresjö reveals a clear zonal pattern (Fig. 4.2:13). Rye and barley were cropped in an inner zone, while oats were cropped in an outer zone at a greater distance from the village core. A corresponding pattern can also be identified in Bjersjöholm. This is a pattern that cannot be explained solely by natural subsoil conditions (cf. Steensberg 1976:113). Rye and barley were generally grown within the three-year rotation system, and on manured lands. With the exception of parts of the East Field in Bjäresjö, the distribution of rye and barley was the same as the distribution of soils with high proportions of mould. These areas represent the core area of the arable, which had long been receiving manure. For practical reasons, it was located near the village core (cf. Ch 4.1.6.2).

The peripheral areas, on the other hand, seldom or never received manure. Moreover, the pollen diagram from lake Bjäresjösjön indicates that oats were a latecomer compared with rye and barley (Ch. 4.2.2). As we saw in Ch. 4.2.6.2 the field *Gallfäladen* and adjacent areas were probably not reclaimed until after 1570 (cf. Persson 1986b). The small field in Bjersjöholm, recently reclaimed in 1698, indicates that reclamation was in progress in this area around 1700. The occurrence of arable sown with oats within the Meadow Field in Bjäresjö (E in Fig. 4.2:13) clearly indicates an expansion of arable at the expense of dry meadows.

The sources from Bjäresjö village give no information on the length of fallow periods. The co-variation of the distribution of rye and barley with the humus-rich soils justifies the assumption that these crops generally followed the proper three-year rotation, that is, two years of grain followed by one year of fallow (cf. Ch 4.1.7.2). As regards oats there is information from Bjersjöholm, where they were sown one year and followed by two

years of fallow. This was probably the general cycle for oats in Bjäresjö too (cf. Persson 1986b).

The general pattern of land use in Bjäresjö/Bjersjöholm is the same as in the villages in the Köpinge area. The only major difference concerns the former commons. Apart from differing physical conditions, there is a difference in how they were reclaimed. In the Köpinge area they had been reclaimed by the farmers in the villages. In Bjäresjö, on the other hand, a large part had been transformed into private land owned by the Bjersjöholm estate (Ch. 4.2.6.2).

4.2.7.3 The agro-ecosystem of Bjäresjö village in the early 18th century

E. Gunilla A. Olsson

Landscape and vegetation

By the end of the 17th century Bjäresjö village was already a treeless agrarian landscape dominated by intensive grain production (LM Bjäresjö 1:1699). The village settlements were grouped around the stone church from the 12th century, close to a small lake situated centrally in the infields (Fig. 4.2:13). The hummocky terrain was covered with grain fields, fallows, and meadows. The land sloping westwards towards the river Svartån was occupied by extensive meadows on peat deposits. Parts of the meadows were flooded each year, and the documents tell of difficulties with hay harvests during rainy summers (LM Bjäresjö 1:1699). The vegetation of the wet meadows was dominated by grasses and sedges. The area of meadowland used by this village was considered large and the peat deposits in the infields also allowed the sale of peat for fuel and turf for fencing purposes. *Iris pseudacorus* (yellow flag) and *Carex caespitosa* (sedge) were frequent in turf-pits (cf. Linnaeus 1751). About one-third of the infields was annually fallowed and used for grazing (Olsson 1988). Vegetation on the fallows (one or two years in duration) were winter and summer annuals: weeds like *Anagallis arvensis* (scarlet pimpernel), *Rumex acetosella* (sheep's sorrel), *Thlaspi arvense* (field penny-cress), and *Papaver rhoeas* (field poppy), but also grasses like *Agropyron repens* (couch-grass), *Holcus lanatus* (Yorkshire fog), and *Agrostis canina* (brown bent-grass). Frequent fallow species in this area were also perennial shrubs like *Rubus caesius* (dewberry). Nitrogen-fixing species like *Ononis repens* (rest-harrow) and the clover species *Trifolium repens* and *T. pratense* were especially common (Linnaeus 1751, cf. Ch. 4.2.2). Barley fields were infested with *Chrysanthemum segetum* (corn marigold), *Avena fatua* (wild oat), and *Neslia paniculata* (ball mustard). Frequent rye weeds were *Melampyrum arvense* (field cow-wheat), *Lithospermum arvense* (corn grom-

LANDSCAPE AND LAND USE IN
BJÄRESJÖ - BJERSJÖHOLM, CA. 1700

Fig. 4.2:13

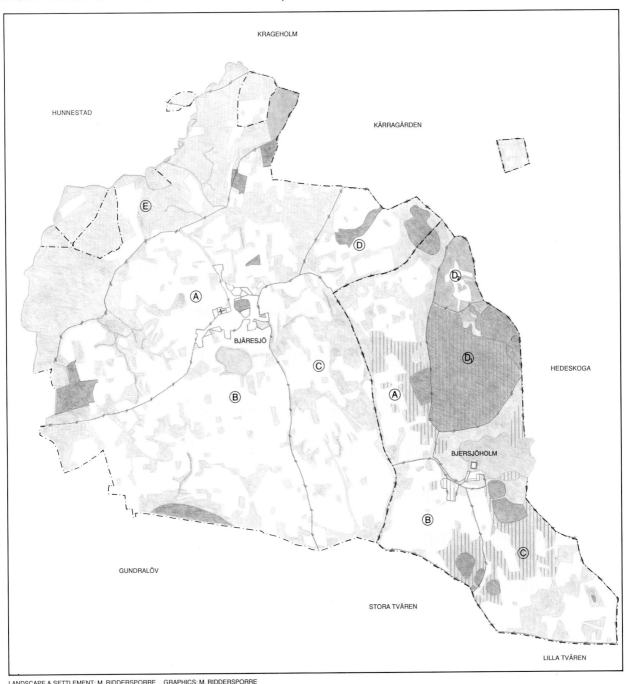

KRAGEHOLM

HUNNESTAD

KÄRRAGÅRDEN

Ⓔ

Ⓓ

Ⓓ₂

Ⓐ

HEDESKOGA

BJÄRESJÖ

Ⓒ

Ⓓ₁

Ⓑ

Ⓐ

BJERSJÖHOLM

Ⓑ

Ⓒ

GUNDRALÖV

STORA TVÄREN

LILLA TVÄREN

LANDSCAPE & SETTLEMENT: M. RIDDERSPORRE   GRAPHICS: M. RIDDERSPORRE

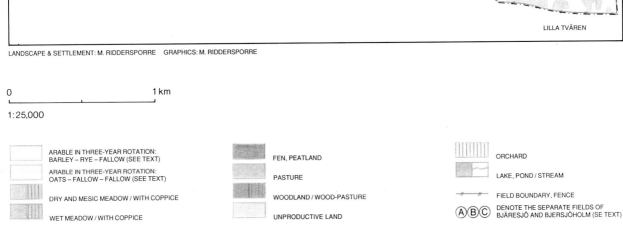

0   1 km

1:25,000

▨	ARABLE IN THREE-YEAR ROTATION: BARLEY – RYE – FALLOW (SEE TEXT)
▨	ARABLE IN THREE-YEAR ROTATION: OATS – FALLOW – FALLOW (SEE TEXT)
▨	DRY AND MESIC MEADOW / WITH COPPICE
▨	WET MEADOW / WITH COPPICE

▨	FEN, PEATLAND
▨	PASTURE
▨	WOODLAND / WOOD-PASTURE
▨	UNPRODUCTIVE LAND

▨	ORCHARD
▨	LAKE, POND / STREAM
—#—#—	FIELD BOUNDARY, FENCE
ABC	DENOTE THE SEPARATE FIELDS OF BJÄRESJÖ AND BJERSJÖHOLM (SE TEXT)

well), *Centaurea cyanus* (cornflower), *Anthemis arvensis* (corn camomile), and *Bromus secalinus* (rye brome). Several of these weeds are today extinct or considered as endangered plants of high conservation value. The only woody vegetation at this time was thorny shrubs of *Rosa* (rose), *Crataegus* (hawthorn), and *Prunus* (blackthorn) growing on small hillocks and collections of boulders which could not be harvested or grazed. Linnaeus, travelling through Bjäresjö on 1 July 1749, described the landscape like this: "The fields looked as if they were beautifully painted and covered by the most exquisite wallpaper. In some places they were bright yellow from *Chrysanthemum* (corn marigold), in other places radiant blue from *Echium* (viper's bugloss), in other places all brown from *Rumex* (sheep's sorrel) and gleaming white with *Anthemis* (corn camomile)".

Ecological aspects of land use and cropping systems

The land, including that in the villages, was owned by large estates. All villages in this region practised three-course rotation, although with modifications. In Bjäresjö village the arable and meadows were organized in five separate common fields. Three of these were used for growing rye and barley in a three-year rotation cycle with one year of fallow. The fourth was reserved for cropping oats, but with two years of fallow. This meant that about 50% of the total arable was used for annual cultivation (cf. Ch. 4.2.7.1). Barley and rye generally received manure every third year according to the fallow periodicity, but oats were never manured (Olsson 1988). Oats were cultivated on soils of less fertility in the north-east of the infields, on the field called Gallfäladen. The name indicates that this area most probably originated from earlier common pasture. According to earlier documents, conversion to arable took place some time during the 17th century (Persson 1986b).

Estimated seed yields for this village were 4.5, 3.7 and 2.8 for barley, rye, and oats respectively (Persson 1986b). Harvest volumes in Bjäresjö reached almost twice the amounts of the villages on sandy soils (Olsson 1988, Ch. 4.1.7). In terms of cultivated area, oats were the predominant crop in Bjäresjö around 1700 (Olsson and Persson 1987).

The hay meadows constituted 40% of the infields (Fig. 4.2:10); 50% of the meadow area was situated in the fifth field, along the river. Additional meadows existed as small meadow plots intermingled among the arable inside the fields. The rotating fallow meant that during the fallow year(s) when the field was grazed, the meadow areas inside the same fields were also grazed and no hay was harvested. Meadow production was highest on the wet meadows close to the river banks. The floodings here supplied water, nutrients, and oxygen to the meadows, thereby helping to maintain a reliable productivity. Maximum yield was 1100 kg ha^{-1} yr^{-1} (Ch. 5.3). Mean annual hay production was approx. 500 kg ha^{-1}, fully representative of the clay till areas.

Fodder areas much smaller than the arable were typical of the three-course rotation areas in southern Sweden in Early Modern Time. In Bjäresjö, 5 ha of the infields were reported as enclosed pasture, but apart from this little area, only the fallows were available for stock grazing (LM Bjäresjö 1:1699). This implies that the amount of livestock was limited, on average amounting to 1–2 cows, a few oxen, and 2–3 horses per farm (Olsson 1988). Mean farm size was 30 ha (Olsson 1988, Ch. 4.2.7.1).

To reduce the demand for raw material for fencing, community pasturing was practised, creating a vast common tract of land between the south field in Bjäresjö and the north field in the village of Gundralöv. Fences were mainly built of soil, giving low walls reinforced by branches and stems of thorny shrubs.

Circulation of nitrogen within the agro-ecosystem

The amount of manure calculated as available nitrogen for the arable soils was approx. 12 kg ha^{-1} yr^{-1} (cf. Olsson 1988, Ch. 5.3). Adding the small amounts of nitrogen from waste material like used roof-thatch (Dodgshon and Olsson 1989) will still give a low nitrogen input compared with the output contained in the harvested crops. The balance resulted in a net annual nitrogen loss of about 10 kg ha^{-1} (Olsson 1988). This deficit can only to a limited extent be explained by supply from the soil organic pool. In clayey soils we can expect considerable annual leaching of nitrogen, approx. 15 kg ha^{-1} (Brink 1983). Obviously, additional nitrogen sources must have existed here. The clayey and undrained soils preserve soil moisture and the lime content connotes a high soil pH value, two factors positively correlated with the activity of nitrogen-fixing bacteria, both free-living and symbiotic with leguminous plants (Hansen and Kyllingbæk 1983). Further, there are several indications of the abundant occurrence of leguminous weeds on fallows and in the arable mentioned above, especially *Trifolium repens* and *Ononis repens* (cf. Olsson 1988). Cultivation of leguminous crops is not recorded from this time, but indications exist of infrequent cultivation of pulses (Dahl 1942).

Bjäresjö, settled in prehistoric time, had by the early 18th century practised 500 years of intensive cropping (cf. Skansjö 1983), albeit with a low yield. The fertile soil here, described today as the third most fertile in Sweden in the soil classification system (Fig. 1.2:16), has acted as an insurance against long-term nutrient exhaustion. Nutrient depletion from the arable fields was probably balanced largely by biological nitrogen fixation (Olsson 1988). By the beginning of the 18th century it seems that the Bjäresjö system had depleted its resources. This was due to the conversion of common pasture into arable land. This limited the number of stock and thus the amount of manure nutrients available (cf. Table 4.4:1, Ch. 5.3). The possibilities of increasing the yield in the absence of artificial fertilizers within the system were thus very small. Ley cultivation would have been one possibility, but more than a century was to pass before this was commonly practised in this area.

4.2.8 The 19th and 20th centuries
Landscape and landholding changes in Bjäresjö since 1800

Tomas Germundsson and
E. Gunilla A. Olsson

The landholding situation in the 19th century

During the 19th century the parish of Bjäresjö was still strongly dominated by the two estates within its boundaries; Ruuthsbo in the west and Bjersjöholm in the east. Throughout the century these estates controlled nearly all the land in Bjäresjö. The only exceptions were one farm in the village of Bjäresjö, belonging to the church, and a few tenant farms in the north-west, belonging to the Krageholm estate situated some kilometres north of the parish. There were in other words no farms owned by their occupiers in the 19th century. This situation meant that the landlords had a very strong influence on the changing production organization in connection with the new achievements in agriculture during the 19th century. Such important improvements as marling, artificial manuring, under-drainage, the introduction of modern crop rotation, together with the development of new crops and implements and mechanization totally changed the prerequisites for agricultural production in the era of industrialization. The utilization of these new features also called for a breakup of the old cultivation systems, and the enclosure movement could be seen as one of many integrated factors in the protracted agrarian revolution taking place in the 19th century. Most of the agricultural innovations were introduced at the landed estates and consequently these units also played a leading role in the introduction of enclosures. In Bjäresjö the estate-owners initiated all the four enclosures carried out in the villages in the parish at the beginning of the 19th century. In essence these processes meant that the landlords created a settlement pattern consisting of either good-sized tenant farms or of smallholdings, where the tenants had labour duties on the estates. As a result, the villages of Bjäresjö, Stora Tvären, and Kärragården were split up, while the village of Gundralöv – adjacent to Ruuthsbo estate – was simply abolished.

The map showing the area in 1815 represents the stage just before these radical changes, and most of the farmsteads are still situated in nucleated villages. The enclosure process had started, however, and there are some moved-out farms in the centre of the parish (Fig. 4.2:14).

Land use, landscape, and vegetation in early modern times

The early 19th century landscape of Bjäresjö (Fig.4.2:14) differs in many details from the scenery in the previous century. However, land use was still determined by the three-course rotation. The large areas used for yearly sowing generated an open landscape dominated by arable fields used for cereal cropping surrounding the nucleated settlement sites. There were no commons or outland and the fallows were used as grazing grounds for livestock. Since the arable in the parish had increased by 20–60% during the past hundred years (Ch.4.2.7.1) the meadow area was correspondingly reduced. The pronounced shortage of wood, not only as building and fencing material but also as fuel, is demonstrated by the widespread marks left by peat-cutting present in almost every small basin or depression between the hummocks. The land survey map of Bjäresjö in 1809 reports totally overexploited peat deposits (LM Bjäresjö 6:1809). The available area for hay harvest was thus at an extreme minimum by the time of the land enclosures.

The formerly wooded meadows at Bjersjöholm lost their trees during the past century and were already treeless in 1791 (LM Bjäresjö 2:1789–91). By the early 19th century only two woodland areas existed within the parish. The wood-pasture, or rather park in the sense used by Rackham (1986), north of Bjersjöholm was a characteristic feature of the large estates and constituted an enclosed area where the tree layer was dominated by wide-crowned old beeches and oaks. Horses and deer were pastured here and consequently there was no closed tree canopy, and grasses and herbs dominated the field layer vegetation. The Svarte Forest, situated close to the outlet of the river Svartån, was a mixed oak-hazel grove, also fenced and belonging to the Ruuthsbo estate. Oak, alder, and hazel were formerly coppiced here and the area was still used as a wooded meadow where hay crops were harvested (LM Bjäresjö 4:1761). The high value of the wood and also of the game is indicated by the presence of two gamekeeper's houses on the borders of the woodland, according to the 1815 map. Besides these woodlands, trees also existed in a planted avenue of elms bordering the entrance road south of the Ruuthsbo estate.

The wet meadows along river Svartån were the largest fodder areas used for hay harvests in the parish. There are no signs in the contemporary documents of any management of water-meadows, that is, irrigation of the grasslands. However, the meadow vegetation could probably benefit from nutrients supplied by the annual natural floodings, as indicated by the presence of recent fluvial deposits along the river. At this time the water flow in the river was considerable, maintaining four watermills downstream from the wet meadows of Bjäresjö.

Gardens did not exist at farmer's settlements in the

THE BJÄRESJÖ AREA
EARLY 19TH CENTURY LANDSCAPE, CA. 1815

Fig. 4.2:14

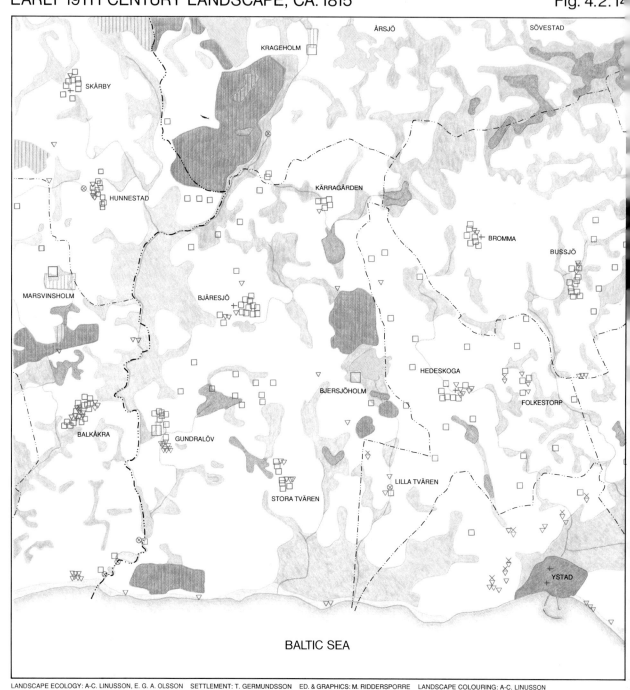

KRAGEHOLM
ÅRSJÖ
SÖVESTAD
SKÅRBY
KÄRRAGÅRDEN
HUNNESTAD
BROMMA
BUSSJÖ
MARSVINSHOLM
BJÄRESJÖ
HEDESKOGA
BJERSJÖHOLM
FOLKESTORP
BALKÅKRA
GUNDRALÖV
LILLA TVÄREN
STORA TVÄREN
YSTAD

BALTIC SEA

LANDSCAPE ECOLOGY: A-C. LINUSSON, E. G. A. OLSSON   SETTLEMENT: T. GERMUNDSSON   ED. & GRAPHICS: M. RIDDERSPORRE   LANDSCAPE COLOURING: A-C. LINUSSON

0   1   2 km

1 : 50,000

OPEN AREAS: ARABLE, FALLOW,
DRY AND MESIC MEADOW, PASTURE

DRY AND MESIC MEADOW
WITH COPPICE

WET MEADOW

OPEN FEN

PASTURE WITH SCATTERED
TREES AND SHRUBS

WOOD-PASTURE

OPEN SAND

LAKE, POND

STREAM

PARK

MANOR

FARM

SMALL HOUSE, COTTAGE

WATERMILL

WINDMILL

TOWN

CHURCH

HUNDRED BOUNDARY

PARISH BOUNDARY (1864)

THE BJÄRESJÖ AREA

ARLY 20TH CENTURY LANDSCAPE, CA. 1915

Fig. 4.2:15

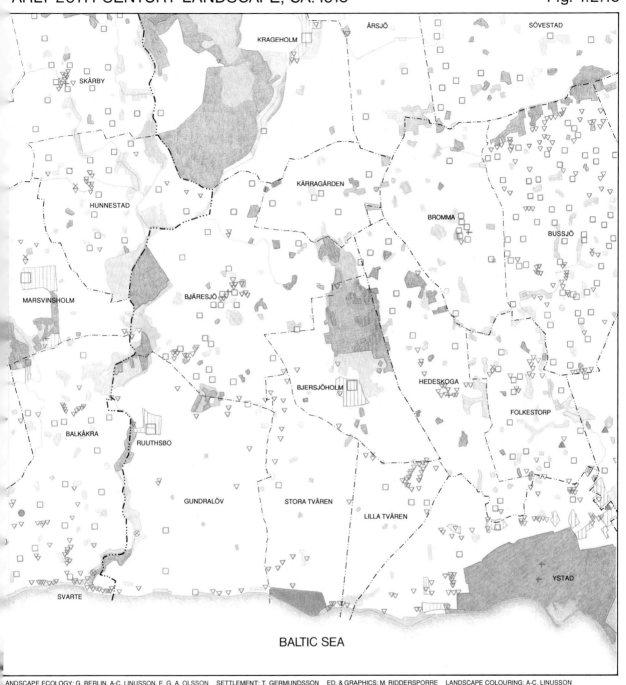

SÖVESTAD

ÅRSJÖ

KRAGEHOLM

SKÅRBY

KÄRRAGÅRDEN

BROMMA

BUSSJÖ

HUNNESTAD

BJÄRESJÖ

MARSVINSHOLM

BJERSJÖHOLM

HEDESKOGA

FOLKESTORP

BALKÅKRA

RUUTHSBO

GUNDRALÖV

STORA TVÄREN

LILLA TVÄREN

YSTAD

SVARTE

BALTIC SEA

LANDSCAPE ECOLOGY: G. BERLIN, A-C. LINUSSON, E. G. A. OLSSON   SETTLEMENT: T. GERMUNDSSON   ED. & GRAPHICS: M. RIDDERSPORRE   LANDSCAPE COLOURING: A-C. LINUSSON

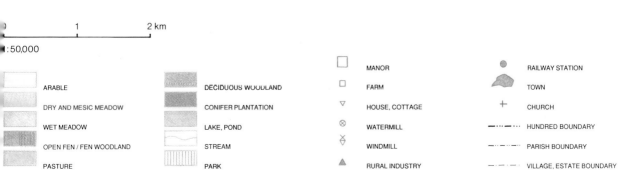

1   2 km

:50,000

ARABLE	DECIDUOUS WOODLAND	MANOR
DRY AND MESIC MEADOW	CONIFER PLANTATION	FARM
WET MEADOW	LAKE, POND	HOUSE, COTTAGE
OPEN FEN / FEN WOODLAND	STREAM	WATERMILL
PASTURE	PARK	WINDMILL

RAILWAY STATION

TOWN

CHURCH

RURAL INDUSTRY

HUNDRED BOUNDARY

PARISH BOUNDARY

VILLAGE, ESTATE BOUNDARY

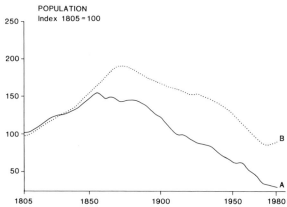

POPULATION
Index 1805 = 100

Fig. 4.2:16. Population in Bjäresjö parish (A) and the Ystad area (B) 1805–1980. Sources: parish registers, official statistics.

early 19th century. At most there were a hop pole and a potato plot. The garden of Ruuthsbo is marked on the map, however, and the description from 1761 (LM Bjäresjö 4:1761) tells of a garden arranged in French Versailles style with hedges of *Buxus* (box), *Ligustrum* (privet) and *Ribes* (currant) bushes, and quarters filled with ornamental flowers.

From the 1830s onwards all the villages had been enclosed. From then on, besides the new settlement pattern, the effect of the reorganization was reflected in changes in farming practices, the size of livestock herds and harvest volumes. The strict three-course rotation was abandoned and succeeded by crop rotation systems including leguminous crops like vetches, contributing positively to the soil nitrogen balance. Soil fertility was also improved by marling. A more elaborate preparation of the soil, as well as equalized seed density and sowing depth as a result of the introduction of agricultural machinery, gave further prerequisites for increased crop yields. Harvest volumes of barley and rye increased by 60% from the beginning of the 18th century to 1855 (Persson 1986, Germundsson and Persson 1987). However, the main direction of agricultural production in Bjäresjö was still the same as in the early 18th century, with cereals as the only important crop. Winter wheat was introduced, but despite its ninefold seed return (Germundsson and Persson 1987) it only accounted for 1% of the arable area in the 1850s. Potatoes were now cropped on roughly 9% of the arable.

Population changes

As regards the population development, Bjäresjö experienced a considerable growth during the first half of the 19th century, although it was more modest than the general trend in Sweden. In 1805 Bjäresjö parish had 672 inhabitants. The number increased to 1045 (56%) in 1855, which also marked the population maximum for the parish. From 1865 onwards, the population has been declining up to the present day.

Compared with the countryside as a whole, and also with the project area in total, this population development was rather limited (Fig. 4.2:16). An increased productivity in agriculture was a necessary condition for the population growth up to the mid 19th century, but compared with other areas with similar natural qualities, it is obvious that the fertile lands of Bjäresjö theoretically could have supported a much bigger population during that time. It is therefore not very fruitful to discuss a direct relationship between population and agricultural production within the parish. Other factors must be considered to explain population changes, and it has been found that one of the most important is the landholding pattern. To a large extent the modest population development in Bjäresjö referred to above can be explained by the landholding situation (Germundsson 1987). Other investigations have shown that, generally, population rise in the 19th century was considerably smaller in areas dominated by big landed estates, and also that population maxima were reached earlier here (Jeansson 1966, Fridlizius 1975a). In these areas the estate-owners could make decisions concerning settlement and farm sizes that had direct influence on the population number.

Some characteristic rationalizations carried out by the estates in accordance with the new economic order during the latter part of the 19th century were to sell off peripheral land – with the aim of investing capital in modernizations – and also to cultivate more land under the estate's own regime, instead of under tenant farms. In Bjäresjö the effect was that a number of tenant farms were closed down, which in turn meant that many people had to move out of the parish during the second half of the 19th century (Fig 4.2:16). Consequently, there was much less partitioning of farms in Bjäresjö than in many other places, and when mechanization and rationalization led to a reduction in the number of people working in agriculture, the effects were especially strong in Bjäresjö as the estates had the greatest possibilities to invest in machines (cf. Germundsson and Möller 1987).

Landscape implications of the mechanization of agriculture

The map of Bjäresjö from 1915 (Fig. 4.2:15) shows the landscape effects of the reorganization of land use and farming practices which occurred successively during the 19th century. This is reflected as a rectilinear landscape, shaped by the new geometry of the landholding units. There was a considerable change in the settlement pattern compared with the early 19th century. In a first phase the enclosure process generally spread out the settlement, as can be seen in Fig. 4.2:15. In the village of Bjäresjö some farms were moved out while others remained on their old sites, but with all the arable land agglomerated into one unit. As mentioned before, the process in the village of Gundralöv, initiated

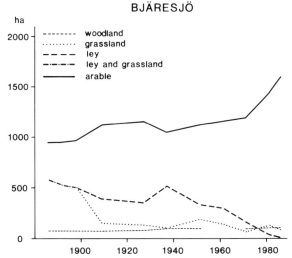

BJÄRESJÖ

Fig. 4.2:17. Land use in Bjäresjö parish during the period 1886–1986: arable, grasslands, and woodlands. Source: Agricultural statistics, unpubl. (1886–1909), and publ. (1909–1986) – see Ch. 2.2.

by the owner of the adjacent Ruuthsbo estate, meant that the whole village was abolished.

In the north and in the south-east, the villages of Kärragården and Stora Tvären were split up and most of the farms were moved out to new sites. In a second phase during the 19th century an "infill" of small units could be noted, especially in the village of Bjäresjö. These smallholdings were created on land not belonging to the estates and they were mostly inhabited by families working both on their own land and as day-labourers, primarily on the estates.

Parallel to the reorganization of land there was an intensification of cropping practices in terms of the proportion of land used for annual cultivation. This is visible in the landscape as reduced fodder areas (Fig. 4.2:17). The former large wet meadows along the river Svartån had been considerably diminished by draining and then cultivation. The only remaining grasslands north-west of Bjäresjö village and east of Ruuthsbo estate are some small fens. Generally there was a strong impact on the hydrology in the whole parish. This is drastically expressed at Bjersjöholm, where the lake has been completely emptied and converted to meadowland. The water level in Bjäresjö lake was probably lowered, as indicated by the invasion of reeds along the shores. Eutrophication of the lakes caused by the use of fertilizers and more intensive management of surrounding arable is indicated by the increased content of a target alga in lake sediments (Ch. 5.1). Regularly spaced over the whole landscape are numerous marl-pits, often with a small open water surface, resulting from the frequent habit of digging calcareous clay from deep soil layers and adding it to the topsoil as a fertilizer.

The woody vegetation was further reduced and

Svarte Forest was cut down. A new plantation of mixed coniferous trees appeared in the south-east close to the outskirts of Ystad. The estate settlements were still surrounded by large parks, and the gardens were now arranged in English style. From the manor buildings avenues radiate in several directions.

The changes in agricultural production implied a greater crop diversity and the introduction of more new crops (Fig. 4.2:18) like sugar-beet which around 1915 was cultivated on 13% of the arable. Winter wheat increased its share of arable to 13%. At the enclosure of Bjäresjö village in 1907 the total grassland area including meadows and natural pastures amounted to 13% of the arable (LM Bjäresjö 15:1907). For the whole parish, on the other hand, the same category reached only 2.2% of the arable and the fallows were almost extinct: 5% in 1909 (Agricultural Statistics 1909). Fodder was obtained to a large extent by culti-

BJÄRESJÖ

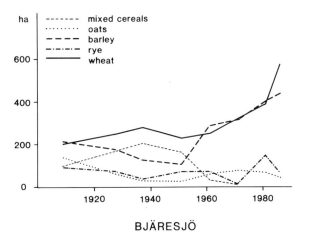

BJÄRESJÖ

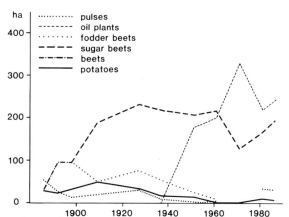

Fig. 4.2:18. Above. Cultivated crops in Bjäresjö parish during the period 1909–1986: grain crops. Source: Agricultural statistics, unpubl. (1886–1909) – see Ch. 2.2. – Below. Cultivated crops in Bjäresjö parish during the period 1886–1986: oil plants, pulses, and root crops. Source: Agricultural statistics, unpubl. (1886–1909), and publ. (1909–1986) – see Ch. 2.2.

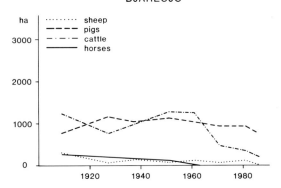

BJÄRESJÖ

ha
.... sheep
---- pigs
-·-·- cattle
—— horses

Fig. 4.2:19. Livestock in Bjäresjö parish during the period 1909–1986. Source: Agricultural statistics, unpubl. (1886–1909), and publ. (1909–1986) – see Ch. 2.2.

vation of leys and vetches on 26% of the total arable (Fig. 4.2:17). Commercial fertilizers were used, although in modest volumes. Oxen as draught-animals had disappeared but there were on average four working-horses per farm (Fig. 4.2:19). This general agricultural development and landscape content was relatively stable until 1945, although mechanization proceeded with an increasing number of tractors accompanied by a reduced number of horses. Cattle-raising for dairy production had approximately the same extent during the 20th century until 1960 when stock decreased significantly (Fig. 4.2:19).

Landholding and cultivation units

The big estates in the parish have continued to exert a strong influence on development. The landholding conditions together with the actual cultivation units during the 20th century are shown in a series of cross-section maps in Fig. 4.2:20.

In short, the picture of 1915 shows a situation where nearly all the land was owned by four big landowners, who thus had a determining influence on land use (Fig. 4.2:20a). Alongside the big units run by the estates, the large number of cultivators meant that the landscape in some areas was strongly varied and small-scale in character (Fig. 4.2:20b). At this time many crofters' holdings – small parcels where the tenants were obliged to do service at the manor owning the land – still existed. To every croft belonged a little piece of land cultivated by the crofter himself.

In 1907 the village of Bjäresjö had undergone a *laga skifte* enclosure. All the land in the village (except for one farm belonging to the church) belonged to Bjersjöholm estate, and it was the estate-owner who ordered the enclosure reform. A number of quite small units were created, and they were later sold off to peasants.

By 1945 the two estates of Ruuthsbo and Bjersjöholm had the same owner, an extraordinary situation where one proprietor controlled almost the whole parish. Another feature, perhaps more typical of its time, was that the Krageholm estate had sold off some of its tenant farms in the north (Fig. 4.2:20c). It was very common during this period that the big estates sold off peripheral land to finance rationalizations on the estates (Möller 1985). Still another change was the introduction of ordinary one-family houses with small gardens in the old village centre in Bjäresjö, inhabited primarily by agricultural workers. This change in settlement reflects the new organization of labour on the estates, where the former tenant cotters had been replaced by wage-labourers.

The cottages had been replaced by tenant farms with about 5–10 ha of arable land, while the rest of the tenant farms had generally grown in size and become fewer in number since 1915. The estates cultivated approximately the same acreage as thirty years earlier, and the picture of the cultivation units had not changed radically during the period (Fig. 4.2:20d). By 1945, half of the cultivation units from 1915 had disappeared, but this reduction took place only at units smaller than 10 ha.

Fig. 4.2:20. a–f. Landholding and cultivation units in Bjäresjö 1915, 1945, and 1984. After Germundsson 1985. Two-way arrow: same owner or farming unit. Arrow pointing out: centre of farming unit lies outside the parish. Arrow pointing in: unit also farms land outside the parish.
a) Landholding 1915. As regards landholding units there is a striking dominance of the big estates. A belonged to Ruuthsbo and B to Bjersjöholm. C was owned by the Krageholm estate situated outside the parish. The area Y was owned by a member of the nobility and was mainly cultivated under the Ruuthsbo estate. The church owned I, and only the small plots D–H and J–V were owned by peasants, or – in the case of the very smallest plots – by other occupational groups. Units J–U were smallholdings created in a *laga skifte* enclosure in 1907.
b) Cultivation units 1915. The arrangement of the cultivation units differed from the landholding situation. The two largest areas, 1 and 2, were cultivated by the estates, while the remaining cultivation units were mostly managed by tenant farmers. These units were principally situated in the north, where most of them belonged to the Krageholm estate outside the parish, and in the south of the parish. The black squares represent crofters' holdings. To every croft belonged a little piece of land cultivated by the crofter himself.
c) Landholding 1945. At this time the two estates of Ruuthsbo and Bjersjöholm had the same owner. Farms D, E and F had been sold off from Krageholm estate. The letter (a) represents ordinary one-family houses with gardens.
d) Cultivation units 1945. The cottages are now replaced by tenant farms, but otherwise the changes since 1915 are fairly slight.
e) Landholding 1984. The two estates have separate owners again. Nearly all the small units have been eliminated.
f) Cultivation units 1984. Note the high correspondence between landholding and cultivation units compared to the situation in 1945 and 1915. Sources: 1915: economic map 1915, with description; 1945: land survey documents, taxation registers, primary census data 1945; 1984: register of landed property, interviews.

210

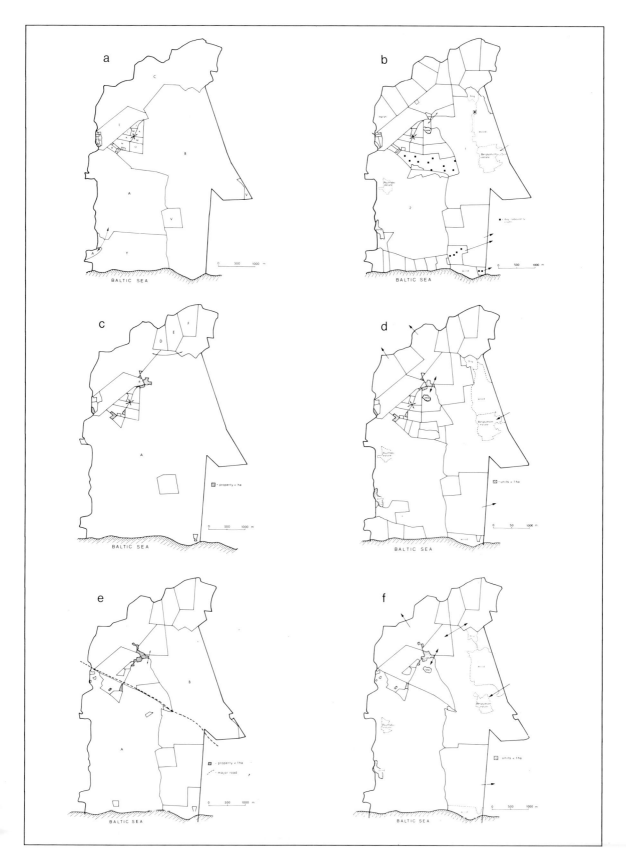

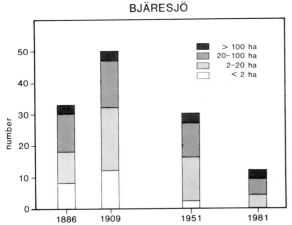

BJÄRESJÖ

> 100 ha
20–100 ha
2–20 ha
< 2 ha

Fig. 4.2:21. Number of cultivation units in Bjäresjö parish 1888–1981, in different size classes. For the year 1981 only units > 2 ha. Source: Agricultural Statistics, unpubl. (1888–1909), and publ. (1909–1981) – see Ch. 2.2.

During the period from 1945 up to the present (1984) land use and landholding have changed considerably. The two estates have once again separate owners, but the most striking change is the elimination of nearly all the small units. It is also notable that there is a greater correspondence between landholding units and cultivation units (Figs 4.2:20e,f and 4.2:21).

Agricultural development in Bjäresjö is in many ways a reflection of the new demands on economic efficiency that have characterized agriculture in Sweden, and other western European countries, in the postwar period (for further discussion, see Ch. 5.12). Mechanization also enabled a single farming family to cultivate larger areas. The estates, on the other hand, generally became smaller in size, when land was sold off. This was the case with Bjersjöholm, while Ruuthsbo had been reduced already before the 19th century. The changes are illustrated in Fig. 4.2:21, showing the number of cultivation units at different points in time.

Rural development since 1945

During the period from 1945 up to the present there have been great changes in agricultural production. Today there is only one dairy farm in the parish and ley cultivation has almost totally ceased (Fig. 4.2:16). However, two categories of animals have increased, namely pigs and chickens (12,300 chickens in 1986), raised in animal factories and fed solely on commercial fodder. An oil-yielding crop, rape, appears for the first time in the agri-statistics from Bjäresjö in 1951 accounting for 12% of the arable (Fig. 4.2:18 below). Interviews with the estate-owners and farmers reveal that Bjäresjö today is an efficient production landscape with high inputs of energy in the form of machines and artificial fertilizers, and also that production is totally concentrated

upon four crops: wheat, barley, rape, and sugar-beet. Compared with the beginning of the present century, crop diversity is again reduced, but to the four main crops must also be added maize and artichokes as main products from Ruuthsbo estate. The agricultural landscape of today is uniform and strongly dominated by the large-scale arable cropping. Rationalizations have also meant that the individual fields have been reduced in number and increased in size (cf. Ihse and Lewan 1986).

Landscape diversity has been greatly impoverished with the continuing removal of physical cultivation obstacles like marl-pits, open ditches, dykes, hedges, and refuges of shrubs and trees in the arable land. Out of 14 existing fen areas in this area in 1914, only one remains in 1987 (Ch. 5.11). A comparison of grassland area, all categories except leys, between the years 1938 and 1987 shows a reduction by 46% (Ch. 5.11), which is high considering that very little remained by 1909 (Fig. 4.2:17). Much of the river Svartån has been directed into culverts, and the formerly impressive wet meadows are reduced to a narrow strip of grassland heavily affected by nitrogen fertilizer. This is reflected in the vegetation by the dominance of plants like *Urtica dioica* (nettle) and *Phalaris arundinacea* (reed-grass). A weak hint of the species content of the former wet meadows is given by a tiny plot of unfertilized, grazed calcareous grassland along some peat-cutting ponds north of the main road between Malmö and Ystad and another plot close to lake Bjäresjösjön (Fig. 4.2:22). Characteristic species are plants like *Dactylorhiza incarnata* (early marsh orchid), *Parnassia palustris* (grass of Parnassus) and *Carex paniculata* (greater tussock sedge). Other small plots of grasslands are either dominated by nitrogen-favoured ruderals and weeds or are overgrown by shrubs in a process of secondary succession since management has ceased. Small-scale spruce plantations have been established on former grasslands with the intention of serving as game refuges, mainly for pheasants. In the former wood-pasture at Bjersjöholm grazing has ceased and the area has now turned into a mixed deciduous forest with beech, elm, ash, and oak. Spruce has been planted here as well, and in the northern part there is a large stone quarry and refuse dump.

As mentioned above, there has been a considerable out-migration from Bjäresjö parish since the 1850s, and by 1910 the population number was back at the 1805 level. The shortage of work in the parish means that this out-migration has continued. Apart from agriculture, there have been virtually no jobs in Bjäresjö this century. Census statistics from 1945 show that there were 192 working people living in the parish at this time; 153 of them were employed, while the rest were farmers and their family members working on their own land, ranging from estate-owners to smallholders. Of the wage-labourers, 73% were working in agriculture, 57% on some of the estates in the parish (Table 4.2:6).

In 1980 only 30% of the economically active population in Bjäresjö were working in agriculture (Table

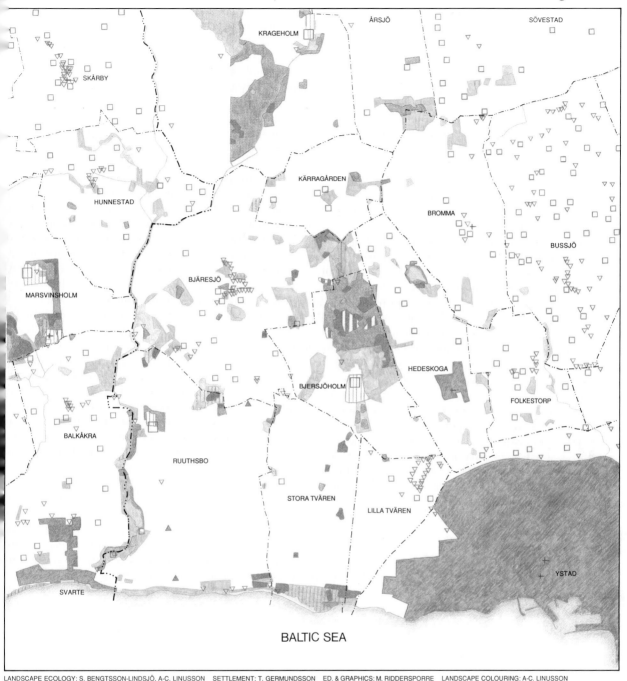

KRAGEHOLM

ÅRSJÖ

SÖVESTAD

SKÅRBY

KÄRRAGÅRDEN

BROMMA

BUSSJÖ

HUNNESTAD

BJÄRESJÖ

MARSVINSHOLM

HEDESKOGA

BJERSJÖHOLM

FOLKESTORP

BALKÅKRA

RUUTHSBO

STORA TVÄREN

LILLA TVÄREN

YSTAD

SVARTE

BALTIC SEA

LANDSCAPE ECOLOGY: S. BENGTSSON-LINDSJÖ, A-C. LINUSSON   SETTLEMENT: T. GERMUNDSSON   ED. & GRAPHICS: M. RIDDERSPORRE   LANDSCAPE COLOURING: A-C. LINUSSON

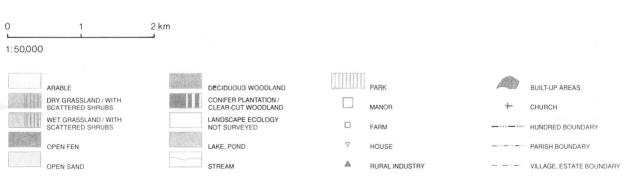

0   1   2 km

1:50,000

ARABLE	DECIDUOUS WOODLAND
DRY GRASSLAND / WITH SCATTERED SHRUBS	CONIFER PLANTATION / CLEAR-CUT WOODLAND
WET GRASSLAND / WITH SCATTERED SHRUBS	LANDSCAPE ECOLOGY NOT SURVEYED
OPEN FEN	LAKE, POND
OPEN SAND	STREAM

PARK	BUILT-UP AREAS
MANOR	CHURCH
FARM	HUNDRED BOUNDARY
HOUSE	PARISH BOUNDARY
RURAL INDUSTRY	VILLAGE, ESTATE BOUNDARY

Tab. 4.2:6. Employees in Bjäresjö parish in 1945.

Place of work	Number	%
Estates	87	(57)
Farms >50 ha	14	(9)
Farms 20–50 ha	11	(7)
Canning factory	3	(2)
Haulage contractor	4	(2)
Schools	4	(3)
Other	11	(3)
		(7)
Outside parish	19	
		(12)
Total	153	(100)

Source: national registration material

4.2:6). In absolute terms the changes are even bigger, and especially on the estates there has been a radical cut in the number of employees. In 1945 around 50 persons worked on Bjersjöholm estate, while the figure in 1984 was 5 – a reduction by 90%. Most of the economically active people today have their work in the town of Ystad, situated only a few kilometres away.

In the estate-dominated parish of Bjäresjö the changes in settlement pattern during the 20th century have not been extensive (Fig. 4.2:22). Compared with the situation in 1915 (Fig. 4.2:15), it is mostly a question of a thinning out of the settlement. Far more important are the functional changes. Keeping in mind the large reduction of cultivation units and the big changes in employment, it is obvious that many houses in the countryside are occupied by people not working in agriculture. It could be said that the physical changes of settlement in the countryside are much slower than the functional ones; former smallholdings and farms are often occupied by commuters or used as summer-houses. Consequently, the distance to towns determines to what degree rural habitats will change function instead of being abandoned when not used as cultivation units any longer (cf. Lewan 1967). The most striking change in settlement since the beginning of the century is the growth of Ystad and the creation of built-up areas in some places. Both Hedeskoga and Svarte mainly consist of modern one-family houses with small gardens, and the people living here mostly work in the town of Ystad.

4.2.9 Conclusions

Marie-José Gaillard, Deborah Olausson and Sten Skansjö

4000–3100 BC, the end of the Mesolithic period

At the end of the Mesolithic period, Bjärsjöholmssjön and the lake at Kumlan were the largest bodies of water in the Bjäresjö area. Bjäresjösjön did not exist at that time. The basin was occupied by an alder carr. The

vegetation around Bjärsjöholmssjön – and probably in the whole area – consisted mainly of elm-ash forest communities with much oak. The wettest areas in the deep hollows between hummocks and along the shores of the lakes and the river Svartån were dominated by alder mixed with ash and by open fens. The well-drained soils, rare in this area, were characterized by lime and oak.

Mesolithic settlement in the Bjäresjö area probably consisted of scattered hunting and gathering family groups who exploited the resources of the forest, the lakes, and the sea, in a pattern of regular seasonal migrations. The shores of the lake at Kumlan were attractive settlement locations. The abundance of hazel in pollen diagrams of this period is surprising and may be interpreted as a reflection of human interference in the forest in the form of clearance fires and coppicing. There is no palaeobotanical indication of cereal cultivation in the area, but small-scale crop tillage in the forest cannot be excluded.

3100–2500 BC, the Early Neolithic period

The major vegetational change at the beginning of the Early Neolithic, around 3100 BC, is shown in the pollen diagrams from the Bjäresjö area, as elsewhere in the project area, by a strong decrease in elm, ash, and lime. Moreover, the Early Neolithic as a whole is characterized by the continued decrease of these trees and the expansion of birch. This speaks undoubtedly for a drastic opening of the forest, but the reason behind this is still widely discussed, and the two major possible causes suggested for the Bjäresjö area are human interference ("landnam") and/or climatic change (dryness). On the basis of the pollen diagram from Bjärsjöholmssjön and sites in Östergötland, Göransson (1987b, 1991) rejects Iversen's "landnam" theory (1941, 1949) and suggests that the forest had been weakened by ecological changes beyond man's control. Dry climate may have been one major influencing factor since the time of the elm decline belongs to a period with low lake levels in southern Sweden (Ch. 5.1). There is indication of low water level in Bjärsjöholmssjön itself (Göransson 1991).

Early Neolithic settlement in the area was characterized by scattered sites of 800 to 1000 m². There are indications of Early Neolithic settlements on narrow headlands on the lake at Kumlan. Little evidence for Early Neolithic settlement was found around Bjärsjöholmssjön, however, which is surprising in view of the pollen-analytical evidence that suggests cultivation of cereals close to the lake and considerable grazing in the area. At the very beginning of the Early Neolithic, cultivation of cereals may have taken place on a very small scale, and then developed into a system of small "wandering arable fields" in hazel coppice woods. Man continued to open the forest by clearance fires throughout the Early Neolithic. The possible ecological catastrophe around 3100 BC and the rarity of healthy trees may have forced people to exploit much larger areas than before.

2500–1800 BC, the Middle Neolithic period

All pollen diagrams from the project area are characterized by a so-called regeneration phase starting ca. 2500 BC, thus at the beginning of the Middle Neolithic period, which consists of the increase of elm, oak, ash, and lime, accompanied by a maximum of hazel. The causes behind the "regeneration" phase may be several. It may be interpreted as a decrease of the human impact on the forest vegetation, and consequently as a regression of the settlement and land-use expansion. However, it may also be the reflection of a concentration of the settlements at particular sites where the expansion continued, which is seen in the coastal Köpinge area for instance (Ch. 4.1.9). Sites situated in more marginal and less attractive areas may have been abandoned and recolonized by forest trees. A further possible cause would be that forests were healed through changing ecological factors such as climatic conditions (Göransson 1991). However, the palaeohydrological evidence speaks for continued dry conditions until ca. 700 BC. Between 2500 and 2000 BC, the lake at Kumlan was completely overgrown by aquatics and reed vegetation. Around 2000 BC, an alder carr occupied the site of the ancient lake.

There is archaeological evidence to support the idea that a change in settlement pattern occurred during the course of the Middle Neolithic at Bjäresjö. While indications of Early Neolithic settlement can be found spread over most of the area, settlement apparently consolidated at the coast during the Middle Neolithic. In the Bjäresjö area we can suggest that the transformation of Kumlan from an open lake to an alder carr might have been one reason for the shift away from this location as the primary focus of settlement. There is evidence in the archaeological record for the use of the ard plough during this period, and livestock apparently made up an important part of the subsistence base. The pollen diagram from Bjärsjöholmssjön points to continuous grazing in the neighbourhood of the site. It may have consisted mainly of forest grazing in coppice woods. Such areas may have been used as common pastures by settlements situated close to the coast. The evidence for cultivation of cereals close to Bjärsjöholmssjön is weak. The fields were possibly concentrated mainly close to the coastal settlements.

1800–800 BC, the Late Neolithic and Early Bronze Age

During the thousand-year period represented by the Late Neolithic and Early Bronze Age, the climatic conditions seem to have been roughly the same as earlier, thus relatively dry until ca. 700 BC. Alder carrs occupied the basin at Kumlan and small hollows between the hummocks, such as that situated south of the present Bjäresjö village (Ch. 5.1).

In the pollen diagram from Bjärsjöholmssjön, as in all pollen diagrams from the project area, a further and definitive decline in elm and lime is dated at ca. 1700 BC, and these trees are very badly represented throughout the Late Neolithic and Early Bronze Age. In parallel, pollen indicators of grazing and cereal cultivation are increasing, which shows that the major vegetational changes during this period were due to increased human impact on the forests.

Signs of gradual but steady increase in human influence on the landscape also show up in the archaeological record as an increase in the frequency of stray finds and graves from the Late Neolithic and Early Bronze Age. We note that once again the area inland from the coast is being settled and exploited. The wealth of archaeological stray finds, as well as several extraordinary Early Bronze Age prestige items from the Bjäresjö area, give a picture of a strong and vigorous society whose knowledge of the area's natural resources allowed the optimal utilization of these resources. Judging by the placement of the barrows, settlement by the end of this period would have been dispersed throughout a large part of the Bjäresjö area, and mobile. Extensive and mobile farming practices were producing an increasingly open landscape.

800 BC–AD 700, the Late Bronze Age and the Iron Age until the Vendel/Viking Age transition period

Around 700 BC, the area was marked by a significant palaeohydrological change leading to a general rise of the water-table, which resulted in the creation of a lake at Bjäresjö (replacing the ancient alder carr) and the flooding of the deepest parts of the alder carr at Kumlan which were replaced by reed vegetation and open fens. These changes are ascribed to a regional climatic shift towards wetter conditions, which is well documented by palaeohydrological studies in South Sweden (Ch. 5.1). The interpretation of pollen diagrams suggests a drastic opening of the landscape around Bjäresjö and Bjärsjöholmssjön at the beginning of the Late Bronze Age, which is a characteristic of the whole project area and is clearly a reflection of increased human impact on the forests. Man undertook new fire clearings of hazel coppice, elm, and lime to gain new areas for cultivation around settlements, such as at Bjäresjö, and for grazing in large common grasslands. However, trees were maintained in certain parts of the area, as around Bjärsjöholmssjön where wooded meadows were probably well developed. Alder carrs, in spite of their regression in certain flooded areas such as those at Kumlan, were still abundant along the river Svartån and in wet bottoms.

There is archaeological evidence for Late Bronze Age settlement activity at four locations in the vicinity of the present village of Bjäresjö. The dramatic change seen in the pollen diagrams is not mirrored by the archaeological evidence, which is instead characterized by continuity in settlement area and settlement intensity. Moreover, the reconstructed resource areas for the four settlements of Bjäresjö are small in relation to the extent of grasslands suggested by the pollen diagrams. The total resource area needed including arable and fodder was at most 40 ha for each settlement. This fact may support the hypothesis that large common pastures in the inland part of the project area were used during the Late Bronze Age by other settlements that lacked winter fodder, such as the coastal ones of the Köpinge area, during a period of crisis resulting from both overexploitation for a long time and climatic deterioration towards cooler and wetter conditions (Ch. 4.1.9). It is proposed that the major part of the open grasslands occupied areas where a concentration of barrows occurred.

The surroundings of the settlement sites at Bjäresjö were probably characterized by a mosaic of small cultivated plots (less than 0.5 ha). Macrofossil analysis shows that barley, wheat, millet, and gold of pleasure were important crops. Hulled barley increased at the expense of spelt wheat, which suggests that manure was needed. The nitrogen need could be covered by the addition of manure and household wastes, or a combination of manuring with short-term coppice rotation among the arable. The few areas with light sandy soils in the neighbourhood of Bjärsjöholmssjön were probably also cultivated with cereals. There is indirect evidence that fishing, stock-raising, trade and bronze-casting were also part of the economic basis of life in the Bjäresjö area. It is also suggested that the shift from mobile to more permanent settlement began during this time.

During the Early Iron Age, drier climatic conditions resulted in a regional lowering of the water-table that led to lower lake levels. Bjäresjösjön was transformed into a very shallow pond with reed vegetation and aquatics. It is hard to judge how this climatic change influenced the vegetational landscape. Alder carrs may have expanded again in earlier flooded areas, and certain open fens may have dried out.

Archaeological evidence for Early Iron Age settlement in the area is unfortunately slight. The trend towards permanent settlement, made possible by the use of manuring, accelerates now, and some form of infield-outland system was probably established in the area by the Roman Iron Age or even earlier. The spread of beech during the Pre-Roman Iron Age may have been favoured in the grazed outlands. Settlement in the Bjäresjö area is apparently less consolidated than we see elsewhere at this time, however, due to the hilly topography here. As in the Köpinge area (Ch. 4.1.9), the transition to the Iron Age does not seem to represent a break, but rather a smooth development towards a more fixed form of settlement with the establishment of permanent fields and hay meadows in an infield/outland system. The pollen diagram from Bjäresjösjön suggests a land use rather like that of the Late Bronze Age.

There are, however, indications of recolonization of open areas by aspen, birch, and hazel during the later part of the Pre-Roman Iron Age and the Roman Iron Age. The one-course rotation system may have been introduced at that time. Moreover, the cultivation of barley increased steadily, particularly during the Migration Period and the first part of the Vendel Period.

That the coastal zone was also densely settled in the Bjäresjö focal area can be seen by cemeteries and habitation sites from Roman Iron Age and Migration Period located here. The economy of the area was integrated in larger spheres of production of for instance iron, and in far-reaching exchange networks.

AD 700–1200, the Late Iron Age (from the Vendel/ Viking Age transition period) and Early Middle Ages

Investigations of past lake-level fluctuations in southern Sweden provide strong indication of a distinct regional climatic change towards wetter conditions around AD 700–800 (Ch. 5.1). As a result of this, the water level of Bjäresjösjön rose, and the shallow pond was transformed into a little lake that reached its highest level between 1200 and 1300. This climatic change caused the flooding of other areas, leading to the occurrence of larger zones of open wetlands and to the disappearance of several alder carrs.

Other drastic changes in the landscape were caused mainly by new land-use practices. The introduction of autumn-sown rye during the Late Vendel Period shows that an important innovation occurred during this time. It is unclear whether this period saw an increase in the number of households within the agglomerated settlement which we presume existed at Bjäresjö from the Late Roman Iron Age. Hemp was also introduced as a cultivated crop and, for that reason, it is obvious that Bjäresjö was an attractive site with its little lake for retting and its peaty, nutrient-rich soils along the lake's outlet suitable for hemp cultivation. During the Viking Age, the cultivation of cereals (rye, barley, and spelt/ emmer wheat) occurred probably in a three-course rotation system on manured fields close to the settlement core. Hemp was grown during the whole Viking Age. Flax and buckwheat were probably introduced as cultivated plants. Arable land of ca. 2–4 ha was harvested each year. The infield/outland system was already well established. Large areas of wet meadows were available at Kumlan. Grazing pressure as a whole was higher than before, and hazel, elm, ash, and lime were the trees showing the largest decrease. Oaks were abundant, and probably common in wood pastures north of Bjersjöholm and in the northern part of the parish. However, the region was certainly more treeless than ever before during prehistoric time.

By the late Viking Age settlement had withdrawn from the coast, moving inland to more secure locations and probably creating pressure on existing settlement inland. Traces of settlement and/or runic stones give us every reason to assume that settlement, probably of village type, existed from this time in the immediate vicinity of the historical villages or on the actual village sites. We have evidence for Viking Age settlement north of Bjäresjösjön, and by the end of the period the greater part of what came to be the medieval village site was occupied. The next settlement was found 2 km southwards near the river Svartån, at the site of the later village of Gundralöv. There is less archaeological information about the Bjersjöholm area, which probably was exploited primarily for grazing.

The Early Middle Ages saw the establishment of a prominent manor complex with a stave church in the village of Bjäresjö. Restructuring in the 12th century gave the manor an even more distinctive character, and in the middle of the 12th century the old stave church was replaced by a Romanesque stone church. The manor was also rebuilt at this time, and the large number of coins from the manor site indicates that its owner was a rich and important person. The major settlement reorganization undergone between 950 and 1200 was accompanied by a probable expansion of the area of cultivated land. The consolidation of the infield/outland system is confirmed by the appearance of juniper and heather in the early medieval landscape. Those species were favoured on the poor leached soils of the permanently grazed outlands, where most of the trees had probably been removed in the course of the Viking Age. Hemp retting was abandoned in Bjäresjösjön during the Early Middle Ages (1000–1200).

1200–1670, High/Late Middle Ages and Early Modern Time

The oldest historically known settlement pattern consisted on the one hand of the two noble manors of Berghusagård/Bjersjöholm and Gundralövsgård, and on the other hand of the villages of Bjäresjö, Gundralöv, Stora Tvären, and Kärragården. With the possible exception of Kärragården, these settlements are all attested in medieval sources from the 14th century onwards. As mentioned earlier, villages like Bjäresjö and Gundralöv, possibly also Stora Tvären, can be traced back to the Viking Age and Early Middle Ages.

During the period 1200–1670, the land use was characterized by the increase in the area cultivated according to a three-course rotation system with rye and barley as dominant crops. Hemp, buckwheat, and flax were cultivated on a small scale. Moreover the organization into an infield/outland system was gradually crystallized and soon resembled the one known from the oldest maps of the parish (around 1700). Juniper was common in the outlands until ca. 1550 when it started to decrease significantly. The surroundings of Bjäresjö village were almost free of trees. The most densely wooded areas were probably situated around Bjärsjöholmssjön, where oak and beech were the dominant trees, and along the river Svartån (alder, hazel, oak). The basin of Kumlan was occupied by open reed swamp and fen vegetation.

The climatic deteriorations known for medieval times in central Europe and northern Scandinavia during the period 1350–1400 and the 17th century could not be demonstrated in the Ystad area and in southern Sweden on the basis of lake-level reconstructions (Ch. 5.1). The general climatic conditions seem to have been relatively wet. Moreover, no significant changes in settlement – expansion or regression – are recorded in written sources from the Late Middle Ages. An interesting fact, however, is that the pollen diagram from Bjäresjösjön indicates a stagnation of the expansion within the three-course rotation system between 1350 and 1450, the classical period of the so-called agrarian crisis of the Late Middle Ages.

However, around 1350 other changes have been observed that may have caused stagnation in the land-use expansion: the manor in Bjäresjö village, with its roots in the Viking Age, was moved to the islet in Bjärsjöholmssjön. In connection with the move, the manor changes its name from Bjäresjögård to Bjersjöholm. The move was made at a time when private castles were growing up in large numbers in Denmark. It can be explained partly as a sign of the growing power of the nobility in relation to a weakened central government, and partly as a result of political unrest: this was a time that simply demanded fortifications and up-to-date defensive sites. Unfortunately there are no pollen data available from Bjärsjöholmssjön concerning historical times.

The archaeological material suggests, furthermore, that the church-village of Bjäresjö may have been replanned and regulated in the Late Middle Ages: in addition to the moving of the manor, an earlier large estate seems to have been replaced by a number of almost equalized farms. From then on, the village structure much resembled what is known from the first surveyor's maps ca. 1700, despite the fact that there was a reduction in the number of farms for most of the villages in Bjäresjö parish during the 16th and 17th centuries. This reduction can largely be explained by equalization caused by the development of the two estates within the parish of Bjäresjö, Bjersjöholm, and Gundralövsgård. At Bjersjöholm the relatively plain 14th century manor was replaced by a splendid castle, largely finished by 1576. The building of this castle coincided with a moderate extension of the lands farmed by the estate. In a Scanian perspective, the area farmed by Bjersjöholm estate was only of medium size, while the grandiose buildings of the manor house were true signs of "the time of noble dominance" as the period 1536–1660 is often referred to in Danish history. Bjäresjö parish around 1670, with its settlement density of 2.7 farms km^{-2}, was very near the average for the entire project area.

The cultivation system of Bjäresjö village is known in detail from 1570. It consisted of a three-field system with the addition of a special annual meadow field. There seems to have been a significant expansion in arable land at the expense of meadowland and pasture (outlands with juniper) from 1570 to about 1700 in Bjäresjö. This is evident from different written sources, and it is also clearly indicated by the pollen diagram from Bjäresjösjön.

1670–1800

This period was characterized by a major development of cultivated land at the expense of pastures. Around 1700, the villages had a greater proportion of arable than the manors. The expansion was greatest on the manor of Bjersjöholm and Bjäresjö village (an increase of about 40%). All the villages were organized according to the three-field system with additional fields. Bjäresjö was a treeless agrarian landscape dominated by intensive grain production. A traditional three-course rotation system was practised on the two manors and the four villages of the parish. The area of meadowland used was relatively large. However, fodder areas were small, and therefore the amount of livestock was limited. Community pasturing was still practised. After 500 years of intensive cropping, nutrient depletion from the arable fields was probably balanced largely by biological nitrogen fixation from leguminous plants on fallows and arable. The outlands with juniper were entirely reclaimed for the cultivation of oats in the course of the 17th century as shown by pollen analysis and historical sources. The wooded meadows of the Bjersjöholm area lost their trees during the 18th century. The area along the river Svartån at Kumlan was still occupied by wetlands.

For this period it is also possible, for the first time in the total investigation period of the project, to get detailed knowledge of the population development of Bjäresjö parish. We observe, in parallel to the expansion of arable land, a population increase by about 100% in the course of the 18th century (from ca. 300 to ca. 670). However, during this period the number of village farms fell owing to equalization, and the area under direct cultivation from the two estates increased. The main cause of the development of the estates must have been a need to increase production in order to compensate for a less favourable market situation in the second half of the 18th century.

As a result of these processes, more people had access to less means of support, and an increasing part of the growing population was directed to a life as cotters: the number of cottages more than doubled during the 18th century. A change in productivity or in the organization of the production seemed inevitable around 1800 in the densely populated parish of Bjäresjö.

1800–1985

During the 19th and 20th centuries, the landscape was mainly transformed and fashioned by the changes in land use and the introduction of new technologies. The major transformations of the landscape were caused by the comprehensive under-drainage during the 19th century and by the enclosure movement (1810–1830). Bjärsjöholmssjön was lowered and replaced by meadowland. The area of wetlands diminished considerably by drainage and subsequent cultivation. Several small swamps and shallow lakes disappeared. By the early 19th century only two woodland areas existed within the parish, one wood-pasture north of Bjersjöholm and the Svarte forest close to the sea shore. However, in the areas that were almost treeless earlier, such as the surroundings of Bjäresjö, trees became more common thanks to plantations along roads and close to the farms. Spruce was also introduced. During the 20th century the Svarte forest was felled, and a new plantation of mixed coniferous trees appeared in the south-east.

The landholding situation during the 19th century was still dominated by the two estates within the parish boundaries: Bjersjöholm and Ruuthsbo (formerly Gundralöv). Most of the innovations associated with the agrarian revolution in the 19th century were introduced by landlords: marling, artificial manuring, under-drainage, new crops, new crop rotation systems, and new machines. Also the *enskifte* enclosures in the four villages were carried out under the auspices of the landowners: Bjäresjö, Stora Tvären, and Kärragården were split up, while Gundralöv village in the immediate vicinity of Ruuthsbo manor was closed down.

The population increase known from the 18th century continued into the first half of the 19th century which experienced a considerable growth: from about 670 inhabitants in 1805 to 1,045 in 1855. The latter date marked a population maximum in Bjäresjö parish: from then on, the population has been declining. In theory, the fertile soils of Bjäresjö could have carried a much bigger population, but the comparatively weak population development here in the 19th century can be explained by the landholding situation, as can the comparatively modest partitioning of farms in the Bjäresjö parish.

At the beginning of the 19th century, land use was still determined by three-course rotation. There was no common or outland, and the fallows were used as grazing grounds for livestock. The available area for hay harvest was at an extreme minimum by the time of the land enclosures. When enclosure was implemented, the strict three-course rotation was abandoned and succeeded by crop rotation systems including leguminous crops, contributing positively to the soil nitrogen balance.

Even in the 20th century the strong influence of the big estates prevailed in Bjäresjö parish: in 1915 only some small plots of the parish area were owned by peasant farmers. By 1945 Bjersjöholm and Ruuthsbo had the same owner who consequently controlled almost the whole parish area. Former tenant cotters had now been replaced by wage labourers living in one-family houses in the village of Bjäresjö. The ordinary tenant farms had become bigger in size but fewer in number since 1915.

The mechanization of agriculture led to changes in agricultural production during the early 20th century. This brought about a greater crop diversity and the introduction of new crops. Fodder was obtained to a large extent by cultivation of leys and vetches. This general agricultural development and landscape content was relatively stable until 1945, although machines increased at the expense of horses.

From 1945 up to the present time the most striking change is the elimination of nearly all the small cultivation units. To a large extent, the modern agricultural development in Bjäresjö can be seen as a response to central governmental political demands for rationalization and economic efficiency. As a consequence, the migration out of Bjäresjö parish, which started in the 1850s and had brought the population in 1910 back to the 1805 level, has continued up to the present day. Especially on the estates there has been a drastic cut in the number of employees. On the other hand, the changes in settlement pattern have not been very extensive in Bjäresjö. Many houses have changed their functions: formerly associated with agriculture, they now have become occupied by commuters or used as summer houses.

Since 1945, because land use and landholding have changed considerably, the agricultural landscape has become progressively uniform and the landscape diversity has been greatly impoverished. Ley cultivation disappeared with the replacement of horses by tractors. Rape was introduced after 1945. Bjäresjö today is an efficient production landscape concentrating on few crops: wheat, barley, rape, and sugar-beet. There are high inputs of energy in the form of machines and artificial fertilizers. In the former wood-pasture of Bjersjöholm, grazing has ceased and the area is now a mixed deciduous forest. The ancient wetlands at Kumlan are occupied by pasture land.

Lake Krageholmssjön surrounded by the woodlands of the Krageholm estate. View towards the southwest. Photo: Ingmar Holmåsen/N, September 1988.

4.3 The Krageholm area

4.3.1 Introduction

Björn E. Berglund

Area and sites studied. The Krageholm area comprises the present parish of Sövestad, which corresponds approximately to the extent of the Krageholm estate. Our studies deal particularly with the areas around lake Krageholmssjön (Figs 4.3:2–5).

The main field study sites are the following:
Palaeoecology: lake Krageholmssjön.
Archaeology: Årsjö village.

Historical and human geographical studies deal with source material from the entire parish of Sövestad and Krageholm estate.

Geomorphology and hydrology. The area is situated within the inner hummocky landscape, which is built up of moraine hills and kame terraces. However, one may regard lake Krageholmssjön as situated in a border zone between the typical hummocky landscape in the east and the slope of the Romeleåsen ridge in the west (see the Romele area, Ch. 4.4.1). The elevation of the Krageholm area varies from about 40 m at the lake and about 70 m in the hills north of the lake and in the north-east of the parish. The distance to the sea is 7–10 km.

Most of the western and southern part of the area is drained towards Krageholmssjön and the river Svartån. The catchment area of the lake extends further westwards on the slope of Romeleåsen (see Ch. 5.2). The northern part of the area is drained northwards through some peatlands towards lake Ellestadssjön and the eastern part is drained towards the river Nybroån. The area east of Sövestad has the character of a hummocky plateau with several fen basins coalescing in the east in Högestads Mosse, today a remnant of a large fenland. Most fens in the Krageholm area were wooded carrs or open fens in the late Holocene, but some are overgrown lake basins, like Vasasjön in the north-east. Most peatlands have been exploited for fuel or cultivated. Dammed water reservoirs have occurred in historical time, e.g. Willvattensjön, east of Krageholm, and Svartskyllesjön (= Vasasjön) in the north-eastern part of Sövestad parish.

Subsoil distribution. The Quaternary deposits are dominated by clayey till, in smaller areas sandy till. Large areas south and north of the lake are built up of glacial clay/silt. The calcium carbonate content is lower than in the Bjäresjö area (approx. 15%). The plateau-like area in the east contains several basins with peat. See map in Fig. 1.2:15 (Daniel 1986, 1989).

Conditions for cultivation. Because of the heavy soils, rarely interrupted by sandy soils, the area may be regarded as less favourable for primitive agriculture. In modern times, however, advanced ploughing techniques and drainage have improved the situation. The agricultural soil quality is rather good (classified as 7–8 on a 10-grade scale, Fig. 1.2:16). The heavy clay sediments (today mainly forested) near the lake have not been attractive for agriculture. These areas and peaty areas have been used for grazing.

4.3.2 Vegetation and landscape through time

Joachim Regnéll

The description of vegetation and landscape through time is part of a larger palaeoecological study of Krageholmssjön and the nearby lake Bussjösjön (Regnéll 1989). That paper contains more detailed information including complete pollen diagrams. In the summary pollen diagram (Fig. 4.3:1) the vegetation history over the last 6000 ^{14}C-years is divided into seven periods.

1) 4050–3150 BC. The landscape of mixed deciduous forest

During the Mesolithic the Krageholm area was covered by forest. Both the pollen diagram and the distribution of subsoil in the area imply that dominant forest types were those that demand soils rich in nutrients.

Alnus (alder) grew in wet and waterlogged places. In areas transitional to drier ground *Fraxinus* (ash) occurred together with *Quercus* (oak) and *Ulmus* (elm). On dry ground *Quercus* and *Tilia* (lime) were dominant. *Corylus* (hazel) probably occurred as a shrub on drier soils and on a variety of soils in open areas such as forest edges and blowdown gaps.

High frequencies of *Corylus* pollen imply that the forest was either rather open, or at least not taller than *Corylus* itself, thus allowing it to be a part of the pollen-producing forest canopy. These same two alternatives have been suggested by Rackham (1980) for interpreting high *Corylus* frequencies in the Mesolithic in Britain.

Göransson (1986, 1987b) proposed that Mesolithic man partly transformed the forest into coppice woods and grasslands by girdling and bush burning, which would have favoured grazing and browsing animals. Göransson also suggested the possibility of small-scale

KRAGEHOLMSSJÖN

POLLEN DIAGRAM

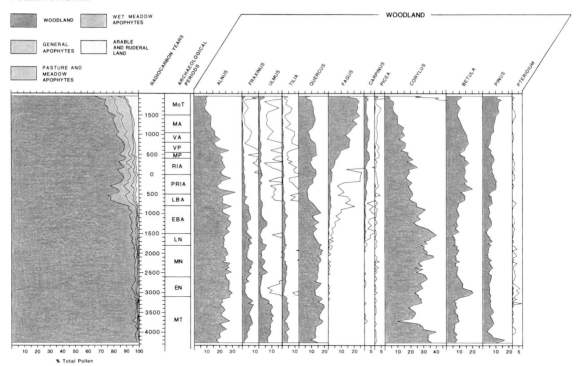

"garden cultivation" of cereals. This contrasts with the more traditional view of the virgin climax forest (e.g. Nilsson 1961, Iversen 1973). Low frequencies of herb pollen, especially cultural indicators, in the Krageholm diagram imply that there was no agriculture in the area during this period.

2) 3150–700 BC. Small-scale agriculture in the forest

The beginning of the period is defined by the elm decline. This event, which is generally considered to be synchronous in north-west Europe, is characterized by rapidly decreasing pollen frequencies of *Ulmus* at about 3150 BP and often also by corresponding decreases in other tree taxa. In the Krageholm diagram the elm decline is accompanied by decreasing frequencies of *Fraxinus, Tilia,* and *Quercus.* Proposed explanations have involved climatic and edaphic changes, interference by man, elm disease, or combinations of these factors (cf. compilation by Birks 1986:49–50). Birks (1986) concludes that "a close, but possibly varying interaction between disease and human interference" is probably behind the elm decline.

After the elm decline came a phase with high frequencies of *Betula* (birch) and *Alnus,* and low frequencies of *Ulmus, Fraxinus,* and *Tilia.* The distinct increase in the pollen frequencies of *Betula* has been recorded from many sites in southern Scandinavia (Iversen 1973, Aaby 1986, Göransson 1986, 1987b) and is often in-

terpreted as an indication of burning. *Populus* (aspen) pollen and *Pteridium* (bracken) spores indicate burning (Uggla 1958, Iversen 1973); both appear in their highest frequencies during this phase of the diagram. The frequencies of charcoal do not show any major rise, however, as might be expected, so there is no direct evidence for a higher frequency of fires in the Krageholm area.

The increasing frequencies of pollen of cultural indicators such as *Artemisia* (wormwood) and *Rumex acetosa/acetosella* (sorrel/sheep's sorrel), together with the first appearance of *Plantago lanceolata* (ribwort plantain) during the Early Neolithic, are the first strong evidence of human impact in the Krageholm diagram. From ca. 3000 BC there seems to be a continuity of stock-raising, probably in combination with the growing of crops. During the Early Neolithic and most of the Middle Neolithic, open areas were probably present on well-drained soils, as indicated by occasional *Juniperus* (juniper) pollen.

After the *Betula* phase, there occur fluctuations between high and low frequencies of *Ulmus, Fraxinus,* and *Tilia* pollen. Low frequencies are not, however, associated with high frequencies of *Betula* pollen, as was the case during the elm decline, but rather with high frequencies of *Corylus.*

Low *Ulmus* frequencies after the elm decline in southern Scandinavia have mostly been attributed to human impact (Berglund 1969, 1985a, Iversen 1973,

222

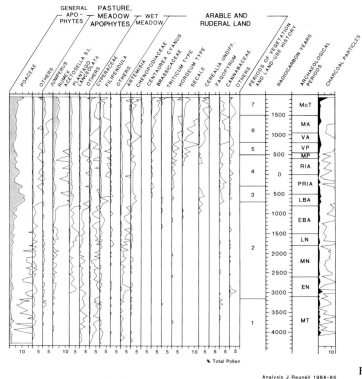

Fig. 4.3:1. Pollen diagram from Krageholmssjön.

Analysis J. Regnéll 1984-85

Aaby 1986). In the Krageholm diagram, however, human impact is not so evident. The frequencies of cultural indicators are continuously low, and the recorded small-scale changes do not correspond to the fluctuations in the tree-pollen curves. Thus, if human impact is also assumed to be behind the tree-pollen fluctuations in the Krageholm diagram, the human disturbances were probably not in the form of land clearance for agriculture and pasture. Possibly the fluctuations reflect changes in a forest managed for leaves, nuts, and acorns (maybe through coppicing) to provide easy sources of nutrients for man, livestock, and game.

Not until the Late Neolithic and Early Bronze Age do rising frequencies of herb pollen indicate an increase in open areas used for agriculture. Also diatom analysis and magnetic studies reveal distinct changes during this period, indicating increased human impact within the lake catchment (Håkansson 1989, Ch. 5.2).

3) 700–200 BC. The pastoral landscape

The cultural landscape changed dramatically during the later part of the Bronze Age. Distinct increases in herb pollen indicate that the landscape was to a large extent open. The vegetation was probably a mosaic with grassland and forest in early successional stages. *Corylus, Betula,* and *Quercus* seem to have been the most abundant trees on dry ground. *Carpinus* (hornbeam) increased as well, and was probably favoured by its ability to colonize abandoned grazing lands (Emanuelsson et al. 1985). High pollen frequencies of *Alnus* indicate that

wet areas were still forested. Low frequencies of Cerealia-type pollen indicate that crops were grown only on small areas. The grazing and browsing of livestock was probably the most important land use in an area-extensive, pastoral economy.

4) 200 BC-AD 500. A period of forest regeneration

The pollen frequencies of *Quercus, Fagus, Carpinus,* and *Ulmus* increase and total herb-pollen frequencies decrease in the part of the diagram which corresponds to Late Pre-Roman Iron Age and Early Roman Iron Age. This pattern represents regeneration of the forest, probably after a decrease in intensity of human impact on the vegetation. From the pollen diagram alone, however, it is difficult to judge the cause of this change. It may have been a regression in the pastoral economy described above, but it may also have been a change towards the more settled economy that characterized the next period.

5) AD 500–800. The central and the marginal cultural landscape

According to the pollen evidence, the composition of the forest changed distinctly during the Migration Period and Vendel Period. *Fagus* together with *Quercus* became the most abundant late successional trees. As the high *Fagus* frequencies are not associated with an increase in total herb pollen, this trend is likely to have resulted from a transition from open grazed forest to a more closed forest in some areas, rather than from intensified grazing of open grasslands containing scattered *Fagus* and *Quercus* trees. This indication of larger areas of closed forest, combined with no major change in the frequencies of cultural indicators, implies a change towards a more settled economy. It is thus suggested that during this period the landscape came to be characterized by an economy of permanent villages, of the type known by archaeologists and historians from at least the Early Viking Age. The landscape was divided into central and marginal areas characterized by different land use and vegetation. Depending on the location within the cultural landscape, the former open forest was turned into either areas of extensively grazed woodland of *Quercus* and *Fagus,* or intensively used fields and meadows. The areas of cultivated lands expanded, and *Secale* (rye) was regularly grown for the first time.

6) AD 800–1500. Forest in decline

During the Viking Age the areas of fields and pasture expanded. The areas of wood-pasture seem to have decreased and the dominance of *Fagus* and *Quercus,* both tolerant of grazing and soils poor in nutrients, is even more pronounced. The more edible nutrient-demanding species, such as *Ulmus, Tilia, Fraxinus,* and *Corylus,* all occur in the lowest frequencies of the last 6000 years.

A distinct rise in the pollen frequencies of *Rumex acetosa/acetosella* is probably related to the increase in arable land. When Linnaeus visited the Krageholm area in 1749 he described *Rumex* as the dominant herb of fallow areas (Linnaeus 1751). Thus, a field-rotation system that included fallow periods may have existed from the Early Viking Age. This is also implied by increasing pollen frequencies of *Secale*. Engelmark (1984) suggests that *Secale* cannot be grown after summer crops without a period of fallow between.

Cannabis sativa (hemp) was commonly cultivated as a fibre plant from the Late Vendel/Early Viking Age until the Early Middle Ages in the Bjäresjö and Bussjösjön areas (Gaillard and Berglund 1988, Regnéll 1989, Ch. 4.2.2). This crop was probably also grown in the Krageholm area, although the pollen frequencies are much lower than in the diagrams of those small lakes. A general decrease in *Alnus* pollen suggests that the alder carrs decreased in total area. Probably these were converted by man to open sedge fens, which were then mowed for winter fodder.

From the Late Middle Ages *Juniperus* became more abundant, indicating pastures on poor soils. This may represent clearance and grazing of the less clayey tills of the western part of the Krageholm area or the Romeleåsen area. However, it may also be a sign of nutrient depletion of existing grazing lands (cf. Ch. 4.2.2). In that case, this edaphic change may have been a consequence of an economy based on settled villages. The production of crops was based largely on nutrients transported from meadows and grazing lands to the fields via manure (Emanuelsson 1988b). In this way, centuries of grazing may result in a less nutrient-demanding vegetation on the grazing lands.

7) AD 1500–1984. The expansion of the fields

During the 16th, 17th, and 18th centuries there was an increase in the cultivation of *Secale* and *Hordeum* (barley). Decreasing pollen frequencies of *Plantago lanceolata* during the same period imply reduced areas of pasture in the Krageholm area.

During the 19th century, fields expanded and pasture areas decreased (Ch. 4.3.9). This change of the cultural landscape is also evident in the pollen diagram as an increase in the frequencies of *Secale* pollen and decrease in *Juniperus* and *Plantago lanceolata*. This shift of grasslands into fields probably also caused the increase in minerogenic matter in lake sediments deposited during this time.

The relative importance of different cereal crops seems to have changed, as indicated by decreasing *Hordeum* and increasing *Secale*. This is possibly related to a change from a field rotation system, to one that allowed the growing of crops in other proportions (Ch. 4.3.9). According to documents in the archive of the Krageholm estate *Picea* (spruce) has been planted since the 1860s in the forests of the estate. There is, however, no major rise in *Picea* in the pollen diagram until the late 19th century.

During the 20th century, according to the pollen diagram and the description by Möller (Ch. 4.3.9), new crops were introduced and the importance of old ones changed. Grass for fodder and sugar-beet *(Beta vulgaris)* have been commonly grown since the early 20th century, and oil-seed rape *(Brassica napus)* was introduced during World War II. In the past hundred years wheat became a more common crop, whereas the cultivation of rye and barley declined. These changes in the cultural landscape are evident in the pollen diagram as changes in the pollen frequencies of Chenopodiaceae, Poaceae, Brassicaceae, *Triticum, Secale,* and *Hordeum*.

4.3.3 The Neolithic
The cultural landscape in the Krageholm area during the Neolithic

Mats Larsson

Stray finds and the cultural landscape

The object of principal interest in the following is an area around Krageholmssjön covering a radius of 1–1.5 km around the lake. The whole of the investigated area lies within the parish of Sövestad.

Compared with Bjäresjö parish, there is very little material from this area in museums and private collections. Only 33 finds from the area are recorded in collections. This small amount of material naturally causes major problems in any discussion of Neolithic settlement in this area.

In 1984 and 1985 and, to a certain extent, 1986, the number of finds was increased by an extensive field survey which took in the whole of Sövestad parish. Only the area around Krageholmssjön will be dealt with here, however.

The finds in the museums and private collections come from three main sites: Skogshuset and Hallarp, to the west of Krageholmssjön, and Snogarp, on lake Ellestadssjön. The latter actually lies outside the area discussed here, but it will be discussed briefly since it is interesting from a number of points of view.

The pointed-butted and thin-butted axes are the items of particular interest which date from the earlier parts of the Neolithic period. The material contains only four such axes. These come from two sites, Skogshuset and Snogarp. These axes are not found on so-called "clean" sites, but are included in large collections of single finds. Also dating from this period is the fragment of a battleaxe, which was found close to the south-eastern part of Ellestadssjön.

There is very little material dating from the earlier part of the Middle Neolithic. Three thin-butted axes have been dated to the Middle Neolithic I–II. The material from the later part of the Funnel Beaker culture, Middle Neolithic IV–V, is rather better represented, however. A total of four thick-butted axes of the Lindø and Store Valby type are represented in the material from Hallarp, Skogshuset, and Snogarp.

The later part of the Middle Neolithic (B) is characterized by the Battleaxe culture and the Pitted Ware culture. In the area with which we are concerned here, Middle Neolithic B is associated only with the Battleaxe culture. The extent of the material is comparable with the later Funnel Beaker culture, and is also represented in the same places. The material is not easy to date accurately, although one of the battleaxes is from the middle part of the Battleaxe culture.

The Late Neolithic gives the impression of having been a period of expansion in this area, at least judging by the number of single finds. A total of 14 artefacts belong to this period and, to a certain extent, to the Early Bronze Age. Late Neolithic implements account for almost half of the total number of single finds from this area. Of the available material, it is the daggers that permit more accurate dating within the period; by far the majority of them belong to the early part of the period. An important number of implements dating from this period were found at Skogshuset, to the west of Krageholmssjön. Other single finds are from Hallarp, Snogarp, and Karlstorp; the latter location is north-east of Krageholmssjön. Neolithic finds are shown on the map in Fig. 4.3:2.

The Neolithic settlement area

The results of the survey of museum collections are combined with the results of the field survey in this section. A total of 13 sites have been recorded in the area around Krageholmssjön and in the area between it and Ellestadssjön. Two of these sites can be dated to the Mesolithic period: Ebbarp, on Ellestadssjön, can be dated to the Maglemose culture, whereas the other site is situated on the southern shore of Krageholmssjön; the latter can be dated to the Ertebølle culture (Larsson and Larsson 1986). The number of sites is relatively small, and field surveys by other members of the project team have failed to produce evidence of any new Neolithic sites (Olausson 1988). The problem with any discussion of settlement in the area is that the majority of the sites have not left any clearly datable material. The following discussion relating to settlement and settlement patterns must, therefore, be regarded as a general model.

Early Neolithic. Mention is made above of the two sites, Skogshuset and Snogarp, where finds of both pointed-butted and thin-butted axes were made. Both may be classified as settlement sites. The former is situated on a light sandy till, whereas the Snogarp region is made up of sand and sandy till. Both sites lie close to watercourses. Skogshuset is situated on a plateau about 500 m from Krageholmssjön and next to a small stream. The actual location of the site at Snogarp is not known, but is assumed to have been in the general area of the now vanished village of Snogarp, by Ellestadssjön. No other Early Neolithic settlements have been found around Krageholmssjön itself. Some evidence of Early Neolithic settlement has been found in the area to the north of Krageholmssjön, on the other hand. Three sites have been found in the vicinity of the Ebbarp estate, south of Ellestadssjön. All contain fragments of thin-butted axes, scrapers, knives, blocks, and waste products. The sites are small, with the distribution of the flint indicating a surface area of approximately 500–1000 m². The soil in this area is lighter, and is interspersed with light sandy till. The sites were not directly on the shore, but at a distance of some 500–600 m from it on slightly raised areas of land. The localization of the sites is the same as that observed in the hinterland of south-west Scania (M. Larsson 1984). A considerable number of Early Neolithic sites have been found around Sturup, about 20 km to the west of the area discussed here; some of these have been investigated (Larsson 1984). The area around the lakes Ellestadssjön and Snogeholmssjön is a part of this same area, with its diverse soil types and considerable ecological variation.

It is doubtful whether these sites existed simultaneously. What is more likely is that they provide the evidence that a small group of people moved about in a small area. Parallels can be drawn with the previously discussed Sturup area, where the sites were situated close to one another in many cases (M. Larsson 1985). One further site dating from this period is worthy of more detailed discussion. In the course of the 1984 field surveys relatively abundant material was found immediately to the west of the Snogarp estate, south-east of Ellestadssjön. This material consists of fragments of polished axes, scrapers, flakes, and waste products. Some of the larger polished fragments are probably from pointed-butted axes. Clearly what we are dealing with here is a site dating from the earliest part of the Early Neolithic.

In summary, it can be stated that the area immediately adjacent to Krageholmssjön constituted a marginal area during the Early Neolithic. This view is also reinforced by the results of the palaeoecological investigations. The pollen diagram for the lake contains evidence of small clearances on light soils. An indication of this is given by the pollen from certain plants and by the presence of juniper (Ch. 4.3.2). The sites which were found are essentially from the area north of the lake. One site, Skogshuset, is situated west of the lake. The sites are localized on light soils next to watercourses. This particular area thus conforms to the situation which existed in the adjacent hummocky landscape (M. Larsson 1984). The pattern of settlement represents a system of base sites; these are estimated at 3–4 in number, which can probably be taken to represent the same number of family-sized groups. The area around Krageholmssjön itself was used primarily for hunting/fishing, an activity which is indicated by the small size of the sites that have been discovered.

Middle Neolithic. Very little evidence of settlement in the area dating from the earlier part of the Middle Neolithic period has been found. Nevertheless, a small number of thin-butted axes found on the Skogshuset site point to a presence in the area. Only with the late Funnel Beaker culture (Middle Neolithic IV–V) is evidence to be found of renewed settlement in the area. This is also in line with the palaeoecological investigations, which clearly point to an increase in the amount of *Plantago lanceolata* (ribwort plantain) on the pollen diagram for the lake (Ch. 4.3.2). This can be taken as an indication of a considerable level of animal husbandry. Three sites have been recorded: Skogshuset, Hallarp, and Snogarp (Ellestadssjön). The Battleaxe culture (Middle Neolithic B) is also represented relatively abundantly in the material found in the area. What is interesting is the very distinct local continuity which can be observed from the late Funnel Beaker culture up to the Battleaxe culture. Compared with the Early Neolithic, it is much more difficult to discuss settlement patterns when no sites were found during the investigation; the only evidence of any sites was in the form of single finds. One possible theory is that the sites were fewer in number, but larger in size, and of a more permanent nature during the late Funnel Beaker culture in particular.

Late Neolithic. It was stated by way of introduction that the material from the single finds contains a high proportion of artefacts from the Late Neolithic. The number of daggers from the Skogshuset, Hallarp, and Snogarp sites is relatively large, which indicates fairly extensive settlement in the area. The pollen diagram for Krageholmssjön, with its increasing values for plant pollen, indicates land clearance in the vicinity of the lake (Ch. 4.3.2).

4.3.4 The Bronze Age
The cultural landscape in the Krageholm area during the Bronze Age

Deborah Olausson

The available evidence for Bronze Age activity in this area points to an extensive, rather than intensive, use of the landscape primarily during the Late Bronze Age.

The Early Bronze Age

Barrows. Study of written sources and early maps shows five *hög* features, representing possible Bronze Age barrows, in the area designated as the focal area. Most of them lie at some distance from the lake and are widely dispersed in the landscape. All are located east of the lake (Fig. 4.3:3).

Single finds, hoards. Ten objects datable to the Early Bronze Age have been found in this area. The find circumstances for two of them indicate that they are hoard-finds: a bronze flanged axe found during the deepening of a ditch between Oretorp and Hersaremölla (YM 126), and a bronze spearhead found during digging in the river Svartån around Hunnestad (Arbman 1948:10). The other stray finds are from a private collection donated to the Museum of National Antiquities in 1906 (SHM 13341): the exact location or circumstances of the finds are not known. The Late Stone Age/Early Bronze Age finds in the collection include nine flint daggers and fragments, four flint projectile points, a fragment of a bronze sword or dagger blade, a bronze spiral bracelet, and bronze smelting debris.

Settlement. Approximately 62 ha (equivalent to 1%) of this area has been subject to surface survey (Larsson and Larsson 1986, Olausson 1988). Survey efforts resulted mostly in indications of settlement datable to the Stone Age and the Late Iron Age (Olausson 1988: Table 1). Evidence for Early Bronze Age settlement consisted of two ploughed-up hearths/cooking pits ^{14}C-dated to this period.

THE KRAGEHOLM AREA
NEOLITHIC SETTLEMENT, CA. 3000 BC – 2000 BC[*]

Fig. 4.3: 2

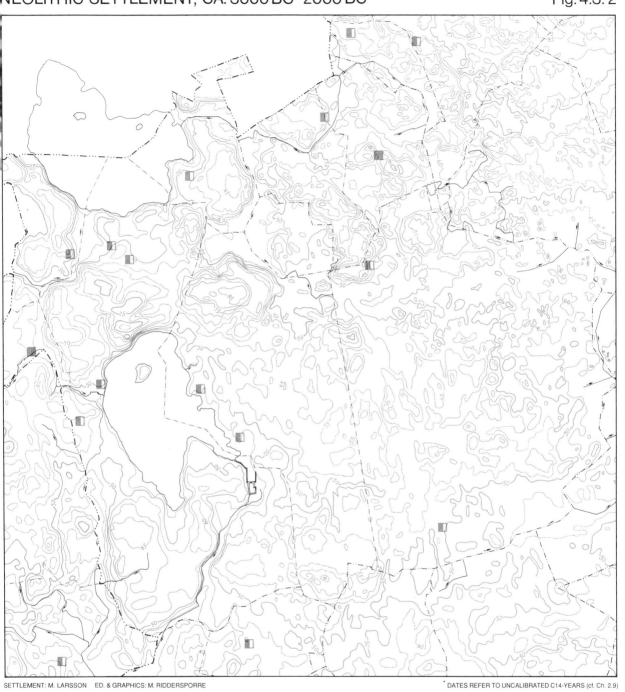

SETTLEMENT: M. LARSSON   ED. & GRAPHICS: M. RIDDERSPORRE

[*] DATES REFER TO UNCALIBRATED C14-YEARS (cf. Ch. 2.9)

0   1   2 km

1:50,000

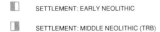 SETTLEMENT: EARLY NEOLITHIC

 SETTLEMENT: MIDDLE NEOLITHIC (TRB)

LAKE / STREAM (CA. 1970)

—50— CONTOUR LINE: 5 M EQUIDISTANT

— ·· — ·· — HUNDRED BOUNDARY (CA. 1970)

— · — · — PARISH BOUNDARY (CA. 1970)

VILLAGE, ESTATE,
SINGLE FARM BOUNDARY (CA. 1970)

The Late Bronze Age

Graves. There are no known Late Bronze Age burials from this area. There is, however, the source-critical problem to be taken into account when trying to evaluate whether the absence of burials reflects the situation during the Bronze Age, or whether it rather reflects the uneven quality of our archaeological data. Late Bronze Age burial consists of cremation burials which can occur either as "flat-earth burials", i.e., under the surface with little or no marking above ground, or as secondary burials sunk into already existing Early Bronze Age barrows. In the first instance, it is difficult to locate such burials by means of surface survey (although ploughed-up cemeteries of the calibre of Svarte or Piledal would probably not be missed; Olausson 1987). In the second case, excavation is necessary to locate such graves. None of the barrows in the region has been excavated.

Single finds, hoards. As was true for the Early Bronze Age, finds of scattered Late Bronze Age objects are known from this area. Three Late Bronze Age bronze socketed axes in the collection of the Museum of National Antiquities come from this area (SHM 13341). The find circumstances of another such axe suggest that it comes from a sacrificial deposit (LUHM 18480).

In addition to these scattered finds, there are two from the area which probably represent hoards (SHM 13224, LUHM 12609).

Settlement. Two of the 33 ploughed-up hearths found during surface survey in this area could be dated to the Late Bronze Age. Five others were radiocarbon-dated to the Late Iron Age (Olausson 1988: Table 1). During excavations near Årsjö village, J. Callmer encountered Late Bronze Age potsherds under a heap of clearance stones which had not been visible above the surface (J. Callmer, pers. comm.).

The Bronze Age cultural landscape

The evidence at hand would seem to indicate an extensive use of the Krageholm region during the Bronze Age, with some change in land use occurring from the Early to the Late Bronze Age. Faced with an absence of direct evidence for settlements, we are forced to rely on approximations based on known graves and single finds. The density of barrows for the area shown in Fig. 4.3:3 (minus the area of the lake) is 1 barrow per 10 km^2 or 0.09 barrow km^{-2} of land area. If we assume that the number of barrows recorded is two-thirds of the number which once existed (a minimum estimate: cf. Ekblad 1981: D7-D8, Kristiansen 1985a:126, Eriksen 1987:19), density increases to 1 barrow per 7 km^2 or 0.14 barrow km^{-2} of land area for the Early Bronze Age. This density is low when compared with that of the outer hummocky zone (Bjäresjö, Ch. 4.2.4) or the coastal zone (Stora Köpinge, Ch. 4.1.4).

Seen in a wider perspective, the Krageholm area in the inner hummocky landscape is a continuation of a zone in which Early Bronze Age barrows are sparse (Fig. 3.3:1). The heavy soils here make this area less suitable for Bronze Age farming than the lighter soils near the coast, but on the other hand, leaf-fodder for cattle and sheep and twigs for goats would have been available in the partly forested landscape of the Early Bronze Age. By analogy with the pattern seen on north-western Zealand (Kristiansen 1978:331) we can interpret the evidence as smaller settlement enclaves subsisting on shifting agriculture with small plots: a continuation of the Late Stone Age subsistence base (Ch. 4.3.3). A mixed economy in which small-scale farming with long fallow periods and extensive land use was combined with livestock grazing freely in the surrounding forests (Kristiansen 1988:75) would be a model which well fits the evidence at hand in the Krageholm area. During the course of the Early Bronze Age, this combination of mobile, dispersed farming and forest grazing led to gradual deforestation of large areas in both the inner and outer hummocky zones ("The Jakobsen Model", Göransson 1987b:42, cf. Kristiansen 1988:80).

The pollen diagrams from the lakes Krageholmssjön and Bussjösjön indicate that the landscape became more open during the course of the Bronze Age here. Arable land was more common during the Late Bronze Age than it had been during the Early Bronze Age, and large parts of the area were covered with grass (Ch. 4.3.2). This is a general pattern which is evident in the Netherlands and the whole of north-western Europe as well (e.g. Welinder 1977:45, Kristiansen 1978:325). The archaeological evidence for human settlement during the Late Bronze Age around Krageholm is, as we have noted, sparse. To a greater extent than during the Early Bronze Age the Krageholm area may have come to be used for summer pasturing and/or hay production for settlements located closer to the coast, rather like the transhumance system (summer pasturing of cows and/or goats; e.g. Nyman 1963, Thyselius 1963) practised in Sweden during the Middle Ages. All available osteological evidence points to a Bronze Age economy in which cattle and sheep or goats were important, which entailed extensive pasture (Strömberg 1954, 1982:155, Lepiksaar 1969, Widholm 1974a, Welinder 1977:154, Jaanusson 1981:21). The question is, of course, whether Bronze Age settlers would have used (been forced to use?) pasture at some distance from their permanent settlements, or whether the settlements were in fact located in the area but have remained more or less "invisible" to us archaeologically. The shieling model does not contradict the former interpretation of the evidence, as a distance of 12 km between summer pastures and farms, as mentioned for Härjedalen in northern Sweden (Thyselius 1963:44), is greater than the distance between the coast and Krageholm. The fact that the intensity of bronze deposition seems to remain constant from the Early to the Late Bronze Age in the

THE KRAGEHOLM AREA
BRONZE AGE, ROMAN IRON AGE, AND VIKING AGE SETTLEMENT

Fig. 4.3:3

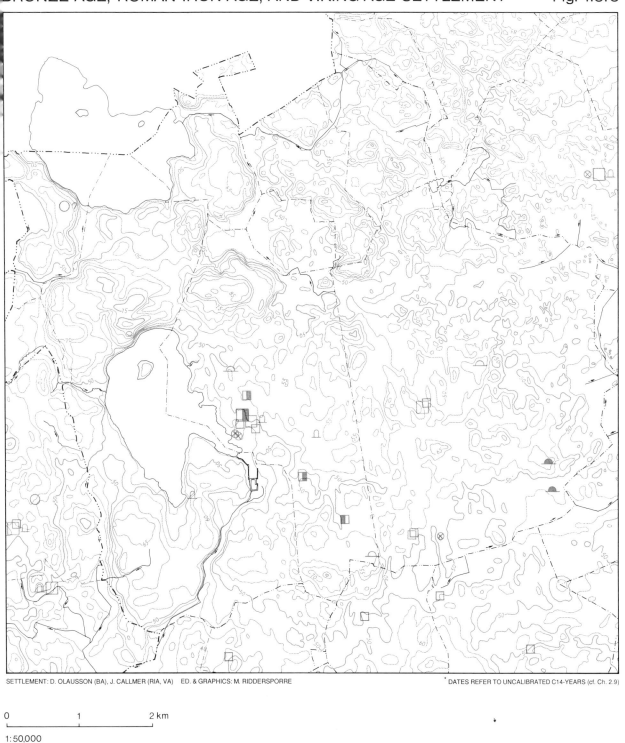

SETTLEMENT: D. Olausson (BA), J. Callmer (RIA, VA)   ED. & GRAPHICS: M. Riddersporre

* DATES REFER TO UNCALIBRATED C14-YEARS (cf. Ch. 2.9)

0   1   2 km

1:50,000

BRONZE AGE, CA. 1500 BC–500 BC[*]

- PERMANENT SETTLEMENT: LATE BRONZE AGE
- UNSPEC. TYPE SETTLEMENT: EARLY / LATE BRONZE AGE
- EARLY BRONZE AGE BARROW: CERTAIN / UNCERTAIN
- HOARD: LATE BRONZE AGE

ROMAN IRON AGE, CA. 0–AD 400

- SETTLEMENT
- BURIAL: > 5 GRAVES
- HOARD

LATE VIKING AGE, CA. AD 1000

- PERMANENT SETTLEMENT
- UNSPEC. TYPE SETTLEMENT
- RUNIC STONE, PICTURE STONE
- MONUMENT INCLUDING RUNIC STONES AND PICTURE STONES
- HOARD

- LAKE / STREAM (CA. 1970)
- ----·---- HUNDRED BOUNDARY (CA. 1970)
- ----- PARISH BOUNDARY (CA. 1970)
- ---·--- VILLAGE, ESTATE, SINGLE FARM BOUNDARY (CA. 1970)
- —50— CONTOUR LINE: 5 M EQUIDISTANT

area means that the area was used, even though we have few concrete indications of settlement (cf. Kristiansen 1978:331). In the absence of more substantial evidence, we are forced to leave the question open as to whether settlement was now consolidated at the coast or was located to any notable extent in the Krageholm area.

4.3.5 The Iron Age
Settlement and landscape in the Krageholm area in the Iron Age

Johan Callmer

Pre-Roman Iron Age

There is little possibility that field survey material and stray finds can be dated to the early part of this period, so our observations must essentially be based on material uncovered by excavations. In the microarea which was intensively studied on the eastern shore of Krageholmssjön, a large area with high phosphate values was excavated extensively. This revealed a few pits, a well, and a badly damaged hearth. On a weakly marked natural terrace on the slope an investigation was made of an area adjacent to two of these structures; this revealed numerous small pits about 10 cm deep. However, the fact that these structures are located in a strictly limited part of the trench and absent from the rest of the area makes it likely that we are dealing with traces of a large long-house, probably with two phases. There is relatively little pottery, but almost all large structures can be dated to a late stage of the Pre-Roman Iron Age. The number of post-holes suggests that the house was rebuilt at least once or twice. Judging by the phosphate values, which appear to indicate that Iron Age habitation sites were localized on heights and slopes, the entire habitation site has been investigated. It is thus not improbable that we have here a habitation site which can function as a model for the character and location of settlement units in the Early Iron Age in the Krageholm area. The close spatial connection with the sites of Late Bronze Age settlement should also be stressed.

It appears probable that Pre-Roman Iron Age agriculture was on a greater scale than that of the Late Bronze Age (Ch. 5.2). Judging by the pollen diagrams, however, the number of animals kept was not greater. The great difficulty in discerning material from the Pre-Roman Iron Age means that the habitation site investigated at Krageholmssjön is the only trace of the Pre-Roman Iron Age in our microarea.

Roman Iron Age

There is fairly good evidence for the Roman Iron Age in the Krageholm area, where over a dozen sites are known. The middle and late stages of the Roman Iron Age in particular appear well represented. The finds – from nine localities – consist largely of stray finds of pottery and beads datable to the period. A gold ring is also known from the area (Strömberg 1963:29). These find localities largely coincide with the parts of the landscape that were utilized during the Bronze Age. It was the same little basins in the hummocky landscape that attracted settlement. In addition, the lakes in the inner zone no doubt continued to play a decisive role in the selection of habitation sites. As a whole, settlement appears to have been concentrated in the south-east. It is uncertain whether the possibly somewhat higher number of points noted for the Roman Iron Age than for the Late Bronze Age, along with the brevity of the period, should be taken as an indication of denser settlement; this uncertainty is due to the fact that traces of the Roman Iron Age consist mainly of very conspicuous finds. On the other hand, the slightly (but undeniably) wider dispersion of the finds in the landscape could be interpreted as an expansion of settlement. Whether this expansion actually occurred in part during the Pre-Roman Iron Age cannot be ascertained (see Fig. 4.3:3).

In those cases where the find material can be closely assigned to a definite area in the landscape, we find distinct heights. In all the six cases which could be studied, these are hills with an area of about a hectare. Only one habitation site from the Roman Iron Age has been (partially) studied within this focal area; it lies within the same south-eastern part of the shore of Krageholmssjön that we have previously discussed in conjunction with settlement in the Late Bronze Age and Pre-Roman Iron Age. The habitation site is on a low hill which, at the original water level, stuck out into the lake as a promontory. The site of the actual habitation on the highest part of the hill had been heavily ploughed up and lacked occupation layers. The ground was stony in places, which made it rather difficult to distinguish traces of structures. It was therefore not considered worthwhile stripping the whole area. The trenches dug on the slopes of the hill were most rewarding. Structures here were better preserved and traces of occupation layers could be noted. In those parts of the habitation site which had preserved occupation layers, rich finds were made of pottery, mainly from the middle of the Roman Iron Age, but probably also with older and younger pottery. Several finds of beads show that people here availed themselves of the exchange contacts that were ultimately linked to the Roman Empire. Slag in considerable quantities revealed the presence of a smithy and possibly also of iron production. Although the habitation site was not fully excavated, it appears clear from its extent that the site consisted of one or two to three farms. By all appearances, the habitation site was retained for a considerable length of time, probably more than a few generations. There was thus a noticeable change in the pattern of settlement towards greater stability in the Roman Iron Age. However, the settlement must be seen as peripheral in relation to the coastal region with its much denser settlement and probably also larger settlement units. The cultural landscape in the Krageholm area in this period began to take on a more distinct division into different zones of exploitation. It is probable that a system of cultivated infields and outland commons was definitively stabilized in the latter part of the Roman Iron Age. After a slight reduction in land use, there was a new increase in the later part of the Roman Iron Age.

Late Iron Age

The Late Iron Age – with the exception of its final phase, is known only very partially in the Krageholm area. Habitation sites from this period are extremely hard to identify on account of the generally scanty and indistinct material. It has not been possible within the framework of our investigations in the Krageholm area to find evidence of more than three habitation sites from various parts of the Late Iron Age in the area by the south-eastern shore of Krageholmssjön. It is possible that one other site in Sövestad parish is indicated by a single find, a bird fibula (8th century) recorded as found in Sövestad (Strömberg 1961, 2:27–28). It appears probable that it did not originate in the area by Krageholmssjön, since it would then have been recorded as coming from Årsjö or Krageholm. In the area by the south-eastern shore of the lake there seems to be evidence of settlement in both the Migration Period and the Vendel Period. As has already been pointed out, the Roman Iron Age habitation site appears to have existed in the same place up to the middle of the period. After that, settlement on this hill ceases. Pottery finds and dated hearths suggest that settlement in the Late Roman Iron Age and Migration Period occupied a number of places near the village site. Material from these habitation sites is restricted to some undecorated sherds of Late Iron Age pottery and two hearths found by field-walking, one of which gave a ^{14}C dating to 1510 ± 100 BP (Olausson 1988). About a kilometre south-east of Krageholmssjön, systematic field surveys have found evidence of settlement from the Migration Period. On a ridge above a stream flowing into the lake basin, traces have been found of hearths and fire-cracked stone. One of the hearths gave a ^{14}C dating of 1410 ± 130 BP. Apart from these observations, we can point to other evidence of continued settlement in the area immediately southeast of Krageholmssjön: this consists of a remarkable hoard found during peat-cutting, with 32 beads, 27 of them amber, the rest of glass, and datable to the 5th century (Strömberg 1961, 2:27). According to local people, the place is identical with the bog north of Krageholm castle.

Some idea of the size and location of the settlement can be obtained by a study of the position of habitation site indications in the terrain, their extent, and the area available for occupation at each site. It then becomes clear that the habitation sites in both cases were rather small. The area can scarcely have exceeded one or two hectares, so there was hardly room for more than a single farm or a restricted number of smaller farms (cf. Tornbjerg 1985). As previously during the Iron Age, the sites chosen for habitation were naturally delimited, generally on quite low hills surrounded by wet bottoms or water.

In the succeeding Vendel Period, settlement in the area probably continued as before, but it has not been possible to find more than two concrete examples of settlement from this time. On a hill above the place where the stream which is the only major feeder flows into the lake, a glass bead from the Vendel Period was found. On the top of the hill the ground has been severely upset since a farm was sited there at a much more recent date. When the farm was abandoned the buildings were demolished and the foundations exploded. It is possible that the hill was the site of a Vendel Period cemetery or habitation. About 500 m north-west of Årsjö village a hearth has been found, dated by ^{14}C to the Vendel Period.

To sum up, the archaeological sources allow the interpretation that settlement in the Late Iron Age had largely the same character as earlier in the Roman Iron Age. Settlements were small, consisting of single farms or a small cluster of farms. The areas of exploitation were strictly distinguished into different zones. Near the farms lay the arable land, possibly still consisting of one-course fields (Regnéll, Ch. 4.3.2). Then came the area where fodder was gathered, and beyond that stretched the wood-pastures. It is impossible to deduce on the basis of the archaeological material whether this agrarian settlement had any economic specialization. In comparison with the more intensively utilized areas near the coast, settlement in the Krageholm area was on the periphery of the region, with relatively large stands of woodland. Settlement in the Late Iron Age can hardly have extended any further to the north-west.

4.3.6 The Viking Age and Early Middle Ages
Viking Age and early medieval settlement and landscape in the Krageholm area

Johan Callmer

The Viking Age

The Viking Age, especially the later phase, is well represented in the Krageholm area. As in the Ystad area as a whole, runic stones are important evidence. From Baldringe village a runic stone is known, probably still in its original site near the historical village (Strömberg 1961:25). From Krageholm there are no less than three runic stones (Strömberg 1961:27). We have no exact information about the original location of one of these stones, but according to a local tradition it was found near the lake, north of the castle. The other stone was found at Möllevången, about 500 m east of Krageholm. The third stone comes from Bokhultsbacken by the southern tip of the western bay; this place probably once marked the boundary between Årsjö and Skårby. All the stones are related to the position of Årsjö village (see Fig. 4.3:3).

Two hoards have also been found from exactly the places just mentioned. Only a hundred metres from the church in Baldringe a 10th century hoard of silver was found (Arbman 1946). Excavations on a small scale on the site also revealed traces of settlement from the same period. Near Årsjö a gold ring from the 10th or 11th century was found by the lake-shore, immediately below the site of the historical village (Hårdh 1976:70). Just outside Sövestad village a silver coin struck for Knut the Great was found (Royal Cabinet of Coins and Medals, Stockholm). About two kilometres south of Sövestad an iron axe of Viking Age type was found (Strömberg 1961:27); the find in itself is scarcely indicative of a habitation site, but when we see it against the background of a whole series of finds within one limited area that was in continuous use since the Late Bronze Age, we have every reason to assume the existence of a Viking Age settlement on this spot (recently Black Earthenware was found on the site). From Borrie we know of sherds of Black Earthenware which can be dated to the late Viking Age or the Early Middle Ages, found in the field immediately south-west of the church (field survey 1987). The find of a Late Viking Age ring-brooch at Allskog farm in the north of Baldringe parish also indicates either a habitation site or a grave (Strömberg 1961:25).

To these observations we can also add the results of excavations on and near the site of the historical village of Årsjö, within the intensive study area. Despite the poor excavating conditions caused by recent disturbances, we were able to demonstrate a considerable extension of settlement in Årsjö and immediately south of the village site marked on the land survey maps of the 18th century. It was here we found concrete signs that we were dealing with a settlement that was much larger than the settlement units assumed for the earlier periods in the Krageholm area. It is very hard to say at present whether this development in settlement can be shown to go back before the Late Viking Age. The excavations on and near the site of Årsjö village turned up some pottery that might suggest a slightly earlier start. This material was found in the western part of the village site and in erosion material in the slope south-west of the village.

The archaeological material thus shows in the case of Årsjö a close connection between, on the one hand, a small area with what appears to be older continuous settlement – although on different habitation sites – and, on the other hand, a habitation site which quickly developed in the Viking Age into a village-like community. We cannot tell whether a manorial building had

already been established at this time near the medieval castle of Krageholm, since practical difficulties made it impossible to undertake any excavations there. In Baldringe and Sövestad as well, historical sources show that settlement at the end of the Viking Age was undoubtedly of a village kind, that is, relatively large agglomerations of several farms, probably between five and ten in both cases. Parts of the Krageholm area thus changed radically as regards the cultural landscape. For example, cereal cultivation increased heavily in the Viking Age, while there was only a relatively slight increase in pasture. In connection with this change some form of rotation was introduced, with the establishment of fields.

It is probable that this development towards bigger settlement units is in part a result of the contraction of an older, more dispersed settlement pattern in the peripheral areas of which the Krageholm area is an example. The settlement that probably still existed in the Late Viking Age about two kilometres south of Sövestad appears to have ceased to exist before or during the Early Middle Ages. The same applies to the indication of Late Viking Age settlement at Allskog, which was not established as an isolated farmstead until the 19th century.

No major changes appear to have affected the picture of settlement in the Middle Ages. Settlement mostly consisted of villages with a large number of farms. The oldest of these appear to be Baldringe, Borrie, Sövestad, and Årsjö. Baldringetorp is possibly a slightly younger foundation (Billberg 1989), but settlement there had probably evolved into a village at a very early stage (12th or 13th century). Very different, however, is the settlement of Bellingaröd, situated between lakes Krageholmssjön and Ellestadssjön; here we seem to have an early medieval foundation which remains throughout the period a very small settlement unit comprising one or two farms on the periphery of the settlement district.

The emergence of the historical villages at this time in areas which are peripheral in relation to the primary settlement districts of the Late Iron Age should also be seen as the result of a gradual increase in population. This increase would have exerted a certain pressure on the expansion of the settlement areas. Although we cannot say with certainty, it is likely that a large-scale abandonment of settlement in the coastal zone in the 10th century (Callmer 1986:200) brought still more pressure to bear on this expansion of settlement in the peripheral areas.

The form of the cultural landscape that was essentially retained in the Krageholm area until the 18th and 19th centuries took shape as far back as the end of the Iron Age. There was probably a slight expansion in the extent of arable land in the Middle Ages, but it is above all development in the area of pasture which is noticeable (cf. Ch. 4.3.2); there was undoubtedly a considerable increase in the middle of the Middle Ages, probably the 13th century.

In early historical times the Krageholm area is in many respects a parallel to the Romele area. It is characterized in the south and south-east by a well developed village landscape with large villages like Årsjö, Sövestad, Baldringe, Baldringetorp, and Borrie. To the north and north-east there is still closed woodland. Only a few isolated farmsteads or double farms interrupted this picture in the Early and High Middle Ages, forming small enclaves of tilled land and pasture. Bellingaröd, between lakes Krageholmssjön and Ellestadssjön, has already been mentioned. There were probably one or two similar settlements north of Sövestad: Fjärshus and possibly Snogarp just south-west of Ellestadssjön may have been such settlements (cf. Ch. 4.3.7).

Towards the end of the Middle Ages and probably in the 16th century in particular, there was a rapid increase in the number of peripheral settlements in the north and north-east of the area. Certainly six new farms were established in the northern half of the Krageholm area. In this we see a clear difference between the state of affairs in Sövestad and Baldringe. Only one farm of this kind was established in Baldringe. It is evident that the two estates of Krageholm and Baldringe – the former owned by a nobleman, the latter by the church – were run according to different principles. Both estates must have taken advantage of the potential of the peripheral areas for animal husbandry.

4.3.7 The Middle Ages and Modern Time until ca. 1670
Settlement, cultivation systems, and property distribution in the Krageholm area

Sten Skansjö

The oldest known historical settlement

Of the settlements in Sövestad parish, the church village of Sövestad is mentioned for the first time in extant written sources as far back as 1085. The village of Årsjö is attested from the mid 14th century, while the first distinct reference to Krageholm manor comes from around 1400. Of the smaller settlements, Bellingaröd and Fjärshus are named in medieval written sources. Early Modern Time saw the coming of several more smaller settlements (see below).

Settlement in the Middle Ages consisted of the manor of Krageholm, two villages, and a number of smaller agglomerations comprising one to three farms. This pattern was retained into Modern Time, although with an increase in the number of smaller clusters and with changes in the structures of the estate and the villages. Sövestad, with its 25 farms in 1670, was one of the largest villages in the Ystad area, while Årsjö, with 13

farms, represented well the average size of villages in the area.

A comparison of settlement density shows that the parish of Sövestad was far below the average, which has been estimated for the entire project area to 2.8 normal farms km^{-2} (Skansjö 1985). Even if we follow contemporary tax assessments and count the demesne of the Krageholm estate as 20 normal farms and include them in this comparison, the settlement density of Sövestad parish was still only 1.3 normal farms km^{-2}. In this respect the parish was one of the most sparsely populated in the whole Ystad area.

There are some, admittedly weak, indications of a few discontinued medieval settlements near the Krageholm estate, probably in Sövestad parish.

In the distribution of the property of the deceased owner of Krageholm, Bonde Jepsen (Thott), in 1470, there is mention of the tenant farms which then belonged to Krageholm manor. These included the whole of Årsjö village with Jepstrup, Getsholmen (in Sövde parish), Fjärshus, and others. The context here suggests that a settlement which disappeared later, Jepstrup, may have existed in the area around Krageholm. It has not been possible to achieve any clarity on this point from the written sources.

From the first half of the 16th century we have records of two localities with names in -*ryd* or -*röd* which indicate forest clearance, namely, Sjöryd (mentioned 1499–1502) and Långaröd (one mention in 1547). These were situated west of Krageholmssjön in an area that was the subject of a dispute between the owners of Krageholm and the village of Skårby, where the archbishop was the main landowner in the Middle Ages. It has not been established whether these names were borne by settlements or just denoted clearings in the forest.

There is no explicit mention of deserted units in the source material from Sövestad parish in the Late Middle Ages.

In Sövestad parish around 1570 there were 60 tithe-paying farms. In the mid 17th century the number recorded is 50. The reduction can probably be explained as a result of the restructuring of the villages in conjunction with the development of the Krageholm estate at this time. This process obviously affected the surrounding villages, where Krageholm was a dominant owner (see below). We see from the 1671 Civil Survey (Jordrevningsprotokoll) that 21 of Sövestad's farms had been equalized and that all 13 of Årsjö's farms were assessed as being of equal size. Another side of the same process, the changes in settlement in the shadow of the expanding estate, is seen in the noticeably large number of cottages found in these two villages: 13 in Sövestad and as many as 17 in Årsjö belonged to the owner of Krageholm in 1671.

Another post-medieval change in settlement affected Sövestad parish. We have noted above the rise in the number of minor settlements – small villages or isolated farmsteads: for example, Snogarp and Vasatorp are first mentioned in written sources in 1583, and Alaröd, Ebbarp, Halarp, and Svartskylle are named from 1632. It is probable that the latter farms were all established at this time (1583–1632) and that, like the corresponding woodland crofts in the parishes of Villie and Skårby, they represent an increase in settlement in the 16th and 17th centuries.

As in the Swedish provinces in Götaland, this development could be interpreted as an effect of a general population growth, as we know from elsewhere in Europe at this time (cf. Ch. 4.4.6). It is possible that this population growth is also reflected in the greater number of cottages, but this should be seen primarily as a result of organizational changes brought about by the increased labour requirements of large-scale estate management (cf. Ch. 5.9, Skansjö 1987a, 1988).

Cultivation according to the oldest written sources

The first written evidence of a three-field system in the church village comes from the latter half of the 16th century, although it is probably much older than our meagre sources suggest.

Based on the diocesan terrier of about 1570, a comparison of tithe records for the various parishes shows that grain production per tithe-paying unit in Sövestad parish was more than 20% below the average for the Ystad area as a whole.

In the absence of old land survey documents, it has not been possible to investigate any changes in the cultivated land in the church village of Sövestad between 1570 and 1700, as was done for the other focal areas (the parishes of Stora Köpinge, Bjäresjö, and Villie) (Skansjö 1988).

Property distribution

What we know of ownership in Sövestad parish shows that it underwent important changes in early historical times. In this respect, it was above all the development of the Krageholm estate in the 16th and 17th centuries which had decisive consequences for the future look of the cultural landscape.

In the Middle Ages the distribution of property was more differentiated than would later be the case. We know, for example, that certain ecclesiastical institutions owned land in the parish. In King Knut's famous deed of gift to the church in Lund (1085) half a hide (*bol*) of land in Sövestad was transferred to the ownership of the church; this had previously been a part of the large body of property which a man called Scora had given to the king to obtain release from outlawry. In the Late Middle Ages the monastery at Dalby owned ten farms in Sövestad village; our first evidence of this comes from the second half of the 14th century. At the same time, various noblemen with estates elsewhere owned scattered property in the parish.

The farms owned by the monastery came into the possession of the Crown at the Reformation. Later these too would join the ranks of tenant farms belonging to the Krageholm estate. This manor is known from around 1400, although it could be a little older (cf. Skansjö et al. 1989). Its first recorded owner was a knight from the Thott family. From estate distributions from 1419 and 1470 we obtain an idea of the tenant farms belonging to Krageholm manor. The core demesne on both occasions consisted of farms in Årsjö (the entire village in 1470), and in Fjärshus, but no farms in the church village seem to have belonged to Krageholm at this stage. However, several tenant farms scattered in the hundreds of Herrestad and Ljunits belonged to this manor.

It was especially in the course of the 16th and 17th centuries that land and tenant farms were concentrated near the manor. In this perspective we can also see the major property dispute of the early 16th century, in which the archbishop – a major landowner in Skårby – and the owners of Krageholm both claimed an area rich in forest west of Krageholmssjön (Sjöryd, etc.). The owner of Krageholm emerged victorious from this struggle. In 1632 the concentration of property meant that the estate had acquired the greater part of the manor parish. At this time the manor had the following number of tenants obliged to do corvée: 23 in Sövestad, 13 in Årsjö, two in Fjärshus, two in Alaröd, two in Svartskylle, and one each in Ebbarp, Halarp, and Snogarp. Apart from these subject peasants within the parish, the manor had a further 23 farms and 14 crofts in the Ystad area with corvée obligations. This gave a total of 82 tenant units providing labour for the estate's lands, and working at transports and building. A well-built Renaissance castle with a barn now succeeded the medieval manor, parts of which were incorporated into the present main building of Krageholm castle (Skansjö et al. 1989).

According to the 1671 Civil Survey, Krageholm owned all 52 farms in Sövestad parish, including the vicarage, since the landowner had acquired the patronage of the church in 1611. In this way a closed estate unit, virtually a state within the state, had developed in Sövestad parish in the course of the 16th and 17th centuries (Skansjö 1988).

4.3.8 The 18th century
Cultivation, population, and social structure in the Krageholm area

Anders Persson

Introduction

The appearance and use of the 18th century cultural landscape in Sövestad parish cannot be described in any depth. The source material does not permit a detailed diachronic analysis of the two villages, the double farms, and the isolated farms. Not only are the main sources – maps with accompanying descriptions – late (from the second half of the 18th century or later), but they are also compiled in a way that does not provide total information about the cultural landscape. However, the lands belonging to the manor of Krageholm can be described at several stages. We can nevertheless obtain a relatively good picture of land use outside the manor. As a whole, the study of Sövestad parish in the 18th century is particularly interesting, since it conveys a picture of a total manor structure with rigorous control over both the use of the landscape and the structure of settlement; this control seems to explain the unusual population development of the parish in the 18th century.

Settlement pattern

The taxation assessment roll of 1700 records a total of 97 settlement units in Sövestad parish. About 80% of the settlement was to be found in the two villages, the remaining 20% being elsewhere. Of these units, 60 were farms and 34 cottages. The high proportion of cottages in the manor parishes (36%) is explained by the seasonal need for labour on the manor, for which the landowners deliberately established a large number of cottages in villages near the manor. At this time the parish was wholly owned by the Piper family of Krageholm manor. Most of the cottages were in the two villages, Årsjö consisting of 15 farms and 15 cottages and the bigger village of Sövestad consisting of 29 farms and 17 cottages. This settlement structure remained largely unchanged during the 18th century. The number of farms fell slightly, from about 60 in 1700 to 56 in 1801, whereas the number of cottages rose from 34 in 1700 to 53 in 1801. The number of farms thus fell somewhat while the number of cottages rose by 55%. The latter figure may be too high, since the increase may not have been so great if some cottages escaped registration in the taxation assessment roll of 1700.

Population

The population of the parish can be estimated at about 550 (index 100) in 1700. By the mid 18th century this had risen to about 770, falling again to 715 at the beginning of the 19th century (index 100–140–130). These figures reflect the development of population outside the manors, whose inhabitants are not included. Sövestad parish, along with Öja parish outside Ystad, are the only rural parishes of the 17 in the project area which had a lower population at the start of the 19th century than in the mid 18th century. Baldringe and Högestad, belonging to the same manor demesne as Krageholm, are the parishes with the smallest population growth in the second half of the 18th century (25% and 10% respectively). For Sweden as a whole the population rose from about 1.76 million to 2.4 million between 1750 and 1810, an increase of about 35% (Guteland et al. 1975). Why did this manor parish show a deviant trend? The answer is probably to be sought in the very particular agrarian structure of the manor of Krageholm.

Agrarian structure and social control

Krageholm manor. A description of the agrarian structure of the manor can be suitably divided into two periods. In the first period, 1700–1764, the estate economy appears to have been largely based on stock-raising, whereas grain production acquired greater importance between 1764 and the end of the century. These conclusions are fairly tentative, since we lack continuous information about cultivation and livestock for this time. The first period saw an increase in the area of rye and barley by some 30% and of oats by about 50%. In comparison with the manor's demesne of 600 ha, however, the area of grain-producing land was small – 15–20% of the total – besides which productivity was low. The landscape around Krageholm was dominated by marshy meadows, pasture, and woodland. In the mid 1760s it is also recorded that the supply of hay was so large every year "that it was able to maintain large studs of horses and cattle, and the stalling of oxen" (Gillberg 1765). Agriculture was intensified at this time, and there was a change from a three-field to a four-field system. In the long run this agrarian reform brought about a significant increase in productivity, thereby making grain cultivation a more important part of the estate's economy (see Ch. 5.9).

A comparison of Årsjö and Sövestad. The primary workforce for the running of the demesne farm was to be found in the two nearby villages east of the manor. Corvée does not appear to have been regulated to any great extent in the 18th century. We may suppose that the peasants were obliged to work whenever the manor demanded (see Ch. 5.9). In the two villages there were in all about 75 tenant farmers and cotters obliged to do corvée.

The village of Årsjö was equalized in 1770, giving each of the 13 farms about 37.5 ha, while the 17 cottages each had about 7.5 ha. The villagers also had turbary rights, with about 1 ha of peatbog for each farm and

0.1–0.4 ha for each cottage. The total area of the village was some 700 ha, or about 100 ha more than the manor lands. The 25 farms in Sövestad village were equalized in 1762; apart from the vicarage, with its 55 ha, the farms were all about 40 ha in size. The 20 cottages in the village each had about 4 ha of arable and meadow. In 1770 the village of Sövestad had an area of about 1125 ha.

The total area cropped with rye and barley in Årsjö was around 100 ha in 1770, but the corresponding figure for oats cannot be stated since it is recorded along with meadow. The area of rye and barley in Sövestad can be estimated at about 340 ha, and that of oats at about 455. The manor's total area of rye and barley in 1764 was around 80 ha, and that of oats about 75. In this perspective, grain cultivation at Krageholm does not appear to have been on a particularly large scale in 1764, since it was no greater than that of a medium-size village of 15 farms (cf. Skansjö 1985:32–33). For Sövestad in 1770 we have an idea of the proportion of arable and meadow: the former was about 70% and the latter 30% of the total area of the village.

A long-term interpretation of ploughed areas 1700–1810. As mentioned earlier, the Krageholm manor is the only settlement unit in the parish for which we have detailed information for the early 18th century. This has necessitated the elaboration of a method for estimating the extent of land ploughed annually in the parish in 1700. This is based on the total *mantal* values for peasant land in the parish. In Villie parish there was about 13 ha of ploughed land per *mantal* unit. The corresponding figure for Stora Köpinge parish was about 19 ha per *mantal*. Calculating from a scale of 12.5–15.0–20.0 (Bjäresjö–Villie–Stora Köpinge), the peasants' land in Sövestad parish should lie somewhere in the top half of the scale. Since the farms in the parish had a total *mantal* value of 24 in 1700, the ploughed peasant land in the parish must have been 300–480 ha, or 0.5–0.9 ha per inhabitant, not counting Krageholm manor. If we add to this the amount of land ploughed by the manor in 1700, about 55 ha, we arrive at a total of 355–535 ha for the whole parish; in other words, 8–12% of the area of the parish was ploughed annually. Consequently, about 10% of the area of Sövestad parish would have been sown with grain at the start of the 18th century. In the parishes on the plain around Ystad, the corresponding figure was over twice as high. There does not appear to have been any great change on the peasant land in the course of the 18th century, but there seems to have been a greater increase on the manor's fields. In the 1760s the amount of ploughed land in the parish can hardly have been more than 15% of the total area. In 1914 there were about 2715 ha of arable fields and gardens in the parish, of which about 10% lay fallow. This means that about 60% of the parish land was tilled that year, or that an estimated 45% of the area of the parish was productive arable land in the period 1760–1910. A slight in-

crease in the amount of arable in the parish thus occurred in the 18th century, but it is quite clear that the extent of this process was especially noticeable in the 19th century. It is against this background that we must see the agrarian development of Sövestad parish in the 18th century.

Relations between the manor and the peasants. It will be evident from what has been said above that the work of cultivating grain on Krageholm manor cannot have been particularly great (cf. Persson 1982:229). If we look only at the 75 units in the two villages (although in fact slightly more units were obliged to do corvée on the estate) this would mean that each peasant was responsible for at most two hectares of arable land. The problem with corvée on the manor was probably not so much the volume as the time when it had to be done; this naturally coincided with the time when the peasants had enough to do on their own farms. It may be noted that the barn-bailiff at Krageholm in 1770, as well as the parish schoolteacher and the fishermen on the manor, lived in a cottage in Årsjö village.

The manor parish of Sövestad in the 1760s generally appears to have been an integrated patriarchal unit. In 1765 there was a newly founded "beautiful brick-built asylum" in which the poor and destitute were maintained for their lifetime, and also a brick building serving as a school for the children of the parish (Gillberg 1765). It is characteristic that schools were founded at an early stage in parishes owned by nobles; this has been attributed to their desire for increased social control of the peasantry (Sandin 1986).

At this time, the 1760s, the system of control manifested itself in a concrete way (map description of Årsjö). In Årsjö the south field was totally treeless while trees and bushes grew in the west and east fields: "alder in wet meadows and fens, some dry meadows and slopes with beech and hazel and many small hornbeams". To protect the woods, relatively large areas of oat and buckwheat fields were turned into meadow, and peasants were forbidden to plough them. In the same way, the map description for Sövestad village in 1762 records that the dry meadow is "for the most part of an upland nature, overgrown with hazel, oak, and thorn bushes, spreading each year since no one dares to clear either large or small woods in defiance of the master's prohibition". The surveyor also notes that "there is woodland here of oak, beech, hazel, and other kinds in the fields". The Piper family of Krageholm thus intervened actively in the village's use of the land, not hesitating to change arable land into meadow to further the growth of forest on the village lands.

Conclusion

The cultural landscape in Sövestad parish clearly bears the stamp of the manor. There were large areas of forest, meadow, and pasture on the demesne lands. In this a change had come about. In the first half of the 18th century animal husbandry was more important on the manor than in the second half of the century, when there was an intensification of grain cultivation. The proportions of arable and meadow (70% to 30%) in the village of Sövestad in the latter half of the 18th century are the same as was found in the villages of the outer hummocky zone and the coastal region at the start of the 18th century. The explanation for this must be the deliberate influence of the manor on the way the village used its land. It should be noted that the pollen diagrams for the 17th and 18th centuries do not suggest any great changes (Ch. 4.3.2). The increase in grain cultivation on the manor in the 18th century, especially the second half, may have been offset by a corresponding decline in the villages. However, we have no directly measurable data to prove that grain cultivation in the villages declined; all we have is the indirect evidence from the map description and the stagnating population figures thereafter. The result is interesting in relation to the testimony of the pollen diagrams.

4.3.9 The 19th and 20th centuries
4.3.9.1 Land and labour on the estate of Krageholm since 1800

Jens Möller

A feudal landscape (ca. 1800)

The estate of Krageholm, as described in Ch. 4.3.8, could be characterized as a "unit of consumption", i.e. most produce stayed within the estate domains. In many respects, 18th century conditions prevailed well into the 19th century. Enclosure took place in the 1830s, but this altered the landscape comparatively little; significant changes did not occur until the estate introduced a new organization of labour in the 1860s (see below).

The organization of the estate can be seen in Fig. 4.3:4. The map shows the situation in the 1760s, and as mentioned this changed only little until the second half of the 19th century. The demesne land was situated immediately south and west of the manor. This was cultivated in a four-year rotation (the expansion from three to four fields occurred in the middle of the 18th century). As can be seen, the three-field system prevailed in the villages. The number of farms per village varied: Sövestad, with 25 farms, was an unusually large settlement; Årsjö had 13 and Kärragården 5 farms. Most of the woods on the estate were situated around the shores of lake Krageholmssjön.

Work at the estate itself was done by the tenant farmers in Årsjö, Sövestad, Kärragården, and half a dozen farms in Bjäresjö and Hedeskoga (not shown on the map), as corvée duties (Sw. *hoveri*). Apart from cultivating their own plots, the farmers had to work on

the manor, in the woods, provide horses and wagons, and of course work the demesne land. The number of employees on the manor was a mere 11 (a hundred years later it was 49), and of these only 6 were employed directly in agriculture. The map shows a *feudal* landscape, with most of the arable land located in the villages, and only a small proportion of demesne land (cf. also Ch. 4.3.8).

The expansion period (1850–1914)

The total acreage at Krageholm during the 19th century varied little from an average of about 3800 ha. In the middle of the century, 54% of the land was arable, 21% meadow (for hay), and 25% woods or pasture (see Table 4.3:1). During the next 60 years this was altered considerably, and at the outbreak of World War I, arable land had risen to 75%. At the same time a mere 4% remained as meadow, and 21% were woods/pasture. The conversion of meadows into ploughed land is also registered in the sediment record of lake Krageholmssjön (cf. Ch. 4.3.2). An increased transport of eroded soils is implied by the higher mineral content of sediments which were deposited during the late 19th century.

The transformation was facilitated by new techniques, such as draining and marling, in which the estate invested large amounts of capital. There is evidence of draining in the late 1850s, which was only ten years after the introduction of the technique to Sweden (Möller 1984). At the same time, a brickyard for making tile pipes was erected. Often the estates undertook marling at the same time as draining, so in many respects the history of draining coincides with that of marling. The

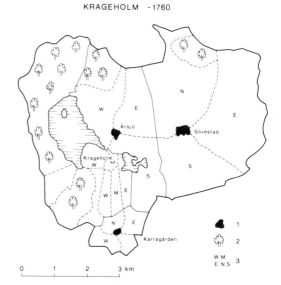

Fig. 4.3:4. The estate of Krageholm ca. 1760. 1. Village, hamlet. 2. Wood. 3. Field. Source: land survey maps. From Möller 1989.

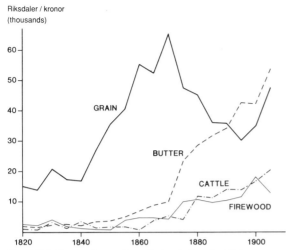

Fig. 4.3:5. Production at Krageholm during the 19th century.
Source: Krageholm records. From Möller 1989.

Ystad region was one of the most marled areas in Scania (Zachrison 1914).

Marl and other primitive fertilizers were superseded by artificial fertilizers from the 1850s onwards. Guano and superphosphate appeared at Krageholm in 1868, which was early in relation to the general adoption in Scania. They were used in large quantities during the 1870s and 1880s. In 1887 Chile saltpetre was introduced, and four years later different kinds of potassium fertilizers.

The technique of flooding meadows, i.e. the irrigation of hay-producing meadows, became widespread during the century (Emanuelsson and Möller 1990). By this method the still remaining meadows could produce up to five times as much fodder, and this was of utmost importance in 19th century Scania, where natural manure was becoming scarce. Krageholm was one of the early adopters of the method. The estate started constructing a system in 1858, and in total about 100 ha were affected. As the maintenance of the flooded meadows involved both high costs and much labour, it was abandoned as soon as cheap artificial fertilizers became available. In the early 1920s – at the latest – the system at Krageholm was abandoned (Zachrison 1922).

Like other large estates in Scania at the time, Krageholm also invested large amounts in the woods. In 1863, the first modern plan for sylviculture was introduced. Before this, the large deciduous woods (mainly beech) had grown with only little management. Felling without reforesting was common, and the remaining woods were in great need of thinning. During the 40 years preceding the plan, only 12 ha of beech had been planted. The first attempts with conifers (pine) were carried out in 1855, but this was a failure and was not continued. The plan of 1863 contained a 120-year circulation for beech; thus a plan for felling and planting up until the 1980s was introduced. In this, it was also de-

cided to replace some beech woods with spruce. In the bogs, alder was to be superseded by ash, as this was more profitable. Heavy grazing had hindered woods from growing, and therefore animals were to be restricted to certain areas. An evaluation was carried out in 1899, in which the spruce cultivation was criticized, mainly because of the poor quality of the timber it produced. Only limited areas of spruce were sown after 1899.

The 19th century, and especially the second half, meant profound changes in production and trade on the estate. Fig. 4.3:5 shows income from different sources at Krageholm during the period 1820–1900. Beginning in the 1840s, income from grain increased and culminated in 1870. Thereafter a reduction can be noted which continued until 1895. The reduction was met by a sharp rise in the sale of animal products, and this became the most important income source beginning in the 1890s. The sale of cattle and woodland produce was not large, but increased somewhat during the century. The evolution at Krageholm was not unique to southern Sweden.

Although grain sales at Krageholm diminished after 1870, the production of grain steadily increased during the century. From 1820 to 1900 the amount produced doubled; from 370 to 760 metric tons per year. Obviously, the change-over to animal products did not mean diminishing harvests, but instead a different use of the grain; towards the end of the century a substantial part of the grain was not sold, but used as fodder for the estate's own cattle.

Skansjö (1987a) has shown that some trade occurred already in the 16th and 17th centuries, but this was probably little compared with what later developed during the second half of the 19th century. Krageholm had a diversified, and over the years changing, network of trade contacts. During the 1820s, 1830s, and 1840s Krageholm sold grain to merchants in Ystad, and malt to the alum works at Andrarum. Grain was also sold to Stockholm, where Count Piper, the owner of Krageholm, resided. Towards the middle of the century grain sales increased, and in 1857 small quantities of milk were also sold to Ystad. The estate changed its trading partners almost every year. Occasionally, grain was conveyed to Malmö as well. Some milk products were carried to Malmö during the 1870s, but the large expansion in animal products came during the 1880s and afterwards. The selling of cattle on a large scale began towards the end of the 1860s, and from 1885 pigs were sold to slaughterhouses, first in Malmö and later in Ystad.

As already mentioned, the labour force more than quadrupled during the 19th century, from 11 to 49 (cf. Möller 1989). Another marked change was the civil status of the male hands. In the early years of the 19th century most were unmarried, but after 1835 over half of the workers were married. This was related to the transition to a capitalistic way of organizing labour. The

Table 4.3:1. Land use on the estate of Krageholm 1851–1911 in hectares (% in brackets).

	Arable	Meadow	Woods	Pasture	Total
1851	2092 (54)	815 (21)	968 (25)		3875 (100)
1875	2823 (74)	168 (4)	734 (19)	117 (3)	3842 (100)
1894	2882 (75)	138 (4)	736 (19)	94 (2)	3850 (100)
1911	2900 (75)	138 (4)	730 (19)	84 (2)	3852 (100)

Source: Krageholm records

outmoded and inefficient corvée system was gradually given up and replaced by a system dependent upon married agricultural labourers *(statare)*, employed on a yearly basis and paid mostly in kind. A prerequisite for employment was that the wives had to carry out milking (without payment). This new landless proletariat grew in size during the 19th century and their barracks became a common feature of the landscape of Scania.

In the 1860s, a completely new organization of farms was established, and this system (some 50 years later) can be seen in Fig. 4.3:6 (cf. Möller 1989). The land in the villages was cultivated by tenant farmers *(åbor)*, cotters with only small acreage *(husmän)*, and crofters *(torpare)*. To the north, east, and south of Sövestad, 25 farms were located, each with 45 ha. There were some farms in Årsjö as well. The tenant farmers had corvée duties throughout the century, but these diminished gradually. Rent in kind was abolished in 1875 after a decade of payments both in kind and in cash. South of Sövestad and north-west of Årsjö about 20 small farms with 10 ha each were located; these were cultivated by cotters. The crofters lived in Sövestad village and in the northern parts of the estate. They had only very small plots (1–2 ha). Apart from their own plots, the cotters and crofters worked the demesne land together with the rising number of agricultural labourers. In the 1860s, 14 houses solely for agricultural labourers were erected at Årsjö (at the site of the former medieval village). This new settlement was one of the largest of its kind in Sweden at this time.

This new organization of production also explains the veritable building boom which began in the 1860s and continued through the rest of the century. Houses for agricultural labourers were also built at Krageholm, Carlstorp, and Ohretorp (Fig. 4.3:6). Schools were founded in Årsjö and Carlstorp, and several houses for retired workers were constructed in Sövestad. Apart from these, a large number of farm buildings suitable for modern agriculture were erected.

In Ch. 4.3.8, Persson shows that population in the parish grew by 30% during the 18th century. Compared with Sweden as a whole this was a modest growth. During the next century Sövestad's population grew by over 70% up until 1880, and has since then experienced a decline. Sövestad, together with several other estate-dominated parishes, experienced a slower population growth with lower population maxima than parishes dominated by freeholders (cf. Ch 5.9).

Intensification and specialization (the 20th century)

In the early 1930s the estate still had about 4000 ha, but during this decade nearly half of the land was sold. It was the Sövestad part which was sold off, mostly to the tenants. This was because the estate needed capital for modernization. For instance, the first combine harvester was purchased in 1939, and the first tractor the year after. The old *statare* system prevailed until 1945. In the 1940s the five Kärragården farms, and in 1987 the remaining farms in Bjäresjö parish were sold. Thus, today the estate comprises less than 2000 ha.

At the beginning of the 1950s there were still over 100 milch cows and 300 pigs, not to mention animals for fattening. The workforce still consisted of at least 30 men. The large stock of animals has now disappeared, and the workforce has declined to a mere seven or eight (compare with the situation in the year 1800!). Tenant farms continue to be sold, and the remaining farms have

KRAGEHOLM 1914

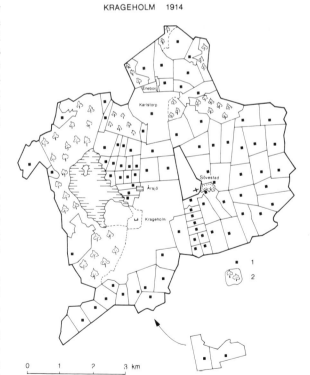

Fig. 4.3:6. The estate of Krageholm in 1914. 1. Tenant farm. 2. Wood. Source: Krageholm records. From Möller 1989.

Table 4.3:2. Use of arable in Sövestad parish in the 20th century

	grain	ley	sugar-beet	oil plants	fallow
1909	51%	28%	1%	–	10%
1927	48%	26%	2%	–	10%
1944	39%	31%	5%	9%*	2%
1966	55%	18%	7%	15%	2%
1981	62%	12%	6%	14%	1%
1987	57%	9%	6%	15%	8%

* This amount is "other crops", but most is probably oil plants.
Sources: Malmöhus läns jordbruk år 1909, Jordbruksräkningen 1927 och 1944, Lantbruksräkningen 1966 och 1981, Lant-
    bruksregistret 1987

been amalgamated. Fig. 4.3:7 shows the estate today; it can clearly be seen that the total area has diminished, but the cultivation units have grown considerably in size. Only seven tenant farms remain, and probably some of the smaller ones today have their last tenant.

Concerning the whole parish of Sövestad during our century, there has been specialization among the crops grown. The area devoted to grain has varied over the years, with the lowest figure in 1944 (39%) and the highest in 1981 (62%). Fallow has decreased from 10% to almost nothing in 1981, but has increased again as a result of government policy (Table 4.3:2). Barley and wheat today make up 89% of all grain. Oil-yielding plants (mainly rape) were first sown on a large scale

during World War II, and today they are a common feature in southern Scania. The area grown as ley (also temporary grazing land) has diminished from 28% to 9%. The increased use of artificial fertilizers during the 20th century has caused a eutrophication of lake Krage-holmssjön. This is shown both by a drastically altered diatom flora in the lake (Håkansson 1989), and by a change towards more organic sediments (Ch. 4.3.2).

The landscape at Krageholm has changed dramat-ically during the last 200 years. The feudal landscape with many smallholdings was superseded by capitalistic farming during the second half of the 19th century. This meant the introduction of drainage and marling, new machines, and a new landless working class. The land-scape became a region of large, drained fields with comparatively few buildings. This evolution has contin-ued in our century. The estate of today is a high-tech-nology food-producing factory, which only allows little room for activities not directly associated with produc-tion. The uniform landscape with huge fields is, of course, a mirror of this evolution.

KRAGEHOLM 1988

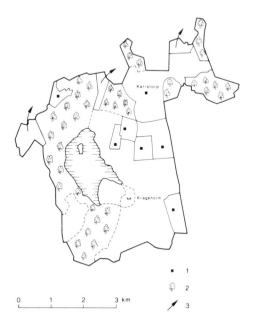

Fig. 4.3:7. The estate of Krageholm in 1988. 1. Tenant farm. 2. Wood. 3. Tenant living outside estate. Source: LRF register (Federation of Swedish Farmers).

4.3.9.2 Natural and semi-natural vegetation in the Krageholm area, 1985

Siv Bengtsson-Lindsjö

The development of the cultural landscape in the Kra-geholm area during the 19th and 20th century has been described by Möller (above). Here I describe the pre-sent (1985) natural and semi-natural vegetation of the area (Fig. 4.3:8). Forest constitutes one-third of the area and is concentrated north, west, and south of lake Krageholmssjön. It is dominated by beech and spruce. Approximately two-thirds of the area is cultivated. The fields are very large and the various crops often border directly on each other. Woods, groves, and clumps of trees are rare, as are stone walls or rows of trees and bushes. However, the monotonous agricultural land-

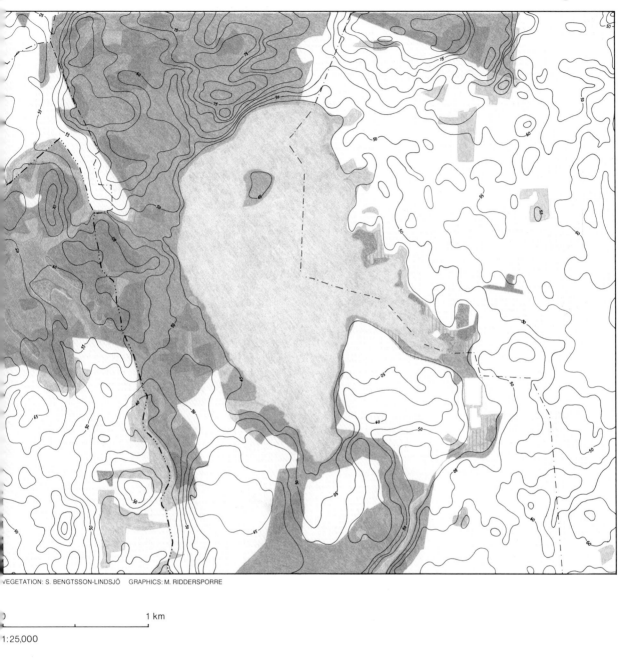

VEGETATION: S. BENGTSSON-LINDSJÖ   GRAPHICS: M. RIDDERSPORRE

0   1 km

1:25,000

	ARABLE, INCLUDING FARMSTEDS AND GARDENS		BROAD-LEAVED WOODLAND INCLUDING BEECH
	KRAGEHOLM CASTLE WITH PARK AND ADJACENT BUILDINGS		OTHER DECIDUOUS WOODLAND
	GRASSLAND / WITH TREES AND SHRUBS		CONIFER PLANTATION
	FEN		LAKE / STREAM

—50— CONTOUR LINE: 5 M EQUIDISTANT

— ··· —··· HUNDRED BOUNDARY

—·—·— VILLAGE, ESTATE BOUNDARY

scape is interrupted by old, planted avenues along the roads leading up to large estates. They often consist of elm (now threatened by Dutch elm disease) as in the beautiful avenues from Krageholm estate to Sövestad and southwards towards Bjäresjö.

The soil is very fertile and nearly all arable land is cultivated. The predominant crops are corn and rape. South of Krageholmssjön, however, a large field differs strikingly from others in the area. It is cultivated, yet sparsely covered with large oak trees. This field was probably once an old pasture connected with the surrounding forest.

Semi-natural vegetation – grasslands including fens – is scarce due to the restricted stock-breeding. The fields east of Krageholmssjön close to the shore and the old pasture along the shore lie 30–40 cm lower than the fields above. This pasture, on the narrow shore, is overgrown with stinging-nettles, favoured by fertilizers leaching from the fields above. Only one pasture, beside Krageholmssjön, is still grazed. Along the southwest side of the lake there were once grazed pastures with scattered big oak trees (Ahlberg et al. 1974). They are not grazed today and are now partly overgrown with bushes and trees. The shore is partly dominated by *Salix* (willow) bushes.

Most pastures in the area receive high loads of commercial fertilizers annually. The result is that the original species composition, consisting of a large group of grassland species, has been replaced by a small number of species tolerant of fertilizers (Ch. 5.11). Mesic and wet grassland are most frequent in the area. Common species on the mesic, fertilized grasslands include *Taraxacum vulgare* (dandelion), *Anthriscus sylvestris* (cow parsley), *Plantago major* (greater plantain), and *Rumex* (sorrel) species. Wet grasslands are usually less fertilized, as illustrated by the meadow along the east side of the river Svartån. Grassland species like *Myosotis* (forget-me-not), *Lychnis flos-cuculi* (ragged robin), and *Glyceria fluitans* (floating sweet-grass) are still found here. Bussjö Mosse in the south-east is another less fertilized pasture (Bengtsson et al. 1986). Dry grassland vegetation is dominated by species like *Agrostis capillaris* (common bent), *Hieracium pilosella* (mouse-ear hawkweed), *Plantago lanceolata* (ribwort plantain), *Rumex acetosella* (sheep's sorrel), and *Silene nutans* (Nottingham catchfly). The mesic pasture on Bussjö Mosse is more affected by fertilizers.

The forest west and north of Krageholmssjön is mainly old forest mixed with younger plantations. The tree species are old beech stands, sometimes mixed with old *Quercus robur* (oak) and *Carpinus betulus* (hornbeam), and there are young stands of planted *Picea abies* (spruce), *Fagus sylvatica* (beech), *Acer pseudoplatanus* (sycamore) or oak. *Alnus* (alder) grows in the wettest parts sometimes mixed with *Fraxinus* (ash) and *Ulmus* (elm). Two small plantations of *Larix* (larch) also exist in the forest.

The topographic variation results in a number of small marshes with birch and alder in the wettest parts. Sedges or tall herbs often dominate the field layer here.

The forest here is, however, mostly of mesic herb type, but not so rich as the forest east and south of the lake. Common species in the field layer are *Mercurialis perennis* (dog's mercury), *Galium odoratum* (woodruff), *Lamium galeobdolon,* and *Milium effusum* (wood millet). The dry herb type is dominated by *Poa nemoralis* (wood meadow-grass), *Lamium galeobdolon, Stellaria holostea* (greater stitchwort), and sometimes *Oxalis acetosella* (wood sorrel). The wet herb type, dominated by *Filipendula ulmaria* (meadowsweet), usually occurs along the shore of the lake and in the brook valleys. North and west of the lake some nutrient-poorer areas also occur: the field layer is dominated by *Deschampsia flexuosa* (wavy hair-grass), or is absent. The field layer is also often missing in the young plantations, especially when planted with dense spruce.

The forest to the south and south-east of Krageholmssjön is the richest part. It consists of old beech stands mixed with oak, elm, hornbeam, and ash. The wettest parts are dominated by ash. A minor area is planted with young oak and there are also a few small areas planted with spruce. The bush layer varies from dense to nearly absent. Common species are beech, elm, ash, and also true bushes like raspberry, elder, blackberry, dog rose, and in some parts hawthorn. Large hazel bushes and blackthorn are remnants from the time when the forest was more open and used for grazing (cf. Möller above). The field layer is dominated by mesic herb species such as those mentioned above, but also *Aegopodium podagraria* (ground elder), *Geranium robertianum* (herb Robert), *Stellaria nemorum* (wood stitchwort), *Cirsium oleraceum* (cabbage thistle), *Campanula latifolia* (giant bellflower), *Allium ursinum* (ramsons), *Festuca gigantea* (giant fescue), *Milium effusum,* and *Carex sylvatica* (wood sedge). In the wetter parts *Filipendula ulmaria* dominates together with *Urticadioica* (nettle), and *Aegopodium podagraria.*

The forest in this area is of special interest, as it shows a comparatively long continuity (cf. Fig. 4.3:4, 4.3:6) as a refuge for wild plants and animals in the otherwise monotonous cultivated landscape.

4.3.10 Conclusions

Johan Callmer and Joachim Regnéll

The Krageholm area in the Late Mesolithic was covered with deciduous forests of varied composition, but dominated by oak, elm, and lime. The forest can hardly have been subject to any human influence, although the conditions favouring hazel may have been deliberately improved.

In the earliest Neolithic times the area already began

to feel the weak effect of a farming population. Data from both pollen analysis and archaeology, however, suggest that human utilization of the Krageholm area for tillage and pasture remained very limited for a long time. The areas of exploited land were tiny enclaves in a large, continuous expanse of forest. Denser settlement on lighter soils was to be found north of Krageholmssjön, and perhaps also west and east of the lake. The area gives the impression of being somewhat on the periphery for the greater part of the Neolithic. To a large extent it was used extensively only for hunting, fishing, gathering, and collecting fodder.

As in so many other eastern parts of south Scandinavia, the Late Neolithic appears to have been an expansive period in the Krageholm area. The number of habitation sites – based on family groups – increased and land use embraced wider areas of pasture. The Krageholm area nevertheless remained weakly exploited throughout the Early Bronze Age. Pollen-analytical data suggest that a smaller number of habitation sites may have existed in the area, but whether or not they represent continuous settlement is unclear. The absence of barrows by Krageholmssjön indicates that we are on the fringe of the settlement system in this period.

The greatest opening of the cultural landscape occurred in the middle phase of the Bronze Age. Open, grazed areas alternating with secondary half-open woodland and scrub characterized the greater part of the landscape. Stock-raising increased in importance during the Bronze Age, reaching a peak in the later years of this period. As a whole, however, the Krageholm area must be viewed as peripheral in relation to settlement in the Ystad area. The number of habitation sites was small. It is possible that the inland zone of which the Krageholm area is part was also used by the coastal population for seasonal grazing. East of the Krageholm area in the south of Sövestad parish there are stronger archaeological indications of a rich settlement.

In the Pre-Roman Iron Age there was a slight regeneration of deciduous forest spreading over former pasture. The change could be interpreted as a greater polarization of the cultural landscape. It is unclear, but scarcely probable, that there was a demographic regression in the area at this time. There is good archaeological evidence of settlement from the Pre-Roman and Roman Iron Age on the eastern shore of the lake. The change in the cultural landscape could be seen as an incipient division into permanent areas for tillage and pasturage in the woodland. The Migration Period and the Vendel Period saw an even greater tendency towards the division of the cultural landscape into infields, outlands, and woodland, and we find a landscape with a structure similar to that of the village communities of historical times. As in the Early Iron Age, the size of the settlement units was relatively modest, but by the end of the period at least, there was a small settlement

consisting of several farms near the historical village site of Årsjö on the eastern shore of the lake. From this time the beech woods begin to be a noticeable feature of the landscape.

In the Viking Age and the Early Middle Ages the areas used for tillage and pasture increased. Since the end of the Viking Age the population in the Krageholm area had been concentrated around the village of Årsjö. It is likely that there was an aristocratic presence as far back as the Viking Age. A field rotation system was introduced in the Early Viking Age. The wetlands, which had hitherto been dominated by alder carrs, were now being increasingly transformed into open sedge fens; these were mowed for winter fodder.

The agrarian crisis of the Middle Ages has left no distinct mark in the Krageholm area on the source material used for scientific analysis. Juniper pastures are a new feature of the cultural landscape from the Late Middle Ages; it is not clear, however, whether they were to be found in the Krageholm area strictly defined, or whether the pollen analysis reflects larger areas of pasture on the Romeleåsen ridge to the north-west. The cultivation system that was introduced in the Late Iron Age remained characteristic of the Krageholm area throughout the Middle Ages as well. It was probably a question then of a three-field system.

It is not certain if there was continuity in the aristocratic element of settlement from the Late Viking Age. What is certain is that the Krageholm estate had been established in the 14th century. The fortified manor at the outflow of the lake and the village of Årsjö are the nucleus of this expansive estate complex from this time.

The estate developed vigorously in the 16th century, and by the 17th century it encompassed the whole of Sövestad parish. In this way a feudal enclave with its own fairly independent economy and society had been formed. The economy of the estate was probably geared early on to animal husbandry, which was favoured by the natural conditions of the area. However, the proportion of land used for grain cultivation increased gradually during the period 1500–1700. Socially the Krageholm area is characterized by the large number of cotters and the far-reaching equalization of the farms. Work on the estate was organized in a corvée system.

When the great restructuring of the Scanian countryside began around 1800, the Krageholm area remained largely unaffected – even retarded – for a long time. The Krageholm estate was a large production and consumption unit isolated from the outside world. This archaic cultural landscape persisted right up to the middle of the 19th century. When the changes began, however, the power of the centrally organized decision-making system manifested itself. Agricultural development proceeded very quickly in the 19th and 20th centuries, with a huge expansion of the area of cultivated land, a reduction in the amount of pasture, new crops, crop sequences, and – not least – large-scale mechanization.

Various measures to increase production, such as drainage and the use of inorganic fertilizers, had a powerful impact at an early stage. The woodlands, which had not been managed to any great extent, were now given considerable attention. The corvée system was gradually abandoned, the work being done by landless labourers who were paid in kind. Parts of the estate were enclosed and settlement was dispersed through the countryside. The cultural landscape thus changed its appearance drastically. The social and cultural community of the village largely vanished.

The tendency towards a greater concentration on grain cultivation has typified agriculture in the Krageholm area up to the present day. The cultivation of oil-yielding plants has come since the end of World War II, while leys and fodder crops have disappeared in conjunction with the total abandonment of stock-raising. In recent decades the estate has closed the last tenant farms and all agriculture has come under the estate. Agriculture today requires only a very small workforce, being highly mechanized.

Fields have been amalgamated into huge monocultural cropping areas. Strips of trees and bushes between the fields have been eliminated. Stone walls have been removed and crushed into macadam for roads. However, since the estate has an interest in game preservation and retaining certain landscape features such as avenues, the monotonization of the landscape around Krageholm has not been total. Forestry in the 20th century has also increased the uniformity of the woodlands.

Throughout the ages studied here, the Krageholm area has been relatively peripheral, albeit to varying extents. The cultural landscape of the area has had a conservative character for long periods; indeed, it has been virtually retarded at times. A special feature of the Krageholm area is the presence of relatively large stands of woodland down the ages. The slower and at times considerably weaker development here is evident when the area is contrasted with the Bjäresjö area immediately to the south. The reason for this is the distal location of the area on the edge of the settlement region that had been occupied since prehistoric times. However, the Krageholm area has also – perhaps by virtue of its slower pace of development – retained a considerable economic potential. In some periods its resources have been successfully mobilized with the aid of various techniques.

The farmstead of Beden on the Romeleåsen ridge. View towards the east. Photo: Ingmar Holmåsen/N, September 1988.

4.4 The Romele area

4.4.1 Introduction

Björn E. Berglund

Areas and sites studied. The Romele area comprises the present parish of Villie, but our studies are concentrated on the northernmost part, primarily the Beden area. Some research also deals with the Romeleåsen ridge west and east of Villie parish. The physicogeographical description here refers mainly to the northern part of Villie parish.

The main field study areas are the following (see map, Figs 4.4:2, 3):

Palaeoecology: The ancient lake Kurarps Mosse (= "the meadows of Beden").

Ecology: The Beden area including Kurarps Mosse.

Archaeology: The Beden area.

Historical and human geographical studies deal with source material from the entire Villie parish and adjacent areas of the Romeleåsen ridge.

Geomorphology and hydrology. The area is situated within the inner hummocky landscape and mainly on the plateau of the Romeleåsen ridge. Also this plateau has an undulating topography caused by moraine hills and kame deposits. The Beden area is situated 90 to 170 m a.s.l. with the ancient lake basin Kurarps Mosse in a depression about 95 m a.s.l. This means that this area has a relatively high relief. The distance to the sea is 12–14 km.

The Beden area forms the watershed between land which is drained southwards, i.e., the major part of Villie parish, and land which is drained northwards into the South Scanian inland depression, the Vomb plain. The area has no major streams.

Besides the fen of the ancient lake basin of Kurarps Mosse, many minor hollows with fens occur. Dammed water reservoirs have been common in historical time, particularly on the south-facing slope of Romeleåsen, from Rydsgård to Knickarp (Möller 1984). Some remnants are still to be found.

Subsoil distribution. The Quaternary deposits of the Romeleåsen ridge are dominated by clayey or sandy tills, around Beden also glaciofluvial deposits with sand and gravel. Except for an area east of Rydsgård, the clay content is much lower in these soils than in the other focal areas. The calcium carbonate content is also lower here (approx. 10–15%). See map in Fig. 1.2:15 (Daniel 1986, 1989).

Conditions for cultivation. The light soils may have been favourable for primitive agriculture. However, these have also been more prone to soil leaching. The agricultural classification confirms the rather poor quality of the soils (classified as 4 to 5 on the 10-grade scale, Fig. 1.2:16). Poor soils, often wooded, and wetlands favoured stock-raising on the Romeleåsen plateau in historical time. The slopes facing south and east have heavy, clayey soils unfavourable for prehistoric cultivation.

4.4.2 Vegetation and landscape through time

Björn E. Berglund and Else Kolstrup

The vegetation history of the Romele area is based on a pollen diagram from Kurarps Mosse covering the time span from 4000 to 1100 BC (Fig. 4.4:1, see Kolstrup 1990). This represents the longest sediment record known until now within the Romele area. The problem of finding a complete lake or peat sequence from the Romeleåsen ridge has been discussed in Ch. 2.1. This means that for the time after 1100 BC the landscape reconstruction cannot rely on local data. However, we assume that the vegetation changes of the Romele area are reflected in the diagram from Krageholmssjön, which also belongs to the inner hummocky landscape (cf. Ch. 4.3.2). Therefore, a tentative, general palaeoecological outline of the development in Kurarp is proposed from the Krageholmssjön diagram for the time from 1100 BC onwards. In the following the description dealing with the time before 1100 BC thus refers to the northern part of the Romele area, meaning the Beden area on the ridge proper, whereas the outline of the following time refers to the Romele area in general. For the time after 1750 we have detailed information of the landscape in written sources (see Ch. 4.4.7, 4.4.8). Six main periods of the landscape development have been distinguished.

1) 4000–3150 BC. The forest landscape and the early human impact

During the Mesolithic the Romele area was forested like the landscape south and east of the ridge. However, there are indications that the forests of the ridge were slightly more open than those closer to the coast, as indicated by more *Betula* (birch), *Pinus* (pine), *Populus* (aspen), *Pteridium* (bracken), and Poaceae (grasses). On dry ground there were forests with *Quercus* (oak), *Tilia* (lime), *Corylus* (hazel), and sometimes with *Pinus* (pine), on damper soils *Ulmus* (elm) and *Fraxinus*

KURARPS MOSSE

POLLEN DIAGRAM

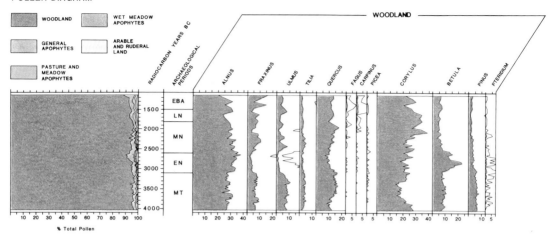

(ash). On wet, peaty ground *Alnus* (alder) was dominant and it seems as if alder was more frequent here than in the outer hummocky landscape. This may be due to the frequently occurring hollows on the moraine plateau of the Romeleåsen ridge.

Regular records of *Juniperus* (juniper) and *Calluna* (heather) together with scattered finds of *Sedum* (stonecrop) and *Rumex* (sorrel) indicate that some open spots occurred. Juniper and heather were probably more frequent here than at the coast.

During the later part of this period, from ca. 3400 BC onwards, there seem to be some minor changes of the forest vegetation. The first pollen grains of *Plantago lanceolata* (ribwort plantain) and *P. major/media* (greater or hoary plantain) appear, and slightly above this level elm starts to decrease and hazel expands. If the findings of *Plantago* are taken as evidence for grazing by domestic animals or mowing, this can support the hypothesis that some activities related to husbandry were carried on in the neighbourhood of the Kurarp lake.

2) 3150–2500 BC. Small-scale farming in a woodland landscape

Also in the Romele area the period starts with an opening of the forest landscape after the elm decline (cf. Ch. 4.2.2). Gradually, the frequencies of *Ulmus, Fraxinus,* and *Tilia* are reduced, followed by an expansion of *Betula* on dry ground and *Alnus* on wet soils. At the same time, herbs and grasses expand indicating more open conditions. Charcoal fragments are slightly more frequent during this period. Together with the expansion of *Pteridium* (bracken) this may indicate that some kind of forest fires were common in the area. Just above the elm decline the first find of *Triticum* type has been made, followed slightly later by additional cereal finds *(Triticum, Hordeum)* and an increase of weeds *(Artemisia* – wormwood, *Urtica* – nettle) together with pasture/

meadow indicators like *Plantago lanceolata*. All this indicates that arable fields and pastures occurred in underwood with a reduced number of standard trees on the Romeleåsen ridge.

3) 2500 BC–ca. 1000 BC. Sporadic farming in a woodland landscape

The period starts with a regeneration of the forests so that elm, ash, lime and oak expand on what had previously been more open land occupied by birch. Around 1500 BC *Carpinus* (hornbeam) and *Fagus* (beech) colonize the area. The herb and grass flora is reduced, but more or less continuous records of pasture as well as field indicators suggest constant but weak human influence in the area (weaker than around the studied lake basins closer to the coast; cf. Ch. 4.2.2). The changing values of the broad-leaved trees, particularly of elm and ash, indicate instability in woodlands, which were probably used as coppiced woods. Around 1800 BC there is an expansion of *Plantago lanceolata* indicating a corresponding increase of pastures/meadows. This may be correlated with the archaeologically documented expansion in the area in the Late Neolithic (Ch. 4.4.3). There seems to be a further expansion of pasture indicators around 1300 BC.

Around 1000 BC the Kurarp lake was filled in and overgrown. This changed the local landscape in the Kurarp-Beden area. The open water disappeared and the landscape became less diverse and possibly less attractive for settlement.

4) Ca. 1000 BC–AD 1000. Marginal woodland grazing

When considering the scanty archaeological information on Bronze Age and Iron Age settlement on the Romeleåsen ridge (Ch. 4.4.4, 4.4.5), we assume that

248

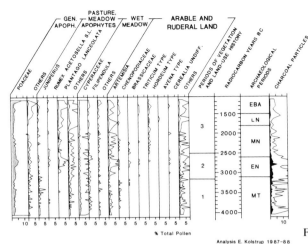

% Total Pollen

Analysis E. Kolstrup 1987-88

Fig. 4.4:1. Pollen diagram from Kurarps mosse.

this area was only utilized extensively for grazing. Probably woodland pastures without permanent settlements covered the area west and north-west of lake Krageholmssjön (cf. Ch. 4.3.4, 4.4.4). It is possible that there was an expansion of such pastures about 1000–700 BC on the ridge as well, but untouched forests may also have been present within this wide area, particularly on wet and damp soils. Such a development has recently been confirmed from the Baldringe area northeast of Krageholmssjön (Lagerås 1991).

5) Ca. AD 1000–1700. Farming expansion in marginal woodlands

The pollen diagrams from Krageholmssjön, as well as from Vasasjön north-east of Krageholm (Hjelmroos unpubl.), show a distinct expansion of fields and pastures during the Viking Age and Early Middle Ages. Archaeological (Ch. 4.4.5) and historical (Ch. 4.4.6) sources do not confirm any colonization in the marginal woodland area on the Romeleåsen ridge during this time. Isolated farms may have been present, but settlement does not seem to have been permanent until the 15th or 16th century. Several of the present isolated farms have roots going back to the 16th century or possibly to the Late Middle Ages. From these indirect sources and the Krageholmssjön diagram it is inferred that the landscape was a woodland landscape also during early historical time, extensively utilized for grazing. The dry ground probably had a half-open woodland character with stands mainly of birch, oak, and beech, and, in more open places, grass-rich vegetation with juniper. Damp soils were probably still forested. The most fertile areas were colonized and cultivated during the Late Middle Ages or the 16th century. The landscape then may have been half-open woodland with "islands" of infields having a different character – arable fields, treeless or sparsely wooded meadows, etc. Gradually the wood-pastures of the outfields were transformed into a woodless, grazed heath.

6) 1700 to the present day. Marginal farming in a woodless pasture landscape

Farming expanded around old and new isolated farms which were surrounded by enclosed infields. The wide outlying pastures were part of an area of common land covering the Romeleåsen ridge. Written sources show that this area had the character of a woodless heath with junipers (Ch. 4.4.7). Probably also alder, sometimes coppiced, occurred on peaty ground. Only the infield areas had a more diverse vegetation including arable fields, treeless meadows and meadows with coppiced alder, oak, and hazel. In addition, small stands of oak and beech may have been important.

This heath-dominated landscape existed until the beginning of this century, when changes in agriculture and economy caused decreased grazing pressure. Large areas of the old common became afforested or overgrown by natural forest successions (Ch. 4.4.8). Former arable fields have in many cases turned into pastures. This means that the landscape today has the character of a forest area with some openings of modern pasture land.

4.4.3 The Neolithic
The cultural landscape in the Romele area during the Neolithic

Mats Larsson

Stray finds and the cultural landscape

The area with which we are concerned here covers the whole of Villie parish. However, the primary area of analysis is concentrated in the northern part of the parish, i.e., the area around the fen of Kurarps Mosse.

From the whole of this area, 58 datable implements have been recorded in museums and private collections. Only a small proportion of these come from around Kurarps Mosse. This area was the subject of a comprehensive field survey in the spring of 1985 and in the autumn/winter of 1986, with the aim of localizing the principal areas of settlement (Larsson and Larsson 1986, Olausson 1988). Seven Stone Age sites were localized in this way. Most of these are to be found in the north of Villie parish.

When discussing the chorological and chronological distribution of the stray finds, the pointed-butted axes from the oldest part of the Early Neolithic period are of particular interest. Eight axes of this kind have been recorded within the whole of this focal area. The precise location where they were found is not stated for most of them, but we know that the number of pointed-butted axes is significantly greater in the north of the area.

Datable finds from the later part of the Early Neolithic and the start of the Middle Neolithic are almost entirely absent. Thin-butted axes are few in number in this area.

A noticeable resurgence in the number of finds can be observed once more only from the later part of the Middle Neolithic (Periods IV–V). This resurgence is even more noticeable if we include Middle Neolithic B (the Battleaxe culture). A similar trend has been observed in the neighbouring Krageholm area (Ch. 4.3.3). It is quite obvious, however, that the real expansion took place during the Late Neolithic. This can be regarded as being generally true of the whole of the outer and inner hummocky zones in the Ystad area. A total of 31 objects dating from this period have been recorded in Villie parish. These account for 55% of all the material. The precise origin of most of the objects has not been established, although the inner hummocky zone is clearly also represented amongst the material.

The Neolithic settlement area

A total of seven sites, including one Mesolithic site, were recorded in the course of the field surveys. A more detailed survey was made, especially of the area around Beden and Kurarps Mosse.

Only one area, Trunnerup, was examined to a minor extent in the autumn of 1985. This area contains evidence of three settlements. The most recent can be dated to the Early Bronze Age, and includes occupation layers and structures. The pottery found there exhibits similarities to the Early Bronze Age material from Denmark (Boas 1983). Another settlement was investigated in the autumn of 1985. This was situated a couple of hundred metres from the aforementioned site. Dating is uncertain, although much of the evidence suggests that it is from the Early Neolithic. The third site is situated immediately to the south of the Early Bronze Age site, on sandy soil. A good deal of surface material has been collected from this site, which is estimated to cover about 1500 m². Finds included fragments of thin-butted axes, scrapers, knives, pottery items, and considerable quantities of flint chippings. Also found was a fragmentary greenstone axe, which can be dated to a late

part of the Early Neolithic (Ebbesen 1984). The site can be dated to the end of the Early Neolithic.

Two sites were found in the area around Kurarps Mosse. One yielded a relatively large quantity of material and can be dated to the Early Neolithic. The other site is more difficult to date, since the quantity of material involved is only small.

In summary, it can be stated that there is sound evidence to indicate Early Neolithic settlement in the area around Kurarps Mosse. With regard to the whole of the Villie parish, a number of Early Neolithic sites have been found there. Just as in the case of the neighbouring areas, a clear relationship exists between the settlement, the lighter soils, the altitude, and watercourses (M. Larsson 1984, Larsson and Larsson 1984, 1986). It is more difficult, however, to discuss the pattern of settlement during the early part of the Middle Neolithic; few finds have been made dating from the Middle Neolithic I–III. It is only during the latter part of the Middle Neolithic (IV–V) and the Battleaxe culture that the area again appears to have been used. The Late Neolithic gives the impression of having been a period of expansion, in Villie parish as elsewhere. A significant number of single finds dating from this period have been recorded. Evidence of settlement during this period has also been found in the area around Kurarps Mosse.

4.4.4 The Bronze Age and Early Iron Age
The cultural landscape in the Romele area during the Bronze Age and Early Iron Age

Deborah Olausson

The scanty material datable to the Bronze Age yields little to substantiate that the increase in activity suggested in Ch. 4.4.3 for the preceding period should have continued into the Bronze Age. On the basis of the evidence available (and commensurate with the pattern seen elsewhere in south-eastern Scandinavia (Kristiansen 1978:334)), the Romele area, 10 km from the coast and with predominantly heavy clay soils, was of marginal importance during the Bronze Age. The difference in the use of the area, heavier during both the preceding Late Neolithic and the succeeding Late Iron Age than during the Bronze Age and Early Iron Age, can be ascribed to changes in the subsistence base and to agricultural practices through time.

The Early Bronze Age

Barrows. Five *hög* features can be located on the early cadastral maps from villages in Villie parish: Torphög, Stehög, Stafhög (Södra Villie 1783), Svenshög (Norra Villie 1791), and Bånghög. However, field inspection

THE ROMELE AREA
NEOLITHIC SETTLEMENT, CA. 3000 BC - 2000 BC[*]

Fig. 4.4:2

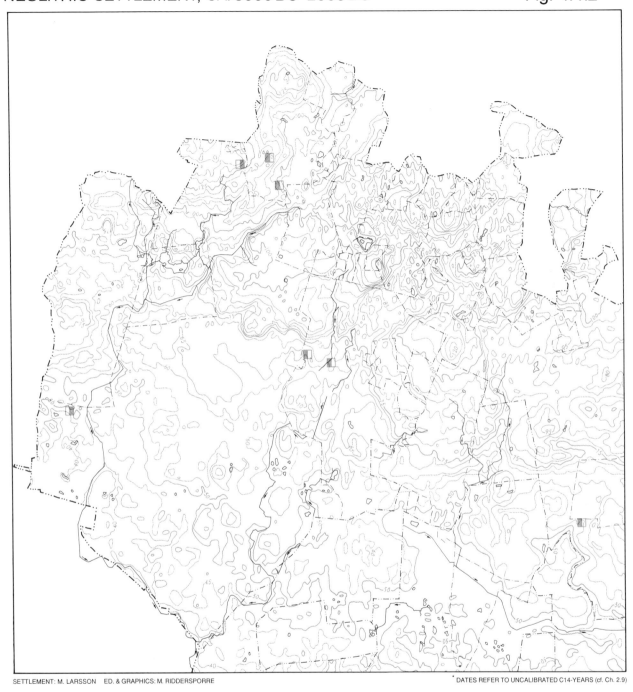

SETTLEMENT: M. LARSSON   ED. & GRAPHICS: M. RIDDERSPORRE

[*] DATES REFER TO UNCALIBRATED C14-YEARS (cf. Ch. 2.9)

0   1   2 km

1:50,000

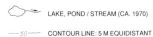

SETTLEMENT: EARLY NEOLITHIC	
SETTLEMENT: MIDDLE NEOLITHIC (TRB)	

LAKE, POND / STREAM (CA. 1970)

—50— CONTOUR LINE: 5 M EQUIDISTANT

—···—···— HUNDRED BOUNDARY (CA. 1970)

—··—··— PARISH BOUNDARY (CA. 1970)

—·—·— VILLAGE, ESTATE, SINGLE FARM BOUNDARY (CA. 1970)

within the project and in connection with the survey for the Register of Ancient Monuments, and place-name research, disqualified all but one of these as Early Bronze Age burials.

Fig. 4.1:4, 4.2:4 shows clearly that Early Bronze Age barrows are concentrated in a band within 5 km of the coast, thinning considerably as one moves further inland and virtually disappearing in the inner hummocky landscape. Against this background, the Romele area is in no way an anomaly.

Single finds. There are no metal finds dating to the Early Bronze Age known from this area. There is of course the possibility that some of the flint artefacts of more general character dated to the Stone Age may in fact originate from the Early Bronze Age (Thrane 1985:146).

Settlement. Surface survey aimed at locating possible settlement was conducted together with subproject B1–2 during the spring of 1985 (Larsson and Larsson 1986:111) and separately during late autumn of 1986 (Olausson 1988). A total of 15.3 ha (less than 1% of the total surface area) was surveyed on these occasions. Although flint flakes and tools were quite abundant, no evidence for Bronze Age settlement was seen during this survey. Several parts of the area were the subject of concentrated survey in December 1986 (Olausson 1988). While heavy concentrations of soot and fire-cracked rocks and 172 g of pottery indicated Iron Age settlement at Norra Villie, no comparable traces of Bronze Age or Early Iron Age activity were found.

During excavation by M. Larsson at Trunnerup, Early Bronze Age settlement remains were found (Ch. 4.4.3). Nothing from the surface survey of this area had indicated settlement later than the Stone Age, which illustrates the difficulty of locating Bronze Age remains by surface survey. However, the absence of Early Bronze Age barrows in the area, even allowing for the possibility that some existed but have been levelled by subsequent agriculture, gives strong support to the contention that the picture gained during surface survey reflects the actual nature of the sparse Early Bronze Age use of this area (see Fig. 4.4:3).

The Trunnerup site (Ch. 4.4.3) should be characterized as a short-term specialized occupation. The limited area, the small amount of finds (approx. 1100 g of pottery) and the lack of evidence for any structures, all point to such an interpretation.

The Late Bronze Age

Graves. There are no known burials from this period in the area.

Single finds. A bronze celt (LUHM 19056) from the collection at the Frankhult farm is the only known Late Bronze Age stray find from Villie parish. Such celts are commonly found as stray finds and need not indicate settlement: in Västergötland, for instance, such celts are not connected with settlement (Bergström 1980:39).

Settlement. Neither surface survey nor the limited excavation carried out in the area has yielded evidence for Late Bronze Age settlement.

The Bronze Age cultural landscape

The Romele area follows the pattern of Bronze Age settlement observable in the other focal areas of the project. As settlement consolidated and became less mobile during this time, preference for specific conditions for settlement location emerged (Jensen 1982:153–4, Thurborg 1985:40, Fagerlund 1985:15ff., T. Larsson 1986:104). As long as preferred locations for settlement were available, less ideal positions were not settled but rather used sporadically for supplementary activities: for temporary grazing, in exploiting some particular resource, on hunting expeditions, and so on.

The nature of the evidence from Romele points to the conclusion that it was used as a marginal area during the Bronze Age. For Stone Age settlement, characterized by mobility, smaller population entities, and more extensive use of resources, the Romele area contained natural resources and potential sites for settlement which fit into the subsistence pattern. Bronze Age farmers and pastoralists, on the other hand, consolidated settlement in locations where both lighter soils (workable by means of an ard) and pasture were within easy reach. The heavy clay soils and more pronounced topography of the inner hummocky landscape were less attractive in the light of this adaptation. The Bronze Age farmers buried their dead near their settlements, located closer to the coast, and exploited the inland for more specialized purposes only.

The Early Iron Age cultural landscape

Concrete evidence, in the form of stray finds, burials, or settlement remains from this period, is almost totally lacking from Villie parish. This would seem to confirm a continuation of the marginal character of this area during the Early Iron Age. Unfortunately, source-critical factors mean that it is difficult to know whether the apparent emptiness of the area during the Early Iron Age is real or whether it is a factor of the difficulties of discovering Early Iron Age sites archaeologically. Iron Age burials are difficult to detect by surface survey. As settlement becomes less mobile, leaving fewer occupation sites, the chances of discovering them will also decrease (L. Nielsen 1982:138).

Apparently, settlement had stabilized nearer the coast, as was seen by the distribution of the Bronze Age remains, and this zone continued to be more heavily populated than areas further inland during the Early Iron Age. The fact that activity in the Romele area was apparently so slight probably means that the agrarian crisis seen in Denmark, in which a gradual deterioration of lighter soils during the course of the Late Bronze Age forced an expansion into forested areas (Kristiansen

THE ROMELE AREA

BRONZE AGE, ROMAN IRON AGE, AND VIKING AGE SETTLEMENT   Fig. 4.4:3

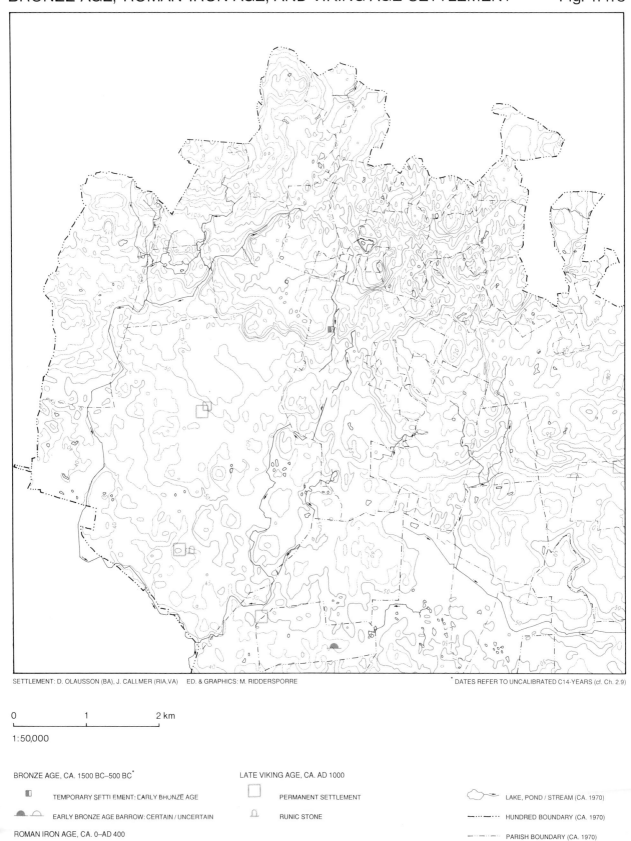

SETTLEMENT: D. OLAUSSON (BA), J. CALLMER (RIA,VA)   ED. & GRAPHICS: M. RIDDERSPORRE   * DATES REFER TO UNCALIBRATED C14-YEARS (cf. Ch. 2.9)

```
0          1          2 km
|_____|_____|
```

1:50,000

BRONZE AGE, CA. 1500 BC–500 BC*

TEMPORARY SETTLEMENT: EARLY BRONZE AGE

EARLY BRONZE AGE BARROW: CERTAIN / UNCERTAIN

ROMAN IRON AGE, CA. 0–AD 400

SETTLEMENT

LATE VIKING AGE, CA. AD 1000

PERMANENT SETTLEMENT

RUNIC STONE

——50—— CONTOUR LINE: 5 M EQUIDISTANT

LAKE, POND / STREAM (CA. 1970)

——·——·—— HUNDRED BOUNDARY (CA. 1970)

——·——·—— PARISH BOUNDARY (CA. 1970)

——·——·—— VILLAGE, ESTATE, SINGLE FARM BOUNDARY (CA. 1970)

1978:337, Hedeager 1988b:176), does not seem to have applied here, this far inland. While the existence of single farmsteads here and there in the Romele area cannot be ruled out, there is no archaeological evidence for the agglomeration of houses with permanent fields and systems of pasture which are evident in Denmark at this time (e.g. Becker 1971, L. Nielsen 1982, Jensen 1982:153, Hvass 1983, 1985). During the Late Iron Age, as agricultural techniques changed and enabled the cultivation of heavier soils, parts of the Romele area once again became a potential area for settlement (see Fig. 4.4:3).

4.4.5 The Late Iron Age
Settlement and landscape in the Romele area in the Late Iron Age

Johan Callmer

In the Late Iron Age and the Early Middle Ages the Romele area was on the periphery of the settlement district. This applies in particular to the central and northern parts of the area. The height above sea level and the noticeably greater relief give a distinctive stamp to the landscape north of a line from around Rydsgård via Norra Kadesjö to the hills north of Gussnava (Fig. 4.4:3). Within this northern part of the area there are no sure indications of settlement from the Early Iron Age, the Late Iron Age, or the Early Middle Ages. However, it must be said that field-walking is extremely difficult in this part of the area. To the west there are large expanses of woodland, and in the east modern farming techniques have brought about considerable erosion of the little hummocks, with the continual transport of material from higher to lower ground. The partial occurrence of stony subsoil has led to even more extreme effects resulting from modern agriculture. Only a few field surveys of this part of the area have been made, but no excavations (see Fig. 4.4:3).

Although it has not been possible to uncover any concrete finds from the period by excavation, dispersed settlement cannot be ruled out. Such an assumption is in fact strengthened by a comparison of the western sections of the Romeleåsen ridge with areas elsewhere showing roughly the same topography and at a similar distance from more continuous settlement (Strömberg 1961, 2:21–22, Nagy 1975, unpubl. excavation at Tullboden, Slimminge parish, by UV-Syd).

South of the line described above, there are larger areas of open landscape. It is also here that we find indications of more continuous settlement in small, limited sectors of the landscape. Of special interest is the habitation site at Norra Villie, discovered in the spring of 1986 by Billberg and Callmer. Within an area measuring about 300 × 200 m, lying partly on the historical

village site, finds of pottery from the Roman Iron Age, the Late Iron Age, and the Early Middle Ages were made in arable land. The material showed somewhat varying tendencies. The early medieval pottery was close to the historical village site, whereas the finds from the Roman Iron Age lay higher up the slope. The location is characteristic. The habitation site is on the southern slope of a low rise above a stream. When Olausson visited the site later in 1986, she found, among other things, a bird fibula from the 8th century. Unfortunately, the distribution of pottery from the Late Iron Age does not enable us to estimate the size of the settlement. By contrast, the settlement from the 11th and 12th centuries was shown to be on a scale that clearly indicates agglomerated settlement. Of great significance is the probability that both Norra and Södra Villie in the immediate vicinity have traces of settlement from the Late Iron Age (a runic stone probably from Södra Villie; see Strömberg 1961, 2:43). This speaks against the assumption that place-names prefixed by Northern and Southern or Eastern and Western are always medieval offshoots of older settlements (cf. Porsmose 1988:238; over ten examples of early settlement in such pairs of localities can be adduced from southern Sweden).

Similar observations to those from Norra Villie were made at Gussnava, which lies on the eastern edge of the map of the Romele area. It was possible to investigate settlement from late prehistoric and early medieval times on similar terrain on the historical village site of Gussnava (runic stone, Strömberg 1961, 2:43). Typical Early Iron Age pottery was also found at Gussnava. Immediately west of this, on the edge of a large gravel quarry which had destroyed a considerable part of the optimum habitation site, a bead from the Migration Period was found in 1988. A further 200 m to the west a similar find was made. Comparable indications are found immediately outside this part of the area, for example at Skårby (runic stone near Gussnava, Strömberg 1961, 2:43) and Borrsjö (Marsvinsholm).

Despite the relatively poor source material from the area, we can note that settlement in the north differed from that in the south. The northern part may at times have had dispersed settlement comprising isolated farms or small clusters without long-term continuity (sherds of Black Earthenware found just north of Trunnerup), but the landscape was mostly uninhabited. However, it is possible that the north was used to some extent to provide fodder and grazing. In the southern part of the area relatively small groups of farms in a delimited part of the landscape appear to show continuity through the Iron Age up to the end of the Viking Age and the Early Middle Ages, when they grew into rather larger settlements consisting of many farms.

In the absence of a pollen diagram for the area it is impossible to ascertain whether there was any economic specialization. It may be pointed out again, however, that there was a plentiful supply of fodder and wood-pasture, which could have led to a greater emphasis on animal husbandry here than in the more central parts of the settlement region. Nor is it possible from the evidence available at present to make any deductions about expansion here in the Late Iron Age and the Early Middle Ages. Some general assumptions about settlement during this period have already been put forward above. A growth and stabilization of settlement in the southern part of the area took place in late prehistoric times and the Early Middle Ages. It is uncertain whether any expansion into the northern part had already begun, but if it had, it can hardly have been more than occasional.

4.4.6 The Middle Ages and Modern Time until ca. 1670
Settlement, cultivation systems, and property distribution in the Romele area

Sten Skansjö

The oldest known historical settlement

Of the settlements in Villie parish, the church village of Södra Villie, along with the villages of Norra Villie, Trunnerup, and Ekarp are attested in written sources from the 15th century. However, this date is merely a rough *terminus ante quem* for the age of the villages, as is clear, for example, from the parish church in the heart of Södra Villie, the origin of which was a Romanesque building of the 12th century. Archaeological observations around the village site of Norra Villie also provide clear evidence of continuity on the spot since at least the Early Middle Ages (Ch. 4.4.5). Of the dispersed settlements, Rydsgård is attested in writing as an isolated farmstead since the latter part of the 15th century, and Beden as a double farm since the 1540s. These units can thus be counted among the settlements known since the Middle Ages, unlike the seven isolated farmsteads, which are first named in the period 1583–1628 and are probably not older. These appear to belong to a younger stage of colonization in the historical development of settlement (see below).

The pattern of settlement known since the Late Middle Ages reflects what is already known from the Late Iron Age and the Early Middle Ages: we have villages in the south of the parish and dispersed settlement in the north. The villages are of two different sizes. In the mid 17th century each of the villages of Villie consisted of 17–18 farms, clearly higher than the number in the average village, whether in Scania as a whole (10 farms per village) or in the Ystad area (13 farms per village). The two villages with names in *-torp* (originally denoting a secondary settlement or outlying farm), Trunnerup and Ekarp, with 10 and 8 farms respectively, were below the average of the Ystad area (Skansjö 1985). It is

not clear from the written sources whether these smaller settlements represent a younger stratum than Södra and Norra Villie. The question could only be answered by archaeological investigation.

According to a comparative calculation of variations in settlement density in the Ystad area around 1670, the parish of Villie (including enclaves of the parishes of Sövde and Slimminge), with 2.1 farms km^{-2}, was well under the average, which was 2.8 farms km^{-2} for the entire project area (Skansjö 1985). This is probably due to the fact that the northern part of Villie parish – the Romele area proper – was still sparsely populated as late as the end of the 17th century (Skansjö 1987b).

Signs of regression in the Late Middle Ages

In comparison with other parts of the Ystad area, the Romele area is clearly seen to be marginal. It was a particularly dynamic area in settlement development, at least from the Late Iron Age and through the rest of historical times. This is especially true of the northern parts of the parishes of Villie and Skårby, but also of the adjoining parts of the parishes of Katslösa and Sjörup. This area appears to have been exploited for settlement and cultivation in times of increased population pressure, only to be abandoned when population trends fell. This interpretation accords with the fact that we can observe here signs of a regression in settlement in the Late Middle Ages.

Written sources mention deserted farms in the villages of Södra Villie and Trunnerup in the period ca. 1450–1500. As far back as the end of the 14th century there is a reference to Rynge in the north of Sjörup parish as a deserted croft *(Rynde öde torp)*, but this farm was reoccupied before 1492. Nearby there were two other deserted medieval settlements: Kadesjö, which is described as unoccupied land *(Kadusse mark)*, from about 1450 at the latest to the start of the 16th century; and *Mattwse mark*, a permanently abandoned settlement, the location of which is unknown, but which probably lay near Kadesjö close to the Romele area.

Settlement expansion in the 16th and 17th centuries

The view of the Romele area as a marginal area colonized and abandoned according to population changes, is also a fruitful approach to the interpretation of changes in settlement in the 16th and 17th centuries. In Villie parish in 1570 there were 60 tithe-paying farms. From around 1650 the recorded figure is 67; this suggests an actual increase in settlement. As mentioned above, the isolated farmstead of Rydsgård is known since the 1470s and the double farm of Beden since the mid 16th century, while seven isolated farmsteads are first mentioned in the period 1583–1628: Frankhult, Gottorp, Grönhult, Klämman, Lindhult, Stora Lönhult, and Lilla Lönhult (Skansjö 1987a, b).

Everything suggests that these farms were newly established in this period, a phenomenon which we also know from the northern parts of Skårby and Sövestad parishes (Skansjö 1987b, 1988). The many place-names in -hult (wood) indicate that we are witnessing the colonization of what was previously continuous forest. As is the case in, for example, the Swedish provinces of Götaland, this can be seen as an effect of a general increase in population, known also from several parts of Europe (Skansjö 1987a).

To complement the observations from the written sources, trial excavations have been undertaken at a few farms in the Romele area (Billberg unpubl.). However, there was no evidence of archaeological material older than the 18th century.

Cultivation according to the oldest written sources

The more or less unoccupied lands which we must assume the northern part of the area to have been in the Iron Age were probably still a feature of the landscape in the High Middle Ages, perhaps chiefly in the form of wide areas of common pasture with areas of woodland. The supposed contraction in settlement in the Late Middle Ages was no doubt accompanied by a somewhat less intensive use of the Romele area proper.

Not until the latter half of the 16th century do we find written evidence (in the Lund diocesan terrier) of a three-field system in the church village of Södra Villie, but this cultivation system must have been used for centuries in the grain-growing villages of the area, particularly Södra and Norra Villie. Compared to the other vicarages in the Ystad area, the one in Södra Villie had especially good meadowland at its disposal, known as the out-meadows (Utängarna).

The dispersed settlement in the north was geared more to animal husbandry than the villages. This is evident from details of the annual rent which the farms had to pay to their owners. Whereas the village farms of Södra and Norra Villie chiefly paid their rent in grain (barley) in the Late Middle Ages, the smaller villages of Trunnerup and Ekarp and the isolated farmstead of Rydsgård paid in cash. The dispersed settlements in the north paid their rent in butter; this applied to the farms of Beden (1547), Stora and Lilla Lönhult (1626), and Lindhult (1626).

A comparison of tithe records from the various parishes in the Ystad area, according to the diocesan terrier from around 1570, shows that the total grain production (rye, barley, oats) per tithe-paying unit in Villie parish was about 20% below the average for the entire project area (Skansjö 1987b). The differences in the productivity of arable land were of course significant within the parish, on account of the disparities in fertility between north and south.

As we have seen, (Ch. 4.1.6.1, 4.2.6.1) from a comparison of the diocesan terrier of 1570 with the oldest land survey descriptions from the early 18th century, there was a noticeable tendency, especially in the vil-

lages on clayey soil in the outer hummocky zone, for the area of tilled land to increase at the expense of meadow. For Villie parish the corresponding increase was slight, but there were opportunities for expansion in the outlands beyond the established villages, where, as we have seen, a number of new settlements arose in the period 1583–1628 (Persson 1986a, b, Persson and Skansjö unpubl.).

Property distribution

Unlike the majority of the parishes in the Ystad area, Villie is not known to have had a nobleman's manor in the Middle Ages. In the Late Middle Ages, however, at least half of the church-village consisted of tenant farms owned by the nobility; the same was true of almost all of Norra Villie and Trunnerup. There were at the same time a number of freeholding peasants (four in Södra Villie, one in Ekarp) and church-owned farms in the parish.

Around 1670 some 59% of the farms in the parish belonged to estates outside the parish. It was not until Rydsgård manor was formed in the 1680s that virtually the whole of Villie parish came under the domination of an estate (see Ch. 4.4.7).

4.4.7 The 18th century
4.4.7.1 Cultivation, population, and social structure in the Romele area

Anders Persson

Introduction

The form and use of the cultural landscape in Villie parish can be described relatively well for the 18th century. There are early maps for the villages of Södra Villie (1698), Norra Villie (1701), Ekarp (1701), and Trunnerup (1750), and for the manor of Rydsgård (1721), all of which lay below the south-western slope of the Romeleåsen ridge. For the smaller settlement units up on the ridge the source material is not as full, and 18th century maps are available only for the double farm of Beden (1750) and the smaller settlement unit of Klämman (1750) (see Fig. 4.4:4).

The picture conveyed by the map descriptions can be supplemented by the Civil Survey assessments of 1670. We can thus obtain a good picture of the conditions for cultivation in the parish around 1700. Moreover, in the period 1800–1815 the four villages in the parish were mapped for the *enskifte* enclosures. Maps were also made around 1720 and 1790 of the manor of Rydsgård, and the cartographic information can be juxtaposed with tax assessment data. In addition, there is relatively good extant demographic information.

The major processes in the parish in this period con-

THE ROMELE AREA

18TH CENTURY LANDSCAPE

Fig. 4.4:4

LANDSCAPE ECOLOGY & SETTLEMENT: A-C. LINUSSON   ED. & GRAPHICS: M. RIDDERSPORRE   LANDSCAPE COLOURING: A-C. LINUSSON

0   1   2 km

1:50,000

ARABLE AND FALLOW LAND	WOOD-PASTURE	VILLAGE, HAMLET	+ CHURCH
DRY AND MESIC MEADOW / WITH COPPICE	UNPRODUCTIVE LAND	MANOR	HUNDRED BOUNDARY
WET MEADOW / WITH COPPICE	NO INFORMATION	SINGLE FARM	PARISH BOUNDARY
FEN, PEATLAND	LAKE, POND	SMALL HOUSE, COTTAGE	VILLAGE AND SINGLE-FARM BOUNDARY
PASTURE / WITH SCATTERED TREES AND SHRUBS	STREAM	WATERMILL	FIELD BOUNDARY

Map labels: BEDEN, RYDSGÅRD, EKARP, NORRA VILLIE, TRUNNERUP, KADESJÖ, GUSSNAVA, SÖDRA VILLIE, RYNGE, VARMLÖSA, NORRA VALLÖSA

cern the manor of Rydsgård, established in the 1680s. The estate undertook a massive cultivation of the outlands at the edge of the woods on the south-western slope of Romeleåsen ridge. There was also a considerable expansion in the arable land farmed by the villages, as is reflected in the increase in population and settlement. Around 1800 tillage was radically transformed by the enclosures in the villages in Villie parish, a process that occurred at roughly the same time here as elsewhere. The next major break came in the mid 19th century, when the village of Norra Villie ceased to exist; the manor replaced it with a large-scale farm named Villiegården.

Settlement

The villages of Södra and Norra Villie had the same number of farms (20), and were thus twice the size of Trunnerup and Ekarp, which had 10 farms each. The villages of Ekarp, Norra Villie, and Trunnerup lay in a line from west to east, to the north of Södra Villie. This village landscape in the south of the parish contained 56 of the parish's 66 farms, or 85% of the total. In the upland northern parts of the parish lay the remaining 15%: the double farm of Beden and 8 isolated farmsteads. These were situated in a wooded landscape, which is clear from the 1671 Civil Survey for the parish, as well as the cartographic evidence from around 1700. Ekarp, Norra Villie, and Trunnerup were thus "on the edge of the woods", an expression that was also used of Rydsgård manor when it was mapped in 1721. The northern parts of the villages of Ekarp and Norra Villie are shown on the 1701 maps as pasture land consisting of "reasonably good grassland with many tussocks and some slopes with a little scattered juniper". On the ridge to the west and east there were also extensive areas of common land, which were not divided among the surrounding landowners – a number of manors – until 1760.

In the parish there were also smaller units (cottages) with little or no farming land. Ten of the twenty units were in Södra Villie village. There were two cottages on the village street in Norra Villie, while Ekarp and Trunnerup each had one in the heart of the village. Just outside the village of Norra Villie lay the cottage of Foglahus, and the cottages at Stråbyhusen were outside the infields of Trunnerup village. Finally, squeezed in between Lilla Lönhult and Stora Lönhult in the northwest corner of the parish was the cottage of Klämman. The proportion of cottages was between 25% and 30% of the total settlement. In order to survive, the cotters must have pursued ancillary occupations or worked for others.

This settlement structure changed during the 18th century. The number of farms increased slightly, from about 65 in 1700 to roughly 70 in 1801, while the number of cottages more than tripled, from 25 in 1700 to 84 in 1801. The proportion of cottages thus increased in the course of the century from 25–30% to around 55%. The increase was mostly made up of new settlement in crofts (about 35 *torp* units) established outside the villages.

Population

The population of the parish can be estimated at around 440 (index 100) in 1700. By the mid 18th century this had risen to around 725, reaching a figure of 1075 at the end of the century (index 100–165–245). These figures reflect the population trend not counting the people on the manor of Rydsgård. Population density in 1700 was roughly 15 people km^{-2}.

The closest equivalents to this figure are to be found in the parishes of Stora Köpinge and Sövestad, with a population density in 1700 of about 14 km^{-2} and 15 km^{-2} respectively. In the parishes of Bjäresjö and Västra Nöbbelöv, by contrast, population density in 1700 was 23 km^{-2} and 24 km^{-2} respectively (Persson 1988b). The lower figures for Sövestad and Villie can in part be explained by the greater proportion of outlands here at the beginning of the 18th century. Villie parish must therefore be considered a relatively sparsely populated parish at this time, despite its large villages. In 1750 the population density had risen to about 25 km^{-2}, and by the end of the century it was 37 km^{-2}. The population increased rapidly in the last quarter of the 18th century, and especially around the turn of the century (1790–1805).

The form and use of the cultural landscape

Rydsgård manor was established in the early 1680s. Within six or seven years it had acquired some 35 farms in the villages in the parish (as well as three farms in Varmlösa village), making about 65% of the village farms in the parish, or about 55% of the total number of farms in the parish. Throughout the 18th century the estate increased its demesnes, and the number of units it owned rose from around 40 in 1700 to roughly 115 in 1800. In the age of corvée this was one of the conditions without which it would not have been possible to carry out the breaking of new land that went on continually during the century. Smaller parts of the lands belonging to Norra Villie and Ekarp were also incorporated into the demesne. The manor's arable lands thus increased, mostly through the breaking of virgin soil, from about 50 ha (index 100) in 1721 to about 220 ha in 1788 (index 100–440). At this time grain cultivation on the manor lands was intensified, the arable being divided for this purpose into eight fields. Just before the end of the century the manor began to grow clover, with around 30 ha being planted with this crop (see Ch. 5.9).

The cultural landscape outside the manor. Details of cultivation systems are to be found in the Civil Survey, dated 6 July 1671. A three-course system dominated in the villages, whereas we learn that "the crofts in the

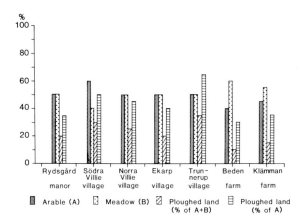

Fig. 4.4:5. Land use in the Romele area in the late 17th and the early 18th centuries.

parish do not have three fields, but some two, some one, which are tilled each year". It is also noted that Ekarp village "has in addition some special grazing for its beasts with Norra Villie village". Trunnerup had, besides its three normal fields, "a common for grazing, and some thorn copses for their young horses". The farms to the north, which the source describes as crofts, had "some woods", and it was noted that "there is no more woodland by these villages, nor any turbary, so the men of the parish must look elsewhere for timber, firewood, and fencing". It may be noted that Varmlösa village, which is actually in Katslösa parish, was a three-course village with a meadow field, and thus had four fields.

Details in the Civil Survey of 1671 can also be used to provide information about the supply of hay in relation to the amount of manured land, that is, the rye and barley fields. The farms in the four villages in the southern part of the parish had on average 1.5–2.5 loads of hay per barrel of rye or barley seed. Up on the ridge, where settlement was dispersed in single or double farms, the volume of hay was between 2.5 and 8.5 loads per barrel of seed corn (Persson 1986a). In Stora Köpinge parish, with its sandier soils, the corresponding figures were much lower (Ch. 4.1.7).

The village farms in Villie parish were, with the exception of the farms in Trunnerup, of different sizes, and usually smaller than the farms in the other three focal parishes. Most farms in 1700 had arable and meadow land covering an area of 10–30 ha. The two farms at Beden were larger, each having about 40 ha of arable and meadow.

As Fig. 4.4:5 shows, arable land made up some 50–60% of the mapped area of the villages. Each year 20–35% of the land was ploughed, which means that 65–80% of the village lands consisted of fallow or meadow. Of the arable land in the villages, an estimated 40–65% was sown each year. It should be stressed that the latter figure of 65% represents a late value from the mid 18th century. The amount of arable sown annually

in relation to the total area of the villages was roughly the same in the villages in Stora Köpinge parish, which had sandier soils. There, however, the amount of ploughed arable was much smaller in relation to the total arable, comprising only about 35–40%. The intensity of cultivation in the villages of Stora Köpinge parish was thus more similar to the northern part of Villie parish, where the arable at Beden in 1750 was 40–45% and the meadow 55–60%, and only 10–15% of this area was ploughed annually. In this part of the parish stock-raising was a more dominant feature than in the farms below the ridge (Persson 1986a). Despite an extensive system of cultivation, however, the yield from seed corn at Beden was almost the same as in the village of Bjäresjö in the parish of the same name (see Olsson and Persson 1987, Ch. 4.4.7.2).

Changes in the area of arable over the period 1700–1800 have been studied for three villages. In Norra Villie the area increased by about 50% and in Ekarp by about 40%. The expansion in Trunnerup was some 40% during the period 1750–1800. By the start of the 19th century, the proportion of arable land in these villages was 70–80%, leaving 20–30% as meadow. Within the three-course system there had thus been a considerable increase in productive lands in the course of the 18th century.

Conclusion

The cultural landscape in Villie parish has been greatly affected by the presence of Rydsgård manor. The manor's own cultivated land was limited in the early 18th century, but a dramatic change took place thereafter, with an almost fivefold increase in the total arable land owned by the manor. From having been scarcely bigger than a couple of farms, the manor grew to a point where its arable lands were as large as the entire village of Norra Villie (arable and meadow together) with its 18 farms. At the same time as this process was in progress, the manor was buying more and more tenant farms and lands near the manor. The arable farmed by the villages nevertheless increased considerably, by an estimated 50%. Judging by changes in population and details about the acreage of arable land in Trunnerup (1750–1800), this process must be dated mostly to the second half of the 18th century. The above account shows that the peasant population increased by about 250% while the amount of arable land in the villages was expanded by some 50%. It is improbable – not to say impossible – that the lands of the isolated farmsteads and the new cottages could have brought new arable land amounting to 200% of the arable acreage of the villages in 1700. Moreover, productivity remained largely unchanged in the investigated villages in the 18th century. An analysis has been made of the evidence from the 1810s for the isolated farmsteads in wooded areas in the north of the parish, but nothing has been found to suggest that there was any radical expansion of arable land there. It thus seems clear that the relation between productive arable land and population in Villie parish changed in such an unfavourable way that this agrarian community acquired a new population group of crofters and day-labourers. At the start of the 19th century these constituted a new economic lower class, below the peasant households.

4.4.7.2 The agro-ecosystem of isolated farmsteads and small villages on the Romeleåsen ridge in the 18th century

E. Gunilla A. Olsson

Landscape and vegetation on the common

The first detailed description of agricultural organization and land use from the Romele area is from 1750; it concerns the two farms in Beden (LM Villie 5:1750, Fig. 4.4:5) and 12 small isolated settlements a few kilometres eastwards (LM Villie 2:1750, Skårby 2:1750, Fig. 4.4:4).

In the first half of the 18th century this area was the south-easterly outpost of the large, formerly continuous area of common outland extending from Malmö all the way to Ystad, a distance of over 80 km. The landscape on the Romeleåsen ridge still bore the mark of the long-term hard grazing pressure. All the surrounding villages had used this area for stock grazing and also for collecting material for fuel and fencing. The area still functioned as the main communication route, crossed by roads and riding tracks (cf. Ch. 3.6.2). Woodlands did not exist here; the only woody vegetation to be found was scattered shrubs of juniper (LM Villie 5:1750). Linnaeus, travelling here in 1749, wrote that blackthorn shrubs everywhere were rare since the farmers collected them to reinforce their fences (Linnaeus 1751). Stone walls were also common here (Persson 1986a), as were fences of coppice twigs of alder, and sometimes also soil dykes were built (LM Skårby 2:1750). The small settlements were all, with the exception of Beden, isolated farmsteads, crofters' holdings, and the land was owned by the surrounding estates. These crofts had been established in the outland fairly recently, most of them after 1550 (Skansjö 1987). By the mid 18th century they formed a more or less continuous belt from Veberöd in the north-west down to the village of Gussnava in the south-east. The crofts were framed by a zone of common pasture constituting a border between them and the villages further down the ridge (see Fig. 3.7:1).

The enclosed infields of the crofts appeared as green and luxuriant islands of varied vegetation. They contained treeless meadows of varying moisture and fertility, as well as coppice meadows with alder, oak, and hazel and meadows – or rather preserved wood lots – with standards of predominantly beech and some oak.

The long-term fallows had their characteristic vegetation of weeds and grasses.

Ecological aspects of land use and cropping systems

The cropping system used in Beden and in the other settlements was a sort of infield-outfield system (cf. Slicher van Bath 1963, Baker and Butlin 1973, Dahl 1978), a field-grass system in Frandsen's sense (Frandsen 1983). The arable fields were permanently enclosed and often, but not always, the meadows were within the same fence. Meadows were harvested annually. Although several separate field plots existed, a regular rotation of cultivation between different fields was very rare. Long fallow periods were used and it was common for every field plot to lie fallow for six years or longer. The long fallow periods meant that only a small part of the arable was cropped, 28–35% for Beden and Klämman in 1750 (Ch. 4.4.7.1, Fig. 4.4:6). The crops grown were rye, barley, oats, and buckwheat, depending on the soil fertility. On the harsh sandy soils, as at Rigårdstorpet, rye and buckwheat were important, but where soils with some humus existed both barley and rye were cultivated (LM Villie 2:1750, Skårby 2:1750). Oats were also cropped and useful since, like buckwheat, they can be cultivated without manure. For all of these tiny crofts there were complaints about the low availability of manure, since the limited meadow area permitted only a small number of stock. The fields of rye and barley were not normally manured more than once every 6–8 years. In Beden, with its large meadows, the manure volume

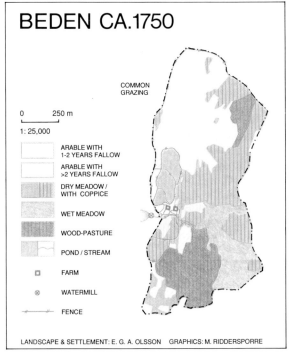

Fig. 4.4:6. Beden. Landscape and land use ca. 1750.

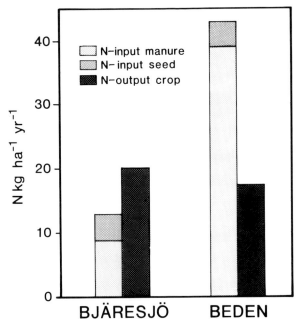

Fig. 4.4:7. Nitrogen budgets for the arable systems in Bjäresjö village 1699 and Beden 1750. Sources: Olsson 1988a, b.

Table 4.4:1. a. Number of livestock per farm in the villages of Beden and Bjäresjö, ca. 1700. The figures are means from two different recordings: 1671 and 1768 (Persson unpubl.) b. Number of farms and number of inhabitants (calculated as adult units) in the same villages.

a. Livestock	Bjäresjö	Beden
horses	2.9	9.4
cattle	3.3	16.5
sheep	4.7	13.8
pigs	2.8	11.5
hens	2.6	3.5

b.	Bjäresjö	Beden
number of farms	20	2
number of inhabitants	75	10

was sufficient, but since long fallow periods were practised here too, much of the arable had the same manure regime as in the crofts. Crop yield in Beden was on average 840 kg grain ha^{-1}, meaning roughly four times the seed input (Persson 1986b, Olsson and Persson 1987). This is similar to Bjäresjö, situated on more fertile soils (Ch. 4.2.7.3). Several of the smallholdings with only very harsh sandy soil sometimes could not even harvest the seed input (LM Skårby 2:1750).

Meadow yield was significantly lower in the Romele area than on the plains (cf. Ch. 5.3). In Beden more than 90% of the meadow area yielded half a load per *tunnland* or still lower, meaning less than 270 kg hay ha^{-1}. This should be compared with meadows in Bjäresjö, where the mean yield was almost double, 500 kg ha^{-1}. However, within the Romele area, Beden had relatively fertile but also larger meadows than other settlements here. Meadow yield as low as 35 kg hay ha^{-1} was reported from several holdings around 1750 (LM Skårby 2:1750). The land surveyor claimed that the meadows were so meagre that there was nothing for the scythe to cut. Moreover, in most of the settlements the meadow grass vegetation was deteriorated by the existence of standard trees of mainly beech, but also oak. These trees and the coppices species were preserved and used only by the landlords. The crofters could sometimes be permitted to use some coppice for fuel and fencing. When the coppice belonged to the farmers, as in Beden, it could be managed to provide fuel and timber, but at the same time it allowed good hay harvests on the ground. Meadow area in Beden was 60% of the infields (cf. Ch. 4.4.7.1). Slightly lower proportions

were valid for the crofts as well. However, since most of these meadows were more or less useless as fodder areas and functioned solely as timber preserves for the landlords, they provided no resources for the farmers.

Stock size in Beden was huge compared with the village of Bjäresjö. In the middle of the 18th century the stock size on each of the two Beden farms is estimated to have been 9 horses, 17 cows, 12 pigs, and 14 sheep (Olsson 1988, Persson unpubl.). This is 4–5 times the number in Bjäresjö, although the size of farms was roughly equal, 35–40 ha (Table 4.4:1). The crofters kept larger numbers of stock than the farms on the plain, although they had small infields. The availability of common pasturage in the outland made this possible.

Circulation of nitrogen within the Beden agro-ecosystem

Focusing on one of the main plant nutrients, nitrogen, and comparing inputs in seed and manure with output contained in the harvested biomass, the crops, it is possible to interpret the balance of this system (cf. Olsson 1988, 1989). Taking into account small cattle size, manure management in the stable and in the arable, we still find a large amount of nitrogen annually added and an amazingly low output in the harvested biomass (Fig. 4.4:7). The relatively large surplus of nitrogen, approx. 25 kg ha^{-1} yr^{-1} (Olsson 1988) could not be consumed by the crops and used for growth improvements. It must have disappeared from the soil system or been made unavailable, for example, by incorporation into soil organic compounds. Some of the surplus nitrogen was leached away from the sandy soils (cf. Gustafson 1982, Brink 1983). Another factor was the use of long fallow periods, 6–8 years. Manure application in the last fallow year was followed by two or three years of cultivation. This probably resulted in inefficient use of added nutrients, since manure has a relatively slow decomposition rate. Some of the nitrogen was leached during the first year of its application on the fallow, and still more of the mineralized nitrogen was utilized by weeds in the following fallow cycle (see Olsson 1988 for further discussion and references).

Ecological aspects of resource use in the Beden system

For the Beden system, the greater availability of manure was not reflected in larger yields, probably because of unfavourable soil conditions. To produce the same amount of crop per cultivated area as in Bjäresjö required twice the field area, taking the extensive fallows into account, and almost four times as much manure (Olsson 1988). There were probably some possibilities of increasing the yield by adopting a more intensive cropping system and at the same time devoting a greater proportion of the annual production to winter crops, which could act as catch crops for the mobile nitrogen in the manure (cf. Gustafson 1985). The environmental resources in the form of meadows and common grazing might have permitted a certain increase in livestock, at least in combination with grazing the fallows. On the other hand, the poor soils at Beden – with low clay content and low pH – had small reserves of organic nitrogen and phosphorus and low capacity to release these nutrients by mineralization. A shortage of phosphorus might have limited the crop production most.

The adoption of a more intensive cropping system required more human labour; this was probably a limiting resource in Beden around 1750. At this time the two farms, totalling 80 ha, had a population of 15 persons, of whom only 7 were adults (Olsson and Persson 1987). Moreover, the Romele area suffered frequent crop failures in the late 17th and early 18th centuries (Persson 1986b). The margins for coping with fluctuations in crop yields were small and set limits to population growth.

4.4.8 The 19th and 20th centuries
Landscape and landholding changes in the Romeleåsen ridge area since 1800

Tomas Germundsson and E. Gunilla A. Olsson

Intensified use of environmental resources in the Romele area during the 19th century

The large outland on the Romeleåsen ridge was divided and enclosed in 1759 and thereby the history of this area as a common resource ceased. In the early 19th century extensive colonization by new settlements occurred here, as shown on the map from 1815 (Fig 4.4:8). In the near vicinity of the Beden farms, six new crofters' settlements had been established since the mid 18th century. A similar increase of settlements occured along the whole of the ridge.

Within the sections of the Romeleåsen ridge belonging to Villie parish the continuous deforestation was complete around 1815 (Fig. 4.4:8). The former wood-pastures of various ecological composition were transformed into arable, into treeless wetlands used for hay-making, or into open pastures. On the other hand, adjacent parts of Romeleåsen in Skårby parish contained considerable quantities of woodlands as well as forested meadows. These differences were related to different landholding conditions, where Beden and surrounding crofter's holdings were owned by the Rydsgård estate (cf. Ch. 4.4.7.1) while the land of the Skårby settlements belonged to other estates with a more liberal policy of land sale and crofter's colonization.

Beden, however, was an exception to the general trend for woodland development during the 19th century. Map records from 1844 show that the infields still contained considerable numbers of young trees. The former coppice meadows had been converted into wood-pastures dominated by young beeches as a response to the loss of formerly primary grazing grounds on the common west of the infields of Beden (cf. Fig. 4.4:8). Also most other meadow areas were reduced by the expansion of arable. Only the wet meadows remained as hay-producing areas.

Settlement pattern

The settlement pattern in the north-western area in the early 19th century was quite different from the rest of the project area. In general the settlements in the Ystad area formed a distinct pattern of nucleated villages, but here in the north-west, single or double farms founded during the late medieval colonization formed a dispersed pattern. These differences are clearly visible in the map in Fig. 4.4:8. The scattered farms can be found in the northern parts of Villie parish and also on the north-western slope of the Romeleåsen ridge. In the eastern part of Villie parish there are also scattered farms, but this is due to the enclosure of Trunnerup village in 1805 – not a result of medieval colonization.

Up to around 1860 there was not much change in the settlement pattern in the northern hilly part of Villie parish; the pattern of single and double farms persisted. In the southern part, however, the village agglomerations were dissolved by *enskifte* enclosures, just as in most of the villages in the plains. Trunnerup was enclosed in 1805, as mentioned above, and the villages of Ekarp, Södra Villie, and Norra Villie were enclosed during the following decades. In Ekarp and Södra Villie the farms were dispersed. In Norra Villie, however, the peasant farms were torn down and replaced by two large tenant farms (*plattgård*, cf. Ch. 5.10), Villiegården and Svenshög. The farms in Norra Villie had previously been family tenant farms, owned by Rydsgård estate. The decision to replace a number of old village farms with a large modern tenant farm was not unique; it was one way for big landowners to rationalize agriculture during the late 19th century (Möller 1985). After the enclosure a large number of farms were divided, and many smallholdings were created.

Up to the 1870s most land in the north-west region was owned by the landed estates. Rydsgård in Villie owned about three-quarters of the land in that parish, while the eastern parts of the Romele area were owned by other estates (Möller 1983). After 1870 the landed estates systematically began to sell off land, and it is interesting to note that, generally, it was on land not owned by estates that the most significant concentration of new settlements occurred at the end of the 19th century. One example is the north-eastern part of Villie parish, where many farms were partitioned on land sold by the estates. This area makes a contrast to the western part of Villie parish, still owned by Rydsgård estate, where very few settlements were allowed during the 19th century (Fig. 4.4:9).

The changes in settlement pattern reflect population growth, and also the new order of labour organization that evolved in the late 19th century. During this period the social position of the labourers on the estates changed from that of crofters, obliged to do corvée on the estates, to a workforce mostly consisting of *statare,* that is, families of landless labourers, living in special houses near the estate (cf. Ch. 5.9). In high seasons this workforce was complemented by day-labourers, mostly living in smallholdings in the countryside. This proletarization of crofters and smallholders is exemplified in the crofter settlement of Rigårdstorp on the eastern part of the Romeleåsen ridge. Here almost all land was laid out for conifer plantation by the landowner, and the crofter was left with a minute plot suitable only for growing potatoes. This forced the family to work elsewhere to survive.

Agricultural changes during the last half of the 19th century

Parallel to this proletarization process, some of the crofts were reorganized and sometimes slightly enlarged in conjunction with the enclosures. Arable land increased continuously at the expense of dry meadows. Also wet meadows and alder coppices were reduced by peat cutting, especially in the eastern parts of the Romele area. As a result of this activity numerous ponds were created. Some of these were drained, however, and left for succession to grasslands to gain back some of the lost fodder area. The increasing shortage of fodder was also reflected by the introduction and use of water meadows, as at Bleckstorp. Management of water meadows was practised all over the Ystad area during the last half of the 19th century as a way of increasing hay biomass production. By intricate systems of successively shallower channels, water from streams or brooks was led out over the meadow areas. Plant nutrients dissolved in the water, or fixed to sedimented clay and soil particles, were used by the meadow plants and thereby increased biomass production. Also the dis-

solved oxygen in the mobile water was beneficial for the plants by promoting microbial activity and by slightly raising soil temperature (Rackham 1986). By 1886, 36% of the meadow area was maintained as water meadows in Villie parish (Olsson and Berlin unpubl.). During the period 1886–1909 natural meadows were reduced by 72% – from 70 to 20 ha – and successively replaced by leys on arable. By 1909 only 2% of the land in Villie was natural grassland; 81% was arable, and 10% consisted of woodlands (Anonymous 1909). During the 19th century most of remaining deciduous woodlands disappeared in the Romele area, but planted conifers were introduced at the end of the century.

Regular two- and three-field rotation systems were never in use in the Romele area, largely because of the settlement patterns with predominantly isolated farmsteads. Various types of field-grass systems (Frandsen 1983) were practised up to the late 18th century, using the common as a large outfield area. This difference from other parts of the Ystad area was clearly reflected both before and after the land enclosures. Land enclosures did not have such dramatic effects on land use or housing location as in districts with village organization.

The general agricultural changes in the Ystad region during the late 19th century – including the introduction of long-term crop rotations without fallow, new crops, and a greater consciousness of the influence of soil nutrient conditions – were valid also for the Romele area. Using the parishes as units for studying agricultural trends in the Ystad region leads to some confusion, especially in Villie parish, since the ecological distinctiveness of the Romeleåsen ridge is obscured. Half of Villie parish is situated on lime-rich clays and similar in ecological characteristics to Bjäresjö while the other, northern part, is dominated by sandy till soils on the slopes of the Romeleåsen ridge. This means that differences in agricultural practices reflected in land use, crops, and stock number in reality were greater than can be shown in summary diagrams for Villie parish (Fig. 4.4:10–12).

By the mid 19th century oats were the dominating crop in Villie, cultivated on 32% of the arable (Germundsson and Persson 1987). This large proportion, compared with other parishes, is probably related to the relatively poor soils in the Romele area, but also to the fodder need of a large stock. Wheat was hardly cultivated at all. Pulses were cropped on 13% of the arable, and in fact it never again became so important here as in the 1850s. Potatoes were cropped on 8% of the arable but never became an important crop in this area (Fig. 4.4:11).

Rural development during the first half of the 20th century

There are considerable differences in agrarian development between the marginal area ("marginal" in a Scanian perspective) on the slope of the Romeleåsen

THE ROMELE AREA

EARLY 19TH CENTURY LANDSCAPE, CA. 1815

Fig. 4.4:8

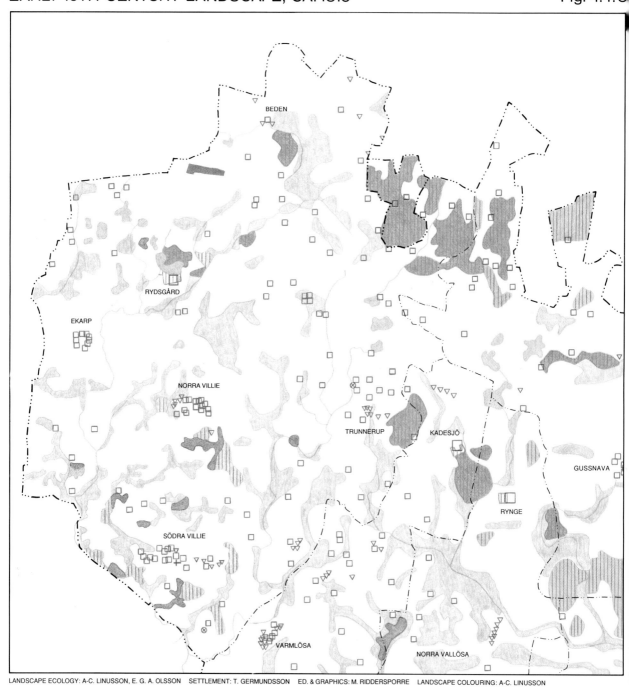

LANDSCAPE ECOLOGY: A-C. LINUSSON, E. G. A. OLSSON   SETTLEMENT: T. GERMUNDSSON   ED. & GRAPHICS: M. RIDDERSPORRE   LANDSCAPE COLOURING: A-C. LINUSSON

0   1   2 km

1:50,000

OPEN AREAS: ARABLE, FALLOW, DRY AND MESIC MEADOW, PASTURE	WOOD-PASTURE	
DRY AND MESIC MEADOW WITH COPPICE	CONIFER PLANTATION	
WET MEADOW / WITH COPPICE	LAKE, POND	MANOR
OPEN FEN	STREAM	FARM
PASTURE WITH SCATTERED TREES AND SHRUBS	PARK	SMALL HOUSE, COTTAGE
		WATERMILL

CHURCH

HUNDRED BOUNDARY

PARISH BOUNDARY (1864)

Place names on map: BEDEN, RYDSGÅRD, EKARP, NORRA VILLIE, TRUNNERUP, KADESJÖ, GUSSNAVA, RYNGE, SÖDRA VILLIE, VARMLÖSA, NORRA VALLÖSA

LANDSCAPE ECOLOGY: G. BERLIN, A-C. LINUSSON, E. G. A. OLSSON   SETTLEMENT: T. GERMUNDSSON   ED. & GRAPHICS: M. RIDDERSPORRE   LANDSCAPE COLOURING: A-C. LINUSSON

0   1   2 km

1:50,000

ARABLE		DECIDUOUS WOODLAND	MANOR		RAILWAY STATION
DRY AND MESIC MEADOW		CONIFER PLANTATION	FARM	+	CHURCH
WET MEADOW		LAKE, POND	HOUSE, COTTAGE		HUNDRED BOUNDARY
OPEN FEN		STREAM	WATERMILL		PARISH BOUNDARY
PASTURE		PARK	WINDMILL		BOUNDARY OF VILLAGE, ESTATE, AND FARM OUTSIDE VILLAGE

BEDEN

EKARP

RYDSGÅRD

NORRA VILLIE

TRUNNERUP

KADESJÖ

GUSSNAVA

RYNGE

SÖDRA VILLIE

VARMLÖSA

NORRA VALLÖSA

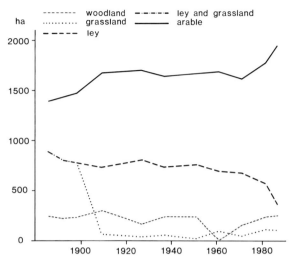

VILLIE

Fig. 4.4:10. Land use in Villie parish during the period 1886–1986: arable, grasslands, and woodlands. Source: Agricultural statistics, unpubl. (1886–1909), and publ. (1909–1986) – see Ch. 2.2.

larger units took place, but in the 1950s this area was still characterized by small-scale farming and many small farms. For this reason, but also because of a more varied topography, the adjustment to modern large-scale agricultural machines has not been on the same scale as on the plains. A number of walls, ditches, rows of trees, and other landscape elements have survived to the present day, contributing to the diversity of the landscape.

Arable reached its maximum in Villie parish in the 1880s and has kept at a stable level throughout the 20th century (Fig. 4.4:10). Among grain crops barley shows a characteristic development, encountered also in other Ystad parishes, with a marked decrease from the turn of the century continuing until 1950 (Fig. 4.4:11). This was balanced by increases in bread wheat and mixed grain

ridge and the high-producing landscape of the plains, as for instance in Bjäresjö.

Comparing the map from 1915 (Fig. 4.4:9) with the one 100 years earlier (Fig. 4.4:8) the most obvious landscape change is the revival of forest. Conifer plantations, mainly spruce, have appeared on the former commons. The map also indicates spontaneous forest successions with deciduous trees (birches) on the commons. This development had several causes. First, there was the general need for wood and fuel. Second, the former commons now were enclosed and divided among many landowners. The use of these land areas for arable was not always successful because of the rather poor soil quality. Third, introduction of new crop rotations with ley cultivation meant that the need for natural grazing grounds decreased. Fourth, the number of sheep decreased from 1880 onwards (Fig. 4.4:12). Sheep were the most effective users of the grazing and browsing areas on the former commons.

Returning to Beden in 1915 (Fig. 4.4:9) wood-pastures with beeches on the former meadows still existed, but conifers had been planted in the northern part. Arable had expanded, especially south of the settlement. The wet meadows had been drained, and mowing was abandoned, replaced by grazing.

The distribution of land on numerous small farms during the late 18th century is characteristic of the ridge area, but different from the other parts of the estate-dominated Ystad region. For example, in 1886 more than 90% of the farms in Villie parish were 20 ha or less in area (Fig. 4.4:13). These small landholdings were often unable to offer the full capacity for agricultural support for their dwellers. Reorganization of land into

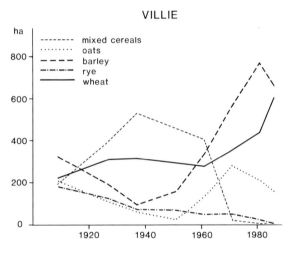

VILLIE

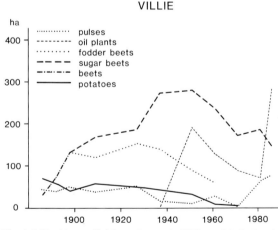

VILLIE

Fig. 4.4:11. Above. Cultivated crops in Villie parish during the period 1909–1986: grain crops. Source: Agricultural statistics, unpubl. (1886–1909), and publ. (1909–1986) – see Ch. 2.2. – Below. Cultivated crops in Villie parish during the period 1886–1986: oil plants, pulses, and root crops. Source: Agricultural statistics, unpubl. (1886–1909), and publ. (1909–1986) – see Ch. 2.2.

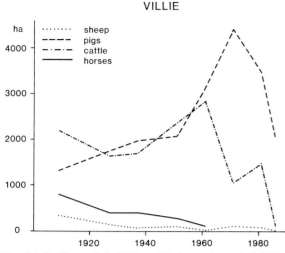

VILLIE

Fig. 4.4:12. Livestock in Villie parish during the period 1909–1986. Source: Agricultural statistics, unpubl. (1886–1909), and publ. (1909–1986) – see Ch. 2.2.

used for stock fodder. After the mid 19th century barley increased again. Rye has never been important in Villie parish after 1910, when it was cultivated on about 20% of the arable (Fig. 4.4:11). Mangel was introduced in the late 19th century and sugar-beet became important from 1910 onwards (Fig. 4.4:11). Villie parish had the largest area devoted to ley cultivation in the Ystad region in the 20th century (Fig. 4.4:11) and the decline occurred later here than in other parishes in the region.

Linked to the extensive ley cultivation has been a relatively large number of stock (Fig. 4.4:12). The number of cattle was almost the same from 1880 to 1955. Pigs exceeded in number the cattle already around 1910, as in most other parishes in this region. However, the emphasis on stock-raising – which has been important here ever since farms were established in the 16th century – was thus maintained during the first half of the 20th century.

Agricultural development after 1945

Changes in the size of agricultural holdings on the ridge since 1915 are shown in Fig. 4.4:14. As a result of amalgamations, many farmsteads have ceased to function as cultivation centres. These farmsteads are today inhabited mostly by people not working in agriculture, by elderly persons or by summer visitors only. It is also clear that very little land is cultivated under tenancy in 1986 (Fig. 4.4:14).Rationalization of the size of farms has occurred, primarily by the purchase of farmland, not by tenancy agreements, something that contrasts with the general Swedish trend. The consequences for the farmers have in many cases been that agriculture is insufficient to provide a living. It is therefore very common that farming is combined with other occupations for all or some of the family members, thus sometimes

turning the small-scale agriculture into more or less a hobby (Erlingsson 1988).

The direction of agrarian production after 1945 agrees with the general tendency for this region. Sugar-beet and the newly introduced oil plants decreased from 1945 onwards. From 1980 the latter experienced a dramatic increase in this parish (Fig. 4.4:11). Ley cultivation did not decrease significantly until after 1980, probably in conjunction with the abandonment of stock-raising also in the smaller units on the Romeleåsen ridge. Barley rose as a share of the arable from 1950, but decreased again from 1970 as a result of the governmental restrictions for Swedish agriculture in the 1960s (cf. Ch. 5.12). Generally all crops have decreased in area after 1970 except wheat and oil plants, and the total arable area has also decreased, but by 1986 it was back to the level of the first half of the 20th century (Fig. 4.4:11).

Changes in agricultural direction after 1960 have implied low land-use intensity either by cattle or racehorse raising or by the abandonment of agricultural land. In 1986 this parish had less than 100 cattle, in fact even less than both Köpinge and Bjäresjö parishes, although they are dominated by arable (Fig. 4.4:12 and Figs 4.1:19 and 4.2:19). The history of Villie parish as part of the Romele common and constituting an important ranging area for stock during more than 700 years (Ch. 3.6.2) has come to an end. Along with the abandoned farming and reduced grazing pressure, arable and grazing grounds are left to spontaneous forest succession. Consequently it is here possible to find considerable areas of semi-natural vegetation, woodlands, half-open pastures, wetlands, and several types of wooded fens.

Returning to Beden around 1945 we find that the two farms here reflect the general tendency. Increased ley cultivation on arable land means that the need for grazing has almost ceased. Hence, the wet grasslands, central in this area, are now partly abandoned, and succes-

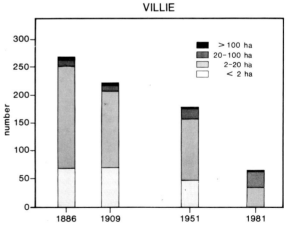

VILLIE

Fig. 4.4:13. Number of cultivation units in Villie parish 1888–1981, in different size classes. For the year 1981 only units > 2 ha were recorded. Source: Agricultural statistics, unpubl. (1888–1909), and publ. (1909–1981) – see Ch. 2.2.

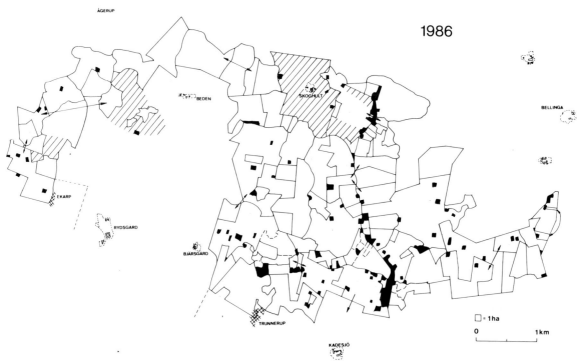

Fig. 4.4:14. Landholding and cultivation units in the northern part of Villie parish 1914 and 1986. Although many amalgamations have occurred during this period, this area on the Romeleåsen ridge east of Beden is still characterized by its small-scale farming. Unfilled areas are units cultivated by their owners. Hatched areas are units cultivated under tenancy. Black areas are non-agricultural units. Sources: economic map 1914, economic map 1982, interviews.

THE ROMELE AREA

LATE 20TH CENTURY SETTLEMENT, CA. 1985

Fig. 4.4:15

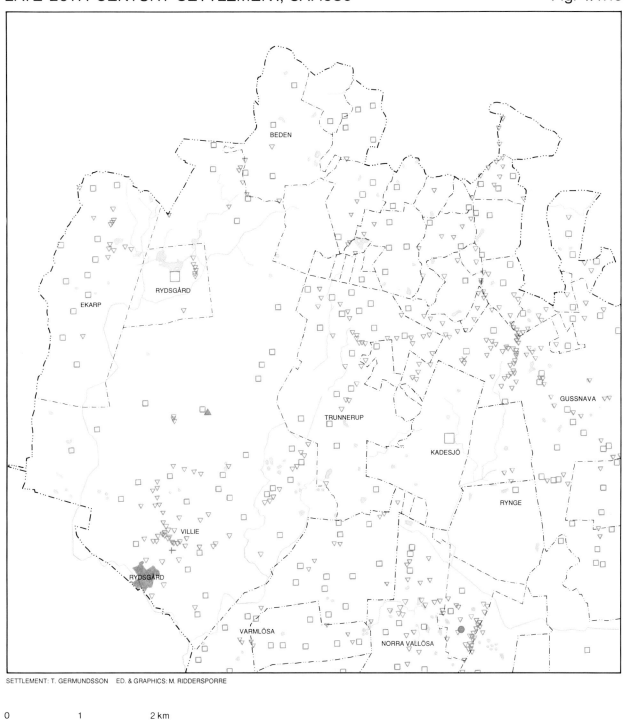

BEDEN

EKARP

RYDSGÅRD

RYDSGÅRD

TRUNNERUP

GUSSNAVA

KADESJÖ

RYNGE

VILLIE

RYDSGÅRD

VARMLÖSA

NORRA VALLÖSA

SETTLEMENT: T. GERMUNDSSON   ED. & GRAPHICS: M. RIDDERSPORRE

0   1   2 km

1:50,000

☐ MANOR	▲ RURAL INDUSTRY	LAKE, POND / STREAM
☐ FARM	● RAILWAY STATION	—··— HUNDRED BOUNDARY
▽ HOUSE	BUILT-UP AREA	—·— PARISH BOUNDARY
WINDMILL	+ CHURCH	— — BOUNDARY OF VILLAGE, ESTATE, AND FARM OUTSIDE VILLAGE

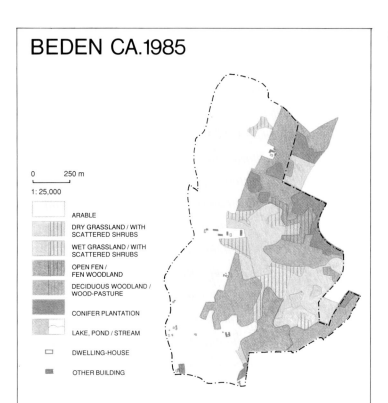

BEDEN CA.1985

```
0          250 m
├─────────────────┤
1: 25,000
```

	ARABLE
	DRY GRASSLAND / WITH SCATTERED SHRUBS
	WET GRASSLAND / WITH SCATTERED SHRUBS
	OPEN FEN / FEN WOODLAND
	DECIDUOUS WOODLAND / WOOD-PASTURE
	CONIFER PLANTATION
	LAKE, POND / STREAM
□	DWELLING-HOUSE
■	OTHER BUILDING

VEGETATION: S. BENGTSSON-LINDSJÖ   SETTLEMENT & GRAPHICS: M. RIDDERSPORRE

Fig. 4.4:16. Beden. Vegetation, ca. 1985.

sion with shrubs like willow, birch, and alder has started (Nilsson and Pröjts 1986). In spite of large ditches established already around 1915, embogging has occurred, and peat-cutting during the 1940s resulted in a number of ponds.

Today most of the former wet meadows are covered by wet deciduous woodland composed of alder and birch (Fig. 4.4:16). Raised water level in some of the ponds has caused further embogging (Bengtsson-Lindsjö unpubl.). Fen areas, reed marshes, sometimes with dense thickets of willows, make the area difficult to transverse on foot. Some areas are still grazed by cattle, but chemical fertilizers have been used on large parts of the pastures. Minute fragments of the vegetation once constituting the hay-meadows here can be traced near some of the ponds where natural grassland plants like several orchids and species like *Lychnis flos-cuculi* (ragged robin) *Geum rivale* (water avens), and *Myosotis scorpioides* (water forget-me-not) are to be found (Bengtsson-Lindsjö, unpubl.). The best preserved historical vegetation in Beden today is the former wood-pastures, although changed by terminated or reduced grazing pressure and selective cutting favouring beeches. Some parts of the former wood-pastures and the former coppice meadows have been replaced by mixed, mesic deciduous forest of oak, ash, elm, beech, bird-cherry, and elder. Hazel and hawthorn are remnants of the former grazed state. Spruce plantations of

considerable size occur in the southern parts of the Beden area.

As for the settlement pattern for the whole north-western area (Fig 4.4:15), it is striking how slight the changes have been since 1915. The patterns established during the enclosure era, and during the expansion that followed up to the population maximum around 1875, have to a great extent remained to the present day. However, the function of the settlements has changed dramatically. Numerous smallholdings, typical of the marginal parts of this area, have ceased to function as cultivation units, as mentioned above. The buildings remain, but the cultivation units have been amalgamated. The result has been a decrease in the number of farms smaller than 20 ha in Villie parish and a rise in the size class 20–100 ha (see Fig. 4.4:13). A feature not visible in the diagram is that the very biggest units have sold off land during this century.

4.4.9 Conclusions

Anders Persson and Björn E. Berglund

There is evidence of sparse Mesolithic occupation of the Romele area. Early Neolithic settlement has also been found, as at Kurarps Mosse. Settlement in this period was localized on lighter soils on high ground near watercourses. The Romele area was evidently utilized in the later part of the Middle Neolithic too, but the Late Neolithic was the most expansive Neolithic period in Villie parish. This kind of expansion is general for the whole of the outer and inner hummocky zones. For the entire Neolithic the landscape was forested, with sporadic open areas around habitation sites, where there was coppiced underwood.

In the Bronze Age the Romele area had a marginal significance in comparison with both earlier and later periods. Although there is evidence of Early Bronze Age settlement in the parish (Trunnerup), everything suggests that the area was little exploited during the period. The reason for this is that Bronze Age settlement was mostly localized on lighter soils and good pasture land. From this point of view, the hummocky landscape was less attractive. The coastal zone was the primary settlement area in the Bronze Age, as is also reflected in the greater number of barrows by the coast. In the Bronze Age the Romele area remained a forest landscape, utilized to only a slight extent as wood-pasture.

In the Early Iron Age there is no evidence of settlement concentration with any fixed system of arable and meadow. Occasional isolated farmsteads may have existed in the area, although there are no finds to prove this. Changes in technology in the Late Iron Age meant that the landscape around the Romeleåsen ridge once again became a potential area of colonization, so the south-western part of the area below the ridge took on greater significance at this time. There was probably growth and stabilization of settlement in late prehistoric and early medieval times. As in earlier periods, the ridge itself appears to have been largely uninhabited; it can be described as marginally located wood-pasture.

In the Late Iron Age and the Middle Ages the greater part of settlement was to be found in the south of Villie parish. The area's marginal position is clear from the low settlement density, which can first be measured in the mid 17th century, and from the fact that settlements here were deserted in the Late Middle Ages. The distal impression is reinforced by the fact that the ridge proper did not really acquire a permanent structure of isolated farmsteads until relatively late, 1580–1630. Unlike most other parishes in the Ystad area, Villie lacked a noble manor before 1680.

The latter half of the 17th century is harder to eval-uate, but there are distinct signs that it was a period of stagnation and that the intensive utilization of the cultural landscape did not get under way until the first half of the 18th century. The 18th century saw an expansion of settlement, population, and cultivated land, but in a way that brought about increased social stratification in the parish. In the latter part of the 18th century there was an increased privatization of hitherto common resources. The greater need for grazing land and timber and firewood in the phase of settlement expansion meant that the landscape of the ridge was gradually changed from a medieval half-open wood-pasture to an open heath with occasional trees or stands of woodland in the 17th and 18th centuries. It was part of an extensive common beyond the settlement areas of the hummocky zone.

In the 19th and 20th centuries the proportion of cultivated land rose from about 20% in 1800 to around 80% in 1910. This happened at the expense of meadow and pasture. The underlying causes were enclosures, changes in technique, and also in ownership, and a result was denser settlement. Now differences between the northern and southern parts of the Romele area have largely disappeared. When artificial manure began to be used there was less need for pasture. The introduction of fodder crops had a similar influence on lands that were now seldom left fallow. In parallel with the expansion of arable land, areas of forest were expanded through new plantations in the 19th century. In the 20th century, agriculture has retained the same area that it had acquired at the end of the 19th century, but there have been important changes in its structure. Above all, bigger fields have been created through amalgamation. The area is nevertheless characterized by relatively small farms with less chance of rationalization. This is also reflected in the more varied cultural landscape of the Romele area, especially in the north, which was once a vast expanse of outland. Here one can find – apart from rationally farmed agricultural land – woodland of different ages, half-open pasture, wetlands, and peatland.

To sum up, we may say that the southern half of the Romele area displays a similar development to the Krageholm area, while the ridge landscape of the north appears to have functioned as a reserve area, used especially intensively in periods when population increased and in periods of technical change.

Ecological Bulletins 41: 00–00. Copenhagen 1991

5 Analysis of changing cultural landscapes – problems and processes

5.1 Palaeohydrological studies and their contribution to palaeoecological and palaeoclimatic reconstructions

Marie-José Gaillard and Gunnar Digerfeldt

The current interest in palaeohydrological studies is mainly due to the possibility of using stratigraphically recorded past lake-level fluctuations as a basis for reconstructing and interpreting palaeoclimatic changes (Digerfeldt 1986). Provided there is unambiguous evidence for a climatic cause, a past lake-level fluctuation indicates a change in humidity and thus a change in precipitation and/or temperature. Unfortunately, palaeohydrological data alone do not allow separation of the effects of changing temperature from those of changing precipitation. The climatic implications of lake-level fluctuations have been discussed by many authors and for several parts of Europe: e.g. Sweden and Scandinavia (Digerfeldt 1972, 1988, Berglund et al. 1983), Scandinavia and central Europe (Gaillard 1985a), Poland (Ralska-Jasiewiczowa and Starkel 1988), and Alpine regions (Magny 1980, 1982, Magny and Olive 1981, Ammann 1982, Joos 1982). In addition, palaeohydrology is an important contribution to our understanding of the palaeoenvironment, as hydrological changes are one of the possible factors that may have influenced human settlement patterns and land use through time (Gaillard 1985a,b).

Studies of lake-level fluctuations were included in the Ystad project in an attempt to answer the following questions (Gaillard 1985b):

1. When did lake-level fluctuations occur in the project area and do they correspond to regionally significant palaeohydrological changes?

2. What are the causes of the recorded lake-level fluctuations – climatic change or human activity?

3. To what extent do past changes in land use and cultural landscape development depend on palaeohydrology and palaeoclimate?

The largest lake in the Ystad area, Krageholmssjön (Fig. 5.1:1), was chosen as the primary site for the reconstruction of lake-level fluctuations during the last 10,000 years (Gaillard 1984). Palaeoecological studies at two other localities, Bjäresjösjön and Kumlan (Fig. 5.1:1), give additional information on palaeohydrology in the Ystad area. Comparison of lake-level changes and land-use development in an intensively exploited area such as that of Bjäresjö village is of particular interest (Gaillard and Berglund 1988). Kumlan is a convenient site for studying the possible influence of palaeohydrology on the past distribution and type of wetlands.

The palaeohydrological investigations were performed according to the methodology developed by Digerfeldt (1986). This consists of the study of transects of marginal sediment cores and of stratigraphical analyses including past changes in (1) the distribution of lake and shore vegetation (plant macrofossil analysis), (2) the sediment composition, and (3) the level of the sediment limit (highest limit for permanent deposition of predominantly organic sediment). The detection of past lake-level changes is, therefore, based on a combination of different types of stratigraphical evidence, in order to obtain positive and convincing correlations between the sedimentary record and lake-level fluctuations. Pollen analysis is used for dating and correlating the cores, and for reconstructing vegetational and land-use history. Fig. 5.1:2 shows, in a simplified way, how water-level changes may affect the distribution of littoral vegetation, sediment composition, and sediment limit, and how these events may be recorded in a sediment core.

In this chapter, all dates are given in uncalibrated ^{14}C years BC/AD (years BP – 1950 = years BC).

Lake-level fluctuations in the Ystad area

The palaeohydrological reconstructions in the Ystad area were obtained from different lines of evidence depending on the sites. At Krageholmssjön, lake-level fluctuations were inferred from sediment-stratigraphical and biostratigraphical (pollen, plant macrofossil, and diatom analyses) investigations of a series of 18 cores in the south-western bay of the lake (Gaillard 1984). Fig. 5.1:3 presents the results from the outermost core analysed for pollen. At Bjäresjösjön, the reconstruction is based mainly on sediment- and pollen-stratigraphical evidence from the central core (Gaillard and Berglund 1988). The investigation of the Kumlan site is still in progress. The interpretation presented in this chapter is, therefore, tentative and relies on sediment-stratigraphical descriptions of a transect of cores across the basin.

Dating lake-level fluctuations

Dating lake-level fluctuations may present several problems. The marginal zone of a lake is often characterized by sediment erosion and redeposition, which makes pollen zonation difficult and ^{14}C datings unreliable. Moreover, if reliable dating turns out to be possible, it is

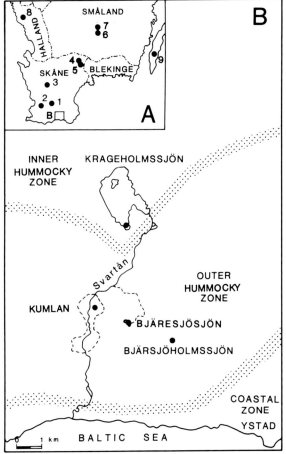

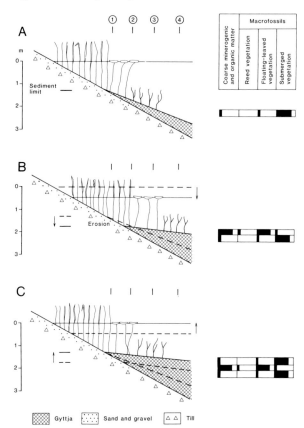

Fig. 5.1:1. A. Sites studied for palaeohydrology in southern
Sweden: (1) Lake Bysjön, Digerfeldt 1988; (2) Torreberga,
Digerfeldt 1971; (3) Ageröds Mosse, Nilsson 1964; (4) Vielången, Farlången, Digerfeldt 1965; (5) Lake Immeln, Digerfeldt 1974; (6) Lake Växjösjön, Digerfeldt 1975b; (7) Lake
Trummen, Digerfeldt 1972; (8) Lake Lyngsjö, Digerfeldt 1976;
(9) Öland, Great Alvar, Königsson 1968.
B. Central part of the Ystad project area with its three geographical zones. The palaeohydrological information is inferred from the studies of Kragehomssjön, Bjäresjösjön, and
Kumlan.

synchroneity of pollen zones for Scania is assumed, the
[14]C dates obtained at Ageröds Mosse for pollen-zone
boundaries (Nilsson 1964) may be used for the Ystad
area. Such a synchroneity is partly demonstrated by [14]C
dates from Bjärsjöholmssjön for the time-span 4050–
550 BC (Göransson 1991, Appendix A). At Krageholmssjön, the age of the recorded water-level lowerings was inferred by interpolation. Therefore, the
closer the record is to a dated level (in this case, a
pollen-zone boundary), the more accurate the age.

At Bjäresjösjön, lake-level fluctuations occurring between 750 BC and AD 750 are ^{14}C-dated. The more
hypothetical water-level lowering during the Middle
Ages is dated by interpolation between [210]Pb- and [14]C-

Fig. 5.1:2. The effects of lake-level fluctuations on the distribution of shore vegetation, the level of the sediment limit and the
sediment composition.
A. The outward spread of the lake vegetation is determined
primarily by water depth, and the level of the sediment limit by
exposure. The content of coarse minerogenic and organic matter in the sediment decreases outwards from the shore. The
schematic diagram on the right shows the macrofossil and
sediment composition in core 3.
B. A lowering in lake level will cause a lowering of the sediment limit and erosion of the marginal sediment. As recorded
in the schematic macrofossil diagram, the lowering will also
result in an outward spread of the lake vegetation and an
increase in the content of coarse matter in core 3.
C. A rise in lake level will produce a rise in the sediment limit,
an inward displacement of the lake vegetation and a decrease
in the content of coarse matter in core 3.

usually for particular levels only – levels easy to recognize in pollen stratigraphy or individual sediment layers
convenient for ^{14}C measurements. Therefore, the age of
events occurring between the dated levels has to be
interpolated. This procedure can involve major errors,
especially in the case of sediment sequences including
obvious sediment-stratigraphical changes or/and hia-
tuses. For instance, a water-level lowering may be indicated by a sandy layer, and it is often impossible to
estimate precisely the period of time the layer represents. Therefore, the ages given for lake-level fluctuations often have to be taken as approximate.

At Krageholmssjön, lake-level fluctuations were
dated by pollen-analytical correlation. The pollen diagram was zoned according to Nilsson (1964). Provided

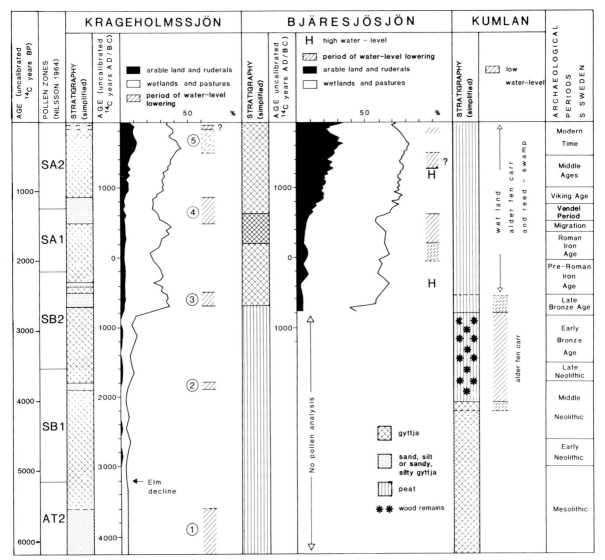

Fig. 5.1:3. Lake-level changes recorded in the Ystad area. The simplified lithostratigraphy, the curve of human impact pollen indicators, and the periods of water-level lowering (hachures) are presented for each site. Note that the lower boundary of a marking corresponds to the beginning, and the upper boundary to the end of a lowering. The markings consequently do not correspond solely to periods of low water level (Digerfeldt 1980, cf. legend to Fig. 5.1:4). Pollen zones according to Nilsson (1964) and archaeological periods in southern Sweden (cf. Ch. 2.9) are given for comparison.

dated levels. Finally, the dating of water-level fluctuations at Kumlan is preliminary and is currently based on a first series of 16 ^{14}C dates from two cores, and on a few pollen analyses.

Lake-level fluctuations and human impact
The palaeohydrological reconstruction obtained in the Ystad area is presented in Fig. 5.1:3.

So far, Krageholmssjön is the only locality where lake-level fluctuations can be reconstructed for the early part of the period investigated in the Ystad project. The study revealed strong indications of five periods of lake-level lowering during the last 6000 years. The earliest one covers pollen zone AT2 until ca. 3500 BC. The second one occurs in the later part of pollen zone SB1 (ca. 1800 BC). These two first low lake levels belong to the major period of lowering at Krageholmssjön during the last 6000 years. This is succeeded by a rise in lake level beginning at the end of zone SB1. The third water-level lowering starts during a land-use expansion phase in zone SB2 that is characteristic of all pollen diagrams from the Ystad area. It may be correlated to the pollen-stratigraphical level SB2c of Ageröds Mosse, which was dated to 895 ± 90 BC, thus to the transition Early/Late Bronze Age (Nilsson 1964). A similar radiocarbon age was obtained for the same event in Bjärsjöholmssjön (Göransson 1991, Appendix A, Ch. 4.2.2). The fourth water-level lowering is dated to the later part of pollen zone SA1 and the earliest part of zone SA2 (AD 450–850). The age of the last lowering is difficult to establish. It may have started either during the 16th century, or much later, during the 19th century. Distinct rises in the lake level occur in the later part of zone SB2, after ca. 500 BC, and in the early part of zone SA2, after ca. AD 900.

A comparison of the reconstructed water-level changes of Krageholmssjön with the curve of human-impact pollen indicators (non-arboreal pollen, NAP, ascribed to arable land, wetlands, and pastures) indicates synchroneity between the start of the water-level lowering during SB2 and the increase in NAP dated to the transition Early/Late Bronze Age (Fig. 5.1:3). The lowering during SA1 is synchronous with a possible decrease or stagnation in land use (low percentages of cultivated plants and grazing indicators). The other periods of lake-level lowering do not correspond to any significant change in land use. However, there is a clear temporal relationship between a rise in lake level (beginning of SA2) and the increase in NAP during the Viking Age.

The transition from alder fen peat to lake sediment (gyttja) in Bjäresjösjön is ^{14}C-dated at 740 ± 60 BC, and indicates a rise in the water-table at that time. Pollen analysis shows successions in the reed vegetation and aquatics (submerged and floating-leaved macrophytes) indicating the start of a lake-level lowering at the beginning of zone SA1, culminating during the later part of SA1, between AD 140 ± 50 and AD 710 ± 45. During that time, the lake was transformed into a very shallow pond with floating-leaved macrophytes and reed-swamp vegetation. Successions in the aquatic pollen taxa suggest a progressive rise in the water-level from ca. AD 710 to ca. AD 1250, when the lake reached its highest level. There is some indication of possible water-level lowering during the 14th and 19th centuries.

The rise in the water-table and the start of lake-sediment accumulation in Bjäresjösjön 740 BC occurred during a period of open landscape with large areas of grazed meadows (Ch. 4.2.2). The establishment of this type of landscape is unfortunately not evidenced in the pollen diagram and may have occurred before 740 BC. However, a distinct decrease in *Alnus* (alder), *Corylus* (hazel), *Quercus* (oak), *Ulmus* (elm), and *Tilia* (lime), and a parallel increase in Poaceae (grasses), between 740 and ca. 550 BC may indicate a transition period in the landscape development at Bjäresjö (Ch. 4.2.2). A possible decrease or stagnation of land use at Bjäresjö during the Roman Iron Age (decrease in percentages of human impact indicators) may be correlated with the start of the water-level lowering during SA1. However, as at Krageholmssjön, the strongest relationship is between the Vendel/Viking Age land-use change and a rise in the lake level.

At Kumlan, a large lake existed during the Early Holocene. According to a series of eight ^{14}C dates from the central core, the lake was completely overgrown and replaced by an alder carr ca. 2050–2250 BC (sharp transition from a coarse detritus gyttja to an alder carr peat). Around 850–550 BC the alder carr was transformed into a reed-swamp (transition from an alder carr peat to a peat rich in *Phragmites* (reed) macro-remains). This event may be concurrent with a general rise in the water-table.

Correlation of lake-level fluctuations in the Ystad area
The palaeohydrological changes described for Krageholmssjön, Bjäresjösjön, and Kumlan show a rather similar history at each site (Fig. 5.1:3).

The discrepancies seen in the record of the lake-level lowerings during zones SB1 and SB2 are probably due to the dating difficulties. The lowering recorded at Krageholmssjön in the later part of zone SB1 may be correlated to the creation of the alder carr at Kumlan (ca. 2050–2250 BC). In all sites there is a tendency towards a progressively higher water level in the later part of zone SB2 around 800–500 BC. At Krageholmssjön, the lowering in SB2 is characterized by two separate sandy layers, the most distinct of which is dated to the middle part of zone SB2. This event has no equivalent in Bjäresjösjön, where a rise in water-table is recorded at 740 BC. Whether the low lake level at Krageholmssjön belongs to the period of decreased humidity between 4800–4500 BC and 900–600 BC demonstrated in southern Sweden (Digerfeldt 1988) or represents a later short event during the long-term change towards progressively increased average humidity after 1000 BC has to remain an open question until better dating and further records from other lakes are available. However, there are good reasons to think that the lowering at Krageholmssjön during zone SB2 occurred earlier than the creation of the lake at Bjäresjö, and that all sites reacted to the same general increase in humidity some time between 800 and 500 BC. The rise in water level indicated at Kumlan around 850–550 BC speaks in favour of a general rise of the water-table in the Bjäresjö-Kumlan area at that time.

The distinct lake-level lowering during SA1 and the beginning of SA2 recorded in Krageholmssjön and Bjäresjösjön (slightly later in the former than in the latter) is most probably a result of the same regional palaeohydrological event. Finally, the rise in the water level dated to AD 710 at Bjäresjösjön is also registered in Krageholmssjön around AD 850.

Regional correlation and synthesis of Holocene lake-level fluctuations in southern Sweden

The palaeohydrological changes established for Krageholmssjön, Bjäresjösjön, and Kumlan can be compared with the reconstruction of Holocene lake-level fluctuations at Bysjön described by Digerfeldt (1988). Bysjön is located in the eastern part of the Vomb plain, 20–25 km NW of the lakes studied in the Ystad area. Particularly favourable stratigraphical conditions made it possible to reconstruct not only the direction, but also the

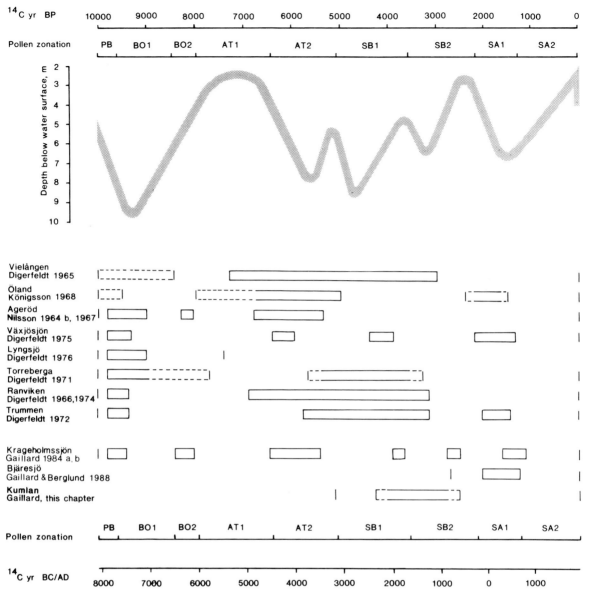

Fig. 5.1:4. Top: Reconstruction of past changes in the sediment limit in lake Bysjön. The relationship between sediment limit and water level in this lake may be assumed to have remained roughly the same throughout the Holocene. Therefore, the reconstructed sediment-limit curve can be used as indication of the approximate magnitude of the recorded lake-level fluctuations. Bottom: Comparison of Holocene lake-level fluctuations recorded in South Swedish lakes. The location of the investigated sites is shown in Fig. 5.1:1. The markings indicate periods of lake-level lowering. The left boundary of a marking corresponds to the beginning, and the right boundary to the end and culmination of a lowering. It should be observed that the markings do not correspond to periods of low lake level. The lake level is relatively high at the beginning of the lowering, and remains relatively low some time after the beginning of the subsequent rise. In some cases (broken line) the lowering is not directly recorded, but indicated by a stratigraphical hiatus. After Digerfeldt 1988, with addition of the present results from the Ystad area – Krageholmssjön, Bjäresjösjön, and Kumlan.

approximate magnitude of the lake-level changes throughout most of the Holocene.

The palaeohydrological reconstruction from Bysjön is shown in Fig. 5.1:4, together with a comparison of the results from most of the recent studies of Holocene lake-level fluctuations in southern Sweden, including investigations in the Ystad area. The precision of the reconstructions and dating certainly differs from one study to the other because of different stratigraphical conditions and methods of investigation. However, there is a fairly good consistency between the major changes. Though additional studies would be desirable in some cases, the regional significance seems to be clearly demonstrated. The synchroneity of changes is obviously not absolute, but the discrepancies may be explained by the existing difficulties in reconstructing and dating precisely the lake-level fluctuations.

The regional significance and synchroneity of the recorded palaeohydrological changes suggest an underlying climatic cause. When considering the studies in southern Sweden, no regional factor other than climate can be proposed as an explanation for the reconstructed palaeohydrological changes.

The comparison of results presented in Fig. 5.1:4 shows three major periods of lake-level lowering during the Holocene, indicating drier climate. A distinct lowering that culminated in Bysjön in the earlier part of zone BO1 – at about 7500–7200 BC – suggests a major period of drier climate during the Early Holocene. After a period of increased humidity, indicated by rising and relatively high lake levels, a second major period with decreasing humidity began in the later part of zone AT1 – at about 4800–4500 BC according to the reconstruction in Bysjön. However, the climate was obviously not constantly drier during this period, since minor lake-level changes were recorded in some of the lakes studied, indicating a fluctuating climate with drier periods alternating with more humid periods. In Bysjön, the decrease in humidity culminated in the earlier part of zone SB1, at about 2900–2600 BC, but the entire period continues until the later part of zone SB2 (ca. 900–600 BC). A third major lake-level lowering, recording a period of distinctly drier climate, occurred in zone SA1 between AD 200 and 800. The separation of this third major period may be somewhat questionable. The lake-level fluctuations recorded in Bysjön during the later part of the Holocene may just as well indicate a long-term change towards progressively increased average humidity, which was interrupted by relatively short-term periods of drier climate. This alternative interpretation is in better accordance with the available peat-stratigraphical evidence from southern Sweden (Nilsson 1935, 1964). The progressive peat growth in raised bogs indicates a major increase in the average humidity throughout the later part of the Holocene, but the occurrence of several regionally significant recurrence surfaces suggests that this change included a series of fluctuations. Recurrence surfaces are peat layers characterized by higher humification and corresponding to areas of slower peat growth in the bog, which may be a result of drier conditions, perhaps climatically controlled (e.g. Aaby and Tauber 1975).

The palaeohydrological changes established for Krageholmssjön, Bjäresjösjön, and Kumlan show a rather good consistency with the reconstructed lake-level curve from Bysjön and the regional comparison presented in Fig. 5.1:4. The water-level lowering recorded at Krageholmssjön in zone AT2 most probably corresponds to the distinct lowering demonstrated in Bysjön during the same time span, which records the beginning of the second major Holocene period of decreased humidity in southern Sweden. Regarding the subsequent

two lowerings recorded at Krageholmssjön in zones SB1 and SB2, it seems reasonable to correlate these with the two lowerings in the same zones in Bysjön, even if the precise dating differs somewhat. This period of low lake levels may have contributed to the rapid overgrowing of the former lake at Kumlan and to the creation of the subsequent alder carr around 2000 BC.

The rise in water-table recorded in both Kumlan and Bjäresjösjön at about 800–500 BC is in good agreement with the lake-level rise shown in Bysjön and several other south Swedish lakes during the same period (Fig. 5.1:4). At Torreberga (30 km NW of the project area), Digerfeldt (1971) showed that the alder carr peat was replaced by a swamp gyttja during the last part of SB2, which indicates the culmination of a rise in the water-table.

The distinct lake-level lowering and subsequent rise recorded in Bysjön during SA1 is also demonstrated in Krageholmssjön and Bjäresjösjön. In the latter lake, the culmination of the lowering is ^{14}C-dated to AD 140–710.

Palaeohydrology, palaeoclimate, and land-use history in the Ystad area – a discussion

Lake-level fluctuations in southern Sweden and their climatic implications

Lake-level fluctuations may result from regional climatic changes, but also from local environmental changes (e.g. lowering of the outflow threshold by erosion, or damming of the outflow by peat growth or landslides) or human activities (e.g. artificial lowering or damming of the threshold, deforestation or reforestation) (Fig. 5.1:5).

As mentioned above, the synchroneity of lake-level fluctuations in a large area, such as southern Sweden, suggests climatically induced palaeohydrological changes. A comparison of the results from the Ystad area with the regional synthesis for southern Sweden (Fig. 5.1:4) shows that the water-level changes recorded at Krageholmssjön, Bjäresjösjön, and Kumlan between 8000 BC and AD 1000 may be correlated with regional palaeohydrological events and hence with climatic changes.

Lake-level fluctuations may therefore be interpreted in terms of changes in the average humidity. Periods of lake-level lowering are attributed to decreased average humidity, whereas periods of rising lake levels are related to increased average humidity. The climatic causes behind the changes in humidity during the Holocene in southern Sweden are discussed in Harrison and Digerfeldt (unpubl.). In addition, the authors show how lake-level fluctuations and, consequently, climatic changes differ from one region to the other (southern Sweden and the Mediterranean region in their example). These results suggest that extrapolation of climatic reconstructions between distant regions may be very ques-

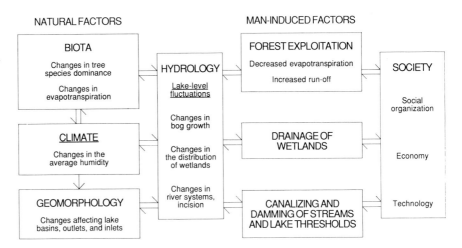

Fig. 5.1:5. Schematic presentation of the numerous relationships between natural and human-induced factors that may influence lake levels. The discussion in the text is restricted mainly to the relationships between climate, lake-level fluctuations (underlined in the figure), and land use.

NATURAL FACTORS

BIOTA
Changes in tree species dominance
Changes in evapotranspiration

CLIMATE
Changes in the average humidity

GEOMORPHOLOGY
Changes affecting lake basins, outlets, and inlets

HYDROLOGY
Lake-level fluctuations
Changes in bog growth
Changes in the distribution of wetlands
Changes in river systems, incision

MAN-INDUCED FACTORS

FOREST EXPLOITATION
Decreased evapotranspiration
Increased run-off

DRAINAGE OF WETLANDS

CANALIZING AND DAMMING OF STREAMS AND LAKE THRESHOLDS

SOCIETY
Social organization
Economy
Technology

tionable and misleading. When lacking other climatic information, the tendency in northern Europe is to use reconstructions obtained from the Holocene history of glacier advances and timber-line changes in northern Scandinavia (Karlén 1988, Kullman 1988, 1989). A comparison of the palaeohydrological events recorded in southern Sweden with the glacier advances demonstrated in northern Scandinavia may be attempted, but should take in account all limits in the interpretation and dating of the events recorded by both methodologies. Climatic changes have not necessarily been similar in southern and northern Scandinavia during all periods of the Holocene. One should therefore be very careful in correlating events and in assuming synchroneity between events.

According to Karlén (1988) the most significant pre-Little Ice Age glacier advances are dated at ca. 7500, 5100–4500, 3200–2800, 2200–1900, and 1500–1100 BP (ca. 5550, 3150–2550, 1250–850, 250–50 BC, AD 450–850). Periods of glacier advances are due to the combination of changes in different climatic factors, for instance decreased temperature and/or increased precipitation, or changes towards more continental climate.

There is no good correlation between periods of glacier advances in northern Scandinavia and periods of rising lake levels (or high lake levels) in southern Sweden during the Early Holocene. For this period, we have to rely entirely on the palaeohydrological data for climatic reconstructions in southern Sweden.

The glacier advances between 1250 and 850 BC, and between AD 450 and 850, may possibly be compared with the general rises in lake levels shown in southern Sweden around 900–600 BC and at ca. AD 800. A severe climatic deterioration around 1000 BC has been demonstrated not only in northern Scandinavia, but also in the Alps, and represents the most severe climatic cooling during the last 10,000 years (Röthlisberger 1986). The tendency towards higher lake levels from ca. 900–600 BC in southern Sweden may therefore be related to this regional climatic event. The lake-level low-

ering recorded at Krageholmssjön during SB2 most probably occurred before this major climatic change.

The general rise in water level registered after AD 740 in the Ystad area and most of the south Swedish lakes is possibly a response to the regional colder and wetter climate indicated by glacier advances in northern Scandinavia around AD 450–850 (Karlén 1988).

The later climatic fluctuations, such as the Little Ice Age (AD 1550–1850) are not recorded by the lake-level fluctuations in southern Sweden. These climatic changes may have been of such a nature that they did not modify the water balance significantly. Dendrochronological investigations show that particularly low summer temperatures prevailed in northern Sweden AD 550, 780–970, 1100–1250, 1350–1400, 1470, 1570–1750, 1810, 1840, 1910 (Bartholin and Karlén 1983, Karlén 1984). From ca. AD 1000, possible decreases in temperature occurring in southern Sweden may have been of little significance for the general water balance during a period of relatively wet climate.

Effects of palaeohydrological and palaeoclimatic changes on the past cultural landscape
The final task of a palaeohydrological study in the context of the Ystad project is to attempt to understand the possible effects of palaeohydrological and palaeoclimatic changes on past land use and landscape development. The complex relationship between palaeohydrology, palaeoclimate, and land-use history was discussed earlier by Gaillard (1985b) (Fig. 5.1:5). The investigations in the Ystad area provide the possibility to compare, at each site, the record of lake-level fluctuations with the reconstructed land-use history, both inferred from the same sediment core. However, the interpretation of such a comparison turns out to be difficult. The following problems should be stressed:

1) The synchroneity of a regionally significant palaeohydrological change and a change in the past land use is an indication that past land use is, in some way, related to palaeoclimate. However, it is difficult to establish

which was the decisive climatic factor – temperature, humidity, or a combination of both.

2) A past lake-level fluctuation indicates possible changes in different palaeoenvironmental factors which could also have influenced land-use development, such as soil (drier or wetter), temperature (higher or lower), precipitation (higher or lower), and the size of the areas available for tillage or grazing (larger or smaller). It is impossible, from the results obtained, to decide which of these factors has been decisive for the observed changes in past land use.

3) The effect of a particular climatic change, towards drier or wetter conditions, may have been different depending on the local palaeoenvironmental conditions and the type of land use prevailing at the time of the change. Therefore, one can only make assumptions concerning possible effects of climatic changes for the development of the cultural landscape in the Ystad area.

There are few indications of a colder or more continental climate around 3000 BC in southern Sweden that would partly explain the elm decline (e.g. Ch. 4.2.2, 3.2.1). There is no distinct palaeohydrological change recorded in the Ystad area for that time. As regards southern Sweden, the best evidence is from Bysjön, where a significant water-level lowering is shown around 5000 BP (ca. 3000 BC). Whether this palaeohydrological change may be related to the elm decline is a question the solution of which would need further evidence of the link between the two types of event in other sites (Göransson 1991). However, the time of the elm decline certainly belongs to the culmination period of increased aridity in South Sweden, which also caused a significant change in the landscape of the Bjäresjö-Kumlan area. The former lake at Kumlan was entirely overgrown around 2000 BC and replaced by swamp forests with alder until ca. 850–550 BC.

The synchroneity of a water-level lowering and land-use change at the beginning of the Late Bronze Age (Krageholmssjön) has been discussed earlier (Gaillard 1985b). Whether the significant shift in the cultural landscape at that time is related to a climatic change has to remain an open question. However, man certainly experienced essential changes in the landscape that may have influenced his land-use practices during the Late Bronze Age. The general rise in the water-table dated at 740 BC at Bjäresjösjön resulted in the creation of new bodies of open water (e.g. the lake at Bjäresjö), and several large areas of swamp forests became flooded and were replaced by fen (e.g. Kumlan). Therefore, some sites became more attractive for settlement (e.g. Bjäresjö) and others less (e.g. Kumlan, where possible grazing and coppicing could not be practised to the same extent as before). Other wooded areas probably had to be opened for grazing and coppicing. This is one of several possible explanations for the major change in the cultural landscape seen in the pollen diagrams of the area (Ch. 4.3.2, 4.2.2, 3.3.1).

The rise in lake levels and the land-use intensification during the Vendel/Viking Age at Krageholmssjön and Bjäresjösjön are clearly synchronous. The general rise in lake levels is most probably due to a regional climatic change towards wetter conditions during the earliest part of pollen zone SA2. Intensive tree-felling during that period could also have produced a rise in the water-table. However, the pollen-analytical evidence indicates a rather open landscape since the Late Bronze Age in the Ystad area. The Vendel/Viking Age is characterized mainly by a change from predominant grazing to more intensive agriculture. Tree-felling does not appear to have been particularly extensive at that time. This would support the assumption of a regional climatically induced rise in lake levels. Bjäresjösjön was a little pond invaded by aquatic vegetation during the Early Iron Age. It became deeper at the end of the Vendel Period and was, therefore, ideal as a retting pit for extracting *Cannabis* (hemp) fibres, which was a common practice in south Sweden from that time until ca. AD 1200. This was probably one of the factors that made Bjäresjösjön particularly attractive from the end of the Vendel Age and during the Viking Age.

The palaeohydrological investigation performed in the Ystad area provides useful information for a more complete description of the palaeoenvironment during the last 6000 years, including climate, hydrology, and extension of lakes and wetlands (see e.g. Ch. 4.2.9). Whether these different environmental factors had a decisive influence on human populations and land use still remains a difficult question to answer.

282

5.2 Erosion and land use

John A. Dearing

Introduction

In Scandinavia and elsewhere in NW Europe soil erosion is not normally a problem, occurring either infrequently, locally, or so slowly as to be imperceptible. Few societies have felt the direct impact of soil erosion through reduced crop yields, or even seen the effects of dramatic erosion in terms of dust storms, gullied slopes, or polluted water supplies. Its unspectacular nature gives the impression that losses of soil are too small to have any effect on the viability of farming either today or in the past; certainly few farmers today actively farm in ways to reduce the soil loss. But the question arises, "how high may an erosion rate become before it poses problems?" and the answer depends largely on how much soil is formed each year. It is clear that soil erosion rates will always pose a potential problem if they exceed natural rates of soil formation. An annual deficit of soil, however small, may produce problems in the long term. And just how long is "the long term" will depend on the depth of soil; the deeper the soil the longer it will take before any cumulative effect is seen. When "the long term" is centuries and millennia, the effect of accumulating soil deficits may not be perceived by the contemporary farming society. Whilst striving to manage an environment which is less fertile and less resilient, they may be ignorant of their inheritance. Future historians may also be ignorant of their inheritance, perhaps arguing that the cause of their less efficient farming system was bad weather, disease, social upheaval, or poor farming methods. The farming society may suffer the unluckiest of fates if by coincidence their period of stewardship witnesses an extreme climatic event when the soil, long-depleted of material and strength, is unable to withstand this higher-than-average pulse of energy. A century before, the same event would have had little effect, but the cumulative depletion finally reaches the point where the soil's resistance to severe erosion is overcome; an environmental threshold is crossed with resultant devastating effects. Whilst that society was not the cause of its downfall, history may record it differently.

In the same way that we have insufficient information to judge past farmers, so today's farming methods are more often than not assessed on the basis of short records with little regard given to the historical perspective. Where visible soil erosion occurs today, is it the present practices of intensive and continuous cultivation that are to blame? Or have past practices dictated that today's farming methods will increase the chance that thresholds are crossed, so that weather conditions which induce erosion will become increasingly less extreme and more frequent? Only by viewing the present and past conditions can we design measures to rectify a problem which has grown over a long period. Where erosion rates are low, but nevertheless leading to net soil losses, we may learn from the past about the consequences of inaction on future landscapes and generations.

One approach to tackle these problems is to obtain soil erosion data over long time-scales. With these data we might be able to answer the following questions:

1. What are the rates of erosion under natural or semi-natural vegetation? – These data give an estimate of the lowest attainable erosion rates which, in an ecosystem at equilibrium, should approximate to the natural rate of soil formation. This allows a true assessment of the net soil losses under agriculture.
2. Under what set of conditions did erosion begin to accelerate from background levels? – This may give us ideas about the mechanisms which trigger erosion.
3. How quickly does the erosion rate return to previous levels or stabilize at a new one? – This tells us about the resilience of the system.
4. Are farming practices today more or less likely to lead to severe erosion, when compared with previous practices? – This should enable a proper assessment to be made of different farming systems; a basis for planning farming in the future.

To answer these questions it is necessary that the records of an environment include soil erosion, vegetation history, and farming history: these records need to be of high time resolution over long time-scales, and covering all the different types of environment within a region. In this context, it is necessary to have records from areas differing in their physical environment (soil type, topography, etc.) and scale (single slopes, whole farms, whole farming regions). An important source of information about past erosion is the sediment which accumulates year by year at the bottom of lakes. For this reason, much of the information about past environments has been obtained for lake-catchments.

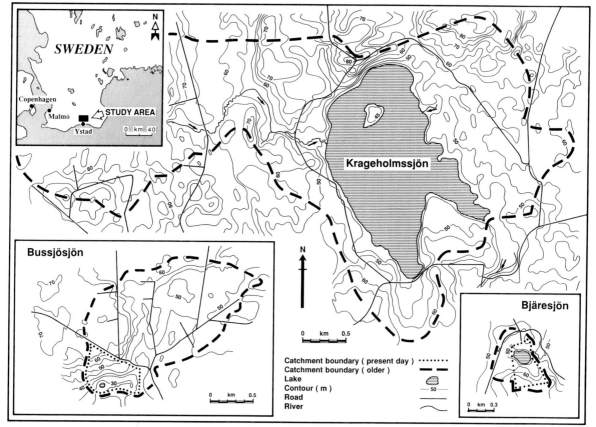

Fig. 5.2:1. Lake-catchments with details of topography and drainage.

Lakes and their catchments in the Ystad area

The Ystad area today is predominantly agricultural, with deep fertile soils and gentle topographies over much of the area, and provides suitable conditions for intensive arable farming. It is a landscape which man has modified for centuries to maximize agricultural output; deep drainage, wetland drainage, forest clearance, field enlargement, and even lake lowerings have all contributed to the present efficient farming systems.

Elsewhere in Scania farmers have witnessed soil degradation in the form of severe wind erosion; northerly areas in the sandy Vomb plain frequently experience dust storms in spring. But in the Ystad area the generally clayey soils are not prone to desiccation and blowing, and up until recent years the loss of soil on slopes by water erosion was not often seen. Soil degradation here was not believed to be a problem.

Recent studies, however, suggest that water erosion has occurred in the past and may well be increasing today. Lines of evidence which suggest this are the accumulations of soil upslope of boundary walls, soil colour indicating subsoil at the surface, observations of rills (small erosional channels) on slopes above fans of sediment, and measurements of rill volume and trapped eroded soil. The period over which evidence for wide-

Table 5.2:1. Site statistics.

	Bjäresjösjön	Krageholmssjön	Bussjösjön	Havgårdssjön
Lake altitude (m)	48	43	24	51
Max. catchment altitude (m)	70	105	70	110
Lake surface area 1983 (ha)	2.32	220	0.86	54.7
Sedimentation area, present (ha)*	1.14	96.25	0.29	no data
Catchment area, present (ha)	15.08	930	106.1	141
Sed. area/catchment area	1:13.2	1:9.66	1:366	1:2.6
Lake/catchment area, present	1:6.5	1:4.2	1:123.4	1:2.6

* Based on area over which more than 1 m of silty/clayey sediment has been deposited.

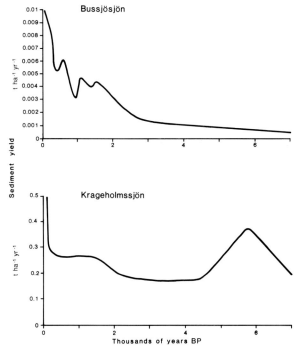

Fig. 5.2:2. Reconstructed sediment yields for Bussjösjön and Krageholmssjön, 0–7000 BP.

were established by correlating pollen diagrams with previously dated (^{14}C) pollen zones (Ch. 2.1), and by making radioisotope analyses (^{210}Pb, ^{137}Cs) directly. These chronologies were transferred to other cores by correlation with either stratigraphic or magnetic variables. Mean accumulation rates across a lake were then calculated and used to estimate total sediment yields from the catchment. The accumulation rates were corrected for organic and carbonate contents and converted to yield by multiplying by the lake area/catchment area ratio (in practice, the sedimentation area/catchment area ratio; Table 5.2:1), and expressed as the annual mass of minerogenic material eroded per hectare of catchment (t ha^{-1} yr^{-1}). In these calculations it was assumed that all the minerogenic material in the lake basin originated in the catchment and that losses

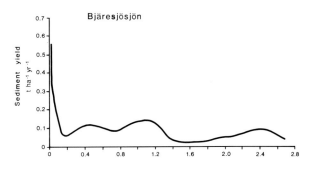

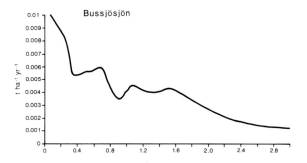

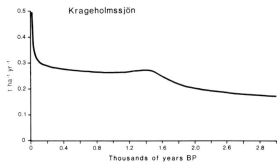

Fig. 5.2:3. Reconstructed sediment yields for Bussjösjön, Bjäresjösjön, and Krageholmssjön, 0–3000 BP.

spread water erosion has come to light has coincided with an intensification in agriculture, and has raised the level of concern about the long-term viability of present farming methods.

This chapter concentrates on the evidence for past and present soil erosion in three areas which drain into the lakes Bussjösjön, Bjäresjösjön, and Krageholmssjön. Fig. 5.2:1 shows the nature of each of these lake-catchment units in terms of their present land use, drainage, relief, and size. Additional statistics for the sites are presented in Table 5.2:1.

Methods

The nature of soil erosion in the area has been assessed in three ways: contemporary monitoring of erosion processes, calculation of historical yields of eroded soil from lake sediment studies, and magnetic analyses of lake sediments to show their source.

Contemporary monitoring of processes has been undertaken at only one site, Bussjösjön. Metal troughs were placed on lower slopes to trap soil washed downslope by sheet erosion, and the total volume of soil eroded by rills was estimated from repeated surveys of the rill systems (Alström and Bergman 1988, 1990).

Lake sediment studies were conducted at all three sites, where a similar approach was adopted. Sediment cores were sampled from a network of positions across the lake, subsampled and analysed for a variety of properties. The ages of the sediment in a single master core

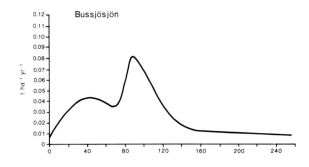

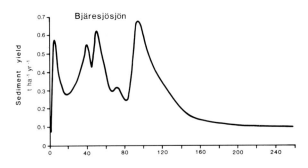

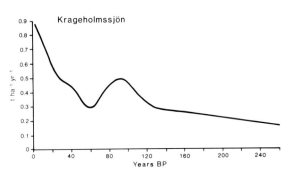

Fig. 5.2:4. Reconstructed sediment yields for Bussjösjön, Bjäresjösjön, and Krageholmssjön, 0–250 BP.

of the lake sediments might be directly compared with similar particle-size ranges in their sources.

In the results that follow, sediment age is expressed in years before the present (BP), which is defined here as the year of sampling between 1983 and 1985: the recent dates post-150 BP are calendar years, older dates are uncalibrated radiocarbon years before 1950 with the addition of the calendar years between 1950 and the sampling year.

Results

Sediment yields and erosion. Past records of sediment yield are shown for three time-spans, 0–7000 BP, 0–3000 BP, and 0–250 BP (Figs 5.2:2–4). Lake sediments at Bjäresjösjön date from only ca. 2800 BP and consequently are not included in the figure. All the curves represent sequences of smoothed mean values covering different time-spans, according to the resolution of the chronologies, and may not always be directly compared in the older periods because of dating errors. It is planned that future work will produce records with greater detail and accuracy.

Fig. 5.2:2 shows that the trend in sediment yield since 4000–3000 BP has been increasing, though the earlier yields at Krageholmssjön are relatively high in contrast to low yields at Bussjösjön; yields at Krageholmssjön are typically two orders of magnitude higher throughout the period. At both sites there is a significant increase between 2500 BP and 1000 BP, from ca. 0.18 to 0.28 t ha^{-1} yr^{-1} at Krageholmssjön, and from ca. 0.002 to 0.0045 t ha^{-1} yr^{-1} at Bussjösjön.

Records from the last 3000 years (Fig. 5.2:3) show generally increasing values up to the present at all three sites, with the most rapid increases occurring during the periods 300–100 BP and 2000–1600 BP (1400–1200 BP at Bjäresjösjön). The record from Krageholmssjön is the least precise, and it is not possible to define the exact timing of these changes. Also there is evidence at Bjäresjösjön for a major water-level lowering during the period 2000–1300 BP, which may have led to a reduction in the amount of sediment reaching the lake. At Bussjösjön an initial rise in yield occurred before 1600 BP, which appears to be at least 400 years earlier than at Bjäresjösjön. At both these sites yield values remain relatively high during the period 1200–400 BP, although minima are recorded at ca. 900 BP at Bussjösjön and at 700 BP at Bjäresjösjön: the recent rises in yield are preceded by periods of relatively low sediment yield at ca. 400 BP at Bussjösjön and ca. 200 BP at Bjäresjösjön. With one exception the Bussjösjön record consistently shows changes occurring 200–400 years before those at Bjäresjösjön. The relatively high yields at 2600–2200 BP in the Bjäresjösjön record are not apparent in the Bussjösjön record, but this may be due to the poor resolution of the data at this time.

Recent changes in sediment yield over the past 250 years are shown in Fig. 5.2:4. Here there are certain

through the outflow have always been insignificant. Diatom silica, atmospheric dust, and eroded lake bank material will have contributed to the minerogenic fraction of the sediment in varying amounts at all times, and together represent an error in these calculations of sediment yield; an overestimation by perhaps 10–20% (cf. Dearing et al. 1987). This overestimation is likely to be compensated to some extent by losses of sediment through the outflow which are estimated to be less than 10%.

The magnetic properties of the sediments were analysed at intervals down several cores from each lake. These properties were then compared with those from soils and eroding sediments in the catchments. Some soils were fractioned in order that magnetic properties

286

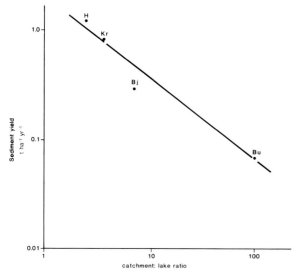

Fig. 5.2:5. A bilogarithmic plot of maximum sediment yield against the ratio catchment area:lake area for Bussjösjön (Bu), Bjäresjösjön (Bj), and Krageholmssjön (Kr) (this paper), and Havgårdssjön (H) (Dearing et al. 1987).

by the highest overall rates of increase in yield and the largest catchment/lake ratio. Indeed, a bilogarithmic plot of maximum sediment yields against catchment/lake ratio (Fig. 5.2:5) approximates to a negative exponential relationship, suggesting that sediment yield increases as catchment size becomes relatively smaller. Conversely, factors of increase in sediment yield within each record decrease as catchment size becomes relatively smaller.

Data for movements of soil on slopes around Bussjösjön (Table 5.2:2) show a wide range of figures ranging from 0.1–30.3 t ha^{-1} over the measuring period (Alström and Bergman 1988). The high levels of erosion occurred in the spring of 1987; the land was bare after winter wheat cultivation during the previous autumn and the soil was desiccated after the prolonged freezing conditions. A rapid thaw with rain led to much of the topsoil becoming mobile as rain and meltwater flowed over the surface above a still-frozen subsurface, removing great quantities of sediment in well-developed rill systems. In the following spring there was little snow cover and no significant thaw, yet rainfall was three times higher than in the previous spring. Rills were weakly developed, few in number, and sediment losses were one-tenth of the previous year's losses.

Sediment sources. The assessment of sediment source is based mainly on a comparison of the results from magnetic analyses of soils and lake sediments. Results of measurements made on soils show that there are magnetic differences between the main soil types, but that these are small in comparison with the differences observed between topsoil and subsoil levels. Topsoils are characterized by having reasonably high levels of very small magnetite minerals, formed within the soil, which may be identified by measuring the frequency-dependent susceptibility value (Xfd). These minerals are not found to any great extent in soil samples taken from depths below 30–50 cm. Other minerals, such as goethite and hematite, are more common in unweathered material from the lower soil horizons below 100 cm and may be identified by another magnetic variable, the isothermal magnetization remaining at a high negative magnetic field (HIRM). Consequently the ratio HIRM/Xfd provides a useful indicator of the depth from which soil material originates, and may be measured on lake sediment samples in order to infer the sediment source:

parallels between the three records. At all sites there is a rapid rise in yield from ca. 160 BP up to 90 BP (ca. 1825 onwards to the end of the last century). This rise is followed by a drop in the values in the period 1905–1925 to levels not much higher than those which existed in the earliest part of the 19th century. Thereafter follows a common rise which peaks ca. 1945 in Bjäresjösjön and Bussjösjön, but which continues to the present (1985) in Krageholmssjön. At both Bjäresjösjön and Bussjösjön the yields have fallen within the past 45 years, although at Bjäresjösjön a further peak is registered in the 1960s.

These broad similarities in trends are in contrast to the absolute values of sediment yield. These vary between the lakes by up to a factor of 200 (Bussjösjön and Krageholmssjön) in the early (ca. 7000 BP) part of the record, reducing to a factor of 7.5 (Bussjösjön and Bjäresjösjön) at the end of the 19th century. The differences between lowest and highest yields within each record also vary between lakes, by factors of 80, 20, and 3 in Bussjösjön, Bjäresjösjön, and Krageholmssjön respectively. The Bussjösjön catchment, while showing the lowest mean sediment yields, is also characterized

Table 5.2:2. Soil losses in fields around Bussjösjön (Alström and Bergman 1988), and summary data for sediment yields calculated from lake sediments (monitoring periods 23 Oct 86–20 Apr 87 and 21 Nov 87–26 Feb 88).

| | 1986/1987 | | 1987/1988 | |
	field B6	field B7	field B6	field B7
Soil losses by sheet erosion (t ha^{-1})	4	16	0.11	0.03
Soil losses by rill erosion (t ha^{-1})	0.4	2720	3	30
Total erosion losses (t)	13	2870	3	30
Erosion losses per hectare (t ha^{-1})	6	303	1	3

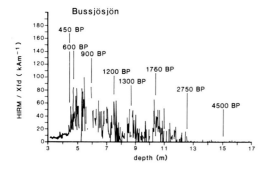

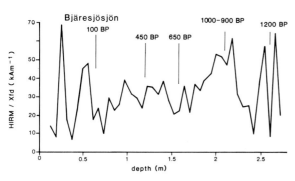

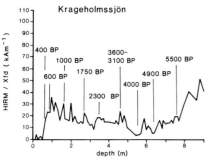

Fig. 5.2:6. Records of the magnetic parameter HIRM/Xfd plotted against sediment depth for Bussjösjön, Bjäresjösjön, and Krageholmssjön. The ratio approximates to the proportions of subsoil and topsoil in the sediment; the ratio increases with increasing proportions of subsoil. Sediment dates are from a combination of radiometric ages and dated pollen zones.

the value of the ratio can be expected to increase as the proportion of sediment from subsoil sources increases. This interpretation is valid for sediments derived from freely-drained and fine-textured soils, but does not apply in situations where sandy or alluvial soils form the major sediment source. Fortunately, all three catchments are dominated by moderately well-drained brown earths developed on clayey deposits. Measurements of HIRM/Xfd were also made on the eroded soil caught in traps at the bottom of slopes at Bussjösjön. Clayey sediment eroded in rills had HIRM/Xfd values generally below 20 kAm⁻¹, coarser sediment and samples of sub-

soil clay had higher values up to 200 kAm⁻¹ (magnetic unit).

Fig. 5.2:6 shows plots of the HIRM/Xfd ratio against depth for central sediment sequences covering the past 6000 years (1500 years at Bjäresjösjön). At Krageholmssjön the sequence begins with high values, suggesting a subsoil source for sediments at this stage. These values decline to reach minima at ca. 4900 BP and ca. 4000 BP before rising to a double peak covering the period ca. 3600–3100 BP. Values remain relatively high until ca. 400 BP, with periods of higher values at ca. 2300 BP, ca. 1750 BP, and 1000–600 BP. After 400 BP the values decline rapidly up to the present day. In some ways the record for Bussjösjön is similar, especially in the recent past. The older sections show the first rise in values to be ca. 2750 BP, somewhat later than in Krageholmssjön, and peak values occur at ca. 1750 BP, ca. 1300 BP, ca. 1200 BP, and 900–600 BP. The values decline from ca. 450 BP and remain low up to the present, except for a short period of high values in the mid 20th century (not shown). At Bjäresjösjön a shorter sequence displays fluctuating values before a peak at ca. 1000–900 BP. Thereafter the record shows declining values up to 200–100 BP, followed by fluctuating values during the 20th century. Overall, the records indicate that subsoil erosion was generally high during the period from 3600–2750 BP to 450–400 BP, with maximum erosion at ca. 1750 BP and ca. 1000 BP. In two of the lakes subsoil material dominates during some periods of the 20th century. Between ca. 1000 BP and the 20th century the subsoil contribution declines at all sites, and at Bjäresjösjön there is a noticeable minimum 700–600 BP.

In terms of the absolute values of the HIRM/Xfd ratio at the peak dated to ca. 1000 BP, Bussjösjön shows the highest, followed by Bjäresjösjön and then Krageholmssjön. This suggests that the soil that was eroded at this time, whilst dominated overall by subsoil, comprised a mixture of topsoil and subsoil in varying proportions in the different catchments; relatively more subsoil from the Bussjösjön catchment and least from the Krageholmssjön catchment.

Discussion

The results show that, for Bussjösjön at least, it is possible to make a direct link between contemporary erosion processes and recent deposits in the lake. At Bussjösjön the uppermost sediments have, on the whole, low HIRM/Xfd values below 20 kAm⁻¹, which suggests that the lake sediment is composed of fine topsoil from surrounding slopes, similar in character to the material which was trapped on the slopes and shown to have low HIRM/Xfd values. However the link between the values of sediment yield, derived from lake sediment, and the soil losses from individual fields during the winters 1986/1987 and 1987/1988 is not so clear. The mean sediment yield from the catchment is calcu-

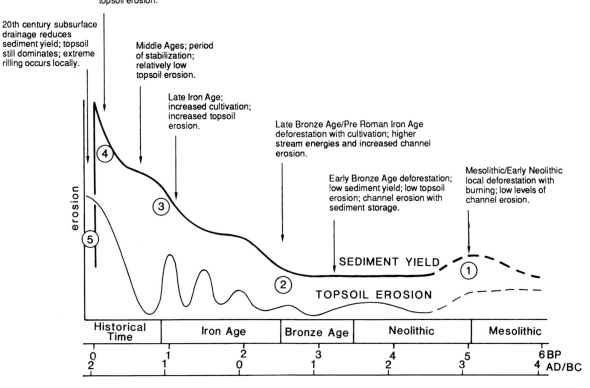

Fig. 5.2:7. A generalized history of land use and erosion in the Ystad area. Circled numbers refer to periods of destabilization. Time-scale based on uncalibrated radiocarbon years.

lated to be <0.05 t ha⁻¹ yr⁻¹ compared with losses from rill and sheet erosion of between 0.1 and 30.3 t ha⁻¹ yr⁻¹ (and these figures assume that erosion outside of the winter/spring period is negligible). This large difference between the two methods of estimating sediment loss could indicate that soil erosion on slopes may be highly variable from year to year, and that the lake sediment-based calculations of mean sediment yield from a catchment may mask this variability. But the relationship between catchment/lake ratio and sediment yield shown in Fig. 5.2:5 suggests, perhaps more significantly, that there are zones of sediment storage in all the catchments; not all the eroded soil reaches the lake. This means that the lake sediment-based records of sediment yield should be interpreted in different ways at each of the sites. The records from relatively small catchments should give values closer to those expected from studies of soil erosion on slopes, while those from relatively large catchments reflect more the changes in mobilization of sediment by streams and the storage of sediment on flood plains and lower slopes. Taking these considerations into account, the records of sediment yield, sediment source, and present-day erosion rates

have been combined with the land-use history of the region to produce a generalized history of erosion and hydrological changes in the Ystad area (Fig. 5.2:7). This diagram provides the basis for discussing the questions posed in the introduction.

The sediment yield data for the earliest times (Figs 5.2:2, 7) show that losses under natural woodland with minimal disturbance were low ca. 7000 BP, about 0.2 t ha⁻¹ yr⁻¹, at Krageholmssjön where sediment yield results most closely reflect soil erosion. This value is similar to results obtained at another lake, Havgårdssjön, where the catchment/lake ratio is even smaller (Dearing et al. 1987). This value approximates to a surface lowering of 0.02 mm yr⁻¹ (assuming a soil bulk density of 1 t m⁻³) which is some five times less than the generally accepted rate of soil formation (Morgan 1986). This suggests that the system as a whole was aggrading at this time, implying that soils were stable, and that sediment transport was dominated by low energy processes. Information about sediment sources indicates that unweathered clays or subsoil were major sources for the eroded sediment, and it is possible to conclude that the eroded material was derived from either stream channel

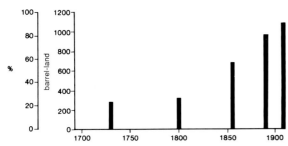

Fig. 5.2:8. Ploughed areas in Bussjö village, 1730–1908 ; area in tunnland as a percentage of total village area (being the same through time). Late 20th century values are similar to those in 1908. Unpublished results of A. Persson.

banks or from the deposits at the lake margin. Since the lake is large, and prone to wind-generated wave erosion of the shore, the possibility of the lake margins contributing a large amount of material to the lake sediments is likely. In contrast, sediment sources at Bussjösjön were dominated by topsoil material, but given the extremely low levels of sediment yield it seems likely that the major process of soil mobilization here was the periodic and natural collapse of the stream banks which included surface soil.

The earliest phase of higher sediment yields, 7000–4500 BP at Krageholmssjön, correlates with a change in sediment source towards an overall greater contribution by topsoil. The pollen record (Ch. 4.3.2) suggests a change in woodland composition at ca. 5100 BP, and there is evidence of burning, with microscopic remains of charcoal preserved in the sediments. At Bussjösjön the sediment yields remain low. What seems likely, though, is that early deforestation could alter the erodibility of the catchment soils, and this locally represented the first period of destabilization (Fig. 5.2:7). Some deforestation also took place at Bussjösjön and it is possible to conclude that its relatively larger catchment was able to store a large proportion of the eroded material. At Krageholmssjön soil was eroded directly into the lake or into small streams whose sediment transport times were short. Interestingly, the small peak in subsoil ca. 4500 BP correlates with lower sediment yields, perhaps suggesting some stabilization of the landscape after the earlier disturbances.

Both Bussjösjön and Krageholmssjön show increases in the proportion of subsoil in eroded sediment during Bronze Age times, though these records do not appear to be synchronous and neither of the sediment yield records shows an increasing trend in response. Deforestation at Bussjösjön and Krageholmssjön (Ch. 4.3.2) suggests a regional expansion phase during the Late Bronze Age, but the evidence suggests that erosion of soil from slopes was less important than an accelerated erosion of subsoil. This would have occurred as a result of streams eroding their channel beds in response to higher peak levels of runoff from a deforested land-

scape, with large areas of sediment storage to account for the relatively stable sediment yield values.

The first rise in sediment yield at both sites is dated to 2800–1500 BP, the Late Bronze Age/Roman Iron Age, indicating a major regional response to human activity or climatic change and correlating with evidence for subsoil as the dominant sediment source. At Krageholmssjön there is again evidence of a major phase of local burning, and it would appear that this period represents the second major destabilization of the catchment soils and deposits (Fig. 5.2:7). The data indicate accelerated channel erosion, and this would be caused by either an appreciably wetter climate or by modification of the hydrological cycle through an intensification of farming activities. A water-level rise at Krageholmssjön, 2800–2000 BP, supports the idea of a wetter climate, but puts in doubt the accuracy of the sediment yield calculations during this period. At Bjäresjösjön the sediment yield values also reach a peak during this period, and there is evidence for a small expansion in arable farming. Together the lines of evidence suggest that cultivation, rather than simply woodland clearance, may have been the crucial factor in causing sediment losses to rise; not so much by causing soil erosion on slopes (though that is likely too), but by accelerating runoff and hence increasing the power of streams to erode their flood plains, to deepen their channels, and to transport sediment.

Peak values in sediment yield occur at ca. 1500 BP at Bussjösjön and Krageholmssjön, and 300 years later at Bjäresjösjön. These dates correspond to the Late Iron Age, and, although low water-levels are recorded at both Bjäresjösjön and Krageholmssjön during the period 1800–1400 BP, at all three sites there is some evidence for a switch in sediment source to one which included substantial proportions of topsoil, presumably by surface erosion processes acting on bare fields. After the Late Iron Age the trend in sediment yield is, for the most part, increasing to the present day at all three sites. And it seems that the Late Iron Age was perhaps the first period in the history of the area when human activity produced irreversible changes in the form of erosion and drainage processes – the third major destabilizing phase (Fig. 5.2:7).

During the period 900–600 BP, the Viking Age and Middle Ages, subsoil makes its greatest contribution to the lake sediments at all the sites, a significant change from the previous period. Sediment yields are no higher, suggesting that hydrological systems had to some extent become stabilized by this time, perhaps achieved by holding more sediment in storage in flood plains and on lower slopes. The pollen records do not indicate any reduction in the area of cultivated land at this time (Ch. 4.2.2, 4.3.2).

After this time the proportion of topsoil entering the lakes increased rapidly. At first sediment yields remained unchanged, but at all three sites there is a dramatic rise in sediment yields from the 18th century up

290

1939   1969   1981

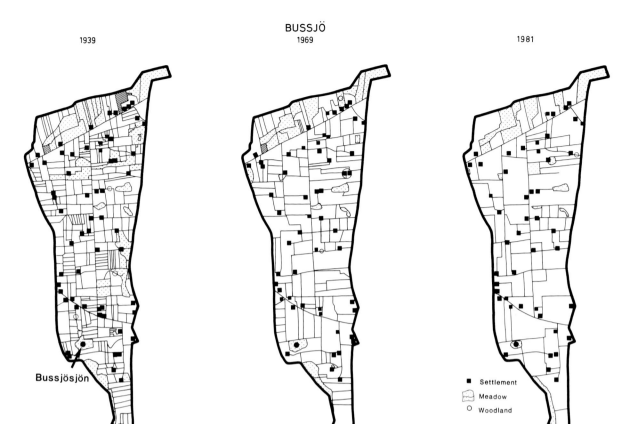

Bussjösjön

■ Settlement

▥ Meadow

○ Woodland

0 _____ 1000m

Fig. 5.2:9. Plans of field boundaries around Bussjösjön 1939, 1969, and 1981. From Ihse 1985.

until the end of the 19th century. During this period sediment yields increased by more than they had done in the previous 6000 years, by between 2 and 10 times over the levels of the 17th century. The rate of increase was greatest during the middle of the last century. The causes of this major increase in erosion are linked to the intensification and growth of farming from the 17th century onwards. Fig. 5.2:8 shows data for the area of ploughed land in Bussjö village during this period; a sixfold growth in population in the parish coupled with the introduction of the five-field system in 1850 combined to increase the annual area of ploughed land from 25% of the village area in 1730 to nearly 90% in 1910 (Persson unpubl.). This change from a "medieval" agricultural system to a more modern one, apparently widespread throughout the Ystad area, appears to have triggered a dramatic phase of soil erosion and may be viewed as the fourth period in which soils were destabilized (Fig. 5.2:7).

As the records approach the present day, the links between land use and erosion become more clear. For instance, it is certain that the practice of under-drainage during the late 19th century and early 20th century was widespread. During the excavation of the drainage

ditches, subsoil was eroded into these lakes, as shown by the peak values of HIRM/Xfd in the upper parts of the Bjäresjösjön record (Fig. 5.2:6). However, the overall effect on erosion was to remove the surface streams that had acted as routeways by which eroded soil could reach the lakes. The response to this is seen in the reduced levels of sediment yield at Bussjösjön and Bjäresjösjön (Fig. 5.2:5). Although this modification of drainage led to a reduction in the amount of sediment leaving the catchments, it is not implied that erosion rates on the slopes were lower.

For recent decades there is a paradox in the data. Sediment yields are apparently low, but there is additional evidence for high rates of soil erosion at the present time in fields around Bussjösjön. Does this mean that the data for slope erosion are atypical or relevant only to the last few years? Why, apparently, is there no record of these measured processes in the lake sediments? What exactly have been the effects of intensive agriculture? Certainly there is also considerable evidence available to show that there has been an intensification in farming during the last 40 years. Fig. 5.2:9 shows the field boundaries in part of Bromma parish around Bussjösjön in 1939, 1969, and 1981 (Ihse

1985). Between these dates hedgerows have been removed, fields have been enlarged, and cropping has become almost continuous with few leys. It seems likely that, at the very least, rills may now develop more readily because of this increase in slope length. The fact that topsoil removed in rills does not appear to reach the lake in any great amount points to short-term sediment storage on the flatter and lower slopes around the lake edge. The stored soil will only move towards the lake when rills develop in those storage zones or when soil movement is assisted by ploughing. It is possible to envisage a time-lag of, perhaps, several decades before the eroded soil reaches the lake. This means that the high erosion rates monitored in the fields are likely to be a recent phenomenon; if they had been as high during the early 20th century it would be expected that they should now be registered in the lake sediments.

The causes of the recent soil erosion are complex: increased slope length through field enlargement, the presence of bare soil during early spring with autumn-grown crops, and the possibility of decreased levels of organic matter in the soil have all combined to alter the previous relationships between farming and the soil. The soil system is now vulnerable to erosion by overland flow and rilling after a rapid thaw, especially when the thaw coincides with rainfall – the fifth period of significant destabilization of the soil system.

Previous destabilization phases have apparently led to new phases of relative stability; periods when farming and soil processes were in a kind of equilibrium. But what of the future? Today's highest erosion rates are over 50 times the rate of soil formation, and many hundreds of times higher than the earliest estimates of sediment yield from forested landscapes at the same site. By NW European standards these rates are very high and seldom exceeded (Morgan 1986). But where similar rates have been measured elsewhere the conditions are much the same. Erosion rates on the thin chalky soils of the South Downs in England frequently exceed 30 t ha^{-1} yr^{-1} and have reached levels of 270 t ha^{-1} yr^{-1}, and here too there has been an intensification of agriculture and a switch to autumn ploughing (Boardman 1990).

Even though the clayey soils in the Ystad area are generally deep and fertile, the high rates of erosion are equivalent to a surface lowering of ca. 3 mm yr^{-1}, which would expose the subsoil, below 50 cm, in less than 200 years. In the past, soil erosion was feared by agricultural societies for its effect on soil fertility, although in the Ystad area this was probably not a frequent problem. The dramatic rise in sediment losses at Bussjösjön during the 18th and 19th centuries was paralleled by a threefold increase in crop productivity. But in modern farming, soil erosion is often symptomatic of different problems: the overall decline in the structural qualities of the soil. Without good structure, natural drainage is reduced, nutrient uptake from fertilizers is inhibited and the soil becomes vulnerable to compaction by animals and farm machinery. Clearly there are certain farmed areas which would greatly benefit from a return to a less intensive agricultural system, with smaller fields and more frequent leys.

Overall the record of erosion in the Ystad area has been one of increase ever since the initial disruption of the natural ecosystem, 7000–6000 BP. At no time have the ecosystems been allowed to recover completely; rather there have been times of destabilization, triggered by human intervention, followed by periods of relative stability as the system reached a different set of equilibrium conditions. It remains to be seen whether or not there have been climatic triggers in the past, but on the evidence for present erosion, climatic conditions have played a secondary role to the factors of slope length and crop husbandry. If a new equilibrium is reached under these conditions it might prove to be one that is far less suitable for the demands made of the soil by today's farmers than it was for their predecessors.

Acknowledgements. Many individuals have contributed to the work summarized here. I am particularly grateful to K. Alström, A. Bergman, F. El-Daoushy, M.-J. Gaillard, R. Grew, M. Ihse, A. Persson, J. Regnéll, and P. Sandgren either for allowing me to use their unpublished results, for assistance in the field, or for comments on this paper. I also thank I. Foster for wide-ranging discussions and comments on an earlier draft.

5.3 Agro-ecosystems from Neolithic time to the present

E. Gunilla A. Olsson

The agro-ecosystem concept

Much of the development and change of the landscape in the Ystad area over the long period of 6000 years is the story of the management, use, and change of agro-ecosystems. Agro-ecosystems are subject to the same processes as natural ecosystems, and they are powered by solar energy for the production of plant biomass. Also, they are always subject to the same physical processes of weathering, erosion, and nutrient circulation. The major difference between natural ecosystems and agro-ecosystems is that the latter are manipulated in various degrees to maximize the production of one aspect of plant biomass – the cultivated crops. The intention is to control both the biological and physical processes, but this is usually at the expense of a reduction in biological diversity. Cultivation seeks to reduce competition for limited resources for the crop plants. A larger part of the solar influx can thereby be directed to human consumption, either indirectly via herbivores or domestic livestock, or via direct human consumption of the crop plants.

Biomass production of both wild and cultivated plants is maintained by the availability of:

* solar energy
* carbon dioxide
* water
* plant nutrients (particularly nitrogen, potassium, and phosphorus)

Carbon dioxide and solar radiation set limits at fixed levels in any geographic location. Radiation limits biomass production both by direct influence on maximum assimilation level during the vegetative season, and because it determines the length of the frost-free season or vegetative period. In the climatic region of our study area, water limitations for biomass production differ greatly from year to year because of regional climatic fluctuations, while the shortage in mineral nutrients is much more constant for plant productivity. In natural ecosystems the circulation of nutrients is relatively balanced. Most of the nutrients in the biomass produced are circulated via the decomposer chain and via additions from stored sources by the weathering processes. In agro-ecosystems the harvest of biomass and the increased leaching will lead to degradation of the soil nutrient pool unless this is balanced by a continuous supply. Generally, water and plant nutrients can be considered the most important limiting factors for crop production in this part of Europe. Agro-ecosystems are generally more or less open systems (cf. Tivy 1990). The degree of recycling and existence of internal regulation processes are a function of the cropping system that is practised.

In contrast to modern systems, pre-industrial agro-ecosystems exchange energy and matter mainly with neighbouring ecosystems and they contain internal feedback systems (Odum 1984). Further, in pre-industrial systems the landscape diversity is generally high. The management methods applied and the spatial scale of their use give rise to vegetation in different successional stages. A consequence of this is high biodiversity on the ecosystem level, and both within and between populations (Oldfield and Alcorn 1987). This high diversity on different scales implies complex ecosystems sharing many features with natural ecosystems in the same region.

Agricultural impact variables

In landscapes where woodlands are the potential natural vegetation, agro-ecosystems are generally created and maintained by a few major factors, although each of them can appear in manifold variations. The principal factors are:

A. Initiating factor
   1. clearing of woodlands
B. Maintaining factors
   1. livestock grazing
   2. fodder collection
   3. cultivation

These factors can be viewed as headings for a large variety of activities and management practices. Thus the term *clearing of woodlands* in this context would mean woodland decline as a result of:
1) spontaneous or intentional fires
2) pests or diseases reducing the tree populations
3) climate change reducing the biological productivity and growth conditions of the tree populations
4) clear-cutting of woodlands

Livestock grazing includes four subvariables:
1) herbivore species: cow, horse, goat, sheep, pig, etc.
2) grazing by single species or a mixture of herbivore species
3) grazing pressure (number of animals per area)
4) time of year and duration of grazing periods (all year round; seasonal; rotating system, etc.)

TOOLS AND TECHNIQUES

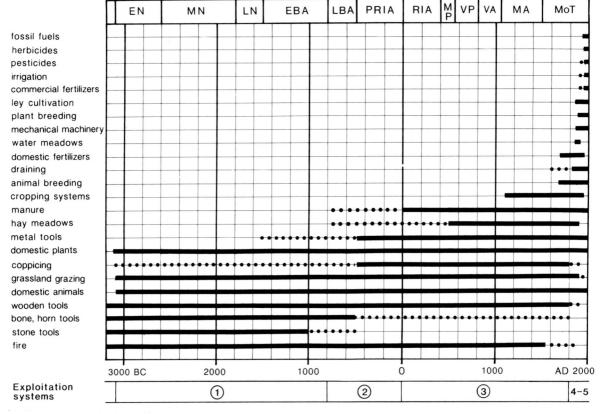

Fig. 5.3:1. The introduction and persistence of tools and techniques of significant importance for the development of agricultural production and for the evolution of the cultural landscape in the Ystad area. Solid line = tool/technique in common use. Broken line = tool/technique in limited use.
Source: Leufvén 1914, Nilsson 1920, Weibull 1923, Johansson 1953, Eskeröd 1973, Barker 1985, and unpubl. archive material.
Exploitation systems: 1 = shifting cultivation; 2 = permanent arable, transition period; 3 = permanent arable, infield-outland system; 4 = permanent arable, external nutrient input; 5 = industrial agriculture. EN = Early Neolithic; MN = Middle Neolithic; LN = Late Neolithic; EBA = Early Bronze Age; LBA = Late Bronze Age; PRIA = Pre-Roman Iron Age; RIA = Roman Iron Age; MP = Migration Period; VP = Vendel Period; VA = Viking Age; MA = Middle Ages; MoT = Modern Time.

Fodder collection is composed of the subvariables:
1) tree leaf and twig harvest
2) coppice management
3) hay harvest

Cultivation, finally, displays almost endless variation:
1) crop species used
2) techniques for seed-bed preparation (different tools etc.)
3) tilling techniques
4) seeding techniques
5) fertilizer/manure techniques (frequency, quality, quantity)
6) draining techniques
7) harvest techniques
8) cropping systems (including fallow regimes etc.)

The factors maintaining the agro-ecosystems can further be classified according to the *exploitation systems* used, e.g. the way of achieving biomass yields for harvests, and maintaining nutrient circulation in the agro-ecosystems. Biological production can be exploited by letting the herbivores (wild or domestic) consume the natural vegetation and then using the animal meat for human consumption (pastoralist economies). Alternatively, biological production can be utilized by managing vegetation as monocultures in arable fields where the primary production is consumed by the humans themselves (arable economies). Further, the *tools* and the *control variables* used to maintain a particular exploitation system must be determined.

In the following, a sequence of outlines of the agro-ecosystems in the Ystad area is presented, from the Neolithic to the present. The theme is composed of the four main factors above (A1–B3). The main exploitation system, along with the control variables, and the

CROP SPECIES DIVERSITY

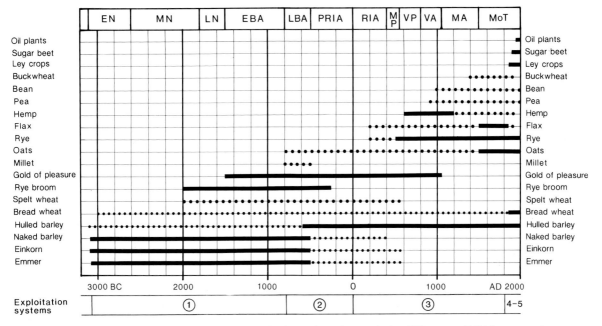

Fig. 5.3:2. Crop species diversity in the Ystad area; introduction and persistence during 6000 years. – Solid line = crop in common use. Broken line = crop in limited use.
Sources: Prehistoric time: Engelmark and Viklund 1991, Gaillard and Berglund 1988 (hemp). Historical time: unpubl. agricultural statistics 1892–1898, Malmöhus läns hushållningssällskap, SCB 1913–1990, Germundsson and Persson 1987. Exploitation systems and archaeological chronology – see Fig. 5.3:1.

techniques and tools for its maintenance are suggested for each period (Figs 5.3:1, 2, 11, 12).

Methods and sources for the history of agriculture

The survey of the agro-ecosystems in the present paper suffers from the lack of continuity in the data. The periods are chosen to represent the main developmental periods, but the time gaps, the data content, and the details differ from period to period. A consistent comparison of the same variables on an equal level of accuracy is not possible.

Prehistoric times. The basis for these periods is the archaeological finds, the sediment cores with pollen data, and other palaeoecological finds. The archaeological finds reflect the different preservation capacity of different materials. The data are fragmentary, showing great differences in terms of information on ecology, technique, and cultural and social conditions. The sediment cores which form the basis for the pollen data often give continuity through time, but there are problems in translating these data into landscape-ecological areas. Besides these three main data sources, analogies with present agro-ecosystems and ecological processes have been used.

Historic times. For the Middle Ages there is sparse documentation of agro-ecosystems. Analogies with similar European conditions have been used, mainly from England. The fact that new agricultural practices were

adopted significantly later in Scandinavia is taken into consideration. The highly accurate cartographic documentation performed during the 17th and 18th centuries in southern Sweden allows detailed analysis of agro-ecosystems on a local level. It is a paradox, though, that for later periods the information is more general and less detailed for the local agro-ecosystems. Even for the present time the statistical data are not available on the level of the local ecosystem, which now is represented by single farms.

Agro-ecosystems in the Ystad area during eight periods

Early and Middle Neolithic farming

In the Early Neolithic, ca. 3100 BC, the pollen diagrams from the Ystad area show a significant decrease in pollen from the predominant forest tree species (mainly lime, ash, and elm) (Ch. 5.13). This decrease, visible throughout southern Scandinavia, has generally been called the *Ulmus* decline, and has been interpreted as a sign of the introduction of agriculture (cf. Iversen 1941). However, several factors coincided to cause a forest decline. A rapid change of climate has been suggested together with the outbreak of a biological pest, *Ulmus* blight (Ch. 5.13). These factors in combination probably contributed to general unstable ecological conditions in the woodland ecosystems. They provided the prerequisites for the initial forest clearings by Neolithic

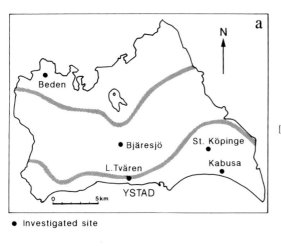

a

N

Beden

Bjäresjö St. Köpinge

L.Tvären Kabusa

YSTAD

0   5km

● Investigated site

RESOURCE AREA LATE BRONZE AGE
Increased above subsistence level

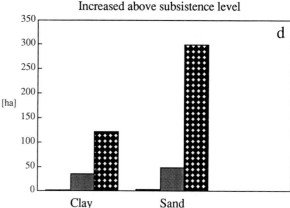

d

[ha]

Clay   Sand

RESOURCE AREA NEOLITHIC AGE

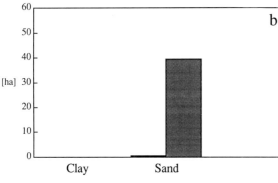

b

[ha]

Clay   Sand

RESOURCE AREA IRON AGE

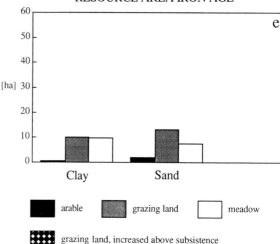

e

[ha]

Clay   Sand

■ arable ▨ grazing land □ meadow

▨ grazing land, increased above subsistence

RESOURCE AREA LATE BRONZE AGE

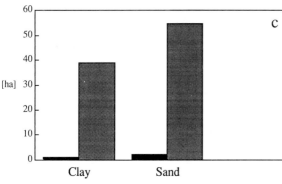

c

[ha]

Clay   Sand

Fig. 5.3:3. Comparison of resource areas used by human populations and their livestock in the Ystad area during prehistoric time.

a. Key map indicating investigated sites.

b. Neolithic Age: Settlements in Kabusa (settlement on sand) in the Early and Middle Neolithic. Calculation for a mean settlement with 7.5 adult humans. Subsistence consumption of livestock and arable crops according to palaeoecological finds.

c. Late Bronze Age: Comparison between settlement on clay (Bjäresjö) and sand (Stora Köpinge). Mean size of each settlement is estimated at 7.5 adult units. Subsistence consumption of livestock and arable crops according to palaeoecological finds.

d. Late Bronze Age: Comparison between settlement on clay (Bjäresjö) and sand (Stora Köpinge). The mean size of each settlement is estimated to 7.5 adult units. Calculation of grassland area needed to maintain the grassland landscape of the barrows. N.B. The area scale differs from that in the other figures.

e. Roman Iron Age: Comparison of resource areas between settlement on clay (Bjäresjö) and sand (Stora Köpinge) needed to maintain 7.5 human adults. Mean settlement size at this time was 30 adults. The actual resource area *per settlement* must be multiplied by 4.

In the figures above the arable figures are derived from different levels of seed return. The winter fodder area is derived from the assumption of hay yield of 750 kg ha^{-1} yr^{-1} for Bjäresjö, and 1000 kg ha^{-1} yr^{-1} for Stora Köpinge. The figures are related to the yield data from the 18th century for those areas. The summer fodder areas are derived from the assumed biomass production of grasslands and wood-pastures of 700 kg ha^{-1} yr^{-1} for Bjäresjö and 500 kg ha^{-1} yr^{-1} for Stora Köpinge. These figures are related to the vegetation data from the pollen diagrams from this period.

human populations, for their cultivation and the grazing needs of their livestock.

The archaeological remains of settlements from the Early Neolithic are all from sandy soils. There is no evidence of cultivation on clayey soils from the Early Neolithic. The excavated habitation site at Kabusa (Fig. 5.3:3a) (Larsson 1987, Ch. 4.1.3) represents a typical Neolithic agro-ecosystem in the Ystad area. This site is located at the transition between three different ecosystems: the terrestrial, sandy soil site with deciduous woodlands, the lacustrine river ecosystem, and the marine shallow sea lagoon off the Baltic coast (Ch. 4.1.3). It is assumed that the human population consisted of a "family" group of roughly 5 adults and 5 children (Ch. 4.1.3). Finds from the settlement also show that the lacustrine and marine ecosystems were used for human nutrition. The human dwellings were moved from time to time, probably when the building material needed renewal (Ch. 4.1.3). These movements were over very short distances, however – only a few hundred metres (Larsson 1987).

The reconstruction of the agro-ecosystem and the archaeological finds (Larsson 1987, Ch. 4.1.3) suggest that very small areas were used for cultivation, usually less than half a hectare. The arable plots were put into long-term fallows and it is reasonable to believe that former habitation sites were used for succeeding cultivation to make use of the accumulated nutrients on such sites. A cropping system with long rotation periods (up to 50 yr) was most probable, perhaps in combination with deciduous coppice management (Göransson 1988). Maintaining coppice among the temporary arable is an effective way to conserve soil nutrients. This agrarian practice of shifting cultivation can be considered as a form of agroforestry. For reconstructing the size of the arable a seed return of 4–6 for einkorn and emmer has been assumed (Ch. 4.1.3.3) which is at least not an underestimation according to long-term cultivation experiments in Rothamsted, England (Rothamsted Experimental Station 1970) and recent experimental cultivation studies (Lüning and Meurers-Balke 1980, Olsson, unpubl.) (Table 5.3:1). Emmer and einkorn were the most important crops during the Early and Middle Neolithic (Engelmark and Viklund 1991), but also important were varieties of barley (Fig. 5.3:2). The small fields were worked with hand implements made of wood and stone. The small size of the arable fields suggests that Neolithic farming had similarities with gardening, and it could well have been possible to weed the crops. Assuming these practices, it is likely that soil erosion and nutrient leaching out of the system were fairly low.

Cattle, sheep, pigs, and dogs were kept as domestic animals. During Neolithic time the climate allowed year-round grazing for the livestock, which had a significant impact on the natural vegetation. A total resource area of about 40 ha is estimated (Fig. 5.3:3b). The creation of semi-natural vegetation, grass-swards, and grazed woodlands had started. The primary factors in this process were the grazing and trampling of the domestic livestock.

The creation of the cultural landscape of the Early Neolithic humans was favoured by biological and climatic factors leading to a decrease in the stability of the woodlands (Ch. 5.13). It was thus possible for the human population to create forest clearings with very simple tools and maintain them by small-scale cultivation and livestock grazing. Primary tools used by Neolithic farmers were (a) fire and (b) hand implements mainly of wood, stone, bone, and horn (Fig. 5.3:1). When environmental conditions changed and affected the regeneration capacity of the woodland tree populations, it probably became difficult for the Neolithic farmers to maintain the semi-open landscape (Ch. 5.13).

Input of matter and energy. Given that cereal seeds for the arable were introduced in a few instances, all the matter and energy needed by the population were withdrawn from the local ecosystem. The settlement and hence the agricultural activities at the Neolithic site at Kabusa lasted for a period of more than 700 yr. Similar farming techniques were applied during the Late Neolithic and Early Bronze Age, mainly on sandy soils in the coastal area and in the outer hummocky landscape.

Limits of the system. Given the assumptions above, it is difficult to find any distinct limitations for the persistence of this agro-ecosystem – as long as the human population remained fairly small. When the population grew, and if people preferred to remain at the same settlement site, it would have been necessary to cultivate larger fields. The availability of uncultivated land, or land that was allowed to lie fallow for about 50 yr, would have been a limiting factor.

Late Bronze Age farming

From this period we have archaeological information about agro-ecosystems on both sandy and clayey soils. The cultivation practice of rotation cropping and long fallows (possibly up to 50 yr) allowed good conditions for nutrient recirculation even on sandy soils. The degree of soil erosion and leaching from grazing and trampling of livestock in the arable areas cannot be estimated (cf. Ch. 5.2). We assume that farming was still combined with coppicing. Bronze Age farmers used mainly the same tools as Neolithic humans for their agricultural activities. The hand implements were supplemented with a wooden ard drawn by humans and/or by oxen. The new metal, bronze, was used mainly for ceremonial activities, worship, and for military purposes (Stenberger et al. 1974).

Crop diversity had increased compared with the Neolithic and, besides einkorn and emmer wheat, and the two varieties of barley, also oats, millet, rye brome, and gold of pleasure *(Camelina sativa)* were cultivated (Engelmark and Hjelmqvist 1991, Engelmark and Viklund 1991, Fig. 5.3:2). Spelt wheat became increasingly important during the later periods of the Bronze Age. A

Table 5.3:1. Comparison of yields from different empirical cultivation studies - unfertilized plots.

Crop	Seed return	Period of study	Study site	Reference
wheat	5.1	1852-1967	clay loam Rothamsted England	Roth. Exp. Station 1969
barley	4.6	1852-1967	clay loam Rothamsted England	Roth. Exp. Station 1970
barley	6.7	1877-1896	sandy loam Bedfordshire England	Cooke 1976
bread wheat	5.0	1877-1896	sandy loam Bedfordshire England	Cooke 1976
einkorn	4.9	1979	loam Rhineland Germany	Lüning and Meurers-Balke 1980
emmer	11.8	1979	loam Rhineland Germany	Lüning and Meurers-Balke 1980
spelt	13.1	1979	loam Rhineland Germany	Lüning and Meurers-Balke 1980
barley	6.5	1979	loam Rhineland Germany	Lüning and Meurers-Balke 1980
barley	6.2	1985-86	boulder clay Scania Sweden	Olsson unpubl.
barley	4.9*	1985-87	boulder clay Scania Sweden	Olsson unpubl.
rye	7.6*	1986-87	boulder clay Scania Sweden	Olsson unpubl.

* = yield depression due to fungus attack in 1987.

reduction in the tree pollen influx and a simultaneous increase in non-arboreal pollen (Ch. 5.13) during the Early Bronze Age indicate a landscape with large grasslands. Whether fires were used to maintain the large grasslands in combination with huge livestock herds is an open question. The increased amount of charcoal particles in pollen-analysed sequences (Ch. 5.13) may be connected with such extensive burning.

The sandy plain in the east, the Köpinge area, was by this time intensively used for farming and foraging. The settlement sites were located so that access to highly productive fodder areas was guaranteed, either in the large wetland of Herrestads Mosse or along the riverbanks. Barrows were scattered over the sandy plain. Since barrows were built to be seen over long distances, the pattern of their distribution indicates that most of the sandy plain must have been covered by grasslands. Maintaining such large grasslands would have required a large animal stock. The estimated grassland area necessary in the Köpinge area to provide fodder for the

stock needed for human subsistence was much smaller than the large areas inferred from the pollen diagrams (Ch. 4.1.2), and estimated to be the environments of the Bronze Age barrows. In fact, five times the number of stock at human subsistence level would be necessary to justify such extensive grasslands (Fig. 5.3:3c,d).

Farming practices on clayey soils were similar, although the resulting agro-ecosystem differed owing to different soil moisture conditions. The investigated site at Bjäresjö (Fig. 5.3:3a) has marked topography compared with the sandy plains in the Köpinge area. The Bronze Age settlements were located on the hilltops where a sandy-clayey soil type is prevalent. The heavier clayey soils in the depressions were still covered by a variety of wet wood-pastures and fen woodland.

The Bjäresjö area consisted at this time of a mosaic of open grasslands, wet and dry wood-pastures, and more or less wooded marshes and fens along the river Svartån (Ch. 4.2.2, 4.2.4). Assuming the same human population per settlement as in Köpinge, and a similar dietary consumption, the total resource area was about 15 ha smaller on the clayey soil in Bjäresjö than on the sand plain in the east (Fig. 5.3:3c). This difference is due to the differences in production potential, such as the more fertile soils in the Bjäresjö area. The harvest yield from the arable was estimated from literature sources on empirical cultivation (Table 5.3:1). In order not to underestimate the areas needed for cultivation, the harvest yields were set slightly lower than could be expected for this soil and climate region.

Fodder production, both in the wood-pastures and in the grasslands, was larger in the Bjäresjö area. However, if we take into account that the people on the sandy soil probably had larger numbers of stock, then the difference in resource area between ecosystems on different soil types in the Bronze Age must have been significant (Fig. 5.3:3c,d).

Comparing land use over time in the sandy soil area (Köpinge) and the boulder clay area (Bjäresjö), we see a difference in the time of duration. The boulder clay areas were apparently first used for cultivation in the Early Bronze Age (Ch. 4.2.4) while cultivation on the sandy soils had started more than a thousand years earlier (cf. Ch. 4.1.3, 5.13). The concentration of Early Bronze Age barrows which can be taken as a measure of land use was about equal in the sandy areas (Köpinge) and in the boulder clay areas around Bjäresjö. It was estimated at about 1.8 barrows km^{-2} and 1.6 barrows km^{-2} in Köpinge and Bjäresjö respectively (Ch. 4.1.4.2). The distribution patterns were different and indicated larger areas of wide-open landscape in Köpinge than the more patchy landscape in Bjäresjö (Ch. 3.3).

There seems to have been a significant overproduction of livestock demanding large areas of grasslands to feed them during the Bronze Age in the Ystad area. A combination of burning and grazing seems to have maintained these grasslands. The reasons for this division of energy and land use can only be surmised. Large numbers of cattle and/or sheep might have been kept for religious and social reasons. During the Late Bronze Age, the permanent arable system was gradually introduced and probably occurred in the area parallel to the shifting cultivation system (see discussion below, and Ch. 3.3).

Input of matter and energy. Matter and energy were withdrawn primarily from the local ecosystem. However, the seed for cultivating the new crops *Camelina*, oats, and barley must have been imported on at least one or a few occasions. The metals copper and zinc as components of the valuable bronze also had to be imported, although the bronze was used mainly for activities not directly related to agriculture.

Limits of the system. Considering the increasing human population, a shortage of "new" land for cultivation, or land fallowed for a considerable time, might have been a limiting factor. The Bronze Age farmer favoured sandy soils, and nutrient depletion of the arable on the sandy soils of the south-eastern plain might have affected crop yields.

Iron Age farming
The Roman Iron Age was characterized by a climatic change towards cooler and wetter conditions. A consequence of this, which is evident from the archaeological finds, was that winter stalling of stock became more common than before, or even necessary. However, archaeological finds give hints that buildings used for stalling stock occurred already during the Bronze Age (Ch. 5.5, Myrdal 1988 with further references). Winter stalling of the livestock demanded more intensive stock management with more intensive and effective fodder collection. This led to a more reliable yield of meat and other animal products (Myrdal 1988). Another important effect was that it was easier to collect and manage manure. There are some indications of the occurrence of manuring of arable already during the Late Bronze Age (cf. Ch. 3.4). However, from the Early Iron Age we also have palaeobotanical evidence of the practice of regular manuring in the Ystad area; this includes the frequent occurrence of seeds from a nitrophilous weed, fat hen *(Chenopodium album)*, in cereal finds (Engelmark 1989). By adding nutrients via manure to the arable it was possible to maintain stationary arable fields. It is uncertain whether the arable yield increased compared with the earlier cultivation practices. Long-term fallowing combined with short-term cultivation can well have the same effect on the nutrient availability for the crops as regular manuring in limited amounts. The time for the breakthrough of permanent arable and manuring can be dated around 800–500 BC in this part of southernmost Scandinavia. This seems to be considerably earlier than for eastern central Sweden, where geographers have assessed the time by the dating of stone fences associated with the arable (Widgren 1983). It is likely, however, that the time from 800 BC to AD 700 was a transition period with combinations of permanent arable fields maintained by manuring and shifting cultivation.

With the establishment of permanent arable based on the addition of manure, winter stalling of livestock, fodder collection by haymaking and leaf harvest, and livestock grazing, the *infield-outland system* was established. This brought about a polarization of land use with a strictly planned division of land use between the infields and the outlands, implying a systematically directed flow of nutrients from the outland grazing areas and the meadows into the arable in the infields. The iron scythe was the most essential tool for obtaining the winter fodder for the livestock. The increased use of iron during the Late Iron Age (from AD 500 onwards) also meant improved tools for fodder collection. The iron scythe with a long blade was a great achievement which is estimated to have doubled the rate of grass harvesting compared with the earlier short scythes (Myrdal 1988). The use of an iron plough-share on the wooden plough was an important improvement for cultivating the arable.

During the Iron Age rye was introduced as a new crop (Fig. 5.3:2). Rye and hulled barley became the predominant crops in the Ystad area. Oats, wheat, and gold of pleasure were also cultivated but to a lesser extent. Other new crops were flax and hemp. The latter displays very high pollen records from Bjäresjö, which is probably a reflection of the habit of using lake Bjäresjösjön for the preparation of the hemp fibres by retting (cf. Ch. 4.2.2). The Neolithic crops (einkorn, emmer, spelt, and naked barley) gradually disappeared during the Iron Age (Fig. 5.3:2).

The pollen diagrams from the Roman Iron Age (ca. 0–AD 500) indicate a slightly decreased human impact on the landscape (Ch. 4.2.2, 5.13). By organizing the fodder collection into summer grazing and the harvesting of leaves and hay for winter needs, smaller total resource areas were needed (Fig. 5.3:3e). Moreover, if we assume a higher proportion of cereals in the human diet (50% compared with 45% in the Late Bronze Age and 30% in the Neolithic) the need for arable areas for subsistence cultivation remained relatively small, an estimated 2 ha for a group of 7–8 humans on sandy soils (seed return of 2–3) and half that size on clay soils (seed return of 4; Fig. 5.3:3; Ch. 4.2.5.2; calculation based on protein need). These figures are slightly lower than the estimates of Enckell et al. (1979) for land during the Roman Iron Age (calculations based on energy need).

Along with the organization of permanent arable fields, the settlements were concentrated in fewer localities than during the Bronze Age. It has been estimated that an average of 4–5 extended "families" or 30–40 persons including children lived in each settlement (Callmer 1991). At the end of the Viking Age such settlements became stationary and thus developed into large villages (Ch. 5.6). Parallel to this there was also an increase in the cultivated areas. Around AD 700 the pollen data indicate an increase in the both the arable areas and the non-forested fodder areas (Ch. 5.13). This could possibly be a reflection of the introduction of crop rotation and the use of the two-field system. Twice the arable figures stated above would then have been needed. The pastoralist system with large stock herds and extensive grasslands finally gave way to an agricultural system with intensive land use of arables, and organization of the nutrient flows from outlands into the infields.

Comparing the agro-ecosystems on sand and clay, the arable area on sand needed to be twice the size of the clay area if a similar human consumption of cereals is assumed (Fig. 5.3:3). The grasslands on the sandy plains yielded lower fodder production than the clay grasslands. The wood-pasture areas on the clays had declined at this time (cf. Ch. 3.5), but since the soil moisture in the clayey area was generally higher than in the sandy region, the clays allowed a higher production both in grasslands and in wood-pastures. However, the production of meadow biomass should have been about equal on sand and clay, or even larger in the sandy area, considering the productive wetlands in depressions (e.g. Herrestads Mosse) and along rivers. The pollen data from the Late Iron Age give further clues to the expansion of meadow management by the decrease of wetland species like *Filipendula ulmaria* (meadowsweet) (Ch. 5.13) which is resistant to low grazing pressure, but sensitive to repeated cutting and hay harvest. At the same time there is an increase of wetlands dominated by low-statured sedge species tolerant to and favoured by haymaking.

The Iron Age was a transformational period with large differences in use of the ecosystems. The earlier periods were similar to the Late Bronze Age, founded on pastoralist economies. The agro-ecosystems of the Late Viking Age with an established infield-outland system and permanent villages resembled the agro-ecosystems of the Middle Ages.

Input of matter and energy. There were no local sources of iron in the Ystad area. As far as we know, the nearest iron sources were in northern Scania and southern Småland, some 150 km north of the Ystad area, which could have been possible import areas. Other matter and energy were withdrawn from the local ecosystem.

Limits of the system. The maintenance of a balanced nutrient circulation of the permanent arable soils would have been fully possible, especially if cropping rotations with fallows were used. However, from the Late Iron Age the rate of soil erosion increased (Ch. 5.2), indicating that nutrient leaching from the arable land would also have been significant. Manure might have been a limiting factor. The import and preparation of the iron implements could have been limiting for individuals, but hardly for the farmers of the Late Iron Age in general.

Agro-ecosystems of the Middle Ages

During the Early Middle Ages the infield-outland system was further developed. All parts of the landscape were used for agriculture (Ch. 3.6) or grazing. Due to local and regional differences in topography, soils, etc., the cultural landscape was further differentiated into increasingly diverse subregions within the Ystad area (Olsson 1989). Woodland grazing by both cattle and pigs was important. A written source on pig fattening in the woodlands of Baldringe in 1611 indicates that about 800 ha in this parish was used for this (Billberg 1989, Olsson 1989). This information also gives a hint about land use in the Early Middle Ages, although pig fattening in woodlands with masting trees (beech, oak, hazel) was probably even more intensive in the Middle Ages. The settlement pattern with an agglomerated village core surrounded by infields and with the outlands on the periphery provided the frame for medieval land use.

The Early Middle Ages (1000–1200) was generally a period of agricultural inventions over much of northern

Table 5.3:2. Medieval yields per measure of seed calculated from 40 manors belonging to the archbishop of Winchester, southern England, during the period 1209-1349 (Titow 1972). Mancorn is a mixture of wheat and rye (Adams 1976). Number of calculations denotes the number of records considered.

Crop Period	No. of calculations	Average yield
Wheat		
1200-49	411	3.8
1250-99	887	3.8
1300-49	1555	3.9
Barley		
1200-49	417	4.4
1250-99	844	3.5
1300-49	1436	3.6
Oats		
1200-49	438	2.6
1250-99	837	2.3
1300-49	1476	2.2
Mancorn		
1200-49	86	3.9
1250-99	183	3.1
1300-49	372	2.9

Europe (Myrdal 1988). The iron-framed spade was an important implement for cultivating new land and thus for expanding the arable areas. The detailed investigation of the medieval farmsteads of Borup in Zealand (Steensberg 1983), showed a total arable of about 50 ha managed in common by 3–4 farms (Porsmose 1988). The configuration of the arable was usually several parallel oblong strips with furrows and ridges. According to Porsmose (1988) the furrows were created with a wheel-plough and maintained using a wheel-ard. It is probable that wheel-ploughs were also used in the eastern part of medieval Denmark, in the Ystad area, at least in the boulder clay areas. However, the wooden ard was the major soil-preparing tool in many parishes in this part of Scania as late as the early 19th century (Bringéus 1964) and it must have been the predominant tool in the Middle Ages. The absence of a mould-board on the ard allowed some root weeds and fragments of the grass-sward to remain in the field and thus exert strong competition with the crop species. The use of furrows and ridges – the only way of draining in this area as late as the mid 18th century (Linnaeus 1751) – was an inefficient method, which further facilitated the growth of weeds.

The invention of cropping systems with regular fallow periods helped to intensify cultivation in the Early Middle Ages. There are indications from both written sources (Skansjö 1983) and palaeoecological finds (Engelmark 1989) that three-course rotation was practised in the Ystad area on boulder clay in the Early Middle Ages. Modifications of three-course rotation were used on the sandy plains as well, while different variations of the one-field or field-grass systems (Frandsen 1983) were used on the Romeleåsen ridge (Dahl 1989). The

long fallow periods (up to 20 yr) and the patchy conditions in the arable fields here contributed to a mosaic landscape with large diversity on both habitat and species level.

Rye and barley were the predominant crops in the three-course villages. Oats played a minor role, although this crop was more important in the northern boulder clay area (Engelmark 1989). Since oats were often used for feeding cattle, this information together with the occurrence of extensive grasslands in Baldringe indicates that animal husbandry was more important here. Flax was cultivated and buckwheat became a common crop on the sandy soils during the Middle Ages (Engelmark 1989, Engelmark and Hjelmqvist 1991, Ch. 4.2.2). Wheat and pulses (beans and peas) appear extremely sparsely in the palaeoecological finds. The cultivation of these crops was probably of little importance during the Middle Ages (Engelmark 1989), and remained so up to the late 19th century (Ch. 3.8.2). This is an interesting contrast to the medieval conditions in English counties with fertile soils. Here wheat cultivation in combination with legumes as a substitute for fallows was totally dominant on the fertile soils, e.g. in Sussex, Leicester, and Norfolk (Slicher van Bath 1963, Brandon 1971, Campbell 1983).

We lack data on crop yield and productivity from the Middle Ages in the Ystad area. Assuming a crop rotation without cultivation of nitrogen-fixing plants but with regular fallows with manure application, the conditions here can be compared with the Winchester estates in southern England. Yield data for cereals from 40 estates have been compiled for a complete series over the period 1209–1349 (Titow 1972). Soil fertility can be assumed to have been equal to the boulder clay area in the Ystad region. Similar crops were cultivated, with an important exception: wheat was important and generally cultivated, while rye was sown just occasionally. The proportion of leguminous crops was low, changing from 1% in the early 14th century to about 8% in 1345 (Titow 1972). Yields of barley and oats decreased significantly during the period from returns per seed of 4.4 to 3.6 for barley and 2.6 to 2.2 for oats (Table 5.3:2). Yields of wheat remained at the same level during the period (Table 5.3:2). According to Titow, the reasons for the low yields and the reduction during the period are:

* deficit of plant nutrients by permanent under-manuring
* climate deterioration

We have little information from the Ystad area about changes in land use in the Early Middle Ages. Up to the late 16th century a relatively "balanced" relationship between fodder areas and arable can be assumed; the fodder areas were considerably larger than the arable (Persson 1986b). This is also indicated by field-names on the 17th century maps suggesting a former land use

Table 5.3:3. Number of farms and number of inhabitants (calculated as adult units) in the villages of Bjäresjö, Beden, Kabusa, and Lilla Tvären. The composite figures for number of farms denote the farms with tenant land (first figure), plus farms without land and animals. Source: Beden and Bjäresjö: Olsson and Persson 1987, Olsson 1988a; Lilla Tvären: Persson 1988; Kabusa: A. Persson unpubl.

	Bjäresjö 1699	Beden 1750	Kabusa 1699	Lilla Tvären 1691
Number of farms	13+7	2	10+3	4+1
Number of inhabitants	75	10	47.5	37

of the arable to have been grazing lands and commons. The historical sources also corroborate such a transformation of land from fodder production to arable during the period from 1570 to 1700 (Persson 1986b).

Input of matter and energy. Iron had to be imported to the area. The nearest source areas would have been the southern parts of Småland where bog-ore deposits existed. Timber for building purposes was becoming scarce and had to be imported. Seeds for the new crop buckwheat occasionally needed to be imported. With the exceptions mentioned here, matter and energy were withdrawn primarily from the local ecosystem.

Limits of the system. Soil nutrient leaching and soil erosion had to be offset by adequate supplies of manure nutrients. Along with an increase of the arable areas, it is possible that a shortage of manure, at least during the latter parts of the Middle Ages, could have been a limiting factor for the agricultural production.

Early Modern Time before the introduction of land enclosures (early 18th century)
Generally, the agricultural practices and cropping systems were still within the framework of the infield-outland system until the early 18th century, although with large variations due to differences in such factors as population density. Four villages were studied to represent the local variation in the agrarian ecosystems of the Ystad region at this time. The methods for using the historical sources and applying the NUE concept have been explained elsewhere (Olsson 1988a, Ch. 2.2). A comparison of the applied cropping systems, the productivity, and the ecological effects on the ecosystems in Bjäresjö and Beden is presented by Olsson (1988b) and Olsson and Persson (1987). The present study uses unpublished data from A. Persson on agrarian conditions in Kabusa.

The four villages (Fig. 5.3:3a) represent the boulder

Fig. 5.3:4. Distribution of land-use categories in the infields in the villages of Bjäresjö, Beden, Kabusa, and Lilla Tvären, ca. 1700 Bjäresjö, Kabusa and Lilla Tvären used three-course rotation. Sources: land survey maps 1691-1750.

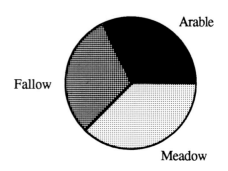

Bjäresjö 1699

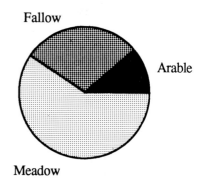

Beden 1750

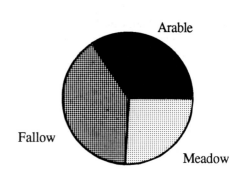

Kabusa 1699

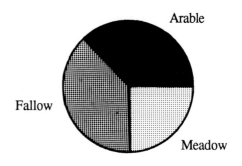

Lilla Tvären 1698

LIVESTOCK

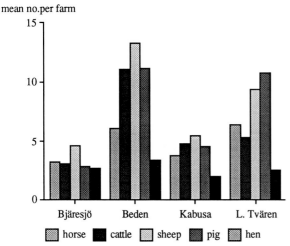

mean no.per farm

Fig. 5.3:5. Numbers of livestock per farm in the villages of Bjäresjö, Beden, Kabusa, and Lilla Tvären, ca. 1700. The figures for Beden are means from two different recordings, 1671 and 1768. Sources: A. Persson unpubl, Olsson 1988.

clay area (Bjäresjö and Lilla Tvären), the sandy plain (Kabusa), and the sandy ridge area (Beden). The numbers of farms and inhabitants in the villages are summarized in Table 5.3:3. The cropping system in Bjäresjö, Lilla Tvären, and Kabusa was three-course rotation, although with large variations between the three sites. In Beden, a variant of a one-field system was practised (Frandsen 1983, Dahl 1989). Considering that most villages and their lands were owned by the demesne or the church, it is obvious that the individual farmers had very little freedom to change land use.

The three three-course villages had much smaller fodder areas than arable fields (Fig. 5.3:4). In the village of Lilla Tvären the meadows were less than 30% of the size of the arable. Commons did not exist in the three-course villages, and the fallowed arables were the primary grazing grounds. On the other hand, in Beden, with its one-field system, the meadows were 40% larger than the arable, and in addition common pastures were available for livestock grazing.

The fodder areas in the different villages (Fig. 5.3:4) reflect livestock numbers, the availability of manure, and thus indirectly the cropping systems. A low number of stock can thus be expected in the three-course villages of Bjäresjö, Lilla Tvären, and Kabusa (Fig. 5.3:5). However, the village of Lilla Tvären is an exception to this prediction. In spite of the minute fodder areas (Fig. 5.3:4), Lilla Tvären maintains about the same number of stock as the stock-raising village of Beden (Fig. 5.3:5).

Hay yield. Hay production from natural meadows was generally low in the 18th century. Yields around 500 kg ha^{-1} yr^{-1} were normal, even in areas on fertile soils (Lindgren 1939). These figures can be compared with recent biomass data for unfertilized semi-natural

meadows ranging from 1000 to 1700 kg ha^{-1} yr^{-1} (data from dry meadows: Sjörs 1954, Persson 1975, Fogelfors and Steen 1982). The generally low 18th century figures must be related to the intensive and long-term exploitation of the grasslands with a combination of haymaking and grazing. Especially when grazing occurred also in spring and autumn, it can be assumed that nutrients were exhausted, with a subsequent decrease in productivity.

Comparing the size of meadow production in the four villages reveals striking differences: the two sites on sandy soils (Kabusa, Beden) had extremely low yields, with means of less than 270 kg ha^{-1} (Fig. 5.3:6). The low figures for Kabusa can be interpreted as a consequence of a combination of the long-term exhaustion of the grasslands here since the Neolithic, more than 4800 yr, and also of the relatively infertile sandy soils here. The low figures for Beden could be related to the coarse soils in the hummocky terrain sensitive to nutrient leaching. Although human cultivation activity with permanent arable started here rather late (during the 16th century), this area had been used since prehistoric time for grazing. The high meadow production for Lilla Tvären is probably related to the fertile boulder clay soil and the existence of wet meadows. However, the great difference in meadow production between Bjäresjö and Lilla Tvären cannot be explained so easily. Possibly the difference can be related to the reorganization of land

BIOMASS PRODUCTION
Arable crops (cereals); hay from meadows

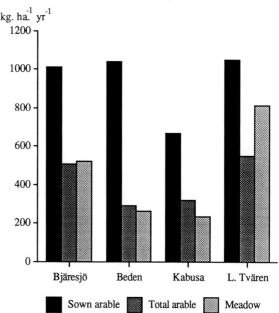

kg. ha^{-1} yr^{-1}

Fig. 5.3:6. Production of cereal crops and hay in the villages of Bjäresjö, Beden, Kabusa, and Lilla Tvären, ca. 1700. Crop production is calculated per sown area and per total arable area. Hay production from infield meadows is calculated as the mean of all meadow types. Sources: Persson 1986a,b, 1988, Persson unpubl.

N-BALANCE: ARABLE FIELDS
N in manures and fertilizers vs N in harvested crops

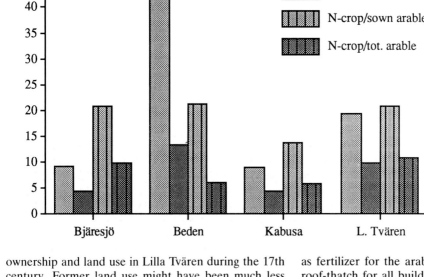

kg N.ha^{-1}.yr^{-1}

Legend:
- N-add/sown arable
- N-add/tot. arable
- N-crop/sown arable
- N-crop/tot. arable

Bjäresjö   Beden   Kabusa   L. Tvären

Fig. 5.3:7. Nitrogen balance in the arable fields in the villages of Bjäresjö, Beden, Kabusa, and Lilla Tvären. Nitrogen output in crop harvest and nitrogen input via fertilizers, manure, and seed.

ownership and land use in Lilla Tvären during the 17th century. Former land use might have been much less intensive here than in Bjäresjö.

Crop yield. The average yields of cereal crops (Fig. 5.3:6) also show interesting differences between the four villages. The two villages on fertile clay soils (Bjäresjö, Lilla Tvären) harvested about 1000 kg ha^{-1} yr^{-1}, while the sandy soil sites reached only 50–60% of this figure (Fig. 5.3:6). On the other hand, the village of Beden, on sandy soil but with a livestock-dominated economy, had an equal crop yield – as long as we compare the harvest per arable sown. Since the fallowed arables must be considered as necessary parts of the nutrient balance system, a relevant comparison should be made on the basis of the total arable area, including fallows. Such a comparison shows that the yield from the two sandy sites, Beden and Kabusa, is almost equal, and just half of the harvest on the clay soils (Fig. 5.3:6). The cost of achieving reasonable yields in Beden is a very large input of manure nutrients. This is possible thanks to the availability of excess manure from the large number of livestock. However, when relating the input of nutrients in Beden and Kabusa to the yields in the two other villages, it seems probable that a more intensive cropping system with shorter fallows could have resulted in larger yields in Beden (Olsson 1988b).

There are no records of fertilizers other than manure for the arable fields. However, we have indications of the use of seaweed in the coastal villages (Linnaeus 1751). From other sites in northern Europe from this time there is information on the recycling of roof-thatch

as fertilizer for the arable. In the present calculations roof-thatch for all buildings in the villages is therefore included, and also seaweed for Kabusa and Lilla Tvären.

Finally, one of the most important factors for the crop yield was competition with weeds. Linnaeus (1751) reports on cereal fields in this area where the weeds were more abundant than the crop plants. The technique of sowing by hand, the heterogeneous soil conditions, and the patchy soil humidity due to the lack of effective draining, provided a favourable situation for many species of arable weeds, many of which profited from the lack of seed cleaning and were permanent companions to the crop seed (Svensson and Wigren 1986).

However, weeds must also be considered as an advantageous factor for crop production. Leguminous (nitrogen-fixing) weed species, such as *Trifolium* spp., *Ononis repens,* etc., contribute nitrogen to the soil system and when they decay, plant-available nitrogen is released in greater amounts than from non-fixing species. The symbiotic nitrogen-fixing bacteria have a pH optimum of 6.5–7.5 (Etherington 1975). Especially on soils with alkaline pH reaction and moist conditions, the nitrogen fixation by the leguminous weeds might have been considerable. Such conditions were characteristic of the boulder clay soils in Bjäresjö and Lilla Tvären and should have brought about a favourable nitrogen balance here.

Nitrogen balance via input of manures and seed versus the output of harvested crops is shown in Fig. 5.3:7. For approximately equal harvest output in the villages, except Kabusa, the input of manure nitrogen was differ-

Table 5.3:4. Agricultural soil improvements and new water-meadows created during the year 1886, and existing remaining water-meadows in 1909. Sources: unpubl. agricultural statistics, Malmöhus läns hushållningssällskap, and Leufvén 1914.

	Herrestad	Ljunits
1886		
Subsoil drainage during the year (ha)	+280	+238
Marling during the year (ha)	+738	+567
Water-meadows during the year (ha)	+108	−122
New water-meadows in Bjäresjö (ha)	+50	
1909		
Total existing water-meadows (ha)	127	98
Total existing water-meadows in Bjäresjö (ha)	0	

ent. It is only in the three-course village of Lilla Tvären that balanced conditions seem to have existed (Fig. 5.3:7). The large manure inputs in the livestock economy of Beden have been discussed above. The low input of manure nitrogen in Bjäresjö was probably offset in part by release from the soil organic component and from leguminous weeds. However, the unbalanced nutrient circulation indirectly caused by the imbalance in distribution of fodder areas relative to the arable led to a continuous decrease in yields in Bjäresjö. At the beginning of 19th century this system was completely disrupted by the imbalance between the nutrient source and nutrient sink areas. A survey of nine different villages in both boulder clay and sandy areas in the Ystad area during the period 1798–1829 showed that all but one village had a meadow area less than half the size of the arable area (Olsson unpubl.). The decreasing crop yields further forced new cultivation on land unsuitable for arable production. In 1825, 42% of the arable area in the village of Stora Köpinge on the eastern sand plain was unsuitable for cultivation (Land survey map 1825). The unbalanced expansion of the arable economy without subsequent maintenance of the livestock led to serious degradation of the soils and decreasing productivity.

Input of matter and energy. Agricultural production was part of a local and regional trade market. The necessary metal implements could be purchased in the regional market. The expansion of the arable fields led to a significant shortage of woodlands for wood, timber, fuel, and fencing material. This situation was most pronounced in the crop economy villages. These products had to be imported from neighbouring ecosystems in the Ystad area or in neighbouring parishes north of the Ystad area. Further, the fodder areas were far too small in the three-course villages in the late 18th century. Either grass had to be purchased as hay from neighbouring parishes, or else the livestock were pastured on grasslands or commons in the parishes on the Romeleåsen ridge or further north of the investigation area.

Limits of the system. It is obvious that during the 18th century the nutrient balance in the agro-ecosystems in the crop economy villages here was severely disrupted. There was a general, acute shortage of manure and other fertilizers. At the same time and intertwined with this problem, the fodder areas, grasslands, and woodlands were seriously insufficient.

Agro-ecosystems of the late 19th century

By the late 19th century, most villages had experienced the land enclosure movement and thus a reorganization of both land property and land use. Conditions for agriculture were significantly different from before the land reform. The farmers became almost independent from their neighbours, free to organize agricultural activities according to their own needs and capacity. The period from 1880 to 1914 was very expansive in all aspects of agriculture. The old cropping systems with two- and three-field rotations and fallows were abandoned and replaced by four-field, five-field, or even longer field rotations. The use of fallows was gradually replaced by the cultivation of fodder crops like clover (*Trifolium* spp.), ley grass, beans, and vetches (*Vicia* spp., *Lathyrus* spp.). The use of ley cultivation was introduced earlier, but it did not become common until this period. The use of semi-natural meadows for hay production, however, was still significant and it was only in the first decades of the 20th century that ley cultivation took over as the main source of winter fodder for livestock (Ch. 4; Figs 4.1:16, 4.2:15, 4.4:9). A special endeavour to increase the yield of meadows was the use of water-meadows (Ch. 3.8.3, 3.9.2). Although extensively used, this practice was restricted to some 30–50 yr during the late 19th century (Table 5.3:4).

Several factors collaborated in the rise of crop production: Firstly, there is the general increase of the arable areas. Comparing the arable used for cereal cultivation in 1850 and 1909, the increase was more than 50% in the western part of the Ystad area (Germundsson and Persson 1987). In some parishes, like Villie, the arable increased almost threefold during this period. The increase of arable in the eastern part had been extensive already before 1850, explaining the more moderate increase between 1850 and 1909. Secondly, soil fertility was improved by two main activities: (1) large-scale draining of the arable fields; (2) improvement of the soil nutrient balance by the addition of fertilizers like marl and limestone.

The large-scale draining of wetlands during the late

Table 5.3:5. Addition of fertilizers to arable land in the Ystad area in 1971 and 1988, kg ha^{-1} yr^{-1}. Source: SCB 1989.

	1971			1988		
	N	P	K	N	P	K
Winter wheat	127	27	50	142	7	15
Rye	70	20	35	92	1	2
Barley	88	23	35	90	4	9
Ley	93	20	35	75	4	11

19th century was probably the most dramatic landscape change ever in this landscape. It transformed both the hydrology and the landscape-ecological conditions in terms of habitat and landscape diversity.

The agrarian and the historical sources give little information on the quantities of fertilizers used in the late 19th century. Fertilizers could be bought, but domestic nutrients predominated. Among these were marl from subsoil deposits frequently used to raise the soil pH. This activity peaked during the 1880s (Zachrison 1914). The addition of ground animal skeletal bones to raise the phosphorus level in the soils became common during the late 19th century, as well as the addition of seaweed (Zachrison 1920). All these fertilizers were known earlier in Scania and Sweden (cf. Linnaeus 1751), but had not been used to any great extent. Contributing to larger nutrient input to the arable was also a greater awareness of the management and storage of manure, along with a general increase in the number of livestock animals, especially milk cows. According to Zachrison (1920), the most important commercial fertilizer in the late 19th century was phosphate, and the root crops were given the largest quantities. The raised yields were also helped by the improved ploughs which enabled more effective and deeper cultivation of the soils (Eskeröd 1973).

The efforts to improve agrarian productivity were rewarded by raised yields (Fig. 5.3:8). The difference in crop yield between the western area (Ljunits) and the eastern part (Herrestad) during the late 19th century was levelled out in the 1909 records (Fig. 5.3:8a). A simultaneous record of purchased fertilizers reveals significantly higher quantities used in Ljunits at this time ("superphosphate" was applied in quantities of approx. 450 and 730 kg ha^{-1}yr^{-1} in Herrestad and Ljunits respectively, Zachrison 1920). The increased levels of phosphorus in combination with effective draining, and the ploughing of the heavy clay soils of Ljunits hundred could possibly explain the equalling of yield levels in the early 20th century (Fig. 5.3:8a).

Another important measure at this time was the increasing availability of improved crop breeds. The establishment of the first seed control institutes in 1879, in two places in Scania (the first in Sweden, Nilsson 1920) made way for the awareness of improved seed crops. The old breeds of cereals were successively replaced by new cultivars from 1890 onwards, with wheat displaying the largest yield increase among the cereals (Fig. 5.3:8b). Bread wheat gradually returned to being an important crop here for the first time since its dominant role in prehistoric cultivation, although then in the form of old cultivars of emmer, spelt, and einkorn (Fig. 5.3:2).

Input of matter and energy. Matter and energy were withdrawn primarily from the regional ecosystems. Organic fertilizers could be purchased on the regional/local market – see above. The import of other commercial fertilizers like phosphate minerals and guano came from far outside the regional systems.

Limits of the system. Manure could be limited locally, but also the availability of fertilizers like marl, and the long-distance imports of "superphosphate" and guano. The extensive increase of arable fields implied new cultivation of land with low productivity and short sustainability. Fertile land for new cultivation became scarce.

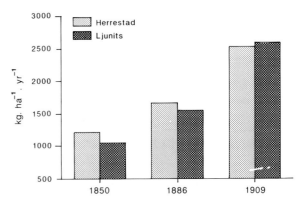

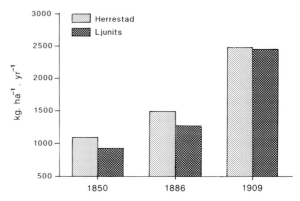

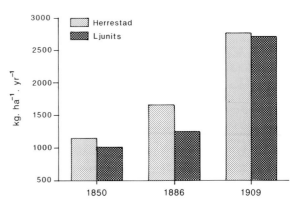

Fig. 5.3:8a. Crop yields in the Ystad area, Ljunits and Herrestad hundreds, 1850-1909.

Fig. 5.3:8. b. Yields of wheat, barley, and rye in the Ystad area 1850–1990, with a comparison with Malmöhus County and Sweden for the period 1910–1990. Source: unpubl. agricultural statistics 1892–1898, Malmöhus läns hushållningssällskap, SCB 1913–1990, Germundsson and Persson 1987.

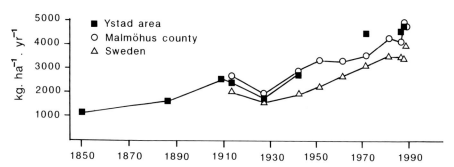

BARLEY 1850-1990

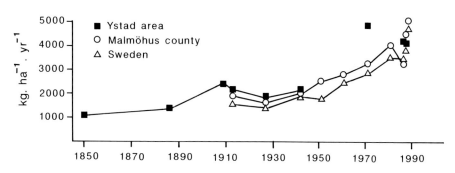

RYE 1850-1990

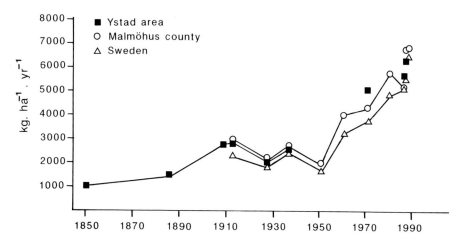

WHEAT 1850-1990

The maintenance of these systems was very labour-intensive. The availability of manpower in the absence of draught animals could have been a limitation, especially for small tenants without access to working animals.

Agro-ecosystems of the mid 20th century

The 1950s were a time of rapid agricultural changes. During about one decade horse power was gradually replaced by tractors powered by (imported) fossil fuels. Imported fertilizers, mainly phosphates and commercial nitrogen compounds along with lime, became available and common to virtually all farmers. Another factor affecting crop production was the increasing availability of chemicals for combating weeds and crop predators, pest and parasites. The input of these chemicals was still at low levels and was not common in the 1950s. How-

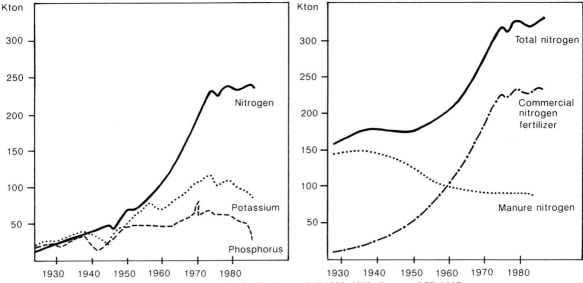

Fig. 5.3:9. Use of fertilizers and manure in Sweden during the period 1930–1980. Source: SCB 1987.

ever, during this decade the direct dependence of the local ecosystems was broken definitively.

Oil crops of rape and turnip rape were first recorded in the agricultural statistics in 1951, reflecting that the earlier cultivation of these crops had been of relatively little importance. A general trend is that the cultivation of barley increased in the whole area, while livestock fodder crops such as mixed grain and oats decreased (Ch. 4: Figs 4.1:18, 4.2:17, 4.4:11). The use of natural fodder areas (semi-natural grasslands) gradually became uncommon, and almost all livestock fodder was obtained from the arable fields by ley and cereal crops. In parishes with a strong orientation to arable production, like Bjäresjö and Köpinge, the ley areas also decreased significantly during the 1950s (Ch. 4: Figs 4.1:16, 4.2:14). The numbers and composition of livestock changed. The decline in the number of horses, replaced by tractors and machinery (at the end of the decade horses were not visible in the agricultural statistics, Ch. 4: Figs 4.1:19, 4.2:18, 4.4:12), was accompanied by a dramatically increased investment in "pig factories", and higher demands on milk production from new breeds of cattle (Nannesson 1914).

Unfortunately, no figures on crop yield from the investigation area are available from the early 1950s. Yield records from Malmöhus County and Sweden as a whole indicate, however, a slight decrease in all crop yields in 1951 compared with the late 1930s (Fig. 5.3:8b). The figures are similar to the records from 1927 (Fig. 5.3:8b). This can possibly be related to the general war-time and post-war shortages of commercial fertilizers and lime. Later, a significant, general increase in all crop yields is evident (Fig. 5.3:8b).

Import of matter and energy. The supplies of commercial fertilizers, chemical pesticides, etc. had to be industrially produced. There is a very high energy cost for

producing nitrogen fertilizers especially. The energy was withdrawn from water power or fossil fuels. The fossil fuel, oil, was also needed for the maintenance of the tractor vehicles and machinery. This energy consumption was covered by the import of energy from distant ecosystems. Fertilizers like turf and marl obtained from the local and regional ecosystems were still important.

Limiting factors. The new implements and the refined draining technique became available because of industrialized production and effective information which forced agricultural rationalizations and raised crop yields. The ongoing soil erosion and nutrient drain could be offset by increasing loads of both manure and commercial fertilizers. However, the tendency towards a gradual increase in the loads of nutrients to watercourses and the sea had started – and thereby the development towards agro-ecosystems without internal recirculation of nutrients. In spite of this the agro-ecosystems of the early 1950s do not seem to have suffered from any distinct limitations – provided the import of nutrients and energy was unhampered.

Agro-ecosystems of the late 20th century

The Ystad area has witnessed the development of large agricultural specialization, accompanied by a significant decrease in ecological diversity. As a generalization, the area can be divided into three categories:

* arable land,
* non-arable land: abandoned grasslands with spontaneous forest succession, or small wood lots,
* urban areas including military and recreation grounds.

In the late 1980s, all crops peak in the yield records. The

Table 5.3:6. Use of pesticides in agriculture in 1979 and 1987/88, percentage of arable area in Malmöhus County and in Sweden as a whole; n.d. = no data available. Sources: SCB 1988, Jordbruksstatistisk årsbok 1988, SCB 1990, Jordbruks-statistisk årsbok 1990.

	Malmöhus County		Sweden	
	1979	1987/88	1979	1987/88
Insecticides	n.d.	51	n.d.	30
Herbicides	n.d.	79	n.d.	8
Fungicides	n.d.	36	n.d.	8
All pesticides	78	82	52	54

investigation area is one of the regions with the highest cereal yields in Sweden. The rate of increase for cereal yields has declined after 1970 (Fig. 5.3:8b). Many factors have contributed to this development, such as increased sensitivity to imbalance in edaphic conditions like water and nutrient availability. The higher crop demands for fertilizer have been counteracted by the environmental need to reduce nutrient leaching to watercourses, which has led to a lower increase of fertilizer additions. Also the modern crop breeds have a successively shorter lifetime for the individual breeds, owing to the rapid development of new pests (Plucknett and Smith 1986).

The cost of these high yields is very high inputs of both pesticides and fertilizers. In 1963 the mean input of N-fertilizer was 92 kg ha$^-$ yr^{-1} in Malmöhus County. This should be compared with a mean of 128 kg ha^{-1} yr^{-1} in 1983 for the same area (Anonymous 1985). Some crops in the investigation area receive still higher loads, for example winter wheat, where the mean for 1988 was 142 kg N ha^{-1} (Table 5.3:5). This is an increase of 11% since 1971, although for Sweden in general the rate of increase for loads of commercial fertilizers has stabilized after the 1970s (Fig. 5.3:9).

In Malmöhus County practically all winter wheat fields are treated with herbicides (99% in 1979 (Naturmiljön i siffror 1986–87), and most of the arable (78%) is also treated with insecticides and fungicides (Table 5.3:6).

Input of matter and energy. The majority of the soil nutrients supplied to the systems are imported either from distant ecosystems, like the phosphorus compounds, or as high-energy-demanding industrial fertilizers. Virtually all the energy for powering the machinery is imported in the form of fossil fuels. None of the crucial supplies to the present agro-ecosystem are withdrawn from the local or regional ecosystems. An exception is manure, but animal husbandry has become rare in this area.

Limiting factors include decreasing soil fertility, resulting from disturbances in the C/N ratio, disturbances in the soil microbial populations due to heavy loads of chemical fertilizers and pesticides, and the depletion of

the humus content in the soil. Pests like fungi and insects develop resistance to pesticides. There are time-lags in the development of new crop breeds. In addition, surface waters and groundwater have been polluted by fertilizer and manure loads deposited on the arable fields.

Ecological effects of the human impact on the agro-ecosystems – general trends

Variables reflecting the agricultural exploitation systems
The crop yield. Several variables can be chosen to reflect the ecological effects of the agricultural activities shown in Fig. 5.3:1. However, the variable which farmers have most sought to affect by their efforts has always been the crop yield. The yield curves of the cereal crops wheat and barley (Fig. 5.3:10) reflect their cultivation history. For wheat the curve consists of emmer and einkorn wheats for the long prehistoric period of ca. 3000–500 BC, while spelt wheat dominated ca. 500 BC–AD 500. Bread wheat was not cultivated (or cultivated to only a very small extent) in the Ystad area before the mid 19th century. During the long prehistoric period when the early wheat varieties were grown, the yield probably fluctuated around a seed return of 4–5 (Table 5.3:1). The dramatic increase in yield did not take place until the late 19th century, and after 1950, with the introduction of the industrial agriculture ex-

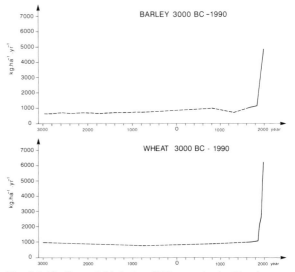

Fig. 5.3:10. Crop yield during 6000 years in the Ystad area. Wheat is here an umbrella term for the prehistoric wheat varieties and the later modern bread wheat – see Fig. 5.3:2. Barley denotes both naked barley and hulled barley – Fig. 5.3:2. The solid line denotes data based on real figures; the broken line denotes estimated figures from models and recent growth experiment. Sources: growth experiments with prehistoric crop varieties (see Table 5.3:1), land survey maps 1691–1890, National Land Survey, Malmö; unpubl. agricultural statistics 1892–1898, Malmöhus läns hushållningssällskap, SCB 1913–1990, Germundsson and Persson 1987, model calculations by Olsson (Ch. 2.2).

LANDSCAPE ECOLOGICAL PERIODS

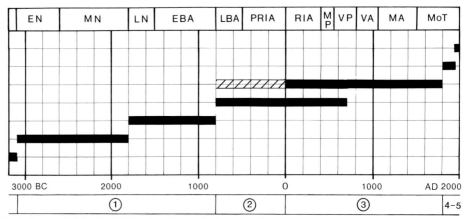

Fig. 5.3:11. Landscape-ecological periods (cf. Ch. 5.13) related to agricultural exploitation systems. Exploitation systems and archaeological chronology – see Fig. 5.3:1. The hatched bar denotes the transition period (exploitation period 2) where shifting cultivation existed parallel to permanent arable.

ploitation system, the increase in yield was exponential (Fig. 5.3:10).

For barley the curve is composed of the two prehistoric varieties naked and hulled barley (Fig. 5.3:10). Hulled barley has been the predominant barley variety during the whole historical period in this area. Existing yield records from the 17th century testify to a yield depression which can be related to a possible nutrient deficit of the arables of the infield-outland system – see above. During the present period of industrial cultivation the seed return can reach 45 (approx. 10,000 kg ha^{-1} yr^{-1}), although means are around 25 (ca. 5000–6000 kg ha^{-1} yr^{-1}) (Fig. 5.3:8b).

The diversity component of the agrarian system. The crop species on which the whole prehistoric cultivation in this area was based were the early wheat forms emmer, einkorn, and spelt, and the barley varieties (Fig. 5.3:2). During the first 3000 years of cultivation there was a relatively slow replacement of species. The turnover rate of species has increased and during the industrial agricultural period the three main cereals barley, rye, and bread wheat remain from the prehistorical periods. Other crops which are important today in this area are of recent origin, introduced during the last hundred years (Fig. 5.3:2).

The adoption of complicated crop rotation systems during the infield-outland period contributed to the overall diversity component. In such systems rotation periods and crops varied according to soil conditions, and fallows of different ages developed diverse vegetation communities; all this encouraged an ecological diversity on both species and habitat level. In this situation we should also include the meadows and commons with vegetation communities shaped by fodder gathering, either haymaking or livestock grazing. These land uses resulted in specific and varied plant and animal communities.

The diversity – on all levels – has probably never been as high as during the existence of the infield-outland system, before it started to become unsettled in the 18th century. The habitat diversity was also accompanied by a high species diversity (cf. Ch. 5.13) and high genetic diversity. The agricultural changes in the industrial agricultural system forced a development where semi-natural vegetation (grasslands and woodlands) expressed as the number of sites has decreased by about 25% during the last 50 yr (Ch. 5.11). The fen area and the number of fen sites have suffered the greatest reduction of all non-arable units during the last 50 yr. This means a reduction in habitat diversity where a reduction of variation along the soil moisture gradient is also involved.

The introduction of nitrogen fertilization of the semi-natural grasslands in this area has decreased the mean plant species richness from about 20 to 11 plant species per m^2 (Ch. 5.11).

The ecological effects of the disruptions of this diversity on the level of the semi-natural grasslands are further discussed in Ch. 5.11.

The landscape-ecological content of the Ystad area. On the basis of the landscape-determining elements, the agricultural exploitation systems can be related to landscape-ecological periods (Fig. 5.3:11). The shift from reliance on the wooded ecosystems during the main part of the prehistoric agricultural period to farming in an open grassland landscape took place during the Bronze Age (Ch. 5.13). The agricultural period of *shifting cultivation* thus contained three landscape-ecological periods; the transition from *wildwoods* to *human-influenced woodlands* and *wood-pastures* to the *pasture landscape* (Fig. 5.3:11).

The *infield-outland* system created a mosaic landscape with increasing differences in vegetation communities between the land-use categories according to their function in the directed flow of nutrients. During the human existence in this area, the maximum habitat diversity was probably reached in the *infield-outland* period (Fig. 5.3:11), before the disruption of the nutrient flows had started.

The two last agricultural periods have created landscapes which are characterized by decreasing habitat and species diversity. The trend from the preceding period has been expanded during the *industrial cultivation period*. The arable farming depending on external fertilizers is reflected by a landscape with decreasing areas of natural and semi-natural vegetation (cf. Ch. 5.11), and a general levelling out of landscape-ecological variation and habitat diversity.

Other ecological variables reflecting the agricultural exploitation systems. The use and persistence of the different agricultural exploitation systems over the long period of 6000 years has led to significant and very influential changes in a number of ecological variables not discussed above. Some of these are discussed in other chapters in this book: erosion (Ch. 5.1), changes in landscape diversity (Ch. 5.11, Ch. 5.13), changes in floral diversity (Ch. 5.11), changes in biological community composition (Ch. 5.13), changes in the frequency of different stages (Ch. 5.13), nutrient deficit in the arable soils (Ch. 6), leaching (Ch. 6). Others have not been investigated within the framework of this project, like changes in faunal diversity, changes in the genetic diversity of crop plants and domestic animals, and local climate changes due to deforestation.

The relationship between the local agro-ecosystem and its neighbouring systems

To compare the efficiency of different farming activities and their degree of exploitation of the environment, it is convenient to use the agro-ecosystem concept. The scale of study of an agro-ecosystem is set by each investigator. It can be a whole region (Zucchetto and Jansson 1979), villages (Dodgshon and Olsson 1988, Olsson 1988b), or single fields (Paustian et al. 1990).

In natural systems, nitrogen and phosphorus generally occur as limiting nutrients. There has recently been a growing interest in the study of the efficiency of uptake, allocation, residence time, and final use of these elements in separate plant individuals. In this context, the term *nutrient use efficiency* (NUE) has been developed (Chapin 1980, Vitousek 1982, Berendse and Aerts 1987).

The NUE concept can be applied to the study of historical agro-ecosystems even though specific field data may be lacking. In such a context nutrient use efficiency is defined as the amount of above-ground crop biomass produced, divided by the amount of mineral nutrient supplied in a given period (Ch. 2.2). It can be written as:

$$NUE = \frac{OUTPUT\text{-}(n)}{INPUT\text{-}(n)}$$

Output: nutrient content in crop biomass recorded as harvest. *Input:* nutrient content in supplied manure and fertilizers.

This approach is used by Olsson in Ch. 4 and in Olsson (1988b), Dodgshon and Olsson (1988) for local agro-ecosystems during historical periods. However, for the local agro-ecosystems of the Ystad area over the period of 6000 yr, the varying accuracy of the quantitative data of nutrient flows makes comparisons less relevant.

Nevertheless, what we can discuss is the relative degree of exchange of nutrients, matter, and energy between the local system and the regional and global systems. We can relate this pattern to the division into agricultural exploitation periods discussed above. The principles for circulation of nutrients and the dependence on other systems are generalized and summarized in Fig. 5.3:12.

Agricultural exploitation systems

We have seen how an area of natural deciduous woodland gradually changed into a cultural landscape along with the increasing impact of the four variables introduced by humans (clearing of woodlands, grazing, fodder collection, cultivation) during the 6000 yr.

The degree of impact and the intensity of exploitation have varied during this long period. The effort to obtain reliable crop harvests and to increase yields has been the recurrent task of agriculture. The various agricultural exploitation systems are characterized by different ways of maintaining the soil-plant nutrient availability in the arable systems. In Fig. 5.3:12, the following six exploitation periods are shown. The tools and techniques used for maintaining the agricultural activity are summarized in Fig. 5.3:1.

1. Shifting cultivation system; Neolithic–Early Bronze Age (3100–800 BC). The introduction of the cultivation of crops often has been called the Neolithic Revolution, or the First Agricultural Revolution. The agricultural techniques used over the longest period are shifting cultivation, the use of simple tools of bone, horn, and stone, and the management of grasslands by livestock grazing (Fig. 5.3:1). For almost 3000 yr these were the main agricultural activities, and they were confined to the local community and the local ecosystems. The arable nutrient availability was maintained by short-term cultivation followed by periods of fallow where the soil nutrients were able to recover by the secondary succession processes with developments into shrub and woodland seres. It is possible that this system also included the management of coppices among the arable, variants of agro-forestry. This system with nutrient and soil-

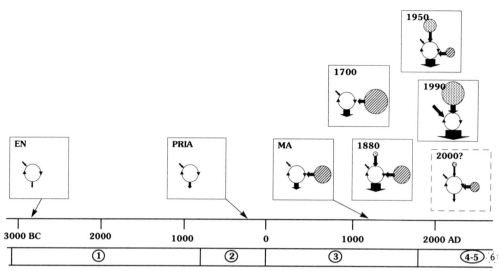

Fig. 5.3:12. Schematic representation of the circulation of nutrients between the local agroecosystem and regional and distant ecosystems during the different time periods and within the different exploitation systems. Exploitation systems and archaeological chronology – see Fig. 5.3:1. This figure also includes exploitation period 6 (sustainable agriculture). The size of the symbols for regional and distant ecosystems denotes their relative importance for the contribution of nutrients to the local agro-ecosystem. The size of the arrows similarly denotes the relative magnitude of the transport of nutrients. An arrow without a source symbol denotes diffuse input from neighbouring ecosystems via leaching and erosion or atmospheric deposition. Open circles refer to nutrients and energy supply from the local ecosystems, hatched circles from regional ecosystems, and dotted circles from foreign areas (imports).

water capture by the tree roots, and nutrient release from decaying tree roots related to the coppicing cycles, could probably allow cultivation at the same spot for long periods, reducing the need for long fallows. The livestock were nourished by year-round grazing and were thus involved in the creation of the semi-natural grasslands. The shifting cultivation system needed large territories for the moving arable fields and for the extensive livestock grazing. Nutrient circulation was predominantly internal. The relatively small losses of nutrients by leaching and soil erosion were offset by the supply from neighbouring ecosystems (Fig. 5.3:12).

2. Permanent arable: transition period from shifting cultivation to infield-outland system, Early Bronze Age–Pre Roman Iron Age (800 BC–0). The introduction of the permanent arable fields based on manuring can be called the Second Agricultural Revolution. This system may have been introduced in this area around 800 BC according to indications from archaeological interpretations of possible byres and sparse archaeological finds from the Ystad area (cf. Ch. 3.3, 4.1.4, 5.5). According to this interpretation, the habit of using permanent arable gradually became more important in the time up to birth of Christ. However, during this period the shifting cultivation was used side by side with the infield-outland system. The bearing principle of this system is treated below. The contribution of energy and matter was maintained within the local ecosystem.

3. Permanent arable system: infield-outland system (0–1800). From about the birth of Christ the infield-out-

land system seems to have been common in the Ystad area. The organization was within the framework of the one-field rotation or grassland system (Frandsen 1983). The vital principle was utilizing certain land-use categories as nutrient sources and the permanent arable land as nutrient sinks. The livestock acted as converters of the nutrients from the outland to the infield arables. By grazing the outlands, and by consuming hay harvests from the mown meadows, the livestock transferred biomass nutrients via manure to the arable. The transition of manure nutrients was accomplished by night folding of the livestock in the infields, or by stalling the livestock at night and in the winter. The system could work as long as there was a balance between the size of the arable land and the nutrient-supporting outlands (Fig. 5.3:12).

During the Middle Ages the degree of exploitation in the infield-outland system had increased by the adoption of new tools and more intensive crop rotation, (e.g. three-course rotation). The agro-ecosystems had an exchange of both energy and nutrients with the neighbouring and regional ecosystems. Such examples are the habit of hiring summer grazing for the young livestock in parishes away from the three-course villages on the most fertile soils, where the fodder-producing areas had been decimated.

The balance between nutrient source and sink areas was disrupted in the Ystad area already in the late 17th century since most agricultural units (farms, villages) in this area appeared with considerably smaller outlands than arable areas. It can be estimated that the sum of meadows and pastures should at a minimum exceed twice the size of arable area in the boulder clay areas and three times or more in the sandy plain areas. Calculations for the sandy heathlands in the Netherlands (Plaggen-soil agriculture) have shown that 3–11 ha of heathland and grasslands were needed to maintain the fertility of 1 ha of arable land (Gimingham and de Smidt 1983).

The dependence of the regional system increased and during the 18th century became an obligate relationship for the three-course villages on the boulder clay areas. The parishes in the northern part of the Ystad area were the nutrient sources for the arable fields in the boulder clay areas. Leaching from the arable increased as a result of the large areas used for tillage and the intensive use of the fallows for grazing domestic pigs and small livestock. The productivity of the arable fields generally decreased in the three-course villages. Several areas of arable were reported to be unsuitable for cultivation in the early 19th century, see above.

During the existence of the infield-outland system several improvements in agricultural technique took place. The use of cropping systems with periods of fallow was able to improve nutrient recovery in the arable areas. The preparation of seed-beds was improved by the use of the plough (Myrdal 1988), which allowed a more effective breaking of the fallow sward, thus favouring crop plants in their competition with arable weeds. During the 800 years when this system was in operation, changes took place gradually, but few were of dramatic enough importance to change the fundamental principles of the system (Fig. 5.3:1). The need for larger harvests to feed the increasing human population was met by expanding the arable at the expense of the grasslands, although only on the dry soils. The general result was decreasing crop harvests due to insufficient nutrient availability (Fig. 5.3:10). The agricultural ecosystems in the framework of the infield-outland system had little exchange of energy and nutrients with other than local systems.

4. Permanent arable: external nutrient import (1800–1950). The agricultural exploitation system during the period 1800–1950 is characterized by the many new techniques, most of which were intended to increase nutrient availability to the arable soils (Fig. 5.3:1). The final collapse of the infield-outland system demanded other ways of ensuring plant nutrients. During a very short period in the 19th century (ca. 1840–1900) water-meadows were used in the area to supply plant nutrients and thereby raise the productivity of the hay meadows. Improved crop rotation and cultivation of nitrogen-fixing crops on the fallows were other methods of raising the agricultural productivity within the framework of the local agro-ecosystem. This period was of short duration; in the Ystad area ca. 1800–1900 (advanced permanent field farming, Emanuelsson 1988). The local agroecosystem was heavily dependent upon exchange of nutrients with neighbouring systems as described above.

Besides, the introduction of supply of fertilizers from different organic and inorganic sources also led to the import of nutrients from distant ecosystems. Examples are the purchase of guano from foreign sources and the import of phosphate minerals from Asia and other remote places (Fig. 5.3:12).

The fairly general introduction around the turn of the century of nitrogen-fixing ley crops cultivated on the arable fields was one of the major characteristics of this period. This development contributed to decrease the economic importance of the semi-natural meadows since the winter fodder could now be grown on the arable land. Parallel to this was the demand to increase the arable by extensive draining of wetlands and even complete draining of shallow lakes. From a landscape-ecological point of view, the levelling out of the hydrological mosaic was one of the most radical changes ever in this landscape.

During the 19th century different types of organic and minerogenic fertilizers were successively introduced, like ground skeletal bones to supply phosphorus compounds to the arable soils, or the addition of marl to supply calcium and potassium. Imported fertilizers also became increasingly available: guano from maritime bird rocks, e.g. in South America, was manufactured as a phosphorus fertilizer. Other imported fertilizers followed and the agro-ecosystems became dependent upon nutrient import from distant ecosystems. The liberation of the agro-ecosystem from the dependence on nutrient supply from local and neighbouring ecosystems was accompanied by the introduction of new tools and techniques like deep ploughing, effective draining, and above all improved crop breeds. These led to higher production, but were also more sensible to weed competition, and patchy soil and micro-climatic conditions.

5. Industrial agriculture (1950–1991?). During the period after 1950 up to the present the agro-ecosystems became more or less totally liberated from the dependence on nutrient supply from local or even regional ecosystems. This development was accompanied by an increased dependence on imported nutrients for the arable soils, and imported fossil fuels to supply power to the heavy agricultural machinery. The pressure was no longer to increase the areas of the arable, but to increase the yield per area. New crops (like oil crops) and a general decrease in ley cultivation in favour of grain crops, and new breeds of the old crops further contributed to the rise in yield. The high production potential of the modern breeds of crops requires increasing loads of fertilizers, and effective monocultures maintained by high loads of pesticides, fungicides, and herbicides. All these chemicals, including the fertilizers, require high energy input for their manufacture which must be considered as further energy dependence for industrial agriculture. With the application of this agricultural system we have entered the Third Agricultural Revolution – industrial agriculture.

The driving force for this development was an increasing import of cheap fossil fuels from distant areas outside the country. The dependence on the regional and neighbouring systems decreased along with the increasing dependence on distant ecosystems (Fig. 5.3:12). In this respect the Ystad area represents the agricultural areas in south Scandinavia with intensive cultivation, and resembles other parts of Europe with highly specialized agriculture, like the Netherlands (Rabbinge 1986). It is interesting to note that the improved modern cereal varieties that are used must be called high-potential rather that high-yielding crops. They are obligately dependent on the high input of nutrients, pesticides, water, etc. for which they have been selected (cf. Evans 1980).

This development has accelerated and today the local agro-ecosystems exist in a complete dependence on the continuous supply of nutrients and energy from distant sources. Along with this development, the degree of internal recirculation has receded and the total losses from the system have increased (Fig. 5.3:12).

6. Sustainable agro-ecosystems: 1991?–? When considering the long period of agricultural exploitation, it is clear that for most of the time the scene has been dominated by systems based on the local agro-ecosystems, with a large degree of internal nutrient circulation, moderate nutrient losses via erosion and leaching, and an exchange of matter and energy with the regional or neighbouring systems (Figs 5.3:1, 5.3:12).

If we wish to predict the agricultural system which has a chance to survive for more than a few hundred years, it seems reasonable that it would resemble the infield-outland system. A system for sustainable agriculture would be similar to the one shown in Fig. 5.3:12. This has a high degree of internal recirculation of nutrients. It contains not only the exchanges with the regional/neighbouring systems, but also input from distant systems, although to a very limited degree. A sustainable system for the future cannot be organized on the basis of a continuous import of matter and energy from distant, and thus energy-consuming transports. The agro-ecosystems of the future would include utilization of modern and specialized technology, although they would be energy- and matter-efficient. Such technology is used along with traditional agricultural methods with the aim of maintaining sustainable agro-ecosystems. If this development arises we will enter the period of the Fourth Agricultural Revolution – sustainable agriculture.

Conclusions

When we look at the agro-ecosystems in our study area in a national perspective in 1990, we find that this area is one of the few which is contemplated for use for crop production also in the future. Along with fundamental changes in the national policies for Swedish agriculture (Ministry of Agriculture 1989) large arable areas are expected to be abandoned as being not economically profitable for agriculture. The remaining agricultural areas will be the ones with fertile soils, favourable land consolidation, and an infrastructure (e.g. ownership, communications, etc.) suitable for industrial crop production. However, the limiting factors for the present agro-ecosystems and the global situation might give rise to other circumstances for future crop production. The problems of environmental pollution (water, air, and soil) partly as a result of high loads of fertilizers and pesticides, airborne pollutants, etc., have contributed to decreasing soil fertility. The monocultures with virtually no or incomplete crop rotation exacerbate decreasing soil fertility. Soil erosion and dust storms in the spring are other problems caused by intensive and large-scale cultivation. The dependence upon high loads of chemicals, and on a continuous supply of new breeds of crop species with changed resistance to coevolving pests all makes the present agro-ecosystems extremely sensitive to disruptions in the external supplies.

On a global scale, human population is expected to increase by 60% during the next 40 years, from 5 billion to 8 billion in 2030 (World Watch Institute 1990). The available areas for food production might increase by 5%, but still the arable area per capita will decrease significantly, from 0.28 ha per person to 0.19 ha per person. Both ecological and ethical considerations suggest that the industrialized and very sensitive agro-ecosystems of this area or in other industrial nations cannot persist in the long term.

The prospects for agro-ecosystems of the future could be based on the principles which governed the agro-ecosystems in this area during the longest period with similar climatic conditions to the present day (Fig. 5.3:1). This is the period of infield-outland cultivation, where the nutrient balance of the agro-ecosystems was maintained mainly by internal circulation and by supplies from the local and/or regional ecosystems. Present knowledge of ecological systems and scientific data and experiences concerning organic agriculture (Granstedt 1990) and low-input systems (Odum 1989) ought to give the prerequisites to counteract nutrient exhaustion and overexploitation. We can envisage the creation of new agricultural systems with combinations of different crops and rotations, agro-forestry systems, and different combinations of food and energy crops. A development of agro-ecosystems which are to a large extent based on local nutrient circulation, and on the use of biological diversity (local ecosystems, vegetation communities, and species adapted to the ecological conditions of the region, etc.) is a way to adopt a global ecological approach, and to contribute to a sustainable agricultural development.

Acknowledgements. Valuable comments on the manuscript were given by W. P. Cramer and N. Malmer. They are kindly acknowledged.

5.4 The introduction and establishment of agriculture

The moving family: aspects of the Early Neolithic in southern Scania

Mats Larsson

Neolithization in southern Scandinavia – background

The problems associated with Neolithization and the form taken by early agriculture have been the subject of intensive discussion in Scandinavia in recent years (Jennbert 1984, 1985, 1987, Zvelebil and Rowley-Conwy 1984, Rowley-Conwy 1985, Berglund 1985a, Larsson 1987, M. Larsson 1987d, Madsen 1987, Nielsen 1987). It is not possible to discuss individual works in detail within the context of this chapter, although the more important trends will be included as a background to developments in the Ystad area.

We can see how larger, more permanent settlements were established within the Ertebølle culture towards the end of the Atlantic period, in about 3600 BC (Andersen 1973). These settlements are localized essentially in ecologically favourable areas, such as fiords, river mouths, and lagoons (Paludan-Müller 1978, Larsson 1984, Andersen 1986). Recent years have seen the discovery of cemeteries in conjunction with a number of dwelling areas in southern Scandinavia, such as Vedbæk and Tybrind Vig in Denmark, and Skateholm in Scania (Albrethsen and Brinch Petersen 1977, Larsson 1984, Andersen 1985).

The growth of what can be described as "formal disposal areas" in conjunction with these settlements has been taken as being indicative of the growth of local territories during the Late Atlantic period. Locally marked groups can be identified during this period (Vang Petersen 1984, Jennbert 1984). Differences in material culture may indicate a desire to emphasize association with a particular group, which in turn may be a sign of economic stress or conflicts (Hodder 1979). Contacts with farming groups are indicated by imported objects such as the "Schuhleistenkeil", a type of axe of Central European origin, which has been found on Ertebølle sites (Fischer 1982). It has been suggested that seed-corn may also have been a barterable commodity, since grain impressions have been identified on Ertebølle sherds (Jennbert 1984).

The Late Ertebølle culture is characterized by increasingly permanent settlement, in which marine resources played an important role. The development towards local groups points to increased social complexity with a certain degree of social differentiation, perhaps based on age – a type of "big man" society.

A knowledge of agriculture was gained from Neolithic groups in northern Germany and Poland (Jennbert 1984). An interesting point, however, is the difference of 1300 years between the establishment of agriculture in central Europe and in southern Scandinavia (Zvelebil and Rowley-Conwy 1986). What are the factors which caused the apparently well-established Ertebølle culture to change? It is natural in this context for interest to be focused on the Ystad area; a model illustrating how this change may have taken place is discussed in the following.

The youngest ^{14}C datings for the Ertebølle culture are between 3200 and 3100 BC, whereas the earliest datings of the Funnel Beaker culture are put around 3100 BC (Madsen and Petersen 1984). In other words, the critical period is around 3200 BC. The fall in the pollen curve for *Ulmus* (elm) occurs in 3200 BC over the whole of NW Europe. This change in the forest ecosystem is now considered to have been caused by a combination of elm disease and climatological changes (Berglund 1985a, Göransson 1986). A series of regressions in sea level occurred at about the same time (Christensen 1982a). These regressions brought about a noticeable change in marine resources (Rowley-Conwy 1984). There is a good deal of evidence to suggest that the principal diet of the Late Ertebølle culture people was made up of fish and molluscs (Jarman et al. 1982). This is supported by measurements of the ^{13}C/^{12}C ratio in skeletons in Mesolithic graves, which are high (Tauber 1982). This indicates a diet associated primarily with marine fish, molluscs, and seals.

An interesting aspect in this context is the settlement which took place in the inland areas simultaneously with the large coastal settlements; this phenomenon can be observed over the whole of the area of distribution of the Ertebølle culture. It can also be seen in a number of pollen diagrams from the Ystad area (Hjelmroos 1985, Håkansson and Kolstrup 1987). All the diagrams provide evidence of clearings in the forest. These have been dated to between 3650 and 3400 BC, in other words, a considerable time before the elm decline. These clearing operations were then followed by culture-indicating plants such as *Plantago lanceolata* (ribwort plantain). It is interesting to note that pollen of the Cerealia type has been identified in a number of cases. These obviously originate from larger varieties of grass, however (Berglund 1985). Nevertheless, pollen grains from both wheat and barley have been identified in Danish material (Kolstrup 1987). Evidence to the effect that cereals were known during the Late Atlantic period is provided by impressions in sherds from the Löddesborg site in western Scania (Jennbert 1984). This shows that the Atlantic forests had been partially transformed into

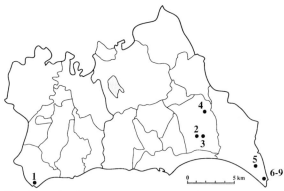

Fig. 5.4:1. Sites from the Neolithic. 1: Mossby, 2: Karlshem, 3: Piledal, 4: Karlsfält, 5: Kabusa and 6–9: Kabusa I–IV.

coppiced woodland. Branches and twigs were burned in small areas, and cereals were then grown there (Göransson 1986). There is no reliable evidence, however, to show that cereals were grown in the Ystad area (Berglund 1985b). It is nevertheless quite clear that the Atlantic forests were influenced by man during the Late Ertebølle culture.

In the period from 3200 to 3100 BC the Ertebølle culture underwent a rapid transformation into the Funnel Beaker culture. The changes in the ecosystem which have been discussed clearly had an influence on the economic basis of the Ertebølle culture. In particular there was less opportunity to utilize marine resources. In my opinion, however, we should not seek an explanation in the ecological factors alone, but also within the social sphere.

The growth of local groups during the Late Atlantic period, such as have been demonstrated in Zealand (Denmark), for example, was associated with the risk of conflicts with other groups (Vang Petersen 1984). The damage observed on the vast majority of skeletons dating from the Late Ertebølle culture can be interpreted as the result of open conflicts. This fact can be paralleled with the previous assumption that society was subdivided by sex and by age. Some of the older men may have acquired control of significant resources. The status of these men was dependent on the number of transactions which they were able to control (Hodder 1982). A move towards a more complex society may have obstructed further development (Madsen 1987). One possibility was to utilize new resource areas, which in this case obviously meant the inland areas with their rich ecological diversity. This may be the situation which is reflected in the pollen diagrams for the Ystad area and also from other areas in southern Scandinavia.

The changes discussed above, both in the ecosystem and in the social sphere, are reflected in the rapid transformation of the Ertebølle culture into the Funnel Beaker culture, and in the adoption of animal husbandry and the growing of cereals.

The Early Funnel Beaker culture in the Ystad area

A distinct change in the pattern of settlement between the Late Ertebølle culture and the Early Funnel Beaker culture has been observed in other thoroughly investigated areas in Scania (M. Larsson 1984). Instead of the large sites on the coast, we now find sites in the inland areas. Altitude, well-drained sandy soil, and proximity to watercourses were decisive factors affecting the choice of settlement locations (Madsen 1982, M. Larsson 1984). The sites are small, rarely exceeding 500–800 m² (Madsen 1982, M. Larsson 1987a). One interesting point is that a number of regional groups are already clearly distinguishable from the earliest part of the Early Neolithic: the Oxie group, the Svenstorp group and the Mossby group in Scania, and the Volling group, the Svaleklint group, and the Oxie group in Denmark (M. Larsson 1984, 1988a, Madsen and Petersen 1984). What we see here is probably the same pattern of development that has been discussed in England, where Mesolithic social territories harmonize closely with Early Neolithic style groups (Bradley 1984).

The oldest Neolithic phase in the Ystad area is characterized by a local group, which has been named the Mossby group after a site at Mossby in Västra Nöbbelöv parish (Larsson and Larsson 1986). A characteristic feature is the wealth of cord ornamentation below the rim of vessels (Fig. 5.4:2). The surface area of the sites was not large – between 800 and 1000 m². One item of some importance in this context is the discovery of a house during the investigations. This was 12 m long and 5 m wide. The roof had been supported by only a single row of posts. The most interesting feature of this site, apart from the construction of the house, is the very early ¹⁴C datings. These put the settlement between 3280 BC and 2965 BC. On average, the site can be dated to approximately 3100 BC. These datings are obviously comparable to the oldest Danish datings (Madsen and Petersen 1984).

Unlike the Oxie and Svenstorp local groups of southwest Scania, the Mossby group does not exhibit the same close association with the Ertebølle culture with regard to flint implements. The flake axe, which is commonly encountered in the other two groups, is absent from the Mossby group. Nevertheless, it is interesting to note that the pointed-butted axe is found in small numbers.

The largest number of sites dating from this oldest part of the Early Neolithic, here called phase 1, are from the Herrestad/Köpinge area. A total of four sites dating from this phase have been investigated in this area. With one exception, Karlsfält, the sites are near the coast. The sites are not, however, located directly on the coast, but in the inner parts of inlets and lagoons. This settlement pattern can be compared with that suggested by Madsen (1982) for eastern Jutland.

A good example of a settlement area of this kind is provided by Kabusa in Stora Köpinge parish. Two sites

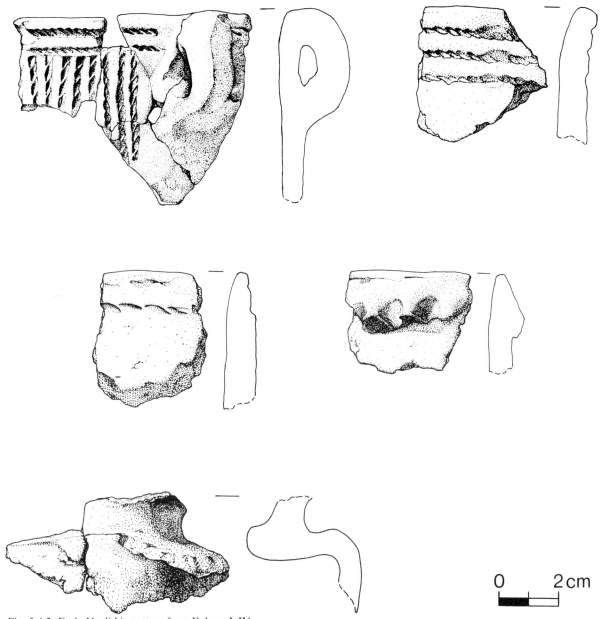

Fig. 5.4:2. Early Neolithic pottery from Kabusa I–IV.

from the Early Neolithic 1 have been investigated in this area (M. Larsson 1987a). The larger of the sites is located on light, sandy soil on a narrow headland projecting into Ingelstorps Mosse. Judging by the size of the occupation layer and the distribution of the finds, the site was small, about 600 m². According to Madsen (1982), the site may be classified as a base site. It is situated in an area where a number of habitats coincide. It is customary for ecologists to describe this as being indicative of the utilization of an "edge effect" (Odum 1971). A settlement in an area of this kind permits the utilization of a number of resources. It is obvious that the lagoon areas, which Kabusa was at that time, have

high productivity (Paludan-Müller 1978). This is a pattern of settlement which has been observed in a number of areas, such as south-west Scania, southern England, and Zealand and Jutland (Madsen 1982, M. Larsson 1984, Barker 1985, Nielsen 1985).

Emmer, einkorn, and naked barley were grown on the light soils around the sites. The results of the macrofossil analyses relate to Mossby, although since the sites and their natural environment are comparable, it is possible to regard the results as being equally applicable to the Kabusa area. Our knowledge of the numbers of cattle, pigs, and goats/sheep is extremely limited for the Early Neolithic 1. This applies to the whole of southern

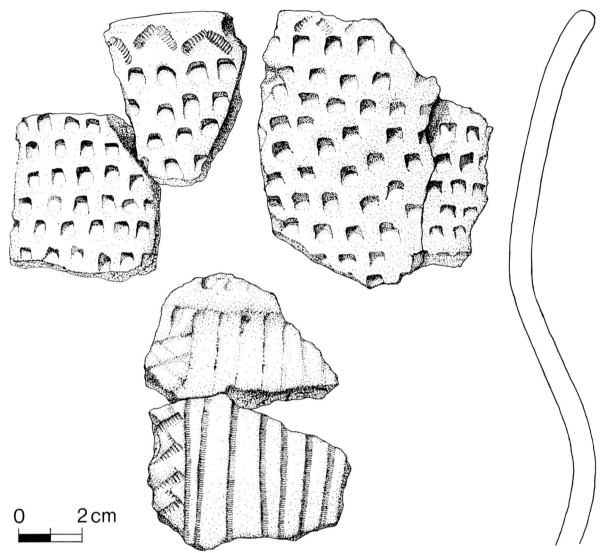

Fig. 5.4:3. Early Middle Neolithic vessel from Kabusa IV.

Scandinavia (Nyegaard 1985). The results of the analyses, in this case also including the Mossby site, nevertheless indicate that all these species were represented. It has been suggested that pigs may well have been the predominant species (Madsen and Jensen 1982). This conclusion is based entirely on an analysis of the settlement pattern. Moist woodland slopes are known to have been good for pig husbandry. Plants such as *Pteridium aquilinum* (bracken) and *Plantago lanceolata* are an important part of the diet of wild pigs (Grigson 1982). In summary, it can be stated that the considerable importance of the pig during the oldest part of the Early Neolithic is supported by a good deal of evidence. This is not clearly documented in the faunal material, although the predominance of pigs to a certain degree in the data relating to the Oxie group from the Björkesåkra 1 site may support the theory (M. Larsson 1984).

The Early Neolithic 1 is characterized in southern Scania by small sites. Those which have been investigated have a surface area of 400–800 m². This has also been confirmed by the field surveys which were made in the Ystad area between 1982 and 1987. A close relationship exists between settlement, sandy soil, and wetlands. The settlement pattern thus resembles that of other areas in southern Scandinavia. The settlement pattern also shows wide dispersion, with large distances between the sites.

Four sites dating from the Early Neolithic 1 have been investigated in the area studied in greatest detail, around Stora Herrestad/Köpinge. These sites are at a distance of a couple of kilometres apart; the exception is Kabusa, where both sites are localized on projecting headlands separated by only a few hundred metres. A pattern of settlement of the kind described above points to a sparse population. Four coeval sites indicate a population in this area of approximately 25–30 persons.

Although the sites are small, the presence of post-built long-houses, as at Mossby, suggests a relatively settled population. This brings us conveniently to the discussion of agriculture and the settlement pattern. The assumption that has been held for many decades is that agriculture in the Early Neolithic was based on the slash-and-burn method (Iversen 1941, 1973, M. Larsson 1985). Over recent years, however, the existence of slash-and-burn has been questioned (Sherratt 1980, Rowley-Conwy 1981, Barker 1985, M. Larsson 1988b). It would probably be much more realistic to imagine a system based on permanently cultivated fields, which were moved at regular intervals within a limited "territory". The question of whether or not a system of cultivation of this kind was combined with the transformation of the forest into coppiced woodland – at least during this part of the Early Neolithic – is the subject of lively debate (Berglund 1985b, Göransson 1986).

The settlement model in the Ystad area is thus one in which the settlements were relatively permanent in nature and were not moved from place to place. A system of this kind also assumes that the fields were fenced in to afford protection against the grazing cattle.

Obviously, a very rapid change in society took place during phase 1 of the Early Neolithic. The process of adaptation to an agricultural economy, under the conditions described here, clearly takes place very rapidly. The very early ^{14}C datings from Mossby indicate that it did not take more than perhaps a couple of generations (50–75 years) for what we now know as the Funnel Beaker culture to develop. What is also interesting is that previous links between different areas, sometimes separated by considerable distances, continue during the early Funnel Beaker culture. This can be observed, for example, in the exchange of amber, copper axes and other copper items, and thin-butted axes (Randsborg 1975, Kristiansen 1982, M. Larsson 1988a). Obvious similarities in the ornamentation on pottery also exist between the Mossby group from southern Scania and the more distant Volling group from Jutland (as defined by Madsen et al. 1984). Copper artefacts have also been found in graves associated with the Volling group (Madsen 1979, 1980). A number of copper axes have been found in the area in southern Scania occupied by the Mossby group (M. Larsson 1985). It is conceivable that an efficient system of exchange existed between Jutland and southern Scania. It has been suggested that the greater the stylistic similarities existing between two areas, the greater the social interaction existing between them (Braun and Plog 1982). Intensive social contacts lead to a certain degree of consistency in, for example, the manner of decorating pottery; however, competition between different groups brings a disposition towards a more highly differentiated material culture (Hodder 1979).

Economic and social change during the later part of the Early Neolithic

Phase 2 of the Early Neolithic is characterized by major changes of both a social and an economic nature. A distinct change in the pattern of settlement between phases 1 and 2 has been reported in work dealing with south-west Scania (M. Larsson 1984, 1985). From having been concentrated in the inner hummocky zone, the pattern of settlement involved a move to the heavier soils in the Malmö area. A clear continuity can be seen in the data from the later part of the Early Neolithic up to an early part of the Middle Neolithic (I-II) (M. Larsson 1984). Is a similar change to be observed in the Ystad area? The evidence provided by the single finds may be a starting-point for the discussion.

Compared with the pointed-butted axes, which are associated in particular with the Oxie group and are encountered above all in the inner hummocky zone, the thin-butted axes are found essentially in the coastal area (Hellerström 1988). This is true not least of the axe-types, which can be dated to the Early Neolithic 2 and the Middle Neolithic I. It is clear that the axe finds indicate a change in the pattern of settlement to sites closer to the coast. The same principal features as in the Malmö area can thus be observed in the Ystad area. This change is particularly noticeable in the area around Öja-Herrestads Mosse and in the Köpinge/Kabusa area. What factors can have influenced this development? There is no doubt that there was impoverishment of the soil in those areas where forest clearance had taken place (Nielsen 1980). Increased erosion has been documented in, for example, Krageholmssjön (Gaillard 1984, Ch. 5.1). The impoverishment of the soil may have been accelerated by the clearance and burning of small areas. The fact that slash-and-burn, when combined with a form of coppice management, had a negative effect on the environment is substantiated from eastern North America, for example, where the first colonists based their agriculture on the slash-and-burn technique. An increase in the population, combined with erosion along the hillsides, brought about a change in the balance between cultivated land and the wooded areas (Otto and Anderson 1982). This may be one of the reasons for the change in the pattern of settlement which can be observed during the Early Neolithic 2.

The archaeological period with which we are concerned is between ca. 2700 and 2600–2500 BC. This has been designated as a regeneration phase by the palaeo-ecologists (Berglund 1985b). A clear regeneration phase can be identified in a number of pollen diagrams from southern Sweden, starting about 2550 BC. This stage has been reinterpreted by Göransson (1986). An examination of a number of pollen diagrams has revealed that the pollen curve for cereals does not fall during the so-called regeneration phase (Göransson 1986). What may be assumed instead is a system of coppiced woodland of various ages on light soils (Bro-wall 1986, Göransson 1986).

Concurrently with a more intensive utilization of the forests, the first signs begin to emerge of the ard plough. A considerable quantity of evidence has been recorded from Denmark, all of which dates back beyond the Middle Neolithic I (Thrane 1982).

We can sum up thus: A number of factors which have changed and influenced the pattern of settlement during the Early Neolithic 2 can be observed: sites close to the coast; more intensive use of the forests; and the in-troduction of the ard. The construction of the first megalithic graves, the dolmens, can also be dated to this period.

Emphasis is placed on these factors in the following discussion; the main area of interest selected is the Herrestad/Köpinge area.

From a starting-point of scattered settlement consist-ing of four identified sites in the Early Neolithic 1, a noticeable increase in the density of settlement occurs during phase 2. The sites are still small and are localized on sandy soils in the vicinity of watercourses. A system of base sites plus sites of a more specialized nature for activities such as hunting and fishing can be observed in the Ystad area during the Early Neolithic 2 and the start of the Middle Neolithic. Various pieces of evidence show that both the overgrown lagoon (the present-day Herrestads Mosse) and the coast were used for these purposes. Small sites with little material remains have been found, and a few of these have been excavated (Tesch 1985, M. Larsson 1988b). A large number of the very small settlements which were found during field surveys in the area can without doubt be described as specialized sites of a more temporary nature, of the same type as in many neighbouring areas (Skaarup 1973, 1985, Strömberg 1978, M. Larsson 1984, Hoika 1987).

Megalithic graves and society

The localization of the megalithic graves on light, sandy soils has been discussed in many contexts (Schirnig 1979, Bakker 1982, Madsen 1982, M. Larsson 1984, Barker 1985, Midgley 1985). A further characteristic feature is the association with rivers, lakes, the sea, or lagoons; the graves are frequently situated on terraces overlooking watercourses (Madsen 1982, Barker 1985). These characteristic features are, of course, equally ap-plicable to southern Scania, where the megalithic graves in both the Hagestad area and the Ystad area are found on sandy soils (Strömberg 1982) (Fig. 5.4:4).

Only a single example of a dolmen is known today in the area around Herrestad/Köpinge; this is known as Trollasten (Strömberg 1968). A study of old maps in-dicates that there were at least three others in the area (Riddersporre 1987, M. Larsson 1987a). Yet another dolmen was discovered in conjunction with the investi-gation of a ploughed-out Bronze Age barrow at Skogs-dala (Jakobsson 1986). A significant feature of the set-tlement in the area discussed here is the central location of the dolmens within certain specific areas. The hab-itation sites are grouped around them, and in a couple of cases hoards of thin-butted axes have also been found (Larsson and Larsson 1984, Hellerström 1988).

The significance of the megalithic graves as territorial markers is frequently referred to in recent archaeolog-ical research (Chapman 1981, Hårdh 1982, Fraser 1983, Renfrew 1984). Chapman (1981), for example, has stressed the connection between formal disposal areas, resources, social groups, and burial rites. Bradley (1984) has suggested that the dolmens are situated in the central areas of the settlement. A picture of this kind has emerged from studies of the localization of the dolmens in Denmark. There is a close relationship be-tween dolmens and highly productive areas such as river mouths, lagoons, and archipelagos (Bekmose 1977). Similar patterns have been obtained in Scania (M. Lars-son 1988b). The interrelationship of resource area, hab-itation site, and megalithic grave is noticeable not least in the Herrestad/Köpinge area. The sites are grouped together within five or six areas: the inner part of Her-restads Mosse; Karlsfält/Fårarp; the Tingshög area; the Kabusa area; the Skogsdala area; and Trollasten. The latter is less distinct, however. There are megalithic graves in four of these areas; none have been found in the Kabusa area or at Tingshög. This may give an idea of how society was structured in the Ystad area during the Early Neolithic 2 and the start of the Middle Neo-lithic. The society of the later period of the Early Neo-lithic was a fundamentally segmentary society, which means that a specific area was utilized permanently by a group of people, the basis for which was the individual site (Renfrew 1984). It is interesting to note that a segmentary society develops above all in conjunction with a system of small settlements during a period of expansion of the society (Service 1962).

Continuous settlement can be seen in many cases from the Early Neolithic 1 up to the late Funnel Beaker culture in the five or six territories which have been identified in the Herrestad/Köpinge area (M. Larsson 1988a). From the Early Neolithic 2 onwards it is pos-sible to assume the permanent utilization of those terri-tories in which agriculture based on coppicing was prac-tised. The primary unit within each area was a family dwelling site. The dolmens served as a burial place for a small number of members of the society, and at the same time served as a central point within an individual territory, thus marking the right of ownership of that specific area.

Conclusions

A model of how Early Neolithic society may have changed over a period of time has been presented above. It includes a number of factors of both an eco-nomic and a social nature. A specific characteristic of

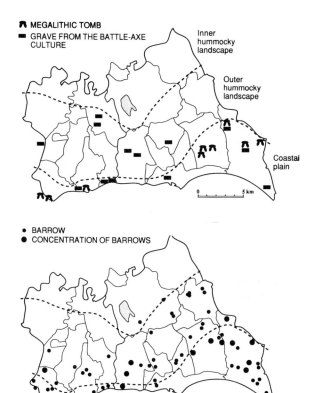

MEGALITHIC TOMB
GRAVE FROM THE BATTLE-AXE CULTURE

Inner hummocky landscape

Outer hummocky landscape

Coastal plain

0 5 km

● BARROW
● CONCENTRATION OF BARROWS

0 5 km

Fig. 5.4:4. Finds of Neolithic graves (above) and graves from the Early Bronze Age (below).

the Herrestad/Köpinge area in particular, when compared with south-west Scania, for instance, is the fact that it has been possible to substantiate the continuity from the Early Neolithic 1 and as far as the Middle Neolithic within a number of the identified group territories. Most of the evidence suggests that the population was small in this area; five or six groups would have provided an estimated population of 40–50 persons during the Early Neolithic 2. Settlement was apparently at its densest during the Middle Neolithic I, after which it either fell or was centralized (M. Larsson 1988b). This may indicate that there was very little opportunity for expansion within the prevailing social framework.

Cultural relationships during the late Middle Neolithic

Lars Larsson

The relationship between the various material cultures during the Middle Neolithic – the Funnel Beaker culture, the Pitted Ware culture and the Battleaxe culture – and the utilization of the landscape by these cultures have been, and continue to be, the subject of heated debate in archaeological research into southern Scandi-

navia. The discussion surrounding this subject has shifted its approach, from concentrating on chronological factors, to examining changes in society from a socio-economic viewpoint. A study of the relationships between the three aforementioned cultural types is thus synonymous with an examination of the structure of society as such during a late part of the Middle Neolithic.

The model put forward by Becker in 1954 to illustrate the relationship between the three cultures pointed to their existence in parallel during a late part of the Middle Neolithic (Becker 1954: Fig. 36, 1960: Fig. 44). Radiocarbon datings pointed clearly to a comparatively lengthy parallel existence in chronological terms between the Funnel Beaker culture and the Pitted Ware culture, corresponding largely to the whole of the Middle Neolithic, whereas datings for the Battleaxe culture produced values which were more recent than those for the Funnel Beaker culture (Malmros and Tauber 1977). This was interpreted as being indicative of a linear development of the Battleaxe culture from the Funnel Beaker culture (Spång et al. 1976: Fig. 1). It should be borne in mind that the radiometric datings came mainly from Jutland, but that in many cases the figures had been accepted as valid for a region made up of south and central Scandinavia (Burenhult 1982:156).

With regard to the relationship between the Funnel Beaker culture and the Pitted Ware culture, the results of a number of investigations have been published in recent years which to a certain extent change our view of these two cultural phenomena. An interest is now being shown in what characterizes these two cultures (Welinder 1978, Nielsen 1979, Becker 1982). This mainly concerns the question of what is to be designated as characteristic of the Pitted Ware culture. This culture, like the Funnel Beaker culture, does not exhibit a comparatively uniform combination of artefacts and style. The additional existence of a noticeable difference, not only in the material culture, but also in the economy, has proved to be associated with considerable interpretation problems (Browall 1986:28ff.). Certain areas are completely dominated by hunting, fishing, and gathering, whilst animal husbandry and arable farming are well represented in other areas.

Clear indications of a mixture of elements from both the Funnel Beaker culture and the Pitted Ware culture have been documented in the area of both Scania and eastern Jutland (Larsson 1982b, 1985, Wincentz Rasmussen 1984:83ff.). The organic finds indicate a mixed economy, with animal husbandry and arable farming being supplemented by hunting and fishing. A major variation in the utilization of the resources of the landscape can be recognized during the final phase, i.e. MN V, even in "pure" Funnel Beaker areas in eastern Denmark (Davidsen 1978:140ff.). This can be appreciated on the one hand from a distinct variation in the position of the sites, and on the other hand from a major variation in the size of the sites. It is not until MN V, for example, that a resource such as molluscs again begins to be used. Even at the sites of finds which are very close to the coast, remains of this resource are absent during the early parts of the Middle Neolithic.

The reasons for this distinctly mixed economy are not entirely clear. The explanation which can be considered the most plausible is that the controlling factors which society was unable to influence contributed to the change. A deterioration in the climate, for example, may have resulted in the increased utilization of those natural resources which were available on land and in the water. Another reason could be that to derive one's livelihood from trapping became socially acceptable in a way it had not previously been. Control mechanisms of a social nature are present in even the most low-technology societies; their importance can have a significant influence on the need to compensate the basal metabolism.

In certain parts of southern Scandinavia, therefore, it is not possible to speak of clearly identified material cultural units. The finds instead point towards cultural assimilation between the existing Funnel Beaker settlement and an expanding Pitted Ware culture. This period, which corresponds to the late Funnel Beaker culture, appears to have been an expansive stage in Scania, as it was in Denmark. The phase was previously often regarded as a degenerate period of decline (Malmer 1986:21). As in the transition between the Early and the Middle Neolithic, when clearly identifiable local characteristics are encountered, which are most clearly reflected in the pottery items (Ebbesen 1979:48ff.), it is possible to observe a world of varying forms during the late Funnel Beaker culture, with regionally limited stylistic features, both in western Scania and on Bornholm (Larsson 1985:31ff., 1986b, 1989b, Nielsen and Nielsen 1985:104ff., Strömberg 1988).

The cultural assimilation between the Funnel Beaker culture and the Pitted Ware culture may have been encouraged by the fact that the two cultures appear to share a common origin. Furthermore, the regional characteristics of the Pitted Ware culture indicate that its structure was such as to permit new elements to be readily accepted. The fusion of these population groups was facilitated by a common basic approach to society.

The problem is one of how the Battleaxe culture is to be related to this acculturation. An examination of the situation in central Sweden fails to reveal any cultural assimilation tendencies between the Pitted Ware culture and the Battleaxe culture during the late Middle Neolithic (Welinder 1978:109f.). The Pitted Ware culture remains a form of society based on hunting and fishing, whereas the Battleaxe culture is supported by a society based on animal husbandry and arable farming. Any displacement of the Pitted Ware sites from the highly productive inner archipelago to a less productive area in the outer archipelago has been interpreted as the result of the extension of its economic sphere of activity by the Battleaxe culture, causing the Pitted Ware groups to be driven out.

There is also no evidence of any tendency towards fusion between the Single Grave culture, which is the west Danish form of the Battleaxe culture, and the coeval complex in eastern Jutland with elements mainly of the Pitted Ware culture, but also of the Funnel Beaker culture (Wincentz Rasmussen 1984:97). At the same time it is difficult in the south Scandinavian Battleaxe complexes to identify any tradition from the Funnel Beaker culture which might reasonably be expected to have existed in the event of a linear change from the late Funnel Beaker culture to the early Battleaxe culture. It is also very important to pay particular attention to the study of relationships, not so much from a supraregional south Scandinavian perspective, but rather as they relate to a number of regions, in order better to appreciate the cultural and economic relationships (Ebbesen 1985:41f.).

With regard to the final phase of the Funnel Beaker culture, the sites in Scania where pottery items of the MN IV-V type have been found have produced many datings ranging between 2420 ± 220 and 2030 ± 90 BC (Larsson 1982a:87ff., Larsson 1989b). The Danish material shows that MN V extends over the period 2340–2160 BC, with an average value of 2270 BC (Tauber 1986:203). The finds which have been made in the west of Denmark, with examples of how elements from the late Funnel Beaker culture were combined with the Pitted Ware culture, exhibit a chronological variation from 2360 to 2000 BC, with an average value of 2150 BC (Tauber 1986b:203).

The radiometric datings from the Battleaxe culture are still few in number, although they nevertheless provide a picture of the chronological relationships of the culture. Four values from sites in the south of Scania vary from 2060 ± 115 to 1860 ± 60 BC (Strömberg 1978b:90, Larsson 1989b). The significance of these values should be considered from the standpoint that they constitute the only collected group of datings carried out on a Swedish site belonging to the Battleaxe culture. Only three ^{14}C datings, from as many graves, had previously been published, with a chronological variation from 2010±120 to 1900±60 BC (Lu-2554) (Larsson 1982a:87, 1988b). All these graves belong to the late Battleaxe culture Period 4 or 5 according to Malmer's classification (Malmer 1962:930, 1975:134).

All datings from the late Battleaxe culture – Periods 3–5 – should accordingly be accommodated within the chronological interval from 2060±115 to 1860±60 BC. Datings from the Under Grave period, a later phase of the Single Grave culture on Jutland, have been allocated a chronological variation of between 2290 and 2040 BC, with an average value of 2170 BC (Tauber 1986b: Fig. 2). Thus, the late Battleaxe culture in southern Sweden should correspond chronologically to the late Single Grave culture.

Radiocarbon measurements are still totally unavailable for the early Battleaxe culture in southern Sweden. A significant number of datings for the early Single Grave culture – the Under Grave period – have provided radiometric values with a range of variation on the central value of 2290–2090 BC, and with an average value of 2210 BC (Tauber 1986b:203). On the basis of our knowledge of the circumstances prevailing during the late Battleaxe culture, there are grounds on which to advance the theory that the early Battleaxe culture extended for approximately the same chronological period as the early Single Grave culture.

The combination of elements from the Funnel Beaker culture and the Pitted Ware culture which has already been identified, referred to here as the Stävie group after a site in western Scania (Larsson 1982a), may have continued to exist for a period which also includes that of the early Battleaxe culture. In eastern Jutland there was an example of the parallel existence of the Battleaxe culture and a cultural form which was an accumulation of the late Funnel Beaker culture and the Pitted Ware culture.

Although the finds dating from the Battleaxe culture are comparatively sparse in terms of ecological facts (Møhl 1962), there is no clear evidence to suggest the existence of any major difference between the forms of occupation of the Battleaxe culture by comparison with those of the Stävie group (Helmquist 1982, 1985, Persson 1982). As has previously been pointed out, on the other hand, significant differences may have existed in the social organization – a factor which may have had major consequences for relations between the societies. Although we have no need to consider alternatives such as a large migration, it cannot be denied that the basis for the social structure of the Battleaxe culture came from outside Scandinavia. Presumably what was involved was the acceptance of a new ideology with its own striking religious and social patterns (Malmer 1962). The Battleaxe culture is characterized by a distinctly individual-oriented structure, with an extremely formalized burial procedure. The continued use of megalithic graves within the Stävie group, which is substantiated not so much by the pottery artefacts as by the forms of the axes, exhibited a more group-related pattern with traditions derived from older phases of the same culture.

Late Middle Neolithic settlement in the Ystad area

Many of the datings referred to above from the period with which we are concerned here come from the sites of finds made within the Ystad area or in its vicinity, on

Fig. 5.4:5. Model for settlement distribution during Early and Middle Neolithic. Legend, A: site from the Funnel Beaker culture, B: sites with elements from Funnel Beaker culture as well as Pitted Ware culture and C: sites from the Battle-Axe culture.

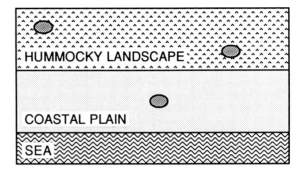

2 700 B C

HUMMOCKY LANDSCAPE

COASTAL PLAIN

SEA

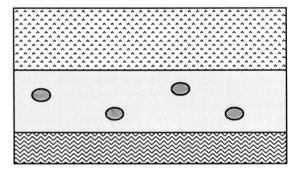

2 400 B C

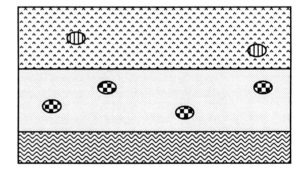

2 100 B C

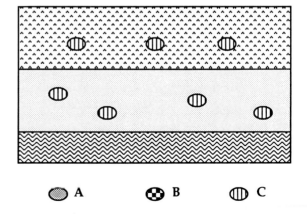

1 900 B C

A   B   C

the basis of which the Stävie group appears to have existed in parallel with the early Battleaxe culture.

The project area contains one site, at Karlsfält, which has been the subject of thorough investigation and which can be included in the Stävie group, in spite of the fact that the Funnel Beaker elements are totally dominant here. Occasional finds have also been made at other sites, which suggests that they are substantiated throughout the whole of the project area. These sites appear to have been of limited extent and were probably widely distributed both within the coastal strip and into the outer hummocky zone. This theory is supported not least by the presence of considerable numbers of late thick-butted axes (Ch. 4.1.3.1). Communities of a more distinct Pitted Ware character may have existed in certain coastal areas (Strömberg 1988:37, Hulthén 1977:126ff.).

Although the number of confirmed sites from the Stävie group has increased noticeably in recent years, traces of settlement dating from an early part of the Battleaxe culture are almost entirely lacking. For the purposes of examining the Battleaxe culture from a point of view which takes in both its older and its more recent phases, the graves are still the only adequate basis for the study of the distribution of the settlements.

As previously proposed by Malmer, the graves are concentrated on the best soils in the coastal regions of Scania (Malmer 1962). If the situation is examined in greater detail, however, in terms of both time and space, then a clear difference emerges. The graves from an early part of the Battleaxe culture are distributed in the most central parts of the best soil areas, which corresponds to the hummocky landscape (Fig. 5.4:4). The distribution of the oldest graves exhibits certain similarities to the distribution of the oldest pointed-butted flint axes (Larsson 1987e: Fig. 2; 1988d, 1989b). During the late Battleaxe culture, however, the majority of graves are situated in areas near the coast. As far as the Ystad area is concerned, the number of graves is very limited. Eight graves are known from a more recent part of the Battleaxe culture. These are distributed both in the hummocky landscape and in the area near the coast.

A well-established and thriving community in parts of Scania, supported by the Stävie group, was able to withstand the influence of the new ideology for a couple of centuries. Consequently, the representatives of those who welcomed the new ideological influences were obliged to look for pasture and arable land outside the established settlements in the hummocky zone. Nevertheless, the Battleaxe culture proved to be a most vigorous form of social organization, as a consequence of which it spread down as far as the coast and incorporated, with more or less resistance, the traditional areas of settlement.

Sites dating from the late Single Grave culture in western Denmark have been documented in a small, but increasing number (Hvass 1977, 1986, Rostholm

1986b, Hansen 1986, Simonsen 1986). One site dating from the late Battleaxe culture was investigated at Kabusa, within the project area (Larsson 1989b). A large, shallow pit was identified here, with post-holes which are presumably the remains of a house structure. Also found were three structures with deposits of what may be presumed to be whole vessels, indicating activity of a religious nature on the site. The finds are extremely limited. The layout of this site is very reminiscent of the sites in Jutland dating from the later Single Grave culture. The surface area of the site is similar to that exhibited by most of the Funnel Beaker sites in the investigated area, although the finds at the latter are considerably more numerous. The limited number of finds may indicate that mobility during the Battleaxe culture was greater than in the preceding period.

One reason why no sites dating from the early Battleaxe culture have been documented may possibly be that they were situated just inside the hummocky landscape, where sites with few finds are very easily destroyed by cultivation of the soil, as a consequence of which they are difficult to observe by field survey. On the other hand, the occasional fragments of axes found on sites in the hummocky zone may indicate that they were inhabited during an early part of the Battleaxe culture.

Late Neolithic settlement in the Ystad area

The complex picture of the relationship between the cultural conditions during Late Middle Neolithic is followed during the Late Neolithic by a more homogeneous material culture. The pattern of distribution of the finds made in the investigated area relating to this period is also more homogeneous from the chorological point of view. The Late Neolithic period is, in fact, well represented both on the coast and inland. The finds which have been made, and in particular the individual artefacts, provide indications not only of a wider distribution than in earlier phases in the hummocky landscape, but also of more intensive settlement in the earlier central parts of the coastal strip (Ch. 4.1.3.2, 4.2.3, 4.3.3, 4.4.3).

Of the investigated sites on the coastal strip, five have revealed traces of house structures. The variation in these constructions points to functional differences between the sites. There are house constructions of up to 20 metres in size (Ch. 4.1.3.2). Some of these have larger or smaller parts of their floor area sunken. This type of house, which has many counterparts both in Scania and Jutland (Strömberg 1971, Callmer 1973, Simonsen 1983:86ff., Björhem and Säfvestad 1989), presumably has its origins in the house constructions of the Battleaxe culture. The other type is represented by houses of up to 40 metres in length, but with the same floor plan as the smaller type (Larsson and Larsson 1984:38ff., 1986:30ff., Nielsen and Nielsen 1985:107ff.). As yet, it has not been possible to identify any subdivision within these house types. What is more, the presence of both types of house on one and the same site has not yet been documented, which further serves to confirm a functional difference. However, no difference has been identified in the sites on which they occur, or in their extent or position. The smaller type of house was in use during an early part of the Early Bronze Age (Boas 1983:99 ff.).

The same distribution of settlement in both the coastal strip and the inner hummocky zone which can be documented for the Late Neolithic also exists during the Early Bronze Age. This is confirmed by the distribution of the barrows which can be observed in zones within both areas (Fig. 5.4:4). Barrows additionally occur sporadically within the previously utilized inner hummocky zone (Larsson 1991b). The steep fall in the number of barrows observed as one moves further from the coast need not in itself entirely reflect a corresponding relationship in the distribution of the population. The struggle for the best soils may have resulted in a disproportionately high number of barrows at the coast in their function as territorial markers. According to the pollen diagrams, however, human activity does not appear to have increased since the Neolithic, as a consequence of which limited settlement in the inland area is reflected on a par with the high number of barrows.

Archaeological and Quaternary biological expansion phases – how closely do they coincide?

The pollen diagrams indicate an increase in human activity during the final part of the Middle Neolithic. In previous papers dealing with the expansion phases during prehistoric times, this second expansion phase was taken to represent the time of major change towards the rapid establishment of a permanent cultural landscape which had a noticeable influence on the environmental picture (Berglund 1969). The findings of new studies have revealed the dating of this period of steeply increased human activity to be incorrect, and that it should instead be moved to the point of transition between the Early and Late Bronze Age. What this means is that the last part of the Middle Neolithic must be reinterpreted, and in particular the Late Neolithic, which was previously regarded as an extremely expansive phase, with regard to the Quaternary biological indications as well.

According to new interpretations of the pollen analyses, the degree of utilization has thus not been found to be any greater than during the expansion phase of an early part of the Neolithic period. Stagnation, or even a regression in the cultural landscape, can be demonstrated between these comparatively weakly marked periods of expansion.

The stagnation could be explained by a source-critical factor, namely, that an established area of coppiced woodland acted as a filter which prevented the distribution of pollen, with an associated decrease in certain species of trees (Göransson 1988). In this case, any increase in the environmental indications influenced by humans could be interpreted as a change associated with a limitation of coppiced woodland. Even plants such as the ribwort plantain, whose spread is caused by the direct presence of livestock, increase concurrently with other indications of a cultivation. This supports the theory that a change in the landscape actually occurred. This has reached a level corresponding to that which existed earlier, notwithstanding the fact that it did not exhibit the same noticeable sequence of changes in view of the existence already of a limited cultural landscape.

5.5 Tradition and change during the Bronze Age and Iron Age. Houses as archaeological sources for the study of changes in the cultural landscape

Sten Tesch

Introduction

The following pages present a discussion of changes in the cultural landscape in the Köpinge area on the basis of a concrete archaeological source – habitation sites with house remains. It has sometimes been claimed that only pollen diagrams reflect the intensity of cultivation (e.g. Malmer 1988:96). It is true that archaeological evidence from habitation sites is primarily a reflection of settlement, but it is also an indirect reflection of tillage and grazing. The agrarian settlement units of farm and village and the pattern of agrarian settlement in different periods can be described as the spatial expression of continuous societal changes in agrarian production and dwelling (Sporrong 1984:61).

Even if the primary causes of the changes are of an ideological or ecological nature, I maintain that these changes can also be indirectly reflected in the evidence from habitation sites. It is therefore important, not least in view of the problems of pollen diagrams as a source, to analyse changes in the cultural landscape with the aid of the indirect evidence provided by the habitation sites. A key role is also played here by the macrofossils found at some of the excavated habitation sites (e.g. Hjelmqvist 1987, Engelmark 1988, Engelmark and Hjelmqvist 1991, Ch. 4.1.4, 4.1.5).

The settled landscape is a part of the cultural landscape. The first long-houses are contemporary with the introduction of agriculture (Björhem and Säfvestad 1989:89f.). An early example has been investigated in the western part of the project area (Larsson and Olausson 1986, M. Larsson 1988, see Fig. 5.5:1). Long-houses continued to be the normal form of agrarian settlement until the end of prehistoric times. The spatial distribution of habitation sites, the size of settlement units, the settlement pattern, and not least the houses themselves bear the stamp of the conservatism and traditionalism of agrarian society. The buildings are shaped according to the cultural milieu and its traditions. It is only when the cultural milieu changes and tradition no longer has a function to fulfil that drastic changes are accepted. The few striking changes that we observe should therefore be reflections of profound changes of the entire society (Näsman 1984, 1987, Lund 1988, Björhem and Säfvestad 1989).

Decisive changes in the cultural milieu which can be studied in changes in the construction of long-houses or in the spatial disposition of habitation sites include the introduction of stalling for livestock, the transition from isolated farmsteads to agglomerated settlements and villages, and in the settlement pattern, with indirect evidence of changes such as the introduction of manuring, the relation between animal husbandry and tillage, the rise of permanent cultivation systems resulting in the regulated medieval village (developments discussed in Ch. 4.1.4, 4.1.5). However, since only a limited dimension of the houses is preserved in the plans and sections drawn by archaeologists (Figs 5.5:1, 2), we must always remember to subject the sources to careful criticism (see also S. Nielsen 1987, Björhem and Säfvestad 1989:89).

Source material

In the Köpinge area – the coastal district – a relatively large number of habitation sites with house remains have been investigated (Ch. 4.1.4). The source material has the advantage of being varied in time and place. It covers all the prehistoric periods and each period is represented by several sites. As regards the chronology of the sites, the scarcity of finds often makes it difficult to date sites more precisely than somewhere within a period of 200 years. It can therefore be hard to determine whether different sites from one period represent a continuous chain or are contemporaneous. At the habitation sites it can also be a problem to see whether all the buildings are remains of a single farm which moved around within one habitation area or an agglomeration of contemporary farms, a village. The variation in the material and many ^{14}C datings have enabled the establishment of a long-house chronology for the Köpinge area, which to a certain extent follows the general development in southern Scandinavia (Fig. 5.5:1), but which also shows regional and local features. It is above all the inner roof-bearing construction that can be used in this context. It develops over several thousand years from a two-aisle structure with a row of posts down the middle to a three-aisle structure where two rows of inner posts support the roof; in the latter type the longitudinal distance between the roof-bearing posts changed through time from four to two metres (Fig. 5.5:3). Towards the end of the prehistoric era there also occur houses with no inner roof-bearing structure, where the walls instead bear the entire weight of the roof. With the passage of time the farms acquire more and more outhouses of different kinds. There is a continuity in that the normal long-houses, disregarding possible functional divisions, are mostly the same size the whole time, around 100 m^2. This pattern is somewhat different from the general development in southern Scandinavia.

We thus have the possibility of investigating whether there is a connection between the changes in cultivation intensity that can be read in the pollen diagrams and the archaeological evidence of changes in the structure of the long-houses, the composition of the farms, and the settlement pattern. The archaeologist's source material – the houses – may also help to provide more precise datings of changes in the cultural landscape, and perhaps explain them to some extent. Yet we must always be conscious of the difficulties of interpreting cause and effect from archaeological material.

Research into settlement archaeology – an overview

Our picture of prehistoric society has long been influenced mainly by the evidence of burials. Grave-finds are the basis of archaeological chronology and are sources for ideology, social structure, and cultural contacts. Before excavations of habitation sites became common, research into settlement archaeology had to rely on the evidence of graves, both visible monuments and grave-finds, as well as stray finds. Previous habitation site investigations in Sweden had either been on a very small scale, often only an indication, or were associated with special settlements and research projects (see Stjernquist 1978). Distribution maps for different categories of finds and graves from different periods have therefore played an important role in a wider context of settlement archaeology (e.g. Malmer 1957, 1962, Strömberg 1961, Stjernquist 1983).

By analysing antiquities from the relatively well preserved grave-material from the Mälaren region, Ambrosiani (1964a) developed an innovative method in settlement archaeology. In recent years, however, it has become obvious that the ploughing up of graves and cemeteries is a problem in totally cultivated regions (e.g. Sporrong 1985b), which means that the analysis of visible graves is a more uncertain instrument for studying changes in the cultural landscape than has previously been assumed.

Ambrosiani also extended the analysis by placing archaeological source material in relation to the oldest land survey documents from the 17th and 18th centuries, which gave rise to fruitful interdisciplinary collaboration between settlement archaeologists and historical geographers. The geographers were also quick to see the significance of archaeological investigations of habitation sites for a long-term perspective on changes in the cultural landscape (e.g. Lindquist 1968, Carlsson 1979).

Declines in the number of known graves in certain periods have often been interpreted as reductions in population and crises instead of new burial customs favouring humble and inconspicuous graves. In recent years, however, the evidence of graves has come to the fore again now that changes in burial customs have also been linked to a discussion of changes in religion and socio-economic conditions (Bennett 1987).

In the past few decades habitation site investigations have become increasingly common. The reasons for this are the growing awareness of the importance of habitation site material and the increase in rescue archaeology as a result of the pressure of exploitation on the cultural landscape. New methods of investigation have also created new possibilities for the investigation of habitation sites.

For Danish settlement archaeology the breakthrough took place back in the early 1960s with Becker's many years of excavation in western Jutland (e.g. Becker 1965, 1983, cf. Näsman 1987, Hvass 1988, Lund 1988). This introduced a new method by which machines quickly exposed large areas down to the substratum. Here post-holes, pits, and the like could be seen as dark patches against the lighter substratum. Entire villages could thus be mapped in a short time. A number of large habitation site were excavated in Jutland, the best known of which are Hodde (Hvass 1985), Vorbasse (Hvass 1987, 1988), Sædding (Stoumann 1978), and Omgård (L. C. Nielsen 1980). On the Danish islands the breakthrough did not come until the 1980s, chiefly as a result of the rescue archaeology of the natural gas project (Näsman 1987).

In Sweden the breakthrough occurred in the 1970s. In many cases, however, it was not a question of a conscious choice but the result of excavations of Late Iron Age cemeteries which turned out to be superposed on earlier habitation sites (e.g. Tesch 1974, 1980, 1983c, Wigren 1978, Bennett 1984).

In Scania it was not until the end of the 1970s that it became common to strip large areas of arable land to discover habitation sites hidden underneath (Tesch 1979a, 1979b, 1980, 1982, 1983b,c, Björhem and Säfvestad 1983, 1989).

Since habitation sites provide a much richer and very different picture of a society than graves and grave-finds, they have radically changed our view of prehistory. Unlike the graves, the habitation sites are shaped by practical and functional considerations, so they are a better reflection of everyday life and material conditions. In particular, habitation sites with house remains prove to be a very suitable source for studying the economy and society of prehistoric communities in a long-term perspective (Näsman 1984, 1987, Lund 1988).

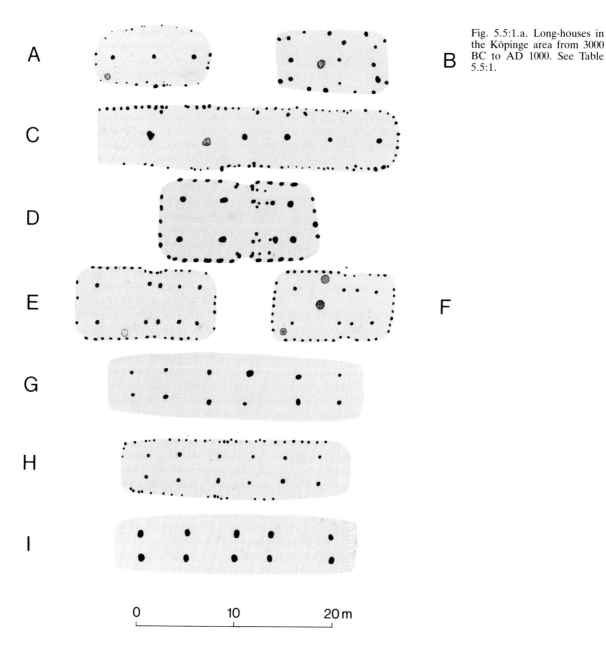

Fig. 5.5:1.a. Long-houses in the Köpinge area from 3000 BC to AD 1000. See Table 5.5:1.

A

B

C

D

E

F

G

H

I

```
0          10          20 m
```

Research into cultural landscape changes – a survey

What then are the underlying causes of the changes in the cultural environment that are reflected in changes in the traditional long-house, the farm, and the settlement pattern?

Research into changes in the prehistoric cultural landscape has mostly been conducted in collaboration between archaeologists, historical geographers, and palaeoecologists. In recent decades there has been discussion of topics such as the existence of generally valid models for the development of the cultural landscape, that there are regular links between several contemporary phenomena in different places throughout Scandi-

navia and northern Europe. The expansion and regression model proposed by Berglund (1969) became important (e.g. Welinder 1975), not least for the conception of the Ystad Project (Ch. 1.1). In recent years, however, it has become increasingly clear that the development of the cultural landscape displays great regional and local differences (Widgren 1983).

When it comes to explaining changes in the cultural landscape, archaeologists and geographers in particular have been heavily influenced by Boserup's (1973) theories of agricultural development and population growth. These theories have also greatly influenced the working hypothesis of the Ystad Project, that there is a link between expansions of cultivation/production and pop-

328

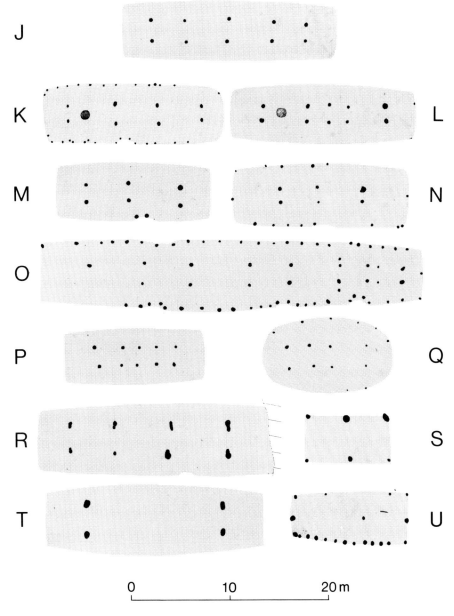

Fig. 5.5:1.b. Long-houses in the Köpinge area from 3000 BC to AD 1000. See Table 5.5:1.

J

K   L

M   N

O

P   Q

R   S

T   U

0   10   20 m

ulation pressure (e.g. Berglund 1984: Fig. 2). Today, however, there is contention about whether it is fruitful to interpret the development solely as a consequence of population pressure and the intensification of agriculture. This discussion has also influenced and changed our working hypothesis in the course of the project (Ch. 1.1). Changes in the cultural landscape appear to be due to several parallel processes which may be mutually interdependent but are not synchronous and have varying impact in different regions (Widgren 1983, 1988).

Stjernquist (1984, 1987:246f., 1989) has presented a simple model which includes general causal connections in changes in the cultural landscape. It is a combination of changes in human activity and ecological changes that causes changes in the landscape. The ecological changes are caused by factors such as climate and occur independently of human activity. Changes in human activity comprise four coordinate factors: technical innovations, changes in economic conditions, in social organization, and in population. The working hypothesis of the Ystad Project has been gradually influenced by this model (Berglund 1986:4f., Ch. 1.1). Stjernquist shows convincingly that the population question cannot be assigned any central significance for changes in the cultural landscape. There is no evidence that people in prehistoric times lived at the ecological limit permitted by their technological knowledge.

Stjernquist considers it unrealistic to believe that a

Table 5.5:1. Measurements and dates for long-houses in the Köpinge area. See Fig. 5.5:1.

Fig.	Property no.	House no.	Length	Width	Distance between roof-bearing posts W-E	Area m²	Dating
A*	Mossby 27:1		12.0	6.0	Two-aisled	66	EN**
B	Köpinge 15:22	VII	11.6	6.0	Two-aisled 4.0/4.0/3.3	66	LN
C*	Herrestad 68:87		30.0	5.8	Two-aisled 5.0?	180	LN/EBA I
D	Köpinge13:14	I	16.0	8.2–7.6	4.0/3.9/3.9/3.7/3.6	127	EBA II-III
E	L. Köpinge 19:1	I	14.0	7.3–7.0	3.6/3.7/3.6/3.6/3.5	98	E-LBA III-IV
F	L. Köpinge 19:1	VIII	12.3	7.0–6.4	3.4/3.3/3.3/3.0	80	LBA IV-V
G	L. Köpinge 19:1	XIII	26.0	6.5	2.2/2.6/2.8/3.0/2.6/2.7	160	PRIA I-II
H	L.Köpinge 6:20	I	23.0	5.0–5.7	2.0/2.5/2.4/2.5/2.3/2.6	125	PRIA III
I	L. Köpinge 6:26	IX	24.0	5.0–5.5	2.3/2.5/2.4/2.3/2.3	125	ERIA
J	Köpinge 15:22	II	21.0	5.0–5.5	2–0/2.1/2.2/1.9/1.6	115	ERIA
K	L. Köpinge 19:1	IV	18.0	5.6–5.8	1.3/1.8/1.9/1.4	105	LRIA
L	L. Köpinge 19:1	IX	18.0	5.0–5.6	1.4/1.5/1.7/1.4/1.5	95	LRIA
M	Köpinge 64:1	III	16.0	5.5–6.0	1.7/1.7/1.7	92	MP-VP
N	Köpinge 64:1	IIA	18.0	6.0	1.3/1.5/1.3	105	MP-VP
O	Köpinge 21:2	III	35.0?	5.0–7.0	- /2.0/2.0/2.0/1.7/1.6/1.5/1.3	210	MP-VP
P	Köpinge 64:1	I	13.0	5.0–5.5	1.8/1.7/1.7/1.7/1.7	70	MP-VP
Q*	L. Köpinge 7:3	III	13.0	5.0–7.0	2.0/2.0/2.0/2.0	78	MP-VP
R	Köpinge 6:26	VIII	26.0?	7.0	2.7/2.8/2.9/3.1	182	VA-EMA
S	Kabusa 28:3	I	8.4	4.3	No internal posts	36	VA-EMA
T	Köpinge 6:26	XIX	22.0	5.0–7.0	2.6/3.0	132	VA-EMA
U	Köpinge 6:26	XVI	12.0	4.0–5.2	Two-aisled	50	VA-EMA

* After Larsson and Larsson 1984, 1986.
** The house is located in the south-west corner of the project area. It has been included in the table to show the entire prehistoric development of the long-house. The other houses are located in a limited part of the Köpinge area in the south-east corner of the project area. For a full account of the houses in the Köpinge area see Ch. 5 and 7 in Callmer et al. (1991).

hypothesis about population growth can be tested using archaeological material. The problems are of both a chronological and a spatial kind. Since it is impossible to reconstruct all the changes in the cultural landscape through time, we must schematize the development in periods of one or two centuries, sometimes even longer. This can make it appear as if there were great, sudden changes. For this reason it is also difficult to compare maps showing the spread of settlement at different times. Anthropological studies of pre-industrial societies have also shown that natural resources are not used to the full, which means that Boserup's theories are not tenable.

We can also reject those explanatory models which presuppose that development was stimulated by crises caused by uncontrolled population growth. It is not the case that population at regular intervals reached a production maximum. Changes in the cultural landscape are instead a result of the will of the local community, of the adaptation of thinking people to changed ecological conditions, although perhaps not always in phase with

each other. In addition, there has been influence from outside in the form of new ideas.

Throughout the period covered by the Ystad Project there has been room for expansion, not least in the hummocky landscape. Because of tradition, the relation to the collectively owned land, the closeness to ancestors' graves, and so on, people more often chose internal expansion rather than external. This can have provoked periodical pressure on resources, perhaps even crises, in the central area, even though the landscape as a whole was not exploited to the full. The hummocky landscape was here used as a buffer zone, but more often extensively than intensively. Privatiza-

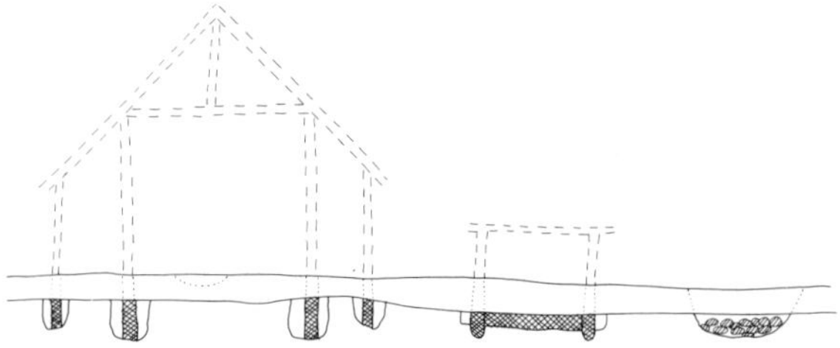

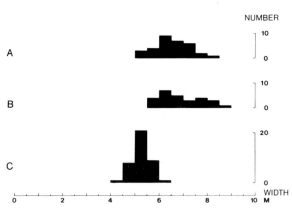

Fig. 5.5:3. The width of southern Scandinavian long-houses of different age. A: Two-aisled house from the Late Middle Neolithic to the Early Bronze Age. B: Three-aisled houses from the Bronze Age. C: Three-aisled houses from the Pre-Roman Iron Age. From Björhem and Säfvestad 1989.

Fig. 5.5:2. Hypothetical section through part of an Iron Age settlement. After S. Nielsen 1987.

tion of arable land, that is, a dissolution of the social-religious bond to the land, can have facilitated external expansion, settlement colonization, as happened in the Ystad area in the Late Iron Age.

Expansion and regression/Stability and instability/ Tradition and change

From a study of pollen diagrams from the introduction of agriculture down to the present day Berglund (most recently 1985, 1986, 1988, Ch. 1.1) has identified six expansion phases, of which three can be described as crucial phases of change in the development of the cultural landscape. Three expansion phases, two of them crucial phases of change (3, 4, 5), fall within the time-span to be discussed here, from the Late Neolithic to the High Middle Ages, covering more than 3000 years. From a correlation diagram from the Ystad area the expansion phases have been dated to 800–400 BC (3), 0–AD 400 (4), and AD 800–1100 (5) (Berglund 1988, cf. Fig. 1.1:2). For the Köpinge area the middle phase should be shifted one or two hundred years back in time (Ch. 4.1.2). There is no room here to discuss what is actually meant by phases of expansion and regression, but they are mostly quantitative measures of the openness of the landscape, the area of land used for pasture in particular (see Ch. 2.9). A regression phase, such as 400–200 BC, can be interpreted as the result of a concentration of settlement and an intensification of production, and thus actually an expansion phase, just as well as it can be seen as a sign of crisis and decreasing population (Ch. 4.1.2).

The archaeological evidence from the Köpinge area suggests an alternative interpretation of expansion and regression phases. An expansion phase, at least its core, is characterized by a stable settlement pattern and some form of equilibrium. The structure and disposition of the long-houses do not undergo any decisive changes. Regression and intermediate phases, on the other hand, appear to be the real phases of change as regards both long-houses and settlement patterns, and hence also the pattern of production. However, it is not a question of changes by leaps but rather an undulating and gradually ascending curve with alternating phases of stability and change (Sporrong 1984), in other words, tradition and change (Fig. 2.9:1). The course of development is cyclic, with phases of stability lasting 200–300 years before the balance is upset. Shifts in the balance ought to be evident in changed settlement patterns, such as agglomerated settlement, and thereby capable of being proved by archaeological research.

The pollen diagrams – or rather their interpretations – involve many problems of source criticism (Berglund 1985:79ff., Ch. 2.1). One must also bear in mind that there are problems in the exact radiocarbon datings of pollen diagrams (Berglund 1988:248).

When working with theoretical models it is also important to remember that, whether they present development in steps or in waves, they are about humans and their activities. Human activities depend on the prevailing cultural milieu and need not always be rational, at least not when seen through our eyes. External circumstances such as climatic changes seem to follow a more or less cyclic course (Aaby 1976), but it is not certain that they always have the same impact on the form of the cultural landscape. Nor is there any clear correlation between climate and the expansion phases discussed here, even if the decisive phases (3 and 5) seem to correspond to the transition to a cold and damp climate with raised groundwater and lake levels (Ch. 3.3.1, 3.4.1, 3.5.1, 4.1.2).

It should also be pointed out that the light soils of the Köpinge area have been sensitive not only to a warm, dry climate but also to a cool, damp one. People therefore usually avoided risking destitution in bad years by not exploiting to anything near the maximum the area's carrying capacity, as it has been estimated for different phases: 3100–400 BC 20 persons km^{-2}, 400 BC-AD 800 50 persons km^{-2}, AD 800–1900 50–200 persons km^{-2} (Emanuelsson 1988).

Expansion phase 3 (800–400 BC)
The pollen diagrams show a dramatic change in land use around 800 BC, "the great leap". Woodland disappeared and the landscape opened up. This change is most evident in the hummocky landscape, however, while the coastal zone had already seen a gradual change in the preceding period (Late Neolithic and Early Bronze Age, 1800–800 BC) to a half-open woodland landscape with much birch and hazel, coppices alternating with open pastures and small arable areas (Ch. 4.1.2). Around 1000 BC there was a radical change in the climate to more maritime conditions with lower summer temperatures and greater humidity. This change is assumed to represent the most severe cooling of the climate in the past 10,000 years (Ch. 5.1). There is thus good reason to examine the archaeological evidence to see if it reveals crucial changes in the time preceding the expansion phase.

The Late Neolithic appears to have seen the establishment of the topographically delimited settlement area discussed above (Ch. 4.1.4), with settlement units consisting of single farms. From the Malmö area at the same time there is evidence that a settlement made similar small moves between three sites located not more than 170 m from each other during a period of some 300 years (Björhem and Säfvestad 1989:123f.).

There is nothing to suggest any great change in this relatively stable settlement pattern in the Early Bronze Age (Ch. 4.1.4.1, Fig. 4.1:4). On the other hand, in period II of the Early Bronze Age we see one of the few real breaks with tradition in the development of the long-house. The decisive change is from the two-aisle long-house to a three-aisle house with two rows of inner posts bearing the roof (Fig. 5.5:1). The shift is synchronous throughout southern Scandinavia (Ethelberg

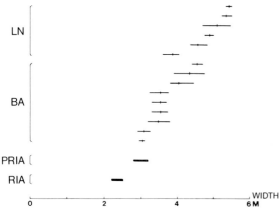

LN

BA

PRIA [
RIA [

WIDTH
0   2   4   6 M

Fig. 5.5:4. The width between the indrawn posts in Late Neolithic houses and the width between the two rows of posts in Bronze Age and Iron Age houses. Maximum, minimum, and mean values. LN (late Neolithic) and BA (Bronze Age) are houses from Fosie IV. PRIA (Pre-Roman Iron AGe) are mean values from houses at Hodde. RIA (Roman Iron AGe) are mean values from houses at Vorbasse. From Björhem and Salvested 1989.

pairs of posts, and there is a tendency to divide the house into functions at the entrance (Ethelberg 1987:164). Long-houses from period II of the Early Bronze Age in the Köpinge area are relatively long, but not enough is preserved for us to say that they had this functional division into dwelling and byre (Fig. 5.5:1). By contrast, from long-houses with ^{14}C datings to period III or the transition to period IV there are many examples of a functional division. Even Late Neolithic long-houses, including one from the project area, appear in many cases to show some division, with the hearth in the west end. These houses are interpreted as chiefly dwelling houses with room for work and storage, not byres, in the smaller western part (Björhem and Säfvestad 1989:98). To sum up, it cannot be entirely ruled out that animals were stalled in the eastern ends of long-houses from period II of the Early Bronze Age onwards. However, the climate was still warm enough for the livestock to be kept outside the year round. The serious cooling of the climate came later in the Early Bronze Age. Stalling may also have been adopted to meet the need for more manure for intensified tillage, although there is no evidence in the Köpinge area for such an intensification before the Late Bronze Age.

The three-aisle house meant a break with tradition which need not require an explanation from the practical and economic point of view but may instead have a social or an ideological/religious background. It is striking that the change in tradition is synchronous with the rise of a new ideology which united much of Europe, with a warrior aristocracy and an economy of prestige goods based on long-distance exchange of bronze and

1987:165, Björhem and Säfvestad 1989:91). It was probably due to functional considerations: a three-aisle structure is, for example, better suited to stalling livestock along the long walls (Näsman 1987:78) and was probably also a more stable structure (Ethelberg 1987:165). The earliest evidence of stalls comes from a small number of houses from Early Bronze Age Jutland (Jensen 1988:160). In long-houses from the Early Bronze Age there are often hearths and cooking pits in the west end of the houses, usually between the last two

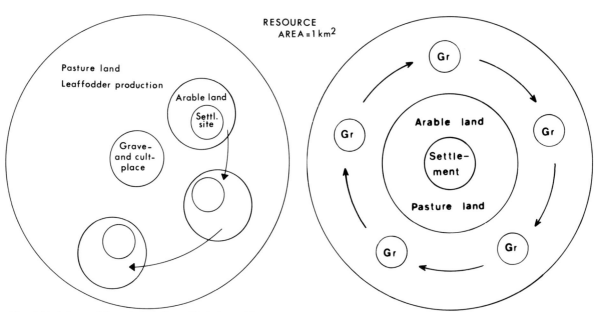

Fig. 5.5:5. The model on the left (from Welinder 1977) can be said to represent the "traditional" interpretation of the territorial behaviour of Bronze Age people, with a burial ground and cult-place in the centre. The model on the right (after Carlsson 1983) is an alternative interpretation with the habitation site and production area in the centre and burial places scattered on the periphery of the territory.

gold (Kristiansen 1987a). This can hardly be described as a stable phase; it was rather a period of expansive change. Livestock may have been a capital for building up a power base through marriage alliances. Keeping much larger herds than were needed to feed the people may have been one way for the chieftains to maintain their power within their own social group. At the same time, a large herd brought security against bad winters, drought, rinderpest, and so on. To ensure the survival of the herd, it may have been the practice to stall some of the livestock (the breeding stock?) over the winter. The use of milk products as food may also have been a relevant factor, in which case the milch cows could have been stalled.

Within the Ystad Project there has been discussion about whether the Bronze Age barrows should be taken to indicate settlement directly (Olausson 1991, Ch. 3.3.2, 4.1.4.2) or indirectly (Ch. 3.3.2, 4.1.4.1). An intensification of land use in Gotland in the Baltic has been linked to a more distinct demarcation of territories, in other words, the rise of distinctive settlement areas. The habitation site and the land used for production were the centre, whereas the burial places were spread out to mark boundaries on the periphery of the territory (Carlsson 1983, 1984, see Fig. 5.5:4). One possibility is that there may nevertheless be a connection between stalling/manuring, suggested by the change in the long-houses, and the building of barrows in the Köpinge area. The Late Bronze Age burials, not only in barrows but also in centrally located cemeteries or scattered graves, suggest a transition to the more differentiated territorial division of the cultural landscape in the Early Iron Age.

Previously (Ch. 4.1.4.1) I have discussed in detail the evidence for the manuring of arable land in the Late Bronze Age in southern Scandinavia and the Köpinge area in particular. The reasoning was based in large measure on analyses of carbonized plant material from the habitation sites. Hulled barley increased noticeably to become the main cereal in the Early Bronze Age. Since hulled barley needs nutrient-rich soils to give an acceptable harvest, this indicates some form of manuring. In the Late Bronze Age people grew more kinds of cereal than in any other period. Since the different crops require different soils and nutrients, the great variety is a sign that agriculture was important and well planned (Engelmark and Hjelmqvist 1991), which reinforces the picture of a stable settlement pattern during the period.

The pollen diagrams show that there was an expansion of sedge and other wetland plants. This supports the hypothesis that former alder carrs were now grazed and mown, providing winter fodder as well (Ch. 4.1.2). In the hummocky landscape, on the other hand, the alder carrs persisted (Ch. 4.3.2).

The houses underwent no major changes, but developed within the framework of tradition. However, towards the end of the period, in the early years of the Pre-Roman Iron Age, the roof-bearing posts in the east end of the house were not placed so close together; instead the span was the same between all the pairs of posts running longitudinally through the house, about 4 m, or roughly the same as in the dwelling section of the Late Bronze Age houses (Fig. 5.5:1). It is difficult to say when this change took place, since there are no houses which can be dated with certainty to the very end of the Late Bronze Age.

Analysis of charcoal from post-holes in various house structures reveals a change around 600 BC. Before this, more alder than oak was used, along with other timber types, but oak dominates after 600 BC. One possible interpretation is that the houses were used for a longer period (Bartholin and Berglund 1991), thus suggesting a more stable settlement pattern. There may also be a connection with the increased span between posts. However, the charcoal from the post-holes does not necessarily come from the standing posts.

Changes in burial practices also suggest a stabilization during the Late Bronze Age. Cremation cemeteries near the dwellings mark the group's own ties to its settlement area, unlike the Early Bronze Age barrows which were a manifestation to the outside world.

I have argued above (Ch. 4.1.4.1) that the much cooler climate towards the end of the period nevertheless led to less stable conditions, caused by factors like soil erosion and leaching. Most camp sites outside the settlement areas can be dated to the later part of the Late Bronze Age, which suggests that people had to look further afield than before for pastures, winter fodder, and so on.

However, the occurrence of stalling/manuring and haymaking indicates that we see in expansion phase 3 the beginning of the changes that led to the next stable phase, expansion phase 4, with its more differentiated farming society.

Expansion phase 4 (0–AD 400)
In the Köpinge area the expansion phase in the Pre-Roman Iron Age can be seen as a continuation of the expansive period that began in the Late Bronze Age. However, the intermediate period began with a woodland regeneration lasting about 200 years (400–200 BC). This can be interpreted in two ways: either as a real decline in human pressure on the landscape (a fall in population) or else a concentration of agricultural land to the immediate vicinity of the habitation site, with fields tilled in single courses surrounded by pasture and meadow. The fact that it was primarily hazel which spread supports both interpretations (Ch. 4.1.2). The same is true of the archaeological evidence from habitation sites, which declines in quantity in this period. The analysed plant material from the sites suggests one-course tillage growing mostly only one crop, hulled barley, which is a drastic change from the Late Bronze

Age cultivation of a great variety of cereals (Engelmark and Hjelmqvist 1991).

From now until AD 300 the pollen diagrams show a continuous heavy human pressure on the cultural landscape. Cereal cropping is proportionately more important than stock-raising compared with the Late Bronze Age (Ch. 4.1.2).

In the Krageholm area, in the hummocky landscape north of the Köpinge area, the same period (300 BC-AD 250) sees reduced human impact on the cultural landscape (Ch. 4.3.2). This could be interpreted as an expression of an increasingly stable production pattern in the central area, with consequent reduced pressure on the resources of the hummocky landscape.

Continuity from the Late Bronze Age is seen in the fact that the long-houses in the Köpinge area do not undergo any major changes during the period. The changes that occur are within the traditional framework (Fig. 5.5:1). In the last part of the Pre-Roman Iron Age the walls become more curved. Around the birth of Christ this is also seen in the roof-bearing structure, since the outer pairs of posts are closer together than the others. The span at either end of the house is often longer (4–6 m) than the span in the middle section. In the Roman Iron Age the roof-bearing posts become sturdier, while the dimensions of the posts in the wall decrease to the extent that they often leave no traces. Houses become gradually longer and narrower during the period. Although the total area remains roughly the same, this change gives room for the stalling of a greater number of animals than in the Late Bronze Age. No stall partitions have survived, however. Where hearths have been preserved they are still in the west end of the house, between the two westernmost pairs of posts.

This ostensibly continuous development in house construction in the Köpinge area is very different from a well-documented area like Jutland. Here there is a decisive break around 500 BC, when the large Bronze Age halls cease to be built. Instead settlement units now consist of small villages of 3–5, later 10–15 little farms (e.g. Lund 1988:158). The long-houses are very small to begin with – 7–12 m long and up to 5 metres wide – and they usually continue to be smaller than those in the Köpinge area throughout the period. The long-houses are also supplemented by smaller houses. The villages moved around within a resource area, in much the same way as in the presumed settlement area at Köpinge.

The rise of clear regional differences in settlement at the transition to the Pre-Roman Iron Age can be a reflection of the collapse of the Bronze Age's international system of prestige goods exchange. However, continuity within each region is clear from the grave-finds (Stjernquist 1985:229, Sørensen 1987:99).

Several sunken-floor huts (weaving huts) from the Köpinge area dated to the latter part of the Pre-Roman Iron Age indicate a differentiation of settlement. This type of settlement has usually been dated to the Late Roman Iron Age at the earliest, yet the tradition has

existed since the Late Neolithic, and recent years' excavations have shown several examples of sunken-floor huts which are older than the Late Roman Iron Age (e.g. Olausson 1989).

A more decisive change is seen in the agglomerated settlements that are attested from the time around the birth of Christ and the transition to expansion phase 4. These "villages" probably did not have a large number of farms. They do, however, appear to remain in the same place for 200–300 years. A stable settlement pattern like this must go hand in hand with some system of infields/outlands with permanently tilled fields. A system like this presupposes a balance between tillage, livestock (manure), pasture, and haymaking. Analyses of carbonized plants from these habitation sites show that mostly hulled barley was cultivated (Engelmark and Hjelmqvist 1991). This long-term exploitation of the poor soils of the Köpinge area reveals the importance of manuring. Even if the climate was favourable, the system bore the seeds of its own destruction in the long run, which led to a radical restructuring of the agricultural system in the Late Roman Iron Age.

Expansion phase 5 (AD 800–1100)

Expansion phase 5 is preceded by a powerful and unstable time of change, known as the crisis of the Migration Period. The change lasted as long as from the 3rd to the 7th century (Näsman 1988:240ff.). The whole of southern Scandinavia saw a change in the structure of agriculture. The development was by no means synchronous, which argues against a deterioration in climate as the direct cause, although it could have contributed to the restructuring in some regions. In the Ystad area the period AD 400–900 is characterized by a cold, damp climate and rising water levels (Ch. 3.5.1). The deterioration in climate thus begins after the restructuring, and the cool, damp weather still prevails when expansion phase 5 begins.

In the Köpinge area expansion phase 5 is preceded by both unstable and stable phases of settlement. In the 3rd century the stable little villages are dissolved and settlement instead takes the form of isolated farms, usually showing no continuity with the sites of the former villages. Nevertheless, there is no major change in the long-houses (Fig. 5.5:1). In other parts of Scandinavia there is evidence of very large houses, interpreted as chieftains' farms; the restructuring is thus seen in connection with a social stratification. There are examples from south-west Scania (Björhem and Säfvestad 1983, Wallin 1987), and in the Köpinge area too there are some indications of a social change of this kind (Ch. 4.1.4.4).

For the Köpinge area the direct cause of the restructuring should be sought in the stable and intensive pattern of production which characterized expansion phase 4. Long-term exploitation of poor soils proved untenable in the end. There was probably not only a restructuring but also a reduction in the number of farms, which is evident from a slight decline in human impact on the landscape around AD 300 (Ch. 4.1.2). Intensive farming in the Roman Empire had a similar result even earlier, in the 2nd century AD, with an impoverishment of the land that is still visible today, and the effect of which can be seen as a decline in cultural landscape indicators in pollen diagrams (Groenman-van Waateringe 1983). The protracted process of change thus suggests that climate was not the decisive factor.

The deterioration of climate in the Ystad area instead seems to coincide with the concentration of settlement that has been archaeologically dated to around AD 400. In the Köpinge area there were probably no more than a few large habitation sites, one in Stora Herrestad parish and one in Stora Köpinge parish. Over much of Europe we see a similar clustering of settlements separated by empty areas (Groenman-van Waateringe 1983).

In the Tingshög area in Stora Köpinge parish excavations have revealed a big, agglomerated, village-like settlement, with such a large number of buildings that it must have consisted of more farms than the little villages of the Early Roman Iron Age (Tesch 1983b: Fig. 19). Settlement is also more differentiated, with large and small long-houses and smaller buildings for various purposes. The settlement concentration, which appears to last for several centuries, has been interpreted as a way of meeting the adverse change in climate. However, there were minor moves within the area. It is probable that this change and stabilization in the settlement pattern can be associated with a change and intensification in the agricultural system to some form of regular crop rotation with fallowing and two or three courses (Tesch 1983b:53ff.). However, it is not until the Viking Age that there is sufficient macrofossil material from habitation sites in the area to prove the occurrence of three-course tillage with regular crop rotation – fallow, autumn-sown rye, spring-sown barley (Engelmark and Hjelmqvist 1991). Around AD 600 there is a chieftain's farm, a house about 40 m long, adjacent to the village.

About AD 700, at the start of expansion phase 5, there is archaeological evidence of colonization and resettlement of former arable lands, a development which is also found in other parts of southern Scandinavia. In the Ystad area as a whole the period AD 700–900 appears to show a vigorous expansion of the cultural landscape, which is especially noticeable in the hummocky zone (Ch. 3.5.1). Expansion phase 5 is interpreted mainly as a significant increase of the area of arable land (Ch. 4.1.2). Both before the expansion and during its opening phase (AD 500–800), the climate was cool and damp, with a spread of wetlands, and thus unfavourable for growing cereals. We must also bear in mind the link between arable and meadow – manure; an expansion of tillage is unthinkable without an expansion of livestock. A three-course system requires a much larger area than a one-course system, and this could be one of the reasons that the expansion is interpreted mainly as an expansion of tillage. At the same time as the expansion is a sign of population growth, one must be aware that there was no market for any great surplus production of grain during this period. In the 11th and 12th centuries it was possible to dispose of the surplus, especially with the dues payable to the crown, the church, and local lords. This was also a period when the climate grew warmer and thus more suitable for growing cereals. However, it was not until the 13th century, when the town of Ystad emerged and foreign trade in everyday commodities became common, that it became profitable to indulge in large-scale grain production (Ch. 4.1.5).

In view of what has been said above, expansion phase 5 could be given an alternative interpretation: that its initial phase was above all an expansion of pasture lands. I have previously put forward the hypothesis that the settlements which colonized the hummocky landscape in the Viking Age took the form of large cattle ranches (Tesch 1983b, 1985b). As well as being in keeping with the power politics of the Viking Age this would have been more suitable in the hilly topography of the hummocky landscape and in the prevailing climate, with the spread of wetlands. According to this view, a new agrarian system with tenant farmers in villages and with cereal cropping assuming a major role did not arise until the Early Middle Ages.

Unfortunately, archaeologically investigated houses from expansion phase 5 are few in number in the Köpinge area, as in the Ystad area as a whole. The tradition of the chieftain's farm in the Tingshög area probably continued. These large houses with their curving walls are typical of the Viking Age in southern Scandinavia, and in their most pronounced form are identical with the type known as Trelleborg houses. The farms are also typically complexes of many houses. Magnates' farms from the Viking Age/Early Middle Ages are attested in Bjäresjö (Ch. 4.2.5.1), in Ilstorp just north of the Ystad area (Tesch 1985b, Ch. 4.1.5.1), and in Kabusa in the Köpinge area (through retrogressive analysis of maps, Riddersporre 1983, 1986). The restructuring of settlement around AD 700 in the Köpinge area meant the abandonment of the light sandy soils. There are also signs of incipient heath formation in the period AD 500–800 (Ch. 4.1.2). Viking Age settlement is normally found close to the medieval village cores, in the tofts (Riddersporre 1988, 1989, Tesch 1988, Ch. 4.1.5.1). It probably consisted of some form of loosely clustered villages, located on the boundary between light and heavy soils, or only on heavy soils. At the end of the expansion phase, around AD 1100, the settlement pattern was stabilized, or rather fixed, through the regulation of the villages (Riddersporre 1988, 1989). It was not until the enclosure movement that settlement was once again dispersed.

Conclusions

The development of houses in the Köpinge area is characterized by continuous changes within the framework of tradition, to a greater extent than in many other areas of southern Scandinavia. In Jutland, for example, the transition from the Bronze Age to the Iron Age was a time when the house-building tradition changed radically, but there is no sign of this break in the Köpinge area.

One of the few real breaks with tradition in the structure of long-houses in the Köpinge area occurred in period II of the Early Bronze Age. However, this change is synchronous throughout southern Scandinavia. Moreover, the change came before "the great leap" in the development of the cultural landscape, which is dated to 800 BC in the pollen diagrams, and the serious change in climate around 1000 BC. What is cause and what is effect? Are the datings correct? As I have said above, we must not only consider ecological and economic causes but also social and ideological factors behind the changes.

A major change in the structure of the long-house is the introduction of a byre section; it is important to date this, since it meant a significant change and stabilization of the agrarian economy. In the absence of concrete testimony in the form of stall partitions, indirect evidence must be used to date the introduction of stalling in the Köpinge area to some time between period II and period IV of the Bronze Age.

The decisive changes in the cultural landscape and breaks with traditional development are easier to see in the archaeological material from habitation sites if we look instead at the composition of farms, the size of settlement units, and the settlement pattern. An important finding above is the connection between changes in the cultural landscape and the recurrent Iron Age alternations between single farms and villages, between dispersed and agglomerated settlement.

It has not been possible to find proof of a clear link between expansion and stability. Nevertheless, expansion phases 3 and 4 can be described as relatively stable. The unstable phases, the times of change, mostly come before or at the very beginning of the expansion phases.

5.6 The process of village formation

Johan Callmer

Research into the development of agglomerated settlement

Since the village is the structural framework for agrarian settlement in many parts of the world, it is of importance for our general knowledge of societal development to establish under what circumstances and in what way village formation occurred in various areas. Research into village formation seeks not only to describe but also to explain the process by defining the (primarily economic and social) factors and combinations of underlying factors.

Research into the village formation process in northwest Europe covers a period stretching from the Late Bronze Age and the Early Iron Age up to the village settlements of the Early Middle Ages (Jankuhn et al. 1977). Development within this vast geographical area can be said to show a number of variants. It is uncertain to what extent these are regionally limited, however, since our knowledge is almost entirely based on a small number of intensively studied little areas (e.g. Becker 1968, 1980a, Becker et al. 1979, Waterbolk 1980, Schmid 1982). Most of these areas are characterized by very light sandy soil, whereas we have only very incomplete knowledge of settlement on heavier soils.

The interest in north-west European research into village formation is largely concentrated on three main themes. The first is the origin of genuine agglomerated settlement, by which is meant here a settlement consisting of more than one household not belonging to the same kin group or family. The difficulties in applying such a definition to archaeological material are naturally great; they will be considered below. Yet these difficulties should not deter us from defining this change as one which involved a significant social transformation.

The second theme concerns the change of the agglomerated communities in the direction of a society with a stable village structure. Individual habitation sites can be shown to undergo phases of growth, maturity, and decay. The pendulum swinging from isolated farmstead to clustered settlements or villages and back appears in varying degree to have been a reality throughout the period in question here. There is also reason to assert that continuity in certain areas is problematic. In a few cases there appear to have been breaks in settlement followed by recolonization (Müller-Wille 1984). In several areas it appears irrefutable that processes of still other kinds are operating: some settlements cease to exist, with their populations being incorporated into a nearby habitation site.

The third theme concerns the formation of the settlement pattern which is encountered in the fiscal accounts and cadastral records of the Late Middle Ages and Early Modern Period, and which is essentially retained in the picture of agrarian settlement seen in the early land survey maps (Grøngaard Jeppesen 1981, Callmer 1986). The transition to this pattern of settlement is closely associated with the question of the fixation of the previously shifting habitation sites. The process by which settlement units ceased to be amalgamated also falls within the scope of this theme. The changes which have taken place in the fixed villages since the Late Middle Ages are confined in principle to evictions in conjunction with the transition to estate farming.

It is only recently that scholars have considered the importance of the landscape in terms of the conditions and the framework (or limitations) it offers for dynamic spatial changes and socio-economic processes. Yet the landscape, together with the social and economic aspects, is of crucial importance for a profound understanding of the transformation of settlement up to the relatively stable pattern of historical times (Kossack et al. 1984).

Important breakthroughs in research into the origin of agglomerated settlements in north-west Europe were made in the 1960s and 1970s. Of particular significance were the results obtained by Becker in his study of a relatively restricted area of western Jutland (e.g. Becker 1968, 1976, 1980a, 1982). He was able to demonstrate that the Late Bronze Age sites in this area could in most cases not be seen as unambiguously agglomerated settlements. They are too small for clusters of farms of village size. Often, however, the sites consist of a few contemporary dwelling houses, which could very well have met the needs of a family group for dwelling, barns, byres, and so on. Later studies appear to provide further support for this view of the structure of settlement in the final phase of the Bronze Age. Only a few observations could be tentatively interpreted as suggesting an occasional settlement of agglomerated type (Jensen 1968).

The change from family community to the cohabitation on one site of several households without close kinship ties appears to be one of the most important social changes in the development of agrarian society in north-west Europe. The village population as a group now becomes a second social identification for the individual, much closer to the family identification than was the case with earlier forms of social organization outside the family group. The changes which can be associated with the development of village-like communities must

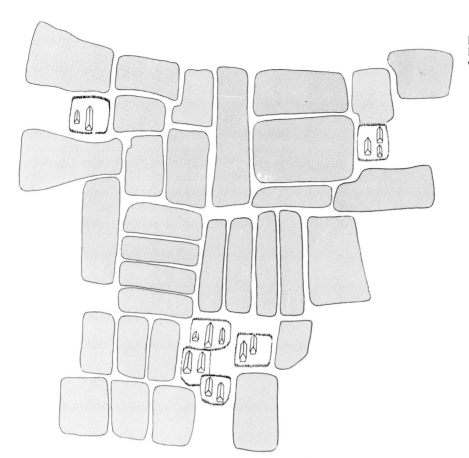

Fig. 5.6:1. Model of Early Iron Age settlement units with arable fields.

also be conceived as a series of transformations of land-use systems.

This social change undeniably requires further detailed study if we are to obtain satisfactory clarity about its course and its causes. In today's research it is often viewed as a rapid and widespread change, which also makes it somewhat amorphous as well as difficult to understand in relation to customary explanations in terms of subsistence technology and demographic changes (Jensen 1979, Hedeager and Kristiansen 1988). However, the course of events can be assumed to have been more differentiated. The relation between social and demographic factors and "pure" innovations in the system of land use is a classic controversy (cf. Stig Sørensen 1987:101). From many viewpoints, however, it must be doubted whether a strict division into, for example, social factors and technical innovations can be made. It can instead be maintained that both are part of an integrated system.

Archaeological research, especially in areas of southern and western Jutland and along the North Sea coast, has clearly shown the Pre-Roman Iron Age to be the period when distinctly agglomerated settlements began to be widespread (Becker 1968, Kossack et al. 1984:193ff.). However, isolated farms, double farms, and more loosely grouped clusters of multiple farms appear to have coexisted to a limited extent (Becker

1980b, Hedeager and Kristiansen 1988:131). Within the framework of contemporary cultivation systems, mostly of the Celtic field type, habitation sites show a certain mobility in space, fluctuating stability in size and probably also in social structure (Fig. 5.6:1).

The state of our knowledge of settlement in the Roman Iron Age has improved considerably thanks to large-scale planned excavations, especially in the North Sea region (Schmid and Zimmermann 1976) and in Jutland (Hvass 1980). It is possible to show a much greater degree of stability and continuity, albeit with considerable restructuring, in the appearance of the individual farms.

The problems of village formation and village transformations in the Late Iron Age have received little attention, and only a few systematic studies and excavation programmes have been conducted. In large parts of north-west Europe there are serious difficulties on account of the sometimes illusory but occasionally real discontinuity in settlement (Waterbolk 1980, Jankuhn 1986). This has also meant that we lack a link with the historically known villages. There is great obscurity about the similarity or dissimilarity of settlement systems in the late prehistoric and early medieval period throughout north-west Europe. Despite considerable research and extensive material, we know little of the development of settlement on the continent in the Early

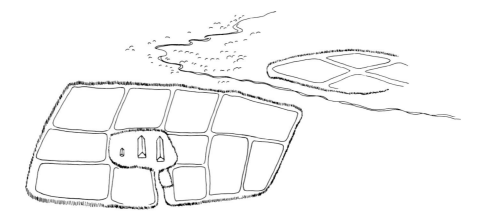

Fig. 5.6:2. Family group settlement model (Late Bronze Age – Early Iron Age).

Middle Ages – particularly the dynamics of change in the individual habitation site or village. It appears as if great changes were still taking place – albeit within a restricted framework – during the developed Middle Ages.

Not all these themes will be dealt with here; the focus will be on the later process of change in the second half of the first millennium AD and the Early Middle Ages which led up to the settlement landscape of historical times, that is, the landscape which remained largely unchanged until our oldest maps. To understand this process, however, it is necessary at times to look back at earlier stages of the development of agglomerated settlement.

Models to describe and explain the process

If we are to extend our current knowledge of the village formation process, we must use the existing material to test dynamic models of the origin, transformation, and dissolution of agglomerated settlements. The need to think in terms of models in this phase of research must also guide the further planning of systematic research programmes geared to various stages of settlement development. These models must be of two kinds. On the one hand, they must be related to changes in the individual components of the sites and their interaction; on the other hand, they must consider the stability or dynamism in relations between different sites. It is of course possible to develop and formulate models of widely varying kinds. In the present state of research it seems essential to begin by applying very simple descriptive models, which can only then be used to a limited extent as explanatory models (Callmer 1986, 1987a).

As we have seen, the agglomerated form of settlement was preceded by settlements based on a family unit, whether a small unit or a kindred in an extended sense. This earliest part of the change towards agglomerated settlement can be seen in strictly evolutionary terms, and it is my view that settlement forms developed towards greater complexity and stability in the course of the Iron Age. It must be emphasized, how-

ever, that during the period of change we must reckon with phenomena such as units based exclusively on a family group existing in parallel with agglomerated settlements. This could be interpreted as regressive development but is better seen as a fluctuation between simple and complex settlement formations. Communities like these, vacillating (within a more or less stable system) between larger and smaller organizational forms, are well known to social anthropologists (cf. Leach 1954).

A family-based habitation site changes to an agglomerated settlement when the population is composed of more than one family group. This change can scarcely be perceived by archaeological methods, as already pointed out; it is not likely to have occurred until the stage where a site has so many farms that it is improbable that it supported a community based exclusively on one kin group. In concrete terms we should reckon that a transition to agglomerated settlement occurs when the number of farms is at least three or four over a lengthy period (cf. Uhlig and Lienau 1972, Müller-Wille 1977).

I shall here present the change of settlement form in a simplified model. This is mainly descriptive, based on the state of archaeological knowledge in south-east Scandinavia (i.e., eastern Denmark, southern Halland, Scania, and parts of Blekinge). The model comprises four stages. It must be stressed that this presentation does not assume that these stages are always consecutive. Different units and different areas can be expected to show different development. In this version I associate settlement forms with certain systems of land use. It is presumed that there is a close link between social organization, settlement form, and land use and that these in turn were linked to social organization.

Stage A (Fig. 5.6:2). In this stage the habitation sites consist of one dwelling house (with outhouses), or a small group of dwelling houses corresponding to the needs of a family group. Agriculture consists of tilling fields that are not systematically manured. The economy is based on animal husbandry, requiring considerable areas for the provision of pasture and fodder (see Ch. 5.4).

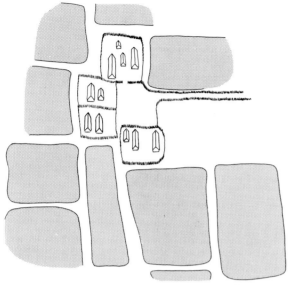

Fig. 5.6:3. Early agglomerated settlement model (Early Iron Age).

Stage B (Fig. 5.6:3). More complex forms of social organization evolve in certain areas, partly under the influence of continental models, and an agricultural system with partly manured Celtic fields develops, probably requiring cooperation between households and family groups. Both factors help to explain the process towards stable agglomerated settlement, but it is impossible to evaluate which was more important. Agglomerated settlements are formed alongside those based on kin groups and more dispersed settlements of single households. The agglomerated settlements are often stable over a period comprising more than a generation, but not stable enough to last more than one or two centuries (Vifot 1936 and later unpubl. obs., Thun 1982:60, Malmö Museum investigations)

Stage C (Fig. 5.6:4). As a consequence of the more scattered pattern of agglomerated settlement, with radically different demands on the spatial disposition of land, people adopt new, more stable cultivation systems. The arable land closest to the settlement is supplied with nutrient and is used to a large extent as permanent fields. Animal husbandry nevertheless retains its primary importance. The settlement form brings considerable stability, and land use allows farmsteads to remain on the same site for a long time (cf. Carlsson 1979, Widgren 1983).

Stage D. The political and economic development of north-west Europe in the second half of the first millennium AD stimulates a social development with more dependence and dominance, which further strengthens the relative importance of certain settlements, and which in some places leads to a concentration of settlement. As part of this process there develops a more stable agriculture with some form of rotation (Callmer

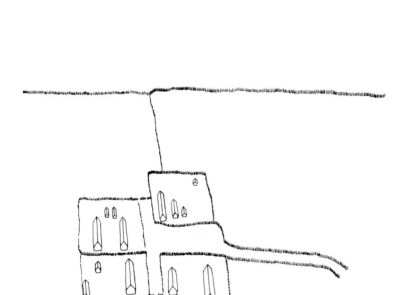

Fig. 5.6:4. Classical agglomerated settlement model (Late Roman Iron Age and later).

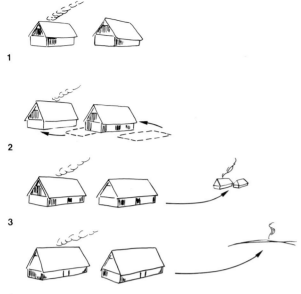

1

2

3

4

Fig. 5.6:5. Variations of mobility of settlement units. 1. Permanence; 2. Minor adjustments; 3. Mobility within the neighbourhood; 4. Mobility over large distances.

1986). Although animal husbandry is still of central importance, the significance of tillage is on the increase.

Models of this kind give us only the simplest outline of the course of development, with little real insight into the pattern of settlement and its dynamism. Let us therefore look at the models that can be used to analyse mobility in the landscape and the interaction of habitation sites. Although they can contribute to a more concrete picture, these models can also be given a greater degree of abstraction and thus be of wider applicability (Callmer 1986:173–175). To begin with, there is the fundamental difference between highly mobile settlements and stable settlements. Between these two extremes there are, of course, different degrees of mobility, which are important for the description of settlement in prehistoric times. Based on concrete observations and theoretical considerations, it is appropriate to set up a limit of 500 m to distinguish between large and small shifts of location, although naturally this limit should not be seen as absolute. We can distinguish four main degrees of mobility (Fig. 5.6:5):

a. permanence in one place
b. minor shifts, with habitation areas overlapping or touching each other
c. mobility in one vicinity, the area where the site's arable land is located – in absolute terms less than 500 m
d. mobility over wider areas.

We must now consider the interaction between settlements when determining the degree of mobility. It is obvious that moves within the first three of these de-

grees do not take people out of the area with which they identify (cf. Malmberg 1969). It is also clear that this mobility can be integrated in the traditional system of land use. Mobility of the fourth degree, however, if it involves ever greater distances within a relatively developed agrarian economy of the kind that existed at this time, would have entailed sweeping changes in living conditions and great strain on the social system and the economy.

To analyse changes from a less stable settlement system presupposing great spatial dispersion to a more stable system, the following models have been elaborated (cf. Callmer 1986) (Fig. 5.6:6):

1. Permanence in one place or minor shifts in one vicinity.
2. A contractive process in which two or more settlements together form a new agglomerated settlement in a new location more than 500 m from the earlier settlement.
3. A contractive process in which one settlement remains in its location, shifts slightly, or is mobile within one small vicinity, while one or more other settlements situated more than 500 m away join it and thus create an agglomerated settlement.
4. Successive moves of an agglomerated settlement over greater distances than 500 m.

Cases 2 and 3 could apply equally to isolated settlements based on a family group or to agglomerated settlements in the initial phase of the process. These models describe a course of events implying either that a settlement remained intact or that it was wholly incorporated into another. It is important to contrast this with patterns showing a course of change in which the agglomerated settlement form is dissolved. Two variants can be distinguished (Fig. 5.6:7):

5. The agglomerated settlement splits up into a number of smaller components moving to new locations at a greater or lesser distance from the original site.
6. The agglomerated settlement likewise splits up but forms new agglomerated settlements in wholly or partly new constellations.

Complex explanatory models for the process of agglomeration leading up to historical settlement are very hard to use at present, since they cannot be tested. This difficulty is associated with the nature of the source material for the study of the village formation process.

Although complex models cannot be applied in our present state of knowledge, it is important to apply conceptual models nevertheless, if only to help to plan future research programmes. Such models should not be monocausal but should integrate demographic, economic, social, religious, and cultural factors (Callmer 1986:173–175). For the present, a tentative discussion of complex causal connections is the best way to proceed. I

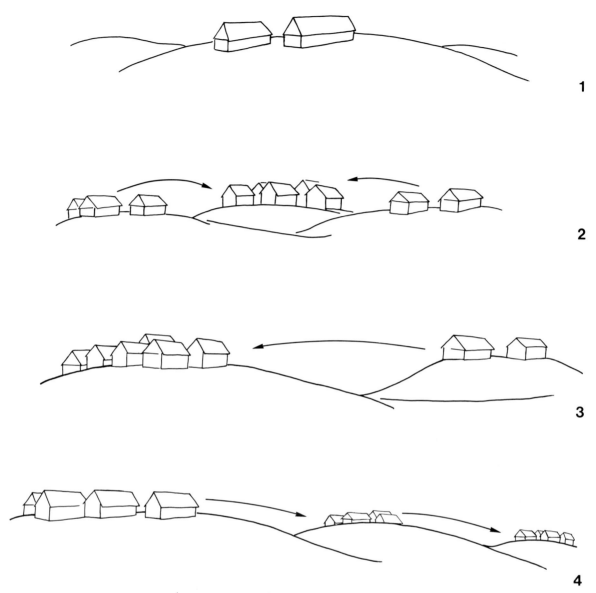

Fig. 5.6:6. Variations of settlement dynamics. 1. Static state or mobility within the neighbourhood; 2. Contractive process with new location; 3. Contractive process with one component remaining static; 4. Successive dislocation of settlement over large distances.

consider below some examples of this kind of explanatory reasoning.

Source material and methodology

When we come to consider source material and research methods, it must be stressed that it is essential to work with positive indications, since it is misleading to rely on negative observations (cf. Jensen 1979, Hedeager and Kristiansen 1988). Since the sources are so meagre, arguments *ex silentio* must be eschewed as far as possible. There are great difficulties in finding evidence of settlement from our period, so excavations on a limited

scale in a potential settlement site cannot be adduced as evidence that there was no settlement in the period.

The most important source material for the study of changes in settlement in the Late Iron Age and the Early Middle Ages is the documentation of habitation site remains and archaeological finds. What this evidence can tell us must be differentiated, however. Excavated habitation sites are particularly significant. After them come habitation site indicators, mostly pottery finds. Graves and finds which can definitely be associated with graves are also important, but cannot be used for the exact localization of habitation sites. When it *has* been possible to determine the relationship between a

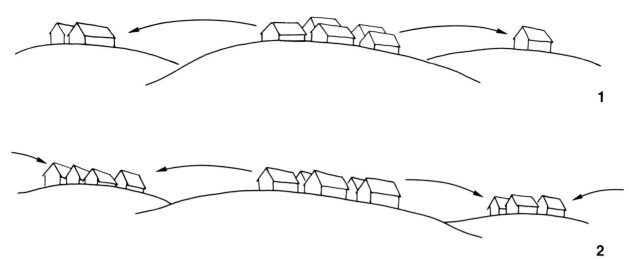

Fig. 5.6:7. Dissolution of settlement units. 1. Dissolution without immediate regrouping; 2. Dissolution and regrouping (new or partly new pattern).

habitation site and a cemetery, it is clear that the cemeteries in the Late Iron Age were in the immediate vicinity of dwelling sites, although the location could be adjusted in individual cases to suit the terrain. On a third level we have the settlement indications of phosphate maps. In conjunction with archaeological finds, these can contribute much to our knowledge of the exact location of sites, and sometimes also their extent. Phosphate indications are important for pointing out suitable areas for intensive field survey and trial digs (Jankuhn 1979, Callmer 1984:64–66).

An important part of the research method is to single out parts of the landscape where we know enough about the location and dating of habitation sites and then test simple dynamic spatial models of change of the kind described above (Jäger 1973, Callmer 1982:146ff.). This approach is basically easy, but it requires much complex consideration since the source material is generally of very uneven quality and often highly incomplete. Of special significance are the large continuous areas where changes can be seen in relation to general features of hydrology, topography, soil composition, and potential vegetation.

When we now turn to look at the process of village formation in south-east Scandinavia, it is necessary to recapitulate in greater depth what has already been said about stage B in the development. Although the actual origin of agglomeration is mostly outside our period, it is important to note here some main features that say something about this form of settlement, which entailed a major fundamental change. Although it is probable – and in some cases demonstrable – that there was highly developed cooperation (fencing and similar communally organized work) between farms and small clusters probably based on kin groups (Widgren 1983:123), when people live together in large agglomerations it involves something quite different. This change meant

that kinship ties as a basic structure now had to coexist with other structures linking different kin groups. The change from one settlement form to the other is probably one of the most important in prehistoric times, as already emphasized. For the agrarian population the next great change was not to come until the dissolution of the village communities by the enclosures in the eighteenth and nineteenth centuries.

The transition to agglomerated settlements meant a radical change in the cultural landscape as well. A dispersed, small-scale form of settlement allows for a more varied landscape. Numerous signs of the transition can be seen in the evidence of vegetation. When agglomerated settlement became the dominant form, there was a sharper polarization between the intensively exploited cultural landscape on the one hand and less heavily utilized areas with less impact on the natural landscape on the other (Hedeager and Kristiansen 1988:171ff., Ch. 4.3.2).

This transition covers a long period in south-east Scandinavia, probably from the Late Bronze Age up to the end of the Pre-Roman Iron Age, and in places even into the Early Roman Iron Age (cf. Tesch 1983:55). It is probable that this long period includes a great many local developments at different paces and with varying degrees of stability. It can be observed that, although agglomerated settlement – or villages – have dominated the plains of south Scandinavia since the Roman Iron Age, there are many variations to be seen in the known sites of this kind, and isolated settlement still existed in certain conditions and in certain regions, especially in peripheral areas (Rønne 1986, Hedeager and Kristiansen 1988:131).

The factors that influenced the transition to agglomerated settlement include demographic growth, changes in agriculture, and social and economic changes. In particular, cultural development on the continent was probably an important model for social organization, land use, and religion (cf. Jensen 1979:222–226, Kossack et al. 1984:246f., Görman 1987).

Let us now proceed to stage C of the village formation process, the rise of agglomerated settlements with relative constancy in one place. As in the previous stage, there was considerable variation between different regions, and perhaps above all between different parts of the same region. Central parts of the region were totally or nearly totally divided into village territories with village cores that gradually became more fixed in relation to one another, with restricted room to move about within the settlement. The settlement landscape can thereby be said to have achieved a certain degree of maturity. The changes that can occur in such conditions are primarily contractive processes, that is, the amalgamation of settlements. Our knowledge of such processes is hardly enough to allow us to follow it with a view to seeing what happened to the component with the greatest human interest – the individual household. However, parts of the process can be followed at habitation site level. The individual habitation sites retained a *potential* degree of mobility. It is indubitable that shifts in the site of dwelling places can be observed for a very long period, covering the whole of the Late Iron Age and the Early Middle Ages, even within the central parts of the settlement region. Mobility appears to have been greater in the less intensively utilized parts of the cultivation landscape and on the periphery (cf. Callmer 1987a).

Known cases of mobility within the settlement area are mostly datable to the middle of this stage, that is, from the end of the Vendel Period to the Early Middle Ages. Our knowledge of settlement in the Migration Period (especially the later phase) and the start of the Vendel Period is extremely slight. It is particularly difficult since there are so few datable artefacts (Nielsen 1985). It is uncertain whether this is due to a sharp decline in the use of pottery (as is known from somewhat later in Norway) and reduced imports of certain characteristic materials. The number of known grave-goods is also very small in southern Scandinavia as a whole. As a consequence of this, it is not possible to follow an unbroken process of settlement change in south-east Scandinavia from the Roman Iron Age to the middle of the Vendel Period. The relatively few indications from the Migration Period and the Early Vendel Period are of considerable interest, however. They show three main connections. First, we can see a link with older settlement locations from the Iron Age. Second, we can see localization in places with evidence of later settlement (Vendel Period and Viking Age). Third, the material reveals some indications which, on the basis of the current view of central and distal settlement areas, must be regarded as peripheral (Ch. 4.4.5). The finds, which include some precious metals, need not be interpreted as hoards deposited in deserted areas, or evidence of hunting in distant woodland, or the

like. This is shown by the corpus of remains and in some cases through excavation. These indications generally suggest a change in the overall settlement structure.

Later periods provide much more material for study. To examine the changes that took place from the Late Vendel Period to the High Middle Ages, we can divide the material into three chronological phases: (1) Late Vendel Age and Early Viking Age up to around 900; (2) Viking Age and the earliest years of the Middle Ages up around 1100; (3) the developed Middle Ages up to about 1350. After the latter date there is a significant decline in material.

Village formation in the Late Iron Age and Middle Ages

Late Vendel Period and Early Viking Age

As we have seen, a crucial question in the village formation process is the connection between prehistoric settlement and the historical settlement pattern. A study based on 78 well-documented indications of settlement from the Late Vendel Period and the Early Viking Age from Scania shows that these are very often closely associated with historical village cores (Callmer 1986:185ff.). In roughly half of these cases the older indications are on or near the actual village core. In some instances the distance from the village is up to 500 m, but in a clear majority of cases the distance is less than 250 m. In almost 40% of the cases the indications are over 500 m (and often further) away from the village core. Of those indications showing no connection with village cores, around half are on or near the coast. Settlement development on the coast in this period is a special case, which will be discussed below. In a few instances it was not possible to reconstruct the oldest historical agrarian settlement well enough on account of early urbanization. A more recent study from the same area based on 102 indications has given largely the same results, but with a higher percentage of indications on or near the historical village cores (Callmer 1987b).

These habitation sites from the early phase are often located on the plain (Fig. 5.6:8). The selection was in large measure determined by the location of large and small watercourses, as well as lakes and wetland areas. The border zones between different landscape types – as on the edge of the plain – were also attractive settlement locations. Habitation sites are often found on light soils (sand and gravel) and medium-heavy soils (light clay and medium clay). Heavy clay was completely avoided, just as it was almost always avoided by medieval villages. It is probable that indications from clay soils are somewhat underrepresented because of the difficulty of finding structures and material here.

It is clear that almost all these habitation site indicators are found in a landscape with a long settlement tradition. In several cases we can talk of well-demarcated settlement areas. It was a distinctive feature of the macrostructure that settlement areas were thus sepa-

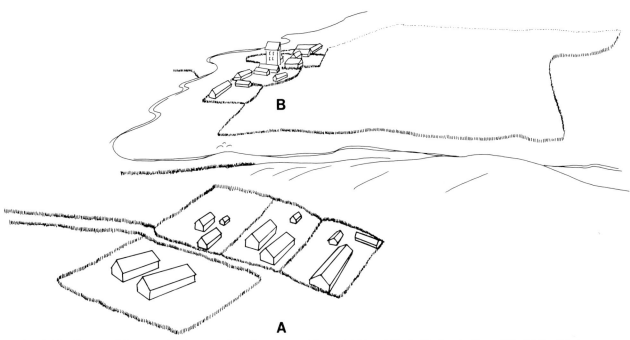

Fig. 5.6:8. Close spatial connection between Late Iron Age settlement (A) and Early Medieval village (B), (Fjelie, Western Scania).

rated from each other, with virtually uninhabited wilderness between them (Jankuhn 1964:268ff., Davies and Vierck 1974:257–259, 272ff.). It has not been possible to obtain detailed evidence of whether large settlement regions had any central site of greater importance, but this is most probable (Uppåkra: Vifot 1936 and later unpubl. observations; Vä: Stjernquist 1941; Thun 1982 and later unpubl. observations). On the other hand, it is now obvious that near the coast there were specialized trading-sites, initially visited only temporarily, but later with a more or less permanent population; their loca-

tion links them closely with larger settlements or regions (Callmer in press).

At this stage there seem to be relatively few changes in settlement. It is probable that there were some shifts, but these would generally have been in the direction of the historical village cores. No cases of amalgamation of settlements can be dated to this period.

The settlement landscape we meet in the Late Vendel Period and the Early Viking Age consists of very large areas with virtually no woodland and with a dense settlement of agglomerated, village-like sites. Agriculture with two- or three-course rotation within the framework

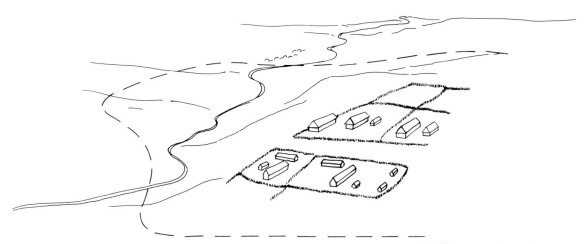

Fig. 5.6:9. Late Viking Age settlement on the Medieval village site (Hjärup, Western Scania). Village area indicated by broken line.

of an infield/outland system develops on the plain, mostly as a result of the interaction of the following three factors (Engelmark 1989): a need for higher yields on account of a new type of estate management with a growing number of non-productive members; a need to use the same arable land for longer periods than before, as central settlement regions were fully claimed and there was little scope for the expansion of pasture land; ideas about three-course farming and rotation reaching Scandinavia from south of the Baltic (cf. Lange 1976). The contractive processes may well have gone hand in hand with the emergence of villages dominated by manors or their predecessors, as already stated.

Late Viking Age and Early Middle Ages

In the phase from the Late Viking Age up to the start of the 12th century, settlement becomes noticeably more like that found in later historical times (Fig. 5.6:9). A selection of 144 well-documented indications of settlement shows that 79% of them are closely associated with historical village cores (Callmer 1985). In fact, three-quarters of these indications are found on or very near the village cores; only two of them are located about 500 m from the villages. When settlement indicators which are found further away from the historical villages are classified in two groups – those found more than 500 m from the historical village cores but less than a kilometre, and those found much further away – there is a very clear distribution: as many as 24 of 26 occur between one and two kilometres from village cores. It is now evident from well-studied areas that an interval of about two kilometres between settlement sites, as is suspected for material from older periods, is now the rule (Callmer 1986:187).

Although habitation sites are often located on or near parts of the historical village core, it is not possible to claim that their location is identical with that of the historical village. There were to be significant (although minor) changes from the settlement picture of this period to the localization of the later villages (cf. Grøngaard Jeppesen 1981, Hedeager et al. 1982).

The settled region in this period witnessed a dramatic change with the abandonment of the settlements which had previously been very common along certain sections of the coast (Holmberg 1977:16, Callmer 1986:200). This change occurred in the course of the 10th century, probably early in the century. It can scarcely be explained by changes in the subsistence economy. On the contrary, the coastal zone must have been fairly resistant to food crises, since alternative resources existed. This change in settlement must have been a reaction to a problem of a different kind. A conceivable explanation is that the coastal population felt threatened by pirate raids.

As before, settlement indications are mostly found in parts of the landscape with a long settlement tradition, and are localized according to the same principles as earlier settlement, for natural reasons. However, it can be said that there is a noticeable rise in the proportion of settlement indications in more peripheral areas. This means that there is now a less sharp division into settlement areas separated by uninhabited wooded tracts.

We know little of central habitations and trading sites in this period. There are observations suggesting that some trading-sites were moved, and there are certain problems in detecting continuity in the central habitation sites which have been conjectured for the preceding period (cf. Callmer in press; continuity in both Vä and Uppåkra is problematic). In this period we also see the earliest urbanization of continental type in southern Scandinavia, with a number of towns which differ considerably from older centres, as regards their location and economic structure, although their central location in important regions places them near the older centres (cf. Redin 1972, Andrén 1980, Thun and Anglert 1984). In addition, towards the end of the period there arise certain new places along the coast with functions for trade and exchange and specialization in fishing.

The settlement changes we can follow on an intermediate level partially come under the heading of colonization and so will be treated below. Contractive processes are still active, although to a lesser extent than in the later part of this period. These changes probably involved one settlement attracting the population of a nearby settlement, who then abandon their own site. A special problem concerns the desertion mentioned above of the coastal zone with its numerous settlements, many of which had been fairly large. This change must have exerted serious pressure on settlement further inland, and it could have meant some growth and reorganization of individual sites. Colonization must have been one of several possible solutions to the problem.

It is more common during this period than before to see settlements being localized on clayey soils. However, it was mainly light clays that were brought under cultivation. It is difficult to arrive at exact details concerning the relationship between soil and settlement here; this would require taking a dense network of samples, since the geological maps are not precise enough to shed light on this question.

The Middle Ages up to ca. 1350

In the last of the three phases defined above, changes continue to move in the direction of known historical settlement. It has been possible to study only 67 well-documented settlement indications (Callmer 1987c). The Romanesque churches have not been included in this material, since they are not necessarily indications of settlement. With very few exceptions, the indications used are located on the historical village cores shown on the oldest land survey maps. (Only three settlements in the archaeological material are located away from any historical villages, and they are at considerable distances from them. In all three cases their names are preserved as field-names.) Although the settlements are in parts of the historical village cores, it is clear that there can be certain differences. The location of most villages had probably not yet been definitively fixed, but there was less extensive change than before. This phase is strikingly similar to the preceding one in the location of settlements as regards landscape and soils. In some cases it can be asserted that villages were shifted to drier sites, but this tendency can scarcely be described as strong; wetlands, lakes, and watercourses continue to be important landscape elements in the localization of settlement.

Finally, it must be underlined that finds from the later part of this period are scarce and the exact dating is uncertain. For this reason it is hardly possible to claim with certainty that the villages did not retain a certain degree of mobility right up to the end of the period (observations from Scania, Blekinge, Halland, and Bornholm; Wienberg 1983).

Stabilization of the village landscape

We have now reached the final phase of village formation, the High and Late Middle Ages, where we meet agglomerated settlements permanently located in one place, with none of the mobility of previous periods. Changes that take place from now on almost exclusively concern the redisposition of agrarian settlement within estates (Ingers 1971–2). There are now often no major changes in the settlement plans of the villages. Minor rearrangements do occur, and some expansion in the form of cottages.

There are several reasons for this stabilization of what had once been dynamic settlement. It is not unlikely that migrations of settlement may have been made for magico-religious reasons (Schlesier 1977), but with the gradual consolidation of Christianity, many heathen practices were abandoned. In the High Middle Ages there were also changes in the agrarian structure, so that it became fairly common for farms in one village to have different owners: the church, the Crown, the nobility. Formerly there had often been only one owner in the village (Ulsig 1968:135ff., Skansjö 1983:271–273, Porsmose 1988:264–265). This composite ownership must have greatly increased the need to establish once and for all the farms' rightful shares of the agricultural land belonging to the village. This would have led to a fixed structure in the spatial division of the village and thus to a fixed permanent location. This tendency may also have been reinforced by more efficient farming techniques (cf. Myrdal 1985, 1990).

In the treatment of the final phase of settlement development above, the term agglomerated settlement has been deliberately replaced by the word village. Strictly speaking, it is only now that we are entitled to claim the existence of true villages with their proper form of organization known from historical sources. Our present knowledge of Late Iron Age settlement in southern Scandinavia allows us to say that settlements which were essentially similar to the historical villages had existed since the Vendel Period at least, and probably long before. With the dating of the introduction of two- or three-course rotation to the Late Vendel Period as suggested here, we can also claim that an internal social organization similar to what we meet later in the historical villages must have existed at least since this time; it is inconceivable that a farming system like this could have been adopted otherwise.

From south-east Scandinavia we have no study to show how one and the same agglomerated settlement changed over the entire period from the Late Iron Age to the Middle Ages. We can nevertheless claim that settlement from the Vendel period consisted to a large extent of clusters of several farms of village type. There was no doubt also dispersed settlement in single or double farms. The multi-farm sites are large, generally covering an area of 3–6 ha. This could mean that six was not an unusual number of farms (Callmer 1986:175). It is very difficult to find traces of tofts in south-east Scandinavia, but in the few cases where they can be detected, they are not noticeably different from those in Jutland (Arkeologi och naturgas i Skåne 1985:28). The houses consist of large long-houses of somewhat varying design, supplemented by barns, byres, and sunken-floor huts. The number of the latter varies greatly, which might indicate economic specialization: numerous sunken-floor huts suggest domestic production of, for example, textiles.

This is the settlement picture that still dominates the agrarian landscape in the 11th century. The houses are built smaller in the 11th century and especially in the 12th, and their design changes from three-aisle to two-aisle constructions (Stenholm 1986, Grøngaard Jeppesen 1983:162–180). It is probably at this time that sunken-floor huts disappear for good. Although farm structure remained largely unchanged, the size of the tofts probably decreased gradually.

Colonization

In the period with which we are concerned – covering the Late Iron Age and the Middle Ages – colonization is an important aspect of settlement development (Callmer 1986:175). The areas colonized were not only on the periphery of the settlement region, but also within it, including land that had been little used or not at all, often poorer farming land. We may thus speak of external and internal colonization. Colonization can have taken essentially two forms (Fig. 5.6:10). The first of these is the movement of an entire agglomerated settlement into the wilderness. This type of colonization cannot be ruled out for our period, but the complexity of the process must be emphasized. Above all, it requires considerable planning and organization. It is particularly important that the pioneering community receives support from established settlements in the initial phase.

Fig. 5.6:10. Colonization models. 1. The establishment of a household outside the settlement area. 2. The establishment of an agglomerated settlement outside the settlement area.

The other type of colonization consists of individual households leaving the existing settlement to found new farms some distance outside it. The first stage of this process leads to very scattered isolated farms. This form of settlement can then lead to agglomerations through contractive processes.

To a certain extent there was both external and internal colonization in the Late Vendel Period and the Early Viking Age, but it is not until the Late Viking Age and the Early Middle Ages that we find numerous examples of colonization, both internal and external in southern Scandinavia. It appears that the colonizing process was often extremely rapid. It took only a few generations for newly founded small units to expand and become villages (cf. Porsmose 1988:234ff.).

There is probably more chance of studying the change from dispersed to agglomerated settlement in this later material than in material from the Early Iron Age. The process is not really comparable, however, since in the Late Iron Age and the Early Middle Ages it takes place following the model of nearby agglomerated settlements.

The process continued with undiminished vigour throughout the Early Middle Ages, when it was mostly a question of external colonization. This settlement expansion leads to a lessening of the distinction between cultivated and uncultivated regions that is claimed to have been so typical of earlier parts of the Late Iron Age. However, certain unoccupied areas retained their wilderness character. After a phase in the 14th century with less expansion – or downright stagnation (as a result of economic overexploitation and the Black Death) – colonization was resumed in the 15th century and above all in the 16th century (Ch. 4.4.6). These new settlements, which were founded on poor farming land, mostly did not develop into agglomerated settlements.

Conclusions – explaining the process

I shall conclude by discussing explanations of the village formation process. It is essential to begin by reiterating that the problem of village formation is connected with three main questions: the origin of the agglomerated settlements and their early settlement patterns; the development of the historical pattern of settlement; and the final fixing of agrarian settlement.

In attempting to explain the origin of agglomerated settlement, primary importance has been attached to causal factors such as population growth, changes in the social system (on various levels of society), new techniques and economic changes, and ecological changes. It is very hard to isolate these individual factors, apart from the purely ecological ones such as changes in climate (cf. Hatt 1937:74ff.). For example, the problem of demographic development cannot be considered without regard to the cultural and social factors (Welinder 1979:26–28). The human-related variables should rather be seen as different facets of the same thing.

Purely ecological factors are not enough to explain the fundamental change – the transition from a family-based form of settlement to one where structures of dependence and hegemony gradually dominate. It is more probable that this development is connected with the new ideas that reached Scandinavia from the continent in the Late Bronze Age and the Early Iron Age. As I have pointed out, the development of agglomerated settlement is a process with an uneven spread over time and space; it lasted almost a thousand years, from the Late Bronze Age to the Roman Iron Age. This is such a protracted period that one is justified in wondering whether it should rather be viewed as a stable phase with a fluctuating degree of organization. This problem, however, is beyond the scope of this study, but the principle is so important for the understanding of future change that it has been reiterated here.

This change in the form of social organization, which is reflected in the tendency towards agglomerated settlement in the plains areas of southern Scandinavia, attained a certain stability and maturity in the Roman Iron Age. The reasons for this stabilization are many. Of great importance are the forms of land use practised in the Pre-Roman Iron Age and the Early Roman Iron Age, which presuppose people working in common, investing great amounts of energy. These forms of land use further reinforce the social organization. The tendency was also strengthened by a very gradual population growth which appears to have been general throughout the Late Bronze Age and Iron Age. It is naïve to view this population growth as a force in itself. It seems preferable to view it as a socially anchored strategy for the solution of certain problems (cf. Hassan 1981:173–175). To keep society on a certain level there had to be a surplus population to compensate for losses through disease, accident, death in battle, and so on. Since these losses struck more or less at random, and people had to be highly prepared, a long-term population surplus was a good answer.

In the Roman Iron Age there was also an increased social stratification, of which the agglomerated settlements are a sign in themselves. An important – perhaps the most important – stimulus for this stratification came from the relations of trade and exchange with the Roman Empire (Lund Hansen 1987). Social models from Rome, and from parts of the continent that were directly and continually under the influence of Roman culture, exerted an ever greater influence on local communities in southern Scandinavia, particularly in the Late Roman Iron Age (specially organized military forces, large habitation sites, cemetery structures), but no real stability was attained (Vifot 1936, Ørsnes 1968, Gebühr 1986). The greater social stratification nevertheless remained a part of the structure and the agglomerated form of settlement continued to dominate the plains. As a consequence of developments in society and land use, arable fields were now concentrated around the village, and land was divided into central and peripheral utilization zones; this division into infields and outlands was to survive in essence until the end of the 18th century. In the earliest phase the dominant form of cultivation was probably in one course, possibly combined with the practising of fallow periods (Hedeager and Kristiansen 1988:155–156, 171ff.).

A process in which settlement units were amalgamated into even larger units went on during the Late Iron Age, doubtless beginning earlier (Stjernquist 1981:21, Callmer 1986, Porsmose 1988:223ff.). This change is probably also due to the social stratification that was increasingly common from the Roman Iron Age onwards. It reflects the emergence of large villages on the plains dominated by "big men". As a result of the amalgamation of settlements, there was often an imbalance in the area immediately surrounding the farmsteads, since there was a shortage of manure for the permanently cultivated fields. It was not possible to

supply this need by expanding the areas of pasture land in the densely populated plains areas. What people tried to do instead was to improve the situation by letting land lie fallow more often. In the Vendel Period and Viking Age there was a transition to more organized rotation of much of the arable land in a system of two or three courses. It may have been important for this process that similar tillage systems were in use south of the Baltic Sea.

The amalgamation process was lengthy, extending over most of the Late Iron Age and into the Early Middle Ages, although it was probably most vigorous in the early part of this period. By the Vendel Period or the Viking Age most of the historical villages in the central plains areas had been formed. Conditions in peripheral areas were very different. Here the process was not completed; it stopped at various stages for a variety of reasons, primarily concerning how peripheral an area was and what the terrain was like. It was in the monotonous types of landscape that the process was carried furthest, whereas varied landscape shows a corresponding variety in settlement forms.

As we have seen, the mobility in individual settlements that can be seen in the Iron Age and the Early Middle Ages is of two kinds. There is the mobility in which people only moved their dwellings a short distance but did not leave their resource area. Then there is the mobility over larger distances, which is generally connected with contractive processes. Short-distance moves of the first type can usually be shown to have occurred at relatively long intervals in the period from the Late Roman Iron Age to the Viking Age. Permanence over a few generations is the rule. Changes like these can scarcely be attributed to problems of subsistence, so they should be seen as answers to other needs. It is more likely that these moves were phases in a cycle with an ideological or – as is most probable – magico-religious basis, or else they were a reaction to extraordinary events (Schlesier 1977). Numerous historical villages can be traced back to the Late Viking Age, and changes at this time are often viewed as explanations for the cessation of mobility. In my opinion, what is interesting is the point when the village ought to have moved next but did not do so. By then, however, Christianity had been essentially integrated into the people's system of norms, and pagan customs had been suppressed; moreover, the character of landowning had changed and land limits had been more sharply defined and subjected to fiscal regulations.

It will be evident from these comments that I claim that changes in settlement structure in the Iron Age and the Early Middle Ages, leading up to the settlement pattern of historical times, were primarily determined by social changes. Especially important are the forms of social organization; when they changed they triggered other economic and technical changes, leading to a dynamic interaction with the surrounding environment.

5.7 The churches in the Ystad area

Barbro Sundnér

Introduction

The medieval churches provide unique source material for the study of various facets of society – economy, settlement, population distribution, social organization, ideology. Most of the surviving churches in Scania were built during the 12th century and are still used as ecclesiastical buildings. Over this long period of use the churches have changed in various ways to suit the new needs that have arisen. These needs in turn have their roots in conditions prevailing in the society of the day.

To exemplify: The sites of medieval churches indicate the spread and concentration of settlement (Fig. 5.7:1). If we compare with the distribution of different soil types, we find that churches are at their densest where the best boulder clay for cultivation is to be found (Fig. 1.2:15). This picture represents a developed stage around 1200, and thus not the entire Middle Ages. If we look at church building through time we can find many variations. Some recent studies in Sweden and Denmark have shown that the oldest stone churches are in fertile arable areas, whereas the later examples were built in less fertile wooded areas (Liepe 1985:230ff., Nyborg 1986, Bonnier 1987:217ff.). On the other hand, the Danish study (Nyborg 1986) shows that only the oldest churches were enlarged during the High Middle Ages; in other words, building activity was once again restricted to the fertile areas.

The scope of church building has often been explained by expansions of settlement for economic and demographic reasons, but it can also show the different functions of the churches or the intentions of the builders (Fernlund 1982, Hurst 1984, Wienberg 1986a). The erection of a church required a certain economic base, but the motivation was not the same at all times. Different factors must therefore be considered, such as the question of who took the initiative to build a church, and why. Although there is only highly sporadic written evidence about churches in Scandinavia, their architecture can reveal something about the builders and the builders' intentions, which varied through time and place (cf. Cinthio 1957, Holmberg 1977, Wienberg 1986b, Bonnier 1987).

Most rural churches are today parish churches, which means that they have some central functions for a certain delimited area. The land was divided into parishes during the Middle Ages. This great organizational change in society was probably not achieved all at once,

and it had different implications at different times and places (Smedberg 1973, Lindquist 1981, Andrén 1985, Redin 1986). (To imagine the dynamic nature of previous periods we need only compare the much more developed social role of the parish in the early 20th century with our own decade, when the church is being increasingly deprived of its administrative and social functions.) There were also churches without parishes; known as chapels of ease, these were generally regarded as dependent on the parish churches, but here too we must see their function in relation to time and place (Anglert 1986).

Church organization played a central role in state formation in the Middle Ages. There are other examples of the political and social functions of the churches. Historical events, for example, have been adduced to explain why some churches were fortified (Boström 1983, Andersson 1983), yet even here the explanations may point rather to social circumstances (Anglert 1984).

The actual work of building churches involved certain basic practical prerequisites: material, expertise, manpower. With the first stone churches a completely new building technique came into use. It was then people learned to use mortar to cement the stones together. Variations and changes in building techniques can also reveal social, economic, and organizational conditions (Sundnér 1982, 1986).

Using churches as historical sources requires dating their building periods. There is a general view – founded both on stylistic comparisons and on archaeological observations, together with occasional written evidence – that the medieval stone churches in Scania were built between about 1050 and 1250, with a concentration in the 12th century. Afterwards, from the 14th to the 16th centuries, they were provided with a range of different additions, such as towers, porches, and vaults. Some had the nave extended to the west, or the chancel extended to the east. There were additions to the churches in the 18th and 19th centuries too, often involving comprehensive rebuilding. Many churches were also demolished in the 19th century and replaced by new buildings. It can be difficult to achieve precision in dating the building periods of the churches, but by studying different features and architectural structures we can arrive at a number of variations which can suggest a chronological background.

Fig. 5.7:1. Medieval churches in Scania, + known sites, (+) sites unknown, □ monasteries. The dotted area shows the extent of the best clayey tills for cultivation (after Ekström 1946), which corresponds to the highest frequency of churches.

The investigation of the churches

The studies of churches in the Ystad area have sought to determine their original plan and structure and the changes they have since undergone. This was done by means of archaeological examination of the masonry (in so far as this was accessible), and by compiling a corpus of earlier descriptions, measurements, and literature concerning the individual churches. In addition, written evidence from the Middle Ages was collected and analysed (Skansjö et al. 1989a). The general aim has been to arrive at an overall picture of the way the churches have varied through time and place, based on the view

that their architecture can give information about the societal phenomena mentioned above. The participants in the church project divided the work and analysis among them so that each had a different angle: Mats Anglert has considered parish formation; Marit Anglert has dealt with the function of the towers; Jes Wienberg has examined demography and churches, and I myself have studied building techniques and chronology. Since the church studies were part of the medieval section of the Ystad project, the analysis has concentrated on the Middle Ages, but later periods have also been treated (Wienberg 1989). The original aim of the project was to

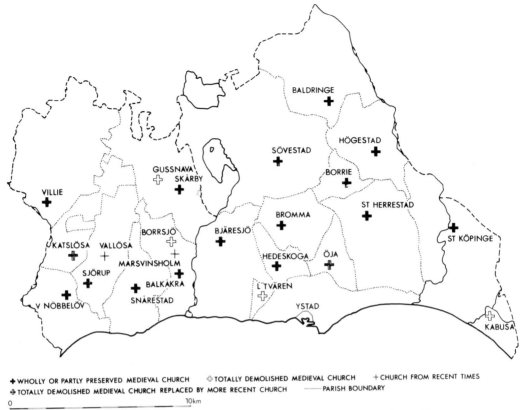

Fig. 5.7:2. Churches in the Ystad area, with present-day parish boundaries. From Mats Anglert 1989a.

+ WHOLLY OR PARTLY PRESERVED MEDIEVAL CHURCH   ⊕ TOTALLY DEMOLISHED MEDIEVAL CHURCH   + CHURCH FROM RECENT TIMES
⊕ TOTALLY DEMOLISHED MEDIEVAL CHURCH REPLACED BY MORE RECENT CHURCH   ········· PARISH BOUNDARY

0   10km

analyse only the countryside, which is why the churches in Ystad have not been examined in any detail.

In Ljunits and Herrestad hundreds there are in all seventeen parishes (Fig. 5.7:2). Most medieval parish churches have survived in varying states of preservation from their first building period. Three churches have been demolished and survive only as ruins. In addition, there were three churches referred to as chapels of ease. Nineteenth century churches were built in two parishes on different sites from the medieval churches, which have thus been preserved.

The variations and changes in the plans of the churches have been grouped in two periods in Fig. 5.7:3. Since most churches are plastered both inside and outside, there is some difficulty in investigating their history and especially in comparing their different building periods in order to arrive at a precise relative chronology. The only visible parts of the masonry may be those parts of the chancel and nave above the secondary vault. The towers can be difficult to study since they are often plastered on the inside too. This is also true of the few surviving porches. The vaults have often been covered with a thick layer of plaster, covering details in the profile of the ribs. In the project it has also been possible to make extensive dendrochronological studies of all the timber assumed to be medieval, chiefly in the roof-trusses. Only four churches have parts of roof-

trusses surviving from their first building period. Two of these have provided exact datings. Some of the other roof-trusses have been shown to be contemporary with the addition of a tower or a vault.

The dating of the churches and their changes is based on a typological classification of building technique and construction. These have been ordered in a relative chronology on the basis of archaeological observations (Sundnér 1989a). Dendrochronological datings and dated wall-paintings can give us exact years for certain groups.

The groupings shown in Fig. 5.7:4 allow us to summarize building activity in the area, grouped in hundred-year periods in Fig. 5.7:5. The columns represent only additions to churches in each period. To this have been added the results of the dendrochronological investigation (Bartholin 1989). Of the 168 samples of timber taken from the medieval churches, it has been possible to date 85; 42 of these have been dated within a ten-year margin, while 43 have been given *terminus post quem* datings. Most of the material consists of secondary roof-truss details and has thus not been ascribable to any specific building period. The surviving parts of medieval roof-trusses nevertheless reveal some building activity and should therefore be compared with what has been assumed on the basis of the archaeological study of the buildings.

352

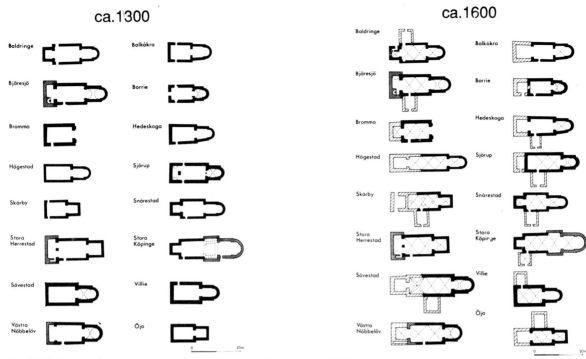

ca.1300 ca.1600

Fig. 5.7:3. Proposed reconstructions of the churches as they were ca. 1300 and ca. 1600. From Mats Anglert 1989a.

Most of the stone churches were built in the period 1050–1250, while vault building prevailed from 1250 to 1450 and tower building between 1450 and 1550. No medieval churches were built after 1350. It was not until the 19th century that new churches were erected. The post-medieval period is dominated by enlargements and

Fig. 5.7:4. Church buildings and their changes, grouped on the basis of construction types.
● Church buildings, ○ extensions, ■ towers, □ tower heightenings, ▲ vaults. Demolished parts are in parentheses. Underlined dates appear on wall paintings; other dates are dendrochronological datings. Arrows mark the churches to which datings refer. I–VIII refers to chronological groups. From Sundnér 1989.
Construction types:
1. Thin walls
2. Thin courses, small stones
3. Sandstone, calcareous tufa, marl: soft stones
4. Inner wall of chancel in line with outer wall of nave
5. Square towers
6. Plinth
7. Limestone
8. Thick walls
9. Thick courses, large stones
10. Rectangular towers
11. Brick
12. Large square towers
13. Vaults of one brick's thickness
14. Three- and three-quarter-profiled groin ribs
15. Rounded ribs along the sides of the vaults
16. Vaults of half brick thickness
17. Square groin ribs
18. Thin walls
19. No courses
20. Ribs on the upper parts of vaults

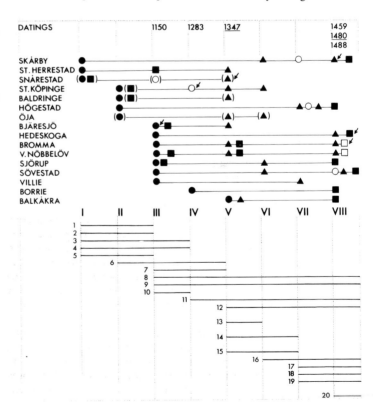

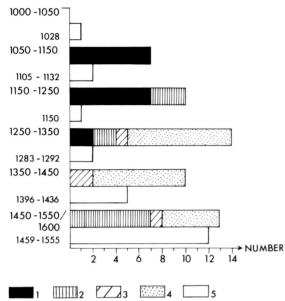

Fig. 5.7:5. Building activity in the Middle Ages and dendro-chronological datings. Apart from the church buildings, only secondary additions are included. 1. Church, 2. Tower, 3. Extension, 4. Cross vault, 5. Dendrochronological dating. From Sundnér 1989.

additions. The greatest total building activity in the Middle Ages was between 1250 and 1350. Activity declined in the following period, increasing again towards the close of the Middle Ages. Stagnation set in after the Reformation, but a new peak in building activity was reached during the 19th century. By contrast, if we look

Fig. 5.7:6. Church murals in Scania, 1250-1350.

at the dated church timber we find a remarkable gap during the vaulting period, the 14th century, when we might expect roof-trusses to have been replaced in conjunction with the building of a vault. This, however, has also been found in other European areas by dendrochronological studies (Baillie 1979). A few datings from the time before 1050, admittedly post quem datings, could be from older wooden churches.

Broadly speaking, this picture can be assumed to agree with the picture for Scania as a whole. In the absence of an overall chronological synthesis for the churches, we may compare with the chronological division of medieval wall-paintings, of which an inventory has been published (Banning 1976–82). Most murals in Scania fall into the two periods 1250–1350 and 1450–1550. A relatively high proportion (22%) of the paintings from the first of these two periods are found in churches in the Ystad area. If we look at the geographical distribution of the paintings from this early period, we see that most of those from the early 14th century are concentrated in the area around Ystad (Fig. 5.7:6). Several of these paintings come from the same workshop, known as the Snårestad group after the lost paintings in Snårestad church, which are dated by an inscription to 1347 (Rydbeck 1943:185ff., Banning 1976–82:4, 38).

Church building from the Middle Ages to the present day

I. 1000–1150 (Fig. 5.7:7a). The first evidence that the new religion, Christianity, had won a foothold in Scandinavia comes from runic monuments. On one of the Jelling stones in Jutland, King Harald Bluetooth Gormsson records that he has converted the Danes to Christianity. The stone is dated to the end of the 10th century. A wooden church was erected adjacent to this monument; it was succeeded by two more wooden churches before the present stone church was built in the 12th century (Krogh 1983).

As at Jelling, we can suspect some connection between the runic monuments in the Ystad area and the first churches. Seven runic stones are found in direct conjunction with the medieval churches (Riddersporre 1989b, Anglert 1989b). The inscriptions reveal that only certain powerful men had reason and need to mark their position by raising such monuments. The first wooden churches, which were erected at roughly the same time as the runic stones, may likewise be viewed as personal manifestations by a magnate class. However, no remains of wooden churches have been found in the Ystad area. Archaeological excavations to discover evidence of any precursors to the stone churches have been undertaken only at Stora Köpinge church. They found no previous wooden church, nor even any older graves (Gustafsson and Redin 1969). On the other hand, there are traces in the church walls at Bjäresjö and Borrie which show that the chancels in the stone churches were

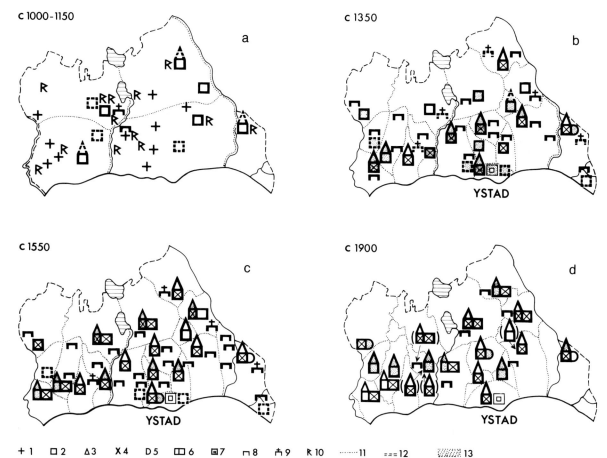

c 1000-1150   a

c 1350   b
YSTAD

c 1550   c
YSTAD

c 1900   d
YSTAD

+ 1   □ 2   △ 3   X 4   D 5   ▥ 6   ▣ 7   ⌐ 8   ♀ 9   ℟ 10   ⋯⋯ 11   === 12   ▨▨▨ 13

Fig. 5.7:7. The churches in the Ystad area and their changes through time and place, together with changes in parish formation (the latter after Mats Anglert 1989b). 1. Site of medieval church, 2. Stone church, 3. Tower, 4. Vault, 5. Eastern extension, 6. Western extension, 7. Monastery, 8. Manor, 9. Archiepiscopal manor, 10. Runic stone, 11. Parish boundary, 12. Uncertain building, 13. Parts built during the period in question.

finished before the naves were started, which might suggest that they were built adjacent to standing wooden churches. Since most churches are plastered, it has not been possible to investigate this possibility at any other churches, with the exception of Stora Herrestad; when the plaster was renewed here the masonry was examined but no signs of any interruption in the building work were revealed. That no traces of previous wooden churches were found at Stora Köpinge or Stora Herrestad does not mean that they did not exist in these localities, since they could have stood on different sites (cf. Cinthio 1980, Anglert 1989b). However, there is reason to suspect that the majority of the stone churches were preceded by wooden churches. According to Adam of Bremen, there were three hundred churches in Scania in the 1060s, and we may assume that most of these were wooden churches (Adam av Bremen 1984:IV, 7).

From written evidence and archaeological finds we know that the first stone church in Denmark was built at Roskilde at the beginning of the 11th century (Krins 1968:6). In Scania the oldest dating so far for a stone

church is the 1060s for the royal church in Dalby. King Knut built a stone church in Lund in the 1080s. King Sven Estridsen, who is assumed to be the builder of Dalby church, also took the initiative in the division of the diocese (Cinthio 1957, 1983). The Crown thereby took an important step towards a great organizational change of medieval society. The first stone churches have been associated with the king and the bishops (Krins 1968). In the Ystad area the church in Skårby is presumed to be the first stone church, closely followed by the church in Stora Herrestad, from around 1100. Unlike the older and contemporary wooden churches, the stone churches can be seen as reflections of some centralized power, with the king and the bishops as initiators. There is evidence that high-ranking magnates were directly associated with both parishes: in Skårby through the imposing runic monument in Hunnestad; in Stora Herrestad through King Knut's great donation in 1085 to the church in Lund.

The two churches differ in several ways. Whereas Skårby church is one of the smallest in the area, Stora Herrestad is the largest. The church in Skårby is built of small field-stones with door surrounds of calcareous tufa, a material that occurs only at Fyledalen on the eastern boundary of Herrestad Hundred. The church in Stora Herrestad, by contrast, is built of an easily worked marly sandstone which is found only in an area between the church-village and Stora Köpinge. The stone has been dressed into squares. The building material and size of the church suggest that it had a different role from the church in Skårby. Perhaps the architecture is also intended to display a social difference – the bishop's church versus the king's church. The building material from Fyledalen is also found in the oldest parts of the cathedral crypt in Lund, which must belong to an earlier stone church, perhaps from the 1080s (E. Cinthio, pers. comm.). This means that when the first stone churches were built, there was already a knowledge of the Fyledalen stone, which was brought to the church in Lund despite the great distance (approx. 65 km), perhaps indicating royal ownership of the quarry in Fyledalen. From the same area came the calcareous tufa for the vault in the cathedral crypt and the entrance hall in Dalby church in the early decades of the 12th century (Sundnér 1984). All this suggests vigorous activity in the area at this time.

With the organization of the diocese came the start of a division into parishes. According to Mats Anglert (1989b), the area of the parishes was based on village boundaries. By starting from conceivable parish churches, he has been able to assume a change in parish formation in the Ystad area in the Middle Ages. In the first half of the 12th century four or five more stone churches were built. It is possible that a chapel mentioned in a written source from 1145, Gussnava, was also built of stone. There has also been a church in Borrsjö, which may have existed at this time. Three of the churches – Baldringe, Stora Köpinge, and Snårestad – had square west towers, which may have been built to demonstrate their status as parish churches. Unfortunately, none of these towers survives. These three churches, along with the churches in Skårby and Stora Herrestad, have been assumed to be the first parish churches whose parishes were demarcated by natural topography. A boundary running from west to east marks the transition from the hummocky landscape of the interior to the flatter coastal zone. The watercourses provide natural north-south boundaries (Fig. 5.7:7a). These oldest stone churches all stand in areas bordering on woodland or other less fertile farming land. They are also along the main roads towards the north-west (Dalby and Lund) and the north-east (Vä). Stora Köpinge, however, is off these roads to the inland, having greater contact with the coast by way of a navigable river.

II. 1150–1350 (Fig. 5.7:7b). Church building in the second half of the 12th century reveals many signs of a change in the social and economic structure. The six new stone churches lie concentrated on the fertile boulder clay in the centre of the area. Most of the churches were built from the beginning as parish churches, as is shown by, among other things, the baptismal fonts preserved from this time. The architecture of the churches now shows greater variation, probably a sign of the varying economic and social circumstances in which they were built. Unlike the earlier churches, many are now richly adorned. Bjäresjö, for example, is the only one of these churches that is built of well-dressed blocks; in addition, the church has such adornments as a frieze of round arcading, a richly profiled portal, and a barrel vault in the chancel. Even the roof-truss is profiled. Moreover, in Bjäresjö there is a clear connection between the construction of the stone church and of the stone house north of the church – the two are built of the same material (Callmer 1986c).' Although there is generally no evidence of an aristocratic presence at this time, the churches reveal that there must have been local magnates, since there were specially marked places in the western part of the naves. None of these survive, but the position of church portals suggests that the magnates' pews were thus distinguished in Bjäresjö, Västra Nöbbelöv, Stora Herrestad, and perhaps also Bromma. Towards the end of the century these special seats were even more clearly manifested by the addition of a west tower, which probably had a private gallery, perhaps also serving as a sepulchral chamber for the patron. The towers in Bjäresjö and Stora Herrestad were particularly imposing. In addition, the tower in Sjörup suggests that there was a manor there as well. The remains of the church in Borrsjö also suggest the existence of a west tower used as a burial chamber. It has generally been assumed that the towers were primarily intended to house the bells, but later research has shown that they cannot have been built for this purpose (Stiesdal 1983). Unlike the earlier towers, which probably reflect a centralized power, these can be seen as a private manifestation by the patron. The Ystad area is dominated by churches which can be presumed to have been built by individual magnates, as is also suggested by the manors attested from later times (Skansjö et al. 1989). The church builders are no longer only kings and bishops, but also other powerful men. It is conceivable that at this time there were also collective building initiatives undertaken by landowning peasants (Nyborg 1979).

Instead of the easily worked stone from Fyledalen, limestone was chosen from the area east of Herrestad Hundred. Contrary to previous practice, stone is now brought from outside the parish. Finished sculptural details can be brought from even further afield. Several fonts and a portal are among the objects made of limestone from Gotland in the Baltic Sea. The stonemasters are no longer tied solely to church building, now being able to work independently.

The 13th century saw the crystallization of two more parishes with the building of churches in Villie and Sövestad. The smallest church in the area was built in

Table 5.7:1. The table shows that parishes with runic stones, towers, and vaults up to 1400 coincide, an expression of the power of magnates from the 11th century down to the mid 14th century (after Mats Anglert 1989b).

	Runic stones	Early medieval tower	Barrel vault in the chancel	Cross vaults in the nave 14th century	Cross vaults in the nave 15th century
Baldringe	X	X		X	
Stora Köpinge	X	X		X	
Snårestad		X		X	
Bjäresjö*	X	X	X	X	
Stora Herrestad	X	X		X	
Västra Nöbbelöv	X	X	X	X	
Sjörup*	X	X	X	X	
Sövestad			X	X	
Bromma				X	
Högestad					X
Villie					X
Öja					X
Skårby	X				X
Hedeskoga					
Balkåkra					
Borrie*		?	X		

* = Recess in the eastern part of the nave

Borrie. The richly sculpted tribune arch and chancel arch seem to have been intended for another church. Borrie Parish was the only one in the 16th century that was held solely by freeholding peasants (Skansjö et al. 1989:107). The architecture and date of the church show that the situation may have been the same during the 13th century.

In the 1280s the old chancel in Stora Köpinge church was replaced by an extension to the east containing a larger chancel with an apse. This change was previously thought to have occurred early in the 13th century, the suggested explanation being that a larger church was planned in association with the market *(köpinge)* but that this was broken off when Ystad was established (Lundberg 1940:124, Tesch 1983d:92f.). Wienberg, however, maintains that the connection between the two places is such that they can have functioned simultaneously for a period (Wienberg 1989). Another explanation is that the extension to the east is an indication of the ecclesiastical function: the archbishop was the dominant landowner in the parish throughout the Middle Ages, and in 1322 he acquired the right of patronage over the church (Skansjö et al. 1989:116, see also Callmer 1984, Ersgård 1988:184f.). By analogy with what we know from England, we may assume that the church had responsibility for the chancel while the congregation were responsible for the nave (Platt 1981:97).

Probably the last church to be built in the Middle Ages was Balkåkra, which may well have replaced the older church at Borrsjö. The town church of Ystad, St Mary's, was built in the early or mid 13th century. St Nicholas' church, which has since disappeared, may have functioned as a chapel of ease even before the founding of the town. Somewhat later a Franciscan friary was established in the town (Tesch 1983c:70f, 1983d.101, Wienberg 1989). For these buildings the new material of brick was introduced, which required a more organized form of production than previous building material. In the rural churches brick was used only for the eastern extension of Stora Köpinge church.

Brick brought with it a new building technique which found special expression in the vault building of the first half of the 14th century. Virtually all the churches in the coastal parishes were vaulted in this period. Some of the vaults have paintings belonging to the Snårestad group, whose work has also been found in areas north-east of Ystad (Fig. 5.7:6). As brick manufacture started in Ystad, it is not unlikely that the continued production of brick for the oldest brick vaults in the rural churches was centralized in the town.

The vaulting of the nave gave the congregation's part of the church a status comparable to the chancel. At the same time, the private function of the tower disappeared. In Västra Nöbbelöv church the first tower was demolished and replaced by a new one. The magnates are still in evidence, however, taking the initiative for the vaulting, but their role is another one (Table 5.7:1). The parishioners are given an active part in the administration of the church by the new system of churchwardens (Nyborg 1979:39). This can be compared with the new manors that were moved out of the villages, underlining the great organizational change that appears to have taken place at the beginning of the 14th century (cf. Skansjö et al. 1989, Reisnert 1989a).

It is probable that the division of the area into parishes was comparable in broad outline to today's, and that it was completed with the formation of the Ystad parishes. The town has two Gotland fonts from the first half of the 14th century. One belongs to St Mary's and the other, which now stands in the Franciscan church, may originally have belonged to St Nicholas' (Sundnér 1969:24f.; these Fröjel fonts, called after a place in Gotland, are to be found in all the coastal towns established in Scania in the 13th century). The churches in

Ystad can thus be assumed to have acquired parish rights only at this time, thereby becoming the last parish churches in the area.

There is some variation in the size of the country churches, but no chronological connection can be detected. That St Mary's church in Ystad is by far the largest can be explained by the emergence of the town (Wienberg 1989).

III. 1350–1550 (Fig. 5.7:7c). A decline in building activity lasting until the mid 15th century was followed by a new upsurge (Fig. 5.7:5). Although vault building still continued until all the churches in the area were vaulted, the building of towers predominated. The new towers were built as belfries with large sound-holes on all four sides. They were generally very big, with spacious floors upstairs. Unlike the older towers, entrance was now directly into the upper floors. In the 15th century towers the ground floors were built with groined vaults and used as chapels, as is shown by surviving paintings in other churches in Denmark. The first floor may have been used for storage. The 16th century towers had no direct access to the church interior; they were probably built expressly for storing the church's tithes. They were wholly owned by the parish in common. As with the towers, the choice of building material also shows that the parishioners were the builders. They selected only the field-stone that was available in the vicinity. People had previously taken stones of one particular size, but now any available size was used. Three churches were extended to the west; they are all in the northern part of the area, where the clearing and cultivation of woodland may have been a contributory factor (Wienberg 1989:257, cf. Billberg 1989:44f).

IV. 1550–1900 (Fig. 5.7:7d). This period has not been studied in any great detail as part of the project, but it can nevertheless be briefly summarized. The 16th and 17th centuries are characterized mainly by a change in the furnishing of the church, with the coming of pews, galleries, pulpits, and organs. Memorial chapels were built in other areas during this period; in the Ystad area there are four churches which were probably given memorial chapels. It is also possible that some of the towers were built at this time. In the first half of the 18th century five churches were enlarged with the building of transepts. Towards the end of the 19th century completely new churches were once again built. Four new churches replaced six medieval churches. Snårestad, Skårby, and Borrie remained without parish churches of their own, although they were all allowed to stand as ruins. Building activity can be seen to shift towards the west of the area. In several cases there is a distinct connection between population growth and church enlargement in this period (Wienberg 1989:258). There are exceptions, however, which can be explained in a way reminiscent of the manifestations of the medieval magnates – the new churches were built partly to give the parishes "a worthy temple" (Fernlund 1982, Wienberg 1989:260). It is also striking that many churches were not enlarged until long after their population had grown. Wienberg believes in this connection that the enlargements of the 15th century could also reflect changes that occurred as early as the 13th and 14th centuries (Wienberg 1989:261f.).

Conclusion

I have discussed some aspects of the way churches can be used as evidence of changes in the cultural landscape. What has not been considered very much is the interior of the church itself. Such things as the furnishings, fixtures, and paintings remain to be studied. Churches are built for the exercise of cults and can as such reveal many of the ideas and concepts of the medieval mentality.

5.8 Ystad and the hinterland

Hans Andersson, Tomas Germundsson, Anders Persson and Sten Skansjö

Herring and farming: the medieval town and its surroundings
The town

In the 1960s a number of coats of arms were uncovered in the Franciscan church in Ystad. Some time in the Middle Ages they had been painted on the walls of the nave and chancel. They are a remarkable and concrete illustration of the town's contacts both with the surrounding countryside and with areas across the Baltic Sea. They represent families associated with large estates around Ystad, burghers from the town itself, and also burgher families with links southwards – and even eastwards – across the Baltic (Raneke 1968). These armorial paintings put Ystad into its geographical and historical context. They show the local and international framework in which the town functioned. Both Ystad's role in the local context and the town's position as an intermediary between the hinterland and the international market will be illuminated in this chapter.

Like several other towns along the south coast of Scania, Ystad emerged in the 13th century. There are signs of activity here before this stage. Archaeological excavations have revealed clay-lined pits associated with herring fishing off the coast (Sw. *lerbottnar*). They have frequently been found along the Scanian coast, especially at localities which grew into towns during the Middle Ages. Their exact function is the subject of debate, but it is evident that they belong to fishing places of some type (Ersgård 1988:41f). According to generally accepted datings, work on the construction of St Mary's Church was begun during the first half of the 13th century. The first true criterion of urbanization, however, is the building of the Franciscan friary in the late 1250s. The town church, St Mary's, underwent considerable renovation and enlargement – including the addition of a vault – at the end of the 13th century or the beginning of the 14th. To this it may be added that the nave of the Franciscan church was built late in the 13th century, followed by the long chancel and a side-chapel in the 14th century, leaving the entire friary complete around 1400. Ystad's formal status as a borough is shown by the naming of town councillors for 1293–1294 (Andrén 1985:187), and the place is termed *villa forensis* in 1303 (Tesch 1983c:12).

Agriculture and herring

The economic role of the town within the area is based on the surrounding countryside and the fishery. An important part of the necessary economic foundation came in all probability from agriculture. Pollen analysis suggests an extension of arable land during the Viking Age (Gaillard and Berglund 1988). This change may be connected with the introduction of a three-field system. In any case it took place before the birth of the town, perhaps as early as the end of the Iron Age (Engelmark 1988, Engelmark and Hjelmqvist 1991). The surplus provided by this new intensive cultivation system was channelled through Ystad. In Scania similar conditions must have applied to other towns, such as Trelleborg (Skansjö 1983:185f.). Comparable agrarian changes and the consequent emergence of towns are known from other places in northern and central Europe at this time. Ystad is thus part of a European pattern in which agrarian development gives an important economic foundation for urbanization in the Early and High Middle Ages.

There are few details of trade in agricultural produce (Kraft 1956:63ff.). By contrast, the role of fishing is seen with greater clarity in the source material. The "pound-toll lists" from the end of the 14th century, giving details of trade with Lübeck, show that the predominant commodity is herring. In trade volume Ystad comes some way behind Malmö and Skanör-Falsterbo, but well before Trelleborg (Weibull 1966:47). Besides this we have a mention of the Lübeck Company's house in Ystad near St Mary's Church (Kraft 1956:62, Tesch 1983c:19).

It is possible that in the Middle Ages the composition of export goods changed from that of the 13th century. There is evidence of trade in oxen in the 15th century from other parts of Denmark; this trade is also attested from the Ystad area, at least in the 17th century (see below). There is no direct evidence from the Middle Ages, but there were alterations in manor structure in the area in the later Middle Ages which may be indicative of changes in farm management (Skansjö 1986). The pollen material from Bjäresjö suggests a greater emphasis on grazing, with an increase in pollen from plants typical of pasture land (Gaillard and Berglund 1988). It is therefore possible that in the late Middle Ages there may have been a surplus in cattle production, which could have been distributed through Ystad.

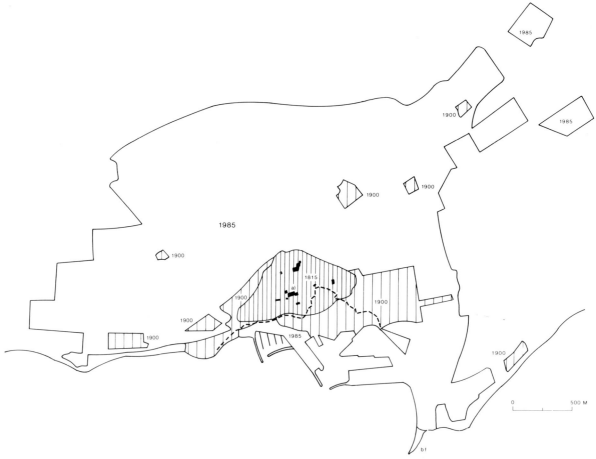

Fig. 5.8:1. The spread of Ystad. The town of Ystad owes its siting to two primary factors: the herring fishery of the twelfth century and the town's role as a trading centre for the agricultural hinterland. The earliest physical remains are clay-lined pits, which are associated with the early herring fishery; these pits are chiefly concentrated around the estuary at the top of the bay which is seen in the reconstructed 13th century coastline (broken line). The building of St. Mary's Church (a) probably began early in the 13th century. The original town may have grown up around the church. The Franciscan friary, probably founded in the 1250s, is situated some 250 m north of the church, and 500 m east of the latter were the remains of St Nicholas' Church, first mentioned in 1452, but of unknown age.

The town had probably grown to its full size within the present town centre by the mid 14th century. This means that the external framework did not change very much during the period from ca. 1350 to 1800. However, it is likely that there were considerable areas of land inside the town which were originally cultivated and only later built on. In the seventeenth century a rampart with ditch and gates was constructed. The map shows the extent of the town in 1815, when the entire network of streets was then of medieval character. The map marks existing buildings from the High Middle Ages (1250–1350) within this area (black).

For Ystad, as for so many other towns, the 19th century was a dynamic time. Trade flourished at the beginning of the century, and in the latter decades the burgeoning of industry changed both the function of the town and its physical structure. Prosperous merchants erected stately houses which contrasted sharply with the older timber-framed housing. Warehouses and factories began to be a feature of the town, at the same time as population grew and many uninhabited sites were occupied. Strictly planned residential areas were created, mostly north and west of the old town centre. In the 1880s the stream, which had flowed through Ystad down all the ages, was drained, and the harbour was extended. The 19th century developments are summarized in the 1900 boundary shown on the map.

In the 20th century it is primarily development since World War II that has changed the town. Up to 1945 the extensions were not so far-reaching, restricted mainly to industrial and military installations and the harbour. After that changes came rapidly, and the enormous expansion in the area of the town in the last four decades can be ascribed to the construction of numerous homes, chiefly single-family houses. In forty years the area of settlement in Ystad has more than trebled, and the little medieval town – which was able to absorb virtually all expansion for nearly seven hundred years – is now deeply embedded in modern-day settlement forms.

This picture of Ystad as a channel for the distribution of agricultural produce, along with fish, finds indirect support in the picture of social and presumably also economic contacts with the hinterland in the 14th and 15th century which we are given by the coats of arms and gravestones in the Franciscan church (Reisnert 1989b: Fig. 12). The families represented here come from an area ranging beyond the borders of Ljunits and Herrestad hundreds, reaching as far eastwards as Simrishamn, over 30 km away, with an occasional offshoot even further afield. What is remarkable is that Simrishamn, along with the other town in the area, Tommarp, do not appear in any way to have affected the extent of Ystad's hinterland. This suggests a ranking order as well as separate functions for the towns. The picture to the west is quite different. Connections are fewer in number, and they do not stretch much more than 10 km. The picture here has probably been influenced by Trelleborg, which followed roughly the same course of urbanization as Ystad (Jacobsson 1982). Even if we question the evidence of this very special and socially restricted source material, it should in any case demonstrate the role Ystad attained as a central locality for south-east Scania (Österlen), from at least the 14th century (Reisnert 1989b).

How did Ystad affect the hinterland?

The rise and presence of a town like Ystad must have affected the hinterland in various ways. Parish organization would have changed when the oldest urban settlement was created on land given up by the parishes. Communications could have been redirected. Supplying the town must have meant changes, for the immediate surroundings at least. The concentration of economic activity, imports and exports, in the town may have facilitated contacts with other markets, and changes in these markets could have influenced production in the hinterland in various ways. Such a process would have been made easier by links between (at least elements of) the burgher population and the aristocrats whose manors were in the countryside. Nor should we underestimate the local trade which developed between rural producers and urban consumers on the regular market days.

The basic pattern of parish division was probably fixed by the mid 12th century. Judging by the oldest maps, when the borough was founded, the area around Ystad was broken away from other parishes – perhaps Hedeskoga and Öja (Mats Anglert 1989) – thus cutting off their contact with the sea. It is also possible to interpret Tvären as the residue of a larger parish, parts of which were claimed by the town (Mats Anglert 1989, Billberg 1989, Persson 1988a).

The rise of the town may have meant the redirection of channels of communication. It is naturally difficult to be certain here. Améen has claimed that a pattern can be seen in the medieval layout of the coastal towns: an original road, leading down to the sea at right angles to the coast, gradually encounters competition from a road parallel to the coast. This pattern can be found in Ystad (Améen 1981, 1984). A similar pattern, albeit with differences in detail, has been demonstrated by Tesch (1983c:60ff., Map 10). Archaeological dating of the roads is not possible, however, but there is much topographical evidence suggesting that the basic framework of Ystad's town plan can go back to an older system of roads leading from the interior towards the shore. Prehistoric communication routes relevant to Ystad have also been reconstructed by Tesch (1983d:74, Fig. 38). There is much to suggest that even in the pre-urban phase there was an established communication structure, and that the rise of Ystad only affected the road system in minor details.

Another interesting question is whether the rise of the town of Ystad brought about any restructuring of the system of localities in the area. Near Ystad lies Stora Köpinge; the name *köpinge* (meaning "market-place") indicates its special character. There is archaeological evidence that it was a trading centre of some kind, with datings up to around 1100 (Tesch 1983d:75ff., Ch.4.1.5.2). Extensive rebuilding of the church chancel and nave was commenced but never completed; these changes have been dated to the end of the 12th century or the beginning of the 13th (Gustafsson and Redin 1969:33). The building work is said to have broken off as a result of the rise of Ystad, which is supposed to have assumed the role previously played by Stora Köpinge.

New dendrochronological datings provide a completely different date for the rebuilding activity, namely 1285 (Bartholin 1988, Sundnér 1988). This makes it difficult to talk of the succession passing from Stora Köpinge to Ystad. The additions to the church at Stora Köpinge are highly unusual for a church outside the boroughs (Wienberg 1988); they show that Stora Köpinge still had special functions in the end of the 13th century.

Ystad and Stora Köpinge may have developed in parallel (or this may have been the intention) during the first phase, the 13th century. Yet development in Stora Köpinge abruptly ceased at the end of the century, perhaps because Ystad had begun to dominate. This may have been due to the different interests in each of the two places. Whereas boroughs were created by the Crown, the *köpinge* localities may have been supported by local magnates, which is the view of Cinthio (1972), or they may have been economic bases for the archbishops, the view of Callmer (1984). Both interpretations see the *köpinge* as representing an economic power centre other than the Crown's. In the Ystad area the town belonged to the king, whereas the hundreds of Vemmenhög, Ljunits, and Herrestad (including Stora Köpinge) were held in fief by the archbishop, at least during the latter part of the 13th century (Kraft 1956:34).

Vault building as an indicator of the extent of the hinterland

As has been said, the end of the 13th century and the first half of the 14th is an important period in the consolidation and development of the town. This is seen above all in the building activity in the town's churches. St Mary's and the Franciscan church were then built in brick. This expansive work may have had consequences for the hinterland – not only for Stora Köpinge – in various ways.

In the churches of the immediate hinterland there is an important parallelism in the development. In the first half of the 14th century the churches in Baldringe, Bjäresjö, Sjörup, Snårestad, Stora Herrestad, Stora Köpinge, and Västra Nöbbelöv were given brick vaults. In a comparative perspective it is relatively uncommon to find such a concentration of vault building in this period. Most of these churches lie in the immediate vicinity of Ystad. There is reason to assume that there is a direct connection between the building activity in the town and in the surrounding parishes. It may have been the case, for example, that brick manufacture for the churches in the town may have been expanded so as to include the nearby rural churches, partly because construction work required a relatively large permanent organization (Sundnér 1989a). Behind this parallel development in the town and its vicinity, marked by vault-building, there must have been a shared economic development (cf. Wienberg 1986a). It has also been assumed that the paintings on the vaults, those known as the Snårestad group, could have come from a workshop in Ystad with contacts in northern Germany (Banning 1976–82:4, 38).

Aristocratic dominance and mercantile growth 1536–1658
Estate building and European economic cycles

The period from the end of the Middle Ages to the beginning of the Swedish era falls entirely within a period in Danish history which is known as "the time of noble dominance". This designation is justified as a covering term for several important processes of change which were typical of the times at various levels in society. Some of these changes contribute in concrete ways to the change of the Ystad area at precisely this time into an agrarian region dominated by the estates of the nobility. It is evident that this transformation also had consequences for Ystad's role in the area, primarily as a channel conveying surplus agricultural produce to an export market. Against the background of a presentation of some of the main relevant features of the period, this and other questions about the town and its function in the area will be discussed here.

The development of the agrarian economy in Europe shows a transition in many places to larger farming units and a more market-oriented agriculture in the period

after 1500. (For the following see Gamrath and Ladewig Petersen 1980:380ff., 403ff., Skansjö 1987 and references there.) An explanation for this can be found in western Europe's sharply increased need for, above all, grain, as a result of general population growth, particularly rapidly swelling towns and expanding armies. Grain was imported to western Europe from the surplus-producing areas of eastern Europe. The interest in introducing large-scale agricultural methods naturally went hand in hand with favourable price trends for agricultural produce in the age of the "price revolution" (from the start of the 16th century to the Thirty Years' War).

It is obvious that Denmark, and hence also Scania, met part of the grain demand of the industrially developed and urbanized countries of western Europe. It even came to occupy a position of rank as regards the export of cattle on the hoof; fatted oxen were an important contribution to the meat consumption of the big towns of the Low Countries and western Germany. For large parts of this period there was something of an economic boom, and Denmark's exports brought in considerable amounts of capital. Profits went primarily to the Crown and the nobility, who totally dominated the production apparatus. The Danish nobility had special privileges giving them a clear advantage to exploit ox-breeding.

European prices and agricultural economic cycles explain a great deal of the development towards large-scale management on the estates of the nobility and the associated building of castles at this time in many parts of Denmark. In Scania this development was especially noticeable in the Ystad area, where the distribution of property was drastically altered in the direction of aristocratic dominance in this period, which also saw the establishment or expansion by nobles of a number of large and medium-sized estates, such as Krageholm, Marsvinsholm, Herrestad, Högestad, and Bjersjöholm (Ch. 5.9). It is obvious that the town of Ystad had important functions for the agrarian hinterland at this stage. Conversely, the development of urban functions can be seen as a reflection of events in the hinterland.

Export and import

As the general European boom affected many other parts of Denmark, we may assume that it influenced activity in the town of Ystad. From the discussion earlier in this chapter it is clear that the burghers of the town in the Middle Ages accounted for the turnover of part of the hinterland produce to an international market, but we have reason to assume that several important nobles directly managed the export of their own produce. What happened in the course of the period under consideration here has the character of outright specialization and division of labour: noble estates produced and merchants conveyed their surplus produce to the international market.

In the wake of the trade boom a number of wealthy merchants crystallized as an upper stratum of Ystad's burgher class. It was these merchants with their plentiful capital who bought up the surplus produced by the nobility, taking care of the export of commodities from the hinterland and meeting the need for imported goods in the area. There could be wide gaps between these merchants and other elements of the town's population, such as the merchants engaged in the local retail trade, craftsmen, and wage-earners.

Both the export and import of goods were subject to customs duties. The surviving customs accounts for Ystad from 1606 onwards give us the first detailed picture of what goods were exported and imported. Studies of this material have shown that the area around Ystad displayed great versatility in agricultural production. (For the following see Wimarson 1918:9ff., 59ff., Kraft 1956:125ff.) The surplus products exported from Ystad included oxen, sometimes also horses, barley, malt, and oats. Fatted oxen and barley were the most important merchandise during the period. Corresponding material from Malmö's hinterland reveals a greater bias towards grain production, while Helsingborg's hinterland specialized in animal husbandry.

Oxen were exported in the spring, when they had been fully fattened. Ystad's position as a legal export harbour for beef cattle was established in a royal charter in 1599. In the 1620s and 1630s the average annual export of oxen from Ystad could be around a thousand head. In normal years it appears that about 1200 head can have been exported, and the highest number known is 1740 head (1646/47). The customs accounts occasionally mention that the oxen in question had been fattened on noble manors near the town. The hinterland alluded to here clearly stretches beyond the boundaries of Ljunits and Herrestad hundreds. Oxen from Svaneholm (23 km west of Ystad), from Övedskloster (31 km to the north), and Ingelstad manor (21 km to the east) were exported through Ystad. Horses were also an important export commodity for the Ystad area. This trade experienced a boom in conjunction with the Thirty Years' War on the continent. For example, the fiscal years 1632/33 and 1633/34 saw the export of a total of some 3800 horses.

Grain export from Ystad was largely in Dutch hands. It was normally most important in the autumn months, but merchants could delay exports from the grain warehouses while waiting to see how prices would develop. Occasional royal regulations affected this export, such as export bans after crop failures. The grain was shipped chiefly to northern Germany (Lübeck, Rostock, and Greifswald), but considerable quantities went to Holland as well. Ystad and Landskrona were the leading grain-exporting towns in Scania after Malmö. According to the customs accounts for 1625–43, the annual exports from Ystad comprised 9400 barrels of barley, 3600 of malt, and 6600 of oats. The figures we have show a marked decline in conjunction with the war

years of 1643–45. Ystad and the hinterland were especially affected by the quartering of Swedish troops in the area during the winter of 1644–45. The town's export business appears otherwise to have flourished during the years when war ravaged the European continent. In this respect Ystad's history in this period is similar to the early 19th century, as we shall see below.

Imports to Ystad are clearly recorded in the customs accounts (Wimarson 1918:55ff.). It was through the intermediacy of the merchants that the people of the town and the hinterland received supplies of necessities such as salt, metal goods, and textiles. Of other imports recorded in the customs accounts we may mention German beer, wine, fish, hops, wood, bricks, glass, tar, and hemp. Most of this came from north German towns, chiefly Lübeck, and from Holland and the British Isles. Grain was also occasionally imported to Ystad: the customs accounts for 1633/34 record the import of some 1600 barrels of rye and 450 of wheat.

Mutual dependence

It should be clear from the above that the mutual dependence of town and hinterland was evident from the period from the end of the Middle Ages to the start of the Swedish era. It is clear, for example, that noblemen in the Ystad hinterland were supplied by the town's merchants with luxuries in the form of textiles, foreign beer, wine, spices, and the like, commodities which gave them a high standard of living for the day. At the same time, a rather aristocratic stratum of merchants emerged as a result of the export boom, mainly through dealing in the production surplus of the noble estates. There was also a mutual dependence between other social strata in the town and the surrounding countryside. This is suggested by the details above concerning the import of essential supplies. It should also be mentioned in this connection that there were two important annual markets in Ystad, one in summer and one in autumn, known at least since the 17th century (Hanssen 1952, 1977:256). These markets provided opportunities for the farmers in the area to sell any surplus and to buy articles from craftsmen and hucksters.

New patterns of trade 1658–ca. 1790
Grain

The most important export products around 1660 were barley, oats, oxen, and horses (Bjurling 1945:26). Exports were traditionally directed towards northern Germany – Travemünde and Lübeck – but Swedish customs policies brought about a partial change in the flow of goods. New rates for customs duties were introduced in 1658. Export goods were reduced in price while imports became much more expensive. The Scanian customs duties for 1669–1715 made the duties on Scanian grain 100–136% more expensive than the corresponding Danish duties. The result was that grain shipping was transformed from a foreign to a national activity. The grain trade took on a pattern which remained largely unchanged during the period in question here. Grain was shipped to towns in eastern Sweden, including Stockholm, but above all to the coastal towns of Blekinge. From there the ships returned laden with timber and other wooden articles to the plains around Ystad and Simrishamn (Bjurling 1945:79ff., 1956:306f.). The accounts from Krageholm manor for 1704–1707 and Gundralöv manor for the 1730s (Landsarkivet, Lund) show the trend of this grain trade clearly. The volume of grain shipped from Ystad varied from year to year depending on the size of the harvest. It is said to have been around 20,000 barrels at the time when Scania came into Swedish hands. In the late 17th century and the first half of the 18th the figures varied between 3000 and 20,000 barrels. The first part of this period saw bad harvests in Scania, with the exception of the 1680s (Persson 1986a). In the latter part of the 18th century there was an increase in the amount of grain shipped, the volume up to 1790 varying between 20,000 and 40,000 barrels, although in 1787 the figure was almost 50,000 barrels (Bjurling 1945:88, 1956:483ff.). Improvements in the grain trade can also be inferred from the increase in luxury craft products in the town in the late 18th century (Bjurling 1956:435, 485), evidently a general phenomenon (Edgren 1987:81). A vigorous expansion in provincial towns at this time was also experienced by tradesmen catering to rural markets (Edgren 1987:79f.). Industrial activity, however, was insignificant during the period in question (Bjurling 1956:412f.).

Oxen

In the late 17th century it was fairly common for Scanian merchants to pay for their oxen to be fattened on manors. During the period 1694–1708 some twenty Ystad burghers leased fattening facilities for oxen in the immediate hinterland of the town. It was also the merchants in the town who managed the trade in oxen between Ystad and northern Germany. A figure that tells us something about the extent of the exports is that around 2450 oxen were shipped out of the town in 1708. After this the export of oxen declined as alum and salt became the dominant export commodities (Bjurling 1941, 1945:40ff., 1956:227). Ystad's burghers were also involved in other ways in operations in the hinterland. Livestock was leased out to farmers in the hinterland. This was evidently a common arrangement, which can be seen as a way of investing capital (Bjurling 1956:213f.). The town merchants thus had direct interests in agrarian production in the hinterland, and indirect involvement through trade in the farmers' produce. Since taxes in Scania were collected in cash, the farmers had to sell produce, primarily grain, in the town. Both promissory notes to merchants and customs rolls allow us to judge the extent of the hinterland, which remained largely the same during the period in

question. It stretched from Trelleborg in the west (which had lost its borough charter in 1619, not to regain it before 1865) towards Simrishamn in the east, and north beyond the hundreds of Ljunits and Herrestad into Färs hundred (Bjurling 1956:233, 235, Hanssen 1952, 1977:253).

Between 1658 and about 1790 Ystad was primarily a channel for the transport of the agricultural produce of the hinterland. The situation is underlined by the fact that the merchants were the leading circle among the burghers. In 1683 as many as 106 of the town's 268 occupied properties (almost 40%) were owned by 40 merchants (Bjurling 1956:195f.).

The town as consumer

Moving on to the town's role as a consumer of the produce of the hinterland, it must be emphasized here that Ystad was in part self-sufficient as regards agricultural produce. On town land and outside the ramparts there were arable fields, meadows, and kitchen gardens, and the rotating sails of several windmills could be seen from the town ramparts. It was reported in 1688 that in 1668 there were over ten windmills on the town lands. In the mid 18th century the people of the town kept about 250 cattle and an estimated 350–400 sheep in the town (Bjurling 1956:371).

At the beginning of the 18th century the population of Ystad was around 2000, rising by the end of the century to some 2500 (Fig. 5.8:2). The combined population of the town and the immediate hinterland, Ljunits and Herrestad hundreds, can be roughly estimated as 6800 in 1700, rising to something like 8900 in 1750 and 11,200 in 1805. The town had a relative share of this combined population amounting to about one-third in 1700, slightly less in 1750, and down to one-quarter in 1805. Over a century the population of the hinterland thus increased in relation to the town from two-thirds to three-quarters (Bjurling 1956:180ff., 339ff., Persson 1988b).

With a virtually stagnant population in the latter half of the 18th century, there can scarcely have been any major change in the situation as regards essential provisions, nor any increase in the flow of trade from the hinterland. At the end of the 18th century Ystad's own fields were probably able to support about 25–35% of the town's population (Bjurling 1956:206). Tax assessments for the town for 1682 and 1688 (Ystads stadsarkiv) show that the town's lands could hardly have supported such a large percentage then. If we nevertheless reckon that 1500 persons required food from outside the town, then each year an estimated 4500 barrels of grain were needed in addition to what the town's fields could provide. Normal rye and barley production from a farm assessed at 1 *mantal* (the fiscal assessment unit) can be estimated to 45–60 barrels (Persson 1988a). In Ljunits and Herrestad hundreds in 1700 there were about 590 farms with a total *mantal*

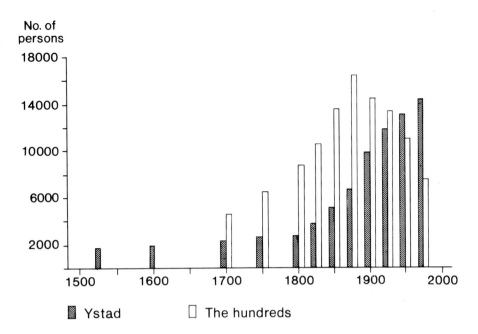

Fig. 5.8:2. Population development for Ystad and the hundreds of Ljunits and Herrestad 1525–1975.

No. of persons

18000

14000

10000

6000

2000

1500   1600   1700   1800   1900   2000

▨ Ystad   ☐ The hundreds

value of 350, not counting the manors. The need for provisions from outside could thus be very roughly calculated to some 20–30% of the total rye and barley production of the hinterland. Despite its modest size, the town therefore depended for its food on quite a large hinterland, which is related to the low rate of cultivation (about 20–30% of the village lands were sowed each year) in conjunction with the low productivity in the villages of the hinterland (Dahl 1946:136, Persson 1987a, 1988a).

Integration and industrialization
Flourishing trade in the early 19th century

It can be said that Ystad experienced a period of prosperity in many respects at the beginning of the 19th century. The foremost reason for this boom is to be found in the advantages the town could exploit in the field of commerce in a time of European unrest. This development came as a contrast to the difficulties experienced by trade at the end of the 18th century, when export bans, a prohibition on distilling, and sky-high customs duties made life hard for the burghers of Ystad. An example of the recession can be seen in the heavy reduction in the town's merchant fleet in the last decades of the 18th century. After the nadir of 1801, however, the trend turned, with the result that for some years during the 1820s Ystad even had the highest tonnage of all the towns of Scania (Morén 1953:202–9).

Trade was thus on the upswing around 1800, when there was also a comparatively high number of craftsmen in Ystad. By contrast, there were few industries. Between 1803 and 1807 there was a sharp increase in the export of food from Ystad. Most of this went to the Swedish possessions in Pomerania, generating consid-

erable income for Ystad's merchants. This lucrative trade ceased when French troops occupied Pomerania, at the same time as the important postal yacht service between Ystad and Stralsund was discontinued. Yet these difficulties were soon compensated by the way relations between Sweden and England developed while Napoleon was applying the Continental System. From 1807 until 1810 Sweden was one of the few European states outside the Continental System, so there was extensive transit trade on the England–Sweden–continent route. These years saw the rise of several of the trading houses in Ystad, which were to maintain their power for so long.

Another reason for the flourishing trade experienced by Ystad at the beginning of the 19th century was the expansion in grain production in the surrounding countryside. Production for sale increased in pace with the growing market (Möller 1985) and reforms such as the *enskifte* enclosures integrated the countryside more and more into a national and international flow of commodities. Merchants in the towns were an indispensable contact between the market and the producers, and the town's economy was "highly dependent on the size and quality of the annual grain harvest in the rich surrounding districts" (Morén 1953:149).

Apart from the purely commercial contacts, there were a few other phenomena which led to changed and intensified links between town and countryside at this time. First, the population of the town increased not only through a birth surplus but also through in-migration, chiefly from the immediate hinterland. Many people from the agrarian lower classes found work in the town, often of a temporary nature, so there was great mobility. Another example of increased integration comes from the commercial contacts between the

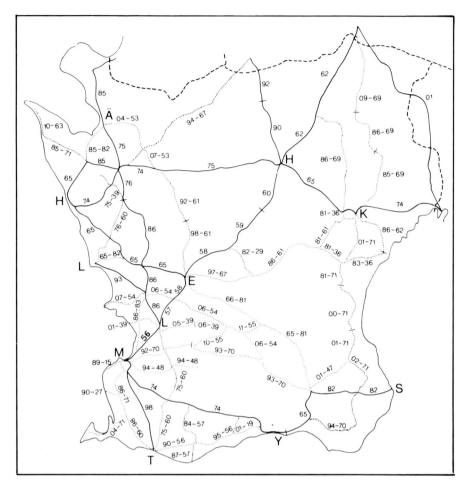

Fig. 5.8:3. The development of the Scanian railway network, 1856–1984. The figures show the years (omitting the centuries) when the lines were opened and when the disused lines (dotted lines) were closed for passenger traffic. The letters mark the towns (Ä = Ängelholm, E = Eslöv, H (east) = Hässleholm, H (west) = Helsingborg, K = Kristianstad,, L (west) = Landskrona, L (east) = Lund, M = Malmö, S = Simrishamn, T = Trelleborg, Y = Ystad). The trunk line to Stockholm from Malmö was opened through Scania between 1856 and 1862, and just a few years later connecting lines to Helsingborg and Landskrona, Kristianstad, and Ystad had been added. Most of the local railways were opened around 1900 (1856–1911) and closed again around 1950 (the first in 1919), in many cases as a result of the sugar-beet transports having been transferred to the roads (Lewan 1985: 143).

town's burghers and the upper stratum of the peasantry. Through their relations with Ystad's merchants the grain-selling peasants came into contact with a new world, with (for them) new things and new habits (Pred 1986:133–134). In addition, they were often economically dependent on the merchants, since many of their purchases were on credit. Yet another example of the integration of town and country is the investment made by wealthy merchants in estates and farms in the district around the town.

When the question of building railways arose in the 1840s, there were advanced plans to let the southern trunk line from Stockholm terminate in Ystad. This plan was not realized, however, and the first railway to Ystad was instead the 1865–66 line connecting Ystad to the main Malmö-Stockholm line at Eslöv. Ystad's position in Scania's regional railway network is shown on Fig. 5.8:3.

The breakthrough of industrialism

In the mid 1860s Ystad had about 6000 inhabitants, making it the smallest town in Malmöhus County. Population grew fairly vigorously throughout the 19th cen-

tury (Fig. 5.8:2), which meant extensive building activity (cf. Fig. 5.8:1). There was a vigorous upsurge in industrialization in the 1870s, a time which was characterized by the gradual transition from craft work to factory or industrial production. About thirty factories are reported as operating in the late 1870s, but it is tricky here to draw a clear line between craft and industry.

Ystad never attained much significance as an industrial town, and its industrial activity was characterized right from the beginning by its proximity to agriculture. In the late 19th century the town's steam-mill accounted for the greatest production, apart from which there was a roller mill and a sugar mill. A highly productive industrial enterprise was the engineering foundry Ystads Gjuteri och Mekaniska Verkstads AB, which made cookers, ovens, tools, machines, mill mechanisms, dairy machines, etc. Through a gradual specialization in agricultural machinery the company was to play a significant role in the modernization of agriculture.

In the field of communications the construction of the railways was the greatest event in the latter half of the 19th century. The opening in 1866 of the Ystad–Eslöv railway linked the town with the growing national communication network. One aim of the line was "to try to transit travellers to the continent via Ystad", another was "to bring the port of Ystad into closer contact with a wider agricultural hinterland" (Améen 1987:178). As earlier in history, this hinterland tended to shift to the east, and in certain places there was commercial competition from the new station villages which arose in the periphery of the hinterland (Swensson 1954:126). When the railway was built it was chiefly with a view to grain exports, but sugar-beet was soon to be the main commodity for transport.

In 1874 the railway line from Malmö to Ystad was opened. It was financed in large part by the estate owners along the line, who thought they could exploit it for the transport of their agricultural produce. The direction of trade was mainly westwards, via the port of Malmö, and industrial and commercial interests in Malmö actively contributed to the financing of the railway. Ystad, on the other hand, had no interest in the line, seeing it instead as a way for the rival port of Malmö to increase its territory (Möller 1987). Ystad was increasingly in the shadow of Malmö and Trelleborg in the field of communications, despite the link with the trunk line and several enlargements of the harbour.

Traditional functions and the loss of the regional position

Ystad's development in the 20th century can be divided into three different phases. The period from the turn of the century to World War I saw an expansion of the town in several respects: population doubled between the 1880s and the 1910s; commerce continued to grow; and settlement expanded. This vigorous period was succeeded by the stagnating trend of the inter-war years. Since World War II, although Ystad has seen increased activity, the town has scarcely regained the position it had up to World War I. It is chiefly Trelleborg and the towns along the Sound which have grown too strong for Ystad in various spheres of competition.

Commerce in Ystad throughout the 20th century has still been characterized by the strong links with the agrarian hinterland, with a consequent sensitivity to economic fluctuations in agriculture. The traditional trading hinterland of the town – to the north, north-east, and east – has been further consolidated during this century, primarily through the coming of the railways. The engineering industry, which is so important for the rest of Sweden, was introduced to Ystad only gradually and on a small scale. The tradition of trading and craft work was hard to break, and "the relatively small and slow changes in commerce in the 20th century must to a large extent have their root cause in this combination of agrarian-centred trading houses and traditional craft. We must also note that the 20th century saw a closure of eastward contacts, first as a result of the Russian Revolution in 1917, later through the reduction in trading contacts with eastern Europe after World War II" (Bunte 1987:222). The westward shift happened early on; around 1910 Trelleborg took over the role of the leading Baltic port. In the 1960s some ferry lines from Ystad were inaugurated, although with varying degrees of success, and it was not until the 1970s that the town was established as the Swedish terminal for the train ferry to Poland – a decision influenced by considerations of regional politics.

Today Ystad has 15,000 inhabitants. After relatively rapid growth in the years immediately following World War II, the curve has levelled off, leaving growth around zero. The area covered by the town has considerably expanded, however (cf. Fig. 5.8:1). Business retains its traditional functions, and modern industrial production is still less important than activities such as trading (mainly in the Baltic region), the processing of primary produce from the surrounding countryside, and the town's role as a centre for local regional trade (Bunte 1987:233).

Ystad and the hinterland – a strong continuity

There is a striking constancy in Ystad's relation to the hinterland from the Middle Ages to the present day. Down the ages the town has functioned as a channel for the agricultural produce of the surrounding countryside. The basis has always been the grain trade (although the evidence from the Middle Ages is scanty), but the cattle trade was probably of importance already at the end of the Middle Ages. The vigorous herring trade of the Middle Ages never attained the same position in later times.

The extent of the hinterland to the north, north-east, and east has also remained strikingly similar down the ages. The picture suggested by the armorial paintings in the Franciscan church is in large part the same in the late 19th century. The changes to the west have been greater, primarily because Trelleborg lost its borough charter for a time during the modern period; Trelleborg's hinterland was then divided between Ystad and Malmö.

Note: The authors of the individual sections are Hans Andersson (Herring and farming: the medieval town and its surroundings); Sten Skansjö (Aristocratic dominance and mercantile growth 1536–1658); Anders Persson (New patterns of trade 1658 – ca. 1790); Tomas Germundsson (Integration and industrialization).

5.9 Landed estates and their landscapes

Tomas Germundsson, Jens Möller, Anders Persson and Sten Skansjö

A characteristic feature of today's cultural landscape is that the Ystad area is situated in one of the most estate-dominated agrarian regions of Scania. Therefore, within the project, an important aim must be to describe and explain the structural changes regarding estate building and forms of large-scale management, and to discuss the impact of these phenomena on the cultural landscape from the Middle Ages until the present day.

From the High Middle Ages to Early Modern Times
Traces of a high medieval manorial system

According to historical research we can expect to find significant changes rather than unbroken continuity in the manorial system during the period from the latter part of the 12th century up to the 16th and 17th centuries. In certain respects the manorial system of the 16th and 17th centuries meant a return to a system resembling that of the High Middle Ages (ca. 1200–1350), whilst, on the other hand, the lay estate organization of the Late Middle Ages (ca. 1350–1536) differed from both these systems.

The Age of the Valdemars (1157–1241) saw the formation of the nobility as one of the estates in Danish society. Written sources from about 1200 reveal that in Denmark, above all in Zealand, some form of large farms (Lat. *mansiones* or *curiae*) existed, resembling the large feudal estates in certain parts of Europe, such as England, France, and parts of Germany. We can find traces of this manorial system in Denmark up to the end of the 14th century. As a rule, an estate of this type has been organized in two main parts, a demesne farm and adjoining smaller holdings. The demesne farms known from Denmark varied in size but in general they seem to have been intended for large-scale farming, although none of the 14th century's manors was on the same scale as the largest demesnes of the 16th and 17th centuries.

In the High Middle Ages many estates were administered by *villici* (Latin for bailiffs, *brytar* in the Scandinavian languages). From about 1300 these manors were referred to in the documents as *curiae, curiae principales* or *curiae villicales*. Attached to these larger farms were much smaller holdings used by *coloni* (tenants with some land to farm) or *inquilini* (almost landless tenants). Both categories had labour duties on the demesne. According to a rather common model for high medieval manorial organization, an entire village could consist of one large *curia principalis* or *villicalis* and a varying number of *coloni* and/or *inquilini* (C. A. Christensen 1964, Ulsig 1968, Lindkvist 1979, cf. Skansjö 1991).

The question is now whether specific observations in the medieval source material concerning the Ystad region correspond with this more general knowledge and, thus, verify the existence of a particular high medieval manorial system. The studies within project have aimed to combine an analysis of the medieval written sources with applicable archaeological evidence and investigations into the oldest extant land survey maps. Even if the combined evidence is still fragmentary it confirms the existence of a manorial system from the High Middle Ages which was later dissolved. We can briefly illustrate this phenomenon by some examples. (For the following, see Skansjö 1987b, 1988, 1991, Skansjö et al. 1989.)

A document from 1387 concerning tenant farms belonging to the Bjersjöholm estate mentions a *curia villicalis* ("brytegård") with 3 *inquilini* ("fästor") in the village of Bromma: "in Brame 1 curia dicta broythegard cum 3 curiis dictis fæsther". This gives a rather unique but distinct example of a small village estate at a time when this high medieval structure was dissolving.

Archaeological observations in Bjäresjö village show that a large farm existed within the village from as early as the 11th century, with continuity up to about 1350 (Callmer 1986, see Ch. 4.2.5.1). At that time the "Bjäresjögård" manor ceased to exist on this site, and it seems to have been replaced by "Bjersjöholm" manor, which was established at some distance from the village (Ch. 4.2.6.1). Although we lack, in the case of Bjäresjö, direct reference in written sources to a *curia principalis* with *coloni* or *inquilini*, we have reason to postulate a manorial structure of the high medieval village estate type. This structure is not only indicated by the village manor, known from archaeological evidence and written sources, including retrogressive observations from the oldest land survey map of 1699 (Ch. 4.2.6.2). We must also consider the distinctive architectonic features of the Bjäresjö church building from the 12th century (see Ch. 5.7). During the Middle Ages there was also a village settlement in Bjäresjö, consisting of holdings with smaller dimensions and representing a lower social stratum than the manor (Callmer 1986) (Ch. 4.2.5.1). In several respects these village units of Bjäresjö must have been dominated by the manor and the church up to at least about 1350.

We have other indications of this estate structure

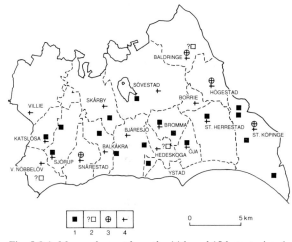

Fig. 5.9:1. Manors known from the 14th and 15th centuries. 1. Manor owned by a nobleman. 2. Possible manor, also owned by a nobleman but with less clear written sources. 3. Manor belonging to the archbishopric of Lund. 4. Parish church. After Skansjö 1987a.

consisting of an entire village settlement connected to a manor. From the early 14th century Borrsjö manor (later known as Marsvinsholm) is known as a residence for a knight, and adjoining the manor was the village of Borrsjö, which around 1550 included twelve tenant farms. There was also an early medieval church in connection with the manor. A similar structure is found in Katslösa, a church village where a stone-built manor for a knight is known since the early 14th century. Evidence from the church-village of Sjörup also reveals a manor with an entire village attached to it. Other indications point to the existence of large manors of the high medieval type also in places like Baldringetorp, Herrestad, Hunnestad, Gundralöv, and Norra Vallösa (Skansjö 1987b, 1988, Skansjö et al. 1989).

The Late Middle Ages

During the latter half of the 14th century the manorial structure of the High Middle Ages dissolved in Denmark. A new estate system developed, characterized by, on the one hand, smaller demesne farms, and, on the other hand, an increasing number of almost equalized tenant farms. Thus, in Denmark and especially in Zealand, many *curiae villicales* were parcelled out to former *inquilini* who were thereby converted into tenant farmers. This process was in many respects parallel to the western European development. In the case of Denmark the dissolution of the high medieval large-scale farming system has been explained in terms of labour shortage due to the effects of the general crisis in the Late Middle Ages.

As another main feature in this period, members of the high nobility created larger estate complexes with many tenant holdings in villages scattered over the country. Within this system the revenues of the estate-owners consisted almost exclusively of dues paid by the tenants (above all land rent in money, corn, or butter). Labour duties were correspondingly of modest dimensions.

Yet another well-known late medieval phenomenon must be considered. In the fifteenth century the lesser nobility ("the gentry") were subjected to a drastic reduction in number due to an inability to fulfil the obligations of an aristocratic way of life. Thus, in many cases manors belonging to the gentry were transferred to the larger estate complexes created by members of the high nobility or by ecclesiastical institutions. This process is often referred to as "the crisis of the gentry" (C. A. Christensen 1964, Ulsig 1968, Dahlerup 1969–70, Gissel et al. 1981).

Many of these general features are reflected in the Ystad area in the Late Middle Ages. For instance, it seems most credible that the Bjäresjögård manor moved out of the village to a new site at lake Bjärsjöholmssjön in connection with a dissolution of an old estate system. It is also probable that a similar development took place in Bromma and in several other previously mentioned villages. The empirical evidence, however, is very fragmentary as we have no explicit record of the sizes of the late medieval lay demesne farms.

Those noble manors in the area which we know from this period were relatively numerous and scattered over the area (Fig. 5.9:1). Out of the 19 manors mentioned in written sources from the Late Middle Ages (1350–1536) only the following six survived into early modern times: Borrsjö, Hunnestad, Bjersjöholm, Gundralöv, Herrestad, and Krageholm. Mentioned in the Late Middle Ages but obviously disappearing as residential manors of the nobility were the manors in Katslösa, Sjörup, Norra Vallösa, Mossby, Baldringetorp, Bromma, Bussjö, Lilla Tvären, Lilla Köpinge, Fårarp, Svenstorp, Kabusa, and Öja.

Some examples may clarify the fact that most of these 19 manors during the Late Middle Ages were incorporated into larger estate complexes belonging either to the high nobility or to the wealthy church institutions. Katslösa manor was transferred to the property of Lund cathedral, and Baldringetorp became part of the possessions of the archbishopric. Sjörup and Mossby manors were integrated into the large and widely spread estates of the Thott and the Oxe families. In a similar way, Lilla Tvären was incorporated into the Bjersjöholm estate, and the Kabusa manor became part of the estate belonging to the Bing family.

In the Late Middle Ages the distribution of ownership in the Ystad area was clearly more diversified than was the case in the latter part of the 17th century (see below). Besides the noble estates, an important feature was the estate complexes belonging to the archbishopric in Lund. At the beginning of the 16th century there were five separate territorial complexes within the area. Apart from the Skårby estate, four of these were

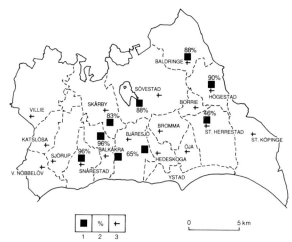

Fig. 5.9:2. Manors ca. 1650. 1. Manor owned by a nobleman.
2. Percentage day-labour tenants within the manorial parishes.
3. Parish church. After Jeppsson 1967, Skansjö 1987a.

administered from a special *curia principalis*, known as
a *skudgård*, which was a comparatively large holding for
a bailiff. Such farms were situated in the villages of
Snårestad, Baldringe, Högestad, and Stora Köpinge.
On average they had access to about 35 ha of arable
land, and they could harvest some eighty wagon-loads
of hay annually. These farms were about three to four
times as big as the ordinary village farms. The amount
of tenant farms belonging to the archbishopric in the
Ystad area can be estimated to represent about 30% of
all the farms in the area at the end of the medieval
period (Skansjö 1987b, 1988).

Estate building in Early Modern Times

In the course of the 16th and 17th centuries many con-
curring processes resulted in a more pronounced estate-
dominated cultural landscape in the Ystad area compared to the Late Middle Ages. The growth of large-
scale farming forms a distinctive and dynamic element
in the cultural landscape of the investigated area from
the 16th century onwards, and much of the later development can be traced back to this period. (For the
following see Weibull 1923, Olsen 1957, Jeppsson 1967,
Skansjö 1985, 1987a,b, 1988, 1991, Porsmose 1987.)

A process not directly visible in the external landscape, but of considerable consequence in the long run,
was a redistribution of ownership in favour of the nobility. This was a general phenomenon in Scania after the
Reformation in the 1530s. Here the large estates owned
by the Catholic church institutions were confiscated by
the Crown and thereafter, in most cases, were redistrib-
uted to representatives of the nobility. By the middle of
the 17th century the nobility owned 54% of about
14,000 farms in the whole of Scania. The trend in the
Ystad area was still more pronounced. By means of
estate exchanges with the Crown, the nobility in the

period 1575–1625 increased its share of tenant farms by
about 30% and ended up with not less than about 80%
of the total number of farms. Now the landowning
nobility also tried to assemble their tenant farmers in
close vicinity to the manors. This was closely connected
to the development of the nobility privileges and the
social organization of production (see below). On the
map (Fig. 5.9:2) we can observe the large quantity of
tenant farmers of the nobility in the manorial parishes.

As previously noted, by the middle of the 17th century only six of the many late medieval noble manors
still existed as such while three additional manors
(Baldringe, Högestad, and Snårestad) had their origin
in the estates previously belonging to the archbishopric
of Lund. Most of these nine manors around 1650 must
be regarded as relatively large production units. Compared to the average size of the peasant farms in the
area, the average manor had access to demesne land
(arable and meadows) 10–11 times as large. Marsvinsholm, Krageholm, and Herrestad, the largest estates,
had lands 16–21 times larger than a normal peasant farm
in the area. In their regional context, these constituted
large-scale farming units of a kind that was new compared to what was common in the Late Middle Ages.

Many of the external changes in the agrarian structure which took place during the 16th and 17th centuries
were connected with this new kind of property distribution and the process of estate building. The erection of
privately owned Renaissance castles (Bjersjöholm, Krageholm, Högestad, and Marsvinsholm) was of course a
striking feature in this context. A fair number of castles
of this kind were built throughout Scania and other
regions in Denmark in this period.

The villages situated close to the manors were affected in one way or another. A fairly mild transformation was to equalize entire villages owned by the nobility, so that every farm became equal with regard to
area, rents, and labour duties. This is known, for example, from the villages of Årsjö near Krageholm manor,
Snårestad at Snårestad manor, Gussnava near Hunnestad manor, Stora Tvären by Bjersjöholm manor, and
Baldringe and Baldringetorp by Baldringe manor.

The most radical change in this connection was the
entire eviction of a village which prevented the expansion of the manorial demesne. This is a phenomenon
well-known in Scania and Denmark west of the Sound
in the period 1500–1700. The most obvious example
from the Ystad area is found in the vicinity of the
Marsvinsholm estate. As previously noted, this manor is
known as a manorial knight's residence under the name
of Borrsjö since the 14th century. The village of Borrsjö
included twelve farms. Around 1600 Borrsjö came into
the hands of the Marsvin family. Otte Marsvin, a prefect of a royal territorial fief *(lensmand)* and one of the
great magnates of his time in southern Scania, had the
farmers evicted and the entire village of Borrsjö demolished. Its cultivated land was added to the demesne
farm. At this time work was begun on the castle edifice,

370

which was completed in the 1640s. In this way a new type of large-scale management was created, consisting of castle buildings, barns, cowhouses, arable fields and meadows, fishing waters and forests. It became the largest estate in the Ystad region with an area of about 500 ha. In addition, the landowner of Marsvinsholm owned and controlled the tenant farms in the neighbourhood, not only in the manorial parish but also in a couple of other neighbouring villages.

To explain phenomena like these, the perspective must be widened far beyond the research area, to Denmark in a national perspective and to northern Europe. In fact, the development reflects the role that our research area, like many other areas in Denmark, played in a wider European context.

An important structural innovation within the development of the agrarian economy in Europe was a change to larger cultivation units and a more market-oriented agricultural production in the period after 1500. This development is usually explained by the large increase in demand for cereals, in particular, because of the rapidly increasing populations in the towns and the growing armies. From the surplus-producing districts in eastern Europe cereals were imported to western Europe. It was in these eastern areas, above all, that the really large estates were created within a system known as "Gutsherrschaft" (cf. below). The interest in large-scale farming was of course closely connected with the favourable development of prices for agricultural products at the time of "the price revolution" from the beginning of the 16th century until the Thirty Years' War.

It is evident that Denmark, and thus also Scania, to a certain extent supplied the grain needed by the industrially developed and urbanized western Europe. Livestock was also exported (see Ch. 5.8). International price levels were substantially higher than prices at home. Denmark's exports yielded substantial amounts of capital, and the profits went primarily to the Crown and the nobility, who were in complete control of the means of production. Obviously, this development gave rise to the structural changes in Denmark in the direction of large-scale farming units, and it is instrumental in explaining the forming of estates and the building of castles on the local level in the Ystad area. But there were other dimensions involved as well.

The 16th and 17th centuries are known as a time of "nation building" in Europe, a period when decentralized administrative structures were replaced by a centralized state with an extensive bureaucracy and substantial military and economic resources. At this time the central power in Denmark was represented by a system of constitutional balance between the King and the *Rigsråd* (the Council of the Realm). This Council was a corporation within the nobility which controlled the King's financial, military, and legislative dispositions. The country was divided into a number of *len* (territorial fiefs) which were governed by prefects of

noble birth, *lensmænd*. They had been entrusted with extensive authority in civilian and military affairs on behalf of the central regime. Danish historiography even refers to this period as "the age of noble dominance" (1536–1660).

Within the nobility an oligarchy now appeared among those families with access to the highest positions in the country as members of the Council of the Realm or as *lensmænd*. This, in turn, implied considerable opportunities to earn private profits which could be used, for instance, to create large estates. The Council aristocracy could also promote and fully exploit the development of the nobility's privileges.

Out of several old and new privileges one important example must be noted. During the first decades of the 16th century, those tenants who lived within a manorial parish became exempt from taxes to the Crown, but instead they had to work on the landowner's estate. For the landed nobility this corvée (Sw. *hoveri*) meant a very favourable social organization of agrarian production. These day-labour tenants could also be used for other than agricultural work, such as transports and, in particular, building. Furthermore they had to pay land rent in kind or cash for their tenant holdings. These rents constituted a substantial source of income for the landowner along with the production of the demesne farms.

As a result of these different processes, an élite was established within the nobility, with unique positions in every respect: economically, politically, socially, and culturally. A physical manifestation of this unique position can be seen when old-fashioned or simple manors were replaced by Renaissance castles. As previously noted, in the Ystad area this was the case in Bjersjöholm, Krageholm, Marsvinsholm, and Högestad. It is evident that the large landowners and castle builders in the area all came from the exclusive group of aristocrats within the Council of the Realm, with close connections to the central government.

Manor management in the period 1658-ca. 1800
New lords on old manors

Scania came into Swedish hands in 1658. The Swedish government regarded the wealth of the Danish-Scanian nobility as a power factor which had to be curbed. Political and economic sanctions were applied to make it easy for Swedes to acquire Scanian estates (Rosén 1944, 1966). The profitable international grain trade, one of the chief sources of income for the manors, largely ceased (see Ch. 5.8). The economy of the manors declined, and the period 1658–1710 saw the manors gradually coming into Swedish hands. This trend is illustrated in figures showing the proportion of noble (*frälse* = tax-exempt) farms in Scania owned by Swedish families. In 1673 Swedes accounted for about 30%, in 1691 about 45%, and in 1708 about 55%. This devel-

Table 5.9:1. Normal harvests of *hartkorn* (rye and barley) on manors and their tenant farms and the relation between them in the hundreds of Ljunits and Herrestad, ca. 1700.

Manor complex	Manor grain (A)	Tenant farm grain (B)	Relation as a percentage (A/(A+B))
Estate complex 1*	1030	870	54
Estate complex 2**	730	540	57
Högestad	400	290	58
Estate complex 3***	370	580	39
Gundralöv	180	300	38
Rydsgård	100	150	40
totals	2810	2730	51

* Marsvinsholm, Hunnestad, and Snårestad
** Krageholm and Baldringe
*** Bjersjöholm and Herrestad

opment was even more noticeable in Ljunits and Herrestad hundreds, as is evident from a comparison of Swedish land ownership in the years 1691 and 1708 (approximate figures for Ljunits Hundred: 90% in 1691 and 95% in 1708; for Herrestad Hundred: 55% in 1691 and 95% in 1708). Around 1710, then, practically all of the noble land in the Ystad area was in Swedish hands (Fabricius 1958:247ff.).

The fundamental structure ca. 1700

We can obtain an idea of the income structure of the manors in the area at this time. As Table 5.9:1 shows, the value of grain production for the ten manors in the area was on average as great as the value of tenancy rentals, which were mostly paid in barley and oats. Large-scale management was more pronounced in estate complexes 1 and 2, and in Högestad manor, which was later incorporated into estate complex 2. It is obvious that in years with bad harvests the income from tenants was more important, in so far as the rents could be collected. Table 5.9:1 presents a picture of a manor structure occupying an intermediate position between the "Grundherrschaft" which was usual in western and southern Germany as well as France, and the "Gut-

sherrschaft" which dominated Germany east of the Elbe (Porskrog Rasmussen et al. 1987:9f.). An important income for the manors in Scania was also the breeding of beef oxen for export, but this activity generally declined in importance during the 18th century (see Ch. 5.8). The total grain production on the manors at the beginning of the 18th century can be estimated to be between 1700 and 2600 barrels of rye and barley and 900–1800 barrels of oats (Table 5.9:2). Rye and barley production was roughly equal to the annual consumption of some 700 people; a fivefold yield and one person's annual consumption equals about three barrels a year (495 l yr^{-1}; 1 barrel = 165 l). This tells us something about the significance of the manors' grain production, which appears mostly to have been shipped via Ystad to towns and regions in eastern Sweden (see Ch. 5.8.3 below). Work on the manors was carried out as corvée by peasants, so the manors, having little need to employ anyone, each had a staff of only 10–20 persons in total.

The 18th century: towards economic and social polarization

The picture presented here is based on a number of sources. A proper estate archive for this period survives only for the estate group of Krageholm-Baldringe-Högestad. Even here, however, the material is sporadic and incomplete for the 18th century. An unequivocal picture can nevertheless be painted. Characteristic of development in Sweden and Scania is a rise in grain prices and a fall in real wage levels for day-labour (Jörberg 1972). This development coincides with agrarian expansion – in the form of markedly increased grain production – on the manors investigated here. This trend is especially characteristic of the second half of the 18th century. A corresponding price trend along with increased grain cultivation was also seen in large parts of Europe at the same time (Slicher van Bath 1963:221ff.), and this period in France has even been described as an agrarian boom (Le Roy Ladurie 1979:201ff.).

At the beginning of the 18th century the majority of the manors (6 out of 10) were in or adjacent to villages.

Table 5.9:2. Grain production (in barrels) on manors in the Ystad area, ca. 1700. Estimates based on different seed returns.

Seed return Manor name	4.0 rye+barley	5.0 rye+barley	6.0 rye+barley	2.0 oats	3.0 oats	4.0 oats
Baldringe	145	182	218	48	72	96
Bjersjöholm	96	120	144	64	96	128
Gundralöv	96	120	144	80	120	160
Herrestad	120	150	180	68	102	136
Högestad	256	320	384	103	154	206
Hunnestad	92	115	138	48	72	96
Krageholm	288	360	432	200	300	400
Marsvinsholm	372	465	558	200	300	400
Rydsgård	76	95	114	10	15	20
Snårestad	176	220	264	60	90	120
Totals	1717	2146	2576	881	1322	1762

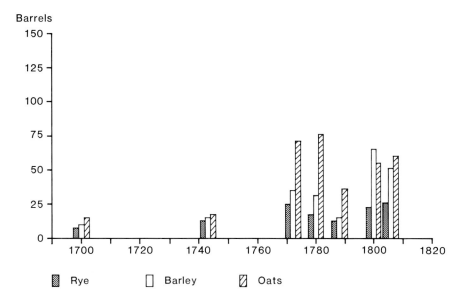

Fig. 5.9:3. Seed-corn on Rydsgård manor 1700–1806. Source: estate inventories from 1772, 1780, and 1806.

In these cases their fields were worked in collaboration with the village. Only the manors of Krageholm, Bjersjöholm, and Marsvinsholm had separate lands (Skansjö 1985:15). To these three can be added the manor of Rydsgård, established in the early 1680s. All the manors at this time operated a three-field system. Breakaways from the village community began in the 1760s, when Gundralöv, Baldringe, and Hunnestad manors were separated from their villages and a four-field system was introduced. In the period 1760–1800 there was a general reform of the cultivation system of the manors. Between 1760 and 1780 the four-field system became increasingly common, while crop rotation using seven or eight fields was applied between 1780 and 1800. From about 1800 we obtain the first reports of clover cultivation. This transformation of estate man-

agement also occurred in Denmark, although a little earlier (Porskrog Rasmussen et al. 1987:25f.).

The manors increased their grain production during the 18th century. Similar trends were revealed by an examination of the four manors in the focal areas, although conditions on Gundralöv manor in Bjäresjö parish cannot be described accurately for the whole of the 18th century, in the absence of sources giving exact land measurements. The greatest change is observed on Rydsgård manor in the Romele area. The manor of Rydsgård extended its lands from some 50 ha early in the 18th century to about 220 ha in 1788. In 1700 an estimated 5–15 barrels each of rye, barley, and oats were sown, compared with 1800, when the figures were 20–25 barrels of rye, 50–65 barrels of barley, and 55–60 barrels of oats (Fig. 5.9:3). It should be noted that the

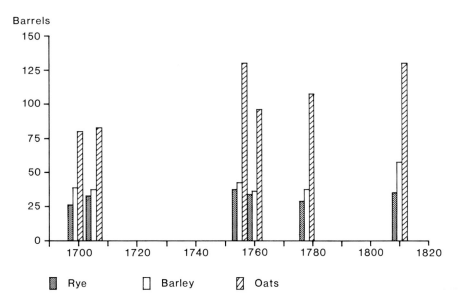

Fig. 5.9:4. Seed-corn on Krageholm manor 1699–1810. Source: farm accounts and statistics.

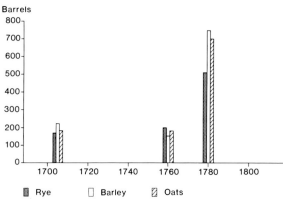

Barrels

Fig. 5.9:5. Seed-corn on Krageholm manor 1705–1760–1780.
Source: farm accounts and statistics.

figures for 1743 and 1788 are minimum values reported for a special state tax. Developments on the other manors were similar, with a particular expansion of oat production at the end of the 18th century. Oats were cultivated on poor, unmanured lands in peripheral locations. The change thus reflects the last stage of the cultivation process. At Bjersjöholm the period 1700–1789 saw an increase in the acreages of rye and barley by about 60% and of oats by 85%. The total area of arable land rose from 75 ha to 220 ha. This expansion occurred within the 270 ha of land owned by the manor itself.

The corresponding figures for changes between 1700 and 1764 in rye and barley cultivation on Krageholm manor were about 35% and 50% respectively, and here too the expansion mostly took place within the lands owned by the manor itself. Krageholm owned some 600 ha, of which the arable land in the period expanded from about 110 to 155 ha. The manor's seed-corn, based on farm accounts and statistics, is shown in Fig. 5.9:4. It is clear that these figures do not convey the same picture of the expansion of cereal fields. The explanation for this is that there was a gradual introduction of a variety of rye which could be sown half as densely as the rye previously in general use. The new rye required a great deal of manuring, but this was repaid by a seed return of 16–20 (Weibull 1923). The values for seed-corn for this rye thus correspond to twice the area of arable land required for ordinary rye. The introduction of the new rye also necessitated the setting up of a new field. As Fig. 5.9:5 shows, harvests on the Krageholm estate increased by 150–400% after 1764, when a new field system was introduced. The increase was chiefly due to a rise in seed return from 4–6 to 15–20 for rye and barley, and a rise for oats from 2 to about 6. The new cultivation techniques at the end of the 18th century thus brought about a significant increase in productivity on the manors in the project area (although for the focal areas there are measurable values only for Krageholm manor). With its extensive outland and pasture and the consequently relatively large-scale stock-breeding in the 18th century, Krageholm was in an ideal situation.

Did increased grain production also mean increased income for the manors? The question cannot be answered without a brief discussion of price formation and possible channels of sale for Scanian grain in the 18th century. During this period Sweden was dependent on imported grain. Cereal price formation in Sweden changed during the 18th century, as great regional differences gave way to more uniform pricing – which should be interpreted as evidence of a more integrated grain market in Sweden at the end of the 18th century (Jörberg 1972). The differences between grain prices in Malmöhus County and the prices paid for Scanian grain in the most usual Swedish markets (Stockholm, Blekinge, and Göteborg and Bohus counties) decreased gradually in the course of the period 1735–1804. This applied in particular to the price of rye and barley. The lower profit level for the grain from the manors at the end of the 18th century was compensated, however, by the fact that the estates at that time were using considerably larger areas for grain production, lands which in addition yielded an estimated 150–300% more grain per unit of square measure compared with harvests before the middle of the 18th century (Persson unpubl.).

The social implications

Everything suggests that the 18th century saw a widening of social divisions in the manorial parishes between well-to-do peasant farmers and cotters, and especially between these two groups and the landed nobility on the manors. The expansion and intensification of operations on the manors of Gundralöv and Rydsgård meant that peasants' lands were swallowed by the manors. In the village of Gundralöv at the start of the 1760s the farms of two estate tenants were demolished when the lands of the manor were broken away from the village and the lands of the two farms were incorporated into the estate. The latter then owned about 155 ha of land, of which 95 ha were defined as arable. In the 18th century it was primarily cottages that increased in number, whereas the number of farms grew only slightly (see Ch. 4.4.7.1). In this setting, with equalized farms able to support relatively prosperous estate tenants, and with poor cotters doing day-labour, the manors came increasingly to appear as the seats of the nobility that they were. Productivity on the estates rose noticeably in the latter half of the 18th century, reaching a level that was not equalled by the peasant farms in the area until the second half of the 19th century (Germundsson and Persson 1987, Persson 1987a). The structure of the estates' economies has not been studied in detail for the latter part of the period under investigation. However, the improved productivity on the manors, combined with their expanded acreage in the late 18th century, brought about a reinforcement of an economy of the "Grundherrschaft" type.

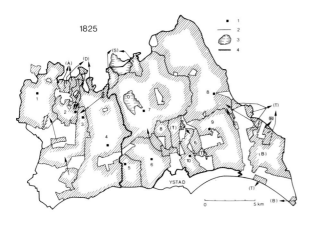

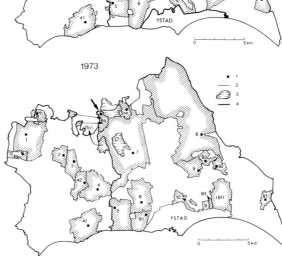

Fig. 5.9:6. Large landed estates in the Ystad region in 1825, 1914, and 1973. Boundaries of estate areas are shaded. The numbers and letters correspond to the different estates (see text).

The blank areas are either freehold or owned by the Crown.
1 = Manor farm, 2 = Estate boundary, 3 = Lake, 4 = Boundary of Ystad region.
From Möller 1983, 1985.

Landholding, population, and the landed estates after 1800

Changes in landholding after 1800

As pointed out above, the Ystad region at the beginning of the 19th century was totally dominated by the landed nobility. Fig. 5.9:6 shows the Ystad region in 1825, 1914, and 1973. The landed estates – i.e. both the demesne land and the land cultivated by tenant farmers – are indicated by shaded areas, and squares mark the manors. Over 80% of the land belongs to the nobility, divided between ten estates, the largest being Marsvinsholm (no. 4 in the figure), Krageholm (7), and Högestad (8). A few estates outside the area also own small parts of land (areas indicated by letters).

Most of the land was cultivated by tenant farmers, who, besides running their own farms, had labour duties on the manor farm. The *enskifte* reform was introduced in 1803, and as early as 1825 most villages in the southern plains were enclosed. An act of 1789 had made it possible for farmers to buy land owned by the nobility, but up to 1825 this had only happened in a few areas in the Ystad region (Möller 1985, 1989).

Half a century later, land owned by the nobility had been severely reduced. Two regions at this time were beginning to emerge: an eastern part still dominated by estates, and a western part where freeholders prevailed. Many of these used to belong to the nobility (cf. above

for the act of 1789). This reduction of the large estates continued and in 1914 quite a number of landed estates had been divided into two or three separate units, and a great deal of land had also been sold off to smallholders. These new domains were often established much earlier as large tenant farms during the second half of the 19th century. Rydsgård (1), Marsvinsholm (4), and Stora Herrestad (9) had all been divided in this way. Marsvinsholm, for instance, had suffered a reduction from around 4000 ha in 1825 to a mere 400 in 1914.

The development during the 20th century can be deduced from the same figure. The reduction begun during the 19th century has continued, but is now slower. The estate of Krageholm (7) has been halved since 1914 (see also Ch. 4.3.9.1), and other reductions have occurred at Öja (10), Carlsfält (92), Beden (11), and Villiegården (12), while Fredriksberg (91) has become a part of the military training ground nearby. Some land was sold out for division into groups of state-subsidized *egnahem* (owner-occupied small farms), especially between 1904 and 1939 (see Ch. 5.12).

The sale of land reveals the 19th century as a period of transition in Swedish agriculture. Population was in an increasing phase; during the century the Swedish population more than doubled. But not only the potential home market grew. When the English Corn Laws were repealed in 1846 the Swedish export market grew considerably. At this point, the southern Swedish es-

Table 5.9:3. Large tenant farms in the Ystad region in the 19th century.

ESTATE Tenant farm	Number of abandoned holdings	Year of creation
RYDSGÅRD		
Vidarp	3	1862-73
Villiegården	7	1856-58
Svenshög	4	1873
Bjergsgård	2	1862
MARSVINSHOLM		
Erikslund	8	1832
Gussnavagården	3	1880s
RYNGE		
Carsborg	9	1837
BJERSJÖHOLM		
Källesjö	8	?
HÖGESTAD		
Högestad Nygård	5	1877
Ljungagården	7	1875-83
Borriegården	6	1869
STORA HERRESTAD		
Fredriksberg	6	(1796) 1822
Carlsfält	6	1822-27
Robertsdal	3	1822-27
Jennyhill	8	1866-75
BOLLERUP		
Köpingsberg	6	1830s
TOTAL	91	

Source: Returns from parish catechetical meetings.

tates started to transport grain in large amounts to the ports, from where it was shipped to England (Fridlizius 1957).

During the 1880s the estates gradually moved towards animal husbandry. The growing number of people living in towns brought a higher demand for animal products. The conversion to animal husbandry was also caused by competition from the USA and Russia, who were able to produce cheap grain. Another factor was that domestic grain prices rose in relation to the international level. Swedish oats went to the growing number of cattle, while barley was used in the breweries (Fridlizius 1985). From the 1880s butter became the main export product from Sweden. Swedish butter was shipped in large quantities to England and Denmark. Technical innovations made this possible, among them the railway and the Laval separator. The railway from Ystad to the port of Malmö was inaugurated in 1874; the line was to a large extent financed by the landowners in the region (Möller 1987).

The large investments affected many sectors: not only new communications but also land reclamation, drainage, tools, and implements. To finance these investments the landlords needed capital. One way of getting this was to sell off land. At the same time as money became available, the estates could in this way get rid of marginal land. Rationalization thus meant smaller

acreage but more intensive agriculture. The division of an inheritance probably also caused reductions in size: when there were many heirs, the one taking over the estate had to buy the others out, for which he needed capital.

This selling of land was thus part of a drive for higher efficiency. Some landlords went one step further by reorganizing their domains. Farmers were replaced by agricultural workers and cotters, and on some estates entire villages were demolished and a new type of large tenant farms up to 200 ha – known as *plattgårdar* – were built in their place (Möller 1986). In some cases, the land previously farmed by smallholders was put under the manor farm itself, and rows of houses for the workers erected nearby. From the landlord's point of view the new system was ideal. His land was efficiently farmed, at the same time as labour became available for the rapidly increasing work on the manor. For the former tenants, however, it was more or less disastrous, as they now either had to move, or give up farming altogether and become labourers.

The creation of *plattgårdar* in Scania probably had its origin in Great Britain. This form of transition to capitalistic farming thus involves large tenant farms where the tenant (or rather: steward) in turn uses wage labour. Similar farms had already been created in England at the beginning of the 18th century (Mingay 1962). The ideas spread to southern Sweden at the beginning of the 19th century, the first examples being on the estate of Engeltofta in north-western Scania

Table 5.9:3 reveals that 17 *plattgårdar* were organized in the 19th century, replacing 91 smaller holdings. This totally remodelled the agricultural landscape of the region. Fig. 5.10:6 gives an example of such a transformation. The village of Gussnava which can be seen on the map from 1812 has totally vanished a hundred years later. No sign of the farmhouses can be seen, the only buildings being the large farms of Erikslund and Gussnavagården which have replaced the old village.

This process seems to have been general in Scania. During the 19th century 143 new large tenant farms were erected in Malmöhus County alone (Möller 1986). They were, not surprisingly, mostly located in the estate-dominated area along the border with Kristianstad County. Not all of these replaced villages or hamlets; 58 villages had been abandoned, and the land of about one-third of these was now cultivated by the manor farms directly. Fridlizius (1979) found that between 1805 and 1865 there was an increase in the share of manorial land in the total area owned by the estates in Scania from 23.8% to 41.8%.

Both the selling of land and the creation of large tenant farms must be seen as methods for the landlords to create modern estates which could produce large quantities of grain and animal products for the expanding market, both overseas and at home.

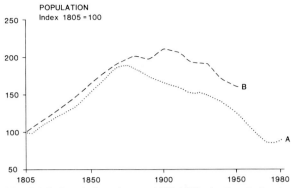

POPULATION
Index 1805 = 100

Fig. 5.9:7. Population changes 1805–1980. A=Population in the investigated area, B=Population in rural districts of Malmöhus County. After Germundsson 1986.

Changes in human population after 1800

One of the most crucial factors of change when studying landscape and settlement is population development. In the general Swedish population development, the period between 1810 and 1880 is identified as a period of strong and steady population increase. In the next phase, 1880–1935, the increase was slower as a result of lower birth rates, while the period after 1935 is characterized by a very small population increase (Hofsten 1986). There are of course regional variations in this pattern, and one factor that has been proved fruitful to analyse in connection with population changes is the difference in landholding (cf. Jaensson 1966, Fridlizius 1979, Germundsson 1987). In the following section the population changes within the estate-dominated Ystad area are discussed in connection with the landholding situation.

In 1805 the Ystad area (outside the town) had around 8700 inhabitants. Five years later a considerable growth in population started, which lasted until about 1870. By then the population had almost doubled (about 16,400), reaching its maximum. During the 1870s, a downward trend set in, which – with the exception of a lull during the first half of the 1920s – lasted unbroken until 1975. Between 1975 and 1980 the population grew slightly, and the number of people living in the project area at the latter point of time was about 7700.

During the growth period up until 1870 the population in the project area increased at about the same rate as in Malmöhus County as a whole. The population decrease in the project area began after 1870, whereas the population in rural areas in Malmöhus County continued to increase until around 1900 (Fig. 5.9:7).

Between 1810 and 1850 the population increased in all the parishes in the Ystad area. A few of them experienced a momentary decline, but none reached its population maximum during this period. Then, around 1850, some of the parishes passed a turning-point and

started the long decline, which in most cases was to last up to the present day. However, the majority of the parishes passed the turning-point during the 1870s. Fig. 5.9:8 shows the relative changes during the years 1810–1850 and 1850–1875. Among other things, it can be seen that the mean point for the whole project area is a result of quite diverse development in the different parishes. In relation to landholding and enclosure the following might be concluded:

One parish *dominated by freeholders* (Nöbbelöv), with early implementation of *enskifte* enclosures, had a population development quite similar to the total figures for the whole project area, and thus to the whole region (Malmöhus County). Early enclosures of this kind were very common in freehold-dominated areas in the plains of southern Scania, and they were generally followed by a strong population growth because of in-migration and high birth rates (Fridlizius 1979).

Several of the *estate-dominated* parishes (Bjäresjö, Stora Herrestad, Hedeskoga, Borrie, Sövestad) had a slower population development with relatively lower population maxima. This is a trend which has been shown to be general for the whole of Malmöhus County (Jaensson 1966), but it is also important to note that some parishes under noble ownership (Öja, Bromma) experienced a strong population increase as a result of *radical enclosure acts* initiated by estate-owners. As there was only one landowner involved in the layout of the cultivation units, these enclosures could be very effectively executed. Here land was generally sold off to farmers *after* the population increase began.

The parishes with a large proportion of *land shifting from estate to freeholders* (Snårestad, Södra Villie, Katslösa, Sjörup, Stora Köpinge) often had quite a strong population increase in connection with the shift in land ownership. Snårestad, for example, is the parish where the highest proportion of land was sold to freeholders during the first half of the 19th century.

In one extreme case (Sjörup), a landowner dividing a domain into *many small holdings* set in motion the fastest population growth in the whole area during the period in question. In 1853 one landowner, possessing a great deal of land in two of the villages in Sjörup, requested a *laga skifte* enclosure. In this process his domain was split up into a great number of smallholdings, obviously with the aim of selling them off. Hence, the action of one man created the opportunity for a "mass" migration into the parish, and the process totally remodelled the landscape; many new houses, gardens, roads, fences, etc. were created, and the arable land was divided into small units. This also demonstrates that there was great pressure for land, and that many people needed a plot to settle on.

Another aspect of importance to landscape change during this period is the increasing number of landowners, as a result of the division of farms in connection with population growth and to the selling of estate land to owner-occupiers. This meant an increasing number

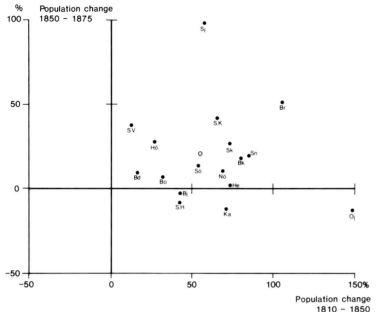

Fig. 5.9:8. Population changes of the parishes in the investigated area, 1810–1850 and 1850–1875. Unfilled circle=mean point for the whole area. After Germundsson 1986. Abbreviations for parishes:
V=Villie
Sj=Sjörup
Sn=Snårestad
Br=Bromma
Bo=Borrie
S.H.=Stora Herrestad
Ka=Katslösa
Sk=Skårby
Sö=Sövestad
He=Hedeskoga
Bd=Baldringe
S.K.=Stora Köpinge
Nö=Nöbbelöv
Bk=Balkåkra
Bj=Bjäresjö
Öj=Öja
Hö=Högestad

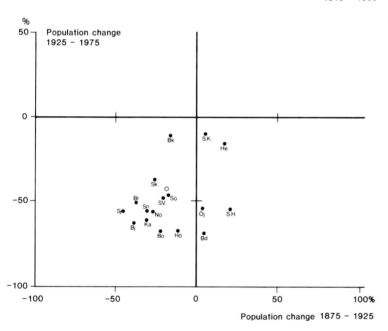

Fig. 5.9:9. Population changes of the parishes in the investigated area, 1875–1925 and 1925–1975. Unfilled circle=mean point for the whole area. After Germundsson 1986 Abbreviations for parishes: see Fig. 5.9:8.

of decision-makers in the landscape, something that created conditions for a more diversified development.

It could also be noted that the four parishes farthest to the left in Fig. 5.9:8 are all situated in the inner hummocky zone of the project area. This indicates that the more marginal land did not experience the same rate of intensification of use as the rest of the area during the first part of the growth period.

As regards the era of decrease, Fig. 5.9:9 shows the development divided into two periods. While five of the parishes had a modest population growth during the first period (1875–1925), all of them experienced a pop-

ulation decline between 1925 and 1975. In absolute terms, most parishes fell to the 1810 population number during the 1950s, and sank below this level thereafter.

If we consider the whole period 1810–1975, it is a striking feature that almost all parishes went through a population decline roughly equivalent to the previous growth. This shows that the new social and geographical structure built up in areas with great population growth was of quite short duration. In times of industrialization and agricultural rationalization, when migration to cities and towns and overseas emigration began, the parishes which were most densely inhabited and with far-reaching subdivision of land did not offer any "positive" alternative to other parishes. The percentage of population which left all parishes after the population maximum was almost the same. The only exception is Stora Köpinge, where an early agricultural industry (sugar-beet) attracted people and settlement from the beginning of this century. In the two other parishes with little decline between 1925 and 1975, Balkåkra and Hedeskoga, the reason is a recent growth of commuters' dwellings in these parishes.

There is no doubt that landholding conditions are important to the question of population development. In the Ystad region it has been shown that areas dominated by landed estates experienced a weaker population development and generally reached their population maximum earlier than areas dominated by freeholders. The reason for this is found to be primarily a question of landholding; in estate-dominated areas, decisions taken by the estate-owner to modernize agriculture did not allow a population increase of the same degree as in areas dominated by freeholders. At the same time it is important to note the exceptions. Sometimes radical land reforms initiated by progressive estate-owners led to considerable in-migration into estate-dominated land in the early 19th century.

Conclusions

The evolution of the landed estates in the Ystad region has not been one of unbroken continuity. Size and structure have varied greatly over time. There were already large farms during the High Middle Ages (ca. 1200–1350), resembling feudal estates on the continent. The estates in the region were often organized as demesne farms with entire villages with smallholdings attached to them.

During the Late Middle Ages (ca. 1350–1536) the demesne farms became smaller and an increasing number of almost equalized farms can be noted. During this period, too, large estate complexes were formed, with tenants paying rents mostly in money or in kind. The complexes were owned either by the nobility or by different church institutions.

In Early Modern Times (1536–1658) the estates began to grow again, which in some respects meant a return to a system resembling that of the High Middle Ages. After the Reformation, the state confiscated church-owned estates, and these subsequently came into the hands of the nobility. The estates gathered together their scattered lands, in order to utilize the privilege of tax exemption for tenants living within the parish; tenants instead had corvée obligations *(hoveri)*. The estates exported grain to western Europe.

In 1658 Scania became Swedish, and this also meant successively more Swedish landowners. Production rose during the 18th century and grain was exported to other parts of Sweden. The corvée system prevailed, which meant small staffs on the manors. The social gap between nobility and peasantry widened as peasant land was put under the manors.

Manorial land diminished in size during the 19th century as the estates sold out large parts of their domains. On the remaining lands, agriculture was intensified with modern techniques. The nobility invested large sums in buildings, machines, and transport facilities. A new organization of labour was introduced which meant that landless labourers replaced tenant farmers and large capitalistic farms were erected. A marked population expansion can be noted from the second half of the 18th century up to the 1880s. In the estate-dominated parishes the landlords could control development, so population growth was slower here.

Note: The authors of the individual sections are Sten Skansjö (From the High Middle Ages to Early Modern Times), Anders Persson (Manor management in the period 1658 – ca. 1800), and Tomas Germundsson and Jens Möller (Landholding, population, and the landed estates after 1800).

5.10 The age of enclosure and improvement in agriculture

Nils Lewan

Introduction

The 19th century stands out, in Sweden as elsewhere in western Europe, as a period of profound change, not only in agriculture but also in the countryside in general. Compared with England or France, however, agricultural change in Sweden was considerably later, the 19th century being the first real period of change (Slicher van Bath 1963, Emanuelsson and Möller 1990). Some of the changes, of importance to the landscape and characteristic of the Ystad region and other parts of southernmost Sweden, will be dealt with below. Interest will concentrate upon enclosure, which was a kind of starting-point for further changes, a number of these changes, and one specific example.

Enclosure

Enclosure or land amalgamation became of great importance from the very beginning of the 19th century. It was, however, preceded by royal decrees from 1749 onwards. These stated that the huge number of land strips of each farm in a village should be reduced to just a few, if the owners asked for a land reorganization. This was a pure amalgamation of land and included no moving of farmsteads. It was called the *storskifte*.

In parts of Sweden, including Scania, this effort came to almost nothing (cf. Dahl 1941, Helmfrid 1961). Other attempts were therefore made, some of them on large landed estates. One such attempt was made by Baron Rutger Macklean, owner of Svaneholm and Skurup with its four tenant villages, immediately west of the Ystad region. Beginning in 1783 Baron Macklean had the open-field system reorganized into some seventy almost square separate pieces of land, each with a farmstead in the middle. The former villages were reduced to places where farm-hands and artisans lived, and within a few years the rural landscape was almost totally changed. Many a tenant farmer opposed the new order. Its efficiency, however, soon became obvious, also helped by Baron Macklean's other measures, viz. new agricultural practices, the organizing of schools, rents to be paid in money instead of by work, and so on. The example set by Macklean was followed by other noblemen, among them Eric Ruuth, the owner of Bussjö. In 1803 a royal decree followed, proclaiming *enskifte* in Scania. This was meant to be a "full enclosure" on the Macklean pattern, and a few years later the decree was extended to the rest of Sweden.

The general causes behind the reform have been discussed elsewhere (Ingers 1948, Helmfrid 1961, Hoppe 1981). As a political radical, Macklean might also have had social reasons, but obviously he was interested in a better economic return. He was well-read in current European agricultural literature, and he had also travelled a great deal. He should have been well aware of the Danish land reforms, originating some years earlier, although not put into general practice until the 1780s. The need to shorten distances between farm and field, and through this and other measures to gain more time for farming, might have been as important to Macklean as the idea of making farmers more independent. Another cause behind the land reform was the obvious need to feed an increasing population. Still another cause, at least according to some people, should have been the insight that the age-old open-field system was in a kind of crisis, arable land having been extended so much that grassland and thus also manure were in short supply (Emanuelsson et al. 1985, Emanuelsson and Möller 1990).

The accomplishment of this land reform depended upon applications from freehold farmers and other landowners, and enclosure took many years to complete. Relatively big villages, a shortage of outland and woodland, an open country, and interested owners helped to speed up the process in much of Scania. Especially in the south and west of the province, most villages implemented the reform within two decades, while much of Sweden lagged far behind, as did parts of the Ystad region (cf. Dahl 1941).

The process of enclosure in this area is interesting from several points of view. It took remarkably many years to complete (Fig. 5.10:1), and the big landowners played an important role. Partly for these reasons the resulting pattern varied between different parts of the region, which has also helped to influence further development. The dominance of a few estate-owners also means that local land reforms might have been implemented without formal registration, thus also being outside our knowledge. As has been discussed earlier, in several cases normal *enskifte* was replaced by the establishment of large tenant farms or *plattgårdar* (cf. Ch. 5.9). From Ch. 4.3.9.1 we also know that inside a single estate one land reform could be replaced by another during the 19th century, this being a sign of new ideas put into practice.

It should also be stressed, however, that neither estate-owners nor ordinary freehold farmers constituted a

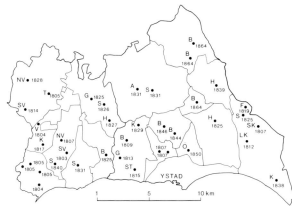

Fig. 5.10:1. Years of enclosure in the Ystad region for different villages and hamlets. In some cases the whole settlement was not enclosed at the same time. Source: LMV archive (National Land Survey Malmöhus County).

homogeneous group with regard to enclosure. We know from the Ystad region that some big landowners were pioneers, like Baron Macklean, with extensive knowledge of modern farming. They endeavoured to reorganize their tenant farming, while others preferred existing practices. The same could be said about common farmers (cf. Dahl 1942, Pred 1986; see also Lewan 1983). From the village of Norra Vallösa there is an example where the *enskifte* was requested by some of the freehold farmers, while the nobleman owning a

number of tenant farms in the village preferred to stick to the old order (Germundsson 1987). The pattern of enclosure in the Ystad region through time followed common rules, already noted by Fridlizius (1979); after pioneering work on some of the estates, villages of freehold farmers became the scenes of early enclosure, while other estates lagged far behind.

In the Ystad area, as elsewhere, two principal systems of enclosure may be discerned (Fig. 5.10:2). The causes of this should not be looked upon as being the whim of the landowner but rather current ideas of efficiency, the opinions of the farmers themselves, and the skill of the land surveyor. The resulting patterns differ, as do the possibilities for future change: such as further subdivision or other kinds of development.

The importance of enclosure was, of course, very much the same as elsewhere, making the single farmer more independent, giving possibilities for new crop systems and other farming practices. As elsewhere, enclosure also became an important prerequisite for other kinds of modernization, such as under-drainage and the use of agricultural machinery. Still another possibility opened by enclosure in the tenant villages of the Ystad region became the selling of tenant farms and other estate-owned land to freehold; without enclosure this would have been almost impossible. Thus even in this respect enclosure was the beginning of important social change.

Summing up, the *enskifte* or enclosure as such meant

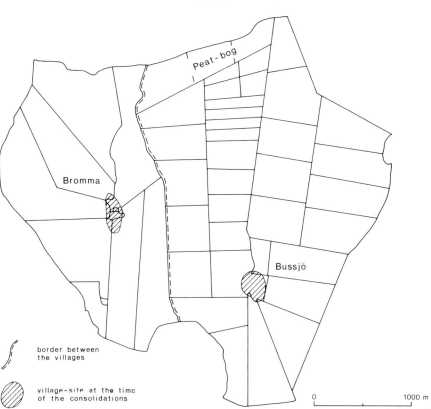

Fig. 5.10:2. The parish of Bromma in 1850 with its two villages. As a result of different consolidation techniques, almost all farmsteads were able to remain at the old village site of Bromma, while most farms had to move out in the case of Bussjö. The checkerboard pattern was very common in this region (cf. Fig. 4.3:8 for villages belonging to Krageholm). Although common elsewhere the star-shaped pattern was used only rarely in the Ystad region, e.g. for the villages of Sjörup and Bjäresjö. From Germundsson 1987.

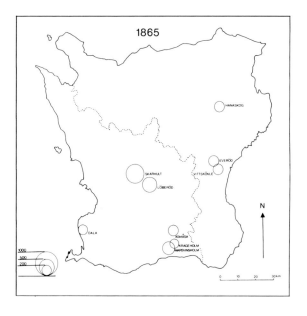

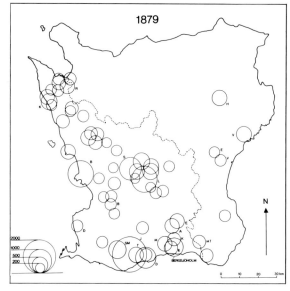

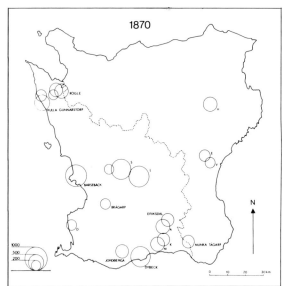

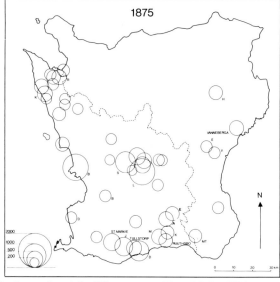

Fig. 5.10:3. Drainage work in Scania including at least 200 ha of land at each unit in different years. From Möller 1984.

immediate and far-reaching changes to people, settlement, and landscape. At the same time it opened essential possibilities for further change. These, however, depended very much upon when and by whom enclosure was accomplished. Some further changes, of importance also to the Ystad region, are discussed below.

A time gap

In most of Scania, including the Ystad region, *enskifte* was implemented well before 1850. This raises the question of whether further changes in agriculture affected the landscape. According to most authors, the middle of the century was a period of transition for Swedish agri-

culture. Before that, any required increase in production was brought about mainly by land reclamation and through the modernization of the cropping system. Later on, other measures played a more important role, among them drainage, marling, and the flooding of meadows. Gradually new farming implements also became more significant.

All this happened before artificial fertilizers were introduced to any extent. A considerable number of measures were adopted in order to reduce the problems arising from the widespread increase of arable land at the expense of meadows, which meant shortage of nutrients. Thus most of the 19th century may be looked upon as a time gap between the old, rather outmoded

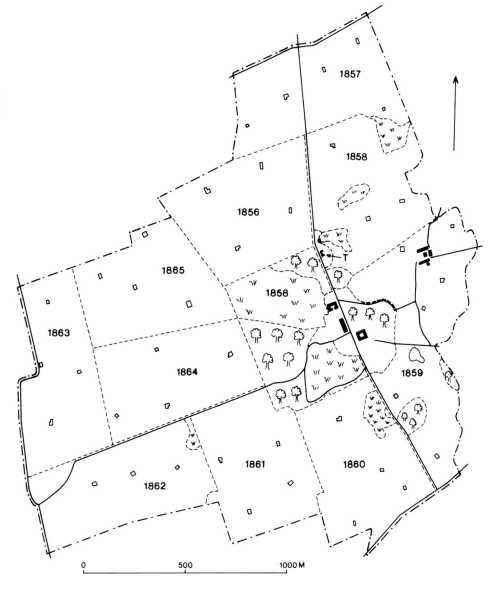

Fig. 5.10:4. Under-drainage and marling at the estate of Charlottenlund, according to maps still kept at the estate. Each year over a ten-year period an area of some 30 ha was under-drained, while at the same time marl was spread from a few pits. From Möller 1984.

three-field system and modern, "industrial" agriculture (cf. Emanuelsson and Möller 1990). As enclosure was relatively early in Scania, where in many areas land other than arable was in short supply, and since enclosure was followed by a rapid increase in population, the time gap became more pronounced here than elsewhere in Sweden. How it was overcome is discussed by Emanuelsson and Möller (1990). Their paper and other sources contribute to the following brief account, covering some of the measures taken to overcome the shortage of grassland and its effects.

Early post-enclosure measures included, as has already been noted, a great increase in the ploughland acreage at the expense of both hay-producing meadows on the infields and grazed commons. Through this, the formerly immobilized nitrogen below the sward was

made available over some years. The turning of the soil by ploughing also made potassium and phosphorus available, although the period of full use of this practice might have differed in length.

Other practices are found as well. Through paring and burning the sward was made loose, dried and burned, and the ashes spread over the bare soil. In Scania this method was much more common than slash and burn, which in the province was mostly applied in the north-east.

Towards the middle of the century, leguminous crops were included as fodder plants, and still later marling was introduced. The marling period covered only some 25 years from about 1860, and the method was widely used in the Ystad region also. As the pH is raised through this use of very calcareous soil, forcing out important nutrients, the method brings advantages only in the short run (Emanuelsson and Möller 1990). As the principal period of marling covered a couple of decades up until the 1880s, it helped to overcome the nutrient problems until artificial fertilizers became more common.

Parallel in time with marling we find the flooding of meadows becoming an important practice. It is known from different parts of Europe and from a long period. In Sweden it was practised both in the north and in the south, although the methods differed.

In Scania it appeared in a number of variants, but the purpose was always the same: to increase the production of grass, mainly for winter fodder but also for grazing. The increase of production was quite substantial, often several times former amounts. The practice, however, also took a good deal of work to get going, and it seems to have been introduced mainly by estate-owners. In southern and western Scania it was in use from around 1850 onwards. The end of the century also saw the end of most systems for the flooding of meadows, as more labour-efficient methods became available. Thus, once again we find a practice being introduced by educated major landowners and only taken into consideration when other less laborious methods were exhausted.

Drainage

During the 19th century covered drains gradually succeeded the old custom of furrow-draining, and the landscape was made much drier than before. Under-drainage has been known in England since the 17th century, with the use of almost the same methods as the Romans, and more widely introduced during the early 18th century (cf. Darby 1964, Phillips 1969). It took another hundred years for this innovation to reach Sweden.

In Scania, it is quite obvious that the big estates played an important role in the history of under-drainage, as reported by Zachrison (1922) and Möller (1984). C. G. Stjernswärd used covered drains at his estate of Engeltofta in north-west Scania, in the very first years of the 19th century, but the practice of underdrainage became widespread only after the middle of the century. By then English machines for the production of tile pipes had been introduced, and they were soon widely used, especially by estate-owners. After a slow beginning, the period 1870–1884 saw some 25% of the total area, or about half the acreage of arable land, being under-drained.

A good picture of the process of drainage in Scania is given in four maps showing places where at least 200 ha have been under-drained (Möller 1984; cf. Fig. 5.10:3). A number of estates may be seen as being innovators, three of them in the region of Ystad, and as late as 1879 the grouping around innovation estates is still visible. In addition to the importance of major landowners comes the role of physical circumstances, drainage of course being most useful on flat land and where soils are heavy. Thus it is only natural that drainage was at first of minor importance in the northern and eastern parts of the province. It is also probable that the south-western freehold farmer part of the province also underwent early draining, although not being noted in these maps, which cover only major operations.

According to Möller (1984) it is obvious that the Ystad region was an area of early drainage. Most of the big estates had their own brick and tile works, and by 1875 drainage is reported to be almost complete on the manor farms, by then also being extended to the land of tenants and freehold farmers. Under-drainage of an estate was usually planned over a period of some years, often with the help of the county drainage engineer. To save labour, the work was often combined with marling, as was the case at the estate of Charlottenlund to the west of Ystad, where both operations were accomplished in parallel over ten years (Fig. 5.10:4). The map also helps to give an idea of the dense and even pattern of marl-pits, with something like one pit for every 8 ha.

Farming implements

It has not been possible to investigate in detail the introduction of new farming implements in the area, but a comparative study has been made of two parishes, covering the period from 1839 to 1900 (cf. Germundsson and Möller 1987).

Using estate inventories, this study has aimed at a comparison of the two westernmost parishes of the Ystad region, namely Villie, which during the period in question was still dominated by the manor of Rydsgård with many small tenant farms, and Nöbbelöv, where freehold farmers prevailed. Although widely used in Sweden as sources in research, estate inventories have their shortcomings. Small or poor estates were not always inventoried, and the inventories might give a more or less antiquated picture of current assets, in this case farming implements. One should also be aware of changes in what would be considered important enough to mention in an inventory. Still, as most inventories are preserved and accessible, they can give an idea about changes over time in property and implements.

Up to around 1850, there seem to have been no real changes in farming implements, the wooden plough and the harrow still being the most important. The "English wheel plough" had, however, been introduced by C. G. Stjernswärd at Engeltofta, helped by imported Scottish makers, and by 1840 it was not uncommon in north-western Scania, i.e., close to Engeltofta. According to available sources, the Rydsgård estate seems to have been in the forefront with modern implements, but at the same time the farmers of Nöbbelöv could have been earlier still to accept iron instead of wooden tools compared with the tenant farmers of Villie. According to Germundsson and Möller (1987), one reason for this might have been that they were freehold farmers. An-

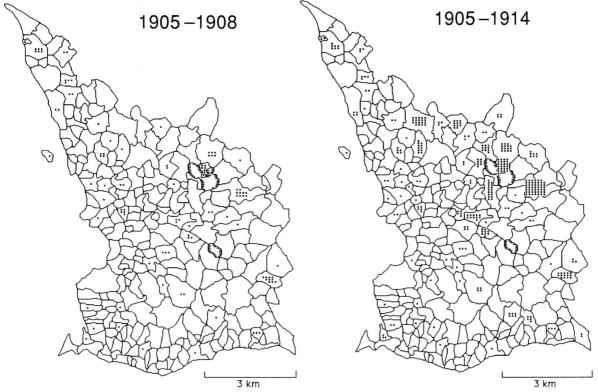

3 km   3 km

Fig. 5.10:5. Loans granted for owner-occupier homes, in Malmöhus County 1905–1914. Each loan indicated by a dot within the parish. During this period most of these homes were introduced in the inner parts of the county and very few in the open plain. There were, however, already small groups in the Ystad region, and more should follow. Source: LLA (Provincial Archives in Lund). See also Germundsson 1989.

other possible reason is that they had more land and heavier soil. On the other hand, as stated by both Gadd (1983) and Germundsson and Möller (1987), ordinary farmers usually felt the need for new implements later than large landowners, and unlike estate-owners they could also rely on the whole family in their farming. But, when time was ripe for change, it seems to have come as fast among ordinary farmers as among gentlemen-farmers.

Denser dispersed settlement

It has already been noted that enclosure resulted in a revolution of the pattern of settlement over the first half of the 19th century, in the Ystad area as in Scania as a whole. Dispersed settlement became the rule all over the agricultural plain in the southern and western parts of the province. This made the appearence of the landscape different from that of many, not to say most, other European regions, where nucleated settlement seems to have been the rule irrespective of the kind of land reform there might have been.

However, the differences between Scania and continental Europe have diminished since then. Behind this there are both general, national, and local causes.

Among the general ones are the advent of the railway and with it the railway villages, and a common tendency of new service facilities to agglomerate. Thus, after enclosure in Sweden and Scania as elsewhere, some villages once more grew into bigger units of settlement, while others never again reached their former size, neither in housing nor in population. As for dispersed, primarily agricultural settlement a kind of increased density can also be observed. It is, however, highly uneven. Mostly it has to do with several reasons of national origin.

After enclosure, when farmers had to move to their new piece of land, many of them sooner or later found out that, if well developed, their land was more than enough for one household or family. As the old governmental restrictions against the division of farms had been abolished, farmers became free to divide their holdings. Many did so in order to provide for the future of their children, or to sell a part of their land for the much needed money to modernize, or for other reasons. This process meant successively more farmsteads and a denser settlement in the landscape. The importance of the process as such has not yet been fully investigated, however, but the effects may be distinguished even in parts of the Ystad region. Thus they

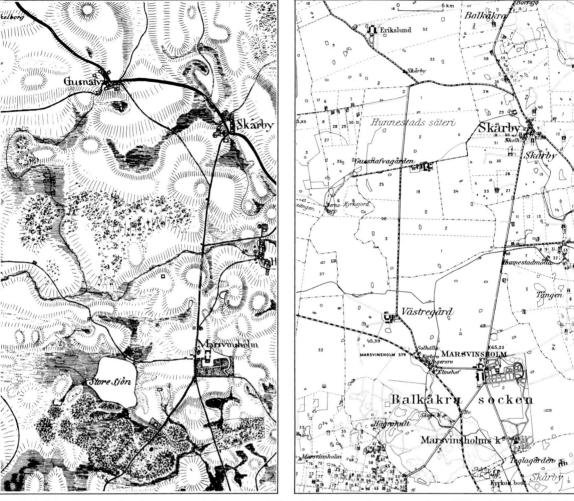

Fig. 5.10:6. Landscape and settlement around the manor of Marsvinsholm according to maps from 1812 and 1913. By 1800 the estate comprised some 4000 ha, which through successive sales have been reduced to about 400 ha in 1980. From Lewan 1988.

seem to have played an important role for the appearance of the landscape in the area of Karlsborg in the northern part of the parish of Sjörup.

Another feature which, however, developed mainly from the beginning of the 20th century, was the establishment of very small freehold farms. The process as such has both foreign and national roots (cf. Wood 1982, Aalen 1986, Smith 1986). When the Swedish parliament took a first decision about small farms in 1904, it was under the influence of the heavy out-migration from rural areas and the interest of major landowners in having a reserve workforce close at hand. The idea was that money should be made available in the form of loans to help people buy a plot of land and there erect their own small farmstead. So far the full history of this own-your-own home movement has not been written, but short accounts are available (cf. Germundsson 1985). Many thousands of new smallholdings were organized before World War II, but the importance varied

very greatly between different parts of Sweden, and within Scania as well. Kristianstad County in the northeast of the province became one of the most important areas for owner-occupier homes, while in Malmöhus County these units are found mainly outside the southwestern plain (cf. Fig. 5.10:5). Land was made available in different ways, often through purchases from bigger units, this being one reason for their location. Thus it became characteristic to find groups of owner-occupier homes, although it also happened that single units were established. As the same drawings were regularly used for all houses, such a group of tiny farmsteads still can appear as a significant feature in the landscape, although most of the units are more than sixty years old, have been rebuilt, and no longer function as farm units. A few groups of them are also found in the Ystad region.

The Marsvinsholm landscape from 1815 to 1915

To illustrate further the changes and circumstances discussed above, the area around Marsvinsholm has been chosen, using maps from 1812 and 1913, reduced to the same scale (Fig. 5.10:6). This area lies in the west central part of the Ystad area. Before the late 19th century it had just one owner, which of course meant that most, not to say all, changes were decided by the owners of Marsvinsholm. By 1800 this estate covered some 4000 hectares, today reduced to a mere 400.

Let me begin by commenting on settlement change. By 1825 enclosure was initiated for the villages of Skårby, Hunnestad, and Gussnava, up to then nucleated settlements. Most farmsteads had to be moved out, and only at Skårby did a kind of nucleus remain, while old Gussnava totally disappeared, being replaced by one big tenant farm, Erikslund. Later on in the 19th century two other major farms according to the *plattgård* system were organized by the estate, viz. Gussnavagården and Västregård, while at Skårby a further subdivision took place. Through these operations the medieval open-field system with all the farms in villages was replaced by a scattered pattern of both major and minor tenant farms and smallholdings (cf. Fig. 5.10:6 B). The number of cottages and smallholdings, however, increased still more in the south-west, where there had previously been no houses at all. This expansion took place outside the domain of the estate, or rather in between three big landholdings, and started around 1870. The habitation was named Ensligheten, i.e. "Solitude", and it must have been founded on the assumption that additional work would be available on the adjacent manor farms; after World War II these were still major places of work for the inhabitants of this neighbourhood.

This is by no means the only rural workers' settlement close to but still outside the boundaries of major estates. On the contrary, it exemplifies a pattern common during the 19th century. It made the inhabitants relatively independent and at the same time put no responsibility upon the estate-owner. Many a cotter had a small piece of land, although rarely sufficient to feed a family.

In the north-west a few other smallholdings can be observed, an example of a pattern common from 1900 onwards being the earliest group of owner-occupied homes in the Ystad region (cf. Ch. 5.9). This settlement was later on vastly extended, when the former *plattgård* of Erikslund and part of the *plattgård* of Gussnavagården were also subdivided into such small-scale freeholdings.

The 19th century saw important changes in the pattern of communication. All over the study area the boundaries of enclosed farms determined the straight course of many a local road, while roads of major importance kept their pre-enclosure courses (cf. Fig. 5.10:6 B). In this case we should look upon the pattern as very much determined by the lord of the manor. Another important addition was the railway; the line between Malmö and Ystad opened in 1874. Its existence and course were decided by a number of financing estate-owners, the most important being the lord of Marsvinsholm (cf. Möller 1987). The embryo of a railway village can be noticed to the west of the manor; it never came to any more, and today the trains rush past it.

Although the 1913 map does not give full information about the landscape, it does allow a number of observations. At the beginning of the 19th century much of the area was still either wetland or woodland, more or less open. The small rivers followed their natural courses, and in the south there was also a shallow lake. Wood and wetland were of course part of the important outland of the villages. Nature limited the possibilities for arable land, and the village sites were close to both dry land and wetland. A hundred years later very little remained of this in many respects medieval landscape.

The only remaining woodland was found in the south-west. Belonging to the manor, it should still have been of interest for timber and hunting, and also as part of a more or less typical estate landscape. The centre of this is the hall, now surrounded by an extended park and garden, and with its modern farm buildings located at a distance, but still conveniently close to the manor house. Still another sign of the importance of the estate is the new church, built in 1867 just a kilometre south of the hall. It was meant to succeed three medieval village churches on the domain of the manor. As it turned out, two of them are still in use today, thus in a way indicating the diminishing power of the nobility.

By 1913 many of the rivers had been turned into straight canals, helping the important under-drainage introduced in the late 19th century. Drained of open water, the former lake was marked as meadowland, presumably of great importance in a landscape rather short of natural grassland, most of which had been turned into arable.

There were, however, also new open waters in the area. On the 1913 map one can see on most fields a number of small square symbols, most of them marking marl-pits. They no longer had their original use but still served as watering places for cattle. Later on they would be regarded as obstacles, especially for modern agricultural machinery, and in many cases they have been done away with, being used for rubbish of different kind. Today the number of marl-pits has been greatly reduced, thus making the landscape much more monotonous than it was at the beginning of our century (cf. Ch. 5.12).

5.11 Landscape patterns and grassland plant species diversity in the 20th century

Siv Bengtsson-Lindsjö, Margareta Ihse and E. Gunilla A. Olsson

Introduction

The landscape in the Ystad area has changed considerably during the long period of agricultural history. Increasing degrees and rates of cultivation have been accompanied by equivalent decreases of natural and semi-natural vegetation areas. The introduction of machinery, fertilizers, crop breeding, pesticides, and other attributes of industrial agriculture have accelerated the landscape changes during the 20th century. This development has resulted in a successive change in the ecological content of remaining biotopes, in terms of distribution of different vegetation types (fragmentation), habitats, and plant and animal species. The semi-natural vegetation, especially all types of grasslands including fenlands, has diminished in the whole Ystad area during the 20th century. Besides being an effect of mechanization, this is also a reflection of the decreased importance of stock-breeding and the replacement of natural fodder-producing areas by ley cultivation. In the south of the Ystad area intensive cereal cropping is traced back to early medieval times (Ch. 4.2.1). At the beginning of the present century most of the land here was used for large-scale grain cropping (Ch. 4.1.8, 4.2.8). However, in the north of the Ystad area land use in a historical perspective has been different, characterized by livestock economies favouring semi-natural grasslands and also woodlands (Olsson 1989a, Ch. 3.6.2, 4.4.7.2).

The aims of the study are:

1. To quantify changes in landscape patterns in the Ystad area during the 20th century as shown in the acreage of natural vegetation types.
2. To investigate the effects of commercial fertilizers on plant species composition and diversity in grasslands on different soil types.

Landscape changes

Visual landscape patterns can be described in terms of three elements: homogeneous patches, linear corridors, and point objects. Different landscape patterns are related to different landscape functions. Several studies have dealt with the relationships between pattern and function (cf. Forman 1984), but only a few (e.g. Agger et al. 1986) have studied changes in pattern over a longer period. Changes in the distribution and size of different vegetation types and land-use categories can be monitored by retrospective inventories of aerial photographs and maps (Ihse 1985). By quantifying the number of vegetation units or land-use categories and their size at any given time, a measure of landscape changes can be obtained.

Fertilizer effect on the diversity of plant species

Generally, one of the most dramatic landscape changes in Sweden during the 20th century has been the large-scale reduction of semi-natural grasslands – fodder areas like meadows and unfertilized pastures on different soil types. Most of the remaining grasslands in the Ystad area receive high loads of commercial fertilizers annually (Bengtsson et al. 1983) and in fact only very small fragments are left unfertilized. At the level of plant species, fertilizer affects the conditions for competition (Grime 1981, Berendse and Elberse 1990) and can bring about changes in the composition of species (Fogelfors and Steen 1982, Vermeer and Berendse 1983, Brunsting and Hei 1985). The composition of grassland plant species communities – the kind of species and their frequency – depends on a number of variables, the most important of which include management methods (like mowing and/or grazing, fertilizer treatment, etc.), the duration and continuity of management. However, the ecological environment around any grassland site also influences the composition of species, since it affects the plants' potential for dispersal, establishment, and survival. One must also consider the direct influence that pesticide sprayed on arable land may have on organisms in adjacent non-arable vegetation.

The combination of changes in landscape pattern and changes in grassland management (such as the introduction of heavy loads of chemical fertilizers and pesticides) has radically changed the conditions for grassland plant communities.

Study sites and methods

Five study areas (4 × 4 km) were selected with the intention of reflecting the different conditions within the Ystad area (Fig. 5.11:1). Their different characteristics are summarized in Table 5.11:1. Two of these areas were used for detailed studies of plant species

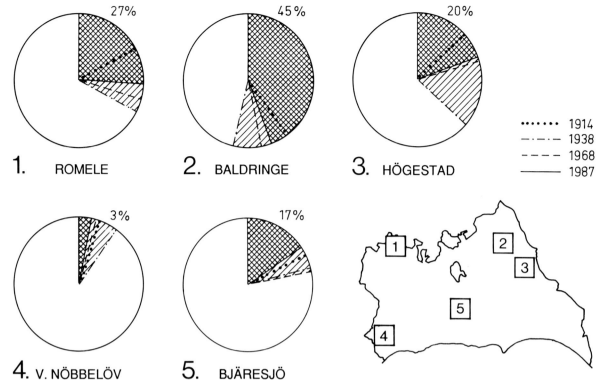

Fig. 5.11:1. Changes in landscape pattern over the period 1914–1987 within study sites. Sector diagrams show the relative distribution of arable (white) and non-arable areas (hatched). Values in percent refer to the situation in 1987.

diversity with respect to soil type and commercial fertilizer.

Landscape changes, with special emphasis on grasslands, were studied by the interpretation of aerial photographs from a 50-yr period (1938–1987). Present-day vegetation (1987) was mapped by means of interpretations of recent aerial photographs and field surveys. This map was used as a reference for consecutive, retrospective mapping of land use and vegetation from the interpretation of black and white aerial photographs. From aerial photos it is possible to recognize areas down to 1 m² and also get information on soil moisture

and vegetation types, land use and management (Ihse and Lewan 1986). Separate maps were constructed for the following years: 1938–39, the oldest available aerial photographs, scale 1:20,000; 1968, aerial photographs, scale 1:30,000; 1987, aerial photographs, scale 1:30,000.

An economic map (scale 1:20,000) from 1914 was used as the oldest available reference.

Units for mapping were:

* Dry and mesic grassland (open or with scattered trees and shrubs)

Table 5.11:1. Characteristics of the five study sites. Landscape types: L = large; M = medium; S = small proportion of woodland grassland, and arable.

Study site	Landscape-type proportion of			Soil type		Soil fertility			Agricultural production		Mean size of arable fields (ha)	
	Wood-land	Grass-land	Arable	Clayey till with lime	Sandy till with clay	9–8	7–6	5	Grain & oil plants	Stock husband-ry	Large >100	Medium Small 50.1–100 <50
	L. M. S.	L. M. S.	L. M. S.									
V. Nöb-belöv	*	*	*	*		*			*			*
Bjäresjö	*	*	*	*		*			*	*		
Högestad	*	*		*	*		*			*	*	
Romele	*	*			*			*	*	*		*
Baldringe	*		*		*		*		*	*		*

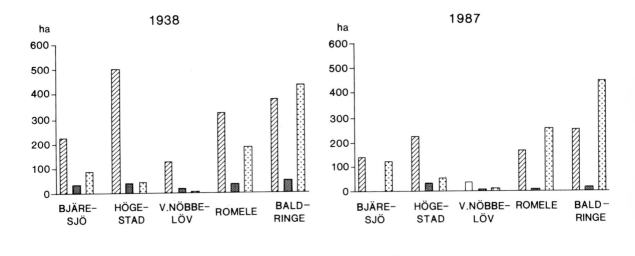

Fig. 5.11.2. Area of semi-natural vegetation units in the study sites, 1938 and 1987, see Fig. 5.11:1.

* Cultivated grassland (15–20-yr-old leys, mostly mesic)
* Wet grassland (open or with scattered trees and shrubs)
* Fen (open or with scattered trees and shrubs)
* Forested marsh
* Woodland
* Open water
* Arable land

The smallest unit on the map is 0.5 ha, where two units are connected, and 0.3 ha for isolated units in arable fields. Areas with a diameter smaller than 30 m were ignored.

Plant species diversity. The study areas of Högestad and Baldringe, which differ in landscape development (Fig. 5.11:1), were used for the investigation of plant species diversity. In each of these, two semi-natural grassland areas (size criterion: site >2 ha) were chosen. The two grasslands in each study area border on each other; one is unfertilized and the other is fertilized (approx. 120 kg N ha^{-1} yr^{-1}, 50 kg K ha^{-1} yr^{-1} and 30 kg P ha^{-1} yr^{-1}). All four sites are grazed during May-September by cattle, and for the Baldringe areas, also by horses. The soil type in the Högestad site is clay with areas of peat, although the latter were excluded from the analysis. The Baldringe site consists of sandy till soil.

Fifty sample squares (size 0.25 m^2) per hectare, organized in a systematic grid, were analysed. The total number of squares was 2 × 200 for Högestad and 2 × 100 for Baldringe. Within each sample square, species frequency and mean vegetation height were recorded. A rough classification of soil moisture (dry, mesic, and wet) was also carried out. For the Högestad site all three soil moisture types were analysed, while for Baldringe only dry and mesic types were considered.

Analysis of plant species data. Vegetation-environment relationships were investigated by canonical correspondence analysis, CCA (Ter Braak and Prentice 1987), with the computer program CANOCO. This program performs a canonical ordination which combines the ordination of vegetation data with multiple regression of the environmental variables related to the ordination axes (for details see Ter Braak 1987). Relative species frequency in each plot was used as an independent variable and the environmental variables were mean vegetation height, soil type (clay soil and sandy soil), the presence or absence of fertilizer, and soil moisture (three classes). The three latter variables were nominal.

Plant species diversity related to fertilizer was investigated by calculating Shannon's diversity index (Pielou 1976):

$$H' = \sum_{i=1}^{S} - p\mathrm{i} \ln p\mathrm{i}$$

where $p\mathrm{i}$ = relative frequency of the ith species and S = species richness.

The computer package DIVCLUS (Rik Leemans, unpubl.) was used for this operation.

In the further analyses, sample squares from wet soil types were excluded. The plant species were classified into four groups according to known preferences: 1) natural grassland species; 2) nitrophiles; 3) ley and weed species; 4) "neutral" species.

Literature references for the classification were: Fogelfors (1979), Fogelfors and Steen (1982), Ellenberg (1984), Olsson (1984), Inventering av ängs- och hagmarker (1987). Frequencies for the species groups were then calculated for the different sites.

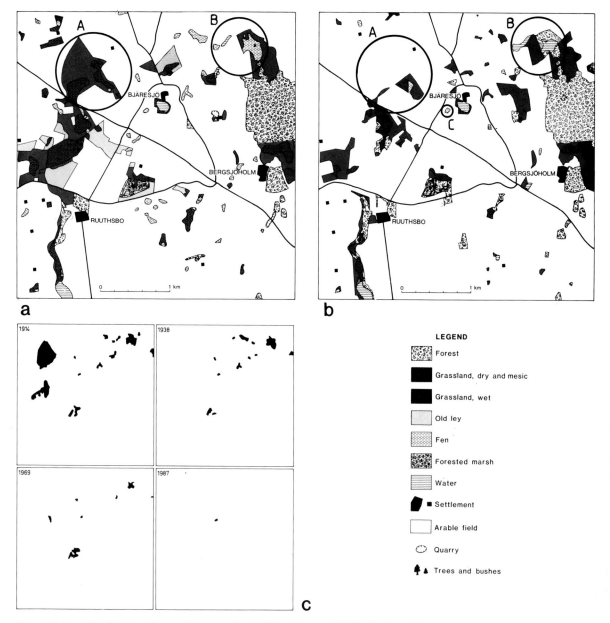

Fig. 5.11:3. a. Distribution of vegetation types in the Bjäresjö area in 1938. See legend.
b. Distribution of vegetation types in the Bjäresjö area in 1987. See legend.
c. Changes in fen area (black) in Bjäresjö during the period 1914–1987.

Results

Landscape changes. The sector diagrams in Fig. 5.11:1 show the distribution of areas of arable and semi-natural vegetation and the overall changes during the period 1938–1987 for the different study sites. A comparison of relative changes reveals that in Bjäresjö, Romele, and Baldringe, 17–25% of non-arable units (grasslands, fens, and forests) have been converted to arable during this period, and as much as 47% and 64% in Högestad and Västra Nöbbelöv respectively. The relatively large non-arable area in Högestad mainly comprises old leys used for grazing. Considering the real size of the areas, the greatest changes have occurred in Högestad and Baldringe, where the areas of non-arable vegetation units that have disappeared are 277 ha and 146 ha respectively (Fig. 5.11:2).

Landscape changes in Bjäresjö (Fig. 5.11:3a,b) are representative of the development in most other study areas. The map from 1914 cannot be used for comparison of areas over time since it is less accurate. However, information on the localization of grasslands in 1914 can

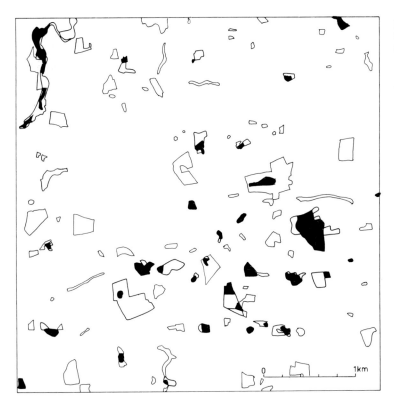

Fig. 5.11:4. Changes in the area of semi-natural grasslands in the Västra Nöbbelöv area during the period 1938–1987.

⬭ Grasslands existing 1938

■ Grasslands existing 1987

be obtained. This map shows that grasslands were mainly localized on organic soils in valleys along streams and rivers (at site A in Fig. 5.11:3a) or adjacent to woodlands (at site B in Fig. 5.11:3b). From 1939 onwards we can distinguish two different trends of change valid for all the study sites, but especially clearly seen in Bjäresjö:

1) Grasslands and fens are drained and converted to arable, cf. A in Fig. 5.11:3a. This development takes place in steps:
* Large fens are drained and converted to wet and mesic grasslands. Connecting grasslands are ploughed.

Table 5.11:2. Non-arable vegetation units (see text) in the five study areas. Changes during the period 1938–1987 in number of subareas and in total size.

Site	1938		1987	
	No. of subareas	Total area	No. of subareas	Total area
Baldringe	119	864	120	718
Bjäresjö	119	358	107	267
Högestad	110	592	91	315
Romele	192	549	125	429
V. Nöbbelöv	139	154	60	55

* Wet grasslands are drained into mesic grasslands and large areas are fragmented into smaller units.
* Mesic grasslands are converted to arable land. Remaining small, wet grasslands are left to forest succession.

2) Grasslands disappear as a result of the invasion of shrubs and trees, as seen at site B in Fig. 5.11:3a. The following happens:
* Open fens are converted to forested marshes or wet forests.
* Trees and shrubs invade surrounding mesic grasslands.
* Small remnants of fens are dammed to create ponds.
* New "cultivated" grasslands, old leys, are created close to the former ones.

Grassland areas have suffered the largest reduction of all the different vegetation units considered in the analysis (Fig. 5.11:2). However, as regards changes in the number of subareas, fens are the big losers. Out of 17 open fens in Bjäresjö in 1914 only one remains in 1987 (Fig. 5.11:3c).

The most extreme landscape changes have occurred in Västra Nöbbelöv. In 1938 all non-arable vegetation units were small and scattered all over the area (Fig. 5.11:4). The reduction of grasslands between 1938 and

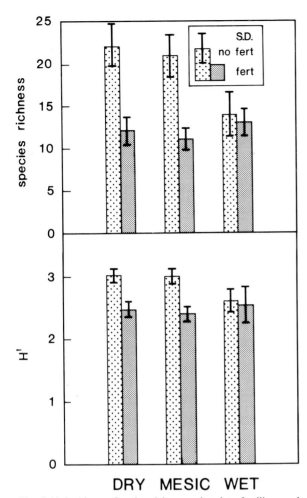

Fig. 5.11:5. Above. Species richness related to fertilizer and soil moisture. Plots on sandy and clayey soils are included. S.D. = standard deviation. Below. Species diversity, H′, Shannon's index, related to fertilizer and soil moisture. Plots on sandy and clayey soils are included.

1987 was very dramatic, 82% (Fig. 5.11:2). The rate of reduction accelerated from 0.3% per year between 1914 and 1957 to 3.1% per year from 1969 to 1987.

The general tendency for all study sites is a decrease in grasslands and fens, both in area and number (Table 5.11:2), and an increase in forested marshes, woodlands, and ponds (Fig. 5.11:2).

Comparing the areas of Högestad and Baldringe (Figs 5.11:1, 2) where the plant species study was carried out, Baldringe has significantly larger woodland areas, while

Table 5.11:3. Mean species number per 0.25 m² related to fertilizer level. Standard deviation in brackets.

Site	Mean no. of species per 0.25 m²	
	No fertilizer	Fertilizer
Högestad, clayey soil	23 (6)	11 (3)
Baldringe, sandy soil	19 (3)	12 (3)

Högestad has larger fen areas (Fig. 5.11:2). Grassland area is about equal, and, like Baldringe, Högestad has 4% of its area protected as nature reserves, containing mainly grasslands and fens. The ecological neighbourhood differs in that Baldringe has a more stable development over the period 1938–1987 with a low reduction in the *number* of subareas (Table 5.11:2) compared with Högestad. The difference in size reduction of non-arable vegetation is even greater (Table 5.11:2). The present surroundings of the Högestad site are dominated by large-scale arable, while the surroundings of the Baldringe site contain considerable areas of woodlands together with arable land.

Effects of fertilizer on species content in pastures. The richness of species, in terms of the mean numbers of species, is significantly larger for unfertilized plots on dry and mesic soils (Fig. 5.11:5a), while there is no significant difference for wet soils. If wet plots are excluded and the four sites are compared, there is a significantly larger number of species in unfertilized plots on clay soil compared with the sandy plots (t = 6.28, p <0.005; Table 5.11:3). However, in a comparison of the fertilized plots on the two soil types, a difference in species is only weakly indicated (T = −2.44, p <0.02; Table 5.11:3). Irrespective of soil type and soil moisture, wet plots excluded, the fertilized plots have approx. 50% less species per 0.25 m² (Table 5.11:2, Fig. 5.11:5a).

Plant species diversity expressed by Shannon's index (Fig. 5.11:5b) gives a combination of species number and dominance (Magurran 1988). Species diversity shows the same trend as species number with significantly larger diversity on unfertilized plots (Fig. 5.11:5b).

In the canonical correspondence analysis (Fig. 5.11:6) fertilizer and soil type are significant variables with the first and the second ordination axes (1st axis: fertilizer: r = 0.57, t = 3.19; 2nd axis: soil type: r = 0.49, t = 2.95; Fig. 5.11:6). In unfertilized grasslands species content is significantly related to soil type (Fig. 5.11:6). Samples from sites with sandy soil and clay soil form two separate clusters, while samples from fertilized sites from both soil types form a third cluster. The effect of fertilizer can thus be interpreted as levelling out the specific species composition related to different soil types, resulting in convergence towards a new plant community (Fig. 5.11:6).

Comparing the qualitative species content (Fig. 5.11:7) the difference between unfertilized and fertilized plots is still greater than the differences in species number. The unfertilized sites have up to 45% of natural grassland species while this group accounts for approx. 10% in fertilized sites. Nitrophiles, ley species, and weeds show the opposite trend, constituting up to 50% of species content on fertilized sites and 10–15% in the unfertilized (Fig. 5.11:7). However, despite different soil types there are no significant differences in the

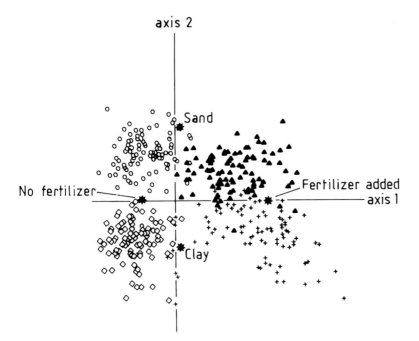

Fig. 5.11:6. Canonical correlation analysis of vegetation sample squares from unfertilized and fertilized pastures. Significant environmental variables are: Fertilizer (nominal variable: no fertilizer, fertilizer added), Soil type (nominal variable: sandy soil, clay soil). The nominal variables are denoted by black stars in the figure. Circles: Baldringe site, unfertilized; Diamonds: Högestad site, unfertilized; Triangles: Baldringe site, fertilized; Crosses: Högestad site, fertilized.

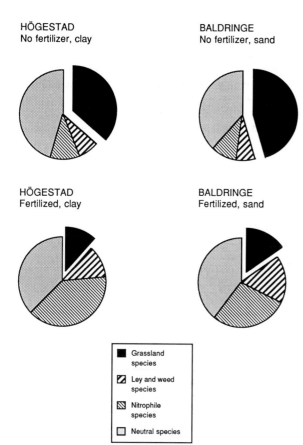

Fig. 5.11:7. Mean percentage contribution of different plant species groups related to fertilizer level.

composition of species groups between the sites when fertilizer level is the same (Fig. 5.11:7). The "fertilized" plant community demonstrated in the ordination (Fig. 5.11:6) is thus dominated by nitrophiles, weeds, and ley species, as seen in Fig. 5.11:7.

Discussion

Landscape pattern. Explanations for changes in landscape pattern must be sought in a combination of several factors: environmental conditions like soil quality and topography; agricultural trends and farming practices; the size of farms and ownership conditions.

Topography and the presence of organic deposits are auto-correlated. Almost every depression in the hummocky landscape was occupied by wet or mesic grassland at the beginning of the century (interpreted from the 1914 map). The grasslands were also located on sandy or clayey till soils close to woodlands. A relationship between soil quality, including moisture, and the localization of the grasslands was thus obvious at the beginning of the century. This relationship becomes increasingly weaker over the investigation period. During the last 30 years grasslands on all soil types have been ploughed at the same rate. In Högestad, where intensive grain production has dominated since early in this century, the soil-grassland relationship was lost back in 1914.

There is no overall relationship between farm size and changes of landscape pattern. Irrespective of farm size, all types of non-arable vegetation are converted to arable at a similar rate in the 20th century (Fig. 5.11:1). However, the starting-point was different (Fig. 5.11:1,

Table 5.11:1). Västra Nöbbelöv was characterized by medium-sized farms and had fewer non-arable units than both large- and small-sized farm areas (Fig. 5.11:1, Table 5.11:1). The influence of agricultural production on landscape pattern is visible as a larger fraction of woodlands in the sample areas of Romele and Baldringe, two areas where animal husbandry is more important than in the other sample areas. This can partly be explained by the fact that stock-raising is here coupled to part-time farming. This allows spontaneous forest succession to occur on former grasslands and farmlands. Another explanation is that 4% of the Baldringe area is protected as nature reserves, including both woodland and grassland.

Landscape changes related to the richness and diversity of species. The outcome of the species analysis with respect to fertilizer shows a marked decrease in the number of species but also a complete change in species composition. The original species assembly, with a large group of grassland species like *Anthoxanthum odoratum* (sweet vernal-grass), *Briza media* (quaking-grass) and *Centaurea jacea* (brown knapweed), are replaced by nitrophiles like *Taraxacum vulgare* (dandelion), *Anthriscus sylvestris* (cow parsley), and *Plantago major* (greater plantain).

Considering the documented landscape changes in the Ystad area, and the present environments of the investigated sites, it is reasonable to interpret the relatively low species richness in the light of these factors. Both the reduction in size of non-arable vegetation areas and the reduced number of subareas during the 20th century would be expected to lower species number by decreasing possibilities for dispersal between different sites (Quinn and Harrison 1988). When the areas of natural grasslands decrease, the edge effect will increase, i.e., the area affected by fertilizer and pesticide sprays increases, and the unaffected area decreases. Species within the "natural grassland" group are sensitive to raised levels of nitrogen (Vermeer 1986) and lose out in competition with fast-growing nitrophilous species. Such species often have rapid height growth and also lateral spread and include species like *Urtica dioica* (stinging-nettle), *Achillea millefolia* (yarrow), and *Deschampsia caespitosa* (tufted hair-grass).

The sites of Baldringe and Högestad display a different development of landscape pattern during the period 1914–1987 (Figs 5.11:1, 2, Table 5.11:3). The reduction of semi-natural vegetation, especially grasslands, was significantly greater in Högestad (Fig. 5.11:5). The Högestad site today forms an island of grassland within a matrix of large-scale arable. The Baldringe site, on the other hand, is surrounded by arable, cultivated grasslands, and woodlands. However, despite a higher landscape diversity around the Baldringe site, neighbouring unfertilized grasslands are as rare here as at Högestad. The greater species richness at Högestad (Table 5.11:2) could probably be ascribed primarily to the differences in soil fertility. A tendency to slightly larger variation in microtopography within the Högestad site possibly also contributes to a greater species richness.

This study has shown that woodland areas have increased over the investigated period, in contrast to other non-arable vegetation units. Small grasslands and fens not economically worth ploughing are left to spontaneous forest succession, or in several cases, are planted with conifers for wildlife purposes. The invasion of woody plants on the former grazed or mowed grasslands implies a substitution of plant species communities where light-demanding, short-statured species, and often highly diverse communities, are replaced by shrubs and woodland species, and less diverse plant communities. This development is especially pronounced where conifer plantations are concerned.

Draining of fens and wetlands (Fig. 5.11:3c) results in changes in landscape hydrology and certainly also in plant species composition. This study has shown that the number of wetland sites has been subjected to the largest reduction during the investigated period, thus indicating that wetland plant communities would be the most reduced both in area and in number of taxa.

Vegetation inventories of the Ystad region have shown that more or less all grasslands, except those protected as nature reserves, are heavily fertilized today (Bengtsson et al. 1985). Commercial fertilizer was not introduced to a significant degree until after 1950. The results from the present study show that about 50% of the grassland species disappear when subjected to fertilizer and that remaining species belong to quite different taxa than in unfertilized sites. Moreover, the present study shows that over the period 1938–1987, 19% (170 ha) of grasslands have disappeared into arable land within the five study areas.

Conclusions

The results of the present study show that:
* during the last 50 years about 20% of the grassland area has become arable within the five studied sample areas
* during the last 50 years 30% of non-arable vegetation has disappeared within the five sample sites
* during the last 50 years the number of fens has suffered the greatest reduction of all non-arable vegetation units
* from the above statement it can be assumed that wetland and fen species are most seriously reduced both in number of taxa within this area, and in plant community area
* during the last 50 years the number of grassland sites has decreased by 25%
* during the last 50 years the number of sites of non-arable vegetation has decreased by 26%
* results from the detailed study of species composition indicate that during the last 35 years roughly 50% of the grassland species have disappeared in this region

* results from the detailed study of plant species indicate that during the last 35 years grassland variability reflecting soil conditions has decreased and homogeneous plant communities dominated by nitrophiles, ley, and weed species characterize the remaining grasslands
* the results of all these changes are a reduced landscape variability, reduced number of habitats, reduced dispersal ability for organisms, reduced plant species diversity in grasslands and fens, reduced accessibility for visitors: fewer dispersal corridors, tracks and roads, and borderlines between different landowners and crops.

Acknowledgements – Unpublished field data and vegetation maps from Baldringe were lent to us by Anders Larsson. Lars-Åke Gustafsson lent us his unpublished flora inventory of the Ystad region. Help with fieldwork was provided by Gudrun Berlin and Elisabeth Åberg. Mimmi Varga and Sigrid Bergfeldt drew the figures. Our thanks are due to all of them.

5.12 The landscape of today and tomorrow

Nils Lewan

Introduction

In this section two principal questions are examined:
– What is left today of yesterday's landscape?
– What will be left tomorrow of today's landscape?
In discussing these, the focus should be on the province of Scania, or rather the plains of southern Sweden, and a number of causes and possibilities will have to be taken into consideration. A few words about changes since the 1950s at a single agricultural unit will serve as an introduction.

In 1950 the Charlottenlund manor farm, some 8 km west of Ystad, comprised 425 ha, all but 45 ha being arable land (cf. Fig. 5.10:4). The ploughland was divided into 36 different fields in a 7–8 year rotation cycle. The animals included 95 milch cows, 135 other cattle, and 27 horses. The permanent workforce comprised 48 persons, while up to another fifty were needed during the most intensive season.

This lively society of yesterday should be compared with the situation today. No animals remain at Charlottenlund, the dairy herd disappeared back in the 1960s. The present workforce includes six persons, with an additional one or two during the harvest. Together they manage 535 ha of arable land, divided into four fields, growing winter wheat and barley, sugar-beet and

peas (with some oats). This means that each single field is more than one square kilometre, and also that a large number of obstacles to the proper use of modern machinery have been removed, including dirt roads, open drains, stone walls, rows of trees, and most of the former marl-pits (cf. Fig. 5.10:4). Quite a number of disused houses, both dwellings and agricultural buildings, have also been pulled down.

These profound and far-reaching changes at Charlottenlund are similar to those at most major farms, everywhere on the Scanian plain, and they are, of course, also general and well-known phenomena in western Europe (cf. Dege 1983, 1987). In the Ystad area, with holdings both small and large, the corresponding changes include all sizes of farms. As the general tendencies are the same all over the open plains of Scania, and in other good agricultural districts in Sweden, figures below (Tab. 5.12:1) have been restricted to the Ystad region (cf. Lewan 1983).

It is clear that arable land has not decreased in area. Meadows and permanent grazing land, however, have been reduced by one-third to about 1400 ha. If rotation grass and ley are added, it can be established that the total agricultural area under more or less permanent cover has decreased from some 7500 ha to around 3500, or from 27% to 13% of the total area.

Soil erosion and other detrimental effects are, however, partly offset by a sharp increase in winter crops of wheat and rye. It should be added that the increase of bare fallow perhaps is temporary, the result of a somewhat panicky political decision in 1987 intended to reduce cereal production. Already in 1988 bare fallow was replaced by cover crops whereby leaching of nutrients should be reduced and not increased.

Other important changes include a general increase in the size of holdings and the decrease of mixed farming. Even if the mean size of a farm in the area has increased only from 20 to 45 ha of arable land, one should be aware that the big units by now farm most of the land. Figures concerning this are published only for larger areas, in this case the open plains of Malmöhus County. Thus holdings with at least 50 ha of arable land by now farm some 65% of the total acreage, compared with around 35% in 1944. The corresponding figures for the smallest farms, i.e. farms with 2–10 ha of arable land, are less than 2% in 1987 compared with about 10% in 1944.

The general expressions of change are well known:

Tab. 5.12:1. Agricultural change in the Ystad area 1944–1987. Sources: Agricultural census returns (Jordbruksräkningen 1944, Lantbruksregistret 1987).

A. Size of holdings (ha arable land), as a percentage of all holdings.

Year	2–10	10–20	20–50	>50	No. of holdings
1944	50	28	18	4	1093
1987	20	20	41	19	482

B. Use of arable land, as a percentage.

Year	grain	oil-plants	sugar-beet	ley, rot. grass	other crops	fallow	Acreage
1944	39	*	13	30	17	1	21600 ha
1987	52	14	12	14	2	6	21600 ha

* included in other crops

C. Number of animals.

Year	cattle	pigs	sheep	poultry	horses
1944	16600	16600	1300	54000	4100
1987	8100	24900	400	40000	*

* no data

SKÖNABÄCK

ÅGERUP

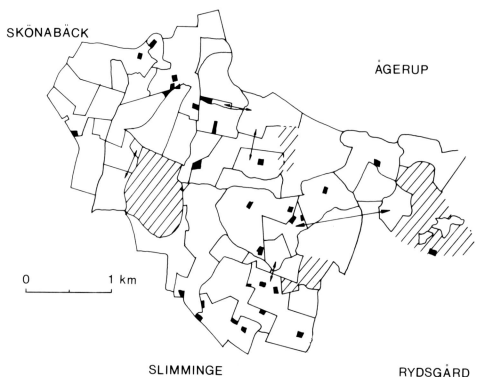

Fig. 5.12:1. Farm units in 1986 in an area just outside the Ystad region, with relatively small units surrounded by the estates of Ågerup, Rydsgård, and Skönabäck. The area lies on the southern edge of the Romeleåsen ridge. Unfilled areas are units cultivated by their owners, hatched areas are units cultivated under tenancy. Black areas are non-agricultural units.

0   1 km

SLIMMINGE

RYDSGÅRD

mechanization and specialization in agriculture and a simplification of the landscape. The prime causes also seem to be both well known and common: national agricultural policy has played an increasing role since World War II, and this has to be borne in mind too, together with non-agricultural reasons for landscape and other rural change.

Swedish agricultural policy

As in many other countries, a national policy for agriculture is by now an old phenomenon in Sweden with its roots in bygone days, through decrees preventing the division of farms up to the early 19th century, by promoting enclosure and encouraging under-draining during the same century, helping people to own their own small farms (*egnahem*) in the early 20th century. The post-war agricultural policy has, however, been very different.

In 1947 the Swedish parliament inaugurated an agricultural policy with decidedly new goals. The central ideas were that food prices should be kept at reasonable levels, and that at the same time ordinary farmers should have incomes equal to those of industrial workers. It was also understood that national self-sufficiency should be maintained, and it was decided that the modernization of the industry was to be furthered with public money.

Major revisions of the policy have since been made in 1967, 1977, 1985, and 1990. The basic ideas have, however, remained the same, although lately evident

threats to the environment have meant that some new measures have been taken, or at least considered.

The overall effect of the policy has been the encouragement of all kinds of modernization: mechanization and specialization, removal of field obstacles, introduction of herbicides and pesticides, increased use of chemical fertilizers, the amalgamation of farms, afforestation of permanent grazing land and less productive arable land, and so forth.

The result in the landscape seems to be almost the same, at least in the open, fertile plains, be it in Sweden or inside the European Community under the auspices of the Common Agricultural Policy. As biological consequences are discussed elsewhere (Ch. 5.11), I will here concentrate upon other effects. Before that, however, other reasons for change ought to be mentioned.

Rural change because of urban influence

With the urban population now an overwhelming majority even in Sweden, one might talk about the agricultural policy as just another urban influence. There are, however, quite other reasons for landscape and other rural change, emanating from outside agriculture. These too are by now in many respects common to western society. One of the most conspicuous is the general urbanization of the countryside, at least within commuting distance of urban centres.

In our context this means that today most of the province of Scania is more or less within reach of people working in urban places but still wishing or needing to

398

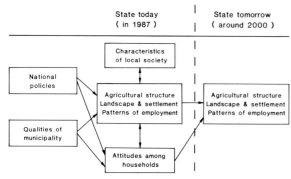

| State today (in 1987) | | State tomorrow (around 2000) |

Fig. 5.12:2. Factors determining tomorrow's occupation and landscape.

reside at a distance from their places of work. Since World War II the increase of rural commuting to central places of work has developed gradually from the western and more thriving parts of the province, so it has not meant very much to the Ystad region until quite recently. Rural areas further west than that went "urban" in the 1950s and 1960s, whereas in our area commuting did not become of great importance until the 1970s. The more obvious manifestations in the landscape are, however, still fairly few, as new building has been confined to just a few places: apart from the town of Ystad and Köpingebro, formerly agricultural villages like Sövestad and Hedeskoga, as well as Svarte, which originated as a fishing village. As in other rural areas, much – not to say most – urbanization has taken place inside already existing settlements, be it in villages, hamlets, or scattered settlements.

In addition to this, even in this region urbanization has many other manifestations. One is permanent use of previously local housing by people connected to urban places; a second is the change of formerly permanent dwellings into second homes, a use found along the seashore as well as inland; a third is the permanent use of houses built as second homes; and a fourth is the secondary use of otherwise disused farm buildings. So far the primary effects of this have been that fewer houses have been pulled down, that the difference between form and function of settlement has increased, and that by now there is considerably more life in the countryside than there otherwise would have been.

Discussing this, one should be aware that development in the future is quite uncertain, and that even in the Ystad area there are important local differences, largely depending upon land ownership and the size of holdings. In this respect the southern slopes of the Romeleåsen ridge together with some adjacent neighbourhoods in the north-western part of the region are obvious areas to investigate, because these areas have had and partly still have a high frequency of small owner-occupied farms, although by now many other people live there as well: what do people in such an area think about recent changes and the threats and possibilities in the future?

Areas of smallholdings since the 1950s

The most interesting area lies to the north-west, on the undulating slopes of the Romeleåsen ridge just outside the Ystad region (cf. Erlingson 1987). Surrounded by big landed estates, there are in this area houses without any agricultural land and a number of relatively small holdings, although their size has increased lately (cf. also Fig. 5.12:1). In the old days the estates were important as places of work, while today commuting is widespread, with only a minority of the households relying totally upon agriculture. Part-time mixed farming is more or less the rule, which, together with the natural circumstances, results in a landscape of great variety, by now also designated by county authorities as an area of outstanding natural beauty. Focusing on the next ten to fifteen years, 40 out of a total of 50 households in the area were interviewed, the informants grouped as follows (cf. Fig. 5.12:2 and Erlingson 1988):

1. *Full-time farmers* (6 households with 25–100 ha arable). No one goes to work away from these farms, of which four have dairy herds, two have pigs. Rationalization and expansion are common, and there seems to be an open attitude to new political signals. In spite of this there is widespread pessimism, and according to interviews, farming in the future may be part-time on a household basis. Decisions in this group will mean much for the future landscape.

2. *Traditional mixed farmers* (5 households with 20–50 ha arable). Work outside the farm is the rule in this group, but even here the future looks bleak, as there seem to be no heirs ready to take over. Presumably, with land included in bigger units, landscape monotony will increase.

3. *Passive part-timers* (13 households with 0–25 ha arable), mostly old-age pensioners, several having left agriculture recently, others growing cereal and rape but only two keeping animals now. Although there seems to be a more widespread optimism in this group, remaining land will probably be included in other units. As the group is one-third of all households, their decisions will not be without impact upon the future landscape.

4. *Commuters* (16 households), a young population with many children. A number of them have moved into the area recently. Backgrounds vary but incomes are fairly high. This is an important group, which might grow in the future. As these people seem to be both optimistic, well informed and well-to-do, they may also move away, if economic circumstances become more difficult, for example if lower tax deductions are granted for journeys to work, etc. Their importance for the future landscape is obvious.

A couple of smaller investigations of areas nearby seem to confirm the impressions in the previous study (cf. Lindelöf 1987, Abramsson 1987). One area developed as a group of owner-occupiers' homes during the inter-war period, while the other study is about an area of smallholdings, partly developed from a 19th century *plattgård* (Oran and Vallösa, respectively). Since World War II the number of farms has decreased with a parallel growth in farm size, the population has grown older, and many houses have been taken over by commuters or as second homes. Although as a result of this the number of houses has remained stable during the last half-century or so, changes seem to be impending, as both people and houses are old, and many former agricultural buildings have remained empty. Most agricultural land is now managed in bigger units, and the function of the settlement has changed very much inside old forms.

Still another study, concerning the agricultural population in three inland districts of Scania, discusses the different strategies of full-time compared with part-time farmers (Jordbruk/landsbygd i utveckling 1986). Basing their conclusions upon several hundred interviews, the authors conclude that both groups are interested in extending their business, that many a part-timer is either on his way into or out of farming, and that full-timers look upon additional work outside their farm as being an important possibility for the future. The obvious conclusion to be drawn here is that future labour demand outside the agricultural industry but inside commuter distance will be crucial for both agriculture and the landscape. Such a demand is small in the Ystad area compared with the much more urbanized western parts of the province.

A changing communication network

As elsewhere, the geographical redistribution of dwellings and areas and places of work has also manifested itself in changes in the communication system. A large number of successive maps have made it possible to plot the changes since the 18th century, thereby also finding the most important periods of change. Obviously these changes also are of importance for the relative accessibility of the area in the future. The general findings as to periods have also been compared with earlier work (cf. Lewan 1968). Looking back over two centuries these periods could be described as below.

Period 1, roughly the 18th century, was still dominated by purely local paths and roads, although a couple of through roads towards Ystad were already to be seen. All roads were heavily dependent upon natural surroundings, avoiding wetland as well as steep hills.

Period 2 witnessed a widespread restructuring of the local network in connection with enclosure. Many of the new roads were straight, following the new boundaries between holdings, while the few main roads retained their former courses (cf. Fig. 5.10:6.). Generally speaking, this period covers the first half of the 19th century.

Period 3 was characterized by a successive adjustment to the railways, which in this area were opened between 1865 and 1901.

Period 4, the early part of our century, includes an important new means of transport, the motor vehicle. Higher speed meant that sharp bends and steep hills had to be smoothed out, although before World War II this was mainly done on a very local level.

Period 5, covering the first decades after World War II, included at least two new tendencies, one being a much more widespread car ownership, a second the closure of local railways and railway stations. The Skivarp-Charlottenlund line was extremely short-lived, operating only between 1901 and 1919. Most of the railway buildings, however, are still in use as private dwellings, while part of the track has been included in one of the main roads. Increasing road traffic also meant that by-passes became important along main routes, which were widened and metalled in one way or other.

Period 6, the last decades, has witnessed still another innovation spreading to our study area, the new through road with limited access but at the same time shortening time distances between major centres. Still, however, most local roads follow the courses laid out before or mainly in connection with enclosure, and most of them are still dirt roads. In addition, one may observe that many field roads have vanished or are in disuse, reflecting changes inside agriculture.

The most general observation to be made is of course that the old local road network, adapted to natural circumstances and serving a rural self-supporting society, has been superimposed by an inter-regional system, mainly serving the needs of major centres.

In the landscape this means that important features have been and also will be changed, also making other changes in the elements of the landscape more interesting, among them the boundaries between holdings, villages, and parishes.

Boundaries and the landscape

The importance of the domain to changes in the landscape has been discussed previously (Ch. 5.10). Continuing this theme, it might be fruitful to discuss different effects of boundaries and of their changes. That can be done at several spatial and functional levels. Here the prime interest will first be the boundaries of parishes and villages, then those of the holdings or farms.

The boundaries of villages and parishes have been looked upon as very stable landscape elements, of importance also to the natural environment. There are, however, signs that the old stability is being weakened. To increase our knowledge, two minor areas have been investigated in southern Scania, one of them in the Ystad area (cf. Nilsson 1987, 1989).

For the parishes and villages investigated so far, it has been established that enclosure in the early part of the 19th century meant defining in detail the older, partly diffuse boundaries. Between then and ca. 1950 no

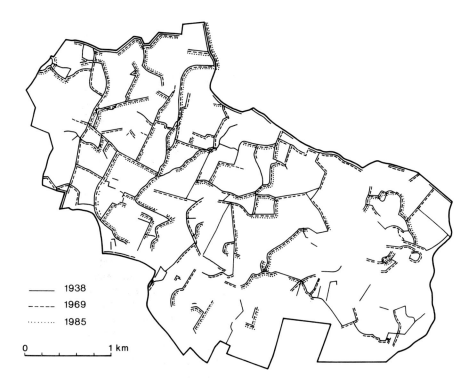

Fig. 5.12:3. Decrease in length of rows of trees and hedgerows in 1938–1985. For location and division into holdings, cf. Fig. 5.12:1. Research: M. Ihse.

——— 1938

- - - - 1969

·········· 1985

0 ⊢——————⊣ 1 km

changes were made, although at least from the turn of the century many holdings comprised land in more than one village or parish. Especially during the last twenty years an increasing number of boundary changes have been made. As new and larger agricultural units are formed with land in more than one village or parish, in many cases their land is also legally brought together. If so, the village boundary in question is regularly changed as well, and nowadays also the parish boundary.

So far we have only witnessed what we suspect to be the beginning of a new tendency, the extent of which cannot yet be evaluated. As said above, these boundary changes are connected with changes in agriculture, but they also have to do with new legal forms and practices. It seems, however, that we are moving towards a pattern of mobile boundaries upon these levels too, i.e., above the level of the single farm units, thus very far from the stable situation during the long period between enclosure and World War II.

In the long run, this might be of importance in several ways. Registrations on the parish level, of people, archaeological finds, and so on, will become difficult to follow. People might also find difficulties in understanding where they belong to. Very often boundaries have meant physical marks in reality, e.g. walls or ditches, rows of trees, etc. When boundaries are removed, such marks are mostly done away with. Changes of boundaries thus also mean the loss of natural habitats, which in the open agricultural plain are already rare phenomena.

Irrespective of boundaries, changes of this kind have

been studied in several cases in Scania. The effects, however, seem to have very much to do with the agricultural units, namely, the size of holdings, amalgamation of units, new farming practices, and the like (cf. Ihse 1984, Ihse and Lewan 1986). In connection with the Ystad Project a study has been made of an area studied also from other points of view (cf. Fig. 5.12:1). Using aerial photographs from 1938, 1969, and 1985, a comparison has been made of a number of different landscape elements, among them hedges and trees along field and farm boundaries (Ihse unpubl.). In 1938 hedges almost everywhere constituted boundaries between farm units, and in most cases they did so also in 1969, as the decrease in total length was limited to some 8 km out of an original 60.5. Since 1969, however, the additional decrease has been 34 km down to a mere 18.5 in 1985. Although it is difficult to visualize a bygone landscape, it is obvious that these undulating slopes of the Romeleåsen ridge have recently undergone both rapid and profound change. With a few exceptions, rows of trees and hedges have been broken down to short sections, if not removed altogether (cf. Fig. 5.12:3). Most longer sections still remaining follow either watercourses or country lanes, while some make up ownership boundaries. Even if land on both sides is used by the same farmer, he cannot, however, remove trees, bushes or hedges, if land is on lease.

This simplification and monotonization of the agricultural landscape is by now a well-known phenomenon, in this case also including draining of ponds and removal

KRAGEHOLM

Fig. 5.12:4. Gardens and parkland on the Krageholm estate during the last three hundred years. A and B = 17th and early 18th century park, respectively, C and D = extension during the 19th century, E = horticulture and orchard, F = grazing land. The park areas reached their maximum area and intensity in the inter-war years. Since around 1950 areas A, C, and F have turned into woodland and shrubs, while B and D have been simplified to open lawns with some scattered trees, and area E is much reduced. From Lewan and Mårtensson 1985.

of coppices and woods. It has been documented all over Europe, from areas as different as Dorset and East Anglia in England, the Paris Basin and Brittany in France, Schleswig-Holstein in Germany, and parts of Denmark. In most cases, however, it is observed in the primary agricultural districts, especially where animal husbandry has been abandoned.

As described before in this chapter, the southern slopes of the Romeleåsen ridge still contain a number of relatively small units, some of them with dairy herds as well. This is also an area of outstanding natural beauty, which has been noted in county documents concerning both nature conservation and the protection of the cultural heritage. If put together, the investigations of landscape change and the strategies for the future among inhabitants suggest that more far-reaching changes are imminent, and that the landscape of an area like this might be just as vulnerable to change as the major farming units upon the real agricultural plain. More work, however, remains to be done on this problem.

The environment of manor and village

When investigating settlement, one also has to contemplate recent changes in the immediate environment of manors and villages. These have been subject to only limited research in the Ystad area, but there is additional knowledge from adjacent parts of the plain (cf. Mårtensson 1985, Lewan and Mårtensson 1985).

The 19th century tendency of the large estates to relocate farm buildings further away from the main building or manor house and to enlarge the parks and gardens has been noted before (Ch. 5.10). With increasing labour costs, the size and elaboration of the manor surroundings peaked in the early 20th century. The last half-century has witnessed a reduction in several ways, of the number of farm buildings because of specialization, of other houses because of reduced staff, of gardens and parkland because of shortage of staff, and so on.

In the project area the early post-war years and the period around 1970 saw most of the reduction and simplification of parks, while vegetable gardens were closed down somewhat earlier. Vast machine-mowed lawns have succeeded the pleasure-gardens, and remote parts of the parkland have been turned into arable or now appear as wilderness, thus being more appropriate for wild animals than before. Additional factors in the change are devastating gales, the advanced age of both parks and avenues, and new economic conditions after changes of ownership, be they through purchase or inheritance. It should also be added that the present owners in most cases declare their intention to maintain both buildings and surroundings as they are today, if this is possible from an economic point of view. The observations from the Ystad area seem to correspond to others made both in other parts of Scania and elsewhere – here the Krageholm estate is shown as an example (cf. Fig. 5.12:4).

At the other end of the settlement scale we find the old village cores, in many cases reduced through enclosure (cf. Ch. 5.10). Nevertheless, there are still in the plain a number of well-maintained villages and hamlets, which from historical, cultural, environmental, and social points of view are well worth protecting. At the same time, however, there are many looming threats and problems. The double question of the historic interest and value of different villages, and of the problems and possibilities concerning their future preservation, has not been examined with regard to the Ystad area but in the district of Trelleborg, just to the west, an area with the same kind of agricultural and natural environment (cf. Mårtensson 1988).

Of about 60 historical villages and hamlets in that district, with roots in the medieval open-field system, one-third were considered to be of particular interest. Today all these villages have distinct remains from the pre-enclosure period; such as a small stream or pond, an obvious green, farmsteads on their old sites, a winding road. On the other hand, none of these places has so far been seriously disturbed by either new buildings or thoroughly modernized houses.

When it comes to the question of future preservation, this can be judged from different viewpoints. Among factors of importance for the future of the village we find land ownership, the actual number of farms in use and their specialization, the amount of money coming in from work outside the farm or village, and the interest of the inhabitants. The understanding and interest shown by authorities, and the suggestions and instructions made by them, also play an important role. Moreover, the agricultural policy and other economic factors are regarded by many owners and inhabitants as prime reasons for change or preservation.

Bearing the Ystad region in mind, one should add that the relative distance from thriving urban centres also plays its role. Proximity may mean that there are more people, functions, and money coming in, while a relative remoteness may help to fend off thorough modernization and extensive new building not adjusted to the existing environment. Obviously, at the same time less money will mean limited possibilities to maintain both buildings and surroundings.

Tomorrow's landscape

The landscape of tomorrow is unknown, and what it will be like depends as much upon yesterday as upon tomorrow.

Contemplating the roots from yesterday, one should also stress that many features in today's landscape are antiquated or obsolete, in spite of thorough modernization and simplification. To take just two examples: most rural buildings are old or very old, many of them not even decently kept; and most trees in gardens and parks, in churchyards and along avenues, are highly mature, planted as they were in the early years of this century or during the 19th century. Thus there is an increasing degree of antiquity in many areas, including parts of the Ystad area. Moreover, in a rural area like this the mean age of the population is also rather high.

If the future partly has its roots in the past, there are at the same time seeds which so far have not even been planted. The probabilities of change also differ between regions. That applies to Europe, to Sweden, and obviously also to the Ystad area. Finally, as the interdependence between different areas has been growing, our region has become more dependent upon western Scania and national decisions, as well as upon international circumstances, like the Common Agricultural Policy of the European Community, and prices of food and raw material outside Europe.

As the objective here is not to give a full view of the probabilities and the possibilities of landscape change in the future, the strategy chosen will be a discussion of a couple of scenarios over the next twenty or so years (cf. Lewan 1986, 1988). Another path, an evaluation of the importance through time of different factors, is left to the following chapter.

Three scenarios might suffice to give an idea of the future landscape:

* *Continued growth*, in economy and farming, together with a continued rationalization and specialization. This should mean vaster open areas, and fewer "obstacles", such as houses and trees, open waters and roads, etc. A resulting landscape that is more monotonous and more susceptible to wind and water erosion will be still more

dominated by major external decisions. The important question is how long such a development can last.

** *A growing bewilderment*, a slow-down because of uncertainty about decisions concerning agriculture and environment, economy and energy. More decisions will be left to the people themselves, and local futures will become more difficult to foresee. A somewhat greater variation in the landscape than above is likely, but still changes will be in the direction of simplification and the like.

*** *A new direction of events* because of more or less immediate decisions as regards agriculture and environment, energy and economy. Such policies might include less use of pesticides and artificial manure, aiming at less intensive agricultural production, a more positive attitude towards part-time farming, and a more restrictive policy towards motor traffic, hampering commuting. Today, in 1989, there are political signs of a less negative attitude towards small-scale farming on a part-time basis, a kind of revival of the idea of the owner-occupier small-farms. The crucial question is of course, if such a policy would just stabilize or even increase population, settlement, and variation in the landscape.

As a general observation, one might argue that in a medium time-span the changes in the landscape will be limited, irrespective of scenario. On the other hand, the possibility of major outside events and of surprises should also be taken into account. So far, however, the most obvious observations concerning the past, and of interest for the future, are that events in a local area become gradually more dependent upon external decisions and events, and that the effect of a single decision or event tends to grow, in both the depth and extent of the area affected. This does not, however, exclude an account of some other factors, changes in which may affect the landscape deeply also in the near future.

The provision of energy will, if and when the Swedish nuclear power stations are closed down, mean a search for other sources. One of the solutions includes an extensive use of wind power, and a large number of windmills are already discussed for the coastal areas of Scania, in our region including groups of windmills both inland and off the coast. Another possibility will be widespread planting of "energy forests", already begun experimentally on a small scale in the area.

Contradictory objectives for agriculture, as regards food and energy production, the natural and cultural environment, and a living rural society as a whole, make an evaluation of the future most uncertain. So does the absence of strict objectives in most of these fields as defined by a political majority. An important detail is an amendment in 1983 to the act on agricultural management, implying the protection of surviving walls and open waters, rows of trees and coppices, and other features in agricultural land, if they do not cause too serious obstacles to farming. This amendment, however, has meant nothing so far to the decisions of the agricultural board of Malmöhus County, which is responsible for enforcing the new law (oral information from the board).

A fresh uncertainty has come with the regulations in our new planning and building law, passed in 1987. According to this legislation, the local responsibility for physical planning also includes the conservation of our natural and cultural heritage. This has up to now been a responsibility of the county authorities, which from now on only have to give advice and to preserve specified national interests.

Conflicting goals in regional policy are still another uncertainty, as today's huge surplus production in agriculture could be restricted in many different ways and regions, in the open plains of southern Sweden, in the intermediate areas, or in the northern half of the country.

A specific factor of great interest not only to the Ystad area but to the whole province of Scania will be the final decision about future road and rail connections between Sweden and Denmark. A road bridge in particular is expected to increase industrial and other interest in the Malmö region. Such an expansion is almost bound to spread further east, also to the Ystad area, and may result in a totally new level of demand upon land resources.

5.13 Landscape-ecological aspects of long-term changes in the Ystad area

Björn E. Berglund, Nils Malmer and Thomas Persson

Ecological processes in a long-term perspective

Landscape is mainly a geographical concept to be seen as a synthesis of geomorphology, bedrock, soil, plants, and animals, where the main structures are formed by topography, hydrology, and vegetation. In a cultural landscape human activity has in one way or another transformed these structures, although much more in vegetation and hydrology than in topography, and far more in urban than in rural areas. In our view, landscape ecology deals with those aspects of the landscape which involve important biological processes: energy flow, synthesis and decomposition of organic matter, and transport of particulates, water, and mineral nutrients among different compartments and of biomass across boundaries, the number of boundaries increasing with the fragmentation of the habitats. We shall restrict our discussion here to the agrarian cultural landscape and exclude urban areas.

The formation of an agrarian landscape means changes brought about by human activity in the structure and species composition of the vegetation in a way that facilitates human use of the productivity (cf. Ch.3). Often the cultural landscape has been contrasted to a natural one in which humans have not caused changes, but such a distinct contrast does not exist. All forces and flows of material – both biological and other – are essentially the same in natural and cultural landscapes. Human influences are not in principle different from other biological and abiotic forces active in creating the structure of the landscape. Only the strength and therefore the effects of the activities are different.

From a biological perspective, landscapes are always patchy (Delcourt et al. 1983). This lack of uniformity is one of the main reasons for the biological variation and diversity in the landscape. In areas with a uniform macroclimate, the varying combinations of topography, hydrology, soil texture, soil acidity, and soil fertility at a specific site always determine much of the large-scale distribution of plants in the landscape. At this scale the patchiness is greater in undulating areas with varying soil depth (e.g., due to tectonized bedrock) than in areas like the Ystad area, which are topographically fairly smoothed and covered by deep sediments and moraines. The small-scale pattern in plant cover is created by biotic interactions (resource utilization, competition, herbivory) among the individual plants of species differing in life strategies, and by the death of individual plants, thus creating gaps that are rapidly colonized by new individual plants.

Large-scale patterns in the vegetation are also created from large-scale disturbances or catastrophes of many kinds (fire, storm-fellings, pests). The regeneration in the plant cover after such events usually takes place as a more or less predictable succession of plants over a long period of time. Forest successions begin with light-demanding, early successional species, and end up with competitive, usually shade-tolerant, late successional species. As different successional stages occur side by side in the landscape, they contribute, together with edaphic conditions, to the large-scale vegetational pattern of the landscape.

In a pollen diagram the proportions of early and late successional tree species will normally remain constant provided that the extension of disturbances and gap formations in the woodland vegetation remains the same or on a low level, and that the pollen source area is not too small. The local fluctuations in the tree population resulting from the woodland successions are then evened out and the structure of the vegetation will appear stable. Increasing disturbances in the woodland vegetation of the pollen source area will always come out in the pollen diagram as an increase in the proportion of early successional tree species, regardless of whether the disturbance is the result of changing abiotic conditions, biotic interferences (e.g. pests), or human activity. The dynamics of populations during such a period of extended disturbances will cause the structure of the vegetation to appear unstable compared with periods with fewer disturbances.

The delicate balance among competitors (Grime 1979) characterizing the late successional stages of the vegetation has undoubtedly changed in several ways during the Holocene period. Small variations in climate, for example, may bring about considerable changes in the balance among competitors in the final stages of succession (Woods and Davis 1989). Decreasing soil fertility has also been suggested as a reason for secular

Table 5.13:1. Characteristic attributes of structural vegetation types in the cultural landscape with special reference to the Ystad area.

Vegetation type	Human activity	Characteristic physiognomy of vegetation structure	Characteristic floristic elements	Characteristic plant strategies (Grime 1979)	Characteristic pollen components
Natural vegetation	Insignificant (hunting, gathering, some forestry)	That of the biotic region, usually stratified	Indigenous plants	Competitors	All components in AP and woodland herb species
Semi-natural vegetation (open or wooded pastures and meadows, field boundaries)	Cutting of trees and shrubs; cattle grazing, haymaking	Low growing and not or only partly stratified	Indigenous plants	Stress-tolerant plants	Apophytes of dry and fresh meadows, general apophytes, and wet meadow apophytes
Vegetation of arable land and urban areas	Disturbance of the soil surface and recurrent destruction of the plant cover	Low growing, not stratified outside urban areas	Introduced cultivated plants and weeds	Ruderals (incl. cultivated plants)	Anthropochores of arable and ruderal land, general apophytes
Potential vegetation	Insignificant (hunting, forestry)	That of the biotic region, usually stratified	Indigenous or introduced plants	Competitors	All components in AP, woodland herb species, some apophytes

changes in forest types (Iversen 1958, 1964, Andersen 1984). There are therefore major limitations in the possibilities for floristic reconstructions of Holocene plant communities. Plants common today have not necessarily always been common, and rare plants not necessarily rare. Pollen-analytical studies have also shown that plant communities without any analogues at present have been widespread (cf. Birks 1981, 1986a, Jacobson et al. 1987).

Vegetation types in a cultural landscape can be divided into four groups (Table 5.13:1). *Natural vegetation* types are dominated by indigenous species in a basic structure determined by the climatic region, the floristic elements present, and the local, edaphic site conditions. *Semi-natural vegetation* types are also dominated by indigenous species, but the stratification and structure have been changed through some human activity. On cultivated fields and other types of *arable land*, in *urban areas*, and in *forests planted with exotic species*, the vegetation is mostly dominated by species introduced to the area, especially as regards plant productivity. Examples of such species are cereals and most weeds.

The development of a cultural landscape means replacement of natural vegetation types with semi-natural ones, and increases in vegetation types characteristic of sites where disturbances in the soil surface have taken place. In a temperate climate, competitors characterize the natural vegetation, stress-tolerators the semi-natural vegetation, and ruderals the cultivated and urban areas (Grime 1979). Plant biomass, particularly that above ground, is greatest in the natural vegetation types (Malmer 1969). Because changes in edaphic conditions usually follow the destruction of natural vegetation types, the secondary successions in cultural landscapes often run toward *potential vegetation types* that are

structurally similar to but floristically different from the natural vegetation types.

The biological diversity of a landscape also depends on its biogeographical history, i.e., on the number of plant and animal species that have spread into the area. The possibilities for introduction of new species were almost certainly much greater during the Late-glacial and Early Holocene periods than later on. A closed and stabilized vegetation cover serves as a barrier to the colonization of new plant species; processes that create gaps in a closed vegetation cover are assumed to facilitate the immigration and establishment of new species in an area (cf. Birks 1986a). Thus, disturbances associated with development of the cultural landscape may allow an increase in the number of species in an area, particularly as human activities actively or passively carry seeds and other types of propagules.

Soil conditions are affected by vegetational and other processes. Over a long-term perspective in a temperate climate, it is generally assumed that the soil processes result in a slow decrease in soil fertility and in more acid soil conditions (Iversen 1958, 1964). Equally important is the accumulation of organic matter, which varies with vegetation type and water regime, continuing until an equilibrium is established between the annual litter formation and the total annual losses of organic matter from the soil. Impeded drainage of the soil or a rise of the subsoil water level can lead to paludification, with an increased accumulation of organic matter and a changed vegetation structure.

The accumulation of organic matter in the cultural landscape is less than in woodlands, more a result of faster decomposition of organic matter than of lower annual productivity and litter formation (Malmer 1969). Because of the lower plant biomass, the recycling of the plant-mineral nutrients is also more rapid in the cultural landscape than in woodlands. Any mechanical disturbance in the soil surface will immediately increase losses by leaching, and may also give rise to soil erosion and redistribution of soil material to topographic basins, particularly when extended over large areas or for long periods. In addition, disturbances in the soil surface will always change the soil profile and increase the mixing of the mineral and organic soil components. The result is usually an increased decay rate in the organic matter. Increased decay will also result from drainage of wet soils.

The potential natural vegetation of the Ystad area is woodland, primarily with broad-leaved, deciduous trees. In a long-term perspective woodland dynamics, including regeneration processes and shifts in dominance among late successional tree species, should therefore be key processes in landscapes not modified by man. The cultural landscape in this area was derived initially from woodlands; without disturbance, secondary vegetational succession could develop towards woodlands again.

Soil fertility does not vary much in the Ystad area, except for the sandy areas of the coastal zone, and the plateau of the Romeleåsen ridge with its less clayey and calcareous soils. This means that variations in water regime and soil wetness would have been the abiotic conditions which, together with vegetation dynamics, contributed most to the variation and diversity of the landscape in the absence of human activities.

Palaeoecological sources for the studies of the ecological development of a landscape, and methods for correlation

In this interpretation of long-term changes in the landscape of the Ystad area we have concentrated on information available from a transect of lake profiles with fairly uniform sediment accumulation records, namely Fårarps Mosse (lake until AD 1650), Bussjösjön (modern lake), Bjärsjöholmssjön (lake until AD 1850 but with disturbed sediments after AD 700); and Krageholmssjön (modern lake). The profile from Bjäresjösjön (lake since 800 BC) has been used as a complement to Bjärsjöholmssjön from about the year 0. For further information about the sites, see Ch. 2.1 and Fig. 2.1:1. Altogether our transect represents a gradient from the sandy, flat coastal area towards the clayey-silty hummocky inland area.

We have used the original pollen data from these sites available both as relative percentages and absolute pollen accumulation rates (PAR or pollen influx) except for Fårarps Mosse where no PAR values are available. Simplified diagrams from the sites are published in Ch. 4 except for Bussjösjön, an additional site discussed separately by Regnéll (1989). In some of the diagrams we have recalculated the pollen counts to compensate for differential pollen production and dispersal (An-

dersen 1970, 1973, 1978). The pollen spectra from the small basins of Fårarps Mosse, Bussjösjön, and Bjärsjöholmssjön are more locally influenced, and therefore have more irregular profiles than the regional pollen spectra from the large lake Krageholmssjön covering a larger area (cf. Prentice 1985, Ch. 2.1). The gradual change of the landscape from an area of woodlands to an almost unwooded open agricultural area has affected the size of the pollen source area (Prentice 1985), particularly during the time from the deforestation around 800 BC. Pollen values in the diagrams in this chapter are expressed as running means of three pollen counts (Figs 5.13:1–6, 8,10,11); the same applies to the charcoal counts (Fig. 5.13:9).

The PAR values should be correlated with tree populations (Aaby 1988, Bennett and Lamb 1988). However, water-level changes and erosion may also affect rates of accumulation of sediment and pollen, and so-called sediment focusing may occur even in the absence of such influence (cf. Birks and Birks 1980). Furthermore, the magnitude of the values is dependent on the size of the lake, i.e. the distance from the accumulation site to the shoreline (cf. Berglund 1973, Jacobson and Bradshaw 1981). Increased erosion influences the PAR through enrichment of all pollen types. This seems to have been the case in the top core parts of all modern lake profiles for the area.

The most uniform sediment sequences of high time-resolution for calculations of the PAR values were derived from the modern lakes Bussjösjön and Krageholmssjön. Also Bjärsjöholmssjön seems to be of reasonably good quality in this respect until 700 BC, although the dating control for the sediments corresponding to the period before 3000 BC is less reliable. Provided that our sequences are not affected by the processes mentioned above, we assume that the PAR values are a good reflection of changes in tree populations. The fact that the general trends are repeated in all three sequences supports the premise that the diagrams really reflect the species ratios among tree populations. For the NAP (non-arboreal pollen) we interpret the PAR values as representing mainly the local open landscape surrounding the lakes, while the corresponding values for trees represent a larger area in the landscape. In general the smaller basins differ in having higher pollen values, particularly for early successional trees and NAP.

The components in the arboreal pollen (AP) have been grouped in the following way:

a) Late successional trees with *Fagus* (beech), *Carpinus* (hornbeam), *Quercus* (oak), *Tilia* (lime), *Ulmus* (elm), and *Fraxinus* (ash) as dominant components. (In some diagrams *Fagus* and *Carpinus* are treated separately.)

b) Early successional trees with *Betula* (birch) and *Corylus* (hazel) as dominant components.

c) Fen-wood trees with *Alnus* (alder) and *Salix* (willow) as dominant components.

Pinus (pine) has been excluded from this grouping because of its sparse occurrence and its noise in pollen spectra caused by dominant long-distance transport (cf. Fig. 5.13:1).

Terrestrial components of the NAP have been grouped with respect to the dependence of the taxa on human activity (anthropochores: species introduced by man; apophytes: indigenous species favoured by man and spread from natural habitats) in the following way. This grouping is used in the pollen diagrams presented in Ch. 4 (Figs 4.1:1, 4.2:1,2, 4.3:1, 4.1:1).

a) Anthropochores (incl. crops and weeds) of arable and ruderal land.

b) Apophytes of dry and fresh semi-natural vegetation types (pastures and meadows).

c) General apophytes of open land, fields, roadsides, etc. (often pollen types which cannot be identified more narrowly than to genus or family and therefore not classified in groups b or d).

d) Apophytes of wet meadows.

e) Woodland herb and fern species (mainly *Pteridium*, *Mercurialis*, *Melampyrum*).

Along with the pollen counts, charcoal particles (>25 µm) were also tallied for all sites except Fårarps Mosse. The results are presented as accumulation rates (particles cm^{-2} yr^{-1}) and plotted against a calendar time-scale (see Ch. 2.9).

The chronology used here is, as in other parts of this volume, the radiocarbon time-scale. However, PAR values are only correct when related to a calendar time scale. Therefore, the PAR values are expressed in calibrated ^{14}C years (according to Stuiver 1986).

The number of taxa identified (to family, genus, or species level) gives an indication of species richness and biological diversity (calculated as richness in species) of the landscape that produced the pollen. The immigration of species was estimated from the number of new taxa found for each 500-yr period on all five sites taken together. Diversity is estimated by a rarefaction analysis in which the diagrams are constructed to show expected numbers of pollen taxa for each level (Birks et al. 1988). Using all taxa in the pollen sum, expected numbers of taxa were calculated for every sample prorated to sample count of 397 grains (the smallest single sample count in the sequence).

Landscape-ecological processes in the Ystad area

Woodland successions

During the Early Holocene, woodland vegetation was present throughout the Ystad area, with the exception of dry hill slopes, spring fens, lakes, and periodically flooded sites along rivers. Today, in contrast, the woodlands are limited to small areas, mainly in the inner hummocky zone (cf. Fig. 3.10:1), where they comprise 2600 ha or 9% of the total area.

Tree population dynamics during the entire Holocene in the Ystad area (Fig. 5.13:1) have generally followed the pattern of development around the southern Baltic (Andersen 1978, 1984, Andersen et al. 1983, Ralska-Jasiewiczowa 1989). On wooded sites with wet soils, *Alnus* and *Salix* have dominated throughout, probably together with some *Betula* and *Fraxinus*. During Atlantic time *Quercus* and *Tilia*, together with some *Ulmus* and *Pinus*, were the dominant late successional trees on the driest soils; *Fraxinus*, together with *Ulmus*, might have been most prominent on places transitional to wet soils (Malmer and Regnéll 1986, Regnéll 1989). Altogether, *Fraxinus* seems to have been the most common tree species for most of the time until the Late Subboreal period, except on well-drained soils (for the distribution of wet soils see Figs 1.2:14, 15).

From Late Subboreal time and onwards, *Fraxinus*, *Quercus*, *Tilia*, and *Ulmus* declined while *Fagus*, and to some extent *Carpinus*, gradually became more prominent among late successional trees on all soil but the wettest soils (Fig. 5.13:1). Pollen grains from *Fagus* and *Carpinus* are found only occasionally in modern lake sediments north of the distribution limits for these tree species in southern Sweden (Prentice 1983). Pollen diagrams with consistent values of 0.5–1.0% for these taxa are suggested to indicate scattered stands in the neighbourhood (Woods and Davis 1989), and frequencies >1.0% indicate regularly occurring stands. For the Ystad area this means that scattered *Fagus* (and *Carpinus*) trees occurred already at 2500 BC, and small stands from 2000 BC. A weak expansion of *Fagus* began at the transition from the Early to the Late Subboreal period (1400 BC); a greater one occurred after 200 BC, resulting in the present strong dominance of beech among late successional trees. The development was similar in Denmark (Andersen et al. 1983, Aaby 1988). Beech has always been rarer, especially on sandy soils, in the coastal zone than in the rest of the Ystad area; in the Bussjö-Fårarp region the first weak expansion was delayed by about 400 years (Fig. 5.13:2).

During this period of increase in *Fagus*, the proportion of *Betula* among the early successional woody species also increased at the expense of *Corylus* (Fig. 5.13:3). In the Bussjö area the increase in *Fagus* is contemporaneous with a distinct decrease in *Corylus*.

A high and increasing proportion of late successional tree species characterized the vegetation on sites with dry soils during the later part of the Atlantic period

KRAGEHOLMSSJÖN. Summary pollen diagram (Regnéll 1989)

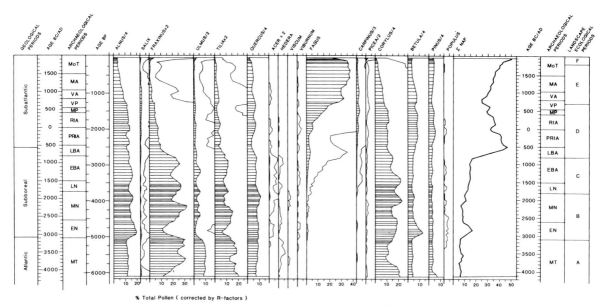

% Total Pollen (corrected by R-factors)

Fig. 5.13:1. Summary pollen diagram from Krageholmssjön, central part. Tree pollen counts corrected by R-factors proposed by Andersen (1970, 1973) which give a more realistic picture of the tree species proportions in the wood vegetation. Calculation sum: recalculated total of terrestrial vascular plants. Each pollen spectrum indicated by a horizontal line. Density of lines reflects depth/time resolution. A complete pollen diagram in Regnéll (1989). Time-scales in radiocarbon years.

WOODLAND VEGETATION

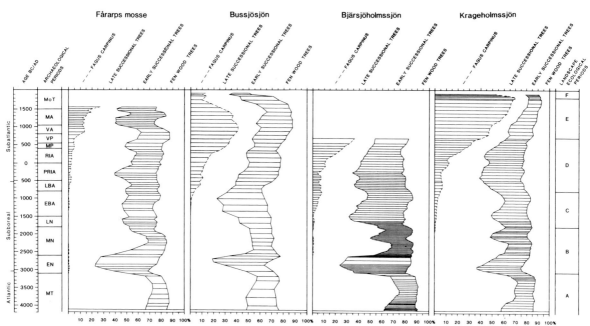

Fig. 5.13:2. Cumulative tree pollen diagrams from four sites along the transect from coast to inland. *Fagus* and *Carpinus*, other late successional tree species, early successional tree species, and fen wood tree species treated separately. Calculation sum: recalculated (Andersen 1970, 1973) total tree pollen. Time-scales in radiocarbon years.

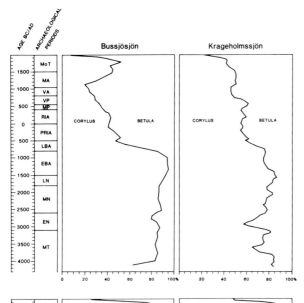

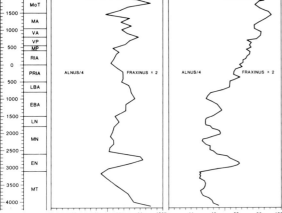

Fig. 5.13:3. The proportions (as percentages) of *Corylus/Betula* and *Alnus/Fraxinus* (values corrected following Andersen 1970, 1973) in the coastal area (Bussjösjön) and the inner hummocky area (Krageholmssjön). Time-scales in radiocarbon years.

eration cycle in the forested areas approached stages in which late successional trees were dominant.

Early successional trees were generally more prominent around Bjärsjöholmssjön and in the Bussjö-Fårarp area than is indicated by the regional diagram from Krageholmssjön (Figs 5.13:2 and 5.13:4). High AP proportions of early successional tree species (in Krageholmssjön about twice the amount of late successional species) were found particularly from 3000–2500 BC and 600–200 BC, in Krageholmssjön also AD 200–400, and in Bjärsjöholmssjön and Bussjösjön/Fårarp also 1700–800 BC. These periods probably followed directly after some kind of strong disturbances in the forested landscape. Their length would have been due to both the duration of the disturbance and the possibilities for late successional species to become dominant. For instance, a closed canopy of *Corylus* is generally more resistant to competing species than a *Betula*-dominated forest (Lindqvist 1938).

Until the Early Subatlantic period tree species characteristic of wet soils were much more abundant at Bussjösjön than at the other sites, and wet-soil species dominate the AP there, even in the most recent samples. This is the influence of the local wetland vegetation. A general feature is that since Late Subboreal time the proportion of *Fraxinus* in AP has decreased more than that of *Alnus* (Figs 5.13:1 and 5.13:3), mainly because of ash woodland being more attractive for meadows than alder fens.

The increasing proportions in AP of *Salix* since the Late Subboreal period (Fig. 5.13:1) may indicate more frequent disturbances and regeneration cycles in the tree cover on wet soils, similar to those on the dry soils, as *Salix* is an early successional species. The increasing proportion of *Betula* could also be a result of such disturbances. From 700 BC and onwards the proportion of wet-soil species decreased (Fig. 5.13:2). Simultaneous increases of Cyperaceae (sedges) and *Filipendula* (meadowsweet) sometimes evident in NAP indicate a change in the vegetation cover on wet soils from woodland to some sort of non-wooded wet meadow or fen (Regnéll 1989; see Fig. 5.13:8).

The spread of non-wooded vegetation
Although the whole Ystad area should be regarded as a potentially wooded area, open, non-wooded areas must always have occurred on a limited scale because of natural events and disturbances. During Atlantic time the percentage of NAP never reached 5% of the AP (Fig. 5.13:5), and accumulation rates were low (Fig. 5.13:4). The replacement of late successional trees by early successional ones on dry woodland sites in the Early Subboreal (beginning 3000 BC) was accompanied by increases in NAP when calculated as both percentages and accumulation rates. Taxa well known to indicate grazing and cultivation occurred together with many true woodland species. Coincident with the re-establishment of late successional trees in the woodland

(Fig. 5.13:2). In the regional Krageholmssjön diagram the accumulation of late successional tree pollen reached about 150% of the accumulation of early successional trees for a period of more than 500 years (Fig. 5.13:4). These proportions changed drastically at the transition to the Subboreal period (3100 BC). Pollen from early successional trees rapidly became dominant in the pollen accumulation for a period of nearly 4000 years. Late successional trees became dominant again from AD 700–800 and onwards, although the relations are not directly comparable because of changes in species. For short periods, however, the proportion of late successional trees was similar to that of early successional species (at 2300–1800 BC, 1200–1000 BC, and around the year 0; cf. Figs 5.13:2 and 5.13:4). These periods might be looked upon as times when the regen-

410

on dry soils after 2500 BC, the NAP taxa returned to the same low figures as during Atlantic time (Fig. 5.13:5). Locally, secondary increases in NAP are seen for brief and apparently not synchronous periods during most of the Subboreal. Those fluctuations, combined with high amounts of early successional trees (particularly at Bjärsjöholmssjön), suggest a period of considerable instability in the forest vegetation.

A small but persistent rise in total NAP percentage occurred 2000–1800 BC at Bjärsjöholmssjön and Krageholmssjön (Fig. 5.13:5). This rise was contemporaneous with a trend of decreasing PAR values from tree pollen (Fig. 5.13:4), but was a few hundred years earlier than the initial expansion of *Fagus* and *Carpinus* among the shade trees on dry soils (Fig. 5.13:1,2). The PAR values for tree species from both wet and dry soils fell by roughly 50% at the same time as the percentage of NAP from open-land areas increased 2–3 times relative to conditions during Atlantic time. Not only tree pollen accumulation rates, but also the proportion in NAP of pollen from woodlands, were reduced (Fig. 5.13:6).

Beginning around 800 BC a distinct second increase in NAP is evident at all the investigated sites, both as percentages and as accumulation rates (Figs 5.13:4 and 5.13:5). There was also a drop in the tree pollen accumulation rate, particularly for the late successional trees. A small drop can also be seen in the woodland NAP, but it is less distinct than the one at 1800 BC (Fig. 5.13:6).

A further but less distinct increase in the total percentage of NAP began around AD 700 (Fig. 5.13:5). In the regional diagram from Krageholmssjön the increase continues until the present, with a short interruption AD 1400–1500. In the local diagrams it was interrupted and even reversed on several occasions. Around AD 700 a drop in AP accumulation rates is also seen in both the species from dry soils and those from wet soils, but the proportion of wet-soil species decreases and that of late successional species increases throughout.

The last 100 years seem to be characterized by increasing accumulation rates for both AP and NAP (Fig. 5.13:4). The increase in AP since the end of the 19th century reveals the reforestation of large areas of low fertility, mainly outside the Ystad area (Lindgren and Malmer 1970). However, because of lowered water level and increased erosion during the 18th and 19th centuries, the pollen accumulation rate may also have been influenced by within-basin disturbances in the sedimentation processes.

Both the increase around 800 BC and that around AD 700 in the NAP were preceded by periods with rather high proportions of late successional tree species in AP (Figs 5.13:2, 4). Corresponding drops in AP start rather distinctly, first among the early successional species and somewhat later among the late successional ones. The first part of the Subatlantic period was characterized by an increasing accumulation rate of AP, including late successional species (Fig. 5.13:4), combined with a decreasing (or at Fårarp fairly constant) percentage of NAP (Fig. 5.13:5). No doubt the wooded area expanded during this period; however the PAR values for the period AD 400–700, for example, in Bussjösjön, might have been affected by changes in sedimentary processes, as both AP and NAP peak at the same time (Fig. 5.13:4).

The spread of non-wooded vegetation and the opening of the landscape was accompanied by an increased sedimentation of charcoal particles in the small basins sampled (but not in the sediments from Krageholmssjön, which represent regional deposition (Fig. 5.13:9)). Contemporaneous with the replacement of the late successional tree species by early successional ones on dry soils and the increase in NAP in Early Subboreal time was an increase in the accumulation rate of charcoal for a period of about 500 years. A second, fairly small rise occurred around 1800 BC, and a third and greater one around 800 BC, both contemporaneous with expansions of semi-natural vegetation (Fig. 5.13:5) and decreases in the tree PAR (Fig. 5.13:4). Around the year 0 the accumulation rate of charcoal decreased, although not to the low values characterizing the woodlands during Atlantic and Early Subboreal time. Around AD 700 there was a new increase and the accumulation rate reached higher values than during the periods before. During the rest of Late Subatlantic time the variation in accumulation rates of charcoal particles seems to be little related to landscape-ecological processes, just as was the case with the PAR values from that period.

The immigration and spread of plant species
Southern Scania is floristically the richest part of Sweden (Lindquist 1966). More than 900 vascular plant species per 1600 km² have been noted from that area in recent inventories. The number of taxa identified in the pollen diagrams is about 150 (excluding aquatic plants, see Ch. 2.1), which is much less than the present total number of plant species.

Samples from the Atlantic period (4500–3000 BC) contained not more than 38% of all the identified pollen taxa (Fig. 5.13:7) During the following period of woodland disturbances and instability, 26 new taxa (19%) were added, most of them herbs. Another 59 (43%) new taxa were recorded from the period with a strong increase in NAP from 700 BC and onwards. The increase in NAP around AD 700–800 was also accompanied by a great number of new taxa. However, for the last millennium only 12 new taxa (9% of all identified) were recorded in the pollen diagrams.

During the development from a wooded landscape towards the present open landscape, great fluctuations in plant populations must have taken place. Some new plants certainly must have colonized the area – real immigrants like crops and the weeds that accompanied them (anthropochores). In most cases however, what appear to be species new to the area may simply have been favoured by the opening of the landscape, and

POLLEN ACCUMULATION RATES

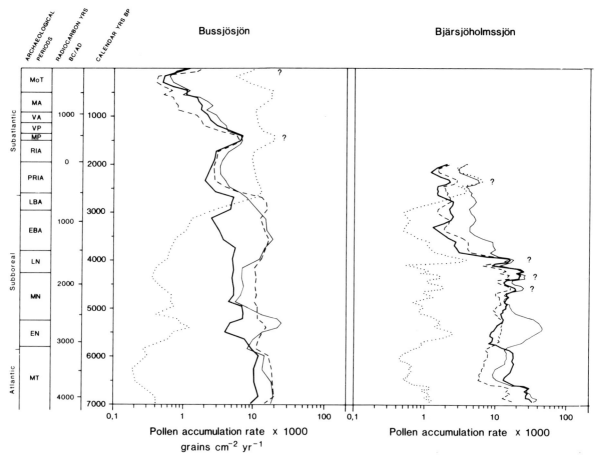

Fig. 5.13:4. Pollen accumulation rates (grains cm^{-2} yr^{-1}, log scale) for the early and late successional tree species, fen wood species and NAP pollen at four sites along the transect from coast to inland. Values plotted against calendar time scale. This time scale is given parallel to a scale based on uncalibrated ^{14}C years. Levels where depositional errors may have influenced the accumulation rates are indicated by ?.

hence spread from seashores, lake and river banks, wood edges, and dry hill slopes (apophytes), for example. In the latter case the apparent change in species richness means merely population expansions. Certainly the number of taxa identified depends to a large extent on the number of pollen grains counted, particularly NAP. Therefore in the present material it is impossible to separate new immigrants to the area from species that already occurred there, but whose populations increased greatly, thereby providing a greater chance to be tallied during the pollen analysis.

Soil conditions
A deforestation and opening of the landscape like that documented in the present pollen diagrams increased erosion as a result of widespread disturbances in the vegetation cover and soil surface. Sedimentation in lake basins increased correspondingly. According to sediment studies by Dearing (Ch. 5.2), the sediment yield

was rather low and constant until the first part of the Late Subboreal period, although there are some indications of increased topsoil erosion during the Early Subboreal (Fig. 5.2:7). At the end of the Late Subboreal, seemingly simultaneously with the strong increase in NAP at about 800 BC, the total sediment yield started to increase, mainly owing to increased channel erosion. This increase in sediment yield continued until the end of the 19th century when it was suddenly much reduced. Topsoil erosion does not seem to have increased as much as channel erosion at this time. However, there is a distinct increase in topsoil erosion at the transition from the Early to the Late Subatlantic period corresponding to the increase in NAP around AD 700. For a short period around AD 1400, topsoil erosion was low; otherwise it has been increasing since then and is high relative to channel erosion even at present. Seen over the whole 6000 years, erosion has changed the pronounced relief of the original moraine landscape to a

412

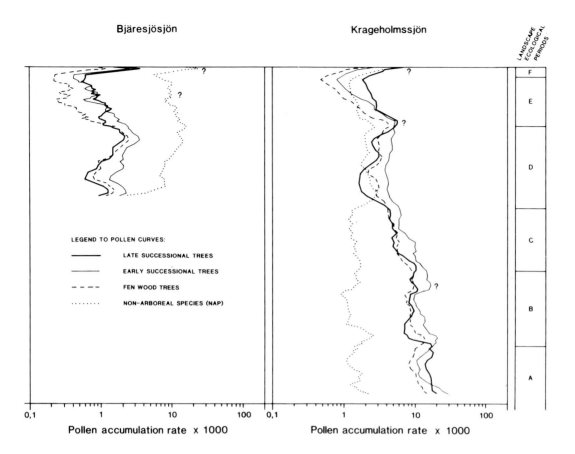

Bjäresjösjön   Krageholmssjön

LANDSCAPE ECOLOGICAL PERIODS

LEGEND TO POLLEN CURVES:

—————— LATE SUCCESSIONAL TREES

——————— EARLY SUCCESSIONAL TREES

— — — — FEN WOOD TREES

.... NON-ARBOREAL SPECIES (NAP)

F
E
D
C
B
A

0,1   1   10   100   0,1   1   10   100

Pollen accumulation rate x 1000   Pollen accumulation rate x 1000

much smoother one in the modern landscape. Organic lake or fen deposits in topographic depressions have often been covered by eroded minerogenic material (Daniel 1977, 1989).

The eutrophication of the lakes indicated by the diatom analysis (Håkansson 1989, unpubl.) must be an effect of this continuous soil erosion. These results do not indicate any general decrease in soil fertility or any increased soil acidity, at least not when the areas with clayey soils outside the sandy coastal zone are considered. Thus, the Ystad area does not seem to have reached the oligocratic phase in the general Holocene ecological development as discussed by Iversen (1958) and Birks (1986a, 1988). The reasons for this maintenance of the soil fertility and soil pH over such a large area and for such a long period could be both the original high fertility of the mineral soil and the erosion continuously exposing new soil to the weathering processes.

The present distribution of organic soils (Figs 1.2:14,

15) reveals areas where an accumulation of organic matter has taken place by either paludification of the ground or infilling of lakes. During Atlantic time there were about 19 lakes with a size >2 ha. Several of them became infilled during the Late Atlantic and Early Subboreal periods, probably because of the low groundwater level prevailing during that period (Digerfeldt 1988, Ch. 5.1). Today only three of the former lakes remain, and the total lake area has been reduced to 220 ha from originally about 500 ha. Some of the earlier lakes have disappeared since the beginning of the 19th century following ditching and drainage of the landscape. However, the greatest infilling with organic matter has taken place at the Baltic coast in the Öja-Herrestad basin, as a result of the combined effects of land upheaval and the development of marsh vegetation in the sheltered bay.

At the end of the Subboreal period large areas around lakes and along rivers became paludified be-

OPEN LAND VEGETATION

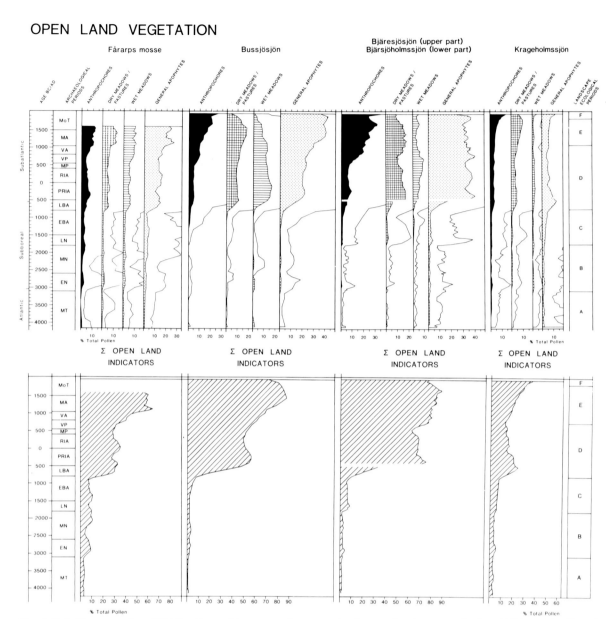

Fig. 5.13:5. The proportion of NAP pollen and fern spores from open land vegetation (groups a–d, cf. p. 408) in total pollen of terrestrial vascular plants in the diagrams from five sites along the transect from coast to inland. Calculation sum: total AP + NAP pollen. Time-scales in radiocarbon years.

cause of a rising groundwater level (Ch. 5.1). In several cases this paludification might have resulted in the development of an alder carr vegetation, usually replaced sooner or later by a wet meadow or sedge fen vegetation (Fig. 5.13:8). The maps from the beginning of the 19th century give strong evidence for how widespread these vegetation types were (Figs 1.2:14, 3.8:1). As far as we know, the accumulation of organic matter has never resulted in the deposition of ombrotrophic *Sphagnum* peat or other types of low-humified organic layers. Such a development in the peatlands has been documented in

many other places on and around the Romeleåsen ridge (Mörnsjö 1968).

Processes in the formation and development of the cultural landscape

Climatic changes
The development of the cultural landscape in the Ystad area is a process which has now been going on for about 6000 years. Stable climatic conditions or an otherwise stable abiotic environment do not exist for such a long period of time. The climatic changes are most evident in

414

WOODLAND HERBS AND FERNS

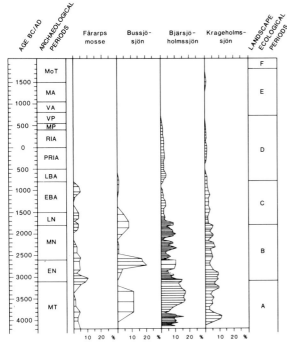

Fig. 5.13:6. Woodland herb pollen and fern spores (group e) as proportion (%) of total NAP in the diagrams from four sites along the transect from coast to inland. Time-scales in radiocarbon years.

the long-term variation in lake-water levels. In lake Bysjön, situated on a sandy plain in central Scania, the water level was much lower than the present one for a long period of the middle Holocene (Digerfeldt 1988). The lowest water level, more than 8 m below the present one, occurred just at the beginning of the Early Subboreal period (3000–2500 BC). Also for the lakes Krageholmssjön, Bjäresjösjön, and Kumlan fluctuations in water level have been recorded (Gaillard 1984, Gaillard and Berglund 1988, Ch. 5.1). A minimum water level occurred at the beginning of the Late Atlantic period (4000 BC), followed by a small rise in water level until the beginning of the Subboreal period when the level fell again with minima around 1700 BC and 700 BC. Both in lake Bysjön and in the Ystad area, the end of the Late Subboreal period was characterized by water levels near the present high one; the water level was much lower than at present near the end of the Early Subatlantic period AD 400–700. Altogether these observations indicate considerably drier water balance than at present, and varying and unstable soil water conditions during a long period from the beginning of the Subboreal period and onwards, and during Early Subatlantic time.

Reconstructions of the variation in water level are founded on observations independent of plant cover. Palynological data, which of course are derived from the plant cover, indicate July mean temperatures about 2°C higher than at present for much of northern and central Europe during the later part of the Atlantic period (Iversen 1944, Huntley and Prentice 1988). Combined with the low lake-water levels observed, this shows that the climate during that period might have been very different from that at present.

The transformation of the natural vegetation

Data from this project suggest that the Late Atlantic period was characterized by fairly stable conditions in the forest vegetation in the Ystad area. The increase in late successional tree species as well as the low and constant values for NAP indicate this. The share of wet-soil species in AP did not change during that period, which indicates that the soil water conditions might have remained the same despite the fact that lake-water levels became lower (Digerfeldt 1988). Small-scale garden cultivation has been postulated (Ch. 4.1.2, 4.2.2, 4.4.2), but if this occurred it affected only small areas in the otherwise wild woods. The transition to the Subboreal period marks a change to much more unstable conditions.

Changes in the woodlands of dry soils starting with the elm decline around 3100 BC were doubtless initiated by a serious ecological disturbance of some kind – an "elm disease", a rapid climatic change, or both in combination – drastically transforming the vegetation over large areas. It has been emphasized that all late successional trees (except *Quercus*) regressed favouring *Betula* and *Corylus* on dry soils and *Alnus* and *Salix* on wet soils. Such a breakup of the wildwoods must have facilitated the spread of human activities like shifting cultivation and cattle-rearing in the area. The pollen diagrams also have clear signs of human activities affecting the vegetation structure, appearing contemporaneously over the whole Ystad area around 3000 BC and lasting for 300–500 years (Fig. 5.13:5). There is an increase in total NAP associated with both cultivated fields and semi-natural vegetation types of dry and fresh soils.

During the following period of about 800 years until 1800 BC, human disturbances of the landscape were local or concentrated at the coast (Fig. 5.13:5). The clearances were evidently not large enough to affect the PAR values of tree pollen. Fairly closed wood vegetation might therefore have been most prominent in the landscape all the time, except in the coastal area.

A more distinct and permanent change in the vegetation structure due to increased human activity is indicated around 1800 BC (Fig. 5.13:5). This change is most apparent in the sites Fårarp and Bjärsjöholmssjön in the coastal and outer hummocky landscape. From that time onwards open areas cleared by man have been a general and widely distributed feature in the landscape of the Ystad area, giving rise not only to a permanently increased proportion of NAP but also drastically reduced tree PAR. The two later distinct rises in the

PLANT TAXA

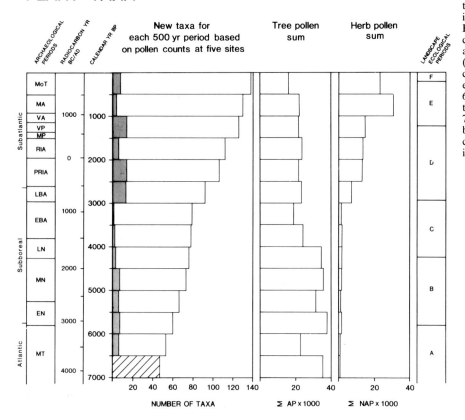

Fig. 5.13:7. The number of terrestrial plant taxa identified at five sites (Få, Bu, BJ+Bj, Kr) since 7000 cal. BP. Values expressed as additional pollen taxa (shaded bars) as well as cumulatively (open bars) for each 500-year period since 6500 cal. BP in comparison to the number of taxa 7000-6500 cal. BP (hatched bar). Absolute number of counted AP and NAP pollen is also given for each period.

proportion of NAP all over the Ystad area, observed around 800 BC and AD 700 (Fig. 5.13:5), indicate that the development of the cultural landscape happened in steps and not as a continuous process.

There is a discrepancy between the increases in percentage and accumulation rates of NAP and corresponding decreases in the woodland NAP and tree pollen accumulation rates (Figs 5.13:4–6). The relatively small increase in NAP around 1800 BC (both as proportion and as accumulation rate) was contemporaneous with the most distinct decrease in accumulation rate of AP from both dry and wet sites, while the much greater increase in the proportion of NAP 1000 years later was contemporaneous with a smaller drop in the tree PAR. We do not believe that either methodological errors or within-basin processes can explain these discrepancies. To some extent a variation in the PAR values could be due to changes in the tree cover from species with high pollen production to species with lower production, but such a replacement cannot explain the discrepancies fully. The reduced tree PAR from 1800–1400 BC at Krageholmssjön and Bjärsjöholmssjön appeared in all three groups of tree species and was combined with the most pronounced reduction of woodland NAP. The most realistic hypothesis is that this resulted from opening of dry grounds near the sampling sites, which greatly reduced the total pollen production in the near vicinity

of the sampled lakes, and thus also lowered the tree PAR and woodland NAP from the nearby areas compared with more distant ones.

It seems most probable that the proportion and accumulation rate of NAP reflect the expansion of the open landscape better than AP. This suggests that the increase of the open areas during the period 1800–1400 BC was considerably less than that from 800 BC and onwards, despite the fact that the reduction in tree PAR was less then. Accepting such an interpretation of the available pollen data, the expansion of the open landscape beginning around 1800 BC took place mainly in the coastal and outer hummocky areas, leaving most of the interior rather untouched. However, from 800 BC onwards the whole Ystad area must have been characterized by an open, sparsely wooded cultural landscape dominated by semi-natural vegetation types and cultivated fields in varying proportions. The periods with reforestation or further expansion of NAP-forming vegetation types, particularly from AD 700 onwards, do not change this general picture.

The increase of wet meadow species in NAP was paralleled by a decrease in *Alnus* in AP (Fig. 5.13:8), particularly evident around 800 BC, but continuing until AD 1400. The accumulation of organic matter in areas of waterlogged soil caused such a development to be common and stratigraphically well documented in sev-

WETLAND VEGETATION

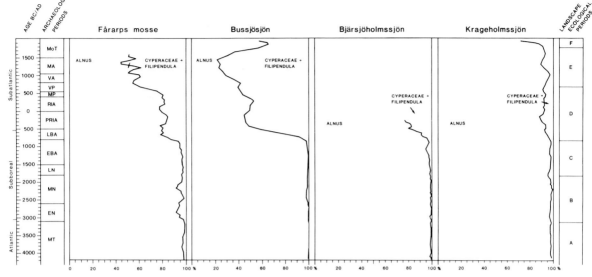

Fig. 5.13:8. The proportion of *Alnus* vs *Filipendula* + Cyperaceae at four sites in the transect from coast to inland. This diagram illustrates the relation between fen woods and non-wooded wetlands. Time-scale in radiocarbon years.

eral areas around the southern Baltic (cf. Mörnsjö 1969, Succow and Lange 1984). Usually it has been attributed to climatic changes and rising water levels. These may also have affected this development in the Ystad area, as the latest part of the Subboreal period was characterized by rising lake-water levels (cf. Ch. 5.1). However, as the initiation of the process around 800 BC is contemporaneous with the deforestation of the areas with dry soils, it must be assumed that this opening up of the fen woods was also strongly influenced by human activities.

We assume a relation between charcoal accumulation rate and clearings (cf. Patterson and Backman 1988). The variation in the deposition of charcoal in the sediment profiles (Fig. 5.13:9) suggests that widespread fires were very important both for the transformation of the woodlands to an open landscape and probably also for the maintenance of that landscape. Through these fires the biomass as well as the amount of energy and organic matter stored in the landscape decreased drastically. Primary production did not necessarily decrease at the same time, however (Malmer 1969), but great amounts of mineral nutrients were released in forms easily available to the plant cover.

Development in the open landscape

The expansion of the open landscape around 1800 BC was concentrated to areas with dry soils where the semi-natural vegetation expanded first (Fig. 5.13:10). The proportion of cultivated land increased more slowly, and partly at the expense of meadows and pastures. Erosion did not increase, however (Ch. 5.2).

Likewise the expansion of the open landscape beginning at 800 BC started with an expansion of the semi-natural vegetation, probably on both dry and wet soils (Figs 5.13:5, 8, 10). An expansion of the cultivated land area must also have occurred, because the total area of open land increased greatly and there are clear signs of a topsoil erosion and increasing sediment yield (Ch. 5.2). Therefore, pastures and meadows with semi-natural vegetation types of dry and wet soils must have dominated the whole landscape for a long time.

Unlike what happened with these Late Subboreal clearances, the openings made before 2000 BC in the unstable woodlands on dry soils did not develop further to become large areas permanently covered with an open, semi-natural vegetation of pastures and meadows (Fig. 5.13:5). Besides woodland herbs, the predominating groups in NAP from this period were anthropochores and general apophytes. Much of these open areas might therefore have been occupied by vegetation characteristic of settlement areas and fields established through shifting cultivation and open for a rapid recolonization by trees and shrubs when abandoned (Fig. 5.13:3). Such a conclusion is confirmed by the constant PAR values and the rise in topsoil erosion indicated in the lake sediments from this period with a dominating natural wildwood vegetation (Figs 5.2:7, 5.13:4).

A distinct increase in the area of cultivated land must have taken place around AD 700–800. The proportion of cultivated plants and weeds of NAP increased at that time, mainly at the expense of species from dry meadows and pastures (Fig. 5.13:10); the topsoil erosion

417

CHARCOAL DUST

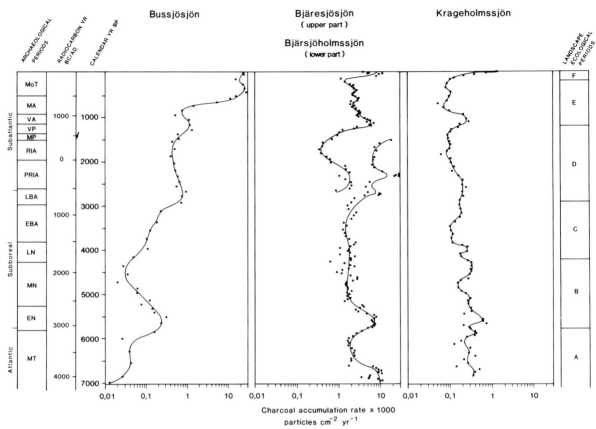

Fig. 5.13:9. The deposition of charcoal particles (> 25 μm) at four sites along the transect from coast to inland. Values expressed as accumulation rates (particles cm^{-2} yr^{-1}, log scale) and plotted against a calendar time-scale.

increased as well (Fig. 5.2:7). This expansion of the cultivated areas, mainly at the expense of semi-natural vegetation, has continued until the present, possibly with the exception of the period around AD 1200–1400, when meadows and pastures expanded in some inland areas (Fig. 5.13:10, Regnéll 1989, Ch. 4.2.2). This process was particularly intense during the last century, as is also demonstrated in the pollen diagrams.

The proportion of taxa from wet or waterlogged soils in NAP was low until 800 BC (Fig. 5.13:5). Variations in the diagrams are best explained by variations in other components of the pollen. We do not believe that open, non-wooded areas existed on wet soils until that time, with the obvious exception of such wetland areas as flooded riversides that were woodless for natural reasons. However, from that date on, non-wooded vegetation types must have expanded on wet soils (Fig. 5.13:8), at about the same rate as the open areas on dry soils, locally even increasing their proportion in the open landscape.

Since about AD 700 the proportion of NAP taxa from wet or waterlogged soils has decreased (Fig. 5.13:10). As the accumulation rate of NAP has been constant or

even increased during the same period (Fig. 5.13:4), such a decrease suggests a real reduction of the area of wet meadows and fens. However, *Filipendula* and Cyperaceae are the dominating NAP components from wet areas. When such areas are exploited for grazing and haymaking not only the production of pollen from wetland species but also their dominance in the vegetation is reduced. *Filipendula* together with tall-growing Cyperaceae species are replaced by a low grass sward forming a type of semi-natural vegetation dominated by other herbs and graminaceous species that produce much less pollen (or pollen included in other groups). It is therefore possible that the reduction of pollen from wet meadow and fen species during the period AD 700–1800 (Fig. 5.13:10) results from a much stronger exploitation of areas with wetter or more waterlogged soils than earlier, and not from reduction of their total area. It is well known that extensive recent drainage has drastically reduced the wet areas in the landscape compared with the natural conditions. This process did not become important in the landscape until AD 1800, and it is badly reflected in the pollen diagrams.

ARABLE LAND / PASTURES / MEADOWS

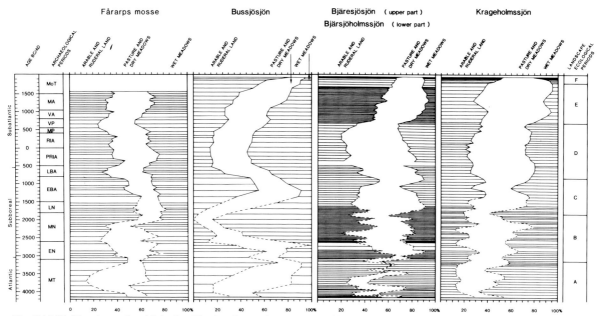

Fig. 5.13:10. The share of arable/ruderal land anthropochores, pasture/dry meadow apophytes and wet meadow apophytes in the NAP from open land vegetation in the diagrams from five localities along the transect from coast to inland. The lower part of the diagrams is based on a low number of pollen counts and therefore difficult to interpret (indicated by broken lines). Calculation sum: NAP of groups a–c (see Fig. 5.13:5). Time-scales in radiocarbon years.

Reforestation and other trends in woodland development

The PAR values for AP remained fairly constant from Atlantic time through the Early Subboreal period, even during disturbances in the wildwood vegetation after the elm decline around 3100 BC and during the first human agricultural activities (Fig. 5.13:4). The decreasing populations of late successional tree species were therefore replaced rapidly by early successional species, both *Betula* and *Corylus* and on wet soils partly by *Alnus*. Around 2300 BC the late successional species seem to have partly reoccupied their former place in the wildwoods again.

The expansions of herb and grass vegetation beginning at 1800 and 800 BC (Fig. 5.13:5) were both followed by regenerations in the woodlands first seen among the early successional trees but with increasing cover of late successional species around 1200–800 BC and 200 BC-AD 500, respectively (Figs 5.13:2, 4). To a minor degree a similar development also followed the expansion around AD 700, as there are some indications of a woodland regeneration for a short period around AD 1300 (Figs 5.13:1, 2). However, the PAR values (Fig. 5.13:4) and the representation of the species in AP show that from AD 1000 and until the present day the early successional species and the tree species on wet soils have been cut back in the landscape.

The predominance of late successional species on dry soils during the last millennium is probably more due to the expansion of the cultivated land and an exclusion of early successional trees from areas with semi-natural vegetation than to a real expansion of the late successional trees. However, late successional trees may have been favoured around the estates for hunting purposes. It is well known that extensive areas have been afforested with *Fagus* during the last 150 years, both inside and outside the Ystad area. The influence of long-distance dispersal on the relative proportions of the tree species in the pollen counts has also continuously increased as the landscape in the Ystad area has been deprived of most of its woods.

The so-called elm decline around 3100 BC represents one major event in the vegetation history of woodlands not only in the Ystad area but over large parts of central and western Europe (Huntley and Birks 1983). In the Ystad area a decline affected not only *Ulmus*, but also *Fraxinus* and *Tilia* (Fig. 5.13:1). However, in a long-term perspective it did not bring about a total change in the balance among late successional trees because a similar type of wood vegetation dominated by *Fraxinus, Quercus, Ulmus,* and *Tilia* seems to have been re-established on the dry soils at least half a millennium later.

There are good reasons (e.g. water-level changes in lakes) to assume widespread and repeated disturbances

419

in the tree cover during the 500-year period after the elm decline. Variation in the soil-water regime, particularly on the sandy soils in the coastal zone, gave rise to a less closed forest canopy and favoured early successional trees, herbs, and grasses. *Quercus*, the tree species dominant on the dry soils, and *Alnus*, dominating on the wettest soils, remained unaffected by the events at the time of the elm decline (Fig. 5.13:1). *Alnus* even increased particularly at the beginning of this period, probably favoured by the regression of *Fraxinus* on damp soils. Further, dense stands of *Corylus*, which sometimes seems to have expanded after the elm decline, may have delayed the re-establishment of the late successional tree vegetation (Lindqvist 1938).

The expansion of *Fagus* and *Carpinus* from 1400 BC and onwards represents another major event in the woodland history around the southern Baltic (Huntley and Birks 1983). During Subatlantic time *Fagus* became the dominant late successional tree in the forests on dry soils in the Ystad area (Figs 5.13:1, 2). Huntley (1988, Huntley et al. 1989) has postulated that climatic changes are responsible for the expansion of *Fagus* and *Carpinus* through Europe. Further, *Fagus* is a tree better adapted than *Ulmus* to such acidic and leached soils as are expected to develop gradually through time (Iversen 1964, Berglund 1966) and its present geographic distribution could well be fitted in with existing climatic gradients (Prentice 1988a). The suggested continuous occurrence of beech stands in the Ystad area during a period of at least 1000 years before the first expansion shows that *Fagus* then had difficulties in invading closed woodlands, just as it seems that *Fagus grandifolia* had in North America before its Subatlantic expansion took place there in native, undisturbed woodlands (Woods and Davis 1989). In Europe, however, the Subatlantic expansion of *Fagus*, just as in the Ystad area, coincided with reforestation periods beginning around 1400/1200 BC and 200 BC following periods with increased disturbances in the vegetation cover from human activities (Ralska-Jasiewiczowa 1964, Berglund 1966, Andersen 1973, Huntley and Birks 1983, Aaby 1986, Regnéll 1989). The spread of both *Fagus* and *Carpinus* in the Ystad area could thus indirectly have been favoured by human activities. There are also studies on recent south Scandinavian vegetation showing *Ulmus* and *Fraxinus* expanding in beech forests on wet, fertile soils (Lindquist 1931, Lindgren 1970, Malmer et al. 1978).

There are fewer reasons to suggest that the expansion of *Salix* and *Betula* at the expense of *Corylus* during Subatlantic time (Figs 5.13:1, 3) may have been climatically determined. *Betula* and *Salix*, with their much lighter seeds, could well be expected to be better adapted to spread in a heavily exploited cultural landscape. Another reason for the decline of *Corylus* could be changes in soil fertility, as *Betula* and *Salix* become more prominent in the early successional stages the more acid and less fertile the soils are. This is the case also when comparing *Alnus* and *Fraxinus*, as *Alnus*

always becomes more prominent than *Fraxinus* on less fertile soils (Fig. 5.13:3). Therefore, the increasing proportions of *Betula* as well as *Salix* and *Alnus* during Subatlantic time may be due partly to a greater share of pollen derived from less fertile, marginal inland/upland areas (the Romeleåsen area) and partly to an increasing restriction of the trees to soils with low fertility. As woody species are excluded from the dry soils in a landscape – as in the Köpinge-Kabusa area in the coastal zone – the proportion of *Betula* and *Alnus* will increase; *Corylus* and *Fraxinus* are less prominent on the wettest soils. All these species but *Corylus* are dispersed mainly by wind or water. Although *Corylus* spread very rapidly in the open landscape of the Early Holocene (Regnéll 1989) it cannot be ruled out that faunal changes (particularly in the avifauna) after the drastic changes in the vegetation during the Late Boreal reduced the possibilities for an efficient dispersal through the landscape not only of *Corylus* but also of *Fagus* and *Quercus*.

Landscape diversity

During Late Atlantic time the variation in the vegetational landscape of the Ystad area mainly reflected the topography and the edaphic conditions, which gave rise to an uneven distribution of tree species, and successional stages, which led to some openings in the cover of wildwoods. In such a landscape all disturbances will serve to increase variation and diversity. Particularly important would be disturbances from human activity, as they are often longer-lasting than natural ones and therefore create better possibilities for the spread of taxa from one point to another. Direct and indirect transports of seeds and plants by man also favour the introduction and spread of new taxa. The increases in number of taxa recorded in the pollen diagrams 3000–2000 BC, around 500 BC, and AD 500–1000 (Fig. 5.13:7) are all closely linked to periods with increasing human activity and to expansion of open landscapes because of these activities.

Rarefaction analysis (Fig. 5.13:11) demonstrates the low diversity in the Late Atlantic wildwoods compared with that in the much more open landscape from 500 BC and onwards. The disturbances by natural events and human activity during the Early Subboreal increased the diversity described in this way. The greatest increase, however, took place during the period after the opening of the landscape in 800 BC. A further but smaller increase took place at the beginning of the Late Subatlantic period. The periods of reforestation do not seem to have reduced the diversity in the vegetation. For the last centuries there are clear indications, particularly in the Bussjösjön diagram, of decreasing diversity in the landscape structure. This is parallel to the findings at Diss Mere in southern England (Peglar et al. 1988, Birks et al. 1988). In both areas it can be explained by the development of a landscape with exten-

VEGETATION DIVERSITY

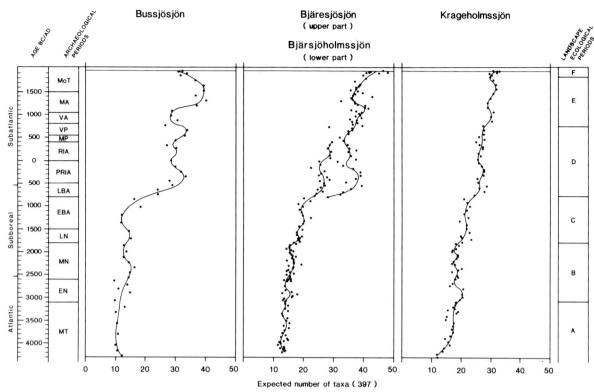

Fig. 5.13:11. Estimated vegetation diversity at three sites according to a rarefaction analysis of each sequence of pollen counts (Birks et al. 1988). Time-scale in radiocarbon years.

sive areas of intensively cultivated fields (Ihse 1985, 1989, Ch. 5.11).

Increasing diversity in the pollen counts (species richness) has been combined with decreasing importance of competitors (Grime 1979) in the vegetation, since competitors dominated the wildwoods during Atlantic time. The expansion of the semi-natural vegetation types during the Late Subboreal period involved an increase of stress-tolerant plants in the landscape. Also the replacement of tall herb and grass vegetation by a low grass sward on wet soils and in fens at the beginning of the Late Subatlantic period increased the importance of stress-tolerant plants. Through the whole development of the cultural landscape ruderal plant species, including crop plants, have increased and today they dominate the landscape.

Landscape-ecological periods in the Ystad area

A. Landscape of wildwoods and woodland stability (before 3100 BC). This period was characterized by wildwoods dominated by late successional tree species over most of the Ystad area. Open areas existed mainly on dry slopes with sandy soils, around spring fens, and along rivers and lake shores. Organic deposits caused by

paludification had a limited distribution. Many lakes existed and contributed to the variation in the wooded landscape. Only a few, small original lake basins from the Early Holocene had been infilled with organic deposits.

There are few indications of human activity in the pollen diagrams. Assumed garden cultivation and pasturing on a small scale (Ch. 4.1.2, 4.2.2, 4.4.2) are difficult to trace. In any case, such agrarian activity must have been local. As a whole, gaps in the vegetation created by man seem to have been insignificant, occurring only near settlements.

B. Landscape of woodland instability and shifting cultivation (3100–1800 BC). This period was characterized by wildwoods and woodlands dominated by early successional trees over most of the Ystad area. Permanently open areas in the landscape were concentrated in areas where tree growth was prevented by sandy soils, a high groundwater level, or regular flooding. However, fluctuations in woodland vegetation as a result of natural events and human activities continuously created gaps in the tree cover, mainly on the dry soils. After a short time these gaps were recolonized by tree species. Because of low water levels in the lakes, several became infilled with organic deposits. There are indications of slightly increased erosion just at the beginning of the period, but erosion was otherwise insignificant during this period.

The pollen diagrams show clear indications of human activity, but it did not affect such large areas of woodlands that the total tree PAR decreased significantly compared with period A. Areas with permanent pastures or meadows seem not to have been widespread. Probably the human activity was a type of shifting cultivation concentrated mainly on dry soils in the coastal area.

C. Landscape of woodlands with areas of semi-natural vegetation (1800–800 BC). This period was characterized by woodlands and wooded pastures over most of the Ystad area. Woodlands dominated the inner hummocky zone while wooded and open semi-natural vegetation types (both pastures and meadows) and cultivated fields were most prominent in the outer hummocky and coastal zones. Early successional trees were more prominent than late successional ones in the woodlands. At the end of the period, however, late successional trees increased, perhaps indicating a greater permanence in the vegetation cover and some expansion of the woodlands. Another group of lakes became infilled, but until the end of the period there are few signs of increasing paludification in the landscape. Soil erosion remained insignificant during this period.

Human activity in the woodlands at this time affected such large areas that the accumulation rate of tree pollen at all the studied localities was reduced. Human activity was confined to dry soils, and was particularly intense in coastal localities. Pastures and meadows were more widespread than cultivated fields. Fires were at least locally more frequent and widespread than in the former wildwoods.

D. Landscape of semi-natural vegetation and woodland areas. (800 BC–AD 700). This period was characterized by an open landscape dominated by semi-natural vegetation types, perhaps partly wooded, over most of the Ystad area. Woodlands covered only limited areas, mainly in the inner hummocky landscape. Alder woods on wet soils became wet meadows and fens. However, around AD 200–500 there are clear indications of woodland expansion over parts of land areas formerly open and cultivated. About 800 BC paludification took place around lakes and the former wetlands expanded because of higher water levels. Organic matter like fen peat began to accumulate over large areas.

The increase of the human agrarian activity at the beginning of this period must have had dramatic effects on the whole landscape. Wooded areas were greatly reduced within a period of less than 200 years and the landscape was then kept open for about 400 to 600 years until some signs of a reforestation can be seen. Burning over wide areas may have been used to open up the

landscape and to keep it open, as high amounts of charcoal are found in sediments from this period (Fig. 5.13:9). Compared with the wide areas of semi-natural pastures, cultivated fields comprised rather small areas.

E. Landscape of semi-natural vegetation and expanding permanent fields (AD 700–1800). This period was characterized by an open landscape where cultivated fields and other types of arable land increased at the expense of pastures and meadows, particularly on the dry soils. Woodlands covered limited areas, although an area of tall woodland comprising as much as 800 ha existed near Krageholmssjön (Olsson 1989a) in the Middle Ages. Late successional trees dominate the tree pollen, probably as a result of an increase in cultivated fields and exclusion of early successional trees from the pastures because of high grazing pressure. Organic matter accumulation continued as a result of a continued paludification on wet soils, but did not give rise to the development of deep peat layers, as happened in many other areas.

The change in landscape starting around AD 700 ought to be looked upon as a result of the introduction of a new agrarian technology and a new organization of land use. An expansion of permanently cultivated land areas is indicated both by an increase of pollen from weeds and crops and by increased topsoil erosion. The wet-soil areas remained as wet meadows or fens, but their vegetation cover changed because of intensified exploitation and management. Around the period AD 1200–1400 there are indications of changes in the agrarian activity in some inland areas. This did not result in a substantial increase in woodlands, however, but mainly in some expansion of pastures and meadows. This period was followed by renewed expansion of arable land in the entire area.

F. Landscape of industrialized agriculture with large-scale, drained and fertilized cultivated fields (from AD 1800). This period is characterized by an open landscape where continuously cultivated fields and arable land dominate over pastures and meadows. Areas of wet soil have been turned into arable land through extensive drainage. The area of woodland remained small and concentrated in the inner hummocky zone. Because all woodlands are less intensively exploited, late successional trees have become increasingly important, and some of the least fertile areas are afforested, partly with species exotic for the area. The hydrological conditions in the landscape are completely changed through the extensive drainage, which caused a lowering of the groundwater level. Accumulation of organic matter does not take place any longer and paludification has been stopped.

This change in the landscape has also taken place as a result of new methods in agriculture and new ideas about how to organize agricultural land use. A very important change is that the agriculture in the Ystad area no longer depends on production within the area of energy and many other resources. Not only fertilizers but also oil for the machines that replaced the horses and oxen are imported into the area to cultivate the fields and improve the production per unit area.

Flow of energy, disturbances, and diversity in the landscape

Changes in the competitive balance among late successional tree species as well as the paludification in the landscape and the formation of organic soils are processes that have been going on in the landscape fairly independently of human activity until the last landscape-ecological period. In that respect the last century is a turning-point, as the present human activity in the Ystad area has had a greater influence on ecological processes in the landscape than ever before. It does not only strongly affect the vegetational landscape, as did, for example, the major deforestation around 800 BC, but also the hydrology and topography of the landscape. The flow of energy and inorganic plant nutrients through the landscape has been much increased, through the increasing use of oil products and fertilizers, particularly during the last decades.

At least until the beginning of the 19th century the pollen counts indicate an increasing diversity of vegetation. However, because of the weak time resolution and the low number of taxa identified, the biological depletion of the landscape during the last century, demonstrated by other methods (see Ch. 5.11), does not appear as distinct in the pollen diagrams as it really is.

Each of the landscape-ecological periods B–F begins with disturbance of the existing systems. In period B the disturbance seems to be the effect of a combination of several natural events and an increasing human activity where the natural events greatly facilitated human activities. At the beginning of each of the following four periods (C–F) it appears that increased or changed human activity led to distinct and rapid transformations of ecological conditions in the landscape. However, this does not exclude the possibility that other, natural events in the landscape, mainly climatically determined, have influenced the processes initiated by man.

The expansions of the open landscape at the beginnings of each of the landscape-ecological periods B–F lasted for a few hundred years. During the latter parts of periods B–D there are clear signs of reforestation and an increasing woodland area, resulting not only in an increase of early successional tree species but later also an increase of late successional ones. Although less evident, signs of such a process could be detected during the last period as well, both around AD 1200–1400 and since AD 1850. An increasing woodland area inevitably must imply a regression in human activity or at least changed land use, with land areas for some reason being abandoned and left open for secondary succession. The development of the cultural vegetational landscape is therefore not only a gradual expansion of semi-natural vegetation and cultivated fields at the expense of woodlands but also long and distinct periods of reforestation with expansion of natural woodland vegetation on both dry and wet soil areas.

The wildwoods dominating the landscape in the Ystad area during Atlantic and Early Subboreal time contained great amounts of organic matter stored as biomass, litter, and soil organic components. Clearances during the Late Subboreal period, probably the one beginning 800 BC in particular, took place during periods as short as a few hundred years and must thus have strongly and rapidly reduced the biomass and soil organic matter. The expanding paludification and infilling of lakes going on at the same time only very marginally counteracted this process. Much more slowly, this process of depleting the landscape of organic matter and stored energy has continued until recent time.

We cannot with any great accuracy reconstruct the primary productivity in the landscape for past periods. However, comparisons with recent plant communities suggest that the overall productivity in the wildwoods was hardly significantly greater than in the pastures and meadows developing from their clearance during the Late Subboreal period. It ought to be noted that only a minor part (20–25%) of the total net productivity in these semi-natural vegetation types could be made available for cattle grazing and haymaking. Until the last few centuries the total biomass production of crops and weeds on cultivated fields may have been lower than on the pastures and meadows with semi-natural vegetation. Today, of course, the total production of biomass on arable land is much higher than it ever has been because of the use of fertilizer. The productivity is most probably significantly greater than it was even in the most productive parts of the wildwoods during Late Atlantic time.

Regional aspects of long-term landscape-ecological changes

The long-term ecological changes may also be compared with other parts of southern Sweden and synthesized by applying the glacial-interglacial cycle of Iversen (1958), modified by Birks (1986a, 1988). Altogether, palaeoecological material is available from different physicogeographical regions mentioned in Ch. 1 (Figs 1.2:2):

a) Lowland area of southern Scandinavia (altitude <100 m a.s.l.) with Cambro-Silurian and Cretaceous bedrock, clayey, calcareous soils, nemoral vegetation (this study)

b) Lowland area of Blekinge (altitude <100 m a.s.l.) with Precambrian bedrock, sandy, non-calcareous soils, transition nemoral/boreo-nemoral vegetation (Berglund 1966)

c) Inland/upland area of north Scania and south Småland (altitude about 200 m a.s.l.) with Precambrian

Table 5.13:2. Main phases in the Holocene ecological development in southern Sweden (cf. Iversen 1958, Birks 1986a, 1988).

Phase	Characteristics	South Scania lowland	Småland-Blekinge inland/upland
Mesocratic	Temperate climate. Fertile brown earths.	7500–2000 BC. Mixed deciduous forest (pine on sandy soils).	7500–2000 BC. Mixed deciduous and coniferous forest
Oligocratic	Temperate climate. Leached soils. Podzols, peat accumulation.	(not represented)	2000 BC–0. Leached forest soils. Expanding mires.
Anthropocratic	Temperate climate. Eroded or fertilized soils. Deforestation, late successional trees dominant.	2000 BC – present. Wood-pastures, open pastures and meadows, arable land – main deforestation 800 BC and AD 700. In the 19th and 20th century the landscape is transformed into intensively exploited monotonous fields.	0 – present. Wood-pastures (expanding spruce), open pastures and meadows, some arable land – main deforestation AD 800 and 1600, and reforestation in the 20th century.

bedrock, sandy, non-calcareous soils, boreo-nemoral vegetation (Digerfeldt 1972, 1974, Thelaus 1989)

d) Inland/upland area of north Småland (altitude >250 m a.s.l.) with Precambrian bedrock, sandy, non-calcareous soils, almost boreal vegetation (Jacobson in prep.)

In this ecological transect, mainly from south to north, we find contrasting bedrock, soils, climate, biota, and human impact between areas a and d with areas b and c being intermediate in human exploitation but edaphically more similar to area d. In Table 5.13:2 we compare the most contrasting areas a and d. Mesocratic and oligocratic phases are defined according to Iversen (1958). Oligocratic is the final phase of an interglacial and its characteristics are not found in the development of south Scania. In the anthropocratic phase, which corresponds to the *Homo sapiens* phase of Birks (1986a), the human exploitation has a dominant impact on the landscape – deforestation leads to a rather open landscape (NAP values > 30–40 %) and early successional trees are as common as late successional trees. A comparison along the transect shows that there is a remarkable difference in the timing and intensity of human impact in southern Sweden. The anthropocratic phase which begins around 2000 BC in the southern Scandinavian lowland was about 1500–2000 years de-

layed in the south Swedish inland. The present study also indicates that the anthropocratic period was delayed in the Scanian inland/upland compared with the Scanian lowland, but the details in this temporal pattern require more study in the future. We can nevertheless conclude that the interactions of climate, soils, and human impact are decisive for spatial and temporal variation in the landscape development of southern Sweden.

Acknowledgements – This chapter is based mainly on pollen diagrams produced within the Ystad project. B. E. Berglund is responsible for the discussions of palaeoecological data, N. Malmer for the ecological approach, and T. Persson for computer calculations and computer graphics. The interpretations are based on joint discussions between the first two authors. We are grateful to our colleagues in the Ystad Project who offered their pollen data for these interpretations, to M.-J. Gaillard, H. Göransson, M. Hjelmroos, and J. Regnéll. M.-J. Gaillard made many valuable comments on the draft. We also express our thanks to G. Skog for assistance with calibration of radiocarbon time-scale and to G. L. Jacobson Jr., Orono, Maine, USA, for an overall discussion of our draft and our conclusions on long-term changes in the environment. We are also grateful to H. J. B. Birks for his generous advice about numerical techniques for the interpretation of past vegetation dynamics.

Ecological Bulletins 41: 00–00. Copenhagen 1991

6 Ecological and social factors behind the landscape changes

Björn E. Berglund, Lars Larsson, Nils Lewan, E. Gunilla A. Olsson and Sten Skansjö

6.1 Cultural landscape periods – patterns and processes

In this chapter we first summarize our results concerning the development of the cultural landscape (Ch. 6.1). A graphical synthesis is presented in Fig. 6.1:1. The time-span of 6000 years is subdivided into seven cultural landscape periods. The boundaries between these mainly follow the landscape-ecological periods as defined in Ch. 5.13. Because the ecological changes roughly correspond to certain important changes in society, the same time periods are also appropriate for describing the social changes, even if other important changes occur in society within these periods, particularly during historical times, that is, during landscape periods 5, 6, and 7 (AD 700 to the present). This chronological summary of our results is followed by a focus on ecological and social factors of importance for explaining changes in the cultural landscape (Ch. 6.2). We here follow the subdivision of background factors as they were classified in Fig. 1.1:2. Finally (Ch. 6.3) we also present our basic ideas about the interaction of environment and social systems, and their influence on landscape changes in the long term, including some critical remarks on our general project hypothesis. We conclude with some reflections about modern society and its future impact on the landscape.

Readers should note that all ages given here refer to the standard radiocarbon time-scale, which has been the normal practice in this volume (see Ch. 2.9). We deliberately refrain from encumbering the text with references to our findings in Ch. 3–5; references are given only when we use data from outside our own area.

1. Virgin woodland until 3000 BC: slight local influence by humans

Hunting-gathering economy. The Late Mesolithic hunting-gathering society is represented by a concentration of settlement in places of high productivity for hunting and fishing, especially near lagoons and estuaries. The extensive size of some habitation sites and the content of the finds indicate that they were base camps occupied by family groups over many generations. The relation between land and sea, which was altered by transgressions, may in certain periods have brought about an increase in fish populations in particular. However, the higher water level meant that sites for camps had to be chosen in higher terrain.

Inland from the coast, the few habitation sites with their scanty finds suggest that the area was of marginal significance for society, probably used for short stays for seasonal hunting and gathering.

Impressions of seeds in Late Mesolithic pottery at a site in western Scania indicate that contacts had been established with farming societies (Jennbert 1984). A limited form of tillage may also have been practised.

Native woodland. Landscape conditions during this period were the starting-point for landscape changes that were to appear in this region during the next 5000 years. Deciduous wildwood ecosystems were governed mainly by internal ecological processes. Local woodland heterogeneity was determined by natural disturbances, including death and uprooting of trees, spontaneous fires, and local soil erosion and vegetation damage from wild fauna. These events involved plant successions with regeneration of forest and field layer species in gaps from fallen trees and in vegetation deflations. The exchange of matter, nutrients, and energy in this ecosystem took place on a local scale and was governed by internal feedback processes. We can consider these woodlands as more or less closed systems with respect to nutrient cycling, with negligible nutrient losses to other systems.

Woodland with broad-leaved trees was the main vegetation type. Open grasslands existed only as narrow strips of shore meadows around lakes and along streams, kept open by natural flooding and by grazing of wild animals. Salt meadows possibly occurred along the seashore, but their distribution was dependent on exposure and flooding as well as grazing. Small open spring fens occurred on some hill slopes.

The human population living here influenced the landscape and the ecosystems to a small extent by their hunting, fishing, and gathering activities. However, around their settlement local influences caused by trampling, small-scale fires, coppicing, and – during the last centuries before 3000 BC – small-scale crop tillage might have occurred. It is reasonable to estimate that human influence on the ecosystems was no greater than that of wild vertebrates.

2. Human-influenced woodland 3000–1800 BC: shifting fields and pastures in unstable woods

The introduction of agriculture about 3000 BC. From a south Scanian perspective, a distinct change is evidenced in both habitation site structure and material culture between Late Mesolithic and Early Neolithic times. This change appears to have occurred in just a few generations. The habitation sites which are known for the early part of the Early Neolithic were located both at the coast and inland, which shows a greater spread of settlement than earlier and just after 2600 BC. The few finds and their relatively limited distribution indicate that the sites were used for short periods and by smaller groups than those from the Late Mesolithic. Both tillage and animal husbandry were practised, and the distribution of settlements suggests that it was these activities which dictated the choice of habitation sites. Tillage was limited to a small area around the site, where individual plots were used in a rotation system. Pigs were among the most important domesticated animals.

Woodland catastrophe and coppice agriculture 3000–2600 BC. The period started with drastic ecological changes, including the opening of the landscape and the recession of the broad-leaved trees. The duration of this phase was 400–500 years. Forest ecosystems were transformed from a rather closed wildwood with tall trees to a more open, patchy underwood with widespread secondary succession of hazel, birch, and alder. We believe that the dominant underlying causes were sudden climatic/hydrological changes along with other interacting factors such as elm disease. A continental climate with dryness and possibly cold winters may have seriously weakened the forest ecosystems. Dryness created gaps in the tree cover. Fires were common and some erosion occurred. There seems to have been some leaching and transport of mineral nutrients to streams and lakes. As a whole it was a landscape with heavy disturbance at least of terrestrial ecosystems.

Early agriculture was indirectly favoured by open patches in widespread underwoods. Land clearance for cultivation of cereals and other crops, and for creating pastures, was easier. It was also easy to do coppicing, which is assumed to have been a fundamental practice for this early agriculture. Mobile cultivation is implied with a rotation interval of more than 30–50 years. The mosaic character of the landscape seems to have allowed the sparse settlement to spread easily in the entire area.

Settlement concentration from about 2600 BC. In a late phase of the Early Neolithic there were considerable changes in society. Settlement in the inland appears to have become sparser, with a consequent shift to the flat area near the coast. From having been fairly dispersed, settlement became concentrated in a few limited areas on the coast, increasing the need to mark territories. This was done by erecting megalithic graves on land of high productivity for both tillage and animal husbandry. Small areas used for settlement during several centuries can be detected around these graves. On the other hand, this phase saw no change in the form of the base camps, which continued to be occupied by small groups. Five or six concentrations of habitation sites can be discerned in the project area. Settlement took the form of single farms which were moved a few tens or hundreds of metres from the earlier location. Old farmyards saturated with waste thus provided new, nutrient-rich cultivation plots. In addition, small habitation sites ap-

peared, probably used for resource utilization during limited parts of the year.

The restructuring of settlement to a few concentrations may have contributed to a spatial restriction of the cultural landscape. The limited extent of cultural landscape which can still be detected inland may be the result of woodland pasturing based on transhumance, that is, movements of livestock between the coast and the inland region.

This settlement pattern in the form of continuity in the same place near the megalithic graves was broken up in a late phase of the Middle Neolithic, when the area outside the concentrations began to be exploited more intensively for base settlement and also for seasonal occupation.

Woodland regeneration and coast-concentrated coppice agriculture 2600–1800 BC. During this period the woodlands seem to have recovered from the earlier crisis – there was an expansion of elm, ash, and lime, together with hazel. Fluctuating frequencies of elm, ash, and lime, together with high frequencies of hazel, indicate dynamic conditions. There seem to have been more unstable conditions than before 3000 BC, particularly at the coast.

The assumed coppice culture with cultivation and stock-raising continued, but agrarian activity seems to have been more concentrated near the coast. This settlement pattern favoured the regeneration of previously utilized marginal woodlands in the hummocky landscape zones. At least in the coastal area there was an expansion of open land around 2200 BC. The small-scale and mobile woodland cultivation implied only a very small loss of mineral nutrients, less than during the preceding period.

3. Wood-pasture landscape 1800–800 BC: gradual deforestation and settlement expansion

Settlement dispersal from about 1800 BC. Settlement during the later part of the Middle Neolithic appears to have become more mobile than before. However, from the Late Neolithic onwards we find bigger houses. The oldest house types differ in design, but most of them have two aisles with one row of posts down the middle. There is evidence of an increased use of the hummocky landscape in this phase.

In a middle phase of the Early Bronze Age we find houses with three aisles and two rows of internal posts. Several features such as isolated farms, settlement locations, and resource utilization agree well with what has been documented for an early stage of the Middle Neolithic. Although houses are of different design, there does not appear to have been any restructuring of settlement on the coastal plain. The same areas were used from the Late Neolithic and during all or most of the Bronze Age. In most cases, these core areas appear to correspond geographically to those of the Middle Neo-

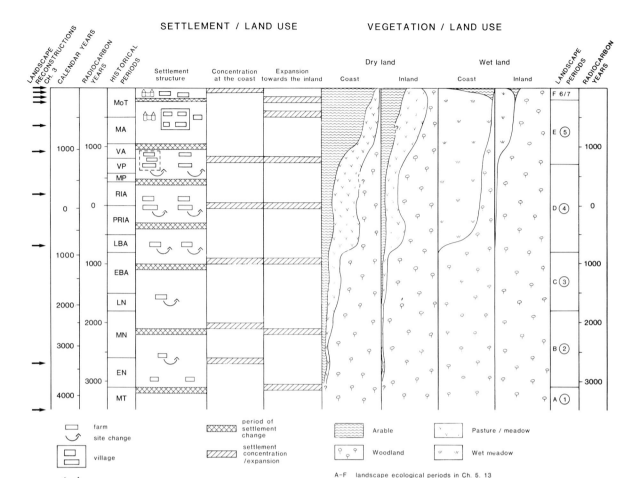

SETTLEMENT / LAND USE **VEGETATION / LAND USE**

Fig. 6.1:1. Synthesis of long-term changes in the landscape during 6000 years. Events are plotted against radiocarbon time-scale, but a calendar time-scale is also given. Landscape reconstructions presented in Ch. 3 are indicated by arrows. Landscape periods refer to the conclusions in Ch. 5.13 and Ch. 6. Settlement/land use is illustrated by symbols according to a simplified model applied in this concluding chapter. Vegetation/land use changes are illustrated by approximate proportions of three main land use categories based on pollen diagrams. Coast refers to the coastal landscape and most of the outer hummocky landscape, inland refers to the inner hummocky landscape.

lithic. No drastic change in population size can be documented, but there was probably a steady increase.

Gradual deforestation with widespread coppice agriculture 1800–800 BC. Around 1800 BC there was a further recession of broad-leaved trees such as elm, ash, and lime, which favoured hazel even more. This meant an opening of woodlands mainly to wooded pastures and some open pastures and meadows. During the later part of the period (1200–800 BC) there was again an expansion of the broad-leaved trees, indicating woodland regeneration and more stable conditions. There is evidence of incipient soil erosion and for the first time also lake eutrophication caused by transport of mineral nutrients from nearby soils.

Settlement was now more widespread. It was prob-

ably a mobile coppice culture as before, taking advantage of regenerating hazel underwoods. In large areas of the coastal plain and parts of the outer hummocky landscape a half-open, mosaic landscape of high diversity developed. It is possible that settlement and land use became more concentrated after 1200 BC.

4. Pasture landscape 800 BC–AD 700: permanent fields in a grassland landscape

Settlement expansion from about 800 BC. In the Middle Bronze Age there were changes which can probably be associated with a restructuring of agriculture. The three-aisled houses may have contained a byre. The form of this functional part of the house becomes increasingly clear in the course of the Late Bronze Age.

The introduction of hulled barley also suggests the start of systematic manuring of the fields. The number of indications of small sites on the periphery of the settlement indicates intensified use of what were previously marginal resources. Coastal fishing is one such resource which may have grown in importance. The increased human impact on the inland can be associated with intensified wood-pasturing in a well-developed transhumance system. The intensification of tillage and animal husbandry was the starting-point for new forms of agriculture in the Early Iron Age.

Deforestation and expanding pasture landscape with wood-pastures in the inland from about 800 BC. The drastic landscape change around 800 BC meant deforestation of large areas in the coastal plains, the outer hummocky landscape, and certain areas of the inner hummocky landscape. Burning seems to have been an important means to open up the landscape to meet the demand for pastures, primarily around permanent settlements, but also in marginal wood-pastures in the upland region. Wetlands were also exploited, probably for hay-cutting and grazing as well as for fuel. This meant a regression of alder and ash woods. Alder woods remained mainly on poorer soils in the inner hummocky area. Increased soil erosion also affected the movement of nutrients into lakes. Eutrophication and sediment accumulation in lakes increased from this time onwards.

Until around 800 BC the agrarian technique with mobile settlement and fields meant a closed nutrient circulation system, but the permanent settlement gradually developing from the Late Bronze Age onwards demanded manuring. Continuous cultivation possibly caused some leaching from the fields, particularly on sandy soils.

Deforestation, visible in the pollen diagrams around 800–500 BC, caused a landscape fragmentation. The mosaic landscape with woods, meadows, pastures, and arable fields favoured a diverse flora and fauna, which persisted until they were devastated by the modern agriculture of the 20th century.

Agglomerated settlement from about 200 BC. There are limited traces of settlement in the Early Iron Age. At the end of the Pre-Roman Iron Age there are more complex farm structures in the shape of a long-house combined with smaller houses such as barns. A greater need to save fodder for the winter meant that the byre section of the long-house had to be bigger than at the end of the Bronze Age.

Settlement can still be distinguished in individual farms, but the concentration of settlement means that there was cooperation in agrarian production between two or more farms. The absence of a fencing system makes it hard to interpret the social organization. Graves containing many more grave-goods than before,

along with the considerable variation in the size of farms, may suggest increased social stratification.

The period 200 BC–AD 200 is characterized by continuity in the same area. We cannot date the origin of an infield/outland system, but it was probably introduced during this period.

Around AD 200 the established farm system appears to break up as previously utilized areas are abandoned. Although the finds for a few centuries onwards are reduced in scope, it is scarcely likely that we are dealing with a radical restructuring of the settlement region. Certain indications suggest that settlement was distributed over a smaller number of settlement units in the form of isolated farms, at the same time as the extent of the settlement region shrank.

The deterioration in climate in the 5th century AD may be connected with a concentration of settlement in just a few places – possibly the establishment of a village organization. New forms of cultivation with fallowing and two-course rotation may have developed as part of this shift.

The limited finds from the Migration Period and afterwards confirm the contraction of the settled region and the decrease in the number of settlements. It is probable that in this period there were farms of varying form and complexity, which may suggest a continued social hierarchy. Evidence of major farms supports this interpretation.

Reforestation of marginal pasture land, from about 200 BC. The concentration of settlement around 200 BC to the coast and some areas of the hummocky landscape meant that marginal areas became forested again. During this period beech and hornbeam expanded in such areas. Pasture land was still dominant until about AD 500. However, at the coast there was an expansion of cereal cultivation after about 200 BC, and the landscape as a whole was more open around 0–AD 200. The landscape became more stable and soil erosion was low during the period 0–AD 500. It is not possible to detect any significant landscape change around AD 200. At least in some areas, like the Bjäresjö area, there seem to have been some land-use changes leading to local reforestation around AD 400–500.

The more intense cultivation – assumed to have been one-field systems at permanent settlements – of the sandy soils at the coast, in contrast to the less favourable silty-clayey soils in the hummocky landscape, caused a polarization of these two areas, more obvious than before. Differences were expressed in vegetation, land use, and nutrient balance.

The time after about AD 600 seems to have been a transitional period to the following one. Pastures were partly reclaimed. Open wetlands expanded again at the expense of fen woods. The gradual expansion of arable fields may also have been due to the two-course rotation system, which demanded larger open areas. Manuring was certainly as important as before. The landscape was characterized by the structure of infields/outlands. The inland area with Romeleåsen was still mainly used as a large wood-pasture.

5. Infield/outland landscape 700–1800: field rotation and technical improvements

Village and population expansion 700–1200. The archaeological source material does not allow us to identify any expansion of settlement units before the 9th century, but according to the pollen diagrams it probably began slightly earlier. The expansion may be connected with the increased importance of both animal husbandry and tillage. Rye and oats are more commonly found.

The archaeological investigations show a close link between habitation sites from the Early Iron Age and those known from the Viking Age. In the Late Viking Age, however, settlement locations on the coast, which were often intensively used in earlier times, were abandoned. This was possibly due to attacks from the sea by hostile groups.

An expansion of settlement in the hummocky landscape is very noticeable in this phase, a result of increased population. Individual persons or families appear to have taken the initiative in this, as indicated by the runic stones. An important feature of the new settlement is the foundation of the precursors of some of the medieval manors.

Through new organizational forms of an administrative nature, coupled with the coming of state formation and Christianity, there arose a need for permanent settlement locations. In the Ystad area there was often a close spatial link between manor and church.

Whereas the settlement pattern had hitherto involved the short-distance movement of habitation sites after a few generations, this practice now ceased. Most of the sites established during the Viking Age therefore correspond to the location of the medieval villages, and the manors in some cases survived as part of a feudal organization. This development was also accompanied by improved forms of cultivation with the introduction of the plough and a changed field system. We can detect the increased importance of cereal cultivation, along with a more distinct trend towards surplus production for sale outside the local area. From the very start of the period there are also indications of trading sites that were used seasonally. However, these places lost their commercial function with the foundation of the town of Ystad.

The consolidation of the infield/outland pattern 700–1200. From AD 600, but mostly distinctly after AD 700, there was again an increased deforestation which is linked to the general village expansion with land use related to the infield/outland system. The land use of the villages and single farms affected all kinds of land, even the wetlands and the marginal woodlands. New clearings and expansion of ploughed land increased soil erosion. Nutrient leaching is demonstrated by increased eutrophication of lakes, particularly around AD 800–1000.

During the Late Iron Age the infield-outland landscape was definitively consolidated. The use of permanent fields where productivity was maintained by the addition of manure had been introduced earlier, but the growing permanent settlements during the Viking Age were both a prerequisite and a driving force for maintaining permanent fields and crop rotation systems. The divergence between the infields and outlands in the development of different biological communities and habitats continued. Implicit in this process was the continuous transport of mineral nutrients from the fodder areas – the outlands – into the infields by the grazing stock, and the subsequent development of heath communities that tolerated the nutrient-poor conditions on the outlands. The use of fire to clear spontaneous invasions of shrubs into the fodder areas, and the ongoing dynamics of interchange between intensively managed and fallowed plots guaranteed a high frequency of small disturbances and small-scale successions vital for maintaining a high species diversity. Landscape characteristics were determined by land use, and secondly, by basic edaphic factors such as hydrology and soil constitution. Due to the scale of land use and its variation, the diversity of the landscapes and habitats was large.

Continuity and fluctuations in the medieval village landscape 1200–1550. After the phase of village formation in the Viking Age and Early Middle Ages, the historically known villages in the area mostly retained the location they were to have in the oldest land survey maps from about 1700. Villages were relatively sparse on the sandy coastal plain and in the inner hummocky zone and nonexistent on the Romeleåsen ridge in the north-west. The outer hummocky zone has been densely settled since the Middle Ages, with just over half of the forty or so villages in the area.

The village consisting of several farms has dominated the settlement pattern in the Ystad area from the Early Middle Ages until the early 19th century. Isolated farmsteads have been few in number. An important feature of settlement was the noblemen's manors, which were distributed in the same way as the villages as regards the geophysical conditions. Often the oldest known manors were linked to villages.

Parish formation was completed in 13th century, and the Romanesque stone churches were built. In the Late Middle Ages certain villages witnessed a dissolution of a high medieval structure with a large farm and smaller tenant units, which was replaced by a pattern with relatively equal farms. In the mid 13th century the town of Ystad was founded, with specific functions for channelling surplus production from the agrarian hinterland.

The development of the cultural landscape in the Early Middle Ages and High Middle Ages was mostly expansive, with a major village-forming process which was paralleled by more intensive, permanent resource utilization. No unambiguous evidence has been found for a major regression in the Late Middle Ages, known from the continent and from several parts of Scandinavia. However, from the more marginal areas in the north and north-west, especially south of the Romeleåsen ridge, we have obtained some indications of regression. It is likely that this part of the area was utilized in periods of population expansion and then abandoned for more fertile regions in periods of population regression.

The medieval land ownership structure was more differentiated than that in the period after about 1600. In the Middle Ages the archbishop of Lund was one of the largest landowners in the area, with a number of manors as centres of a considerable holding of tenant farms. The church also manifested itself physically in the form of a few chapels and many parish churches. The crown was not an important landowner in the area. Evidence of local lords is found as far back as the Viking Age, with large farms and runic stones, and from the Early Middle Ages we find indications of such landowners in several church buildings. From the 13th century there is a fully developed nobility with manors and tenant farms, although these are not well attested until the Late Middle Ages. Only a few freeholding farmers are evidenced in sources from the Middle Ages and the 16th century. The production surplus of the major landowners, and what the tenant farmers had left after their own consumption, taxes, tithes, and other duties, could be sold via the market provided by the town of Ystad. The agrarian surplus of the area, along with the herring fishery, were essential conditions for the high medieval urbanization of Ystad.

Village landscape in ecological balance from about 1200. The landscape in the Middle Ages was at the height of its diversity on all levels: landscape pattern, habitat diversity, species and genetic diversity.

Most woodlands were transformed into a dynamic system of grassland coppices maintained for the purpose of producing hay, firewood, and a wide range of wood products. Few if any woodlands were kept purely for timber. We have indications that land use during this period allowed a proper balance between wooded areas, grasslands, and cultivated fields in order to maintain the productivity of the arable. There are some indications of reforestation around the 14th century in the inner hummocky landscape.

With the introduction of the three-field system, evident at least as far back as the High Middle Ages, cereal production was increased. This system, however, demanded larger arable areas and also regular input of manure. The possibility of keeping cattle, especially oxen, on pasturage in the large commons away from the big cereal fields yielded enough manure to allow a concentration on cereal cropping both in the boulder clay

areas and on the sandy plains. However, soil erosion and nutrient leaching increased as a result of this practice and were probably substantial in the sandy areas.

Settlement expansion and pronounced estate formation from about 1550. The settlement pattern was dominated by the agglomerated villages continuing from the Middle Ages. In the mid 17th century no less than 93% of the farms were in villages, with only 7% as dispersed settlement concentrated in the inner hummocky landscape. To this we may add nine noble manors of varying size, some of them dominating the landscape with magnificent castles surrounded by huge areas of farming land. Settlement density in the Ystad area was far above the average for Scania, and in a comparative perspective it was well representative of a central settlement region in southern Scandinavia. Within the area there were distinct differences between the sparsely populated parishes in the north and the outer hummocky zone and the coast.

This period saw an estate-forming process which was to be of decisive importance for the future. Whereas the noble manors known from the Late Middle Ages were relatively small as farming units, most of the manors around 1650 were relatively large units in their regional context.

A process which had significant long-term consequences was a redistribution of the ownership structure in favour of the nobility. In the mid 17th century the nobility owned over 80% of the farms in the area, a figure which can be compared with 54% for the whole of Scania and 45% for Denmark west of the Sound. The large estates produced both cereals and live oxen for export via Ystad.

Another process at the same time was an expansive settlement development in the form of colonization of the Romeleåsen ridge by small, new single farms. These farms were now established as permanent agrarian settlements in an area which had previously mostly been used only on an extensive basis; we have no real evidence of any permanent settlement on the ridge during the Middle Ages.

Up to the middle of the 18th century the three-field system was still the normal one for cultivation in the open landscape of the village region, and rye and barley were cultivated on land that was regularly manured. Oats were grown on the villages' sandy soils or more peripheral lands. In the period from the mid 16th century up to about 1700 the arable lands of the villages increased. This expansive development continued in the period 1700–1800, when it can be measured at between 40% and 50% of the farmed land. On the estates' demesne lands there is evidence of an even greater increase. With a gradual abandonment of the three-field system from the late 18th century in favour of a multi-field system, the estates obtained increased productivity as well. In the same period the population of the area virtually doubled. The peasant population thereby increased more than the cereal-growing area per individual.

Intensive exploitation of the infield-outland landscape, 1550–1800. During the 16th century, arable fields were allowed to expand at the expense of the fodder areas, both in the boulder clay areas and on the sandy plain in the east. The colonization and subsequent enclosures of the commons on the Romeleåsen ridge during the 16th-18th centuries reduced the availability of grassland resources and indirectly of manure for arable areas in the large villages in the outer hummocky zone. The general tendency of increasing the arable disrupted the balance between nutrient source areas and nutrient sink areas. With continuously smaller fodder areas and subsequently decreasing potential for feeding stock, the nutrient deficit of the arable was rising, resulting in reduced productivity in the large three-course rotation villages, where mean harvests reached just 3–4 times the seed input.

The use and maintenance of coppices, which were so important during the Middle Ages, decreased after ca. 1500, which meant that grassland areas were also decreasing. Coppices did exist, however, and in some parts of the Ystad area they were significant. Standard trees and wood lots and parks for game hunting were favoured only at the large manors and castles. The landscape became gradually more deforested. The acute shortage of wood was reflected in the use of peat for fuel. Fallow fields used as meagre grazing grounds were not allowed to recover. The result in the late 18th century was a highly diverse landscape, although strongly affected by nutrient deficits, especially on the arable-fallow system.

6. Arable agrarian landscape 1800–1950: crop rotation, industrialization, and international trade

Enclosure, agricultural innovations, and population increase during the 19th century. The decades around 1800 were dominated by enclosure and population growth, and important changes took place in land use and settlement. Capitalist farming began to be of importance, first on the big landed estates. In many respects, however, this was still a period of traditional subsistence agriculture.

Population increase became important during this period, still more so later on in the century, but already including a strong growth of unpropertied groups. General causes such as peace, potatoes, and small-pox vaccination cannot be denied as influential. On the other hand, radical land enclosure and more efficient land use opened possibilities for a division of farms, and thus also for more people to live in the area. Such divisions were most frequent where estate land had been sold to common farmers. There, too, population increase was at its fastest.

The area was still dominated by a number of big landlords, whose decisions were of crucial importance. From that time onwards innovations came mostly through a few of these landlords. During the period the effects of their influence showed most in the patterns of settlement and land use. Some early enclosures took place at the estates, with a subsequent scattering of settlement, while the establishment of large tenant farms had to do with ideas about the best organization of economy and labour on big landed units.

The impact of industrialization during the first half of the 20th century. In the second half of the 19th century population maxima were reached in almost every parish, the whole region passing a maximum around 1875. The increase up to then was almost totally a result of circumstances within agriculture, while the changes during the following period are less easily interpreted. Outside forces, including international economy and technology, began to play a more decisive role. This, of course, applies also to the use of land for different purposes, for division of land, etc.

Commercial agriculture now became the rule, and even if mixed farming still dominated, the relative importance of animal husbandry and field crops for sale was to an extent determined by the international market, especially England. This in turn led to quite a number of other changes, most of them still within agriculture.

To mention only some of them, one should be aware of the importance to the landscape of under-drainage and of the flooding of meadows, of marling, and of artificial fertilizers, along with the successive introduction of new farming implements. Early forms of agro-cooperation developed, local dairies being among the first. Other rural industries followed, like sugar-beet factories, brick and tile works, machine shops, and distilleries.

Another important factor was the modernization of transport, the steam railway being the major innovation, its locating force increased by the railway stations. With the new liberty to establish shops and all kinds of trades even outside towns and cities, the railway villages became the new growth points of rural areas.

Still another sign of a re-evaluation of work and independence during this period was the establishment of small owner-occupied farms (Sw. egnahem).

The whole period from the later 19th century to World War II could be looked upon as a period of more or less unbroken change. At the end of the period, however, the area as a whole was still almost as rural as at the beginning.

Intensified cultivation with reclamation of grassland and wetlands, accelerated leaching and erosion since the beginning of the 19th century. The infield/outland system was beginning to break up in the early 19th century. Many factors influenced this process. The increasing transformation of grassland into arable had now reached such large dimensions that many villages has extremely small fodder areas. This in turn necessitated a reduced number of stock and thus led to insufficient manure for the arable. Increasing population meant far-reaching division of the land in the villages, with consequent huge practical problems in land management.

A more rational distribution of the individual land plots for the farmers led to intensified cultivation techniques. From the mid 19th century fodder was cultivated on arable leys, thus continuously replacing the natural grassland fodder. Crop rotation with elements of nitrogen-fixing crops became increasingly spread during the first half of the 19th century, a process that paralleled the disappearance of fallows. During the latter half of the 19th century new knowledge of the importance of mineral nutrients for arable production led to a widespread use of a variety of fertilizers as a complement to manure in this area. Marl was dug from pits in the Ystad area and phosphorus was added by using skeletal bones from slaughtered stock. Commercial fertilizers were introduced here during the 19th century, although they did not become important until the end of the period. The dramatically increased arable area which relied more and more on mineral fertilizers was more susceptible to nutrient leaching than ever before, and also to an increased rate of soil erosion. At the end of the period the majority of the agro-ecosystems had experienced a development to "open" systems with substantial net nutrient losses.

Of the new techniques introduced during this period, the most important from the landscape-ecological point of view was large-scale draining of wetlands. At the start of the period wetlands with diverse characteristics were among the dominating habitats, having about the same magnitude as the arable land. By 1950 only very small areas of these wetlands remained. Even shallow lakes and ponds were emptied during the later part of the 19th century – a process that not only changed the landscape hydrology but also reduced the ecological diversity of the flora and fauna by eliminating wetland biotopes. Grassland ecosystems created by grazing and mowing activities critical for the survival and establishment of the grassland species still existed here, but they were confined to ever shrinking areas.

Fires were intentionally prevented in this period, and thus became of minor ecological importance, although fire was still used to clear the pastures of non-palatable thorny shrubs. The impact of browsing and coppicing also decreased along with the adaptation of new fodder cultivation techniques. In the mid 20th century these activities were of little importance as landscape-forming processes. In addition, the importance of secondary successions, such as recoveries after fire, on fallow land, and so on, decreased during the period; this feature also led to reduced biological diversity.

A factor increasing ecological diversity among the anthropochorous species was greater trade and transport of agrarian goods. The influx of diaspores via long-distance dispersal brought about the establishment of new plant species, most of them weeds and ruderals, and also of some animal species like the wild rabbit.

7. The landscape of modern technology since around 1950

Industrial agriculture and urbanization, increasing trade and environmental concern. In the Ystad area, as elsewhere, the 20th century has meant profound changes in population numbers and distribution, and in possibilities for work, in technology and land use, in settlement and landscape. Population has by now definitely become a dependent factor, while advance in technology may be looked upon as a more decisive factor than before. New agricultural technology has meant the growth of farm units, specialization, and further mechanization resulting in bigger fields, fewer farms, and fewer people. While there has been a fast concentration of other rural industries and services, private car ownership has opened urban jobs for rural people, and physical and hidden urbanization has spread throughout the area. This would not have been the case without widespread prosperity, the roots of which cannot be specified here.

Sweden's national agricultural policy has without doubt been a crucial factor, perhaps the most crucial. Although its details have varied over the decades, the overall goals have remained the same: cheap food for consumers and decent incomes for producers. These goals have furthered changes in the agricultural industry of great importance to the landscape. Up to now, other political ambitions, intended to further the natural or cultural environment or centred upon the future of the rural population, have played a very limited role in our area.

As a whole, outside forces now almost totally decide the destiny of these rural areas. International circumstances are even more important than during the 19th century, especially as they apply to increasing agricultural overproduction and the rural social policies within the European Community, even though Sweden does not belong to it.

While noticing all these outside forces, one should be aware that purely local decisions are still important. Various households and population groups use rather different time horizons in their decisions and actions, and these in turn have very different importance for the future of the landscape. At one end of such a time-scale we find the big landed estates, together with advanced ordinary farmers, while the other end is dominated by old-age pensioners, temporary settlers, and perhaps also by politicians, with a much more limited time perspective. Looking somewhat forward, the destiny of the landscape might still, with some exceptions, be decided by today's forces. Overproduction still seems bound to increase, and so does the importance of national and international agricultural policies. This might be somewhat moderated by various environmental initiatives. At the same time, other needs and external decisions may foster increasing changes in the landscape.

The industrial large-scale landscape. The most important landscape changes after 1950 are due to the amplification of earlier trends in the following processes:
* the drainage of wetlands
* the removal of semi-natural vegetation
* the management of agricultural ecosystems on the basis of inputs of commercial products – fertilizers and pesticides
* cropping routines adjusted to rapid industrial processing of crops

The result of these processes is a monotonous agrarian landscape. In certain areas the transformation has been total, so that the fields completely cover the landscape. Where patches of grassland exist they are not managed and the grassland communities are deteriorating. Generally they are heavily affected by fertilizers, as reflected, for example, in the considerable decrease in the mean number of plant species. Crop diversity has also decreased, and crops like sugar-beet are grown without any regular crop rotation. Most farms specialize in cultivation of cereals or root crops and it is rare for this to be combined with stock-raising. Never before has diversity on all levels been so low and the variation so exhausted. This landscape has never been so dry and devoid of surface waters. However, a tendency to polarization in development between the plains and the Romeleåsen ridge is once again visible. Several farms in the ridge area have been abandoned since around 1950 and some of these areas are now sites for spontaneous forest succession. There are now refuge habitats here for some species which were driven away from their former grassland and wood-lot habitats on the plains. From the point of view of conserving biodiversity, favourable processes are secondary successions and vegetation dynamics reflected in small-scale processes like infrequent management. Such processes were common in the infield-outland landscape and appear again in this degenerating cultural landscape where grazing and haymaking are no longer economically viable.

During this recent period the intensively cultivated plains were turned into sensitive agro-ecosystems heavily dependent on large-scale inputs of commercial resources – and characterized by open-ended flows of nutrients out of the systems. Water pollution problems from high loads of nitrates and ammonia in the rivers and groundwater are among these effects. Regular "sandstorms" during the early spring on the sandy eastern plain are another sign of the vulnerability of these systems.

6.2 Factors behind changes in society and environment

Social factors behind long-term changes

Population. Of the factors which contributed to the introduction and establishment of agriculture, the population pressure is not regarded as significant because of the limited number of inhabitants during the Neolithic. The tendency to settlement concentration detected in the Middle Neolithic may have been the result of a faster population increase which can be followed during the Bronze Age as well. In the Roman Iron Age an additional increase in population growth may have contributed to the expansion, which was temporarily checked in part of the Migration Period and Vendel Period. Several factors of both local and super-regional kinds led to an expansion of settlement in the inner hummocky landscape during the Viking Age.

The expansive period in the cultural landscape identified in the Early Middle Ages and the High Middle Ages is generally considered to have been characterized by a heavy population increase, a widespread European phenomenon with noticeable repercussions in south Scandinavia. The population variable can be regarded as an essential condition for a number of developments: the formation of the villages and with it the intensified resource utilization of the historical settlement region, the increase in number of parish churches, and the urbanization at Ystad. The general European population decline in the Late Middle Ages as a result of demographic catastrophes is not clearly recorded in the Ystad area, but we can see some signs of stagnation or regression in settlement and cultivation.

In the 16th and the early 17th centuries the late medieval decline was followed by a general population rise in Europe, including Scandinavia. During this period we see an expansion of settlement in the marginal parts of the Ystad area, at the same time as an expansion in the arable land farmed by the villages.

In the period 1700–1800 the population of the area doubled, which may have necessitated the contemporary increase in the area used for cereal cultivation. This expansion did not fully meet the needs created by the population increase, which points forward to the need for the agrarian reforms of the 19th century. This century witnessed a rapid increase in population, despite heavy out-migration, followed by stagnation since around 1910. Rural population peaked by 1880, however, and since then it has been reduced by two thirds. Today only 25% of the population lives in truly rural areas, and only about 10% of the total labour force is employed in the primary sector (food production), compared with 60% and 40% respectively in 1900. Urban

living in Ystad, and urban occupation in the whole area, now dominate the population picture, although still not the landscape.

By 1800, and also during much of the subsequent century, population still seems to have been an independent factor, while in our century it has to a greater degree become a dependent one. Still, the inauguration of owner-occupied smallholdings during the first decades of our century seems to indicate population as a deciding force.

Population pressure has undoubtedly been an essential condition behind all phases of expansion in settlement and cultivation – within the limits of the social system. Conversely, decline in population has repeatedly been a cause of regression in the cultural landscape. Generally speaking, the population factor has changed from being an independent one to become a dependent one. In our century, other societal factors have become more important (Fig. 6.2:1).

Social organization. Stationary settlement can be documented for the Late Mesolithic near coastal areas, which were highly productive for fishing. This settlement was broken into smaller units in the Early Neolithic. A concentration of settlement during the transition from the Early to the Middle Neolithic may have provided the impetus for the first hierarchically based social structure, as exemplified by the introduction of monumental megalithic graves combined with the social values that apparently developed around them.

The hierarchical structure was accentuated still more in the Late Neolithic and the Early Bronze Age, when it is noticeable in the grave-finds. This hierarchy cannot be traced in the remains of buildings until the Roman Iron Age, when large farms begin to appear, subsequently to develop into major structures in the Late Iron Age and the Viking Age/Early Middle Ages. The appearance of a warrior aristocracy in the Late Roman Iron Age and its development towards more permanent organizational forms appears to have been of great importance for societal changes in the latter part of prehistoric times. This development included increased contact with the outside world in the form of organized trade.

The state formation process in the Early Middle Ages and the establishment of the Christian church meant that the area was incorporated in a system of new administrative divisions: hundreds and parishes. The fundamental structures of medieval society had concrete manifestations in the landscape, and an important factor behind several changes in the landscape was *land ownership* and the attendant influence that was exerted

POPULATION

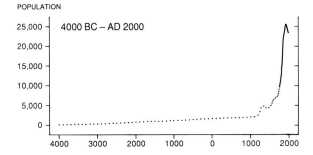

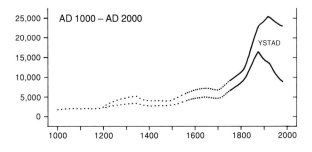

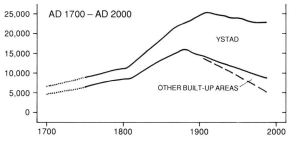

Fig. 6.2:1. Population diagram for the Ystad area. For the time span 4000 BC to AD 1000 given values are an estimate based on interpretations of prehistoric material (Olausson 1991). For the time span AD 1000-1550 the population figures represent an estimation based on general trends in Northern Europe, especially in Southern Scandinavia, adjusted in accordance with local contemporary sources. For the period 1550-1750 the figures represent a calculation based on analyses of local contemporary fiscal source material, from ca. 1700 combined with demographic data. Values for the period 1750-1985 are based on demographic statistics. Accuracy of population values indicated by different lines. Compilation by L. Larsson, S. Skansjö, A. Persson, and N. Lewan.

by secular magnates and ecclesiastical institutions. This is reflected, for instance, in the way the manors varied through time in size, location in relation to the villages, and degree of fortification. Several church buildings also bear the external marks of having been associated with the secular nobility, and changes in the structure of the villages in the Late Middle Ages can often be explained by the owners' preferences as regards cultivation.

A clear example of how the social structure must be considered as a decisive explanatory factor in connection with changes in the cultural landscape can be found in the period from the mid 16th century to the mid 17th

century. Power relations in society are clearly suggested by the designation of the period 1536–1660 as "the time of noble dominance" in Denmark: socially, economically, politically, and culturally, certain strata of the nobility took on a special position. This was manifest in the cultural landscape, above all in the castles that were built. The development of the nobility's privileges created a social organization that favoured tillage (especially the corvée obligations) and stimulated an increase in the estates' tillage of their demesne lands.

The fact that over the last two centuries local self-contained societies have given way to national and international ones indicates both the relocation of decision-making and the decreasing importance of distance. This has been so even in local communities and within single farming units. The latter might today include more than one farm, sometimes a considerable distance apart. The diminished importance of distance is also marked by the increasing commuter ranges. Environmental forces may, however, even in a near future, mean that distance once again will count more.

Land ownership has been of continuing importance, although the number of owners has varied. Since World War II an increasing number of outside influences has reduced the role of the landowner, as more outside restrictions have come into existence.

By 1800 almost all decisions of any importance rested with something like a dozen estate owners, who had to decide about the pros and cons of enclosure, made possible in royal decrees, about modernization of farming practices in the light of possibilities for corn export, and so on. Around 1900 the number of decision-makers had increased greatly, now including both more owner-farmers and many smallholders. As most of them also farmed for sale, they were dependent upon both buyers of produce and intermediaries in places like Ystad and Malmö. In the 1980s the number of decision-making farmers is again reduced, but each of them controls a much larger area. The latitude of decisions has, however, been greatly diminished, as both international competition and national decisions concerning agriculture and other matters have increased in relative importance.

Local authorities meant almost nothing around 1800, apart from the "village assembly", and rather little around 1900. They have, however, increased in importance lately, especially since the many local parishes were amalgamated through a process covering the period 1951–1974. Since 1974 most of the area is in one single municipality, which has to organize physical planning through general as well as detailed plans. A general structure plan from 1990 should, among other things, include areas of natural and cultural interest, delimit them, and specify measures for their protection.

By 1800 the influence of national authorities came mostly through various decrees, giving options and possibilities, e.g., enclosure decrees and decrees concerning the organization of entailed estates. By 1900 the influence was still mostly of the same kind, among other things concerning customs duties and financial aid to smallholders. By 1990 acts of parliament, national decrees, and recommendations cover a vast number of interests, giving the local people much less freedom of action, whether concerning farming or the preservation of areas of outstanding natural beauty or parts of the cultural heritage.

Through time social organization has been a most important factor. It has also become more complicated, as local and independent decisions have been supplemented by decisions at several higher levels of society, successively leaving less room for the individual or the purely local society to decide about changes in settlement and land use.

Economy. The relation between the earlier hunting/fishing and gathering economy and the introduced agricultural economy appears to have been subject to several changes in prehistoric times. In periods such as the middle part of the Middle Neolithic, the Late Bronze Age, and the Viking Age, fishing may have been an important part of the economy.

For considerable parts of prehistoric times animal husbandry appears to have been of greater economic and social importance than was tillage. Not until the Iron Age can we detect any large increase in tillage. An organizational form corresponding to transhumance may have combined the utilization of both the coastal region and the hummocky landscape inland. The Viking Age definitely saw the establishment of the trend that resulted in the predominance of tillage.

Economic surplus can be recognized archaeologically in various periods of prehistory. The construction of large graves like megalithic tombs during an early part of the Middle Neolithic or burial mounds during the Early Bronze Age are examples of this. The introduction of bronze was combined with an increase in the number of artefacts of foreign production. During the Late Iron Age the area was involved in a larger economic system.

Historians in general speak of an economic upswing in the Early Middle Ages or High Middle Ages, associated with increased productivity. Whether this was ultimately due to an increase in population requiring techniques to raise productivity, or to technical development gradually leading to population growth is still debated. From this time on, the peasants had to hand over parts of their surplus production as taxes to the crown and as tithes to the church. The possibility of trading in surplus produce, combined with the high medieval urbanization, further stimulated increased production.

The European trade cycle for much of the 16th and 17th centuries, in the wake of the price revolution and increased demand for agricultural produce in urbanized parts of north-west Europe, had clear repercussions in Denmark – and hence for the cultural landscape in the

Ystad area. In this connection we note again the interest in intensified farming on the estates.

Just before 1700, cereal cultivation in the Ystad area was not on a very great scale, due partly to bad times for agriculture in the period 1658–1700 and partly to the ravages of the Scanian war, which directly affected the area in 1676–78. In the first decades of the 18th century, however, the farming economy improved and as commercial agriculture spread all over the area during the 19th century, both national economic forces and international trade have become of primary interest, also to the individual farmer. Local and regional competition for labour and international competition concerning exports have become some of the major forces behind agricultural specialization in the area, also helped by the import of fertilizers, oil, and machinery. This has in turn forced a reduction in the number of farm units. The economic need for bigger machinery has, as elsewhere in cereal farming, played a great role in both the visual and the biological simplification of the landscape.

As a factor of importance, economy has meant different things. Favourable trade conditions have stimulated productivity and surplus production, and thus expansion in cultivation. Efficient use of resources has always had an impact on the landscape, but with growing force and depth, especially since World War II. Time has also meant interdependence between areas in successively widening geographical regions.

Technology. Contacts with the outside world were the overwhelming factor in the introduction of tillage and animal husbandry. The form of tillage practised in the Early Neolithic was extensive in character because of the need for uncultivated areas and long fallow periods. There were limited possibilities of using nutrients stored just under the surface layers in the soil. Improved methods of cultivation, such as the introduction of ard ploughing and the creation of coppiced woods, made it easier to establish lasting settlements in well-delimited regions. Changes in livestock farming, with the winter stalling of the animals in the Late Bronze Age, also had consequences for tillage, as manure facilitated the creation of more permanent arable fields. The introduction in the Iron Age of new crops like rye and hemp and implements like the plough are important factors in societal changes.

Important conditions for the upswing in the High Middle Ages were provided by several improvements in farming techniques. It appears reasonable to say that it was in this period that innovations such as the wheeled plough and other implements made their real breakthrough, and the three-field system, although possibly established earlier, was now fully developed and generally adopted in the village region.

Nevertheless, technical factors scarcely appear to have been decisive during the expansion period in the 16th century. In the late 18th century, a period with population growth and a rising number of cottages in the villages, we have evidence that the estates were abolishing the three-field system and introducing the more productive multi-field system.

Throughout the last two hundred years technology has gained in importance, promoted by the national agricultural policy and by the inauguration of agricultural societies, by general scientific progress and international influence. Lately it has also been furthered by agricultural specialization towards cereal farming. As elsewhere, revolutions in transport have been of the utmost importance.

In the last decade of the 20th century it is – for the Ystad area, as for Sweden in general, and for much of western Europe – arguable to what proportions technology and economy, agricultural and environmental policy are the prime forces behind agricultural change, and thereby also behind changes in landscape and settlement over most of the area.

Technological innovations have made possible the support of increasing populations during several periods in the past. Through time technology has gained in importance, and today it seems to be the most important societal factor in our area.

Ecological factors behind long-term changes

We summarize here those ecological factors which play an important role for society. However, we are not able to evaluate their importance for society during different times and situations. We separate these into two categories: the natural ones and the human-induced ones. Each category is further divided into four groups: climate, hydrology, soils, and biota. The relative changes are illustrated in diagrams, which show the main trends of changes (Fig. 6.2:2). It has to be emphasized that for the natural factors these figures illustrate the natural (potential) trends even when human activities have disturbed the development. These diagrams have a weak resolution in time as well as in the amplitude of the changes. They give an impression of slow environmental changes. However, it is quite probable that rapid changes occurred, particularly in the climate, and, if so, they were of vital importance for the people in the area.

Climate. The *mean summer temperature* was about 2°C higher than today during the Late Mesolithic, Neolithic and Early Bronze Age, possibly with periods of very warm summers during the Early Neolithic. Precipitation during this period was probably lower and the winters colder than today, more characteristic of a continental climate. The growing season was longer than today, at least until the mid Bronze Age. Temperature decreased in parallel with increased *humidity,* particularly during the Bronze Age. From this time onwards there is a tendency to increased *oceanity* which lasted until historical time, except for some periods of warmer summers and lower humidity, particularly during the Roman Iron Age and the period about AD 900–1200 (the Medieval

Warm Epoch). After 1200 there was a period of fluctuating conditions until around 1350, when the climate suddenly became harsher for a long period lasting until about 1850 (the Little Ice Age). However, fairly mild conditions prevailed around 1450–1580. During the years 1300–1450 and 1580–1700, extreme climatic conditions were particularly common. Temperature changed gradually towards milder conditions during the 19th century, and a particularly warm climate characterizes the period 1920–1950 (Alexandersson and Eriksson 1989). Since then there has been a cooling trend in Scandinavia, which is an anomaly when compared with the global warming trend (Houghton and Woodwell 1989). The amplitude of the mean summer temperature has been about 1°C during the whole of historical times. The small climatic contrast between coast and inland/ upland seems to have been of minor importance for agrarian activities through time. There is no reason to believe that local deforestation changed the climate noticeably. If this occurred, it was mainly during the Late Bronze Age because of increased albedo and wind exposure. The temperature and oceanity curve is based on the general palaeoclimatic trend in NW Europe (Ch. 3), the humidity curve is based on our studies of lake-level changes (Ch. 5.1).

The agrarian economy has certainly been affected by the climate, particularly by short-term fluctuations, that is, periods of severe winters, cool and wet summers. This may have been one of the causes of some settlement regression in the Late Middle Ages. Prehistoric changes in the agrarian economy were also probably, to some extent, caused by climatic changes. The growing need for pasture land in the mid Bronze Age may be traced in the change from warm and dry to cool and wet growing seasons. Finally, expanding agriculture since the mid 19th century has certainly been favoured by a warmer climate.

Hydrology. Natural changes of groundwater and surface-water level are a direct result of changes in humidity, although deforestation has at least locally caused a groundwater-level rise because of changed evaporation (Moore 1986). The changes of surface water, measured as *lake area,* are to some extent related to the lake-level changes, but more to lake infilling processes. The Mesolithic landscape was juvenile in the sense of having many lakes which contributed to the landscape diversity. The number of lakes at that time in our area was about 19 with an area >2 ha and a total surface area of about 480 ha. Today the number of natural lakes is 3 with an area of about 220 ha. A rise of the groundwater level around 1000 BC caused the creation of at least one new lake (Bjäresjösjön) and flooding of some wetlands (e.g. Kumlan).

At the coast, *transgressions/regressions of the Baltic Sea* led to the formation of lagoonal bays (with fluctuating water depth), which became infilled and overgrown as early as about 1500 BC. The earlier trans-

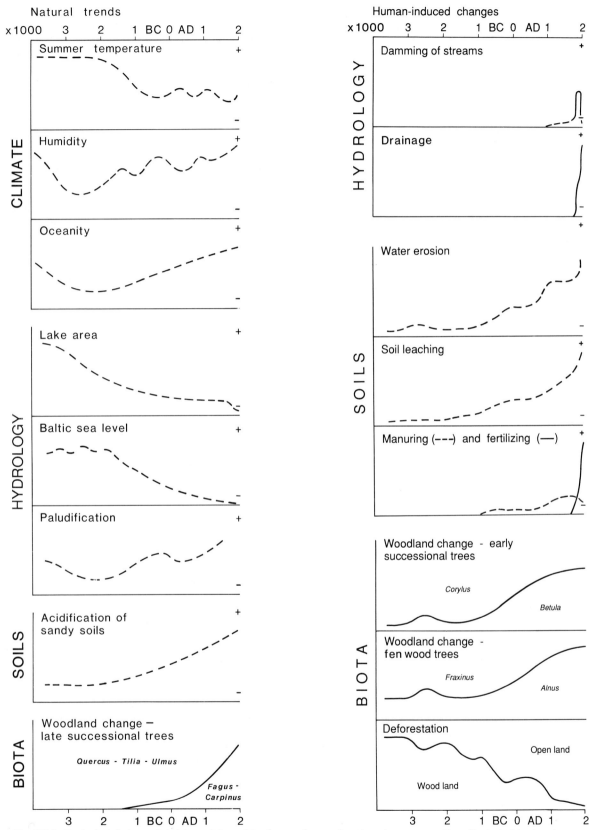

Fig. 6.2:2. Ecological factors related to society and landscape changes in a long-term perspective, illustrated by trend curves showing the relative changes through time. Higher precision in these curves indicated by continuous lines.

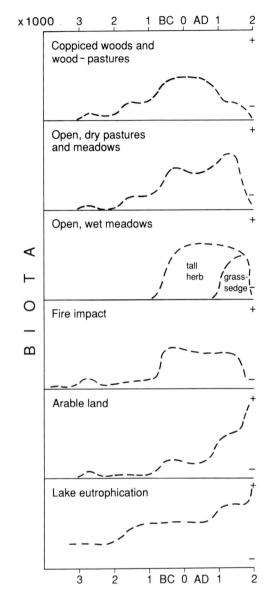

Time is continued paludification, but this is counteracted by drainage and reclamation of land.

Human-induced hydrological changes have been important, but mainly during historical times. Deforestation is associated with increased surface run-off and thus leads to an increase in river discharge. The local climate and microclimate also became drier. Locally small basins may have been embogged or flooded because of deforestation. Already in the Middle Ages people started to create *ponds* for livestock and human needs in the villages and at castles. Later on when peat-cutting became common, ponds and smaller bodies of water often resulted from exploited peat reserves, in both infields and outlands. Numerous water-filled marl-pits were created during the 19th century. *Damming* of rivers and brooks was common from the Middle Ages for powering watermills. The habit during the 18th and 19th centuries of damming streams to create water-meadows to improve meadow production lasted just less than 100 years and decreased rapidly after 1900, when various types of fertilizer became available. *Draining* of arable land and fodder areas was relatively uncommon before 1800. Great effort was expended to drain the landscape in the 19th century. Large-scale draining of arable land and of many wetlands became common to provide new land for cultivation. Even lakes like Bjärsjöholmssjön were tapped and drained for cultivation purposes. Before 1850 open ditches predominated but around that time there began a change to under-drainage. Tile pipes were first introduced on the big estates, but other owners soon followed. Since 1950 this trend has accelerated, and in 1990 it is difficult to find any site in the Ystad area, including most pasture land, which has not been under-drained.

In prehistoric time a raised groundwater level as a consequence of increased humidity was a disadvantage for cultivation, as in the Bronze Age and possibly in the High Middle Ages. In Modern Time it was possible to manipulate the local hydrological conditions in a favourable way by damming/ponding or by draining, always with the aim of increasing agrarian production.

Soils and soil nutrients. There is a general trend of soil leaching, *acidification,* throughout an interglacial like the Holocene. However, in an area with calcareous, clayey to silty soils like the Ystad area the soils are well buffered against acidification. An exception may have been the sandy soils of the Köpinge area and the less clayey soils of the Romeleåsen ridge. Deforestation and development of pastures have favoured an accelerating *soil leaching* which runs parallel to the human-induced *water erosion.* Under natural conditions there is almost no water erosion to be expected in a woodland landscape in south Scandinavia. But deforestation, grazing, and cultivation favour erosion, which is traced in the accelerating sediment load of the lake basins. The first weak increase occured during the Early Neolithic, followed by increased erosion in the Late Bronze Age,

gressions may also have caused a slightly raised groundwater level inland from the coast. The pattern of streams did not change much during this period; brooks and small rivers meandered across the woodland landscape, but deforestation is assumed to have caused increased run-off.

Embogging or *paludification* is dependent on humidity affecting the water balance. In this landscape we believe that the natural fen type has always been minerotrophic fens, mainly fen woods, but along hill slopes and lake shores with groundwater leakage also open sedge/herb-rich fens. Increased humidity probably caused a lateral expansion of at least wooded fens. This trend was also favoured by deforestation, at least from the Late Bronze Age onwards, which probably caused a general expansion of peat. The natural trend of Modern

more distinct from the Iron Age – correlated with expanding arable land and the introduction of deep ploughing techniques – again accelerating during the last 200 years because of expanded ploughed areas. Soil nutrient balance is a ruling factor for the maintenance of productivity and carrying capacity of the agro-ecosystems. Loss of nutrients in cultivated areas was compensated by *manuring,* possibly from as far back as the Late Bronze Age onwards when the settlements and the arable fields became permanent. Rotation systems practised from the Late Iron Age onwards were a further tool to prevent nutrient depletion. Land use in villages with an infield-outland system depended on the transport of nutrients from outlands into infields and within the infields. This system developed during the Iron Age and was fully consolidated during the Middle Ages. Until ca. 1500 this system functioned in a reasonable balance, at least in terms of relatively small changes of the relation between arable fields and fodder areas. However, when the demand for increased crop yields grew, the infield-outland system became increasingly unbalanced. The increase of the arable at the expense of the fodder areas led to nutrient deficit also in the arable. Too small fodder areas did not permit large enough numbers of livestock to provide enough manure to maintain the nutrient balance of the arable. The result was that crop yields decreased, which in turn required still larger areas of arable to give food for the human population. In regions where the availability of fodder was never limited, the nutrient balance of the arable soils could still be a strong restrictive factor. The application of a cropping system with long fallow periods due to a shortage of human labour in areas with light soils could have led to large nutrient losses. Unbalanced soil nutrient conditions, especially as regards nitrogen and phosphorus, were obviously crucial factors for the landscape development in terms of the distribution of arable and fodder areas. At the same time, it was also the main limiting factor for the maintenance – and the increase – of agrarian productivity before 1800.

At the beginning of the 19th century the nutrient conditions of the arable were poor, especially in the three-course villages. With the introduction of the *iron plough* in the 19th century both decomposition and the nutrient circulation rate accelerated and leaching increased. The increase of the ploughed areas after the enclosures, and later the introduction of deep-ploughing techniques, also hastened nutrient losses. During the 19th century various kinds of *organic and minerogenic fertilizers* were introduced to compensate for the losses. Such fertilizers were marl originating from fossil calcareous lake sediments and quarried below clay soil layers – see above. Phosphorus was obtained by spreading ground animal bones. However, it was not until around 1900 that *commercial fertilizers* became available, although used in modest quantities. The high loads used after 1950, with peaks during the mid 1970s, enabled the highest harvest yields in history in the area.

This was only possible either through the import of minerogenic substances from distant ecosystems and nations, or by the import of fossil fuels for the domestic industrial production of fertilizers. This has led to increasing environmental problems with leaching of nitrogen and phosphorus into rivers and streams and eventually to the groundwater.

Altogether the mobilization of soil nutrients for increasing agrarian production seems to have been a key factor behind cultural landscape changes in the past: from the mid Bronze Age (ca. 800 BC) onwards when manuring was practised on arable fields at permanent settlements, during the Iron Age when villages with infield/outland systems were organized, during the Iron Age and Middle Ages when rotation systems were developed, during the 19th century when new techniques like deep-ploughing and the flooding of meadows were applied together with marling, followed by the import of commercial fertilizers (cf. Emanuelsson 1988).

Vegetation and land use. The natural trend in the *woodland development* has been a gradual replacement of oak-lime-elm by beech-hornbeam on dry soils. This may be regarded as a partly climatically conditioned change among the late successional trees. Partly human-caused is the change among the early successional trees from hazel to birch, which resulted from processes such as soil leaching, but mainly from the displacement of half-open wood pastures from the fertile landscape near the coast (with more hazel) to the less fertile inland landscape (with more birch). Human exploitation caused a stepwise *deforestation* from the Neolithic, and the remaining woodlands were transformed by human activities into *coppiced woods and wood-pastures.* Dry areas were transformed in a gradual process of grazing and mowing into *open, dry pastures and meadows.* Damp soils (in depressions and on slopes) with fen woods of alder and moist woods of ash were exploited for fodder production from the Late Bronze Age onwards. These woodlands were transformed into *open, wet meadows.* More alder than ash remained, particularly on poor marginal soils. Tall herb meadows were replaced to a great extent by more exploited grass-sedge meadows after AD 700. Also of importance was *fire* – as a tool to open new land and to keep it open. Areas affected by fire seem to have increased from the Late Bronze Age onwards with a pronounced maximum during the Early Middle Ages. *Arable land* increased more or less parallel with areas used for fodder production. Distinct expansion is noticed for the Late Iron Age and Early Middle Ages and since the 16th century. The biota of the lacustrine ecosystems were also affected by land use. Diatom stratigraphies in the lake sediments are strong evidence for increased *eutrophication* which runs parallel to deforestation and expansion of pastures, meadows, and arable land in each catchment (H. Håkansson 1989, unpubl.).

The maintenance of the agro-ecosystems within the

infield-outland system determined the framework for various land-use practices with direct impact on the landscape. Such land uses can be divided into two classes: management of semi-natural vegetation, meaning grazing lands, meadows, woodlands, and coppice areas (1–3 below); management of cultivated fields (4 below).

1) Management of grazing lands. – The grazing and browsing by the livestock was certainly the most important factor for the management and development of pastures. However, recurrent clearing of the grasslands from shrubs and the like by fire also contributed significantly to the development of grassland and heathland communities with high species diversity. The long-term continuity of the grassland management is in itself a powerful landscape-forming factor.

2) Management of meadows. – Dry and wet meadows were maintained for hay harvests from the Late Bronze Age as a response to climatic change and the resulting need for winter stalling of the livestock. Management with mowing and post-harvest grazing favoured species-rich plant communities. Dry meadows, being easier to harvest, were more susceptible to yield depressions due to nutrient depletion by continuous harvests, as demonstrated for the 18th century. All types of mown meadows, of which many existed into the late 19th century, were sites of high species diversity for all levels of organisms.

3) Management of woodlands and coppice areas. – The deliberate management of the woodlands started in the Neolithic, and by the beginning of historical times virtually all woodlands in the Ystad area were managed. This management took the form of:
1) grazing by livestock, including pigs, and later from the 16th century by introduced deer; development of wood-pastures
2) the cutting of timber trees, in both wood-pastures and coppices
3) the development of coppices
4) the harvest of nuts, grass sward, or peat, mostly in wood-pastures.

The diverse management forms of the wooded areas all implied a substantial change in the characteristics of the ecosystem. The ground-level flora and fauna reached a higher species diversity because of the grazing and trampling of animals, and the greater entry of light from the reduced shrub layer compared with ungrazed woods. Timber harvest affected the age distribution of woodland trees. The management of coppice grassland implied higher biomass production and greater species diversity than in surrounding woodlands.

4) Cultivation and cropping practices. – Five practices can be distinguished as having the greatest influence on the landscape:

1) seed-bed preparation: techniques and tools for arding and ploughing
2) nutrient additions: manure and fertilizers (discussed above)
3) cropping and fallow cycles
4) management of fallows
5) drainage (discussed above)

All these practices have affected the agro-ecosystems since the introduction of permanent arable. At the beginning of the 19th century the wooden ard of similar construction to the prehistoric ones was still the chief implement for seed-bed preparation, not only in the Ystad area but also in most of Scandinavia (Bringéus 1964). The absence of a mould-board on the ards allowed both root weeds and grass sward partly to remain in the field and thus exert strong competition with the crop plants. The system of ridges and furrows was an imperfect form of drainage.

The predominant cropping system on the boulder clay soils from the Early Middle Ages was three-course rotation. The fallows were the primary grazing grounds here, and in order to preserve the fallow weeds as grazing plants, the fallow lands were not worked regularly. On areas outside the boulder clay soils, fallow periods of 10–20 years prevailed, allowing the development of grassland and shrub communities. This land use guaranteed the permanent existence of a broad variety of successional stages contributing to habitat diversity.

Since the early 19th century the cultural landscape has undergone general simplification and rationalization in terms of land use and vegetation. The enclosures during the 19th century led to the breakdown of the infield-outland system. This stimulated reorganizations of cropping practices and later the introduction of large-scale mechanization. The result has been the *disintegration* of the manifold land-use practices involving the *management of semi-natural vegetation*, leaving the remaining forms of land use associated solely with the *management of cultivated fields or forest plantations*. The ecological result of this is a broad spectrum of environmental problems associated with: (1) the fragmentation of ecosystems; (2) decreasing landscape, habitat, and species diversity; (3) threats to species populations; (4) chemical pollution by residues of pesticides; (5) increased nutrient leaching.

Each landscape-ecological change as identified in this synthesis is associated with changes in land use. The underlying causes are complex, but the driving force has been the demand for increased agrarian production by exploiting more land and/or by increasing areal production. The complexity involves new techniques, organization, mobilization of new land, crop choice, the balance between cultivation and pasturing, and so on. The main ecological factor behind all changes seems to be the soil nutrient balance.

6.3 Interaction of environmental and social factors behind landscape changes

We have defined cultural landscape periods and given dates for *changes in society and the environment*. This is an attempt to describe the landscape development in a traditional systematic way, but we have to bear in mind that the development has been gradual. There has in most cases been a response in the environment to changes in society – qualitative as well as quantitative changes of land use will have an effect on the landscape. It is more difficult to judge whether society has responded to independent environmental changes. We are convinced, however, that society and environment have interacted in the long-term perspective.

In the case of the transitions Early/Late Bronze Age around 800 BC, corresponding to the transition between landscape periods 3 and 4, there is a discrepancy between the palaeological and archaeological evidence. Our pollen-analytical studies indicate a remarkable deforestation in the whole area with an expanding pasture landscape also in the inland. Archaeological material demonstrated fairly intense land use in the coastal area without any expansion of settlement towards the inland. The explanation for the opening of the inland woods is therefore to be sought in wood-pasturing in a well-developed transhumance system.

Otherwise there is a general pattern of agreement: most of our cultural landscape periods end with a concentration of settlement leading to woodland regeneration, later followed by a new expansion of settlement to marginal areas, leading to deforestation and expansion of pastures and arable land (Fig. 6.1:1). This is obvious for period transitions 1/2 ca. 3000 BC with the first settlement expansion, 3/4 ca. 800 BC, and 4/5 ca. AD 700. Other periods of concentration have been identified within the landscape periods like the beginning of the Middle Neolithic ca. 2600 BC, the transition Pre-Roman/Roman Iron Age 200 BC–AD 200, and to some extent the Late Middle Ages. In this way we find support for the hypothesis – modified from the one in Ch. 1.1 – that the development of the cultural landscape has passed through periods of areal expansion followed by periods of concentration – the latter characterized by inner expansion, or exceptionally by stagnation/regression.

We believe that, in general, landscape changes, as described for the transitions between the periods identified, are caused by changes in society. Environmental factors like climate and hydrology mainly have a long-term effect on the landscape. However, short-term, drastic changes of climate may have caused a changed land-use strategy. Such environmental changes were general and they certainly influenced society in a wider region like south Scandinavia and caused the dispersal of ideas of importance to land-use changes.

This leads to a discussion of *ecological and social crises*. For an agrarian society there seems to have been a surplus in this region of natural resources throughout the past. The region is large enough to include areas for hunting and fishing as well as areas for cultivation and ample pasture land. At least in prehistoric time the society was flexible and adapted easily to ecological changes. The most sudden change was around the so-called "elm decline" when a natural deforestation made it easier and more favourable to clear land for cultivation. Other natural changes of climate and soils seem to have been gradual. Internal, cultural factors kept the population at a suitable level without overexploitation. And in a long-term perspective with increasing population it has been possible to increase the production by improving agrarian technique, organization, and so on together with land mobilization. Several factors have interacted in this development, but possibly the availability of soil nutrients has been the crucial factor. This emphasizes the close relation between environmental conditions and human adaptation, of importance also for the future prospects of mankind. This also means that our project hypothesis is oversimplified (Ch. 1.1). Interaction of society and environment is too complex to be described in a model like this (see Fig. 1.1:2). Human beings have always taken advantage of the spectrum of available possibilities in new situations.

In agrarian society man started to change nature more actively than before, which means that it became more difficult to see the consequences beyond each generation of individuals. However, it seems as if tradition and experience in combination with cultural ideas made it possible to *adapt to new situations*. The synchrony in the observed landscape changes in south Scandinavia makes it more probable that cultural factors were more important than ecological restrictions in causing these changes. Social factors may for some periods be the explanation for local overexploitation leading to some kind of crisis. This has been discussed for the Bronze Age when the society seems to have produced more cattle and sheep than were needed for its own consumption. The reason in this case may be traced in status behaviour.

The *socio-economic interrelation with areas* situated outside the Ystad area grows with time. Certainly there was trade with foreign areas already in Neolithic time, but it is more obvious during the Bronze Age and Iron Age. We know that the Roman Empire had an economic impact on south Scandinavia because of trade exchange. During historical time the area became more influenced by central Europe. This is particularly obvious for Modern Time – from the 16th century onwards the area became increasingly integrated with the agrarian market of Europe. During the last two centuries there are also parallels, especially since World War II, with the increasing mechanization and the accompanying development of the large-scale agrarian landscape.

Several important *decisions* for society and the landscape have moved from the local tribe to the national and international community. When decisions are taken outside the area the local society has lost its independence, which is obvious for the last century. When the agrarian economy is ruled by international decisions that force farmers to increase their efficiency in order to make a profit, this has negative consequences for the preservation of a rich cultural landscape.

Human society has passed through several unstable periods which may be named crises. They do not seem to have been caused by environmental factors but by social factors indirectly leading to overexploitation, nutrient depletion, etc. In all these cases society has adapted to the new situation by changing the organization and the economy or by applying a new technology, which improved the overall carrying capacity.

Our Ystad area has now become part of the global scene. We seem to be in an unstable period after the 20th century boom of industrialization. Human activity has led to serious environmental problems which are threatening human survival. In the past it was difficult for man to see the consequences of his various actions. Nowadays we know that modern production, traffic, and the like, with emissions into the atmosphere and the oceans, will seriously affect the future environment and even the global climate. This negative development of modern expansion has to be controlled if serious consequences are to be avoided. A radical change of society and its organization, economy, and technology will be needed for a sustainable development. This requires a consciousness and a common will to adapt to the restrictions set by the local ecosystems where nutrient recirculation must be established again. Man has to determine and respect environmental limits to growth – this is the message from ecologists to politicians.

Altogether this means that, in the future, national and international decisions will have a decisive influence on the settlement landscape in the Ystad area, where local decisions will play a minor role.

Ecological Bulletins 41: 00–00. Copenhagen 1991

7 Retrospect and prospect

7.1 Retrospect on the project. Views expressed by project collaborators

Ed. Björn E. Berglund

Aims, working hypothesis, and project results

Studies of long-term changes in the landscape, where human activities are one of the main driving factors, demand an interdisciplinary approach. This means adopting a joint working hypothesis, utilizing different sources, applying a variety of analytical methods, and organizing collaboration between several disciplines and many scientists. Being a large interdisciplinary enterprise, the Ystad Project seems to have been one of the more successful projects of this kind, thanks to a period of long discussions about practical and theoretical questions before the project started, and to the acceptance of a simple hypothesis among the collaborators. Crises during the long project period have occurred, and most of them can be traced back to moments when the theoretical background was less obvious for some project members.

The general aims of the project were very ambitious, but perhaps too general. Some project members also felt that the working hypothesis was too general. The interdisciplinary theoretical basis was good on the whole, but the common theoretical questions were not penetrated in sufficient depth. They should have been discussed in small groups throughout the project period. Therefore, the programme and the aims were too weak, at least at the beginning. In a long-term project like this it is also difficult to consider new theoretical ideas developed during the time.

The restructuring of the project after 2–3 years, with a concentration of study on focal areas, strengthened the project, but still several theoretical questions were rarely discussed until the concluding stage. We arranged several constructive workshops about the regional results from the focal areas, but we did not find time for joint concluding seminars as a background for the conclusions formulated in Ch. 6. They are based mainly on the text in Ch. 3, 4, and 5 and discussions in the small author team. We also neglected to discuss the applications of the project results – for the management of the natural landscape, the cultural landscape and the ancient monuments. These problems have been penetrated by some individuals, and we shall present some thoughts later on (Ch. 7.3, 7.4).

One view of the research approach is that we have been working traditionally in a descriptive-inductive way instead of a hypothetico-deductive way, which is nowadays normal for ecologists. Now, when dealing with the past environment, it is hard to test hypotheses by experiments, because the conditions today are very different from prehistoric time at least. To some extent the plant ecological conclusions are based on experiments or knowledge from analogous environments. Otherwise we had to analyse the past in the normal way: documentation of pieces of the society/environment, bringing this information together by applying a holistic view, always testing different working hypotheses.

Intradisciplinary versus interdisciplinary research

An interdisciplinary project like this has to apply theories and methods available within each discipline. It may be difficult for the project members to take advantage of the progress of knowledge within the "mother subject". Therefore the approach may be out of date in the final stage of such a long-term project. In the meantime, interdisciplinary methods and analyses are maturing. In the case of the Ystad Project it may be that the results are of minor importance for each participating discipline, but the value is to be found in the interdisciplinary, holistic view of the interaction between society and landscape through time.

Another problem is the load of disciplinary tradition which each scientist brings to this collaborative work. This means that key terms related to the cultural landscape are used with different meanings. However, in this project we have noticed that personal contacts have meant an improved understanding and development of a uniform terminology (see Ch. 2.9).

Organization and research strategy

The organization with subject-related subprojects and their leaders forming a project board has in general worked smoothly. Thanks to generous financial support it has been possible to fund subprojects when the research programme has been modified and new needs have appeared. From a scientific point of view the subdivision of the social history research following the time-scale (see Fig. 1.1:4) has been a disadvantage for the interdisciplinary discussions and joint efforts. It was compensated for by the concentrated studies in focal areas where scholars from all disciplines were involved (see results presented in Ch. 4). However, this work in selected areas was not planned in detail from the beginning, and the interdisciplinary discussions started rather late. We also regret that interdisciplinary workshops around such key sites were not organized until the summing up of the results started during the final year. However, during the last 4–5 years seminar series were arranged around results from the project with the participation of project members as well as students and colleagues from our institutes. We have lacked some disciplines in our research, e.g. ethnology, animal ecology, etc., but some restrictions had to be made.

The organization as a whole was a decision from the planning years and was never changed. Possibly this made the project more rigid than necessary. This organization should have been questioned and partly changed when the key site studies were initiated, from year 3 onwards. However, the subprojects were given large freedom for their own initiatives.

As regards the personnel, it was an advantage for those scholars employed from the beginning that they had a contract for 5–6 years. However, with the Swedish university system this also meant that young scholars have to find work outside the universities after a postdoctorate research period of about four years. Therefore we lost some collaborators during the project, and this discontinuity of employment has been a serious disadvantage (affecting palaeoecology, ecology, medieval archaeology).

The main leader of a project like this cannot act as an efficient leader when having this commitment parallel with the leadership of his own department. Only during the final year was it possible to finance a half-time leave for the work with this project synthesis. The main authors and members of the editorial board suffered from the same difficulties. Particularly in the final stage of writing the conclusions – one year after the formal closing of the project – it was very difficult to gather author teams for discussions and polishing of various text versions. These problems will be evident also during the compilation of the popular book based on this synthesis.

Collaboration

Interdisciplinary collaboration has been stimulated through this project, e.g. between archaeologists and palaeoecologists/ecologists, historians and geographers, ecologists and historians/geographers, medieval archaeologists and ecologists/dendrochronologists, etc. However, this collaboration should have been deeper – in the field and in indoor discussions. The joint seminars were meant to stimulate discussions and to build bridges between subjects. However, lack of time and other causes often meant a low frequency of representation at these seminars. Perhaps smaller workshops for a few scholars would have been more attractive. Interdisciplinary collaboration was sometimes difficult because of bad time coordination. Palaeoecological research

should have been given higher priority during the first two years as a background for selecting archaeological excavation sites, for example.

In general, the project collaborators have also increased their contacts with colleagues dealing with related problems in other areas of Sweden or abroad. Individual international collaboration has increased throughout the duration of the project. We have also demonstrated our results at several international field symposia visiting the Ystad area, resulting in constructive discussions.

Educational aspects

The Ystad Project has been stimulating for students on graduate as well as undergraduate level, particularly as a demonstration project related to the cultural landscape. This has improved the understanding of the interaction between society and landscape. It has also favoured interdisciplinary studies for analysis of the cultural landscape. During 1989 a research course was organized at the Department of Social and Economic Geography with the topic of cultural landscape analysis. A new research profile with this topic has now been established at this institute. During the summer of 1989 a Nordic field course on ecological analysis of the cultural landscape was organized by the Department of Plant Ecology (sponsored by the Nordic Council for Ecology).

The information to a wider audience through lectures, excursions, exhibitions, radio and television programmes, etc., has been appreciated. We are convinced that this has increased the understanding of the historical perspective of the landscape. The local inhabitants are also eager to see a popular, well illustrated version of this book.

Publishing

Planning and writing up results for the final publications should have started earlier. This is a general experience of large projects, so also for the Ystad Project. We have also used an additional year after the six-year period of the project. During this year several authors have left the university for other jobs and they have had great difficulties finding time for their writing. The archaeological monographs have been written partly in parallel with this synthesis, but partly after the synthesis chapters. This is of course a disadvantage.

The publishing of primary results is always a matter of discussion among us. The archaeologists publish monographs following an old and well-functioning tradition, the natural scientists restrict their publishing of primary data to international journal papers besides the chapters in this synthesis. In some cases raw data are made available also in institute report series.

7.2 Prospect on future research into the cultural landscape. Views expressed by project collaborators

Ed. Björn E. Berglund

Research gaps

Although this project has been well supplied with money and manpower, we find gaps in our material. There are several reasons for this, such as lack of source material, lack of time or personal capacity. The following research gaps have been mentioned:

- Improved absolute chronology in the sediment profiles based on accelerator technique (see Ch. 2.1).
- Continuous palaeoecological profiles from the Romeleåsen ridge (see Ch. 4.4.2).
- Mire-stratigraphical studies with accurate chronology for understanding palaeohydrological changes and their impact on the landscape (see Ch. 3).
- Extended archaeological studies of the inner hummocky landscape to test the hypothesis about prehistoric transhumance (see Ch. 4.3, 4.4, 5.5).
- Detailed investigations of ecological processes in the agro-ecosystems of the Middle Ages (see Ch. 5.3).
- Specific investigations of ecological processes related to the introduction of commercial fertilizer around 1900 (see Ch. 5.3).
- Detailed palaeoecological analysis of a basin within an area of continuous and dense population (see Ch. 2.1).
- Detailed analysis of town/hinterland interaction (see Ch. 5.8).
- Applied research, particularly for managing the cultural environment (see Ch. 7.3).
- Methodological improvements (see below).

New projects

The Ystad Project with its aims, results, interdisciplinary collaboration and holistic view of the landscape has stimulated new project initiatives, some of which may be mentioned here:
- Methodological improvements for interpretation and quantification of vegetation/land use from pollen diagrams (B. E. Berglund, M.-J. Gaillard, and others).
- Vegetation and land-use changes in village landscapes during historical time = calibration of data (B. E. Berglund, J. Regnéll, A. Persson, and others).
- Regional studies of long-term changes of the cultural landscape (B. E. Berglund and others).
- Prehistoric island settlement in the lakes of southern Sweden (L. Larsson).
- Methodological improvements related to retrogressive analysis of land survey documents (M. Riddersporre).
- The cultural landscape during the Middle Ages (H. Andersson, E. G. A. Olsson, and others).
- Building material of churches and the landscape (B. Sundnér).
- Man and landscape on the landed estates in southern Sweden during the agrarian revolution (J. Möller)
- The Swedish "own your own home" movement in the early 20th century (T. Germundsson)
- The future landscape – the importance of the big landed estates (L. and N. Lewan)

Educational aspects

Studies of different aspects of the cultural landscape will increase on the graduate level, if possible with an interdisciplinary approach. This will also have an influence on the educational contents on the undergraduate level, particularly in human geography. The international contacts will also stimulate education and research.

Continued organized collaboration

The Ystad project has acted as a centre for collaborative research on the cultural landscape at Lund University. This organization will now be dissolved. Instead minor research groups will be linked to new projects. Some colleagues have proposed the creation of a centre for continued collaboration in research and education on the cultural landscape in the past, present, and future. It depends on individuals, department representatives, and financial bodies to choose the way for creative research into the cultural landscape with a national and international perspective. Organization ought to be flexible and a formal centre may not be the best way for creative research collaboration.

7.3 The Ystad Project and the conservation of the cultural environment

Hans Andersson

The material studied by researchers on the Ystad Project consists to a large extent of physical remains of earlier periods, whether visible or invisible in the landscape. The results presented in this volume help us to understand the landscape we see and its component parts, in both a spatial and a chronological perspective. There is here a direct link between the research findings and the work of ensuring that the historical content of the landscape is preserved, the management of the cultural environment (Sw. *kulturmiljövård*). This link is expressly stated as one of the goals of the project, the results of which are supposed to provide a holistic scientific foundation on which to base tomorrow's conservation of nature and historical monuments, and with which to evaluate the way resources are used today.

Modern development has meant that the countryside and the towns have changed to the extent that all traces of the culture of former times are in danger of disappearing, and with them the contextual meaning of the landscape. There is legislation in this sphere which offers differing degrees of protection to historical monuments (far from all old buildings enjoy the same strong protection as ancient monuments and churches), but this legislation is geared to individual structures and sites. What we need is an approach that puts more emphasis on complete contexts in the countryside or the towns, so that cultural monuments in the widest sense are preserved in such a way that they reveal their role in the landscape. Recent general legislation – the Natural Resources Act and the Planning and Building Act of 1987 takes some account of this approach: for instance, areas of national interest for their natural or cultural value are to be demarcated. To champion such an approach in a situation of radical change requires not least a profound knowledge of cultural landscape processes in the widest sense. A greater emphasis on composite environments also involves knowledge which differs in part from that required for the conservation of single features.

The management of the cultural environment comprises three tasks. First, it must go hand in hand with town and country planning to ensure that the cultural landscape is preserved, with its significant milieux, ancient monuments, and historic buildings. Second, it must document whatever disappears, functioning in

other words as society's memory. Third, it must spread knowledge and consciousness of the importance of the physical remains.

To struggle for these objectives in the tough world of physical planning, with all its competing interests, those engaged in conservation must possess all-round knowledge or have the opportunity to obtain the necessary knowledge from relevant research. It is not always possible to meet this demand when the necessary knowledge is not available. This applies not least to the new situation described here, with its stronger contextual emphasis. Yet this creates important challenges for research. A holistic view of the landscape needs more interdisciplinary cooperation, a greater willingness to study more complex connections between natural preconditions and social circumstances, to arrive at a better understanding of the appearance of the landscape and its variation through time and space. These demands are especially important when the laws on natural resources have to define what is meant by "national interest".

It is in such a light that we must review the Ystad Project. This gives reason to draw attention to concrete results which can be directly translated into practical activity in the three tasks specified above. But the project can also inspire more general theoretical and methodological reflections about how the experiences can be applied to other areas and other conditions.

Highly important material of a concrete nature has been obtained by the reconstructions of the landscape at various points in time (Ch. 3). This allows a discussion of various phenomena in their context and an analysis of certain connections in the landscape. It also gives us the opportunity of assessing the value of individual fragments that remain from different times. For example, we have within the project area something that we now know is worth preserving – the area around Baldringe in the north-east, with its landscape reflecting a structure that was much more common in the Ystad area in the past.

There are areas where studies have shown strong continuity within fairly clear boundaries. The settlement locations may have changed, but they can be followed from early prehistoric times down to modern days. An example of this is Bjäresjö (Ch. 4.2). An area like this can serve as a reference and should therefore be protected from excessive change.

Another example in this respect comes from the expressions of power which can still be seen in the landscape. Thanks to the work of the project, we now know that a hierarchical social structure had been established in the area by the Late Iron Age at least. Clear traces of this are seen in the manors and churches, which were once closely associated, and which can sometimes still be seen in the same place, as at Stora Herrestad, even though their exterior forms have changed. In other cases we can only observe this association by excavation, but there is reason to assume that the manor moved to a new location (e.g. Bjäresjö-Bjersjöholm). Visible manifestations of social status also include runic stones, although relatively few of these still stand on their original sites. The churches have occasionally retained their displays of power, in the form of towers or remains of galleries for the patron's family.

Of course, a similar discussion can be conducted for later periods. The estates, the villages, the railways, the industrial buildings are part of the cultural landscape of their times. The project has produced material which puts these phenomena in their historical and geographical context. The analyses carried out have been more detailed and precise than those for earlier periods as a result of the greater availability of source material. Nevertheless, surprisingly little may survive of even the 19th century landscape in an area like the Ystad area. It is not least the villages which have suffered in recent decades from the transformation of agriculture. This means that the aspects of preservation must pay as much attention to recent material as to that from early times, and equally close analysis is required in the field.

Another task, not wholly separate from the preceding one, is the responsibility of documentation. Berta Stjernquist has pointed out the importance of the combination of work in the project and the material uncovered by rescue archaeology. She takes as an example the area around the river Svartån, where there had been vigorous exploitation, necessitating many rescue excavations. On the other hand, archaeologists had had little chance to process the material. Now, however, they have been able to incorporate these excavations into the research of the project and to supplement them, thereby giving a very detailed picture of prehistoric times. It is crucial from the point of view of the cultural environment that documentation has not been confined to strictly archaeological material, but that scientific evidence has also been incorporated (Stjernquist 1989). To this we may add the importance which the new research situation can have for future investigations connected with the conservation of the cultural environment. Rescue excavations in the future can tackle clearly defined problems in a way which has not hitherto been possible.

Another area where the problem of documentation has been prominent is the churches. These are complex buildings, where changes are largely dependent on changes in the world around. It is therefore important to study the building history of the churches. This is best done when the plaster is removed to expose the masonry. This means that it is particularly important to take advantage of the occasions when churches undergo extensive restoration. Churches are protected buildings, and since the authorities responsible for conservation are closely involved they have a particular responsibility to facilitate research. The church investigations of the Ystad Project have shown how the evidence that is revealed can be important for our understanding not only of the churches, but also of the landscape in a broad sense (Andersson and Anglert 1989, Ch. 5.7).

Yet another point should be considered. Although the techniques of natural science are becoming an important part of the archaeological investigation of the cultural landscape, it can be argued that scientific resources should be used even more in the documentation of a vanishing landscape. The current legislation is not strong enough to ensure this, but the question deserves to be raised.

The third task concerns information and the dissemination of knowledge. Here too the project ought to achieve something through the research that has been pursued and presented in various ways. The workers on the project have already presented results to the public in various contexts – lectures, popular articles, exhibitions – but this is not enough. The results must become an integral part of the information for the conservation of the cultural environment, readily available to politicians and public alike in the Ystad area. This is not least important just now, when the new legislation on planning and building is to result in physical structure plans, in which the interests of the cultural environment are to be marked.

But it is not only in this perspective we must see the significance of the dissemination of the project's results. A shared concern of both research and conservation is to create an understanding of the importance of this kind of knowledge. Without this understanding it is impossible to do effective work in cultural conservation.

Despite these direct results and the connections that exist between the project and the conservation of the cultural environment, there are still vital problems waiting to be solved. We need a more profound discussion about how to select environments and how to analyse them. An attempt at such an analysis has been made (Andersson 1989), but this work has to be developed. A discussion of this kind will have to focus on the relation between research and its application to the conservation of the cultural environment. Moreover, the Ystad Project, with its link between nature and culture, can clearly be an important starting-point in a more general perspective.

7.4 The Ystad Project and nature conservation

Nils Malmer

It is generally accepted today that all town and country planning and all exploitation of natural resources such as land and water should be done with regard for the maintenance of nature's productive capacity. Of course, this idea is not new. For centuries there has been legislation to protect, for example, fruit-bearing trees and game species, and to prevent the overexploitation of woodland and maintain forestry production. Nowadays, however, with the growth and diversity of technical methods of exploitation, there is an urgent need to consider the long-term preservation of natural resources.

Compared with older forms of legislation intended to safeguard natural resources, the idea of nature conservation and the awareness of the need to protect nature for its own sake is a relatively recent phenomenon; it is largely a product of the present century. Nature conservation can also be seen as part of an active form of community planning, since it seeks to limit the exploitation of natural resources, using legislation and other means, for reasons other than merely the maintenance of nature's productive capacity.

The protection of the landscape afforded by the legislation on nature conservation in Sweden today is motivated by cultural-scientific, social, and welfare considerations.

The whole of the studied landscape-ecological development (Ch. 5.13) gives us a valuable basis for tackling problems related to the economical use and conservation of natural resources in areas dominated by modern agriculture.

Development towards today's highly exploited agricultural landscape can be said to have consisted of two phases. The first spans the period from the introduction of agriculture up to the Iron Age, by which time virtually the whole landscape bore the stamp of agricultural activities. The second phase, from the Iron Age to the present day, is characterized above all by constantly increasing farming intensity which has been made possible by technical development and – during the last century – the use of auxiliary energy.

The first phase mostly affected the structure of vegetation, primarily by changing the character of the primeval forests and by greatly expanding the area of pastoral grassland. In other words, the diversity of the landscape increased considerably, and several new, human-influenced landscape types took shape. Hydrology was probably only slightly affected. There was almost certainly a reduction in biomass, but it is reasonable to assume that primary production remained largely unchanged. Numerous plant and animal populations were probably able to expand, while others may have declined. Occasional new species may have immigrated, but it is unlikely that any species was threatened with extinction as a result of this development.

The largely continuous increase in the intensity of farming in the second phase, from the Iron Age to the present day, has found expression in many different ways in the vegetational landscape. From the ecological point of view, three developments are particularly striking: the heavy increase in the area of regularly ploughed land, at the expense of natural grassland; the now almost total drainage of all wetlands; and the increased supply of plant nutrients through the use of industrially produced fertilizers. Each of these practices in itself means a reduction in diversity, gradually impoverishing and eliminating not just the original landscape types but also the human-influenced ones. The hydrology of the landscape in particular is undergoing a radical change. The vast majority of the original plant and animal populations have seen drastic reductions as regards both their geographical distribution and the number of individuals.

The intensive exploitation of the agricultural landscape has accelerated since World War II as a direct consequence of modern farming technology, which is regulated by agricultural policies established by governmental decisions. The lake sediments studied in the Ystad area reveal a much faster sedimentation rate than in previous periods, and they consist to a large extent of terrestrial, allogenic matter. Although sedimentation in the past few decades has decreased significantly owing to under-drainage, the results show a sharp increase in water erosion in the area, especially in later historical times. It is beyond all doubt that this increased erosion of the surface layer is connected with modern agricultural techniques. The result is a gradual impoverishment of farming land and is therefore incompatible with the economical use of natural resources. Modern agriculture has been criticized in recent years for leaching nitrogen and phosphorus, polluting our waters. Although it has not attracted the same attention, the rapid rise in erosion is a further example of the misuse of natural resources. Production may not have declined noticeably, or else it has been offset by, for example, increased doses of mineral nutrients. Since erosion is a slow process, it takes some time before the damage can be seen, but when it eventually makes itself felt, it will be irreparable.

In the first decades of the nature conservation movement especially, there was a strong tendency to direct activity towards what was viewed as survivals of the original natural landscape. Given a narrow outlook like this, the Ystad area has very little to offer today, nor has it had for a long time. However, the oldest publications about nature conservation in Scania also emphasized the need to protect various types of human-influenced landscape, such as grasslands of the kind that were probably created during the first phase of agricultural development, and which can contain a rich flora and fauna of spontaneously occurring species. In other words, the early conservationists were fully aware of the significance of these environments, although they may not have known that they were created by human activity and require special forms of land use for their survival.

For today's nature conservation these landscape types reflecting older land use are of the utmost interest, along with all original natural types, of course. The studies in the Ystad area show that what could be called original natural environments were probably virtually eradicated by the start of the Middle Ages, if not earlier. The landscape was already greatly changed even then. Today the Ystad area has retained very few environments preserving the varieties of cultural landscape which were typical of much of the area up to the mid 19th century.

When farming began to be intensified in the 19th century, it was precisely these older human-influenced landscape types which were most heavily affected. This was one reason why nature conservationists espoused these environments at an early stage. Nevertheless, the reduction of the area and distribution of grassland has continued since World War II, constituting as much as 20–30% in a fifty-year period. Grassland in wet, often fen-like areas, has been especially affected. On top of this comes the reduction that has resulted from the large-scale elimination of roadsides and ditch-banks, with their often similar vegetation.

Not as noticeable, but biologically at least as serious, is that the plant and animal content of the human-influenced landscape types has been steadily impoverished. Even seemingly minor changes in land use, such as the change from mowing to grazing, changes in the total grazing pressure, or the use of commercial fertilizers, have been shown to lead to dramatic changes in species composition. Two particularly striking changes are the reduction by half of the number of plant species and the increase in nitrophilous species, both in terms of the number of species per unit of square measure and the proportion of the biomass. It is obvious that the use of commercial fertilizers and thereby the increase of mineral nutrients in the soil results in a long-term change in the balance between the different species in the plant cover, and that competitive species which are better able to exploit the new minerals are favoured at the expense of other species. Vegetational diversity is thus being radically reduced, which is very serious from the point of view of nature conservation.

Biologically there is a great difference between arable land, which is regularly ploughed and fertilized, and so-called natural meadow and the kind of woodland which is not managed by methods like those used in agriculture. In the former case the vegetation consists of a few cultivated species along with a number of weed species. In the latter case there is a much greater variety thanks to a higher frequency of indigenous species. The landscape-ecological maps and the studies of grasslands in the area clearly show how the total area available for indigenous species has shrunk in the past hundred years. Furthermore, the areas still available to them are now much smaller and separated by greater distances. No study has been made of how many species have disappeared from the area, but it is clear that the number is fairly high. It is also obvious that there have been drastic reductions in the populations of virtually all indigenous species. This in turn probably means that in the coming decades more native species will vanish from the area. This extinction process is hastened not only by the declining population sizes but also by the fragmentation of biotopes.

The small-scale biological variation in the landscape, of which the variation in species composition and richness is an example, is not evident from the historical maps, nor is it easy to represent in landscape-ecological maps. This aspect tends to be overlooked in general planning work, which is often geared more to the look of the landscape than to its content. The results of the Ystad Project clearly show the importance of taking these aspects into consideration, since both the look and the content of the landscape must be preserved. When the conservation of historic areas seeks to recreate the original environments as far as possible, there should also be an effort to restore their biological content.

As for the social, welfare, and recreational aspects of nature conservation, the development in the Ystad area shows how modern farming techniques severely hamper accessibility almost everywhere, making it harder for people to move freely in the countryside and to find recreation there. This is happening at a time when the ongoing urbanization of society makes it even more important that land is available for the exercise of *allemansrätt,* the right of public access on which Sweden prides itself. Yet the landscape-ecological maps show a gradual decrease in the amount of accessible land, in pace with but not on account of urbanization. This development indicates yet another problem which must be considered in the planning of nature conservation. Purely social and welfare considerations are also invoked by conservationists: people must have easy access to nature and recreation.

Ecological Bulletins 41: 457–497. Copenhagen 1991

Collections, archives, and maps

Archaeological sources

– Collections in Statens Historiska Museum (Museum of National Antiquities) in Stockholm.

– Collections and excavation reports in Lunds Universitets Historiska Museum (Lund University Historical Museum).

– Collections in Ystads Museum (provincial museum in Ystad).

– Excavation material and reports at Riksantikvarieämbetet UV-Syd (Central Board of National Antiquities, Archaeological Excavations Department, Southern Sweden) in Lund.

– Notes and reports in Riksantikvarieämbetet, Antikvariskt Topografiskt Arkiv (Central Board of National Antiquities, Topographical Antiquarian Archives) in Stockholm.

– Maps and reports produced by the Survey of Ancient Monuments, conducted by Riksantikvarieämbetets Fornminnesavdelning (Central Board of National Antiquities, Archaeological Heritage Department, Documentation Division); a systematic inventory and documentation of cultural resources and historic areas. The documentation material for the Ystad area is available at the Central Board of National Antiquities in Stockholm and at Länsstyrelsens Kulturmiljöenhet (Department of Cultural Heritage, County Administrative Board, Malmöhus County) in Malmö.

– Reports and restoration programs at *Skånes Hembygdsförbund* in Lund.

Historical sources

– Cadastral records, terriers, accounts, etc. in Rigsarkivet (Danish National Archives), Copenhagen, in Riksarkivet (Swedish National Archives), Stockholm, and in Kammarkollegiet (National Judicial Board for Public Lands and Funds), Stockholm.

– Cadastral records, terriers, accounts, ecclesiastical records, estate records, etc. at Landsarkivet i Lund (Provincial Archives in Lund) and at Domkapitlet i Lund (Cathedral Chapter in Lund).

– Estate records in Lunds Universitetsbibliotek (University Library, Lund).

– Place-name records at Dialekt- och ortnamnsarkivet (Institute of Dialect and Place-Name Research) in Lund and at Institut for navneforskning (Institute of Place-Name Research) in Copenhagen.

– Land survey documents at Lantmäteriverket (National Land Survey, Central Office in Gävle and County Office in Malmö). See also Maps (below).

– Letter collection, microfilm archive at Historiska institutionen (Department of History), Lund University.

– Agrostatistical records at Landsarkivet i Lund (Pro-

vincial Archives in Lund), at Lantbruksnämnden i Malmö (Agricultural Board of Malmöhus County, Malmö), and at Statistiska Centralbyrån (Statistics Sweden), Stockholm.

– Drainage records at Kungl. Skogs- och Lantbruksakademien (Royal Academy of Agriculture and Forestry), Stockholm.

– Parish registers, taxation registers, register of landed property at Länsstyrelsen (County Administrative Board), Malmö.

– Primary census data (population records at Statistiska Centralbyrån (Statistics Sweden), Stockholm.

Maps

Publication dates refer to sheets covering the Ystad area

– Gerhard Buhrmann's map of Scania from 1684. Krigsarkivet (Armed Forces Archives) in Stockholm.

– Land survey documents (maps and descriptions) from the late 17th century to the 19th century: large-scale maps showing the contemporary layout of villages and single farms (*geometriska arealavmätningar*) or the plans for reallotment programmes (*storskifte, enskifte*, and *laga skifte*). Länsstyrelsens Lantmäterienhet (National Land Survey, County Office) in Malmö and Lantmäteriverket (National Land Survey, Central Office) in Gävle.

– *Skånska rekognosceringskartan* 1812–1820. Handdrawn map of Scania at a scale of 1:20,000. The original sheets and descriptions are kept in Krigsarkivet (Armed Forces Archives) in Stockholm. The map sheets published at a scale of 1:30,000 by Lantmäteriverket (National Land Survey) 1986.

– *Generalstabens topografiska karta*, scale 1:100,000. Official topographic map. First published in 1864. Partly revised version 1907.

– *Ekonomisk karta över Malmöhus län*, scale 1:20,000. Economic map published 1914–1916. Maps and descriptions published by Malmöhus Läns Hushållningssällskap (Agricultural Society of Malmöhus County).

– Land survey documents (maps and descriptions) from the 19th and 20th centuries showing the development of ownership on individual farms and estates subsequent to the reallotment programmes of the early 19th century. Länsstyrelsens Lantmäterienhet (National Land Survey, County Office) in Malmö.

– *Topografisk karta över Sverige*, scale 1: 50,000. Official topographic map. Third edition, October 1982. Published by Lantmäteriverket (National Land Survey) in Gävle.

– *Ekonomisk karta över Sverige*, scale 1:10,000. Official economic map. Edition 1972/73. Published by Lantmäteriverket (National Land Survey) in Gävle.

Abbreviations of archives, museums, and authorities

Abbreviation	Swedish name	English name
ATA	Se RAÄ	See RAÄ
DAL	Dialekt- och ortnamnsarkivet i Lund	Institute of Dialect and Place-Name Research in Lund
	Hushållningssällskapet – Malmöhus län	Agricultural Society – Malmöhus County
KRA	Krigsarkivet	Royal Military Archives
KSLA	Kungliga Skogs- og Lantbruksakademien	Royal Academy of Agriculture and Forestry
LLA	Landsarkivet i Lund	Provincial Archives in Lund
LMV	Lantmäteriverket – i Gävle – vid Länsstyrelsen i Malmöhus län	National Land Survey – Central Office, Gävle – County Office at the Malmöhus County Administrative Bord
LRF	Lantbrukarnas Riksförbund	Federation of Swedish Farmers
	Lantbruksnämnden – Malmöhus län	Agricultural Board – Malmöhus County
LUHM	Lunds universitets historiska museum	Lund University Historical Museum
	Länsstyrelsen i Malmöhus län – Kulturmiljöenheten	Malmöhus County Administrative Board – Department of Cultural Heritage
MM	Malmö Museer	Malmö Museums
RA	Riksarkivet	National Archives
RAÄ ATA UV-Syd	Riksantikvarieämbetet – Antikvariskt topografiskt arkiv – Fornminnesavdelningen, – Fornminnesinventeringen – Undersökningsverksamheten Syd	Central Board of National Antiquities – Topographical Antiquarian Archives – Archaeological Heritage Department – Documentation Division – Archaeological Excavations Department, Southern Sweden
SCB	Statistiska Centralbyrån	Statistics Sweden
SGU	Sveriges Geologiska Undersökning	Swedish Geological Survey
SHM	Statens Historiska Museum	Museum of National Antiquities
UV-Syd	See RAÄ	See RAÄ
YM	Ystads Museum	Ystad Museum

References

Aaby, B. 1975. Cykliske klimavariationer de sidste 7500 år påvist ved undersøgelser af højmoser og marine transgressionsfaser. – Danmarks Geologiske Undersøgelse, Årbog 1974: 91–104.
- 1976. Cyclic climatic variations over the past 5500 years reflected in raised bogs. – Nature, Lond. 263: 281–284.
- 1983. Forest development, soil genesis and human activity illustrated by pollen and hypha analysis of two neighbouring podzols in Draved Forest, Denmark. – Dan. Geol. Unders. II: 114, København.
- 1986. Trees as anthropogenic indicators in regional pollen diagrams from eastern Denmark. – In: Behre, K.-E. (ed.), Anthropogenic Indicators in Pollen Diagrams. A. A. Balkema, Rotterdam, pp. 73–93.
- 1988. The cultural landscape as reflected in percentage and influx pollen diagrams from two Danish ombrotrophic mires. – In: Birks, H. H., Birks, H. J. B., Kaland, P. E. and Moe, D. (eds), The Cultural Landscape – Past, Present and Future. Cambridge Univ. Press, pp. 209–228.
- and Tauber, H. 1975. Rates of peat formation in relation to degree of humification and local environment, as shown by studies of a raised bog in Denmark. – Boreas 4: 1–17.
Aalen, F. H. A. 1983. Perspectives on the Irish landscape in prehistory and history. – BAR Brit. Ser. 116: 357–377.
- 1986. The rehousing of rural labourers in Ireland under the Labourers (Ireland) Acts, 1883–1919. – J. Hist. Geogr. 12/3: 287–306.
Aarup Jensen, J. 1973. Bopladsen Myrhøj. 3 hustomter med klokkebægerkeramik. – Kuml 1972: 61–122.
Adam av Bremen. 1984. Historien om Hamburgstiftet och dess biskopar. – Ed. Samfundet Pro fide et christianismo, Stockholm.
Adams, I. H. 1976. Agrarian landscape terms. A glossary for historical geography. – Spec. publ. 9. London Inst. of British Geographers.
Addicott, J. F., Aho, J. M., Antolin, M. F., Padilla, D. K., Richardson, J. S. and Soluk, D. A. 1988. Ecological neighborhoods: scaling environmental patterns. – Oikos 49: 340–346.
Agger, D., Brandt, J., Byrmak, E., Jenssen, S., Ursin, M. 1986. Udviklingen i agerlandets småbiotoper i Østdanmark. – Roskilde universitet. Forskarrapport 48. 545 pp.
Ahlberg, E. et al. 1974. Naturvårdsundersökningar inom Krageholmsområdet. – Växtekologiska institutionen, Lunds Universitet.
Ahlén, I. and Tjernberg, M. 1988. Hotade och sällsynta ryggradsdjur i Sverige. – Sveriges Natur 1988/2: 33–42.
Albrethsen, S. E. and Brinch Petersen, E. 1977. Excavation of a Mesolithic Cemetery at Vedbæk, Denmark. – Acta Archaeologica XLVII: 5–28.
Alexandersson, H. and Eriksson, B. 1989. Climate fluctuations in Sweden 1860–1987 – SMHI Reports, Meteorology and Hydrology 58, Swedish Meteorological and Hydrological Institute, Norrköping.
Alström, K. and Bergman, A. 1990. Water erosion on arable land in southern Sweden. – In: Boardman, J., Foster, I. D. L. and Dearing, J. A. (eds), Soil erosin on agricultural land. Wiley, Chichester, pp. 107–118.
Ambrosiani, B. 1964. Fornlämningar och bebyggelse. Studier i Attundalands och Södertörns förhistoria. – Uppsala.
- 1984. Settlement expansion – settlement contraction: A question of war, plague, ecology or climate.– In: Mörner, N. A. and Karlén, W. (eds), Climatic changes on a yearly to millennial basis. D. Reidel, pp. 241–247.
Améen, L. 1981. Stadslandskapets förändringsrisker. – Svensk Geografisk Årsbok 1981: 11–18.
- 1984. Stadsplaner som källa. Bebyggelsehistorisk tidskrift 7: 44–56.
- 1987. Transport och trafik. – In: Jacobsson, B. (ed.), Ystad

under nittonhundratalet. Ystads historia. Del IV, 1914–1971. Ystad.
Ammann, B. 1982. Säkulare Seespiegelschwankungen: wo, wie, wann, warum? – Mitt. Natf. Ges. Bern NF 39: 97–106.
Andersen, K., Jørgensen, S. and Richter, J. 1982. Maglemose hytterne ved Ulkestrup Lyng. – Nordiske Fortidsminder, Bind 7. København.
Andersen, S. H. 1973. Overgangen fra ældre stenalder i Sydskandinavien set fra en mesolitisk synsvinkel. – In: Simonsen, P. and Stamsø-Munch, G. (eds), Bonde Veideman. Bofast-ikke bofast i nordisk forhistorie. Tromsø Museums Skrifter XIV, pp. 26–44.
- 1985 Tybrind Vig. A preliminary report on Submerged Ertebølle site on the West coast of Fyn. – J. Danish Archaeol. 4: 52–67.
- and Johansen, E. 1986. Ertebølle Revisited. – J. Danish Archaeol. 5: 31–61.
Andersen, S. T. 1970. Relative pollen productivity of North European trees, and correction factors for tree pollen spectra. – Danmarks Geologiske Undersøgelse II, 96.
- 1973. The differential pollen productivity of trees and its significance for the interpretation of a pollen diagram from a forested region. – Birks, H. J. B. and West, R. G. (eds), Quaternary Plant Ecology. Blackwell, Oxford.
- 1978. Local and regional vegetation development in eastern Denmark in the Holocene. – Danmarks Geologiske Undersøgelse Årbog 1976: 5–27.
- 1984. Forests at Løvenholm, Djursland, Denmark, at present and in the past. – Kongelige Danske Vid. Selskab, Biol. Skr. 24:1. København.
- Aaby, B. and Odgaard, B. B. 1983. Current studies in vegetational history at the Geological Survey of Denmark. – Danish Archaeol. 2: 184–196.
Andersson, Hilbert. 1959. Parzellierung und Gemengelage. Studien über die älteren Kulturlandschaft in Schonen. – Meddelanden från Geografiska Institutionen vid Stockholms Högskola 122. Gleerups, Lund.
Andersson, Hans and Anglert, M. (eds) 1989. By, huvudgård och kyrka. Studier i Ystadsområdets medeltid. – Lund Studies in Medieval Archaeology 5. Almqvist & Wiksell, Stockholm.
Andersson, K. 1983. Kalmarkustens kyrkor under tidig medeltid. – Hikuin 9. Højbjerg.
Andersson, L. and Appelqvist, T. 1987. Ljunglav och almlav, indikatorer på värdefull lövskog. – Sv. Bot. Tidskr. 81: 185–194.
Andrén, A. 1980. Lund. Medeltidsstaden, Rapport 26. Stockholm.
- 1985. Den urbana scenen. Städer och samhälle i det medeltida Danmark. – Acta Archaeol. Lund., Series in 8⁰: 13 Bonn/Malmö.
Anglert, Marit 1984. Vem försvarade vad? Några reflexioner kring de s k försvarskyrkorna. – Meta 1984: 3–4.
Anglert, Mats 1986. Medeltida kapell i Skåne – en första sammanställning. – Medeltiden och arkeologin. Lund Studies in Medieval Archaeology 1. Lund.
- 1989 a. Kyrkorna. Inledning. – In: Andersson, H. and Anglert, M. (eds), By, huvudgård och kyrka. Studier i Ystadområdet medeltid. Lund Studies in Medieval Archaeology 5. Almqvist & Wiksell, Stockholm, pp. 167–181.
- 1989 b. Den kyrkliga organisationen under äldre medeltid. – In: Andersson, H. and Anglert, M. (eds), By, huvudgård och kyrka. Studier i Ystadsområdets medeltid. – Lund Studies in Medieval Archaeology 5. Almqvist & Wiksell, Stockholm, pp. 221–242.
Anonymous. 1978. Livsmedelstabeller. – Statens Livsmedelsverk
Anonymous. 1983. Näringslära för högskolan. – Esselte Studium.

Anonymous. 1985. Handelsgödsel, stallgödsel och kalk i jordbruket – långa tidsserier. – Statistiska meddelanden. Na 15 SM 8501.

Anonymous. 1986. Jordbruksstatistiks årbok 1986. – Statistiska Centralbyrån. Stockholm.

Anonymous. 1987 Inventering av ängs- och hagmarker. – Statens Naturvårdsverk, Solna.

Anonymous. 1987. Naturmiljön i siffror. Miljöstatistisk årsbok 1986–87. – Statistiska Centralbyrån. Stockholm.

Anonymous. 1988. Jordbruksstatistisk årsbok 1988. – Statistiska Centralbyrån. Stockholm.

Anonymous. 1989. Jordbruksstatistisk årsbok 1989. – Statistiska Centralbyrån. Stockholm.

Anonymous. 1990. Jordbruksstatistisk årsbok 1990. – Statistiska Centralbyrån. Stockholm.

Arbman, H. 1946. The Baldringe hoard. – Meddelanden från Lunds universitets historiska museum. 1946: 99–136.

– 1948. Två nyfunna praktvapen från bronsåldern. – Skånes hembygdsförbunds årsbok 1948: 1–11.

– 1962. Vikingarna. – Bonniers, Stockholm.

Arkeologi i Sverige 1982–1983. – Riksantikvarieämbetet och statens historiska museer. – Rapport RAÄ 1985:5. Stockholm.

Arkeologi i Sverige 1984. Riksantikvarieämbetet och statens historiska museer – Rapport RAÄ 1986:2. Stockholm.

Arkeologi i Sverige 1985. – Riksantikvarieämbetet och statens historiska museer – Rapport RAÄ 1987:1. Stockholm.

Armstrong, P. 1975. The changing landscape. The history and ecology of man's impact on the face of East Anglia. – Terence & Dalton, Lavenham, Suffolk, UK.

Arrhenius, H. 1934. Fosfathalten i skånska jordar. – Sveriges Geologiska Undersökning Ser. C 383. Stockholm.

Arup, E. 1926. Danmarks historie. Bd. I. – København.

Baillie, M. G. L. 1979. Some observations on gaps in tree-ring chronologies. – Symp. archaeol. sci. 4–7 January 1978.

Baker, A. R. H. and Butlin, R. A. (eds) 1973. Studies of field systems in the British Isles. – Cambridge.

Bakker, J. A. 1982. TRB settlement patterns on the Dutch sandy soils. – Analecta Praehistoria Leidensia XV: 87–123.

Banning, K. (ed.) 1976–82. A Catalogue of Wall-Paintings in the Churches of Medieval Denmark 1100–1600. Scania, Halland, Blekinge. Vol. I–IV. – Akad. forlag, Copenhagen.

Barker, G. 1985. Prehistoric farming in Europe. – Cambridge Univ. Press. Cambridge.

Bartholin, T. 1978. Alvastra pile dwelling: tree studies. The dating and the landscape. – Fornvännen 73: 213–219.

– 1989. Dendrokronologiske undersøgelser af Ystadområdets kirker. – In: Andersson, H. and Anglert, M (eds), By, huvudgård och kyrka. Studier i Ystadsområdets medeltid. Lund Studies in Medieval Archaeology 5. Stockholm, pp. 211–219.

– and Karlén, W. 1983. Dendrokronologi i Lappland AD 436–1981. – Dendrokronologiska Sällskapet, Meddelande 5: 3–16.

– and Berglund, B. E. 1991. Reconstruction of the prehistoric landscape. – In: Callmer, J., Larsson, L. and Stjernquist, B. (eds). The Archaeology of the Cultural Landscape. Field Work and Research in a South Swedish Rural Region. Acta Archaeol. Lund., Series in 4⁰. Lund.

– Berglund, B. E. and Malmer, N. 1981. Vegetation and environment in the Gårdlösa area during the Iron Age. – In: Stjernquist, B. (ed.), Gårdlösa. An Iron Age community in its natural and social setting. Kung. Humanistiska Vetenskapssamfundet i Lund 75: 45–53.

Becker, C. J. 1947. Mosefundne lerkar. – Aarbøger 1947: 5–318.

– 1953. Stenaldergrav fra Gabøl med stridsøkser af jysk og svensk type. – Kuml 1953: 155–164.

– 1954. Die mittel-neolitischen Kulturen in Südskandinavien. – Acta Archaeologica XXV: 49–150.

– 1960. Stendyngegrave fra mellem-neolitisk tid. – Aarbøger 1959: 1–90.

– 1965. Ein früheisenzeitliches Dorf bei Gröntoft, Westjütland. Vorbericht über die Ausgrabungen 1961–63. – Acta Archaeol. 42, Köpenhamn, pp. 209–222.

– 1968. Zum Problem der ältesten eisenzeitlichen Dörfer in Jütland. – Studien zur europäischen Vor- und Frühgeschichte (Festschrift für H. Jankuhn). – Hildesheim, pp. 74–82.

– 1971. Früheisenzeitliche Dörfer bei Gröntoft, Westjütland. 3. Vorbericht: Die Ausgrabungen 1967–68. – Acta Archaeol. 42: 79–110.

– 1976. Problemer omkring de tidlige jernaldersbyer i Jylland belyst af udgravningerne ved Grøntoft. – Finska fornminnesföreningen Iskos 1: 55–58.

– 1980 a. Bebyggelsesformer i Danmarks Yngre broncealder. – In: Thrane, H. (ed.), Broncealderbebyggelse i Norden. Skrifter fra historisk institut 28: 127–141.

– 1980 b. Ein Einzelhof aus der jüngere vorrömische Eisenzeit in Westjütland. – Offa 37: 59–62.

– 1982a. Om grubekeramisk kultur i Danmark. – Korte bidrag til en lang diskussion (1950–80). – Aarbøger 1980: 13–33.

– 1982b. Siedlungen der Bronzezeit und der vorrömischen Eisenzeit in Dänemark. – Offa 39: 53–71.

– 1983. Enkeltgård og landsby i Danmarks ældre jernalder. – In: Thrane, H. and Gröngaard Jeppesen, T. (eds), Gårdens udvikling fra jernalderen til nyere tid, Skrifter fra Historisk institut nr 31, Odense universitet, Odense, pp. 5–16.

– 1986. Nørre Sandegård og Snogebæk. En boplads og en gravplads på Bornholm med fund fra stridsøksekultur. – In: Adamsen, C. and Ebbesen, K. (red), Stridsøksetid i Sydskandinavien. Arkæologiske Skrifter 1, København, pp. 22–32.

– 1990. Nørre Sandegård. Arkæologiske undersøgelser på Bornholm 1948–1952. – Historisk-filosofiske Skrifter 13. Copenhagen.

Behre, K. E. 1981. The interpretation of anthropogenic indicators in pollen diagrams. – Pollen et Spores 23,2: 225–245.

– 1988. The role of man in European vegetation history. – In: Huntley, B. and Webb, T. III (eds), Vegetation history. Kluwer, pp. 633–672.

Bekmose, J. 1977. Megalitgrave og megalitbygder. – Antikvariske Studier I. København, pp. 47–66.

Bengtsson, G. and Kristiansson, S. 1958. Gödsling och kalkning. – LTs förlag. 6th ed. Stockholm.

Bengtsson, S., Regnéll, G. and B. Risinger. 1986. Översiktliga vegetationsbeskrivningar av områdena kring Ystad. – En arbetsrapport från Kulturlandskapet under 6000 år. Dept of Plant Ecology, Lund University.

Bennett, A. 1984. Karleby och Gärtuna. Bebyggelse och gravar från bronsålder och järnålder i Östertälje socken Södermanland. – Riksantikvarieämbetet Rapport UV 1984:29.

– 1987. Graven, religiös och social symbol. Strukturer i folkvandringstidens gravskick i Mälarområdet. – Theses and Papers in North European Archaeology 18. Stockholms universitet.

Bennett, K. D. 1988. Post-glacial vegetation history: Ecological considerations. – In: Huntley, B. and Webb, T. III (eds), Vegetation history. Kluwer, pp. 699–724.

– and Lamb, H. F. 1988. Holocene pollen sequences as a record of competitive interactions among tree populations. – Trends Ecol. Evol. 3: 141–144.

Benthien, B. 1963. Karten zur Entwicklungsgeschichte des vollgenossenschaftlichen Dorfes Stresow (Kreis Greifswald). – Geogr. Ber. 26: 1–9.

Berendse, F. 1985. The effect of grazing on the outcome of competition between plant species with different nutrient requirements. – Oikos 44: 35–39.

– and Aerts, R. 1987. Nitrogen-use-efficiency: a biologically meaningful definition? – Funct. Ecol. 1: 293–296.

– and Elberse, W. T. 1990. Competition and nutrient availability in heathland and grassland ecosystems. – In: Grace, J. B. and Tilman, D. (eds), Perspectives on plant competition. Academic Press., San Diego, pp. 93–116.

Bergendorff, C. and Emanuelsson, U. 1982. Skottskogen – en försummad del av vårt kulturlandskap. – Svensk Bot. Tidskr. 76: 92–100.

Bergeron, T., Fries, M., Moberg, C.-A. and Ström, F. 1956. Fimbulvinter. – Fornvännen 1956: 1–18.

Berglund, B. E. 1966. Late-Quaternary vegetation in eastern Blekinge, south-eastern Sweden. II. Post-Glacial time. – Opera Bot. 12,2: 1–190.

– 1969. Vegetation and human influence in South Scandinavia during Prehistoric time. – In: Berglund, B. E. (ed.), Impact of man on the Scandinavian landscape during the Late Post-Glacial. – Oikos Suppl. 12: 9–28.

– 1973. Pollen dispersal and deposition in an area of southeastern Sweden – some preliminary results. – In: Birks, H. J. B. and West, R. G. (eds), Quaternary Plant Ecology. Blackwell, Oxford, pp. 117–129.

– 1984. Kulturlandskapet under 6000 år – den teoretiska bakgrunden for Ystadsprojektet. – In: Teoretiska studier om kulturlandskapets utveckling. Rapport från ett symposium 10-11 november 1983. – Riksbankens Jubileumsfond, HSFR och NFR, pp. 20–26.

– 1985 a. Early agriculture in Scandinavia: Research problems related to pollen analytical studies. – Norw. Archaeol. Rev. 18, No. 1–2: 77–105.

– 1985 b. Neolitiska expansioner och regressioner i Ystadområdet. – In: Regnéll, G. (ed.), Kulturlandskapet – dess framväxt och förändring – Dept. of Plant Ecology, Lund Univ., pp. 4–49.

– 1986 a. The cultural landscape in a long-term perspective. Methods and theories behind the research on land-use and landscape dynamics. – Striae 24: 79–87.

– 1986 b (ed.). Handbook of Holocene palaeoecology and palaeohydrology. – Wiley, Chichester.

– 1986 c. Det sydsvenska kulturlandskapets förändringar under 6000 år – Ystadsprojektet. – In: Det Sydsvenska kulturlandskapets förändring under 6000 år. Ystadsprojektet. En utställning om människan och miljön ur ett tvärvetenskapligt perspektiv. Utställningskatalog, pp. 3–9. Lunds universitets historiska museum, Report series no, 26.

– 1988. The cultural landscape during 6000 years in South Sweden – An interdisciplinary project. – In: Birks, H. H., Birks, H. J. B., Kaland, P. E. and Moe, D. (eds), The Cultural Landscape – Past, Present and Future. Cambridge Univ. Press, pp. 241–254.

– and Ralska-Jasiewiczowa, M. 1986. Pollen analyses and pollen diagrams. – In: Berglund, B. E. (ed.), Handbook of Holocene palaeoecology and palaeohydrology. Wiley, Chichester, pp. 455–484.

– , Aaby, B., Digerfeldt, G., Fredskild, B., Huttunen, P., Hyvärinen, H., Kaland, P. E., Moe, D. and Vasari, Y. 1983. Palaeoclimatic changes in Scandinavia and on Greenland – a tentative correlation based on lake and bog stratigraphical studies. – Quatern. Stud. Poland 4: 27–44.

– , Emanuelsson, U., Persson, S. and Persson, T. 1986. Pollen/vegetation relationships in grazed and mowed plant communities of South Sweden. – In: Behre, K. E. (ed.), Anthropogenic indicators in pollen diagrams. Balkema, Rotterdam, pp. 37–57.

Bergström, E. 1980. Produktion och samhällsförändring. Bronsålder och äldre järnålder – ett västsvenskt exempel. – Institution för arkeologi, Göteborgs universitet, Göteborg.

Billberg, I. 1989. Inland – kustland. Bebyggelse och resursutnyttjande för två medeltida byar inom Ystadsområdet. – In: Andersson, H. and Anglert, M. (red.) By, huvudgård och kyrka. Studier i Ystadsområdets medeltid. Lund Studies in Medieval Archaeology 5. Almqvist & Wiksell, Stockholm, pp. 15–48.

Birks, H. H., Birks, H. J. B., Kaland, P. E. and Moe, D. 1988. The cultural landscape. Past, present and future. – Cambridge Univ. Press.

Birks, H. J. B. 1981. The use of pollen analysis in the reconstruction of past climate: a review. – In: Wigley, T. M. L., Ingram, M. J. and Farmer, G. (eds), Climate and History. Cambridge Univ. Press, pp. 111–138.

– 1986 a. Late-Quaternary biotic changes on terrestrial and lacustrine environments, with particular reference to northwest Europe. – In: Berglund, B. E. (ed.), Handbook of Holocene Palaeoecology and Palaeohydrology. – Wiley, pp. 3–65.

– 1986 b. Numerical zonation, comparison and correlation of Quaternary pollen-stratigraphical data. – In: Berglund, B. E. (ed.), Handbook of Holocene palaeoecology and palaeohydrology. Wiley, Chichester, pp. 743–774.

– 1988. Long-term ecological change in the British uplands. – British Ecological Society, Spec. Publ. 7 of The British Ecological Society (eds Usher, M. B. and Thompson, D. B. A), pp. 37–56.

– and Berglund, B. E. 1979. Holocene pollen stratigraphy of southern Sweden: a reappraisal using numerical methods. – Boreas 8: 257–279.

– and Birks, H. H. 1980. Quaternary palaeoecology. – Arnold, London.

– and Gordon, A. D. 1985. Numerical methods in Quaternary pollen analysis. – Academic Press, London.

– and Moe, D. 1986. Comments on early agriculture in Scandinavia. – Norw. Archaeol. Rev. 19: 39–43.

– , Line, J. M. and Persson, T. 1988. Quantitative estimation of human impact on cultural landscape development. – In: Birks, H. H., Birks, H. J. B., Kaland, P. E. and Moe, D. (eds), The cultural landscape. Past, present and future. Cambridge Univ. Press, Cambridge, pp. 229–240.

Bjurling, O. 1940. Om oxstallningen och exporten av stalloxar från Skåne under årtiondena före och efter 1700. – Scandia, vol. XIII, 1941 pp. 257–276.

– 1945. Skånes utrikessjöfart 1660–1720. En studie i Skånes handelssjöfart. – CWK Gleerup. Lund.

– 1956. Ystad stads historia 1658–1792. – In: Kraft, S. and Bjurling, O. Ystads historia. Del 1. Ystads kommun. Ystad, pp. 163–495.

Björhem, N. 1986. Plats- och områdeskontinuitet? – Ett tolkningsproblem belyst utifrån ett boplatsmaterial från sydvästra Skåne. – In: Det 4. nordiske bronsealder-symposium på Isegran 1984. Varia 12, Universitetets Oldsakssamling, Oslo, pp. 88–107.

– and Säfvestad, U. 1983. Fosie IV. En långdragen historia. – Ale 1983/1: 3–29.

– and Säfvestad, U. 1987. Stenåldershus. Rekonstruktion av ett 4000 år gammalt hus. – Rapport nr. 2, Malmö museer, Stadsantikvariska avdelningen. Malmö museum, Malmö.

– and Säfvestad, U. 1989. Fosie IV. Byggnadstradition och bosättningsmönster under senneolitikum. – Malmöfynd 5, Malmö museum, Malmö.

– , Säfvestad, U. and Tesch, S. 1982. Var vikingen backstusittare? – META, Medeltidsarkeologisk tidskrift 1982/4: 32–35.

Boardman, J. 1990. Soil erosion on the South Downs: a review. – In: Boardman, J., Foster, I. D. L. and Dearing, J. A. (eds), Soil erosion on agricultural land. Wiley, Chichester, pp. 87–106.

– and Robinson, D. A. 1985. Soil erosion, climatic vagary and agricultural change on the Downs around Lewes and Brighton, autumn 1982. – Appl. Geogr. 5, 243–258.

Boas, N. A. 1983. Egehøj. A Settlement from the Early Bronze Age in East Jutland. – J. Danish Archaeol. 2: 90–101.

Bodén, A. E. 1977. Boplatslämningar. Balkåkra 18:31. Balkåkra sn, Skåne. – Rapport, Riksantikvarieämbetet, Stockholm.

Bondorff, K. A. 1939. Forelæsninger over landbrugets jorddyrkning. II. Gødbubgsklreb, – Den Kgl veterinær- og Landbohøjskole. København.

Bonnier, A. C. 1987. Kyrkorna berättar. Upplands Kyrkor 1250–1350. – Upplands fornminnesförenings tidskrift 51. Uppsala.

Borman, F. H. and Likens, G. E. 1979. The nutrient cycles of an ecosystem. – Sci. Am. 223, 4: 92–101.

Boserup, E. 1973. Jordbruksutveckling och befolkningstillväxt. – Lund.

Boström, R. 1983. Ölands medeltida kyrktorn. – Hikuin 9, Højbjerg.

Bottema, S. 1988. Back to nature? Objectives of nature management in view of archaeological research. – In: Bierma, M., Harsema, O. H. and van Zeist, W. (eds), Archeologie en Landschap. Groningen Univ., Groningen.

Bradley, R. 1981. "Various styles of urn" – cemeteries and settlement in southern England c. 1400–1000 bc. – In: Chapman, R., Kinnes, I. and Randsborg, K. (eds), The archaeology of death. Cambridge Univ. Press, Cambridge, pp. 93–104.

– 1984. The social foundations of prehistoric Britain: Themes and variations in the archaeology of power. – Longman. London.

Bradshaw, R. H. W. 1988. Spatially-precise studies of forest dynamics. – In: Huntley, B. and Webb, T. III (eds), Vegetation history. Kluwer, pp. 725–751.

– and Webb, T. III 1985. Relationships between contemporary pollen and vegetation data from Wisconsin and Michigan, USA. – Ecology 66: 721–737.

– , Coxon, P., Greig, J. R. A. and Hall, A. R. 1981. New fossil evidence for the past cultivation and processing of hemp (Cannabis sativa L.) in Eastern England. – New Phytol. 89: 503–510.

Brandon, P. F. 1971. Demesne arable farming in coastal Sussex during the later Middle Ages. – Agricult. Hist. Rev. 19: 113–134.

Braun, D. P. and Plog, S. 1982. Evolution of "Tribal" social networks: Theory and prehistoric North American evidence. – Am. Antiquity 47: 504–526.

Bringéus, N.-A. 1963. Brännodling. En historisk-etnologisk undersökning. – Skr. fr. Folklivsarkivet i Lund 6. 1833 pp.

– 1964. Tradition och förändring i 1800-talets skånska lanthushållning. – Kristianstads läns hushållningssällskap 1814–1964. Kristianstad.

Brink, N. 1983. Gödselanvändningens miljöproblem. – Ekohydrologi 14: 15–19.

Browall, H. 1986. Alvastra pålbyggnad. Social och ekonomisk bas. – Theses and Papers in North European Archaeology 15. Stockholm.

Brunsting, A. M. H. and Heil, G. W. 1985. The role of nutrients in the interaction between a herbivorous beetle and some competing plant species in heathlands. – Oikos 44: 23–26.

Bruzelius, N. G. 1869. Antiquarisk Beskrifning öfver Bjeresjö eller Bergsjö Socken (Bierghusa 1383) i Herrestads härad, Malmöhus län, Skåne. – Samlingar till Skånes historia, fornkunskap och beskrifning 1868–69: 153–158.

Bunte, R. 1977. Fiske och fiskelägen i Skåne. – Skånsk kust. Skånes Hembygdsförbunds Årsbok 1976. Kristianstad.

– 1987. Näringsliv. – In: Jacobsson, B. (ed.), Ystad under nittonhundratalet. Ystads historia. Del IV, 1914–1971. Ystad.

Burenhult, G. 1982. Arkeologi i Sverige 1. Fångstfolk och herdar. – Höganäs.

– 1983. Arkeologi i Sverige 2. Bönder och bronsgjutare. – Höganäs.

– 1984. Arkeologi i Sverige 3. Samhällsbyggare och handelsmän. – Höganäs.

Callmer, J. 1973. Preliminary report on a complex of buildings from the Late Neolithic-Early Bronze Age at Norrvidinge, Scania. – Meddelanden från Lunds universitets historiska museum 1971–1972: 120–143.

– 1980. Topographical notes on some Scanian Viking Period and Early Medieval hoards. – Meddelanden från Lunds universitets historiska museum 1979–80: 132–149.

– 1982. Production site and market area. Some notes on fieldwork in progress. – Meddelanden från Lunds universitets historiska museum 135–165.

– 1984. Recent work att Åhus. Problems and observations. – Offa 41: 63–75.

– 1986 a. Bebyggelse och landskap under sen järnålder och äldsta medeltid. Delprojekt B4. – In: Det sydsvenska kulturlandskapets förändring under 6000 år. Ystadprojektet. En utställning om människan och miljön ur ett tvärvetenskapligt perspektiv. – Inst. of Archaeol., Univ. of Lund, Report Series 26: 41–49.

– 1986 b. To stay or to move. Some aspects of the settlement dynamics in Southern Scandinavia in the seventh to the twelfth centuries A.D. with special reference to the province of Scania, Southern Sweden. – Meddelanden från Lunds universitets historiska museum 1985–1986: 167–208.

– 1986 c. Ett adelsnäste i Bjäresjö. – Våra härader. Ljunits och Herrestads hembygdsförening 19. Ystad.

– 1987. Iron Age and Early Medieval settlement development in southern Scandinavia: some contemporary and future research perspectives. – In: Burenhult, G. et al. (eds), Theoretical approaches to artefacts, settlement and society. Studies in honour of Mats. P. Malmer. Brit. Archaeol Rep. 15 366 (II). Oxford, pp. 429–444.

– 1991 a. Village excavations in the hummocky landscape: Gussnava and Skårby. – In: Callmer, J., Larsson, L. and Stjernquist, B. (eds) 1991. The archaeology of the cultural landscape. Field work and research in a South Swedish rural region. – Acta Archaeol. Lund., Series in 4°. Almqvist & Wiksell, Stockholm.

– 1991 b. The Bjäresjö-investigation. – In: Callmer, J., Larsson, L. and Stjernquist, B. (eds) 1991. The archaeology of the cultural landscape. Field work and research in a South Swedish rural region. – Acta Archaeol. Lund., Series in 4°. Almqvist & Wiksell, Stockholm.

– 1991 c. Late Iron Age localities and coastal networks related to supraregional exchange and trade in Southern Sweden. – In: Redin, L. (ed.), Frühmittelalter im Ostseegebiet. Stockholm.

– , Larsson, L. and Stjernquist, B. (eds) 1991. The archaeology of the cultural landscape. Field work and research in a South Swedish rural region. Acta Archaeologica Lundensia. Series in 4°. Almqvist & Wiksell, Stockholm.

Campbell, B. M. S. 1983. Agricultural production in Medieval England: some evidence from Eastern Norfolk. – Econ. Hist. Rev. 36: 26–46.

Campbell, Å. 1928. Skånska bygder under förra hälften av 1700-talet. Etnografisk studie över den skånska allmogens äldre odlingar, hägnader och byggnader. – Lundequistska bokhandeln, Uppsala.

– 1933. Vångalaget i Skåne – en boskapsskötselns samfällighet. – Skånes Hembygdsförbunds Årsbok 1933: 11–33.

Carlsson, D. 1979. Kulturlandskapets utveckling på Gotland. – Kulturgeografiska Institutionen, Stockholms universitet, Meddelande B 49.

– 1983. Bronsåldern – tiden för kulturlandskapets territoriella framväxt och etablering på Gotland. Ett försök till förklaringsmodell med utgångspunkt från gotländska förhållanden. – In: Struktur och förändring i bronsålderns samhälle. Rapport från det tredje nordiska symposiet för bronsåldersforskning i Lund 23–25 april 1982. Inst. of Archaeol., Lund Univ., Report series no. 17, pp. 23–36.

– 1984. Revirhävdande och gränsdragning som förklaringsfaktorer av kulturlandskapets rumsliga struktur. – In: Teoretiska studier om kulturlandskapets utveckling. Rapport från ett symposium 10–11 november 1983. – Riksbankens Jubileumsfond, HSFR och NFR, pp. 39–48.

– 1987. Äldre hamnar – ett hotat kulturarv. – Fornvännen 1987/1: 7–18.

Chambers, F. M., Kelly, R. S. and Price, S.-M. 1988. Development of the Late-Prehistoric cultural landscape in upland Arduwy, North-west Wales. – In: Birks, H. H., Birks, H. J. B., Kaland, P. E. and Moe, D. (eds), The cultural landscape. Past, present and future. Cambridge Univ. Press, Cambridge, pp. 333–348.

Chapin, F. S., III. 1980. The mineral nutrition of wild plants. – Ann. Rev. Ecol. Syst. 11: 233–260.

Chapman, R. 1981. The emergence of formal disposal areas and the "problem" of megalithic tombs in prehistoric Europe. – In: Chapman, R., Kinnes, I. and Randsborg, K. (eds), The archaeology of death. Cambridge Univ. Press, Cambridge, pp. 71–82.

Christensen, C. 1982 a. Havniveauændringer 5500–2500 f.Kr. i Vedbækområdet, NØ-Sjælland. – Dansk geologisk Forening, Årsskrift 1981: 91–107.

– 1982 b. Stenalderfjorden og Vedbækbopladserne. Havspejlets svingninger 5500–2500 f. Kr. – Nationalmuseets Arbejdsmark: 169–180.

Christensen, C. A. 1964. Ændringerne i landsbyens økonomiske struktur i det 14. og 15. aarhundrede. – Historisk Tidsskrift 12. r. 1: 257–348.

Christophersen, A. 1982. Drengs, thegns, landmen and kings. – Meddelanden från Lunds universitets historiska museum 1981–1982: 115–134.

– 1984. Landsbygrundleggelse og torpdannelse i Baldringe under tidlig middelalder. – En arbetsrapport från Kulturlandskapet under 6000 år. Dept. of Plant Ecology, Lunds University.

Cinthio, E. 1957. Lunds domkyrka under romansk tid. – Acta Archaeol. Lund., Series in 8^0 nr. 1. Bonn.

– 1972. Variationsmuster in dem frühmittelalterlichen Städtewesen Schonens. – In: Hinz, H. (ed.), Kiel Papers. 72. Frühe Städte im westlichen Ostseeraum. Symposium des Sonderforschungsbereich 17 "Skandinaviens- und Ostseeraumforschung" Christian-Albrechts-Universität. Kiel, pp. 57–64.

– 1975. Köping och stad i det tidigmedeltida Skåne. – Ale: 1–10.

– 1983. Dalby kungsgård. – Kungl Vitterhets Historie och Antikvits Akademiens Årsbok 1983. Stockholm.

Cinthio, H. 1980. The Löddeköpinge Investigation III. The early medieval cemetery. – Meddelanden från Lunds universitets historiska museum 1979–80. Lund.

Cook, C. W., Harris, L. E. and Young, M. C. 1967. Botanical and nutritive content of diets of cattle and sheep under single and common use on mountain range. – J. Anim. Sci. 26: 1169–1174.

Cooke, G. W. 1976. Long-term fertilizer experiments in England: the significance of their results for agricultural science and for practical farming. – Ann. Agron. 27: 503–536.

Coupland, R. T. (ed.). 1979. Grassland ecosystems of the world: analysis of grasslands and their uses. – Int. Biol. Progr. 18, Cambridge.

Cullberg, C. 1968. On artifact analysis. – Acta Archaeol. Lund. 4:7.

Dahl, S. 1940. En översiktskarta över Skånes byar vid mitten av 1600-talet. – Svensk Geografisk Årsbok 1940: 26–31.

– 1941. Storskiftets och enskiftets genomförande i Skåne. – Scandia 14: 86–97. Stockholm.

– 1942 a. Torna och Bara. Studier i Skånes bebyggelse och näringsgeografi före 1860. – Meddelanden från Lunds Universitets Geografiska Institution Avh. 6. Lund.

– 1942 b. Skånska herrgårdar under medeltidens senare del och omkring 1700. Ett bidrag till det skånska herrgårdslandskapets geografi. – Svensk Geografisk Årsbok 18: 282–289.

– 1968. Vångalag i Skåne. – Ymer 88: 72–86.

– 1975. Tvåsäde i Skåne. – Svensk Geografisk Årsbok 51: 35–43.

– 1978. Two Scanian types of two-field system. – Geogr. Pol. 38: 37–40.

– 1989. Studier i äldre skånska odlingssystem. – Meddelanden serie B 69. Kulturgeografiska institutionen Stockholms universitet.

Dahlerup, T. 1969–70. Lavadelns krise i dansk senmiddelalder. – Historisk Tidsskrift 12. r. 4: 1–43.

Dam Kofoed, A. and Nemming, O. 1976. Askov 1894: fertilizers and manure on sandy and loamy soils. – Ann. Agron. 27: 583–610.

Daniel, E. 1977. Beskrivning till jordartskartan Trelleborg NO. – Sveriges Geologiska Undersökning Ser. Ae, 33.

– 1986. Beskrivning till jordartskartorna Tomelilla SO/Simrishamn SV, Ystad NO/Örnahusen NV. – Sveriges Geologiska Undersökning Ser. Ae, 65–66.

– 1989. Beskrivning till jordartskartorna Tomelilla SV, Ystad NV. – Sveriges Geologiska Undersökning Ser. Ae, 99–100.

Davidsen, K. 1977. Relativ kronologi i mellemneolitisk tid. En diskussion af C. J. Beckers kronologisystem på baggrund af nye og gamle stratigrafiske fund. – Aarbøger 1975: 42–77.

– 1978. The final TRB Culture in Denmark. – Arkæologiske Studier V. København.

Davies, W. and Vierck, H. 1974. The context of Tribal Hidage: social aggregates and settlement patterns. – Frühmittelalterliche Studien VIII: 223–293.

Davis, M. B. 1969. Palynology and environmental history during the Quaternary period. – Am. Sci. 57: 317–332.

– 1981. Quaternary history and the stability of forest communities. – In: West, D. C., Shugart, H. H. and Botkin, D. B. (eds), Forest succession, concepts and application. Springer, New York, pp. 132–153.

Dearing, J. A. 1986. Core correlation and total sediment influx. – In: Berglund, B. E. (ed.), Handbook of Holocene palaeoecology and palaeohydrology. Wiley, Chichester, pp. 247–270.

– , Håkansson, H., Liedberg-Jönsson, B., Persson, A., Skansjö, S., Widholm, D. and El-Daoushy, F. 1987. Lake sediments used to quantify the erosional response to land use change in southern Sweden. – Oikos 50: 60–78.

Dege, E. 1983. Knoop – ein Gutsbetrieb am Kieler Stadtrand. – Kieler Geogr. Schr. 58: 169–186.

– 1987. Die ostholsteinische Gutslandschaft. – Schleswig-Holstein: Sammlung Geogr. Führer 15.

Dehn, T. 1985. Gravpladsen Karensdal. – In: Hvass, S. (ed.), Hodde. Et vestjysk landsbysamfund fra ældre jernalder. Arkæologiske studier, Vol. VII, Akademisk Forlag, København, pp. 195–200.

Delcourt, H. R., Delcourt, P. A. and Webb, T. 1983. Dynamic plant ecology: the spectrum of vegetational change in space and time. – Quatern. Res. 19: 265–271.

Delcourt, P. A. and Delcourt, H. R. 1987. Late-Quaternary dynamics of temperate forests: applications of palaeoecology to issues of global environmental change. – Quatern. Sci. Rev. 6: 129–146.

Digerfeldt, G. 1965. Vielången och Farlången. En utvecklingshistorisk insjöundersökning. – Skånes Natur 52: 162–183.

– 1971. The Post-glacial development of the ancient lake of Torreberga, Scania, South Sweden. – Geol. Fören. Stockh. Förh. 93: 601–624.

– 1972. The Post-glacial development of Lake Trummen. Regional vegetation history, water-level changes and palaeolimnology. – Folia Limnol. Scand. 16.

– 1974. The Post-glacial development of the Ranviken bay in Lake Immeln. I. The history of the regional vegetation and II. The water-level changes. – Geol. Fören. Stockh. Förh. 96: 3–32.

– 1975 a. A standard profile for Littorina transgressions in western Skåne, South Sweden. – Boreas 4: 125–142.

– 1975 b. Post-glacial water-level changes in Lake Växjösjön, central southern Sweden. – Geol. Fören. Stockh. Förh. 97: 167–173.

- 1976. A Pre-Boreal water-level change in Lake Lyngsjö, central Halland. – Geol. Fören. Stockh. Förh. 98: 329–336.
- 1986. Studies on past lake-level fluctuations. – In: Berglund, B. E. (ed.), Handbook of Holocene Palaeoecology and Palaeohydrology. Wiley, Chichester, pp. 127–143.
- 1988. Reconstruction and regional correlation of Holocene lake-level fluctuations in Lake Bysjön, South Sweden. – Boreas 17: 162–182.
- and Welinder, S. 1988. The prehistoric cultural landscape in South-West-Sweden. – Acta Archaeol. 58: 127–136.
Dodgshon, R. A. and Olsson, E. G. 1988. Productivity and nutrient use in eighteenth-century Scottish Highland townships. – Geogr. Ann. 70B: 39–51.
Draiby, B. 1985. Fragstrup – en boplads fra yngre bronzealder i Vesthimmerland. – Aarbøger for nordisk Oldkyndighed og Historie 1984: 127–212.
Ebbesen, K. 1975. Die jüngere Trichterbecherkultur auf den dänischen Inseln. – Arkæologiske Studier II. København.
- 1979. Stordyssen i Vedsted. Studier over tragtbaegerkulturen i Sønderjylland. – Arkaeologiske Studier, vol. II.
- 1984. Tragtbaegerkulturens grønstensøkser. – KUML 1984: 113–143.
- and Mahler, D. 1980. Virum. Et tidigenolitisk bopladsfund. – Aarbøger 1979: 11–61.
Edgren, L. 1987. Lärling – gesäll – mästare: hantverk och hantverkare i Malmö 1750–1847. – Dialogos, Lund.
Ekblad, K. 1981. Attributes – ancient monuments – regions. – In: Moberg, C.-A. (ed.), Similar finds? Similar interpretations? – Dept. of Archaeology, Univ. of Gothenburg, pp. D1–D39.
Ekström, G. 1946. Jordartskarta över Skåne. – Sveriges Geologiska Undersökning. (Map).
- 1950. Skånes åkerjordsområden. – Socker 6,3: 53–61.
Ellenberg, H. 1979. Zeigerwerte der Gefässflanzen Mitteleuropas. – Scripto Geobotanica IX.
Ellison, A. 1981. Towards a socioeconomic model for the Middle Bronze Age in southern England. – In: Hodder, I. et al. (eds), Pattern of the Past. Cambridge Univ. Press, Cambridge, pp. 413–438.
Emanuelsson, U. 1987. Översikt över det nordiska kulturlandskapet. – Nordiska Ministerrådet, Miljörapport 1987, 6: 13–52.
- 1988 a. A model describing the development of the cultural landscape. – In: Birks, H. H., Birks, H. J. B., Kaland, P. E. and Moe, D. (eds), The cultural landscape. Past, present and future. Cambridge Univ. Press, Cambridge, pp. 111-121.
- 1988 b. Skånes vegetationshistoria. – Svensk Geografisk Årsbok 63: 70–93.
- and Bergendorff, C. 1983. Skånes natur vid 1800-talets början – en växtekologisk utvärdering av den skånska rekognosceringskartan. – Ale 1983/4: 18–40.
- and Möller, J. 1990. Flooding in Scania. A method to overcome the deficiency of nutrients in agriculture during the 19th century. – Agr. Hist. Rev. 38: 127–148.
- , Bergendorff, C., Carlsson, B., Lewan, N. and Nordell, O. 1985: Det skånska kulturlandskapet. – Signum, Lund.
Enckell, P. H., Königsson, E.-S. and Königsson, L.-K. 1979. Ecological instability of a Roman Iron Age human community. – Oikos 33: 328–349.
Engelmark, R. 1985. Ogräs för tolkning av agrara system. – In: Regnéll, G. (ed.), Kulturlandskapet – dess framväxt och förändring. Dept. of Plant Ecology, Lund Univ.
- 1988a. Bjäresjö 19:17. Analys av förkolnat växtmaterial. – Report, Umeå.
- 1988 b. Analyser av förkolnat växtmaterial från St. Köpinge och St. Herrestads socknar i Skåne. – Report, Umeå.
- 1989. Makrofossilanalys från L. Tvären och Baldringetorp. Preliminära resultat. – In: Andersson, H. and Anglert, M. (eds), By, huvudgård och kyrka. Studier i Ystadsområdets medeltid. Lund Studies in Medieval Archaeology 5. Almqvist & Wiksell, Stockholm, pp. 49–50.

- and Viklund, K. 1990 (1991). Makrofossilanalys av växtrester belyser lantbrukets utformning och historia. – Bebyggelsehistorisk Tidskrift 18/90.
- and Hjelmqvist, H. 1991. Farming and land use in the Köpingebro area. – In: Callmer, J., Larsson, L. and Stjernquist, B. (eds), The archaeology of the cultural landscape. field work and research in a South Swedish rural region. Acta Archaeol. Lund., Ser. in 4°. Almqvist & Wiksell, Stockholm.
Eriksen, P. 1987. Kampen for gravhøjene. – Skalk 1987/6: 18–27.
Erlingsson, K. 1988. I går – idag – i morgon. En analys av ett område på södra Romeleåsen och dess förutsättningar för framtiden. – Dept. of Social and Economic Geography, Lund Univ.
Ersgård, L. 1988. "Vår marknad i Skåne". Bebyggelse, handel och urbanisering i Skanör och Falsterbo under medeltiden. – Lund Studies in Medieval Archaeology 4. Lund.
Eskeröd, A. 1973. Jordbruk under femtusen år – redskapen och maskinerna. LT. Borås.
Ethelberg, P. 1987. Bronze Age Houses at Höjgård South Jutland. – In: Kristiansen, K. and Nielsen, P. O. (eds), J. Danish Archaeol. 5, 1986, Odense, pp. 152–167.
Etherington, J. R. 1975. Environment and plant ecology. – Wiley, London.
Evans, L. T. 1980. The natural history of crop yield. – Am. Sci. 68: 388–397.
Fabricius, K. 1958. Skaanes Overgang fra Danmark til Sverige 4. – Rosenkilde & Bagger, Copenhagen.
Fagerlund, D. 1985. Ekologiska och topografiska faktorer som utgångspunkt vid sökandet efter bronsålderns boplatser. – Mimeographed, Dept. of Archaeology, Stockholm.
Fernlund, S. 1982. "Ett Herranom värdigt Tempel". Kyrkorivningar och kyrkobyggen i Skåne 1812–1912. – Lund.
Fisher, A. 1982. Trade in Danubian Shaft-Hole Axes and the Introduction of Neolithic Economy in Denmark. – J. Danish Archaeol. 1: 7–12.
Fitter, A. H. and Hay, R. K. M. 1983. Environmental physiology of plants. – Academic Press. London.
Fitter, F. and Fitter, A. 1974. The Wild Flowers of Britain and Northern Europe. – William Collins, Glasgow.
Fogelfors, H. 1979. Changes in the flora of farmland. Arable land with special regard to chemical weed control. – Dept of Ecology and Nature Conservancy. Swedish Agricultural Univ. Uppsala. Report 5.
- and Steen, E. 1982. Vegetationsförändringar under ett kvartsekel i landskapsvårdsförsök i Uppsalatrakten. – Statens Naturvårdsverk. SNV pm 1623.
Forman, R. 1984. Landscape function as determined by structure. – IALE Proc. from the first international seminar. 5: 4–14.
Fox, H. S. A. 1984. Some ecological dimensions on medieval field systems. – In: Biddick, K. (ed.), Archeological Approaches to Medieval Europe. Studies in Medieval Culture, XVIII. pp. 119–159. Western Michigan Univ.
- 1986. The alleged transformation from two-field to three-field systems in medieval England. – Economic History Review 39(4): 526–548.
- 1988. Social relations and ecological relationships in agrarian change: An example from medieval and early modern England. – Geografiska Annaler B(1): 105–115.
Forman, R. T. T. 1990. Ecologically sustainable landscapes: the role of spatial configuration. – In: Zonneveld, I. S. and Forman, R. T. T. (eds), Changing landscapes: an ecological perspective. Springer, New York, pp. 261–278.
Frandsen, K. E. 1983. Vang og tægt. Studier over dyrkningssystemer og agrarstruktur i Danmarks landsbyer 1682–83. – Esbjerg.
- 1988. The field systems of southern Scandinavia in the 17th century. A comparative analysis. – Geografiska Annaler 70 B(1): 117–121.

Fraser, D. 1983. Land and society in Neolithic Orkney. – BAR, British Series 117, Oxford.

Freij, H. 1977. "Balkåkratrumman" – i solkultens tjänst? – Fornvännen 1977/3–4: 129–134.

Fridlizius, G. 1957. Swedish Corn Export in the Free Trade Era. – Lund.

– 1975 a. Some New Aspects on Swedish Population Growth. I. A Study at County Level. – Economy and History XVIII:1: 3–33.

– 1975 b. Some New Aspects on Swedish Population Growth. II. A Study at Parish Level. – Economy and History XVIII:2: 126–154.

– 1979. Population, Economy and Property Rights. – Economy and History 22: 3–37.

– 1985. Från spannmålstunnor till smördrittlar. – In: Bjurling, O. (ed.), Malmö stads historia, del IV. Malmö, pp. 395–522.

Furingsten, A. 1985. Samhällsförändringar i ett långtidsperspektiv. – GOTARC Series B. No. 1, Gothenburg University, Gothenburg.

Gadd, C.-J. 1983. Järn och potatis. Jordbruk, teknik och social omvandling i Skaraborgs län 1750–1860. – Göteborg.

Gaillard, M.-J. 1984. A palaeohydrological study of Krageholmssjön (Scania, South Sweden). Regional vegetation history and water level changes. – LUNDQUA Report 25.

– 1985 a. Postglacial palaeoclimatic changes in Scandinavia and central Europe. A tentative correlation on studies of lake-level fluctuations. – Ecol. Mediterranea 11: 159–175.

– 1985 b. Palaeohydrology, palaeoclimate and cultural landscape – a palaeohydrological study in the context of the Ystad project. – In: Regnéll, G. (ed.), Kulturlandskapet – dess framväxt och förändring. Dept. of Plant Ecology, Lund University, pp. 15–24.

– and Berglund, B. E. 1988. Land-use history during the last 2700 years in the area of Bjäresjö, S. Sweden. – In: Birks, H. H., Birks, H. J., Kaland, P. E. and Moe, D. (eds) The cultural landscape – past, present and future. Cambridge Univ. Press, pp. 409–428.

– , Göransson, H., Håkansson, H. and Lemdahl, G. 1988. The palaeoenvironment at Skateholm-Järavallen (Southern Sweden) during Atlantic and Early Subboreal time on the basis of pollen–, macrofossil–, diatom– and insect analyses. – In: Larsson, L. (ed.), The Skateholm project. I. Man and environment. Kungl. Humanistiska Vetenskapssamfundet i Lund, 79: 52–55.

– , Dearing, J. A., El-Daoushy, F., Enell, M. and Håkansson, H. 1991. A multidisciplinary study of Lake Bjäresjö (S Sweden): land-use history, soil erosion, lake trophy and lake-level fluctuations during the last 3000 years. Hydrobiologia 214: 107–114.

– , Berglund, B. E., Göransson, H. and Regnéll, J. in prep. Guideline for the presentation of pollen diagrams as a tool for the interpretation of past cultural landscape and land-use, with some examples from South Sweden. – LUNDQUA Report.

Gamrath, H. and Ladewig Petersen, E. 1980. Perioden 1559–1648. – In: Christensen, A. E. et al. (eds), Danmarks historie 2. Gyldendal, Copenhagen, pp. 359–555.

– , Dearing, J. A., El-Daoushy, F., Enell, M. and Håkansson, H., in press. A late Holocene record of land-use history, soil erosion, lake trophy and lake-level fluctuations at Bjäresjösjön (South Sweden). Journal of paleolimnology.

Garner, H. V. and Dyke, G. V. 1969. The Broadbalk yields. – Rothamsted Experimental Station, Report for 1968, Part 2, pp. 26–49.

Gebühr, M. 1986. Ursachen für den Siedlungsabbruch auf Fünen im 5. Jahrhundert n. Chr. Studien zu Voraussetzungen und Motiven für Wanderbewegungen im westlichen Ostseegebiet. – Hamburg.

Germundsson, T. 1984. Kulturlandskapets förändring. Ett diskussionsinlägg med tonvikten på 1800- och 1900-talen. – En arbetsrapport från Kulturlandskapet under 6000 år. Dept. of Plant Ecology, Lund University.

– 1985. Jordbruksegnahem i Sydskåne. – Ale 1985/22: 24–32.

– 1986. Population, landownership and the landscape. changes in the Ystad-region in southernmost Sweden during the 19th and 20th centuries. – Espace Population Soc. 1986-III: 126–130.

– 1987. Population, landholding and the landscape. – En arbetsrapport från Kulturlandskapet under 6000 år. Dept. of Plant Ecology, Lund University.

– and Möller, J. 1987 a. Åkerns redskap i förändring. En studie av två sydskånska socknar under 1800-talet. – Ale 1987/1: 14–26.

– and Möller, J. 1987 b. Redskap och maskiner i Nöbbelöv och Villie. En longitudinell studie av två sydskånska socknar under 1800-talet. – Technical report from the project The cultural landscape during 6000 years. Department of Human Geography. University of Lund.

– and Persson, A. 1987. Odling och djurhållning i Ystadsområdet på 1850-talet. En arbetsrapport från Kulturlandskapet under 6000 år. Dept. of Plant Ecology. Lund University.

Gillberg, J. L. 1765 (1980). Historisk, Oeconomisk och Geographisk Beskrifning Öfwer Malmöhus Lähn uti hertigdömmet Skåne. – Walter Ekstrand Bokförlag (1980), Lund.

Gimingham, C. H. and Smidt de, J. T. 1983. Heaths as natural and semi-natural vegetation. – In: Holzner, H., Werger, M. J. A. and Ikusima, I. (eds), Man's impact on vegetation. W. Junk publishers, The Hague, pp. 185–199.

Gissel, S., Jutikkala, E., Österberg, E., Sandnes, J., and Teitsson, B. 1981. Desertion and land colonization in the Nordic countries c. 1300–1600. Almqvist & Wiksell Intern., Stockholm.

Glob, P. V. 1945. Studier over jysk Enkeltgravskultur. – Aarbøger 1944: 1–283.

– 1971. Højarnas folk. – Natur och kultur, Stockholm.

Goldstein, L. 1981. One-dimensional archaeology and multidimensional people: spatial organisation and mortuary analysis. – In: Chapman, R., Kinnes, I. and Randsborg, K. (eds), The archaeology of death. Cambridge Univ. Press, Cambridge, pp. 53–70.

Grahn, B. and Hansson, A. (eds) 1961. Handledning om gödselmedel och kalk. – Gödsel- och Kalkindustriernas Samarbetsorganisation. Stockholm.

Gransteds, A. 1990. Fallstudier av kväveförsörjning i alternativ odling. (Case studies on nitrogen supply in alternative farming) – Alternative Agriculture 4. Swedish University of Agricultural Sciences.

Grigson, C. 1982. Porridge and pannage: pig husbandry in Neolithic England. – In: Bell, M. and Limbrey, S. (eds), Archaeological aspects of woodland ecology. BAR, British Series 146, pp. 297–315.

Grime, J. P. 1979. Plant strategies and vegetation processes. – Wiley, New York.

Groenman-van Waateringe, W. 1983 a. The disastrous effect of the Roman occupation. – In: Brandt, R. and Slofstra, J. (eds), Roman and native in the Low Countries, spheres of interaction. BAR International Series 184.

– 1983 b. The early agricultural utilization of the Irish landscape: the last word on the elm decline? – BAR, British Series 116: 217–232. Oxford.

– 1988. New trends in palynoarchaeology in Northwest Europe or the frantic search for local pollen data. – BAR Intern. Ser. 416: 1–19.

Grove, J. M. 1988. The Little Ice Age. – Methuen, London.

Gräslund, B. 1979. Efterskrift. – In: Stenberger, M. (ed.), Det forntida Sverige. Almqvist & Wiksell, Stockholm, pp. 871-912.

Grøngaard-Jeppesen, T. 1981. Middelalderlandsbyens opståen. Kontinuitet og brud i den fynske agrarbebyggelse mellem yngre jernalder og tidlig middelalder. – Fynske Studier, IX, Odense.

– 1983. Fynske gårdsundersøgelser med udblik til resten af landet. – Gårdens utvikling fra jernalder til nyere tid. Odense.

Gustafson, A. 1982. Växtnäringsförluster från åkermark i Sverige. – Ekohydrologi 11: 19–27.

– 1985. Växtnäringsläckage och motåtgärder. – Ekohydrologi 20: 44–59.

Gustafsson, E. and Redin, L. 1969. Stora Köpinge kyrka. – Ale 1969/2: 32–34.

Guteland, G. (ed.) 1975. Ett folks biografi. – Liber Förlag, Vällingby.

Göransson, H. 1982. The utilization of the forests in North West Europe during Early and Middle Neolithic. – Second Nordic Conference on the Application of Scientific Methods in Archaeology. PACT 7: 207–221. Strasbourg.

– 1984. Pollen analytical investigations in the Sligo Area. – In: Burenhult, G. (ed.), The Archaeology of Carrowmore. Theses and Papers in North-European Archaeology 14: 154–193.

– 1986. Man and the forests of nemoral broad-leaved trees during the Stone Age. – In: Königsson, L.-K. (ed.), Nordic Late Quaternary biology and ecology, Striae 24: 143–152.

– 1987 a. Comments on early agriculture in Scandinavia. On arguing in a circle. – Norw. Arch. Rev. 20, 1: 43–45.

– 1987 b. Neolithic man and the forest environment around Alvastra pile dwelling. – Theses and Papers in North-European Archaeology 20, Lund Univ. Press, Stockholm.

– 1988 a. Comments on remodelling the Neolithic in Southern Norway. On pollen analytical myths. – Norw. Arch. Rev. 21,1: 33–37.

– 1988 b. Pollen analytical Investigations at Skateholm, Southern Sweden. – In: Larsson, L. (ed.), The Skateholm Project. I. Man and environment. Acta Regiae Societatis Humaniorum Litterarum Lundensis LXXIX. Almqvist & Wiksell, Stockholm, pp. 27–33.

– 1989. Kommentar till "Konstanter och variabler i det förhistoriska samhället". – Fornvännen 84: 43–47.

– 1991. Vegetation and man around Lake Bjärsjöholmssjön, during prehistoric time. – LUNDQUA Report 31.

Görman, M. 1987. Nordiskt och keltiskt. Sydskandinavisk religion under yngre bronsålder och keltisk järnålder. – Lund.

Hafsten, U. 1981. Palaeo-ecological evidence of a climatic shift at the end of the Roman Iron Age. – Striae 14: 58–61.

Hallberg, G. 1975. Ljunits härad. – Skånes ortnamn. Serie A. Bebyggelsenamn, Del II. Lund.

– 1989. De skånska namnen på -husa. – In: Studia Onomastica. Festskrift till Thorsten Andersson 23 februari 1989. Almqvist & Wiksell, Stockholm, pp. 115–125.

Hannerberg, D. 1958. Skånska bolskiften. Veberöd. – Svensk Geografisk Årsbok 34: 7–49.

– 1960. Schonische "Bolskiften". – Lund Studies in Geography Ser. B. 20. Lund.

Hansen, F. 1924. De arkeologiska fynden vid Svarte fiskeläge. – Fornvännen 1923: 119–163, 229–242.

– 1938. Skånska bronsåldershögar. – Gleerupska universitetsbokhandeln, Lund.

Hansen, J. and Kyllingbæk, A. (eds) 1983. Kvælstof og planteproduktion. – Statens Planteavlsforsøg. Beretning nr. S 1669. København.

Hansen, M. 1986. Enkeltgravskulturens bopladsfund fra Vesthimmerland og Ribeområdet. – In: Adamsen, C. and Ebbesen, K. (red.), Stridsøksetid i Sydskandinavien. Arkæologiske Skrifter 1. København, pp. 286–291.

Hansen, V. 1976. The medieval dispersal of rural settlement in Denmark as a function of distance from primary nucleations. – In: Buchanan, R. H., Butlin, R. A. and McCourgt, D. (eds), Fields, farms and settlement in Europe. Ulster Folk and Transport Museum, Belfast, pp. 53–79.

Hanssen, B. 1952 (1977). Österlen: en studie över social-antropologiska sammanhang under 1600- och 1700-talen i sydöstra Skåne. – LTs förlag, Ystad.

Harrison, S. D. and Digerfeldt, G. (manuscript) European lakes as palaeoclimatic indicators. – In: Berglund, B. E., Birks, H. J. B. and Ralska-Jasiewiczowa, M. (eds). Palaeoecological Events In Europe During The Last 15000 Years – Patterns, Processes, And Problems.

Hassan, F. 1981. Demographic archaeology. – New York.

Hasselmo et al. 1987. Från tidigmedeltida centralorter till högmedeltida städer i Östsverige. – In: Andrae, T., Hasselmo, M. and Lamm, K. (eds), 7000 år på 20 år, Arkeologiska undersökningar i Mellansverige, Riksantikvarieämbetet UV, Stockholm, pp. 201–326.

Hedeager, L. 1988 a. Jernalderens landbrug. – In: Bjørn, C. (ed.), Det danske landbrugs historie I. Oldtid og Middelalder, Odense, pp. 146–164

– 1988 b. Oldtid o. 4000 f.Kr. – 1000 e.Kr./Jernalderen. – In: Bjørn, C. (ed.), Det danske landbrugs historie I, Oldtid og middelalder, Odense, pp. 109–203.

– and Kristiansen, K. 1985. Arkaeologi Leksikon. – Politiken, Köpenhamn.

– and Kristiansen, K. 1988. Oldtid o. 4000 f.Kr. – 1000 e.Kr. – In: Bjørn, C. (red.). Det danske landbrugs historie I. Oldtid og Middelalder, Odense, pp. 13–203.

– , Poulsen, B. and Tornbjerg, S. Å. 1982. Land og byundersøgelse af Østersøkeramikken på Stevns. – Hikuin 8: 125–148.

Heitz-Weniger, A. 1976. Zum Problem des mittelholozänen Ulmenabfalls im Gebiet des Zürichsees (Schweiz). – Bauhinia 5: 215–229.

Helbæk, H. 1952. Preserved apples and Panicum in the prehistoric site at Nørre Sandegaard in Bornholm. – Acta Archaeol. 23: 107–115.

Hellerström, S. 1988. Ystadbygden. En studie av lösfynd och bosättning under neolitikum. Seminar paper. – Institute of Archaeology, Univ. of Lund.

Helmfrid, S. 1961. The Storskifte, Enskifte and Laga Skifte in Sweden. – Geografiska Annaler 42: 114–129. Stockholm.

– 1962. Östergötland "Västanstång". Studien über die ältere Agrarlandschaft und ihre Genese. – Geografiska Annaler 44, 1–2.

– 1966 (1985). Europeiska kulturlandskap: en forskningsöversikt. – Kulturgeografiska institutionen, Stockholms universitet.

– 1980. Kulturlandskapet och forskningen. – In: Människan, kulturlandskapet och framtiden. Kungl. Vitterhets Historie och Antikvitets Akademien. Konferenser 4, pp. 9–15.

Hernek, R. 1989. Den spetsnackiga yxan av flinta. – Fornvännen 1988: 216–223.

Hinsch, E. 1956. Yngre steinalders stridsøksekulturer i Norge. – Bergen.

Hjelmqvist, H. 1955. Die älteste Geschichte der Kulturpflanzen in Schweden. – Almqvist & Wiksell, Stockholm.

– 1979. Beiträge zur Kenntnis der prähistorischen Nutzpflanzen in Schweden. – Opera Bot. 47.

– 1982. Economic Plants from a Middle Neolithic Site in Scania. – In: Larsson, L., A Causewayed Enclosure and a Site with Valby Pottery at Stävie, Western Scania. – Papers of the Archaeological Inst., Univ. of Lund 1981–1982: 108–113.

– 1985. Economic plants from two Stone Age settlements in southernmost Scania. – In: Larsson, L. (ed.), Karlsfält. A Settlement from the Early and Late Funnel Beaker Culture in Southern Scania, Sweden. – Acta Archaeol. 54: 57–63.

– 1987. Några ekonomiväxter från Sydskånes forn- och medeltid. – Report, Lund.

Hjelmroos, M. 1985. Vegetational History of Fårarps mosse, South Scania, in the Early Subboreal. – In: Larsson, L. (ed.), Karlsfält. A settlement from the Early and Late Funnel Beaker Culture in Southern Scania, Sweden. – Acta Archaeol. 54: 45–50.

Hodder, I. 1979. Economic and social stress and material culture patterning. – Am. Antiquity 44, No. 3: 446–454.

– 1982. Symbols in action. – Cambridge Univ. Press, Cambridge.

Hofsten, E. 1986. Svensk befolkningshistoria. – Kristianstad.

Hoika, J. 1987. Das Mittelneolithikum in Nordostholstein. – Offa-Bücher, Band 61. Karl Wachholtz Verlag, Neumünster.

Holmberg, R. 1977. Den skånska Öresundskustens medeltid. – Acta Archaeol. Lund., Series in 8⁰, no. 11.

Hoppe, G. 1981. Enclosure in Sweden. Background and Consequences. – Kulturgeografiskt seminarium 79/9. Dept Human Geography, Stockholm.

Hubbard, C. E. 1978. Grasses. – Penguin Books, Harmondsworth, Middlesex, England.

Huntley, B. 1988. Europe. – In: Huntley, B. and Webb, T. III (eds), Vegetation History, Kluwer 341–383.

– and Birks, H. J. B. 1983. An atlas of past and present pollen maps of Europe: 0–13000 years ago. – Cambridge Univ. Press.

– and Prentice, I. C. 1988. July temperatures in Europe from pollen data, 6000 years before present. – Science 241: 687–690.

–, Bartlein, P. J. and Prentice, I. C. 1989. Climatic control of the distribution and abundance of beech (*Fagus* L.) in Europe and North America. – J. Biogeogr. 16. 551–560.

Hurst, J. G. 1984. The Wharram Research Project: Results to 1983. – Medieval Archaeol. vol. XXVIII.

Husdjursskötsel. 1919. Avel af svin, får, getter, fjäderfän och bin. – Skrifter utgivna av de skånska hushållningssällskapen med anledning af deras hundraårsjubileum år 1914. Lund.

Hvass, S. 1977. A house of the Single-grave Culture excavated at Vorbasse in Central Jutland. – Acta Archaeol. 48: 219–232.

– 1979. Fra Hodde til Vorbasse, linier i jernalderens bebyggelsebilde. – In: Thrane, H. (ed.), Fra jernalder til middelalder, Skrifter fra Historisk institut, nr. 27. Odense Universitet, Odense, pp. 31–44.

– 1980. Die Struktur einer Siedlung der Zeit von Christi Geburt bis ins 5. Jahrhundert nach Christus. Ausgrabungen in Vorbasse, Jütland, Dänemark. – Studien zur Sachsenforschung 2: 161–180.

– 1983. Vorbasse. The development of a settlement through the first millenium A.D. – J. Danish Archaeol. 2: 127–136.

– 1985. Hodde. Et vestjysk landsbysamfund fra ældre jernalder. – Arkæologiske studier, Vol. VII, Akademisk Forlag, København.

– 1986. En boplads fra enkeltgravskulturen i Vorbasse. – In: Adamsen, C. and Ebbesen, K. (red.), Stridsøksetid i Sydskandinavien. Arkæologiske Skrifter 1, København, pp. 323–335.

– 1987. Status over Vorbasseundersøgelsen og dens resultater i forhold til jernalderens bebyggelse i Midt- og Østjylland. – In: Thrane, H. (ed.), Diakrone bebyggelsesundersøgelser, Skrifter fra Historisk institut nr. 34, Odense universitet, Odense, pp. 49–58.

– 1988. Jernalderens bebyggelse. – In: Mortensen, P. and Rasmussen, B. M. (eds), Fra Stamme til Stat i Danmark 1, Jernalderens stammesamfund. Jysk Arkæologisk Selskabs Skrifter XXII, pp. 53–92.

Hyenstrand, Å. 1976. Bronsåldershyddor i Mälarområdet. – ISKOS 1, Helsinki, pp. 45–50.

– 1979. Ancient Monuments and Prehistoric Society. – Central Board of National Antiquities, Stockholm.

Håkansson, H. 1984. Diatom analysis of profile XVII from the south-western bay of Krageholmssjön (Scania, South Sweden). – Appendix in: Gaillard, M.-J., A palaeohydrological study of Krageholmssjön (Scania, South Sweden). Regional vegetation history and water-level changes, LUNDQUA Report 25.

– 1989. Diatom succession during Middle and Late Holocene time in Lake Krageholmssjön, southern Sweden. Nova Hedw. 48, 1–2: 143–166.

– and Kolstrup, E. 1987. Early and middle Holocene developments in Herrestads mosse (Scania, South Sweden). I. Diatom analysis and vegetational development. – LUNDQUA Report 28.

Håkansson, S. 1979. University of Lund radiocarbon dates XII. – Radiocarbon 21/3: 384–404.

– 1986. University of Lund radiocarbon dates XIX. – Radiocarbon 28/3: 1111–1132.

Hårdh, B. 1976. Wikingerzeitliche Depotfunde aus Südschweden. (Katalog und Tafeln, Probleme und Analysen). – Lund.

– 1982. The megalithic grave area around the Lödde-Kävlinge river. – Papers of the Archaeological Inst., Univers. of Lund 1981–1982: 26–48.

– 1986. Ceramic decoration and social organization. Regional variations seen in material from South West Swedish Passage Graves. – Scripta Minora 1985–1986: 1.

– 1989. Pattern of deposition and settlement. Studies on the megalithic tombs of west Scania. – Scripta Minora 1988–1989: 2. Lund.

Ihse, M. 1985. Skåne – kulturlandskap i förvandling. – Kulturminnesvård 5/1985: 3–11.

– 1989. Air photo interpretation and computer cartography – tools for studying the cultural landscape. – In: Birks et al. (eds), The cultural landscape. Past, present and future, – Cambridge Univ. Press, Cambridge, pp. 153–63.

– and Lewan, N. 1986. Odlingslandskapets förändringar på Svenstorp studerade i flygbilder från 1940-talet och framåt. – Ale. 2/1986: 1–17.

Ingers, E. 1948. Bonden i svensk historia II. – Stockholm.

Ingers, I. 1971–72. Nedlagda byar och gårdar i Skåne, 1–2. – Ale 1971:3, 1972:1 29–44, 19–34.

Inventering av ängs- och hagmarker. 1987. – Statens Naturvårdsverk, Solna.

Iversen, J. 1941 (reprinted 1964). Landnam i Danmarks Stenalder (Land occupation in Denmark's Stone Age). – Danm. Geol. Unders. 2/66.

– 1944. *Viscum, Hedera* and *Ilex* as climate indicators. – Geologiska Föreningens i Stockholm Förhandlingar 66: 463–483.

– 1949. The influence of prehistoric man on vegetation. – Danm. Geol. Unders. 4/3.

– 1958. The bearing of glacial and interglacial epochs on the formation and extinction of plant taxa. – Uppsala Universitets Årsskrift 6: 210–215.

– 1964. Retrogressive vegetational succession in the postglacial – J. Ecol. 52 (Suppl.): 59–70.

– 1969. Retrospective development of a forest ecosystem demonstrated by pollen diagrams from fossil mor. – Oikos, Suppl. 12: 35–49.

– 1973. The development of Denmark's nature since the last glacial. – Danmarks Geologiske Undersøgelse 5/7C.

Jaanusson, H. 1981. Hallunda. A study of pottery from a Late Bronze Age settlement in central Sweden. – The Museum of National Antiquities, Studies 1, Stockholm.

Jacobson, G. L. Jr and Bradshaw, R. H. W. 1981. The selection of sites for paleovegetational studies. – Quatern. Res. 16: 80–96.

–, Webb, T. III and Grimm, E. C. 1987. Patterns and rates of vegetation change during the deglaciation of eastern North America. – In: Ruddiman, W. F. and Wright, H. E. Jr. North America and Adjacent Oceans During the Last Deglaciation. Geol. Soc. Am., The Geology of North America, K-3, pp. 277–288.

Jacobsson, B. 1982. Trelleborg. – Riksantikvarieämbetet. Rapport. Medeltidsstaden 38. Stockholm.

– 1986. The Skogsdala dolmen. – Papers of the Archaeological Inst. Univ. of Lund 1985–1986: 84–114.

– 1987. Oden. Ett vikingakvarter i staden Trelleborg. – Ale 1987/2: 1–10.

Jacobsson, B. (ed.) 1987. Ystad under nittonhundratalet. – Ystad stads historia del 4, 1914–1971. Ystad.

Jankuhn, H. 1964. Die römische Kaiserzeit und die Völkerwanderungszeit. – In: Klose, O. (ed.), Geschichte Schleswig-Holsteins 2. Kiel, pp. 251–416.
– 1979. Siedlungsarchäologie als Forschungsmethode. – In: Jankuhn, H. and Wenskus, R. Geschichtswissenschaft und Archäologie. Vorträge und Forschungen Bd XXII: 19–44.
– 1986. Die eisenzeitlichen Wurzeln unserer mittelalterlichen Dörfer. – Kunde NF 3: 93–101.
– et al. (Hrsg v.) 1977. Das Dorf der Eisenzeit und des frühen Mittelalters. – Abhandlungen der Akademie der Wissenschaften in Göttingen. Philologisch-historische Klasse. Dritte Folge Nr 101. Göttingen.
Jansson, I. 1978. Ett rembeslag av orientalisk typ funnet på Island. Vikingatidens orientaliska bälten och deras eurasiska sammanhang. – Tor 17: 383–420.
Jansson, S. L. 1956. Stallgödseln, dess egenskaper, vård och användning. – In: Handbok om växtnäring. Del III. Markvård. Gödsel- och Kalkindustriernas Samarbetsorganisation. Stockholm.
Janzon, G. 1986. Stridsyxekultur och metallurgisk know-how. – In: Adamsen, C. and Ebbesen, K. (red.), Stridsøksetid i Sydskandinavien. Arkæologiske Skrifter 1, København, pp. 126–137.
Jarman, M. R., Bailey, G, N. and Jarman, H. N. 1982. Early European agriculture. – Cambridge Univ. Press, Cambridge.
Jeansson, N. R. 1966. De stora domänerna i Malmöhus län omkring år 1914. – Svensk Geografisk Årsbok 42: 42–51.
Jennbert, K. 1984. Den produktiva gåvan. Tradition och innovation i Sydskandinavien för omkring 5300 år sedan. – Acta Archaeol. Lundensia. Series in 4°. 16.
– 1985. Neolithisation – a Scanian perspective. – J. Danish Archaeol. 4: 196–197.
– 1987. Neolithisation processes in the Nordic Area. – Swedish Archaeol. 1981–1985, pp. 21–35.
Jensen, J. 1968. Voldtoftefundet. Bopladsproblemer i yngre bronzealder i Danmark. – Aarbøger for Nordisk oldkyndighed og historie 1967: 91–154.
– 1979 a. Bronzealderen 1: Skovlandets folk. – Danmarkshistorien, Viborg.
– 1979 b. Oldtidens samfund. Tiden indtil år 800. – Dansk socialhistorie bd. 1. København.
– 1981. Et rigdomscenter fra yngre bronzealder på Sjælland. – Aarbøger for Nordisk oldkyndighed og historie 1981: 48–96.
– 1982. The Prehistory of Denmark. – New York.
– 1988. Bronze Age research in Denmark 1970–85. – J. Danish Archaeol. 6: 155–174.
Jensen, S. P. 1988. Landbruget 1860–1914 – produktion af teknologi. – In: Björn, C. (ed.), Det danske landbrugs historie III. Odense, pp. 243–351.
Jeppsson, G. 1967. Veckodagsfriheten i Skåne under 1500- och 1600-talen. – Unpubl. thesis. Dept. of History, Lund Univ.
Johansen, E. 1986. Tre bosättelser fra sen enkeltgravskultur/tidlig senneolitikum ved Solbjerg, Østhimmerland. – In: Adamsen, C. and Ebbesen, K. (red.), Stridsøksetid i Sydskandinavien. Arkæologiske Skrifter 1. København, pp. 280–285.
Johansson, I. (ed.) 1953. Husdjursraserna. – LTs förlag.
Johansson, S. et al. 1991. Bjersjöholm. Storgods och slott. – Skånsk senmedeltid och renässans 13. Vetenskapssocieteten i Lund.
Jonsson, L. 1988. The vertebrate faunal remains from the Late Atlantic settlement Skateholm in Scania, South Sweden. – In: Larsson, L. (ed.), The Skateholm Project. I. Man and environment. Acta Regiae Societatis Humaniorum Litterarum Lundensis LXXIX. Almqvist & Wiksell, Stockholm, pp. 56–87.
– 1989. Rapport, osteologisk analys av boplatsmaterial från St. Köpingeområdet. – Unpubl. report.
Joos, M. 1982. Swiss Midland-lakes and climatic changes. – In:

Harding, A. (ed.), Climatic Change in Late Pre-History. Edinburgh Univ. Press, pp. 44–51.
Jordbruk/Landsbygd i utveckling. 1986. – Länsstyrelsen i Malmöhus län.
Jäger, H. 1973. Altlandschaftforschung. – Hoops Reallexikon der germanischen Altertumswissenschaft (2. Aufl.) Bd I: 225–233.
Jörberg, L. 1972. A history of prices in Sweden 1732–1914 1–2. – C. W. K. Gleerup, Lund.
Jørgensen, B. 1984. Danske marknavns alder. Onomastisk datering. – Fortid og Nutid 31(4): 259–269.
Jørgensen, E. 1977. Hagebrogård – Vroue – Koldkur. Neolithische Gräberfelder aus Nordwest-Jütland. – Arkæologiske Studier IV. København.
– 1981. Gravhusenes problem. – Skalk 1981/3: 3–9.
Jørgensen, G. 1982. Korn fra Sarup. Med nogle bemærkninger om agerbruget i yngre stenalder i Danmark. – Kuml 1981: 221–231.
Jørgensen, M. S. 1982. To jyske bronzealderveje – og en ny metode til arkæologisk opmåling. – Nationalmuseets Arbejdsmark 1982: 142–152.
Kardell, L., Dahlén, R. and Andersson, B. 1980. Svedjebruk förr och nu. – Sveriges Lantbruksuniversitet. Avd. för landskapsvård, Rapport 20.
Karlén, W. 1984. Dendrochronology, mass balance and glacier front fluctuations in northern Sweden. – In: Mörner, N.-A. and Karlén, W. (eds), Climatic changes on a yearly to millennial basis. Reidel, pp. 263–271.
– 1988. Scandinavian glacial and climatic fluctuations during the Holocene. – Quatern. Sci. Rev. 7: 199–209.
Kempfner-Jørgensen, L. and Watt, M. 1985. Settlement Sites with Middle Neolithic Houses at Grødby, Bornholm. – J. Danish Archaeol. 4: 87–100.
Kieffler-Olsen, J. 1987. Vorbasse by. – META, Medeltidsarkeologisk tidskrift 1987/1–2: 51–57.
Kolstrup, E. 1987. Tidligt landbrug. – Skalk 1987/5: 9–12.
– 1990. Early and middle Holocene vegetation development in Kurarp (Scania, South Sweden). – Rev. Palaeobot. Palynol. 63: 233–257.
Kossack, G. et al. (Hrsgg. v.) 1984. Archäologische und naturwissenschaftliche Forschungen an Siedlungen im deutschen Küstengebiet vom 5. Jahrhundert v. Chr. bis zum 11. Jahrhundert n. Chr. – Weinheim.
Kraft, S. 1956. Den danska tiden. – In: Kraft, S. and Bjurling, O. Ystads historia. Del 1. Ystad, pp. 9–160.
Krins, H. 1968. Die frühen Steinkirchen Dänemarks. – Hamburg.
Kristiansen, K. 1975. Bebyggelsens relation til den sociale og økonomiske struktur i Danmarks yngre bronzealder – oplæg til en analyse. – Kontaktstencil 10, Oslo, pp. 79–106.
– 1978. Bebyggelse, erhvervsstrategi og arealudnyttelse i Danmarks bronzealder. – Fortid og nutid 27/3: 320–345.
– 1982. The formation of tribal systems in later European prehistory: Northern Europe, 4000-500 B. C. – In: Renfrew, C., Rowlands, M. and Seagraves, B. (eds), Theory and explanation in archaeology. Academic Press. New York, pp. 241–280.
– 1985 a. Early Bronze Age burial finds. – In: Kristiansen, K. (ed.), Archaeological formation processes. Nationalmuseets forlag, København, pp. 116–128.
– 1985 b. Post-depositional formation processes and the archaeological record. – In: Kristiansen, K. (ed.), Archaeological formation processes. Nationalmuseets forlag, København, pp. 7–11.
– 1987 a. From stone to bronze – the evolution of social complexity in Northern Europe 2300–1200 BC. – In: Brumfiel, E. M. and Earl, T. K. (eds). Specialization, exchange, and complex societies. Cambridge Univ. Press, pp. 30–51.
– 1987 b. Ideologi og samfund i Danmarks bronzealder. – In: Det 4. nordiske bronsealder-symposium på Isegran 1984, Varia 12, Universitetets Oldsakssamling, Oslo, pp. 144–155.

– 1988. Oldtid o. 4000 f.Kr. – 1000 e.Kr./Sten- og bronzealder. – In: Bjørn, C. (ed.), Det danske landbrugs historie I, Oldtid og middelalder, Odense, pp. 21–107.

Krogh, K. 1983. The royal viking-age monuments at Jelling in the light of recent archaeological excavations. – Acta Archaeol. 53: 183–216.

Krok, O. B. N. and Almquist, S. 1984. Svensk Flora, fanerogamer och ormbunksväxter. – Esselte Herzogs, Uppsala.

Kullman, L. 1988. Holocene history of the forest-alpine tundra ecotone in the Scandes Mountains (central Sweden). – New Phytol. 108: 101–110.

– 1989. Tree-limit history during the Holocene in the Scandes Mountains, Sweden, inferred from subfossil wood. – Rev. Palaeobot. Palynol. 58: 163–171.

Königsson, L.-K. 1968. The Holocene history of the Great Alvar of Öland. – Acta Phytogeogr. Suecica 55.

Körber-Grohne, u. 1987. Nutzpflanzen in Deutschland. – Konrad Theiss Verlag.

Lagerås, P. 1991. Kontinuitet i utnyttjandet av Baldringes utmarker. En pollenanalytisk studie i Skogshejdan, Skåne. – Examination paper in geology 33. Lund University.

Lamb, H. H. 1977. Climate: Present, past and future. – Methuen, London.

– 1982. Climate, history and the modern world. – Methuen, London.

– 1984. Climate and history in northern Europe and elsewhere. – In: Mörner, N.-A. and Karlén, W. (eds), Climatic changes on a yearly to millennial basis. Reidel, Dordrecht, pp. 225–240.

Lange, E. 1976. Grundlagen und Entwicklungstendenzen der frühgeschichtlichen Agrarproduktion aus botanischer Sicht. – Z. Archäol. 10: 75–120.

Larsson, L. 1978. Ageröd I:B – Ageröd I:B. A study of Early Atlantic settlement in Scania. – Acta Archaeol. Lund., Series in 4°. 12. Almqvist & Wiksell, Stockholm.

– 1980. Some aspects of the Kongemose Culture of Southern Sweden. – Papers of the Archaeological Inst., Univ. of Lund 1979–1980: 5–22.

– 1982 a. A causewayed enclosure and a site with Valby Pottery at Stävie, Western Scania. – Papers of the Archaeol. Inst., Univ. of Lund 1981–1982: 65–107.

– 1982 b. Segebro. En tidigatlantisk boplats vid Sege ås mynning. – Malmöfynd 4, Malmö museum, Malmö.

– 1983 a. Ageröd V. An Early Atlantic bog site in Central Scania. – Acta Archaeol. Lund. 8°. 12. Almqvist & Wiksell, Stockholm.

– 1983 b. Mesolithic Settlement on the Sea Floor in the Strait of Öresund. – In: Masters P. M. and Flemming, N. C. (eds), Quaternary coastlines and marine archaeology. Towards the prehistory of land bridges and continental shelves. Academic Press, New York, pp. 283–301.

– 1984 a. Bronsvagnen från Hedeskoga. – Våra härader 17: 22–28.

– 1984 b. Gräberfelder und Siedlungen des Spätmesolithikums bei Skateholm, Südschonen, Schweden. – Archäologisches Korrespondenzblatt 14, Mainz: 123–130.

– 1984 c. Spår efter Ystadtraktens äldsta innevånare. – Ystadiana 1984. Ystad Fornminnesförening skrifter XXIX: 97–105.

– 1984 d. Stenålder i hägn och ram. – Ale 1984/1: 1–18.

– 1984 e. The Skateholm Project. A Late Mesolithic settlement and cemetery complex at a Southern Swedish Bay. – Papers of the Archaeol. Inst., Univ. of Lund 1983–1984: 5–38.

– 1985. Karlsfält. A Settlement from the Early and Late Funnel Beaker Culture in Southern Scania, Sweden. – Acta Archaeol. 54: 3–44.

– 1986. Archaeological survey mapping. – In: Berglund, B. E. (ed.), Handbook of Holocene Palaeoecology and Palaeoohydrology. Wiley, Chichester, pp. 219–228.

– 1987 a. Hus över stenåldersgrav. En gravanläggning från Ullstorp 11, Ullstorps sn. – Ystadiana 1987: 59–75.

– 1987 b. Some aspects of cultural relationship and ecological conditions during the Late Mesolithic and Early Neolithic. – In: Burenhult, G., Carlsson, A., Hyenstrand, Å. and Sjøvold, T. (eds), Theoretical approaches to artefacts, settlement and society. Studies in honour of Mats P. Malmer. BAR, International Series 366, pp. 165–176.

– 1988 a. Aspects of exchange in Mesolithic societies. – In: Hårdh, B., Larsson, L., Olausson, D. and Petré, R. (eds), Trade and exchange in prehistory. Studies in honour of Berta Stjernquist. Acta Archaeol. Lund. Series in 8°. 16, Almqvist & Wiksell, Stockholm, pp. 25–32.

– 1988 b. Mortuary building above Stone Age grave. A Grave from the Battle Axe culture at Ullstorp, Southern Scania, Sweden. – Papers of the Archaeol. Inst., Univ. of Lund 1987–1988: 81–98.

– 1988 c. The Skateholm Project. Late Mesolithic settlement at a South Swedish lagoon. – In: Larsson, L. (ed.), The Skateholm Project. I. Man and environment. Acta Regiae Societatis Humaniorum Litterarum LXXIX. Almqvist & Wiksell, Stockholm, pp. 9–19.

– 1988 d. The use of the landscape during the Mesolithic and Neolithic in southern Sweden. – Archeologie en Landscap. Bijdragen aan het gelijknamige symposium gehouden op 19 en 20 oktober 1987, ter gelegenheid van het afscheid van H. T. Waterbolk. Groningen, pp. 31–48.

– 1989 a. Big dog and poor man. Mortuary practices in Mesolithic Societies in Southern Sweden. – In: Larsson, T. B. and Lundmark, H. (eds), Approaches to Swedish Prehistory. A spectrum of problems and perspectives in contemporary research. BAR, International Series 500, Oxford, pp. 211–223.

– 1989 b. Långåker. En boplats från en bondestenålder i förvandling. – Gamle Trelleborg Årsbok 1989: 6–20.

– 1989 c. En kökkenmödding långt från havets strand. – Populär Arkeologi 1989/4: 12–16.

– 1989 d. Boplatser, bebyggelse och bygder. – In: Larsson, L. (ed.), Stridsyxekultur i Sydskandinavien. Univ. of Lund, Inst. of Archaeol. Report Series 36: 53–76.

– 1990. Sydskånsk öbösättning Boplatslämningar från jägarstenålder och sen vikingatid – tidig medeltid vid Ellestadssjön. – Ale 1989/4.

– 1991. Settlement and environment during Middle and Late Neolithic. – In: Callmer, J., Larsson, L and Stjernquist, B. (eds), The archaeology of the cultural landscape. Field work and research in a South Swedish rural region. Acta Archaeol. Lund, Series in 4°. Almqvist & Wiksell, Stockholm.

– and Larsson, M. 1984. Flintyxor, skoskav och massor av stolphål. – Ystadiana 29: 9–95.

– and Larsson, M 1986. Stenåldersbebyggelse i Ystadsområdet. En presentation av fältverksamhet och bearbetning hösten 1984 - våren 1986. – Ystadiana 31: 9–78.

Larsson, M. 1984. Tidigneolitikum i Sydvästskåne. Kronologi och bosättningsmönster. – Acta Archaeol. Lund, Series in 4°. 17. Almqvist & Wiksell, Stockholm.

– 1985. The early neolithic Funnel Beaker culture in SW Scania, Sweden. Economic and social change 3000–2500 BC. – BAR, International Series 264, Oxford.

– 1986. Bredasten – An early Ertebølle site with a dwelling structure in South Scania. – Papers of the Archaeol. Inst., Univ. of Lund 1985–1986: 25–49.

– 1987 a. Människor vid en havsvik. Stenåldersboplatser vid Kabusa, St. Köpinge sn. – Ystadiana 32: 13–58.

– 1987 b. Förändringar i bosättningsmönster under neolitikum i Ystad området. Ekonomiska och sociala faktorer. – Det dolda kulturlandskapet. Metodkonferens 1985. Riksantikvarieämbetet Rapport 1987/2: 47–69.

– 1987 c. Gravplats, boplats, åker. Ett exempel på kulturlandskapets utnyttjande kring Köpingebro i sydligaste Skåne. – Ale 1987/3: 1–14.

– 1987 d. Neolithization in Scania – A Funnel Beaker perspective. – J. Danish Archaeol. 5: 244–247.

– 1988 a. Exchange and society in the early neolithic Funnel Beaker society in Scania, Sweden. – In: Hårdh, B., Larsson, L., Olausson, D. and Petré, R. (eds), Trade and exchange in prehistory. Studies in honour of Berta Stjernquist. Acta Archaeol. Lund, Series in 8° 16. Almqvist & Wiksell, Stockholm, pp. 49–58.

– 1988 b. Megaliths and society. The development of social territories in the south Scanian TRB culture. – Papers of the Archaeol. Inst., Univ. of Lund 1987–1988: 19–39.

– and Olausson, D. 1986. Bönder och handelsmän under vikingatiden vid Mossby i sydligaste Skåne. – Ystadiana 31 79–104.

– in press, Lokale Gruppen des Frühneolithikums in Südschonen, Schweden. – In: Hoika, J. (ed.), Frühe Trichterbecherkultur. Offabücher.

Larsson, T. B. 1986. The Bronze Age metalwork in southern Sweden. Aspects of social and spatial organization 1800–500 B.C. – Archaeology and environment 6. Dept. of Archaeol., Univ. of Umeå.

Larsson, U.-K. 1973. De svenska fynden av flinteggade benspetsar. Seminar paper. – Inst. of Archaeol., Univ. of Lund.

Le Roy Ladurie, E. 1979. The territory of the historian. – Univ. of Chicago Press, Chicago.

Leach, E. 1954. Political systems of Highland Burma. A study of Kachin social structure. – London.

Lepiksaar, J. 1969. Knochenfunde aus den bronzezeitlichen Siedlungen von Hötofta. – In: Stjernquist, B. (ed.), Beiträge zum Studium von bronzezeitlichen Siedlungen. Acta Archaeol. Lund., Series in 8⁰. No. 8. CWK Gleerups Förlag, Lund, pp. 174–207.

Leufvén, G. 1914. Malmöhus läns jordbruk år 1909. Statistisk undersökning. – Skrifter utgivna av de skånska hushållningssällskpem vid deras hundraårsjubileum år 1914.

Lewan, N. 1963. Från godsby till kolchos. – Geografiska Notiser 3: 112–114.

– 1967. Landsbebyggelse i förvandling. En studie av utvecklingen i Skåne sedan 1910 med särskild hänsyn till arbetstillfällenas omfördelning. – Lund.

– 1968. Från markväg och stig till motorväg. – Svensk Geografisk Årsbok 44: 79–89.

– 1982. Om skånska rekognosceringskartan. – Ale 1: 14–24.

– 1983 a. Två revolutioner i Hofterup – skifte och tätortstillväxt. – Kävlingebygden nu och då 6: 35–46.

– 1983 b. Ystadsområdet idag. Iakttagelser – intervjuer – intryck. – En arbetsrapport från Kulturlandskapet under 6000 år. Dept. of Plant Ecology, Lund Univ.

– 1985. Kulturlandskapet och det moderna samhället. – In: Emanuelsson, U. et al. Det skånska kulturlandskapet. Signum, Lund, pp. 139–152.

– 1986. Jordbrukspolitikens landskap. – Årsboken Ymer 182–193.

– 1988. Contradictory planning. Against and in favour of rural change in Sweden. – Geo-Regards No. 15, Neuchâtel.

– 1988. Kring Marsvinsholm förr och nu. – Geografisk Orientering 18/3: 344–349.

– and Mårtensson, E. 1985. Slottsparker i södra Skåne under 1900-talet. Förminskning, förenkling, förfall. – Byggnadskultur 4: 11–15.

Liepe, A. 1985. Medeltida kyrkobygge i Värend. – Kronobergsboken 1984–85. Växjö.

Lindahl, A. 1976. Ystad Sandskog FL 43. En stenåldersboplats. Seminar paper. – Inst. of Archaeol., Univ. of Lund.

Lindgren, G. 1939. Falbygden och dess närmaste omgivning vid 1600-talets mitt. En kulturgeografisk studie. – Geographica 6.

Lindgren, L. 1970. Beech forest vegetation in Sweden – a survey. – Botaniska Notiser 123: 401–424.

– och Malmer, N. 1970. Skogsutvecklingen i Skåne under fyra decennier. – Skånes Natur 57, 3: 2–8.

Lindkvist, Th. 1979. Landborna i Norden under äldre medeltid. – Studica Historica Upsaliensis 110, Uppsala.

Lindquist, B. 1931. Den skandinaviska bokskogens biologi. (The ecology of the Scandinavian beech-woods). – Svenska Skogsvårdsföreningens Tidskrift 1931: 177–532.

– 1938. Dalby Söderskog, en skånsk lövskog i forntid och nutid. – Acta Phytogeogr. Suecia 10.

– 1966. Vegetationsregioner och floraelement. – In: Atlas över Sverige. Generalstabens Litografiska Anstalt, Stockholm, pp. 43–44.

Lindquist, S.-O. 1968. Det förhistoriska kulturlandskapet i östra Östergötland. – Acta. Univ. Stockholm. Studies in North-European Archaeology 2. Stockholm.

– 1981. Sockenbildning på Gotland. En korologisk studie. – Gotländskt arkiv 53.

Lindroth, B. 1987. Ystadsprojektet – diskussion kring osteologiskt material från några medeltida landsbyar. – META, Medeltidsarkeologisk tidskrift 1987/3: 33–40.

Linkole, K. 1916. Studien über die Einfluss auf die Flora in den Gegenden nördlich vom Ladogasee. I. – Acta Soc. Fauna Flora Fennica 45,1:

Linnaeus, C. 1751. Carl Linnaeus Skånska Resa år 1749. – von Sydow, C.-O. (ed.) 1977. Wahlström & Widstrand, Stockholm.

Liversage D. and Singh, P. K. 1985. A comparison of two Neolithic flint industries. – J. Danish Archaeol. 4: 70–78.

Lomborg, E. 1973. Die Flintdolche Dänemarks. Studien über Chronologie und Kulturbeziehungen des südskandinavischen Spätneolithikums. – Nordiske Fortidsminder, Serie B, Bind 1.

Lund, C. 1979. Ett märkligt fornfynd från Balkåkra. – Våra härader 12: 18–23.

Lund, J. 1988. Jernalderens bebyggelse i Jylland. – In: Näsman, U. and Lund, J. (ed.), Folkevandringstiden i Norden. En krisetid mellem ældre og yngre jernalder, Viborg, pp. 139–168.

Lund Hansen, U. 1974. Mellem-neolitiske jordgrave fra Vindinge på Sjælland. – Aarbøger 1972: 5–70.

– 1987. Römischer Import im Norden. Warenaustausch zwischen dem Römischen Reich und dem freien Germanien während der Kaiserzeit unter besonderer Berücksichtigung Nordeuropas. – Kopenhagen.

Lundberg, E. 1940. Byggnadskonsten i Sverige under medeltiden 1000–1400. – Stockholm.

Lundborg, L. 1971. Den förhistoriska forskningen i Halland under åren 1961–1971, en översikt. – Halland 1971/54: 5–21.

Lundmark, H. 1986. Vad säger oss bronsföremålen? En diskussion om kopplingen mellan föremålsanalys och samhällsanalys. – Varia 12, Oslo, pp. 32–41.

Lüning, J. 1980. Getreideanbau ohne Düngung. – Archäologisches Korrespondenzblatt 10, Heft 2: 36–51.

– and Meurers-Balke, J. 1980. Experimenteller Getreidenanbau im Hambacher Forst, Gemeinde Elsdorf, Kr. Bergheim/Rheinland. – Bonner Jahrbücher 180: 305–344.

Lönnberg, E. 1929. En stenåldersboplats i Ystads Sandskog. – Meddelanden från Ystad Fornminnesförening. Ystad.

Madsen, H. J. 1971. To dobbeltgrave fra jysk enkeltgravskultur. – Kuml 1970: 127–149.

Madsen, P. K. Kontinuitet eller brud. Linier i det sidste tiårs bebyggelsehistoriske middelalderforskning i Danmark. – META, Medeltidsarkeologisk tidskrift 1985/2–3: 51–67.

Madsen, T. 1979. Earthen long barrows and timber structures: Aspects of the Early Neolithic mortuary practice in Denmark. – Proc. Prehis. Soc. 45: 301–321.

– 1980. En tidligneolitisk langhøj ved Rude, Østjylland. – Kuml 1979: 74–109.

– 1982. Settlement systems of early agricultural societies of East Jutland, Denmark. A regional study of change. – J. Anthropol. Archaeol. 1: 197–236.

– 1987. Where did all the hunters go? – J. Danish Archaeol. 5: 229–239.

– and Jensen, H. J. 1982. Settlement and land use in Early

Neolithic Denmark. – Analecta Praehistorica Leidensia XV: 63–86.

– and Petersen, J. E. 1984. Tidlig-neolitiske anlæg ved Mosegården, Østjylland. Regionale og kronologiske forskelle i tidligneolitikum. – Kuml 1982–1983: 61–120.

Magny, M. 1980. Fluctuations lacustres et paléoclimatologie post-glaciaire. – Bull. Ass. française pour l'Etude du Quaternaire 1–2: 57–60.

– 1982. Atlantic and Subboreal: dampness and dryness? – In: Harding, A. (ed.), Climatic change in late pre-history. Edinburgh Univ. Press, pp. 33–43.

– and Olive, P. 1981. Origine climatique des variations du niveau de Lac Léman au cours de l'Holocène. La crise de 1700 à 700 ans BC. – Arch. suisses d'anthropologie générale 45: 159–169.

Magurran, A. 1988. Ecological diversity and its measurement. – Croom Helm, London.

Mahler, D. 1985. Ragnesminde. A Germanic-Early Viking Age House-Site in Eastern Sjælland. – J. Danish Archaeol. 4: 164–167.

Malmer, M. P. 1957. Pleionbegreppets betydelse för studiet av förhistoriska innovationsförlopp. – Finska Fornminnesföreningens Tidskrift 58: 160–184.

– 1962. Jungneolithische Studien. – Acta Archaeol. Lund, Series in 8°. 2. Lund.

– 1975. Stridsyxekulturen i Sverige och Norge. – Liber läromedel, Lund.

– 1986. Vad kan vetas om stridsyxekulturens kronologiska och sociala förhållanden? Teoretiska betraktelser. – In: Adamsen, C. and Ebbesen, K. (red.), Stridsøksetid i Sydskandinavien. Arkæologiske Skrifter 1: 7–21.

– 1988. Konstanter och variabler i det förhistoriska samhället. – Fornvännen 1988: 88–97.

– and Bartholin, T. 1983. Pælebygning. – Skalk 1983/4: 18–27.

Malmer, N. 1969. Organic matter and cycling of minerals in virgin and present ecosystems. – Oikos Suppl. 12: 79–86.

– and Regnéll, G. 1984. Mapping past and present vegetation. – In: Berglund, B. E. (ed.), Handbook of Holocene palaeoecology and palaeohydrology. Wiley, Chichester, pp. 203–258.

–, Lindgren, L. och Persson, S. 1978. Vegetational succession in a South Swedish deciduous wood. – Vegetatio 36,1: 17–29.

Malmros, C. and Tauber, H. 1977. Kolstof-14 dateringer af dansk enkeltgravskultur. – Aarbøger for Nordisk oldtidskyndighed og historie 1975: 78–95.

Mangerud, J., Andersen, S. T., Berglund, B. E. and Donner, J. 1974. Quaternary stratigraphy of Norden, a proposal for terminology and classification. – Boreas 3: 109–126.

Mathiassen, T. 1948. Studier over Vestjyllands oldtidsbebyggelse. – Nationalmuseets skrifter. Arkæologisk-Historisk Række, III, Nordisk Forlag, København.

– 1959. Nordvestsjællands Oldtidsbebyggelse. – Nationalmuseets Skrifter. Arkæologisk-Historisk Række VII.

McNeely, J. A. et al. 1988, 1990. Conserving the world's biodiversity. – IUCN publications, Gland, Switzerland.

Mengel, K. and Kirky, E. A. 1981. Principles of plant nutrition, (3d ed.) – Worblaufern, Bern.

Midgeley, M. 1985. The origin and function of the North European earthen long barrows. – BAR, International Series 259.

Mingay, G. E. 1962. The size of farms in the eighteenth century. – Economic Hist. Rev., 2nd series 14: 469–488.

Molloy, K. and O'Connell, M. 1987. The nature of the vegetational changes at about 5000 B. P. with particular reference to the elm decline: fresh evidence from Connemara, western Ireland. – New Phytol. 106: 203–220.

Montelius, O. 1874. Bronsvagnen från Ystad. – Kungl. Vitterhets Historie och Antikvitets Akademiens Månadsblad 1874/13: 4–10.

– 1889. Ett bronskärl funnet vid Bjersjöholm i Skåne. –

Kungl. Vitterhets Historie och Antikvitets Akademiens Månadsblad 1889: 125–140.

– 1917. Minnen från vår forntid. – Norstedt, Stockholm.

Moore, P. D. 1984. Hampstead Heath clue to historical decline of elms. – Nature, Lond. 1984.

– 1986. Hydrological changes in mires. – In: Berglund, B. E. (ed.), Handbook of Holocene palaeoecology and palaeohydrology. Wiley, Chichester, pp. 91–107.

Morén F. W., 1953. Ystads historia. Del 2. 1792–1862. – Ystad.

Morgan, R. P. C. 1986. Soil erosion and conservation. – Longman, England.

Müller-Wille, M. 1977. Bäuerlich Siedlungen der Bronze- und Eisenzeit in den Nordseegebieten. – In: Jankuhn, H. (ed.), Das Dorf der Eisenzeit und des frühen Mittelalters. Abhandlungen der Akademie der Wissenschaften in Göttingen. Philologisch-historische Klasse. Dritte Folge Nr 101. Göttingen, pp. 153–218.

– 1984. Siedlungsarchäologische Untersuchungen in Nordwestdeutschland und benachbarten Gebieten. – Perspective on archaeological theory and method. Inst. of Archaeol., Univ. of Lund. Report Series 20: 63–101.

Myrdal, J. 1984. Elisenhof och järnålderns boskapsskötsel i Nordvästeuropa. – Fornvännen 1984/2: 73–92.

– 1985. Medeltidens åkerbruk. Agrarteknik i Sverige ca 1000 till 1520. – Nordiska Museets Handlingar 105, Stockholm.

– 1988. Agrarteknik och samhälle under två tusen år. – In: Näsman, U. and Lund, J. (eds), Folkevandringstiden i Norden. En krisetid mellem ældre og yngre jernalder. Aarhus Universitetsforlag, pp. 197–226.

– 1989. Jordbruk och jordägande. En aspekt av sambandet mellan agrarteknik och samhällsutveckling i äldre medeltid. – In: Andrén, A. (red.), Medeltidens födelse. (Symposier på Krapperups borg 1). Lund.

Mårtensson, E. 1985. Parkanläggningar i Ljunits och Herrestads härader under 1900-talet. – En arbetsrapport från Kulturlandskapet under 6000 år. Dept. of Plant Ecology, Lund University.

Möhl, U. 1962. Übersicht über Knochenfunde aus Gräbern der schwedisch-norwegischen Streitaxtkultur. – In: Malmer, M. Jungneolitische Studien. Acta Archaeologica Lundensia 8: 2, Lund, pp. 883–910.

Möller, J. 1979. Den skånska diagonalen. En studie över gränsen mellan Kristianstads och Malmöhus län. – Mimeographed. Dept. of Social and Economic Geography, Lund Univ.

– 1983. Gods, gård och landskap. Domänstruktur och kulturlandskap i Ljunits och Herrestads härader, Malmöhus län, under 1800- och 1900-talen. – Rapporter och Notiser 73, Dept. of Social and Economic Geography, Lund Univ.

– 1984. Dikning i Skåne. – Ale. 1984/2: 14–28.

– 1985. The landed estate and the landscape. Landownership and the changing landscape of southern Sweden during the 19th and 20th centuries. – Geografiska Annaler 67 B: 45–52.

– 1986. Den skånska plattgården. Exempel på agrar stordrift i södra Sverige. – Ymer 106: 161–170.

– 1987. The landed estate and the railway. The introduction of a new means of transport in southern Sweden. – J. Transport Hist., 3rd Series 8: 147–163.

– 1989. Godsen och den agrara revolutionen. Arbetsorganisation, domänstruktur och kulturlandskap på skånska gods under 1800-talet. – Lund Univ. Press.

Mörnsjö, T. 1968. Studies on vegetation and development of a peatland in Scania, South Sweden. – Opera Botanica 24.

Nagmér, R. 1983. Balkåkra. Gravar och boplatslämningar från sten- och bronsåldern. – Riksantikvarieämbetets Rapport UV 1983: 11.

– 1987. Balkåkra 1:32 m fl, Balkåkra sn, Skåne. – Rapport, Riksantikvarieämbetet, Stockholm.

Nannesson, L. 1914. Skånes nötkreatursskötsel från 1800-talets början till nuvarande tid. – Skrifter utgivna av de skånska hushållningssällskapen vid deras hundraårsjubileum år 1914. Lund.

Nelson, H. 1935. Den skånska rekognosceringskartan 1812, 1815–1820. Ett märkligt svenskt kartverk. – Svensk Geografisk Årsbok 11: 191–207.

Neumann, H. 1963. Trækonstruktioner over grave fra stenaldrens slutning. – Haderslev amts museum 10.

Nielsen, A.-L. 1987. Hus i Mälardalen, från yngre bronsålder – romersk järnålder. Försök till funktionsanalys. – Stockholms universitet, arkeologi särskilt nordeuropeisk (Mimeographed).

Nielsen, F. O. and Nielsen, P. O. 1985. Middle and Late Neolithic houses at Limensgård, Bornholm. – J. Danish Archaeol. 4: 101–114.

– and Nielsen, P. O. 1986. En boplads med hustomter fra mellem- og senneolitikum ved Limensgård, Bornholm. – In: Adamsen, C. and Ebbesen, K. (red.), Stridsøksetid i Sydskandinavien. Arkæologiske Skrifter: 175–193.

Nielsen, H. 1978. Det tilfældige fundstofs anvendelse i bebyggelsesarkæologien med Østsjælland som eksempel. – In: Thrane, H. (ed.), Bebyggelseshistorisk Metode og Teknik. Skrifter fra historisk institut, Odense universitet 23: 88–93.

– 1979. Jernalderfund og stednavnetyper – en sammenligning af fynske og sjællandske forhold. – In: Thrane, H. (ed.). Fra jernalder til middelalder. Skrifter fra Historisk institut, Odense universitet 27: 87–98.

Nielsen, L. C. 1980. Omgård. A Settlement from the Late Iron Age and the Viking Period in West Jutland. – Acta Archaeol. 50, 1979: 173–208.

– 1982. Vestjyske gårde og landsbyer fra bronze- og jernalder. – Nationalmuseets Arbejdsmark 1982: 131–141.

Nielsen, P. O. 1977. Die Flintbeile der Frühen Trichterbecherkultur in Dänemark. – Acta Archaeologica 48: 61–138.

– 1979. De tyknakkede flintøksers kronologi. – Aarbøger for Nordisk oldkyndighed og historie 1977: 5–71.

– 1981. Stenalderen 2. Bondestenalderen. – Danmarkshistorien.

– 1985. De første bønder. Nye fund fra den tidligste tragtbægerkultur ved Sigersted. – Aarbøger for Nordisk oldkyndighed og historie 1984: 96–126.

– 1987. The beginning of the Neolithic – Assimilation or complex change? – J. Danish Archaeol. 5: 241–243.

Nielsen, S. 1980. Det forhistoriske landbrug. – Antikvariske Studier IV: 103–127.

– 1984. Karby-udgravningen på Mors med nogle bemærkninger om den keramiske udvikling i yngre jernalder. – Aarbøger for Nordisk oldkyndighed og historie 1985: 260–281.

– 1987. Huller i jorden. – In: Danmarks længste udgravning. Arkæologi på naturgassens vej 1979–86. Nationalmuseet og de danske naturgasselskaber, pp. 87–93.

Nielsen, V. 1984. Prehistoric field boundaries in Eastern Denmark. – J. Danish Archaeol. 3: 135–164.

Nilsson, A. 1987. Gränser i upplösning. En studie av by- och sockengränsernas förändring i Alstads församling. – Ale 1987/4: 20–32.

Nilsson, A. 1920. Växtodling och växtförädling. – Skrifter utgivna av de skånska hushållningssällskapen vid deras hundraårsjubileum år 1914. Lund.

Nilsson, H. 1983. Boplats 2. En stenåldersboplats i Ystad Sandskog. – Seminar paper, Inst. of archaeol. Univ. of Lund.

Nilsson, T. 1935. Die pollenanalytische Zonengliederung der spät- und postglazialen Bildungen Schonens. – Geol. Fören. Stockh. Förh. 57: 385–562.

– 1961. Ein neues Standardpollendiagramm aus Bjärsjöholmssjön. – Lunds Univ. Årsskrift N. F. 2, 56, 18.

– 1964. Standardpollendiagramme und C14-Datierungen aus dem Ageröds mosse in mittleren Schonen. – Lunds Univ. Årsskr. N.F. 2, 59, 7.

Nordiska Ministerrådet, 1984. Naturgeografisk regionindelning av Norden. – Nordiska Ministerrådet, Köpenhamn.

Noring, A. 1836. Ett och annat rörande Boskaps- och Ladugårdsskötseln. – Berlings, Lund.

– 1841–42. Handbok i husdjursskötseln. Del 1–3 – Berlings, Lund.

Nyborg, E. 1979. Enkeltmænd og fællesskaber i organiseringen af det romanske sognekirkebyggeri. – Strejflys over Danmarks bygningskultur. Festskrift til Harald Langberg. København.

– 1986. Kirke-sognedannelse-bebyggelse. – Hikuin 12. Højbjerg.

Nyegaard, G. 1985. Faunalevn fra yngre stenalder på øerne syd for Fyn. – In: Skaarup, J., Yngre stenalder på øerne syd for Fyn. Meddelelser fra Langelands Museum, Rudkøbing, pp. 426–455.

Nyman, A. 1963. Den svenska fäboden – ålder, uppkomst, utbredning. – In: Lidman, H. (ed.), Fäbodar. LTs Förlag, Kristianstad, pp. 15–50.

Näsman, U. 1984. Husforskning i Norden. En skiss av situationen för yngre järnåldern, vikingatiden och medeltiden. – In: Liedgren, L. and Widgren, M. (eds), Gård och kulturlandskap under järnåldern, Kulturgeografiskt seminarium 2/84, Stockholms universitet, pp. 79–108.

– 1987. Hus, landsby, bebyggelse. – In: Danmarks længste udgravning. Arkæologi på naturgassens vej 1979–86. København, pp. 69–86.

– 1988. Den folkvandringstida krisen. – In: Näsman, U. and Lund, J. (eds), Folkevandringstiden i Norden. En krisetid mellem ældre og yngre jernalder, Viborg, pp. 227–255.

Odgaard, B. 1985. A pollen analytical investigation of a Bronze Age and Pre Roman Iron Age soil profile from Grøntoft, Western Jutland. – J. Danish Archaeol. 4: 121–128.

Odum, E. P. 1971. Fundamentals of ecology. – Saunders, Philadelphia.

– 1984. Properties of agroecosystems. – In: Lowrance, R., Stinner, B. R. and House, G. J. (eds), Agricultural ecosystems – unifying concepts. Wiley, New York.

– 1989. Input management of production systems. – Science 243: 177–182.

Ohlsson, T. 1976. The Löddeköpinge investigation. The settlement at Vikhögsvägen. – Meddelanden från Lunds universitets historiska museum 1979–80, Lund, pp. 59–161.

Olausson, D. 1987. Piledal and Svarte. A Comparison between two Late Bronze Age cemeteries in Scania. – Acta Archaeol. 1986/57: 121–152.

– 1988. Where have all the settlements gone? Field survey methods for locating Bronze and Iron Age settlements in a cultivated landscape. – Papers of the Archaeol. Inst. Univ. of Lund 7/1987–88: 99–112.

– 1991. The archaeological work in the western part of the project area – research goals, methods, and results. – In: Callmer, J., Larsson, L. and Stjernquist, B. (eds), The Archaeology of the cultural landscape. Field work and research in a South Swedish rural region. Acta Archaeol. Lund., Series in 4⁰. Lund.

–, Carlie, L., Engelmark, R. and Hulthén, B. unpubl. Grävningsrapport, Bjäresjö 19:17, Bjäresjö sn, Skåne. – Dept of Archaeology, Lund University.

Olausson, M. 1989. Ett uppländskt grophus från äldre järnåldern. – In: Arrhenius, B. (ed.), Laborativ Arkeologi 3: 35–56. Arkeologiska Forskningslaboratoriet, Stockholms Universitet.

Oldeberg, A. 1927. Ett smedfynd i Ystad från yngre bronsåldern. – Fornvännen 22/1927: 107–121.

– 1952. Studien über die schwedische Bootaxtkultur. – Stockholm.

Oldfield, M. and Alcorn, J. 1987. Conservation of traditional agroecosystems. – BioScience 37: 199–208.

Olsen, G. 1957. Hovedgård og bondegård: studier i stordriftens udvikling i Danmark i tiden 1525–1774. – Landbohistoriske Skrifter 1, Copenhagen.

Olsson, E. G. 1984. Old field forest succession in the Swedish

west coast archipelago. – Dept of Plant Ecology, Univ. of Lund, Sweden, Doctoral dissertation.
– 1988 a. Nutrient use efficiency as a tool for evaluating pre-industrial agroecosystems. – Geografiska Annaler 70B: 69–73.
– 1988 b. Nutrient use and productivity for different cropping systems in south Sweden during the 18th century. – In: Birks, H. H., Birks, H. J. B. Kaland, P. E., Moe, D. (eds), The cultural landscape – past, present and future. Cambridge Univ. Press, pp. 123–137.
– 1988 c. Productivity and nutrient use in eighteenth-century Scottish townships. Geografiska Annaler 70B: 39–51.
– 1989 a. Odlingslandskapet i Ystadsområdet under medeltiden. Rekonstruktion av landskapstyper och vegetation under tidigt 1300-tal. – In: Andersson, H. and Anglert, M. (eds), 1989. By, huvudgård och kyrka. Studier i Ystadsområdets medeltid. Lund Studies in Medieval Archaeology 5. Almqvist & Wiksell, Stockholm, pp. 51–55.
– 1989 b. Åkerbruket i Ystadsområdet under medeltiden. – In: Andersson, H. and Anglert, M. (eds) 1989. By, huvudgård och kyrka. Studier i Ystadsområdets medeltid. Lund Studies in Medieval Archaeology 5. Almqvist & Wiksell, Stockholm, pp. 57–60.
– and Persson, A. 1987. Skånejordbruk på 1700-talet – varför så olika på slätten och på åsen? – Forskning och Framsteg 7/1987, pp. 12–18.
Olsson, I. U. 1986. Radiometric dating. – In: Berglund, B. E. (ed.), Handbook of Holocene palaeoecology and palaeohydrology. Wiley, Chichester, pp. 273–312.
Osvald, H. 1962. Vallodling och växtföljder. Uppkomst och utveckling i Sverige. – Natur och Kultur, Uppsala.
Otto, J. S. and Anderson, N. E. 1982. Slash-and-burn cultivation in the highlands south: Problems in comparative agricultural history. – Comparative studies in society and history vol. 24(1): 131–149.
Paludan-Müller, C. 1978. High-Atlantic food gathering in Northwestern Zealand, Ecological conditions and spatial representation. – In: Kristiansen, K. and Paludan-Müller, C. (eds), New Directions in Scandinavian Archaeology. Nationalmuseet, Copenhagen, pp. 120-157.
Patterson III, W. A. and Backman, A. E. 1988. Fire and disease history of forest. – In: Huntley, B. and Webb, T. III (eds), Vegetation History. Kluwer Academic Publishers, pp. 603–632.
Paustian, K., Andrén, O., Clarholm, M., Hansson, A. C., Johansson, G., Lagerlöf, J., Lindberg, T., Pettersson, R. and Sohlenius, B. 1990. Carbon and nitrogen budgets of four agro-ecosystems with annual and perinnial crops, with and without N fertilization. – J. Appl. Ecol. 27: 60–84.
Pearson, G. W., Pilcher, J. R., Baillie, M. G. L., Corbett, D. M. and Qua, F. 1986. High precision ^{14}C measurements of Irish oaks to show the natural ^{14}C variations from AD 1840 to 5210 BC. Radiocarbon 28, 2B, 911–932.
Peet, R. K., Glenn-Lewin, D. C. and Walker Wolf, J. 1983. Prediction of man's impact on plant species diversity. – In: Holzner, H., Werger, M. J. A. and Ikusima, I. (eds), Man's impact on vegetation. Junk, The Hague. pp. 41–54.
Peglar, S. M., Fritz, S. C. and Birks, H. J. B. 1989. Vegetation and landuse history at Diss, Norfolk, U.K. – J. Ecol. 77: 203–222.
Pequart, M. and Pequart, S.-J. 1954. Hoëdic. Deuxième station-nécropole du Mésolithique cotier armorican. – De Sikkel, Anvers.
– Pequart, S.-J., Boule, M. and Vallois, H. 1937. Téviec. Station-nécropole Mésolithique du Morbihan. – Archives de L'Institut de Paléontologie Humaine, Memoires 18.
Persson, A. 1981. Svedalabygden från svensktidens början till skiftena (1658–ca 1800) – In: Svedala genom tiderna. Trelleborg, pp. 177–303.
– 1986 a. Villie socken c:a 1700. Bebyggelse – markanvändning – social struktur. – En arbetsrapport från Kulturlandskapet under 6000 år. Dept. of Plant Ecology, Lund Univ.

– 1986 b. Agrar struktur i Bjäresjö by 1570 och 1699. En trendalys utförd med ADB-stöd. – En arbetsrapport från Kulturlandskapet under 6000 år. Dept. of Plant Ecology, Lund Univ.
– 1987 a. Aspects on land use in Bussjö village in Bromma Parish and Herrestad Hundred in Scania ca 1670–1910. – En arbetsrapport från Kulturlandskapet under 6000 år. Dept. of Plant Ecology, Lunds Univ.
– 1987 b. Jordrevningen 1670–1672 och Adlerstens indelningsverk 1691–1693. – In: Haskå, G. (ed.), Skånsk släktforskning: Jubileumsskrift 1937–1987. Lund, pp. 65–77.
– 1988 a. Agrar struktur i Lilla Tvärens by c:a 1700. – En arbetsrapport från Kulturlandskapet under 6000 år. Dept. of Plant Ecology, Lund Univ.
– 1988 b. Befolkningsförhållanden i Ljunits och Herrestads härader c:a 1700–1805. – En arbetsrapport från Kulturlandskapet under 6000 år. Dept. of Plant Ecology, Lund Univ.
– 1989. Levnadsvillkor på skånska gatehus c:a 1700–1800: strukturer och utvecklingsdrag. – In: Huse og husmaend i fortid, nutid og fremtid. Odense University Studies in History and Social Science.
Persson, C. B. 1987. Bebyggelse. – In: Jacobsson, B. (ed.), Ystad under nittonhundratalet. Ystads stads historia del IV, 1914–1971, Ystad, pp. 168–169.
Persson, H. 1975. Deciduous woodland at Andersby, Eastern Sweden: field-layer and below-ground production. – Acta Phytogeographica Suecica 62. Uppsala.
Persson, O. 1982. An osteological analysis of some bones from a settlement at Stävie 4:1. – In: Larsson, L., A Causewayed Enclosure and a site with Valby pottery at Stävie, western Scania. – Papers of the Archaeol. Inst., Univ. of Lund 1981–1982: 114.
Petré, B. 1984. Indikationer om bebyggelseintensitet, samhällsstruktur och ekonomi i östra Mälarområdet under yngre bronsålder. – In: Hyenstrand, Å. (ed.), Bronsåldersforskning. Stockholm Archaeological Reports No. 17, Univ. of Stockholm, pp. 52–61.
Pickett, S. T. A. and White, P. S. 1985. The ecology of natural disturbance and patch dynamics. – Academic Press, Orlando, FL.
– , Kolasa, J., Armesto, J. J. and Collins, S. L. 1989. The ecological concept of disturbance and its expression at various hierarchical levels. – Oikos 54: 129–136.
Pielou, E. 1975. Ecological diversity. – Wiley, New York.
Platt, C. 1981. The parish churches of medieval England. – London.
Plucknett, D. L. and Smith, N. J. H. 1986. Sustaining agricultural yields. – BioScience 36: 30–45.
Porskrog Rasmussen, C. et al. (eds) 1987. Det danske godssystem – udvikling og afvikling 1500–1919. – Forlaget Historia, Århus.
Porsmose, E. 1987. De fynske landsbyers historie – i dyrkningsfællesskabets tid. – Odense University Studies in History and Social Sciences 109. Odense Universitetsforlag, Odense.
– 1988. Middelalder o. 1000–1536. – In: Bjørn, C. (ed.), Det danske landbrugs historie I. Oldtid og middelalder, Odense, pp. 205–417.
Post, L. 1946. The prospect for pollen analysis in the study of the Earth's climatic history. – New Phytol. 45: 193–217.
Pott, R. 1986. Der pollenanalytische Nachweis extensiver Waldbewirtschaftungen in den Haubergen des Siegerlandes. – In: Behre, K.-E. (ed), Anthropogenic Indicators in pollen diagrams. Balkema, Rotterdam, pp. 125-134.
– 1988. Entstehung von Vegetationstypen und Pflanzengesellschaften unter dem Einfluss des Menschen. – Düsseldorfer Geobot. Kolloq. 5: 27–54.
Poulsen, B. 1988. Land-By-Marked. To økonomiske landskaber i 1400-tallets Slesvig. – Dansk Centralbibliotek for Sydslesvig. Flensborgs studieafd. Flensborg.
Poulsen, J. 1980. Om arealudnyttelsen i bronzealderen: nogle

praktiske synspunkter og nogle synspunkter om praksis. – In: Thrane, H. (ed.). Bronzealderbebyggelse i Norden. Skrifter fra historisk institut, Odense universitet. pp. 142–164.

Pratt, R. M., Putman, R. J. Ekins, J. R. and Edwards P. J. 1986. Use of habitat by free-ranging cattle and ponies in the New Forest, southern England. – J. Appl. Ecol. 23: 539–557.

Pred, A. 1987. Place, practice and structure. Social and spatial transformation in southern Sweden: 1750–1850. – Polity, Cambridge.

Prentice, I. C. 1983. Pollen mapping of regional vegetation patterns in south and central Sweden. – J. Biogeogr. 10: 441–454.

– 1985. Pollen representation, source area and basin size: toward a unified theory of analysis. – Quatern. Res. 23: 76–86.

– 1986 a. Multivariate methods of data analysis. – In: Berglund B. E. (ed.), Handbook of Holocene palaeoecology and palaeohydrology. Wiley, Chichester, pp. 775–797.

– 1986 b. Vegetation response to past climatic variation. – Vegetatio 67: 131–141.

– 1988 a. Palaeoecology and plant population dynamics. – Trends Ecol. Evol. 3: 343–345.

– 1988 b. Records of vegetation in time and space: the principles of pollen analysis. – In: Huntley, B. and Webb, T. III (eds), Vegetation history. Kluwer.

Påhlsson, I. 1981. Cannabis sativa in Dalarna. – In: Königsson, L. K. and Paabo, K. (eds). Florilegium Florinis Dedicatum Striae 14: 79–82.

Quinn, J. F. and Harrison, S. 1988. Effects of habitat fragmentation and isolation on species richness: evidence from biogeographic patterns. – Oecologia (Berl.) 75: 132–140.

Rabbinge, R. 1986. The bridge function of crop ecology. – Netherland. J. Agr. Sci. 3: 239–251.

Rackham, O. 1980. Ancient woodland, its history, vegetation and uses in England. – Edward Arnold, London.

– 1984. The forest: woodland and wood-pasture in medieval England. – In: Beddick, K. (ed.), Archeological approaches to medieval Europe. Studies in Medieval Culture, XVIII. Western Michigan Univ. pp. 70–105.

– 1986. The history of the countryside. – Dent & Sons, London.

Ralska-Jasiewiczowa, M. 1964. Correlation between the Holocene history in the Carpinus betulus and prehistoric settlement in North Poland. – Acta Palaeobot. 29: 461–468.

– 1986. Palaeohydrological changes in the temperate zone in the last 15000 years. Subproject B. Lake and mire environments. Project catalogue for Europe. – Dept. of Quaternary Geology, Lund Univ.

– and Starkel, L. 1988. Record of the hydrological changes during the Holocene in the lake, mire and fluvial deposits of Poland. – Folia Quatern. 57: 91–127.

Ralska-Jasiewiczowa, R. (ed.) 1989. Environmental changes recorded in lakes and mires of Poland during the last 13000 years. Part three. – Acta Palaeobotanica 29,2: 1–120.

Randsborg, K. 1975. Social dimensions of Early Neolithic Denmark. – Proc. Prehist. Soc. 41: 105–118.

– 1986. Woman in prehistory: The Danish example. – Acta Archaeol. 55: 143–154.

Raneke, J. 1968. Vapenmålningarna i Ystads gråbrödrakloster. – Ale 1968/3: 16–30.

Rasmussen, P. 1988. Løvfoding af husdyr i Stenalderen. En 40 år gammel teori vurderet gennem nye undersøgelser. – In: Madsen, T. (ed.), Bag Moesgårds maske. Kultur og samfund i fortid og nutid. Aarhus Universitetsforlag, pp. 187–192.

Redin, L. 1972. Tommarp. – Skånes Hembygdsförbund. Årsbok 1972: 61–80.

– 1986. Stad, befolkning og bebyggelse i Halland under medeltid – problem i en generalisering. – Medeltiden och

arkeologin. Lund Studies in Medieval Archaeology 1. Lund.

Regnéll, J. 1989. Vegetation and land-use during 6000 years. Palaeoecology of the cultural landscape at two lake sites in Skåne, South Sweden. – LUNDQUA Thesis 27, Lund.

Reiners, W. A. 1986. Complementary models for ecosystems. – Am. Nat. 127: 59–73.

Reisnert, A. 1989 a. Huvudgårdar i Ystadsområdet. Några medeltidsarkeologiska synpunkter. – In: Andersson, H. and Anglert, M. (eds), By, huvudgård och kyrka. Studier i Ystadsområdets medeltid. – Lund Studies in Medieval Archaeology 5. Almqvist & Wiksell, Stockholm, pp. 145–149.

– 1989 b. Ystads franciskanerkloster och aristokratin. – In: Andersson, H. and Anglert, M. (eds), By, huvudgård och kyrka. Studier i Ystadsområdets medeltid. – Lund Studies in Medieval Archaeology 5. Almqvist & Wiksell, Stockholm, pp. 151–163.

– 1989 c. Bjäresjö socken: Arkeologiska iakttagelser. – In: Andersson, H. and Anglert, M. (eds), By, huvudgård och kyrka. Studier i Ystadsområdets medeltid. Lund Studies in Medieval Archaeology 5. Almqvist & Wiksell, Stockholm, pp. 102–107.

Renfrew, C. 1983. Introduction: The megalith builders of western Europe. – In: Renfrew, C. (ed.), The megalithic monuments of western Europe. Thames and Hudson, London, pp. 8–18.

– 1984. Approaches to social archaeology. – Edinburgh Univ. Press, Edinburgh.

Retzius, A. J. 1796. Utkast til de hufvudsakeligaste grunder för ängsskötseln. – Diss. Resp. C. G. Lundberg. Lund.

Riddersporre, M. 1983. Kabusa. Exempel på retrospektiv kulturgeografi i Sydskåne. – Mimeographed (seminar paper). Dept. of Social and Economic Geography, Univ. of Lund.

– 1985. Lantmäterihandlingar som källmaterial vid retrospektiva bebyggelse- och kulturlandskapshistoriska studier. – In: Land og by. Beretning fra symposium i Skanör 1984, pp. 103–114.

– 1986. Fragment av medeltidens kulturlandskap – lantmäterihandlingar som källa till studiet av samhällsförändringar. – Ymer 106: 52–62.

– 1987. Retrogressiv analys av äldre lantmäterihandlingar. – In: Det dolda kulturlandskapet. Metodkonferens 1985, Riksantikvarieämbetet och Statens Historiska Museer Rapport 1987:2, Stockholm, pp. 19–46.

– 1988. Settlement Site – Village Site. Analysis of the toftstructure in some Medieval villages and its relation to Late Iron Age settlements. A preliminary report and some tentative ideas based on Scanian examples. – Geografiska Annaler B (I): 75–85.

– 1989 a. Bjäresjö socken: Lantmäterihandlingar; Stora Herrestads socken: Lantmäterihandlingar. – In: Andersson, H. and Anglert, M. (eds), By, huvudgård och kyrka. Studier i Ystadsområdets medeltid. Lund Studies in Medieval Archaeology 5. Almqvist & Wiksell, Stockholm, pp. 100–102, 115.

– 1989 b. Lantmäterihandlingar, runstenar och huvudgårdar. Några kommentater och spekulationer i ett lokalt geografiskt perspektiv. – In: Andersson, H. and Anglert, M. (eds), By, huvudgård och kyrka. Studier i Ystadsområdets medeltid. Lund Studies in Medieval Archaeology 5. Almqvist & Wiksell, Stockholm, pp. 135–144.

– 1991. Eighteenth century landscape and archaeology. Retrogressive analysis based on land survey documents from the 17th and 18th centuries – an interdisciplinary approach with archaeological aspects. – In: Callmer, J., Larsson, L. and Stjernquist, B. (eds), The archaeology of the cultural landscape. Field work and research in a South Swedish rural region. Acta Archaeol. Lund., Series 4⁰. Gleerups, Lund.

Ring II, C. H., Nicholson, R. A. and Launchbaugh. 1985. Vegetational traits of patch-grazed rangeland in west-central Kansas. – J. Range Manage. 38: 51–55.

Rosborn, S. 1984. Hököpinge – Pile. Medeltida centralbygd i sydväst-Skåne. – In: Pugna forensis – Arkeologiska undersökningar kring Foteviken, Skåne 1981–83. Länsstyrelsen i Malmöhus län.

Rosén, J. 1944 a. Sjättepenningen och den svenska godspolitiken i Skåne på 1600-talet. – Scandia 15: 41–85.

– 1944 b. Skånska privilegie- och reduktionsfrågor 1658–1686. – Skrifter utgivna av Kungl. Humanistiska Vetenskapssamfundet i Lund 38, Lund.

Rostholm, H. 1986 a. Kornaftryk fra enkeltgravskulturen. – In: Adamsen, C. and Ebbesen, K. (red.), Stridsøksetid i Sydskandinavien. Arkæologiske Skrifter: 230–237.

– 1986 b. Lustrup og andre bopladsfund fra Herning-egnen. – In: Adamsen, C. and Ebbesen, K. (ed.), Stridsøksetid i Sydskandinavien. Arkæologiske skrifter: 301–317.

Rothamsted Experimental Station. 1969. The Broadbalk wheat experiment. – Rothamsted Experimental Station. Report for 1968, part 2. Harpenden.

– 1970. Details of the classical and long-term experiments up to 1967. – Harpenden.

Rowley-Conwy, P. 1979. Forkullet korn fra Lindebjerg. En boplads fra ældre bronzealder. – Kuml 1978: 159–171.

– 1981. Slash and burn in the temperate European Neolithic. – In: Mercer, R. (ed.), Farming practice in British Prehistory, Edinburgh, pp. 85–96.

– 1984. Mellemneolitisk økonomi i Danmark og Sydengland. – Kuml 1984: 77–111.

– 1985 a. The origin of agriculture in Denmark. A review of some theories. – J. Danish Archaeol. 4: 188–195.

– 1985 b. The Single Grave (Corded Ware) economy at Kalvø. – J. Danish Archaeol. 4: 79–86.

Rydbeck, M. 1943. Valvslagning och kalkmålningar i skånska kyrkor. – Skrifter utgivna av Kungl. Humanistiska Vetenskapssamfundet i Lund 35, Lund.

Rønne, P. 1986. Gård på vandring. – Skalk 1984-22: 11–14.

– 1987. Torstorp/Nörreby. Recent excavations and discoveries. – J. Danish Archaeol. 5: 264.

Röthlisberger, F. 1986. 10000 Jahre Gletschergeschichte der Erde. – Sauerländer AG, Aarau.

Sandgren, P. and Björck, S. 1988. Late Weichselian and Holocene palaeomagnetic studies in South Sweden. – Geographia Polonica 55: 129–140.

Sandin, B. 1986. Hemmet, gatan, fabriken eller skolan. – Arkiv, Lund.

SCB Report, 1990. Statistiska meddelanden 1990. – Statistiska centralbyrån, Stockholm.

Schirnig, H. 1979. Siedlungsräume der Trichterbecherkultur am Beispiel des Landkreises Uelzen. – In: Schirnig, H. (ed.), Grosssteingräber in Niedersachsen. Verlagsbuchhandlung A. Lax, Hildesheim, pp. 223–228.

Schlesier, E. 1977. Etnologische Aspekte zu Begriff »Dorf«. – In: Jankuhn, H. (ed.), Das Dorf der Eisenzeit und des frühen Mittelalters. Abhandlungen der Akademie der Wissenschaften in Göttingen. Philologisch-historische Klasse. Dritte Folge Nr 101. Göttingen, pp. 81–85.

Schmid, P. 1982. Ländliche Siedlungen der vorrömischen Eisenzeit bis Völkerwanderungszeit im niedersächsischen Küstengebiet. – Offa 39: 73–96.

– and Zimmermann, H. 1976. Flögeln – Zur Struktur einer Siedlung des 1. bis 5. Jahrhunderts n. Chr. im Küstengebiet der südlichen Nordsee. – Probleme der Küstenforschung 11: 1–77.

Schrøder, H. 1985. Nitrogen losses from Danish agriculture – trends and consequences. – Agriculture, ecosystems and environment 14: 279–289.

Seebohm, M. E. 1927. The evolution of the English farm. – George Allen & Unwin, London.

Service, E. R. 1962. Primitive social organization. – Random House, Chicago.

Sherratt, A. 1980. Water, soil and seasonality in early cereal cultivation. – World Archaeol. 11: 313–341.

– 1985. Plough and pastoralism: aspects of the secondary products revolution. – In: Hodder, I., Isaac, G. and Hammond, N. (eds), Pattern of the Past. Studies in honour of David Clarke. Cambridge Univ. Press, pp. 261–305.

Simonsen, J. 1983. A Late Neolithic house site at Tastum, Northwestern Jutland. – J. Danish Archaeol. 2: 81–89.

– 1986. Nogle nordvestjyske bopladsfund fra enkeltgravskulturen og deres topografi. – In: Adamsen, C. and Ebbesen, K. (red.), Stridsøksetid i Sydskandinavien. Arkæologiske Skrifter: 292–300.

Sjöbeck, M. 1964. Simremarken, Krattmarken och Sommarängen. – Skånes Natur 51: 140–167.

– 1965. Några markhistoriska miljöundersökningar i Skåne. – In: Skånes Natur, Årsbok 1965, Lund.

– 1973. Det sydsvenska landskapets historia och vård. – Föreningen Landskronatrakten natur VI.

Sjörs, H. 1954. Slåtterängar i Grangärde Finnmark. – Acta Phytogeographica Suecica 43. Uppsala.

– 1963. Amphi-Atlantic zonation, Nemoral to Arctic. – In: Löve, A. and Löve, D. (eds), North Atlantic biota and their history. Oxford.

– 1965. Forest regions. – Acta Phytogeogr. Suecica 50.

Skaarup, J. 1973. Hesselø-Sølager. Jagdstationen der südskandinavischen Trichterbecherkultur. – Arkæologiske Studier I. København.

– 1985. Stenalderen på øerne syd for Fyn. – Meddelelser fra Langelands Museum, Rudkøbing.

Skamby Madsen, J. 1984. En regionalundersøgelse af Hads herreds bebyggelse i yngre stenalder. – Fortid Nutid XXXI/3: 169–182.

Skansjö, S. 1983. Söderslätt genom 600 år. Bebyggelse och odling i äldre historisk tid. – Skånsk senmedeltid och renässans 11. Vetenskapssocieteten i Lund.

– 1985. Jordägare och bebyggelsemönster i Ystadsområdet vid mitten av 1600-talet. – En arbetsrapport från Kulturlandskapet under 6000 år. Dept. of Plant Ecology, Lund Univ.

– 1986. Från vikingatida stormansgård till renässansslott: några huvuddrag i Bjersjöholms äldre historia. – En arbetsrapport från Kulturlandskapet under 6000 år. Dept. of Plant Ecology, Lunds Univ.

– 1987 a. Estate building and settlement changes in S. Scania c. 1500–1650 in a European perspective. – In: Nitz, H.-J., (ed.), The medieval and early-modern rural landscape of Europe under the impact of the commercial economy, Göttingen, pp. 105–114.

– 1987 b. Aspekter på Ystadsområdet under medeltiden. Historisk kasuistik med kommentarer. A. Ljunits härad. – En arbetsrapport från Kulturlandskapet under 6000 år. Dept. of Plant Ecology, Lund Univ.

– 1988. Aspekter på Ystadsområdet under medeltiden: historisk kasuistik med kommentarer. B. Herrestads härad. – En arbetsrapport från Kulturlandskapet under 6000 år. Dept. of Plant Ecology, Lund Univ.

– 1991. Från vikingatida stormansgård till renässansslott. Bjersjöholm under äldre historisk tid. – In: Johansson, S. et al., Bjersjöholm. Storgods och slott. Skånsk senmedeltid och renässans 13. Vetenskapssocieteten i Lund.

– , Riddersporre, M. and Reisnert, A. 1989. Huvudgårdar i källmaterialet. – In: Andersson, H. and Anglert, M. (eds), By, huvudgård och kyrka. Studier i Ystadsområdets medeltid. Lund Studies in Medieval Archaeology 5. Almqvist & Wiksell, Stockholm, pp. 71–133.

Skov, T. 1982. A Late Neolithic house site with Bell Beaker pottery at Stendis, Northwestern Jutland. – J. Danish Archaeol. 1: 39–44.

Slicher van Bath, B. H. 1963. The agrarian history of Western Europe 500–1850. – Edward Arnold Ltd, Frome and London.

Smedberg, G. 1973. Nordens första kyrkor. En kyrkorättslig studie. – Biblioteca Theologiæ Practicæ 32. Uppsala.

Smit, J. G. 1986. Ländliche Neusiedlung in Mitteleuropa vom Ende des 19. Jahrhunderts bis zur Gegenwart als nationalpolitisches Instrument: Ziele, zeitgenössische Stellungsnahme und Ergebnisse. – Erdkunde 40/3: 165–174.

Solomon, A. M., West, D. C. and Solomon, J. A. 1981. Simulating the role of climate change and species immigration in forest succession. – In: West, D. C., Shugart, H. H. and Bitkin, D. B. (eds), Forest succession: Concepts and application. Springer, New York.

Sporrong, U. 1970. Jordbruk och landskapsbild. – Gleerups, Lund.

– 1983. Individen, samhället och kulturlandskapet. – In: Roeck Hansen, B. and Sporrong, U., Rapporter från Barknåreprojektet. Kulturgeografiska institutionen, Stockholms Universitet. Kulturgeografiskt seminarium 1983.1: 18–26.

– 1984. Människa – landskap – förändring. Några teoretiska utgångspunkter vid kulturlandskapsstudier. – In: Teoretiska studier om kulturlandskapets utveckling. Rapport från ett symposium 10–11 november 1983. – Riksbankens Jubileumsfond, HSFR och NFR, pp. 60–82.

– 1985 a. Mälarbygd. Agrar bebyggelse och odling ur ett historiskt-geografiskt perspektiv. – Kulturgeografiska institutionen, Stockholms Universitet, Medd. B61.

– 1985 b. The destruction by cultivation of ancient monuments documented on cadastral maps. – In: Archaeology and Environment 4. In honorem Evert Baudou. Dept. of Archaeology, Umeå Univ., pp. 163–170.

Spång, K., Welinder, S. and Wyszomirska, B. 1976. The introduction of the Neolithic Stone Age into the Baltic Area. – In: de Laet, S. (ed.), Acculturation and continuity in Atlantic Europe mainly during the Neolithic period and the Bronze Age. Brügge.

Steen, E. 1980. Dynamics and production of semi-natural grassland vegetation in Fennoscandia in relation to grazing management. – Acta Phytogeogr. Suecica 68: 153–156.

Steensberg, A. 1976. Store Vallby and Borup: Two case studies in the history of Danish settlement. – In: Sawyer, P. H. (ed.), Medieval settlement. Continuity and change. Edward Arnold, London, pp. 94–113.

– 1983. Borup AD 700–1400. A deserted settlement and its fields in South Zealand, Denmark. – Copenhagen.

Steineck, S. 1984. Stallgödselproduktion. – Kursmaterial vid kurser anordnade av Statens Lantbrukskemiska Laboratorium 1984.

Stenberger, M., Christiansson, H. and Malmer, B. 1974. Konsten i Sverige. Forntiden och den första kristna tiden. – Almqvist & Wiksell.

Stenholm, L. 1981. Lerbottnar till ny belysning. – Ale 1981/2: 17–30.

– 1986. Önnerup – en skånsk by mellan två revolutioner. – In: Medeltiden och arkeologin. Festskrift till Erik Cinthio, Lund, pp. 73–86.

Stiesdahl, H. 1978. To middelalderlige huse på Stevns. – In: Arkitekturstudier tilegnede Hans Henrik Engquist. København.

– 1980. Gård og kirke. Sønder Jernløse-fundet belyst ved andre sjællandske eksempler. – Årbøger for Nordisk Oldkyndighed og Historie 1980: 166–172.

– 1981. Types of public and private fortifications in Denmark. – In: Skyum-Nielsen, N. and Lund, N. (eds), Danish medieval history 1, New currents. Museum Tusculanum Press, Copenhagen, pp. 207–220.

– 1983. Tidlige sjællandske og lollandske vesttårne. – Kirkens bygning og brug. Studier tilegnet Elna Møller. København.

Stig Sørensen, M.-L. 1987. Material order and cultural classification: the role of bronze objects in the transition from Bronze Age to Iron Age in Scandinavia. – In: Hodder, I. (ed.). The archaeology of contextual meanings. Cambridge, pp. 99–101.

Stjernquist, B. 1948. A grave of the Boat Axe folk. – Meddelanden från Lunds universitets historiska museum 1948: 162–164.

– 1951. Vä under järnåldern. – Lund.

– 1955. Simris. On cultural connections of Scania in the Ro-

man Iron Age. – Acta Archaeol. Lund., Series in 4⁰, No. 2. Lund.

– 1961. Simris II. Bronze Age problems in the light of the Simris excavation. – Acta Archaeol. Lund., Series in 4⁰. No. 5. Lund.

– 1969. Beiträge zum Studium von bronzezeitlichen Siedlungen. – Acta Archaeol. Lund. Series in 8⁰. No. 8. Lund.

– 1978. Approaches to settlement archaeology in Sweden. – In: World Archaeology, Vol. 9, No. 3, Landscape archaeology. London, pp. 251–264.

– 1981. Arkeologisk forskning om den agrara bebyggelsen i Skåne vid vikingatidens slut – källäge och problemställningar. – Bebyggelsehistorisk tidskrift 2: 17–25.

– 1983. Gravarna som källa till kunskap om yngre bronsålderns bebyggelse i Skåne. – In: Struktur och förändring i bronsålderns samhälle. Inst. of Archaeol., Lund Univ., Report Series, No. 17, Lund, pp. 130–140.

– 1984. Teoretiska utgångspunkter för studiet av kulturlandskapets förändringar. – In: Teoretiska studier om kulturlandskapets utveckling. Rapport från ett symposium 10–11 november 1983. Kulturgeografiska institutionen, Stockholms Universitet.

– 1985. Approaches to the problem of settlement patterns in eastern Scania in the first millennium BC. – In: Settlement and society: aspects of West European prehistory in the first millennium BC, pp. 223–228.

– 1987. Die siedlungsarchäologische Forschung, Generelle Probleme und anschauliche Beispiele aus Südschweden. – In: Hässler, H.-J. (ed.), Studien zur Sachsenforschung, Band 6, Hildesheim, pp. 237–250.

– 1988. On the Iron Age Settlement at Östra Torp and the pattern of settlement in Skåne during the Iron Age. – In: Meddelanden från Lunds universitets historiska museum Vol. 7, 1987–88, Lund, pp. 125–141.

– 1991. Introduction. – In: Callmer, J., Larsson, L. and Stjernquist, B. (eds) The archaeology of the cultural landscape. Field work and research in a South Swedish rural region. Acta Archaeol. Lund., Series in 4⁰. Lund.

Stoumann, I. 1978. De der blev hjemme. En vikingelandsby ved Esbjerg. – Esbjerg.

Strömberg, M. 1954. Bronzeitliche Wohnplätze in Schonen. – Meddelanden från Lunds Universitets historiska museum 1954: 295–380.

– 1961. Römerzeitliche Brandbestattungen in Schonen. – Meddelanden från Lunds universitets historiska museum 1960: 43–72.

– 1961. Untersuchungen zur jüngeren Eisenzeit – Wikingerzeit. – Acta Archaeol. Lund., Series in 4⁰, No. 4. Lund.

– 1963. Järnåldersguld i Skåne – Lund.

– 1968. Der Dolmen Trollasten in St. Köpinge, Schonen. – Acta Archaeol. Lund. 8:7. Lund.

– 1971. Senneolitiska huslämningar i Skåne. – Fornvännen 1971/4: 237–254.

– 1975 a. Bronsålder på Österlen. – Berlingska Boktryckeriet, Lund.

– 1975 b. Studien zu einem Gräberfeld in Löderup. – Acta Archaeol. Lund., Ser. in 8⁰ No. 10, Lund.

– 1976. Kontinuitätsprobleme in der Siedlungsarchäologie Südostschonens. – In: DeLaet, S. J. (ed.), Acculturation and continuity in Atlantic Europe mainly during the Neolithic period and the Bronze Age. Dissertationes Archaeologicae Gandenses, Ghent, pp. 251–256.

– 1978 a. En kustby i Ystad före stadens tillkomst. – Ystadiana 1978: 11–101.

– 1978 b. Three Neolithic sites. A local seriation? – Papers of the Archaeol. Inst. Univ. of Lund 1977–1978: 68–97.

– 1980a. The Hagestad investigation – a project analysis. – Papers of the Archaeological Institute, University of Lund 1979–1980: 47–60.

– 1980b. Var kustbon fiskare eller bonde? – Ystadiana XXV, Ystad, pp. 6–30.

– 1981. Vattenförsörjning och verksamhet i forntidsbyn. – In: Ystadiana XXVI, Ystad, pp. 21–60.

– 1982. Ingelstorp. Zur Siedlungsentwicklung eines südschwedischen Dorfes. – Acta Archaeol. Lund, Series in 4° 14. C. W. K. Gleerup, Lund.

– 1985. Jägare flintsmed bonde järnsmed i Gislöv. – Simrishamn.

– 1986. Signs of Mesolithic occupation in South-East Scania. – Papers of the Archaeol. Inst. Lund Univ. 1985–1986: 52–83.

– 1988 a. A complex hunting and production area. Problems associated with a group of Neolithic sites to the South of Hagestad. – Papers of the Archaeol. Inst., Lund Univ. 1987–1988: 53–80.

– 1988 b. Från bågskytt till medeltidsbonde. – Kulturnämnden i Ystad, Ystad.

Stuiver, M. (ed.) 1986. Radiocarbon. Calibration issue. – Radiocarbon 28, 2B. Am. J. Sci.

Stumman Hansen, S. 1984. Aspekter omkring relationen mellem landbrug og bebyggelse i Danmarks yngre bronzealder og ældre jernalder. – In: Thrane, H. (ed.), Dansk landbrug i oldtid og middelalder, Skrifter fra Historisk institut nr. 32, Odense universitet, Odense, pp. 41–50.

Succow, M. and Lange, E. 1984. The mire types of the German Democratic Republic. – In: Moore, P. D. (ed.), European Mires. Academic Press, London, pp. 149–175.

Sundnér, B. 1969. Två "Fröjelfuntar" i Ystad. – Ystads fornminnesförenings årsskrift XIV. Ystad.

– 1982. Maglarp. En tegelkyrka som historiskt källmaterial. – Acta Archaeol. Lund., Series in 4°. No. 15. Bonn.

– 1984. Från stenbrott till stenkyrka – presentation av ett planerat projekt. – META, Medeltidsarkeologisk tidskrift 84: 3–4.

– 1986. Kan murningstekniken datera medeltidskyrkorna? – Medeltiden och arkeologin. Festskrift till Erik Cinthio. Lund Studies in Medieval Archaeology 1. Lund.

– 1989 a. Medeltida byggnadsteknik. En kronologisk studie av Ystadsområdet kyrkor. – In: Andersson, H. and Anglert, M. (eds), By, huvudgård och kyrka. Studier i Ystadsområdets medeltid. Lund Studies in Medieval Archaeology 5. Almqvist & Wiksell, Stockholm, pp. 183–210.

– (ed.), 1989 b. Dendrokronologi och medeltida kyrkor. Rapport från ett symposium i Lund 11–13 april 1988. – Lund Studies in Medieval Archaeology 6. Almqvist & Wiksell, Stockholm.

Svensson, M. 1986. Trattbägarboplatsen "Hindby Mosse" – aspekter på dess struktur och funktion. – Elbogen 16:3: 97–125.

Svensson, R. and Wigren, M. 1986. A survey of the history, biology and preservation of some retreating synanthropic plants. – Acta Universitatis Upsaliensis, Symbolae Botanicae Upsaliensis. 25:4. Uppsala.

Swensson, G. 1954. Ystads historia. Del III. 1863–1914. – Ystad.

Szalay, J. 1987. Osteological analysis of Piledal and Svarte. The cremated bone material from two late Bronze Age cemeteries in Scania. – In: Olausson, D. 1987, Acta Archeol. 57: 148–150 (1986).

Sörensen, M. L. S. 1987. Material order and cultural classification: the role of bronze objects in the transition from Bronze Age to Iron Age in Scandinavia. – In: Hodder, I. (ed.), The archaeology of contextual meanings, Cambridge, pp. 90–101.

Tauber, H. 1965. Differential pollen dispersion and the interpretation of pollen diagrams. – Danmarks Geologiske Undersøgelse IV:4,4.

– 1967. Investigations of the mode of pollen transfer in forested areas. – Rev. Palaeobot. Palynol. 3: 277–286.

– 1977. Investigations of aerial pollen transport in forested areas. – Dansk Bot. Arkiv. 32.

– 1982. Carbon-13 evidence for the diet of prehistoric humans in Denmark. – PACT 7 pt.1, Conseil de L'Europe, Strasbourg, pp. 235–239.

– 1986 a. Analysis of stable isotopes in prehistoric populations. – Mitt. Berliner Anthropol., Ethnol. Urgeschichte, Band 7, pp. 31–38.

– 1986 b. C14 dateringer af enkeltgravskultur og grubekeramisk kultur i Danmark. – In: Adamsen, C. and Ebbesen, K. (red.), Stridsøksetid i Sydskandinavien. Arkæologiske Skrifter: 196–204.

Ter Braak and Prentice, I. C.1987. CANOCO – a Fortran program for canonical ordination by partial detrended canonical correspondence analysis, principal components analysis and redundancy analysis. Ver. 2. – TNO Institute of Applied Computer Science, Statistics Dept, Wageningen, The Netherlands.

– C. J. F. and Prentice, I. C. 1988. A theory of gradient analysis. – Adv. Ecol. Res. 18: 271–317.

Tesch, S. 1974. Arkeologisk undersökning 1969. Del av fornlämningsområde 20, Ekshammar, Kungsängens sn (fd Stockholms Näs sn), Uppland. – Riksantikvarieämbetet UV rapport 1974 B 10.

– 1979 a. Bronsåldershus i St. Köpinge. – Ale 1979/1: 27–29.

– 1979 b. Forntidens bopålar. – Ale 1979/3: 1–10.

– 1980. Ett par bronsåldersmiljöer med huslämningar i Skåne och Södermanland. – In: Thrane, H. (ed.), Broncealderbebyggelse i Norden, Skrifter fra Historisk institut, nr. 28, Odense universitet, Odense, pp. 83–101.

– 1981. Provundersökning Balkåkra 1:81 m fl (Svarte), Balkåkra sn, Skåne. – Riksantikvarieämbetet Rapport, Stockholm.

– 1982 a. Fosfatkarteringsmetoden i sydskånsk fullåkersbygd. – META, Medeltidsarkeologisk tidskrift 1982/2: 25–37.

– 1982 b. Det stora sillafisket och Ystads äldsta stadsplan. – Ale 1982/3: 5–29.

– 1983 a. Ystad I. – Riksantikvarieämbetet Rapport. Medeltidsstaden 44. Stockholm.

– 1983 b. Ystad II. En omlandsstudie. – Riksantikvarieämbetet Rapport, Medeltidsstaden 45, Stockholm.

– 1983 c. Förändringar i bosättningsmönster och gårdsstruktur under yngre järnålder/tidig medeltid – ett par exempel från Skåne och Södermanland. – In: Thrane, H. and Grøngaard Jeppesen, T. (eds), Gårdens udvikling fra jernalderen til nyere tid..., Skrifter fra Historisk institut nr. 31, Odense universitet, Odense, pp. 100–132.

– 1983 d. Ett par bronsåldersmiljöer med huslämningar i Skåne och Södermanland. – In: Arkeologi i Sverige 1980. Riksantikvarieämbetet och statens historiska museer. Rapport 1983:3. pp. 31–50. Stockholm.

– 1985 a. Vad gömmer åkern? – en fråga om det arkeologiska materialets representativitet. – In: Regnéll, G. (ed.), Kulturlandskapet – dess framväxt och förändring, Symposium sept. 1984, Kulturlandskapet under 6000 år, Lund, pp. 55–64.

– 1985 b. "Vilda Östern". Bebyggelsekolonisation, landsbybildning och urbanisering i Ystadområdet under vikingatid och äldre medeltid. – Land og by 3, Beretning fra symposium i Skanör 1984, pp. 75–96.

– 1985 c. Klostergatan i Ystad under Medeltiden – vad en arkeologisk undersøkning i Kv. Yngve Norra avslöjade. – Riksantikvarieämbetet Rapport UV 1985: 14.

– 1988. En slättbygd vid kusten under brons- och järnålder – arkeologiska undersökningar inom ramen för Ystadsprojektet (B3). – In: Ystadiana 33: 189–230.

– 1991. Ch. 5. The archaeological work in the eastern part of the project area, research goals, methods and results. Ch. 7. House, farmstead and settlement structure during the Bronze and Iron Age. – In: Callmer, J., Larsson, L. and Stjernquist, B. (eds), The archaeology of the cultural landscape. Field work and research in a South Swedish rural region. Acta Archaeol. Lund., Series in 4°. Lund.

– and Wihlborg, A. 1981. Medeltidsstaden Ystad. – Riksantikvarieämbetet UV-Syds skriftserie 2, Lund.

–, Widholm, D. and Wihlborg, A. 1980. Forntidens bopålar. Stora Köpinge-bygden under femtusen år. – Riksantikvarieämbetet UV-Syds skriftserie nr. 1. Lund.

Thelaus, M. 1989. Late Quaternary vegetation history and palaeohydrology of the Sandsjön-Årshult area, southwestern Sweden. – LUNDQUA Thesis 26, Lund.

Thrane, H. 1965. Dänische Funde fremder Bronzegefässer der jüngeren Bronzezeit (Periode IV). – Acta Archaeol. 36: 157–207.

– 1980. Nogle tanker om yngre broncealders bebyggelse på Sydvestfyn. – In: Thrane, H. (ed.), Broncealderbebyggelse i Norden. Odense Univ. Press, Odense, pp. 165–173.

– 1982. Dyrkningsspor fra yngre stenalder i Danmark. Om yngre stenalders bebyggelseshistorie. – Skrifter fra Historisk Institut, Odense universitet 30: 20–28.

– 1984 a. Lusehøj ved Voldtofte. – Fynske Studier XIII, Odense.

– 1984 b. Vidnesbyrd om landbrug i broncealderen. – In: Thrane, H. (ed.), Dansk landbrug i oldtid og middelalder, Skrifter fra Historisk institut, Odense universitet, Odense, 32: 7–17.

– 1985. Bronze Age settlements. – In: Kristiansen, K. (ed.), Archaeological formation processes. Nationalmuseets forlag, København, pp. 142–151.

– 1987. Sydvästfynsundersögelsen. – In: Thrane, H. (ed.), Diakrone bebyggelsesundersögelser, Skrifter fra Historisk institut, Odense universitet, Odense, 34: 3–24.

Thun, E. 1982. Vä, järnålder, medeltid. – Rapport. Örebro (Mimeographed).

– and Anglert, M. 1984. Vä. – Medeltidsstaden, Rapport 57, Stockholm.

Thurborg, M. 1985. Bebyggelse och ekonomi i Kävlingeområdet under bronsåldern. – Mimeographed, Dept. of Archaeol., Lund.

Thurstone, J. M., Williams, E. D. and Johnstone, A. E. 1976. Modern developments in experiment on permanent grassland started in 1956: effects of fertilizers and lime on botanical composition and crop analyses. – Ann. Agron. 27: 1043–1082.

Thyselius, T. T. 1963. Fäbodvall. – Rabén & Sjögren, Stockholm.

Tilley, C. 1982. An assessment of the Scanian battle-axe tradition: towards a social perspective. – Scripta Minora 1981–1982: 1. Almqvist & Wiksell, Stockholm.

– 1984. Ideology and the legitimation of power in the middle neolithic of southern Sweden. – In: Miller, D. and Tilley, C. (eds), Ideology, power and prehistory. CUP, Cambridge, pp. 131–154.

Titow, J. Z. 1972. Winchester yields. A study in medieval agricultural productivity. – Cambridge Univ. Press, Cambridge.

Tivy, J. 1990. Agricultural ecology. – Langmans Scientific and Technical, New York.

Tornbjerg, S. A. 1985. Bellingegård, a late Iron Age settlement at Køge, east Zealand. – J. Danish Archaeol. 4: 147–156.

Trass, H. and Malmer, N. 1973. North European approaches to classification. – In: Whittaker, R. H. (ed.), Handbook of vegetation science 5, pp. 529–574.

Troels-Smith, J. 1960. Ivy, misteltoe and elm. Climate indicators – fodder plants. – Danmarks Geologiske Undersøgelse 4, 4: 1–32.

– 1984. Stall-feeding and field-manuring in Switzerland about 6000 years ago. – In: Lerche, G., Fenton, A. and Steensberg, A. (eds), Tools Tillage 5,1: 13–25.

Tüxen, R. 1956. Die heutige potentielle Vegetation als Gegenstand der Vegetationskartierung. – Angew. Pflanzensoziol. 13: 5–55. Stolzenau.

Uggla, E. 1958. Skogsbrandfält i Muddus Nationalpark. – Acta Phytogeogr. Suecica. 41.

Uhlig, H. and Lienau, C. 1972. Die Siedlungen des ländlichen Raumes. – Materialen zur Terminologie der Agrarlandschaft 2. Bonn.

Ulsig, E. 1968. Danske adelsgodser i middelalderen. – Skrifter udgivet af Det Historiske Institut ved Københavns Universitet 2, Copenhagen.

Ullén, I. 1988 a. Apalle – vardagsliv i en bronsåldersby. – In: Forntid i vägen. Sagt hänt meddelat nr. 3 1988, Statens historiska museer, Stockholm, pp. 23–27.

– 1988 b. Bronsåldersbyn i Apalle. – In: Populär arkeologi nr. 4 1988: 18–21.

Urban, D. N., O'Neill, R. V. and Shugart Jr, H. H. 1987. Landscape ecology. – BioScience 119–127.

Van Dyne, G. M. and Meyer, J. H. 1964. Forage intake by cattle and sheep on dry annual range. – J. Anim. Sci. 23: 1108–1115.

–, Brockington, N. R., Szocs, Z., Duek, J. and Ribic, C. A. 1980. Large herbivore sub-system. – In: Grasslands, systems analysis and man. IBP 19. Cambridge Univ. Press, Cambridge. pp. 269–558.

Vang Petersen, P. 1984. Chronological and regional variation in the Late Mesolithic of Eastern Denmark. – J. Danish Archaeol. 3: 7–18.

Vedsted, J. 1986. Fortidsminder og kulturlandskap. – Djurslands Museum, Grenaa.

Vemming Hansen, P. and Madsen, B. 1983. Flint Axe Manufacture in the Neolithic. An experimental investigation of a flint axe manufacture site at Hastrup Vænget, East Zealand. – J. Danish Archaeol. 2: 43–59.

Vermeer, H. J. G. 1986. The effect of nutrients on shoot biomass and species composition of wetland and hayfield communities. – Acta Oecologia 7: 31–41.

– and Berendse, F. 1983. The relationship between nutrient availability, shoot biomass and species richness in grasslands and wetland communities. – Vegetatio 53: 121–126.

Vifot, B. M. 1936. Järnåldersboplatsen vid Uppåkra. – Meddelanden från Lunds universitets historiska museum 1936: 97–139.

Vitousek, P. 1982. Nutrient cycling and nutrient use efficiency. – Am. Nat. 119: 553–572.

Vorting, H. C. 1973. Et usædvanligt enkeltgravsanlæg ved Veldbæk. – Mark Montre 1973: 5–9.

Vuorela, I. 1986. Palynological and historical evidence of slash-and-burn cultivation in South Finland. – In: Behre, K.-E. (ed.), Anthropogenic indicators in pollen diagrams. Balkema, Rotterdam, pp. 53–64.

Wallin, C. 1951. Gravskick och gravtraditioner i sydöstra Skåne. – Ystad.

Wallin, L. 1986. Nyby-Kabusvägen. – Ale 1986/2: 31–32.

– 1987. Järnålder i Önskvala. – Ale 1987/2: 31–32.

Waterbolk, H. T. 1979. Siedlungskontinuität im Küstengebiet der Nordsee zwischen Rhein und Elbe. – Probleme Küstenforschung 13: 1–21.

– 1980. Hoe oud zijn de Drentse dorpen? Problemen van nederzettingscontinuiteit in Drente van de bronstijd tot de middeleeuwen. – Westerheem 20: 90–98.

Watt, M. 1985 a. Det antikvariske arbejde. – Bornholms Museum 1984–1985: 45–76.

– 1985 b. En gravplads fra sen vikingatid ved Runegård, Åker. – Bornholms Museum 1984–1985: 77–100.

Watts, W. A. 1961. Post-Atlantic forests in Ireland. – Proc. Linn. Soc. Lond. 172: 33–38.

Webb, T. III 1986. Is vegetation in equilibrium with climate? How to interpret Late Quaternary pollen data. – Vegetatio 67: 75–91.

Weibull, C. 1966. Lübecks sjöfart och handel på de nordiska rikena 1368 och 1398–1400. – Scandia 1966/1: 1–123.

Weibull, C. G. 1923. Skånska jordbrukets historia intill 1800-talets början: skrifter utgivna av de skånska hushållningssällskapen med anledning av deras hundraårsjubileum år 1914. – C. W. K. Gleerup, Lund.

Welinder, S. 1971. Tidigpostglacialt mesoliticum i Skåne. – Acta Archaeol. Lund, Series in 8° M: 1.

- 1975. Prehistoric agriculture in eastern Middle Sweden. – Acta Archaeol. Lund., Series in 8° Minore 4.
- 1977. Ekonomiska processer i förhistorisk expansion. – Acta Archaeol. Lund., Series in 8° Minore 7.
- 1978. The Acculturation of the Pitted Ware Culture in Eastern Sweden. – Papers of the Archaeol. Inst., Univ. of Lund 1977–1978: 98–110.
- 1979. Prehistoric demography. – Acta Archaeol. Lund. Series in 8° Minore 8.
- 1982. Varför teori? – Fornvännen 1982/2: 140–141.
- 1983. The ecology of long-term change. – Acta Archaeol. Lund., Series in 8° Minore 9.
- 1984. A systematic approach to understanding long-term change in a culture landscape. – In: Kristiansen, K. (ed.), Settlement and economy in later Scandinavian prehistory. BAR Int. Ser. 211, 1–25.
Wester, E. 1960. Några skånska byar enligt lantmäteriakterne. – Svensk Geografisk Årsbok 36: 162–180.
Whittle, A. 1985. Neolithic Europe: A survey. – Cambridge Univ. Press, Cambridge.
Widgren, M. 1983. Settlement and farming systems in the early Iron Age. A study of fossil agrarian landscapes in Östergötland, Sweden. – Acta Univ. Stockholm. Stockholm Studies in human Geography 3. – Almqvist & Wiksell, Stockholm.
- 1984. Kulturlandskapsförändringar under första årtusendet. – In: Teoretiska Studier om Kulturlandskapets utveckling. Rapport från ett symposium 10–11 november 1983. Riksbankens Jubileumsfond, HSFR och NFR, pp. 96–116.
- 1987. Bebyggelse och kulturlandskap i Östergötland under järnåldern. – In: Petré, B. (ed.), Bebyggelsearkeologiska exempel, Arkeologiska rapporter och meddelanden från institutionen för arkeologi vid Stockholms universitet nr. 19, Stockholm, pp. 37–43.
- 1988. Om skillnader och likheter mellan regioner. Några kommentarer till 3 figurer. – In: Näsman, U. and Lund, J. (eds), Folkevandringstiden i Norden. En krisetid mellem ældre og yngre jernalder, Viborg, pp. 273–285.
Widholm, D. 1974 a. En bronsåldersboplats vid Kvarnby. Undersökning 1969. – Kring Malmöhus 4: 43–89.
- 1974 b. Nya fornfynd i Sydsverige. – Ale 1974/3: 1–10.
- 1980. Problems concerning Bronze Age settlements in southern Sweden. – Meddelanden från Lunds universitets historiska museum 1979–1980, Lund, pp. 29–46.
Wienberg, J. 1983. På sporet af Bornholms middelalderlige gårde? – Bornholms Museum 1983: 49–58.
- 1986 a. Gotiske kirkehvælvinger – et økonomisk perspektiv. – In: Medeltiden och arkeologin. Festskrift till Erik Cinthio. Lund Studies in Medieval Archaeology 1.
- 1986 b. Bornholms kirker i den ældre middelalder. – Hikuin 12. Højbjerg.
- 1988. Dendrokronologi og kildekritik – om dateringen av Store Köpinge kirkes romanske ombygning. – META, Medeltidsarkeologisk tidskrift 1988/1–2: 70–82.
- 1989. Kirker og befolkning i Ystadsområdet. – In: Andersson, H. and Anglert, M. (eds), By, huvudgård och kyrka. Studier i Ystadsområdets medeltid. – Lund Studies in Medieval Archaeology 5. Almqvist & Wiksell, Stockholm, pp. 243–264.
Wigren, S. 1987. Sörmländsk bronsåldersbygd. En studie av tidiga centrumbildningar daterade med termoluminescens. – Theses and Papers in North-European Archaeology 16, Edsbruk.

Wihlborg, A. 1976. Bronsåldershög, nyupptäckt. St. Köpinge, St. Köpinge sn, Skåne. – Riksantikvarieämbetets Rapport 1976, B:40.
- (ed.) 1985. Arkeologi och naturgas i Skåne. – Riksantikvarieämbetet UV-Syds skriftserie nr. 6. Lund.
Wilcox, B. A. and Murphy, D. D. 1985. Conservation strategy: the effects of fragmentation on extinction. – Am. Nat. 125: 879–887.
Willerding, U. 1986. Aussagen von Pollenanalyse und Makrorestanalyse zu Fragen der frühen Landnutzung. – In: Behre, K.-E. (ed.), Anthropogenic indicators in pollen diagrams. A. A. Balkema, Rotterdam, pp. 135–152.
Wimarson, N. 1918. Ystad mot slutet av den danska tiden: några anteckningar rörande Ystads administrativa, ekonomiska och merkantila förhållanden under förra hälften av 1600-talet. – Skrifter utgivna av Ystads Fornminnesförening 2, Ystad.
Wincentz Rasmussen, L. 1984. Kainsbakke A47. A settlement Structure from the Pitted Ware Culture. – J. Danish Archaeol. 3: 83–98.
Wood, B. A. 1982. The development of the Nottinghamshire County Council smallholdings estate 1907–1980. – East Midland Geographer 8: 25–35.
Woods, J. 1989. Limits to exploitation of the renewable resources of the ocean. – Symposium II "Renewable Resources of our Planet". Academia Europaea meeting, London, (in press).
Woods, K. D. and Davis, M. B. 1989. Palaeoecology of range limits: beech in the upper peninsula of Michigan. – Ecology 70: 681–696.
World Watch Institute. 1990. State of the world 1990. Tillståndet i världen 1990. – Naturskyddsföreningen, Naturvårdsverket. Stockholm.
Wright, H. E. Jr. 1984. Sensitivity and response time of natural systems to climatic change in the Late Quaternary. – Quatern. Sci. Rev. 3: 91–131.
Wyszomirska, B. 1986. The Nymölla Project. A Middle Neolithic settlement and burial complex in Nymölla, North-East Scania. – Papers of the Archaeol. Inst., Univ. of Lund 1985–1986: 115–138.
Zachrison, A. 1914. Gödsling och jordförbättring i Skåne från 1800-talets början till nuvarande tid. – Lund.
- 1920. Åkerbruksredskap och jordens bearbetning i Skåne 1800–1914. – C. W. K. Gleerups förlag, Lund.
- 1922. Nyodling, torrläggning och bevattning i Skåne 1800–1914. – Lund.
Zucchetto, J. and Jansson, A. M. 1979. Total energy analysis of Gotland's agriculture: a northern temperate zone case study. – Agro-Ecosystems 5: 329–344.
Zvelebil, M. and Rowley-Conwy, P. 1984. Transition to farming in Northern Europe: A hunter-gatherer perspective. – Norw. Archaeol. Rev. 17, No. 2: 104–128.
Ørsnes, M. 1970. Der Moorfund von Ejsbøl bei Hadersleben und die Deutungsprobleme der grossen nordgermanischen Waffenopferfunde. – In: Jankuhn, H. (ed.), Vorgeschichtliche Heiligtümer und Opferplätze in Mittel- und Nordeuropa. Abhandlungen der Akademie der Wissenschaften in Göttingen. Philologisch-historische Klasse 7A. Göttingen, pp. 172–187.
Åkerblom, F. 1981. Historiska anteckningar om Sveriges nötkreatursavel. – N. J. Gumperts Bokhandel. Göteborg.

Papers from the project

Compiled by Gösta Regnéll

As far as possible, a complete list of papers from the project is given. Papers at the moment (spring 1991) in press are included, but papers still in preparation are omitted. Abtracts of papers presented at conferences have been omitted.

Alström, K. 1990. In: Dearing et al. 1990.

Andersson, H. 1989a. Medeltiden och kulturminnesvården. In: Andersson, H. and Anglert, M. (eds), pp. 267–277.

– 1989b. Byn, huvudgården och kyrkan. In: Andersson, H. and Anglert, M. (eds), pp. 281–288.

Andersson, H. and Anglert M. (eds) 1989. By, huvudgård och kyrka. Studier i Ystadsområdets medeltid. – Lund Studies in Medieval Archaeology 5. Almqvist & Wiksell, Stockholm. 305 pp.

Anglert, M. 1989. Den kyrkliga organisationen under äldre medeltid. In: Andersson, H. and Anglert, M. (eds) pp. 221–242.

Bartholin, T. 1989. Dendrokronologiske undersögelser af Ystadsområdets kirker. In: Andersson, H. and Anglert, M. (eds), pp. 211–219.

– and Berglund, B.E. 1991. Reconstruction of the prehistoric landscape. In: Callmer et al. (eds).

Bengtsson, S., Regnéll, G. and Risinger, B. (1984) 1986. Översiktliga vegetationsbeskrivningar av områden kring Ystad. Andra, utökade upplagan. – En arbetsrapport från Kulturlandskapet under 6000 år. Lunds universitet. 27 pp.

Berglund, B.E. 1983. Paleoekologiskt och ekologiskt källmaterial. In: Berglund, B.E., et al., pp. 94–97.

– 1984. Kulturlandskapet under 6000 år – den teoretiska bakgrunden för Ystadsprojektet. In: Teoretiska studier om kulturlandskapets utveckling. Rapport från ett symposium 10–11 november 1983. – Riksbankens jubileumsfond, HSFR och NFR. pp. 20–26.

– 1985a. Det sydsvenska kulturlandskapets förändringar under 6000 år – en presentation av Ystadsprojektet. In: Kulturlandskapet – dess framväxt och förändring. Symposium sept. 1984. (ed. G. Regnéll) – Växtekologiska inst., Lunds univ., pp. 3–8.

– 1985b. Neolitiska expansioner och regressioner i Ystadsområdet. In: Kulturlandskapet – dess framväxt och förändring. Symposium sept. 1984. (ed. G. Regnéll) – Växtekologiska inst., Lunds univ., pp. 41–48.

– 1985c. Early agriculture in Scandinavia: Research problems related to pollen-analytical studies. – Norw. Arch. Rev. 18: 77–105.

– 1986. The cultural landscape in a long-term perspective. Methods and theories behind the research on landuse and landscape dynamics. – Striae 24: 79–87.

– 1987. In: Gaillard, M.-J. and Berglund, B.E. 1987.

– 1988a. The cultural landscape during 6000 years in south Sweden – an interdisciplinary study. – In: Birks, H.H., Birks, H.J.B., Kaland, P.E. and Moe, D. (eds), The cultural landscape – past, present and future. – Cambridge Univ. Press, pp. 241–254.

– 1988b. In: Gaillard, M.-J. and Berglund, B.E. 1988.

– 1991. In: Bartholin, T. and Berglund, B.E. 1991.

– Digerfeldt, G., Gaillard, M.-J., Hjelmroos, M., and Regnéll, J. 1986. In: Det sydsvenska kulturlandskapets förändringar under 6000 år – Ystadsprojektet. En utställning om människan och miljön ur ett tvärvetenskapligt perspektiv. – University of Lund, Institute of Archaeology, Report series. 26., Lund. pp. 3–26.

– and Engelmark, R. 1987. ... där skogen får vika för odlad bygd ... – Forskning och Framsteg 22(4): 12–21.

– and Regnéll, G. 1987. Presentation av Ystadsprojektet. In: Det dolda kulturlandskapet. Metodkonferens 1985. – Riksantikvarieämbetet och Statens historiska muséer. Rapport RAÄ 1987(2): 1–10.

– and Stjernquist, B. 1981. Ystadsprojektet – det sydsvenska kulturlandskapets förändringar under 6000 år. – Luleälvssymposiet 1–3 juni 1981. Skrifter från Luleälvsprojektet 1: 161–182, Umeå.

– Stjernquist, B., Skansjö, S., Persson, A. and Lewan, N. 1983. Ystadprojektet. En översikt av källäget för en landskaps- och samhällshistorisk analys av en sydskånsk bygd. – Rapporter från Barknåreprojektet 1. Individen, samhället och kulturlandskapet. Symposium i Stockholm. 7–9 juni 1982. (ed. B. Roeck Hansen and U. Sporrong) Kulturgeografiska inst., Stockholms univ. pp. 91–112.

Bergman, A. 1990. In: Dearing et al.

Billberg, I. 1985. Medeltidsarkeologiska studier i Baldringeområdet. In: Kulturlandskapet – dess framväxt och förändring. Symposium sept. 1984. (ed. G. Regnéll) – Växtekologiska inst., Lunds univ. 1985. pp. 71–74.

– 1986. Platskontinuitet och bebyggelseutveckling på landsbygden under medeltiden. In: Det sydsvenska kulturlandskapets förändringar under 6000 år – Ystadsprojektet. En utställning om människan och miljön ur ett tvärvetenskapligt perspektiv. – University of Lund, Institute of Archaeology, Report series. 26., Lund. pp. 50–54.

– 1989. Inland – kustland. Bebyggelse och resursutnyttjande för två medeltida byar inom Ystadsprojektet. In: Andersson, H. and Anglert, M. (eds), pp. 15–48.

Callmer, J. 1985. Agrar bebyggelse i backlandskapet vid övergången till historisk tid. In: Kulturlandskapet – dess framväxt och förändring. Symposium sept. 1984. (ed. G. Regnéll) – Växtekologiska inst., Lunds univ., pp. 65–70.

– 1986a. Bebyggelse och landskap under sen järnålder och äldsta medeltid. In: Det sydsvenska kulturlandskapets förändringar under 6000 år – Ystadsprojektet. En utställning om människan och miljön ur ett tvärvetenskapligt perspektiv. – University of Lund, Institute of Archaeology, Report series. 26: 41–49.

– 1986b. Ett adelsnäste i Bjäresjö. – Våra härader. 19. Ystad. pp. 4–7.

– 1986c. To stay or to move. Some aspects of the settlement dynamics in Southern Scandinavia in the seventh to twelfth centuries A.D. with special reference to the province of Scania, Southern Sweden. – Meddelanden från Lunds universitets historiska museum, 1985–86, NS Vol. 6. Lund. pp. 167–208.

– 1987. In: Gaillard, M.-J., Callmer, J., Dearing, J., Regnéll, J. and Sandgren, P.

– 1991a. An Iron Age Cemetery at Bellinga. In: Callmer et al. (eds).

– 1991b. Erosion as an archaeological source-critical problem. In: Callmer et al. (eds).

– 1991c. From Bronze Age dispersed settlement to historical village in the Krageholm area. In: Callmer et al. (eds).

– 1991d. The Bjäresjö-investigation. In: Callmer et al. (eds).
– 1991e. Village excavations in the hummocky landscape: Gussnava and Skårby. B4. In: Callmer et al. (eds).
– , Larsson, L. and Stjernquist, B. (eds) 1991. The Archaeology of the Cultural Landscape. Field work and research in a South Swedish rural region. – Acta Archeol. Lund, Series 4o. Almqvist & Wiksell, Stockholm.
Christophersen, A. 1984. Landsbygrunnleggelse og torpdannelse i Baldringe under tidlig middelalder. – En arbetsrapport från Kulturlandskapet under 6000 år. Lunds universitet. 25 pp.
The cultural landscape during 6000 years. Report from the project Dec 1983. – En arbetsrapport från Kulturlandskapet under 6000 år. Lunds universitet. 29 pp.
Dearing, J.A. 1987a. In: Gaillard et al.
– 1987b. In: Regnéll et al.
– 1990a. In: Gaillard et al.
– , Alström, K., Bergman, A., Regnéll, J. and Sandgren, P. 1990. Recent and Long-term Records of Soil Erosion from Southern Sweden. In: Boardman, J., Foster, I.D.L. and Dearing, J.A. (eds). Soil erosion on agricultural Land. – John Wiley and Sons, Chichester. 173–192.
Digerfeldt, G. 1986. In: Berglund et al.
Dodgshon, R.A. and Olsson, E.G. 1988. Productivity and nutrient use in eighteenth-century Scottish Highland townships. – Geografiska annaler 70 B (1): 39–51.
El-Daoushy, F. 1990. In: Gaillard et al.
Enell, M. 1990. In: Gaillard et al.
Emanuelsson, U. and Möller, J. 1990. Flooding in Scania. A method to overcome the deficiency of nutrients in agriculture during the 19th century. – Agricultural History Review 38.
Engelmark, R. 1987. In: Berglund, B.E. and Engelmark, R. 1987.
– 1989. Makrofossilanalys från Lilla Tvären och Baldringetorp. Preliminära resultat. Appendix (to Billberg 1989). In: Andersson, H. and Anglert, M. (eds), pp. 49–50.
Engelmark, R. and Hjelmqvist, H. 1991. Farming and landuse in the Köpingebro area. In: Callmer et al. (eds).
Exkursionguide 1984–09-22. – Kulturlandskapet under 6000 år. Ett projekt stött av riksbankens jubileumsfond. Lunds universitet. 31 pp.
Exkursionguide 1986–06-16. – Kulturlandskapet under 6000 år. Ett projekt stött av riksbankens jubileumsfond. Lunds universitet. 42 pp.
Gaillard, M.-J. 1983. Water-level changes and climate: A palaeohydrological study of lake Krageholmssjön (Scania, southern Sweden). – Abstracts of papers Severn 1983 (eds Barber and Gregory). Dept. Geography, Univ. of Southampton. 92–94.
– 1984a. Water-level changes, climate and human impact: A palaeohydrological study of Krageholm Lake (Scania, southern Sweden). – In: Mörner, N.A. and Karlén, W. (eds), Climatic changes on a yearly to millennial basis. pp. 147–154. – D. Reidel Publishing Company.
– 1984b. A palaeohydrological study of Krageholmssjön (Scania, South Sweden). Regional vegetation history and water-level changes. – LUNDQUA Report 25: 1–40.
– 1985a. Palaeohydrology, palaeoclimate and cultural landscape – a palaeohydrological study in the context of the Ystad project. In: Kulturlandskapet – dess framväxt och förändring. Symposium sept. 1984. (ed. G. Regnéll) – Växtekologiska inst., Lunds univ., pp. 15–24.
– 1985b. Postglacial palaeoclimatic changes in Scandinavia and central Europe. A tentative correlation based on studies of lake level fluctuations. – Ecologia Mediterranea 11 (1): 159–175.
– 1986. In: Berglund et al.
– 1987. In: Regnéll et al.
– 1988. Palaeohydrological changes and land-use history during the last 6000 years in southern Skåne, Ystad area (Scania, Sweden). In: Binzer, K. et al. (eds): 18. Nordiske

Geologiske Vintermøde, København 1988. Abstracts. – Danmarks Geologiske Undersøgelse. pp. 128–129.
– 1990. Paleohydrologi och bosättningshistoria – forntida vattenståndsförändringar och deras betydelse för kulturlandskapet. – Bebyggelsehistorisk tidskrift. 19.
– and Berglund, B.E. 1987. Land-use history during the last 2700 years in the area of Bjäresjö, S Sweden. – En arbetsrapport från Kulturlandskapet under 6000 år. Lunds universitet. 36 pp.
– and Berglund, B.E. 1988. Land-use history during the last 2700 years in the area of Bjäresjö, southern Sweden. – In: Birks, H.H., Birks, H.J.B., Kaland, P.E. and Moe, D. (eds), The cultural landscape – past, present and future. – Cambridge Univ. Press, pp. 409–428.
– , Dearing, J., Regnéll, J. and Sandgren, P. 1987. Bjäresjö. In: Digerfeldt, G. (ed.), IGCP 158 – Palaeohydrological changes in the temperate zone in the last 15000 years. Symposium at Höör, Sweden. 18–26 May 1987. Excursion guide. LUNDQUA Report 26. 50–60.
– , El-Daoushy, F., Enell, M. and Håkansson, H. 1990. A multidisciplinary study of the lake Bjäresjösjön (S. Sweden): land-use history, soil erosion, lake trophy and lake-level fluctuations during the last 3000 years. – Hydrobiologia 214: 107–114.
Germundsson, T. 1984a. Egnahem som småbruk. – Kulturgeogr. inst., Lunds universitet, stencilerad sem.uppsats. 66 pp.
– 1984b. Kulturlandskapets förändring. Ett diskussionsinlägg med tonvikten på 1800- och 1900-talen. – En arbetsrapport från Kulturlandskapet under 6000 år. Lunds universitet. 9 pp.
– 1985a. Brukare och markägare i Bjäresjö sedan 1915. En studie av förändringar i en sydskånsk församling. – Rapporter och Notiser 79. Inst. f. Kulturgeogr. och Ekon. geogr., Lund. 18 pp.
– 1985b. Jordbruksegnahem i Sydskåne. – Ale 1985(2): 24–32.
– 1986a. 1800-talet och 1900-talet. In: Det sydsvenska kulturlandskapets förändringar under 6000 år – Ystadsprojektet. En utställning om människan och miljön ur ett tvärvetenskapligt perspektiv. – University of Lund, Institute of Archaeology, Report series. 26: 63–65.
– 1986b. Population, landownership and the landscape. Changes in the Ystad-region in southernmost Sweden during the 19th and 20th centuries. – Revue Espace – Populations – Sociétés 1986(3): 126–130.
– 1987. Population, landholding and the landscape. – En arbetsrapport från Kulturlandskapet under 6000 år. Lunds universitet. 33 pp.
– 1989. Den svenska egnahemsrörelsen och småbruksbildningen i Skåne. In: Huse og husmaend i fortid, nutid og fremtid – småbrugets udbredelse og vilkor i Norden. Beretning fra 12. beyggelsehistoriske symposium ved Odense universitet. ed. Damgaard, H. and Möller, P.G. – Odense Universitetsforlag.
– , Lewan, N. and Möller, J. 1985. Ystadsprojektet och domänstrukturen. – Bebyggelsehistorisk tidskrift 1985(9): 20–32.
– and Möller, J. 1987a. Redskap och maskiner i Nöbbelöv och Villie – en longitudinell studie av två sydskånska socknar under 1800-talet. – En arbetsrapport från Kulturlandskapet under 6000 år. Lunds universitet. 28 pp.
– and Möller, J. 1987b. Åkerns redskap i förändring. En studie av två sydskånska socknar under 1800-talet. – Ale 1987(1): 14–26.
– and Persson, A. 1987. Odling och djurhållning i Ystadsområdet på 1800-talet. – En arbetsrapport från Kulturlandskapet under 6000 år. Lunds universitet. 124 pp.
Göransson, H. 1991. Vegetation and man around Lake Bjärsjöholmssjön during prehistoric time. – LUNDQUA Report. 44 pp.
Hjelmqvist, H. 1985. Economic Plants from Two Stone Age

Settlements in Southernmost Scania. In: Larsson, L. Karlsfält. A Settlement from the Early and Late Funnel Beaker
Culture in Southern Scania, Sweden. – Acta Archaeologica. Vol. 54: 57–63.
– 1991. In: Engelmark, R. and Hjelmqvist, H.
Hjelmroos, M. 1985a. Vegetational history of Fårarps mosse,
South Scania, in the Early Subboreal. – Acta Archaeologica 55.
– 1985b. Mänsklig miljöpåverkan runt Fårarps mosse, Skåne.
Preliminära resultat från en paleoekologisk undersökning.
In: Kulturlandskapet – dess framväxt och förändring. Symposium sept. 1984. (ed. G. Regnéll) – Växtekologiska inst.,
Lunds univ. 1985. pp. 25–32.
– 1986. In: Berglund et al.
Hulthén, B. 1985a. Keramiken från Karlsfält. En teknologisk
studie. In: Kulturlandskapet – dess framväxt och förändring. Symposium sept. 1984. (ed. G. Regnéll) – Växtekologiska inst., Lunds univ. 1985. pp. 39–40.
– 1985b. The pottery from Karlsfält. In: Larsson, L. Karlsfält. A Settlement from the Early and Late Funnel Beaker
Culture in Southern Scania, Sweden. – Acta Archaeologica. Vol. 54: 50–57.
Håkansson, H. 1987. In: Regnéll et al.
– 1990. In: Gaillard et al.
Håkansson, H. and Kolstrup, E. 1987. Early and Middle Holocene developments in Herrestads Mosse (Scania, South
Sweden). Part I. Diatom analysis and vegetational deve-
lopment. Lundqua Report 28. 28 pp.
Ihse, M. and Lewan, N. 1986. Odlingslandskapets föränd-
ringar på Svenstorp studerade i flygbilder från 1940-talet
och framåt. – Ale 1986(2): 1–17.
Jonsson, L. 1985. Analysis of the Fauna from the Karlsfält
Settlement, St. Herrestad Parish, Scania. In: Larsson, L.
Karlsfält. A Settlement from the Early and Late Funnel
Beaker Culture in Southern Scania, Sweden. – Acta
Archaeologica. Vol. 54: 64.
– 1986. Animal bones from Bredasten – preliminary results.
Appendix to: Larsson, M. 1986 (see below). pp. 50–51.
– 1991. Osteological analyses of the site finds from the Kö-
pingebro area. In: Callmer et al. (eds).
Knutsson, H. 1985. Functional analysis of flint tools from
Karlsfält, Southern Scania. In: Larsson, L. Karlsfält. A
Settlement from the Early and Late Funnel Beaker Culture
in Southern Scania, Sweden. – Acta Archaeologica. Vol.
54: 64–71.
Kolstrup, E. 1987. In: Håkansson, H. and Kolstrup, E.
– 1990. Early and middle Holocene vegetational development in Kurarp (Scania, South Sweden). – Review of Pa-
laeobotany and Palynology 63: 233–257.
Larsson, L. 1982. Stenåldern efterlyses. – Våra härader.
(Ystad) 1982: 17–26.
– 1984a. Kulturlandskapet under neolitisk tid. Några aspekter på ekologiska förutsättningar och kulturella förhållanden. – En arbetsrapport från Kulturlandskapet under 6000
år. Lunds universitet. 24 pp.
– 1984b. Spår efter Ystadtraktens äldsta innevånare. – Ystadiana. Ystads fornminnesförenings skrift. 97–105.
– 1985. Karlsfält. A settlement from the Early and Late Funnel Beaker culture in Southern Scania, Sweden. – Acta
Archaeologica Vol. 54: 3–57.
– 1986a. Archaeological survey mapping. In: Berglund, B.E.
(ed) Handbook of Holocene palaeoecology and palaeohydrology. – Wiley. pp. 219–228.
– 1986b. Skåne under sen trattbägarkultur. In: Adamsen, C.
and Ebbesten, K. (eds). Stridsøksetid i Sydskandinavien. –
Arkæologiske Skrifter Nr. 1: 146–155.
– 1987a. Ett gott liv på stenåldern. – Forskning och Framsteg
22(2): 32–38.
– 1987b. En fremmed fugl. Våra härader. (Ystad) : 23–28.
– 1987c. Hus över stenåldersgrav. En grav från båtyxekultur
vid Ullstorp, södra Skåne. – Ystadiana. Ystads fornminnesförenings skrift. 32: 59–75.

– 1987d. Skolästkilar och trädklyvning. Några funderingar
kring ett mellanskånskt fynd från jägarstenåldern. – Frosta-
bygden pp. 5–10.
– 1987e. Some aspects of cultural relationship and ecological
conditions during the Late Mesolithic and Early Neolithic.
– In: Burenhult, G., Carlsson, A., Hyenstrand, Å. and
Sjövold, T. Theoretical Approaches to Artefacts, Settlement and Society. – BAR International Series 366: 165–
176.
– 1988a. Mortuary Building above Stone Age Grave. A
Grave from the Battle Axe Culture at Ullstorp, Southern
Scania, Sweden. – Papers of the Archaeological Institute,
University of Lund 1987–1988. New Series Vol. 7: 81–98.
– 1988b. The use of the landscape during the Mesolithic and
Neolithic in Southern Sweden. – In: van Zeist, W. (ed.)
Archeologie en Landschap. Bijdragen aan het gelijknamige
symposium gehouden op 19 en 20 oktober 1987, ter gelegenheid van het afscheid van H.T. Waterbolk. Groningen.
pp. 31–48.
– 1989a. Boplatser, bebyggelse och bygder. Stridsyxekultur i
södra Skåne. In: Larsson, L. (ed.) Stridsyxekultur i Sydskandinavien. – University of Lund, Institute of Archaeology, Report Series 36: 53–76.
– 1989b. Brandopfer. Der frühneolitische Fundplatz Svartskylle im Südlichen Schonen, Sweden. – Acta Archaeologica. Vol 59: 143–153.
– 1989c. Brännoffer. En tidigneolitisk fyndplats med brända
flintyxor från södra Skåne. In: Larsson, L. and Wyszomirska, B. (eds) Arkeologi och religion. – University of Lund,
Institute of Archaeology, Report Series 34: 89–97.
– 1989d. Långaker. En boplats från en bondestenålder i förvandling. – Gamla Trelleborg Årsbok pp. 6–20.
– 1990a. Vikingatida öbosättning. – Våra härader 1990. Ljunits och Herrestads hembygdsförening 23: 9–13.
– 1990b. Sydskånsk öbosättning. Boplatslämningar från jägarstenålder och sen vikingatid-tidig medeltid vid Ellestadssjön. – Ale. Historisk tidskrift för Skåneland 1989(4):
1–11.
– 1991a. Final remarks. The Ystad project and archaeological
research. In: Callmer et al. (eds).
– 1991b. Settlement and environment during a late part of the
Middle Neolithic and Late Neolithic. B2. In: Callmer et al.
(eds).
– in press. Eine Küstensiedlung der frühen Trichterbecherkultur bei Rävgrav in Südschonen, Schweden. In: Hoika, J.
(ed.) Frühe Trichterbecherkultur in Nordeuropa. Schleswig.
– and Larsson, M. 1984. Flintyxor, skoskav och massor av
stolphål. Resultaten av arkeologisk inventering och under-
sökningsverksamhet under de senaste två åren i Ystadsområdet. – Ystadiana. Ystads fornminnesförenings skrift.
XXIX: 9–95.
– and Larsson, M. 1986. Stenåldersbebyggelse i Ystadsområdet. En presentation av fältverksamhet och bearbetning hösten 1984-våren 1986. – Ystadiana. Ystads fornminnesförenings skrift. XXXI: 9–78.
Larsson, M. 1984. In: Larsson, L. and Larsson, M.
– 1985a. Neolitikum i Ystadsområdet. In: Kulturlandskapet –
dess framväxt och förändring. Symposium sept. 1984. (ed.
G. Regnéll) – Växtekologiska inst., Lunds univ., pp. 33–38.
– 1985b. The Early Neolithic Funnel Beaker Culture in
South-west Scania, Sweden. – British Archaeological Reports 264: 1–146.
– 1986a. Tidiga bönder i Ystadsområdet. In: Det sydsvenska
kulturlandskapets förändringar under 6000 år – Ystadsprojektet. En utställning om människan och miljön ur ett tvärvetenskapligt perspektiv. – University of Lund, Institute of
Archaeology, Report series. 26: 32–35.
– 1986b. Bredasten – An early Ertebölle site with a dwelling
structure in South Scania. – Meddelanden från Lunds Universitets historiska museum 1985–1986: 25–49.
– 1986c. In: Larsson, L. and Larsson, M.

– 1987a. Gravplats, boplats, åker. Ett exempel på kulturlandskapets utnyttjande kring Köpingebro i sydligaste Skåne. – Ale 1987(3): 1–14.
– 1987b. Neolithization in Scania – A Funnel Beaker perspective. – Journal of Danish Archaeology. Vol 5: 244–247.
– 1987c. Människor vid en havsvik. Stenåldersboplatser vid Kabusa, St. Köpinge socken. – Ystadiana. Ystads fornminnesförenings skrift. 32: 13–58.
– 1987d. From hunters to farmers. The evolution of the Early Neolithic TRB culture in Scania, Sweden. – The Neolithic and Early Bronze Age in the Chelmno Land. The materials from the international symposium, 11–13 XI 1986, Torun (ed. T. Wislanski). Torun 1987: 323–348.
– 1987e. Förändringar i bosättningsmönster under neolitikum i Ystadsområdet. Ekonomiska och sociala faktorer. In: Det dolda kulturlandskapet. Metodkonferens 1985. – Riksantikvarieämbetet och Statens historiska muséer. Rapport RAÄ 1987(2): 47–68.
– 1988a. Exchange and society in the Early Neolithic in Scania, Sweden. In: Trade and exchange in prehistory. Studies in honour of Berta Stjernquist. – Acta Archaeologica Lundensia series in 8o. N: o 16: 49–59.
– 1988b. Megaliths and Society. The development of social territories in the South Scanian TRB Culture. – Meddelanden från Lunds universitets historiska museum. 1987–1988. 9: 19–39.
– 1988c. Stenåldersbondens hus. – Populär Arkeologi 1988 (4): 10–13.
– 1989. Stenåldersjägare i Ystads sandskog. – Ystadiana. Ystads fornminnesförenings skrift. XXXIV: 21–28.
– 1991. The Funnel Beaker culture during Early and Middle Neolithic. Economic and social changes. B1. In: Callmer et al. (eds).
– in press. Lokale Gruppen des Frühneolithikums Südschonens. Symposiumbericht. In: Hoika, J. (ed.) Frühe Trichterbecherkultur in Nordeuropa. Schleswig.
– in press. Settlement sites from the Early and Middle Neolithic at Kabusa, southern Scania. In: Middle Neolithic TRB culture in north and central Europe. Poznan.
– and Olausson, D. 1986. Bönder och handelsmän under vikingatiden vid Mossby i sydligaste Skåne. – Ystadiana. Ystads fornminnesförenings skrift. XXXI: 79–104.
– and Olausson, D. 1991. Field survey: methods and problems. In: Callmer et al. (eds).
– and Olsson, E.G. in press. Stenåldersböndernas landskap i Sydskåne. – Forskning och Framsteg.
Lewan, N. 1983a. Slätten som natur- och kulturresurs – en översikt. – Sv. Geogr. Årsbok 59: 16–33. (Abstract in English).
– 1983b. Ystadsområdet idag. Iakttagelser – intervjuer – intryck. – En arbetsrapport från Kulturlandskapet under 6000 år. Lunds universitet. 9 pp.
– 1983c. Kulturgeografiskt källmaterial efter 1800, särskilt kartmaterialet. – In: Berglund et al. pp. 111–112.
– 1985a. Mackleanskiftet 200 år. – Sv. Geogr. Årsbok 61: 253–254.
– 1985b. Jordbrukspolitikens landskap. In: Kulturlandskapet – dess framväxt och förändring. Symposium sept. 1984. (ed. G. Regnéll) – Växtekologiska inst., Lunds univ., pp. 97–100.
– 1985c. Kulturlandskapet och det moderna samhället. – In: U. Emanuelsson m fl: Det skånska kulturlandskapet. pp. 139–152.
– 1985d. In: Germundsson et al.
– 1986a. Jordbrukspolitikens landskap. – Ymer 106: 182–193.
– 1986b. Vad kan forskningen ge kulturminnesvården: Ystadsprojektet. – RAÄ: s konferens Landskaps- och bebyggelseförändringar under 1800-talet. (Lidingö, sept. 1985) pp. 14–22.
– 1986c. Ystadsprojektet – det långa perspektivet och det kortare. In: Symposium nov 1985. – Skrifter fra Historisk Institut, Odense Universitet. 35: 54–63.
– 1986d. Historisk atlas, som visar utvecklingen 1815–1915-1985 i ett tjugotal områden i och omkring tätorter i sydvästra Skåne. (textförf.). – Berlings, Arlöv 52 pp.
– 1986e. In: Ihse, M. and Lewan, N.
– 1987a. Ystadsprojektet som regionstudium. – Ymer 107: 138–144.
– 1987b. Le paysage rural suédois. Remarques générales et l'example de la région d'Ystad. – Hommes et Terres du Nord 1987(2): 101–107.
– 1988a. Kring Marsvinsholm förr och nu. – Geografisk Orientering. Brenderup. 3: 344–349.
– 1988b. Contradictory planning – against and in favour of rural change in Sweden. – Geo Regards. Neuchatel. No 15: 149–157.
– and Mårtensson, E. 1985. Slottsparker i södra Skåne under 1900-talet. Förminskning, förenkling, förfall. – Byggnadskultur. 4: 11–15.
– and Möller, J. 1985. Byarna i Lundatrakten förr och nu. – Årsredovisning för Lundabygdens sparbank 1984. Arlöv. pp. 1–27.
Lindersson, H. and Thuning, B. 1986. Stratigrafi och överplöjning i Bussjöområdet. – Examensarbeten i geologi vid Lunds universitet. 15. 65 pp.
Malmer, N. 1984. Förändringskrafterna i kulturlandskapet ur naturvetenskaplig synvinkel. In: Teoretiska studier om kulturlandskapets utveckling. Rapport från ett symposium 10–11 november 1983. – Riksbankens jubileumsfond, HSFR och NFR. pp. 49–59.
– and Regnéll, G. 1986. Mapping present and past vegetation. – In: Berglund, B.E. (ed) Handbook of Holocene palaeoecology and palaeohydrology. Wiley, pp. 203–218.
Müller-Wille, M. 1984. Siedlungsarchäologische Forschungsprojekte in Schweden. Ein Bericht. – Praehistorische Zeitschrift 59(2): 145–187.
Mårtensson, E. 1984. Förenkling och förfall, en studie av skånska godsmiljöer under 1900-talet. – Ale 1984(4): 8–15.
– 1985a. Parkanläggningar i Ljunits och Herrestads härader under 1900-talet. – En arbetsrapport från Kulturlandskapet under 6000 år. Lunds universitet. 28 pp.
– 1985b. In: Lewan, N. and Mårtensson, E.
Möller, J. 1983. Gods, gård och landskap. Domänstruktur och kulturlandskap i Ljunits och Herrestads härader, Malmöhus län, under 1800- och 1900-talen. – Inst. f. kulturgeografi och ekon. geografi. Rapporter och Notiser 73. 56 pp. (Summary in English).
– 1984. Dikning i Skåne. – Ale 1984(2): 14–28.
– 1985a. Domänstruktur och landskapsutveckling. In: Kulturlandskapet – dess framväxt och förändring. Symposium sept. 1984. (ed. G. Regnéll) – Växtekologiska inst., Lunds univ., pp. 91–96.
– 1985b. The landed estate and the landscape: Landownership and the changing landscape of southern Sweden during the 19th and 20th centuries. – Geogr. Annaler 67B 1985(1): 45–52.
– 1985c. In: Germundsson et al.
– 1985d. In: Lewan, N. and Möller, J.
– 1986. Den skånska plattgården – exempel på agrar stordrift i södra Sverige. – Ymer 106: 161–170.
– 1987a. The landed estate and the railway. The introduction of a new means of transport in Southern Sweden. – The Journal of Transport History. 3rd series, vol. 8: 147–163.
– 1987b. In: Germundsson, T. and Möller, J. 1987a.
– 1987c. In: Germundsson, T. and Möller, J. 1987b.
– 1989a. Godsen och den agrara revolutionen. Arbetsorganisation, domänstruktur och kulturlandskap på skånska gods under 1800-talet. Lund. – Lund University Press. 153 pp.
– 1989b. Godsorganisation i förändring: statsystemets uppkomst i Skåne. – Bebyggelsehistorisk tidskrift. 17: 43–58.
– 1990. In: Emanuelsson, U. and Möller, J.
Nilsson, H. 1983. Boplats 2. En stenåldersboplats i Ystads Sandskog. 60-poängsuppsats i arkeologi. – Institutionen för arkeologi, Lunds universitet, HT 1983. (Stencil) 40 pp.

Nilsson, M. and Pröjts, J. 1986. Markhistorisk studie över Bedens fuktängar. – En arbetsrapport från Kulturlandskapet under 6000 år. Lunds universitet. 22 pp.

Olausson, D. 1986a. Kulturlandskapet under brons- och järnålder inom den västliga delen av Ljunits och Herrestads härader. – In: Det sydsvenska kulturlandskapets förändringar under 6000 år – Ystadsprojektet. En utställning om människan och miljön ur ett tvärvetenskapligt perspektiv. – University of Lund, Institute of Archaeology, Report series 26: 36–40.
– 1986b. In: Larsson, M. and Olausson D.
– 1987. Piledal and Svarte. A Comparison between two Late Bronze Age Cemeteries in Scania. – Acta Archaeologica. København Vol. 57/1986: 121–152.
– 1988a. The cultural landscape during the Bronze and Early Iron Ages in the districts of Ljunits and western Herrestad – a microstudy in the project "The cultural landscape during 6000 years". – Finska Fornminnesföreningens ISKOS. 7: 291–298.
– 1988b. Where have all the settlements gone? Field survey methods for locating Bronze and Iron Age settlements in a cultivated landscape. – Meddelanden från Lunds universitets historiska museum 1987–1988 (Papers of the Archaeological Institute, University of Lund 1987–1988). New Series Vol. 7: 99–112.
– 1991a. The archaeological work in the western part of the project area – research goals, methods and results. In: Callmer et al. (eds).
– 1991b. In: Larsson, M. and Olausson, D.
– in press. Comparison and contrast. Looking for evidence for regional groups in two late Bronze Age cemeteries. Proceedings, 5th Nordic Bronze Age Symposium 1987.

Olsson, E.G. 1985. Resursutnyttjande inom Bjäresjö by under 1700-talet. In: Kulturlandskapet – dess framväxt och förändring. Symposium sept. 1984. (ed. G. Regnéll) – Växtekologiska inst., Lunds univ. 1985. pp. 87–90.
– 1988a. Nutrient use and productivity for different cropping systems in South Sweden during the 18th century. – In: Birks, H.H., Birks, H.J.B., Kaland, P.E. and Moe, D. (eds), The cultural landscape – past, present and future. – Cambridge Univ. Press, pp. 123–138.
– 1988b. Nutrient use efficiency as a tool for evaluating pre-industrial agroecosystems. – Geografiska Annaler 70 B (1): 69–73.
– 1988c. In: Dodgshon, R.A. and Olsson, E.G.
– 1989a. Odlingslandskapet i Ystadsområdet under medeltiden. Rekonstruktion av landskapstyper och vegetation kring tidigt 1300-tal In: Andersson, H. and Anglert, M. (eds), pp. 51–55.
– 1989b. Åkerbruket i Ystadsområdet under medeltiden. In: Andersson, H. and Anglert, M. (eds), pp. 57–60.
– in press. In: Larsson, M. and Olsson, E.G. (in press).
– och Persson, A. 1987. Skånejordbruk på 1700-talet: Varför så olika på åsen och på slätten? – Forskning och framsteg 22(7): 12–18.

Persson, A. 1983. Historiskt källmaterial från 1600- och 1700-talen, särskilt den skånska tiondekommissionens handlingar 1681–1682. In: Berglund pp. 109–110.
– 1985. Teori, metod, problem: erfarenheter från en historisk undersökning av Bjäresjö socken ca 1690–1800. – In: Kulturlandskapet – dess framväxt och förändring. Symposium sept. 1984. (ed. G. Regnéll) – Växtekologiska inst., Lunds univ., pp. 81–86.
– 1986a. Villie socken c: a 1700. Bebyggelse – markanvändning – social struktur. – En arbetsrapport från Kulturlandskapet under 6000 år. Lunds universitet. 62 pp.
– 1986b. Agrar struktur i Bjäresjö by 1570 och 1699. En trendanalys utförd med ADB-stöd. – En arbetsrapport från Kulturlandskapet under 6000 år. Lunds universitet. 64 pp.
– 1986c. Kulturlandskapet från sent 1600-tal till skiftena under 1800-talet. – In: Det sydsvenska kulturlandskapets förändringar under 6000 år – Ystadsprojektet. En utställning

om människan och miljön ur ett tvärvetenskapligt perspektiv. – University of Lund, Institute of Archaeology, Report series. 26: 59–62.
– 1987a. Jordrevningen 1670–1672 och Adlerstens indelningsverk 1691–1693. In: Skånsk släktforskning. Jubileumsskrift 1937–87. – Skånes Genealogiska Förening. pp. 65–77.
– 1987b. Aspects on land use in Bussjö Village in Bromma parish and Herrestad hundred in Scania c: a 1670–1910. – En arbetsrapport från Kulturlandskapet under 6000 år. Lunds universitet. 20 pp.
– 1987c. In: Olsson, E.G. and Persson, A.
– 1987d. In: Germundsson, T. and Persson, A.
– 1988a. Agrar struktur i Lilla Tvärens by c: a 1700. – En arbetsrapport från Kulturlandskapet under 6000 år. Lunds universitet. 13 pp.
– 1988b. Jordbruk på Romeleåsen i äldre tider. – Skånes Natur 1988(4): 281–287.
– 1989. Levnadsvillkor på skånska gatuhus cirka 1700–1800: strukturer och utvecklingsdrag. In: Huse og husmænd i fortid, nutid og fremtid – småbrugets udbredelse og vilkor i Norden. Beretning fra 12. bebyggelseshistoriske symposium ved Odense universitet. (Ed. Damgaard, H. and Møller, P.G.). Odense Universitetsforlag.

Pröjts, J. 1986. In: Nilsson, M. and Pröjts, J.

Regnéll, G. 1985. Vegetationsrekonstruktioner – syfte och möjligheter. In: Kulturlandskapet – dess framväxt och förändring. Symposium sept. 1984. (ed. G. Regnéll) – Växtekologiska inst., Lunds univ., pp. 49–54.
– 1986a. Landskapets vegetation förr och nu. In: Det sydsvenska kulturlandskapets förändringar under 6000 år – Ystadsprojektet. En utställning om människan och miljön ur ett tvärvetenskapligt perspektiv. – University of Lund, Institute of Archaeology, Report series. 26: 27–31.
– (1984) 1986b. In: Bengtsson et al.
– 1986c. In: Malmer, N. and Regnéll, G.
– 1987a. Växtekologiska rekonstruktioner. In: Det dolda kulturlandskapet. Metodkonferens 1985. – Riksantikvarieämbetet och Statens historiska muséer. Rapport RAÄ 1987 (2): 11–18.
– 1987b. In: Berglund, B.E. and Regnéll. G.

Regnéll, J. 1984. Regional vegetationsutveckling i Ystadsområdet. Preliminär redovisning. – En arbetsrapport från Kulturlandskapet under 6000 år. Lunds universitet. 30 pp.
– 1985. Odlingslandskapets utveckling kring Krageholmssjön i södra Skåne. Preliminär redovisning av en kvartärgeologisk undersökning. In: Kulturlandskapet – dess framväxt och förändring. Symposium sept. 1984. (ed. G. Regnéll) – Växtekologiska inst., Lunds univ., pp. 9–14.
– 1986. In: Berglund et al.
– 1987. In: Gaillard et al.
– 1989. Vegetation and land use during 6000 years. Palaeoecology of the cultural landscape at two lake sites in southern Skåne, Sweden. – Lundqua Thesis 27. 61 pp.
– 1990. In: Dearing et al.

Regnéll, J., Gaillard, M.-J., Håkansson, H. and Dearing, J. 1987. Krageholmssjön. – In: Digerfeldt, G. (ed.): IGCP 158 – Palaeohydrological changes in the temperate zone in the last 15000 years. Symposium at Höör, Sweden. 18–26 May 1987. Excursion guide. Lundqua Report 26: 39–49.

Reisnert, A. 1989a. Borgar och huvudgårdar i Danmark och Skåne. Utveckling och forskningshistorik. In: Andersson, H. and Anglert, M. (eds), pp. 65–70.
– 1989b. Huvudgårdar i Ystadsområdet. Några medeltidsarkeologiska synpunkter. In: Andersson, H. and Anglert, M. (eds), pp. 145–149.
– 1989c. Ystads franciskanerkloster och aristokratin. In: Andersson, H. and Anglert, M. (eds), pp. 151–163.
– 1989d. In: Skansjö et al.

Riddersporre, M. 1983. Kabusa. Exempel på retrospektiv kulturgeografi i Sydskåne. 60-poängsuppsats i kulturgeografi. – Institutionen för kulturgeografi och ekonomisk geografi vid Lunds universitet. HT 1983. 35 pp, 4 bilagor. (Stencil).

– 1985. Lantmäterihandlingar som källmaterial vid retrospektiva bebyggelse- och kulturlandskapshistoriska studier. – Land og by i middelalderen 3. Symposium Skanör 1984, pp. 103–115.
– 1986. Fragment av medeltidens kulturlandskap – lantmäterihandlingar som källa till studiet av samhällsförändringar. – Ymer. 106: 52–62.
– 1987. Retrogressiv analys av äldre lantmäterihandlingar. In: Det dolda kulturlandskapet. Metodkonferens 1985. – Riksantikvarieämbetet och Statens historiska muséer. Rapport RAÄ 1987(2): 19–46.
– 1988. Settlement site – village site. Analysis of the toft-structure in some medieval villages and its relation to Late Iron Age settlements. A preliminary report and some tentative ideas based on Scanian examples. – Geografiska Annaler 70 B (1): 75–85.
– 1989a. Lantmäterihandlingar, runstenar och huvudgårdar. Några kommentarer och spekulationer i ett lokalt geografiskt perspektiv. – In: Andersson, H. and Anglert, M. (eds), pp. 135–144.
– 1989b. In: Skansjö et al.
– 1990. Settlement site – village site. Analysis of the toft-structure in some medieval villages and its relation to late Iron Age settlements. A preliminary report and some tentative ideas based on Scanian examples. – In: Sporrong, U. (ed.) The transformation of rural society, economy and landscape. Papers from the 1987 meeting of The Permanent European Conference for the Study of the Rural Landscape. – Meddelanden serie B 71. Department of Human Geography, Stockholm University, pp. 85–95.
– 1991a. Eighteenth century landscape and archaeology. In: Callmer et al. (eds).
– 1991b. Sugar-beet cropping and aerial photography in the service of archoeology. – In: Callmer et al. (eds).
Risinger, B. (1984) 1986. In: Bengtsson et al.
Sandgren, P. 1987. In: Gaillard et al.
– 1990. In: Dearing et al.
Skansjö, S. 1980. PM angående ett planerat kulturlandskapsprojekt i Sydskåne. – Julittaseminariet 9–10 juni 1980 kring kulturlandskapsforskningens metodfrågor, anordnat av Kungl. Vitterhets-, Historie- och Antikvitetsakademien. pp. 32–34.
– 1982. Kring kulturlandskapet vid mitten av 1600-talet. Tabellsammanställning utifrån decimantjordeboken 1651. – Kulturlandskapet under 6000 år. Seminarium 1982–11-19. Lunds universitet. 7 pp.
– 1983. Historiskt källmaterial från medeltiden och 1500-talet. In: Berglund et al. pp. 106–108.
– 1985a. Godsbildning, bebyggelseutveckling och resursutnyttjande 1550–1680. Problem och exempel. In: Kulturlandskapet – dess framväxt och förändring. Symposium sept. 1984. (ed. G. Regnéll) – Växtekologiska inst., Lunds univ., pp. 75–80.
– 1985b. Jordägare och bebyggelsemönster i Ystadsområdet vid mitten av 1600-talet. – En arbetsrapport från Kulturlandskapet under 6000 år. Lunds universitet. 53 pp.
– 1985c. Estate building and settlement changes in S. Scania c. 1500–1650 in a European perspective. – En arbetsrapport från Kulturlandskapet under 6000 år. Lunds universitet. 20 pp.
– 1986a. 6000 Jahre schonische Kulturlandschaft: das Ystad-Projekt. In: Schonen als Brennpunkt skandinavischer Geschichte. Zusammenfassungen der Vorträge zum gleichnamigen Colloquium von 26. und 27. April 1985. Zentrum für Nordische Studien (ZNS), Kiel. pp. 9–13.
– 1986b. Från vikingatida stormansgård till renässansslott. Några huvuddrag i Bjersjöholms äldsta historia. – En arbetsrapport från Kulturlandskapet under 6000 år. Lunds universitet. 82 pp.

– 1986c. Storgodsdrift och bebyggelseexpansion från 1500-talets mitt. – In: Det sydsvenska kulturlandskapets förändringar under 6000 år – Ystadsprojektet. En utställning om människan och miljön ur ett tvärvetenskapligt perspektiv. – University of Lund, Institute of Archaeology, Report series. 26: 55–58.
– 1987a. Estate building and settlement changes in southern Scania c. 1500–1650 in a European perspective. – In: H.-J. Nitz (ed.) The medieval and early-modern rural landscape of Europe under the impact of the commercial economy. – Göttingen. pp. 105–114.
– 1987b. Aspekter på Ystadsområdet under medeltiden. Historisk kasuistik med kommentarer. A. Ljunits härad. – En arbetsrapport från Kulturlandskapet under 6000 år. Lunds universitet. 53 pp.
– 1988. Aspekter på Ystadsområdet under medeltiden. Historisk kasuistik med kommentarer. B. Herrestads härad. – En arbetsrapport från Kulturlandskapet under 6000 år. Lunds universitet. 89 pp.
– 1991. Från vikingatida stormansgård till renässansslott. Bjersjöholm under äldre historisk tid. – In: Johansson, S. et al., Bjersjöholm. Storgods och slott. Skånsk senmedeltid och renässans 13. Vetenskapssocieteten i Lund.
– Riddersporre, M. and Reisnert, A. 1989. Huvudgårdarna i källmaterialet. – In: Andersson, H. and Anglert, M. (eds) pp. 71–133.
Stjernquist, B. 1981. In: Berglund, B.E. and Stjernquist, B.
– 1983. Arkeologiskt källmaterial. – In: Berglund et al. pp. 98–105.
– 1984. Teoretiska utgångspunkter för studiet av kulturlandskapets förändringar. In: Teoretiska studier om kulturlandskapets utveckling. Rapport från ett symposium 10–11 november 1983. – Riksbankens jubileumsfond, HSFR och NFR. pp. 83–95.
– 1985. Approaches to the problem of settlement patterns in eastern Scania in the first millennium B.C. – In: Settlement and Society: aspects of West European prehistory in the first millennium B.C. (ed. T.C. Champion and J.V.S. Megaw) – Leicester, pp. 223–238.
– 1987. Die siedlungsarchäologische Forschung. Generelle Probleme und anschauliche Beispiele aus Südschweden. – Studien zur Sachsenforschung. (Hildesheim) Bd 6: 237–250.
– 1991. The importance of the Ystad project for contemporary land planning. In: Callmer et al. (eds).
Sundnér, B. 1989. Medeltida byggnadsteknik. En kronologisk studie av Ystadsområdets kyrkor. – In: Andersson, H. and Anglert, M. (eds) 1989. pp. 183–210.
Tesch, S. 1985a. Vad gömmer åkern? En fråga om det arkeologiska materialets representativitet. In: Kulturlandskapet – dess framväxt och förändring. Symposium sept. 1984. (ed. G. Regnéll) – Växtekologiska inst., Lunds univ. pp. 49–54.
– 1985b. "Vilda Östern" – Bebyggelsekolonisation, landsbybildning och urbanisering i Ystadsområdet under vikingatid och äldre medeltid. – Land og by i middelalderen 3. Symposium Skanör 1984. pp. 75–96.
– 1988. En slättbygd vid kusten under brons- och järnålder – arkeologiska undersökningar inom ramen för Ystadsprojektet (B3). – Ystadiana. Ystads fornminnesförenings skrift 33: 189–230.
– 1991a. House, farmstead and settlement structure during the Bronze and Iron Age. In: Callmer et al. (eds).
– 1991b. The archaeological work in the eastern part of the project area – research goals, methods and results. In: Callmer et al. (eds).
Thuning, B. 1986. In: Lindersson, H. and Thuning, B.
Wienberg, J. 1989. Kirkene og befolkningen i Ystadsområdet. – In: Andersson, H. and Anglert, M. (eds), pp. 243–264.

Authors and project leaders

Hans Andersson, professor of Medieval Archaeology at Lund University. Leader of subproject B 5 (1987–), main editor of the monograph presenting the results of B 5. – Dept of Archaeology, Krafts torg 1, S-22350 Lund.

Anders Andrén, docent in Medieval Archaeology at Lund University. Leader of subproject B5 (1986–87). – Dept of Archaeology, Krafts torg 1, S-22350 Lund.

Siv Bengtsson-Lindsjö, graduate student in Plant Ecology at Lund University. Research in subproject A 5. – Dept of Plant Ecology, Östra Vallgatan 14, S-22361 Lund.

Björn E. Berglund, professor of Quaternary Geology at Lund University. Leader of subprojects A 1–2, main coordinator of the Ystad project. – Dept of Quaternary Geology, Tornavägen 13, S-22363 Lund.

Johan Callmer, docent in Archaeology at Lund University. Leader of subproject B 4. Co-editor of the monograph for B 1–4. – Dept of Archaeology, Krafts torg 1, S-22350 Lund.

Eric Cinthio, professor emeritus of Medieval Archaeology at Lund University. Leader of subproject B 5 (–1986). – Dept of Archaeology, Krafts torg 1, S-22350 Lund.

Alan Crozier, translator, linguistic consultant to the Ystad project. – Skatteberga, Pl. 36, S-24017 Södra Sandby.

John Dearing, Dr, lecturer in Physical Geography at Coventry Polytechnic, England. Leader of subproject A 4. – Dept of Geography, Coventry Polytechnic, Priory Street, Coventry CV1 5FB, England.

Gunnar Digerfeldt, professor of Quaternary Geology at Lund University. Co-leader of subproject A 3. – Dept of Quaternary Geology, Tornavägen 13, S-22363 Lund.

Marie-José Gaillard, docent in Quaternary Geology at Lund University. Co-leader of subproject A 3, Research in A2–3. Dept of Quaternary Geology, Tornavägen 13, S-22363 Lund.

Tomas Germundsson, graduate student in Human Geography at Lund University. Research in subprojects B 8–9. – Dept of Social and Economic Geography, Sölvegatan 13, S-22362 Lund.

Hans Göransson, fil.dr in Quaternary Geology at Lund University. Research in subproject A 1–2. – Dept of Quaternary Geology, Tornavägen 13, S-22363 Lund.

Mervi Hjelmroos, docent in Quaternary Geology, researcher at Palynology Laboratory, Stockholm. Research in subproject A 1–2. – Palynology Laboratory, Museum of Natural History, S-10405 Stockholm.

Margareta Ihse, docent in Physical Geography at Stockholm University. Research in subprojects A 5 and B 9. – Dept of Physical Geography, S-10691 Stockholm.

Else Kolstrup, Dr of Quaternary Palaeoecology, consulting palaeoecologist in Sønderborg, Denmark. Research in subprojects A 1–2. – Mosevej 12, Blans, DK-6400 Sønderborg, Denmark.

Lars Larsson, professor of Archaeology at Lund University. Leader of subproject B 2, co-editor of the monograph for B 1–4, member of the editorial committee for the project synthesis. – Dept of Archaeology, Krafts torg 1, S-22350 Lund.

Mats Larsson, fil.dr in Archaeology at Lund University. Leader of subproject B 1. – Central Board of National Antiquities, Branch office in Linköping, Järnvägsgatan 8, S-58222 Linköping.

Nils Lewan, docent in Human Geography at Lund University. Leader of subprojects B 8–9, member of the editorial committee for the project synthesis. – Dept of Social and Economic Geography, Sölvegatan 13, S-22362 Lund.

Anna-Carin Linusson, graduate student in Plant Ecology, research in subproject A 5. – Dept of Plant Ecology, Östra Vallgatan 14, S-22361 Lund.

Nils Malmer, professor of Plant Ecology at Lund University. Co-leader of subproject A 5. – Dept of Plant Ecology, Östra Vallgatan 14, S-22361 Lund.

Jens Möller, fil.dr in Human Geography at Lund University. Research in subproject B 8. – Dept of Social and Economic Geography, Sölvegatan 13, S-22362 Lund.

Deborah Olausson, docent in Archaeology at Lund University. Research in subproject B 3. – Dept of Archaeology, Krafts torg 1, S-22350 Lund.

E. Gunilla A. Olsson, fil.dr in Plant Ecology at Lund University. Leader of subproject A 5. – Dept of Botany, Univ. of Trondheim, N-7005 Trondheim-Dragvoll, Norway.

Anders Persson, graduate student in History at Lund University, archivist at the Provincial Archives in Lund. Leader of subproject B 7, research in B6–7. – Mossavägen 10, S-24017 Södra Sandby.

Thomas Persson, research engineer, responsible for data processing related to subprojects A 1–3. – Dept of Quaternary Geology, Tornavägen 13, S-22363 Lund.

Gösta Regnéll, fil.dr in Plant Ecology at Lund University, main project secretary (–1988). Research in subproject A 5. – Dept of Plant Ecology, Östra Vallgatan 14, S-22361 Lund.

Joachim Regnéll, fil. dr in Quaternary Geology at Lund University. Research in subprojects A 1–2. – Dept of Quaternary Geology, Tornavägen 13, S-22363 Lund.

Mats Riddersporre, graduate student in Human Geography, project secretary (1988–), managing and map editor of the project synthesis. Research in subprojects B 3, 5–7. – Dept of Social and Economic Geography, Sölvegatan 13, S-22362 Lund.

Sten Skansjö, docent in History at Lund University, leader of subproject B 6, research in B5–6, member of the editorial committee for the project synthesis. – Dept of History, Magle Lilla Kyrkogata 9A, S-22351 Lund.

Berta Stjernquist, professor emeritus in Archaeology at Lund University. Co-leader of subproject B 3, co-editor of the monograph for B 1–4. – Dept of Archaeology, Krafts torg 1, S-22350 Lund.

Barbro Sundnér, fil.dr in Medieval Archaeology at Lund University. Research in subproject B 5. – Dept of Archaeology, Krafts torg 1, S-22350 Lund.

Sten Tesch, graduate student in Medieval Archaeology at Lund University, Head of Sigtuna Museum. Co-leader of subproject B 3. – Sigtuna Museum, Stora gatan 55, S-19300 Sigtuna.

Appendix A. Chronology of the pollen diagrams from the Ystad area

Marie-José Gaillard, Björn E. Berglund, Hans Göransson, Mervi Hjelmroos, Else Kolstrup and Joachim Regnéll

Introduction

Dating the pollen stratigraphies from the Ystad area has been a major task for the palynologists working within the framework of the Ystad Project. It was essential to establish relevant time-scales for the palaeoecological changes inferred from palaeolimnological studies, so that comparison of these changes in time and space was possible.

The palaeoecological reconstructions in the project area are based primarily on investigations of lake sediments. Peat deposits represent only short time-spans of the studied period (the last 6000 years BP). Unfortunately, in a region with carbonate-rich bedrock, ^{14}C dates from lake sediments tend to be affected by the hard water effect (Olsson 1986). Lake sediments from some of the sites in the project area were tested by ^{14}C dating, and apart from a few exceptions, they provided dates that are too old (Regnéll 1989, Kolstrup 1990, Göransson 1991). Therefore, the dating had to rely primarily on other methods. The ^{210}Pb and ^{137}Cs dating methods were used for recent lake sediments, while pollen-stratigraphical correlation was applied for lake sediments older than 200 years (Gaillard 1984, Hjelmroos 1985, Håkansson and Kolstrup 1987, Regnéll 1989, Kolstrup 1990, Göransson 1991, Gaillard et al. in press). In the following the applications of the various methods are outlined and discussed.

Pollen-stratigraphical correlation

Individual pollen stratigraphies of the Ystad area were correlated to the ^{14}C-dated standard pollen sequence from Ageröds Mosse (Fig. A: 1), following the definitions of Nilsson (1964) for pollen-zone boundaries and pollen-stratigraphical key levels. These levels are listed in Table 1 with their respective ^{14}C ages in Ageröds Mosse together with notes on the pollen-stratigraphical characteristics.

The feasibility of correlation of south Scanian pollen stratigraphies with the one from Ageröds Mosse has been discussed earlier by the authors of the present appendix (e.g. Gaillard 1984, Regnéll 1989, Göransson 1991). This correlation proved to be difficult in the case of pollen-zone boundaries defined by changes in the *Fagus* and *Carpinus* curves, such as the boundaries SB1/SB2 and SB2/SA1. As a matter of fact, *Fagus* and *Carpinus* are poorly represented in south Scanian pollen diagrams during the SB2 pollen zone, in particular in the outer hummocky and coastal zones of the project area (Fig. A:1), for instance in Bjäresjösjön and Bussjösjön (Gaillard and Berglund 1988, Regnéll 1989, Ch. 5.13),

which makes the identification of key levels such as the empirical limit of *Fagus* (first regular occurrence of a taxon with low percentages, Nilsson 1964) very difficult. Therefore, for such levels, other regional pollen-stratigraphical characteristics had to be used (Regnéll 1989).

In certain cases, additional pollen-stratigraphical key levels have been described and used for the establishment of individual chronologies, as in the case of Bjärsjöholmssjön (Göransson 1991) (Table A:1). The key levels SB1 b, e, and g of Ageröds Mosse proved to occur in most pollen diagrams from southern Sweden and to be synchronous (Göransson 1991). Moreover, the original AT2/SB1 boundary of Nilsson (1964), thus the sharp decrease of *Ulmus* dated ca. 5000 BP in south Sweden (Göransson 1991), was used as a key level in Bjärsjöholmssjön. The start of the elm decline dated at 5150 BP in south Sweden was used as a key level and also as a new zone boundary AT2/SB1 in all sites except Bjärsjöholmssjön.

Pollen-stratigraphical key levels were also detected for historical time and compared with ^{210}Pb dates (see below). In the lake sediments of the last 200 yr (Krageholmssjön, Bjäresjösjön, Bussjösjön), there are at least two key levels that are of interest for south Scanian pollen stratigraphies. A distinct increase of Brassicaceae is attributed to the significant development of rape cultivation after the Second World War. This level is dated to ca. AD 1945 by the ^{210}Pb method. Similarly, the increase of *Picea* pollen is a regular feature in the recent lake sediments and it is correlated with the start of systematic spruce plantations in Scania in the middle of the 19th century. The *Picea* increase is dated by the ^{210}PB method at ca. AD 1890 at Krageholmssjön, and AD 1840 at Bjäresjösjön. This level may not be perfectly synchronous for the whole area but it gives a good estimate of the levels representing the second part of the 19th century. Moreover, interpolation of the ^{210}Pb chronology with the calibrated ^{14}C chronology at Bjäresjösjön (see below) provided a relevant age of ca. AD 1700 for a pollen-stratigraphical level characterized by minimum values of juniper after a distinct decrease. This level could be correlated with the total reclamation of juniper outland in the Bjäresjö area as known from historical sources (Persson 1986b).

Radiocarbon dates

All radiocarbon dates obtained from the sites shown in Fig. A:1 are listed in Table A:2. These dates are also presented in Figs A:2–8.

At Krageholmssjön, ^{14}C dating was attempted for the

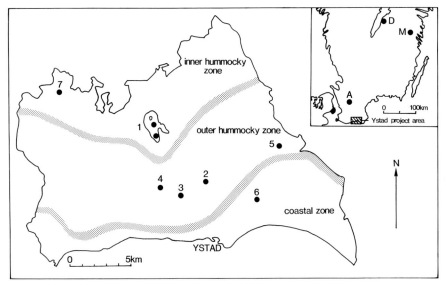

Fig. A:1. Location of the Ystad Project area with the main sites investigated for pollen stratigraphy and palaeoecological reconstructions, the reference site of Ageröds Mosse, and two additional sites mentioned in the text.
1. Krageholmssjön.
2. Bussjösjön.
3. Bjärsjöholmssjön.
4. Bjäresjösjön.
5. Fårarps Mosse.
6. Herrestads Mosse.
7. Kurarp.
A. Ageröds Mosse.
D. Dags Mosse.
M. Mabo Mosse.

Table A:1. List of the pollen-stratigraphical key levels (Nilsson 1964) used to date the pollen stratigraphies from the Ystad area by pollen-stratigraphical correlation. Age in uncalibrated ¹⁴C years of the key levels at Ageröds Mosse and at other sites in southern Sweden (*), in particular Dags Mosse (Östergötland) and Mabo Mosse (NE Småland) (Fig. A:1) (Göransson 1991). Ages indicated by + were obtained at Ageröds Mosse by interpolation of dated levels close above or close below a pollen-zone boundary (Nilsson 1964). (?) = ¹⁴C date from Bjärsjöholmssjön which should be considered with caution (Göransson 1991).

Pollen-zones boundaries and pollen-stratigraphical key levels (Nilsson 1964)	Age in Ageröds Mosse * other sites (uncalibrated ¹⁴C years BP)	Notes on pollen-stratigraphical characteristics (Nilsson 1964, Gaillard 1984, Regnéll 1989, Göransson 1991)
SA1/SA2	1250±85	*Fagus* increase, large *Alnus* decrease
SA1 c	1645±95	*Fagus* and *Carpinus* increase, *Corylus* decrease
SB2/SA1	2140±85	small but distinct *Fagus* increase, *Ulmus, Quercus* and/or *Fraxinus* increase
SB2c	(ca. 2750)+	*Corylus, Quercus, Ulmus, Tilia* decrease, *Betula,* Poaceae increase
immediately below SB2c	2845±90	
SB1/SB2	3560±60	empirical limit of *Fagus* and *Carpinus, Betula* increase, *Corylus* minimum between 2 maxima
SB1 b	* 3700±60	*Tilia* and *Ulmus* reach low values after a decrease. *Corylus* starts to increase. Poaceae, *Plantago lanceolata, Rumex* increase
SB1 c	4000±90	*Corylus* increase, *Ulmus, Fraxinus* minima
SB1 e	* 4450±60	*Quercus* has a maximun, and *Betula* a minimum. *Ulmus* has very low values, just below its new increase
SB1 f	4510±80	*Corylus* peak, low *Betula* values
SB1 g	* 4750±60	absolute *Ulmus* minimum
AT2/SB1 (Nilsson 1964, Göransson 1991)	5090±80 * 5020±60	sharp *Ulmus* decrease (due to a sediment hiatus, Göransson 1991)
AT2/SB1 (this volume)	* 5150±50	start of *Ulmus* decrease
AT2 c (dated only in Bjärsjöholmssjön)	* 5700±70 ?	*Corylus* decreases to 20%
immediately above AT2 f	6170±120	
AT2 f	no date	*Corylus* maximum, *Ulmus* increase
AT1/AT2	(ca. 6350)+	*Corylus* increase, *Betula, Pinus* decrease, *Quercus* increase above AT1/AT2
immediately below AT1/AT2	6750±95	

Table A:2. List of the ^{14}C dates obtained from lake sediments and peat deposits at the sites investigated in the Ystad area. These dates were published first by Regnéll (1989, Krageholmssjön), Göransson (1991, Bjärsjöholmssjön), Håkansson (1986, Bjäresjö-sjön), Håkansson (1979, Herrestads Mosse), and Kolstrup (1990, Kurarp).

Site and sediment type FDG fine detritus gyttja CDG coarse detritus gyttja AG algae gyttja P peat		Level (cm)	Date (uncalibrated ^{14}C years BP) * error	Error (years) + too old − too young		Possible cause
Krageholmssjön	FDG	680-685	*6130±70	+	1000	hard water effect
Bjärsjöholmssjön	CDG	75-85	*2480±50	+	290	id.
Bjärsjöholmssjön	CDG	100-105	*2530±50	+	180	id.
Bjärsjöholmssjön	CDG	137-143	2690±50	±	0	–
Bjärsjöholmssjön	AG	207-213	3550±60	±	0	–
Bjärsjöholmssjön	AG	232-238	3700±60	±	0	–
Bjärsjöholmssjön	FDG	377-383	*4710±60	+	220	hard water effect
Bjärsjöholmssjön	AG	402-408	*4790±60	+	240	id.
Bjärsjöholmssjön	AG	412-418	4790±60	±	0	–
Bjärsjöholmssjön	AG	463-465	*5260±60	+	220	hard water effect
Bjärsjöholmssjön	AG	471-476	*5320±50	+	220	id.
Bjärsjöholmssjön	FDG	525-530	*5920±70	+	220	id.
Bjärsjöholmssjön	AG	600-605	*6180±70	+	220	id.
Bjäresjösjön	CDG	265-268	1240±45	±	0	–
Bjäresjösjön	CDG	270-273	1310±45	±	0	–
Bjäresjösjön	CDG	302-305	1810±50	±	0	–
Bjäresjösjön	CDG	357-360	*2760±50	+	100	hard water effect
Bjäresjösjön	P	361-364	2690±60	±	0	possible contamination by the overlying lake sediments
Bjäresjösjön	P	369-375	2680±50	±	0	–
Herrestads Mosse	P	45	3590±60	±	0	–
Herrestads Mosse	FDG	120	*5050±65	+	400	reservoir effect and
Herrestads Mosse	FDG	150	*5600±70	+	600	hard water effect
Herrestads Mosse	FDG	200	*6090±70	≥	500	id.
Herrestads Mosse	FDG	250	*6350±75	≥	500	id.
Herrestads Mosse	FDG	305	*7590±80	≥	1100	id.
Herrestads Mosse	P	315	*7680±80	≥	1200	id.
Kurarp	CDG	96-99	3140±50	? or	(-400) ±0	contamination with younger roots?
Kurarp	FDG	345-348	*5780±70	+	500	hard water effect

level of the elm decline. This pollen-stratigraphical level is regarded as synchronous throughout north-west Europe (Nilsson 1964, Huntley and Birks 1983) and has been dated to ca. 5150 BP (uncalibrated ^{14}C years) in several sites in southern Sweden (Göransson 1987b). The sample from Krageholmssjön provided an age of 6130 ± 70 BP, thus 1000 years older than the expected age. This date is clearly affected by contamination with ^{14}C-depleted carbonate from the carbonate-rich catchment subsoils (Regnéll 1989).

A single date from Herrestads Mosse, the one performed on the peat deposit, agrees well with the pollen stratigraphy and the dates obtained for a similar level at Ageröds Mosse. The ages obtained from the gyttja are 500–1200 years too old, which is probably the result of combined hard water effect and reservoir effect in a brackish environment (Håkansson and Kolstrup 1987).

Two test samples were selected for ^{14}C dating of the Kurarp sequence. A level just below the elm decline provided a date of 5780 ± 70 BP, thus about 500 years too old, and is, therefore, affected by the hard water effect (Kolstrup 1990). The second dated level (3140 ± 50 BP) is from the upper coarse detritus gyttja deposit,

close above a pollen-stratigraphical level correlated by Kolstrup (1990) to the pollen-zone boundary SB1/SB2 dated at 3560 BP in Ageröds Mosse (Nilsson 1964). If this correlation is correct, the date would be too young, probably as a result of contamination by fine roots from younger vegetation (Kolstrup 1990). An alternative definition of the boundary SB1/SB2 in Kurarp has been proposed by Berglund and Kolstrup (Ch. 4.4.2) at the presumed empirical limits of *Fagus* and *Carpinus* at 125 cm. This second interpretation suggests that the ^{14}C date from the peat sample might be correct or only slightly too young (Fig. A:8).

At Bjäresjösjön, ^{14}C dating was performed on coarse detritus gyttja with particularly low calcium carbonate content and on peat deposits. The ^{14}C dates obtained on the coarse detritus gyttja between 265 and 305 cm (Table A:2) are in good agreement with the expected age of the pollen-stratigraphical level SA1/SA2, which suggests that they are not affected significantly by the hard water effect. The pollen-stratigraphical level SB2/SA1 is very difficult to determine precisely in the Bjäresjösjön sequence. Therefore, we rely on the ^{14}C date instead (Gaillard et al. in press). If we consider that the [14]C

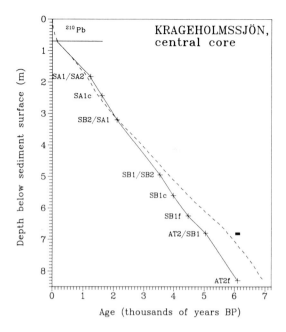

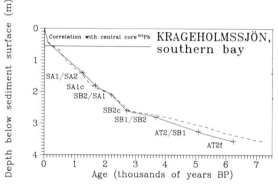

Fig. A:2. Above. Time/depth curves (uncalibrated –, and calibrated - - -) for the central core of Krageholmssjön. (+) ^{14}C dates inferred from Ageröds Mosse (Nilsson 1964) for correlated pollen-stratigraphical levels. (■) ^{14}C dates from Krageholmssjön central core. BP refers to years before 1950 except the upper part where ^{210}Pb dates are in calendar years before 1983. See the text for detailed explanations.
Below. As above for the core from the south-western bay of Krageholmssjön.

dates of the peat are correct, the date of 2760 ± 50 (coarse detritus gyttja) appears to be slightly too old.

The dates from the gyttja sequence of Bjärsjöholmssjön exhibit a more or less systematic error of + 220 ^{14}C years (too old dates, Göransson 1991). However, a few samples of algae gyttja provided expected ages (Table A:2). When taking account of the hard water effect, these dates confirm the synchronism of Nilsson's pollen-stratigraphical levels in Ageröds Mosse and Bjärsjöholmssjön. Moreover, the synchronism of several key levels has been demonstrated for a larger area of Southern Sweden, from Scania to Östergötland (Göransson 1987b, 1991).

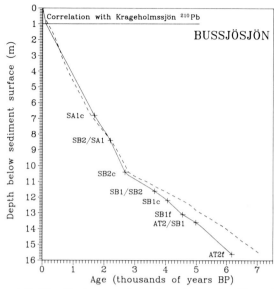

Fig. A:3. Time/dept curves (uncalibrated –, and calibrated - - -) for Bussjösjön. See legend to Fig. A: 2. BP refers to years before 1950 except in the upper part correlated to the ^{210}Pb-dated pollen stratigraphy from Krageholmssjön (calendar years before 1983).

Therefore, in view of the ^{14}C dates obtained at Bjäresjösjön and Bjärsjöholmssjön, pollen-stratigraphical correlation of the pollen sequences of the Ystad area with that of Ageröds Mosse is justified. In addition, the important pollen-stratigraphical key level SB2 c, tentatively dated at 2750 BP in Ageröds Mosse by interpolation of two ^{14}C dates, has obtained a convincing ^{14}C age of 2680 ± 50 and 2690 ± 50 BP at Bjäresjösjön and Bjärsjöholmssjön respectively.

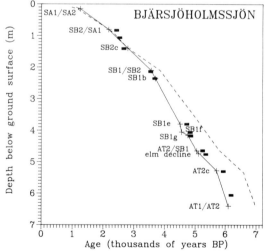

Fig. A:4. Time/depth curves (uncalibrated –, and calibrated - - -) for Bjärsjöholmssjön. See legend to Fig. A: 2. BP refers to years before 1950. (■) ^{14}C dates from Bjärsjöholmssjön deposits (Göransson 1991). The uncalibrated time/depth curve follows the Ageröds Mosse chronology (see text for more explanations).

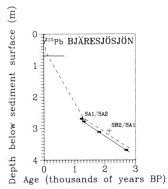

Fig. A:5. Time/depth curves (uncalibrated –, and calibrated
- - -) for Bjäresjösjön. See legend to Fig. A: 2. BP refers to
years before 1950 for the uncalibrated curve, and to calendar
years before 1983 for the calibrated curve. (■) ^{14}C dates obtained from core C3 (Gaillard and Berglund 1988). The time/
depth curves are based on the ^{210}Pb and ^{14}C datings from core
C3 (see text for more explanations).

The ^{210}Pb chronologies

^{210}Pb dating was performed on the upper sequence of
two sites, Krageholmssjön (Regnéll 1989) and Bjäresjösjön (Gaillard et al. in press). The series from Krageholmssjön provided 16 dates between 2.5 and 170 calendar years before 1983 for the upper 70 cm of sediments.
In addition, two levels from this core were dated by the
^{137}Cs method, and the obtained ages were comparable
to the ^{210}Pb dates. At Bjäresjösjön, two cores (C3 and
C4) were dated and provided comparable series of ages.
Core C3 (from the central part of the lake) provided a
series of 24 dates between 4 and 149 calendar years
before 1983 for the upper 70 cm of sediments. Comparison with historical records suggests that the ^{210}Pb chronologies from Krageholmssjön and Bjäresjösjön are relevant (see above).

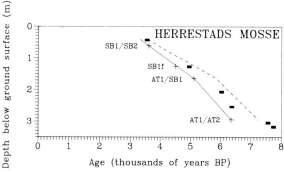

Fig. A:7. Time/depth curves (uncalibrated –, and calibrated
- - -) for Herrestads Mosse. See legend to Fig. A: 2. BP refers
to years before 1950.

Uncalibrated and calibrated chronologies

The uncalibrated chronologies (Figs A:2–8) were established on the basis of interpolation between the ^{14}C
dates (years BP) of pollen-stratigraphical key levels (Table A:2) for the time-span 6350 BP (AT1/AT2) to 1250
BP (SA1/SA2) (Figs A:2–4 and 6–8). For Krageholmssjön and Bussjösjön (Figs A:2 and 4), the chronology of
the upper part of the sequences was based on (a) the
upper ^{210}Pb chronology (calendar years) and (b) interpolation between the ^{14}C chronology and the oldest
^{210}Pb date (in calendar years before 1983) (Regnéll
1989). We are aware that the latter is not the most
correct procedure, but it was regarded as sufficient for
providing a good estimate of the age of the levels when
the analyses have relatively low resolution (one analysis
every 10 to 40 cm). The deviation between uncalibrated
and calibrated chronologies for the last 2000 years is so
small that it may be ignored in such cases. A similar
method was adopted at Fårarps Mosse (Fig. A:6),
where the chronology of the upper part of the profile
relies on the interpolation between the uncalibrated ^{14}C

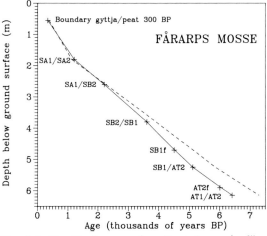

Fig. A:6. Time/depth curves (uncalibrated –, and calibrated
- - -) for Fårarps Mosse. See legend to Fig. A: 2. BP refers to
years before 1950.

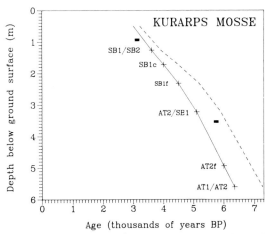

Fig. A:8. Time/depth curves (uncalibrated –, and calibrated
- - -) for Kurarp. See legend to Fig. A: 2. BP refers to years
before 1950.

date of the SA1/SA2 boundary and the assumed time for infilling of the ancient lake (ca. 300 BP, estimated from a series of old maps). The time-scale published by Hjelmroos (1986) has been revised later (this volume) according to the time/depth curve presented in Fig. A:6.

These uncalibrated chronologies have been used for the time-scales of all pollen diagrams presented in Ch. 4, except that of Bjäresjösjön (see below).

In Ch. 5.13, calibrated time-scales were used in a few diagrams to ensure the best possible estimate of (a) the influx of pollen and charcoal particles (number of pollen grains/charcoal deposited per unit area and time) (Figs 5.13:4 and 5.13:11) and (b) the number of new pollen types per 500-year unit (Fig. 5.13:7). For Krageholmssjön, Bjärsjöholmssjön, and Bussjösjön, these calibrated time-scales were obtained by "translation" of the uncalibrated ^{14}C dates of each level (as inferred from the uncalibrated time/depth curves of Figs A:2 (above), 3, and 4) into calibrated ages according to the calibration curves of Pearson et al. (1986). When several alternative calibrated dates were given by the calibration curve, a single date was chosen in order to obtain as smooth a curve as possible. The calibrated time/depth curves of Fårarps Mosse, Herrestads Mosse, and Kurarp (Figs A:6–8) were constructed following a simpler procedure. The uncalibrated ^{14}C dates of the key levels only were translated into calibrated years, and the calibrated time/depth curves were obtained by interpolation of these few calibrated ages.

In the case of Bjäresjösjön, the main part of the profile belongs to Nilsson's pollen zone SA2, where no dated pollen-stratigraphical level could be used for the construction of a time-scale. Because the resolution of the analyses was high (one analysis every 5 cm or one analysis every 2.5 to 30 yr) it was desirable to obtain the best time-scale possible, and a different procedure was applied. The four relevant ^{14}C dates from the lower part of the profile (Table A: 2) were translated into calendar years before 1983. Interpolation of these calibrated dates produced a calibrated time/depth curve which was in turn interpolated with the time/depth curve based on ^{210}Pb dates (calendar years before 1983). In this way, the constructed calibrated time/depth curve provided the best possible estimate of the age of each analysed level (Gaillard et al. in press). This time-scale was used for all the Bjäresjösjön diagrams presented in Ch. 4 and 5.1.3.

Additional comments on individual chronologies

Krageholmssjön. Fig. A:2 (above) shows (1) the time/depth curve for the central core as it was published by Regnéll (1989), which is a combination of calendar years before 1983 for the upper 70 cm and of uncalibrated ^{14}C dates for the remaining part of the sequence, and (2) the time/depth curve based on calibrated years BP.

The chronology of the core from the south-western bay of Krageholmssjön (Fig. A:2 (below)) is based on a time/depth curve relying on (a) interpolation of pollen-stratigraphical levels dated at Ageröds Mosse for the period 6170–2150 BP (uncalibrated ^{14}C years), (b) interpolation of ^{210}Pb dates inferred from the central core for correlated pollen- and lithostratigraphical levels (increase in *Picea* and significant increase in calcium carbonate), and (c) interpolation between the ^{14}C time/depth curve and the ^{210}Pb time/depth curve. This time-scale was used to date lake-level fluctuations in Krageholmssjön and in the diagrams presented in Ch. 5.1. It differs from the time-scale published earlier by Gaillard (1985), which was later revised after discussion with the co-authors of this section.

Bussjösjön. The chronology of Bussjösjön was established in the same way as that of Krageholmssjön (central core) with the exception of the upper part of the core where no ^{210}Pb dates were available. For this section the ^{210}Pb date of the increase in Brassicaceae and in CaCO$_3$ content was inferred from the central core of Krageholmssjön and interpolated with (a) the uncalibrated time/depth curve between 6170 and 2150 BP and (b) the surface level of AD 1983. Fig. A:4 presents the time/depth curve as it has been published in Regnéll (1989), and the time/depth curve using calibrated ages.

Bjäresjösjön. The lake-sediment sequence from Bjäresjösjön covers only the last 2700 yr BP (uncalibrated). Therefore, no more than three pollen-stratigraphical levels dated at Ageröds Mosse could be used, SB2/SA1, SA1 c, and SA1/SA2. However, because of the predominantly local origin of the pollen (Gaillard and Berglund 1988), correlation with the pollen diagrams from Ageröds Mosse and other sites in the Ystad area was problematic, particularly in view of the very low frequencies of *Fagus* and *Carpinus*. Therefore, the time/depth curve between 2690 BP and 1240 BP (uncalibrated years) (Fig. A:5) relies entirely on the relevant ^{14}C dates obtained from peat and coarse detritus gyttja (Table A:2). Note that one of the dates is in good agreement with the tentative position of the pollen-stratigraphical level SA1/SA2, which is the least questionable, and characterized by a distinct and large decrease of *Alnus* (see Table A:1). Moreover, the increase of Poaceae at the bottom of the sequence may be correlated to the regional pollen-stratigraphical key level SB2 c, dated at 2690 BP at Bjärsjöholmssjön (Göransson 1991). Therefore, the ^{14}C date obtained in the peat lying immediately below the lake sediments of Bjäresjösjön is in good agreement with the date from Bjärsjöholmssjön. The remaining part of the time/depth curve is based on interpolation between the younger calibrated ^{14}C date and the ^{210}Pb chronology (in calendar years) of the upper 70 cm (Fig. A:5).

Conclusions

1. The lake sediments of the Ystad area are in general not adequate for ^{14}C dating because they are contaminated by older carbon from the carbonate-rich bedrock characterizing the region (the hard water effect, Olsson 1986). ^{14}C dating of lake sediments from the Ystad area provided too old dates with errors of ca. ±200 to ±1000 yr. A few dates from coarse detritus gyttja (Bjäresjösjön) and algae gyttja (Bjärsjöholmssjön) may be regarded as correct.

2. The assumption that the pollen-stratigraphical key levels defined by Nilsson (1964) at Ageröds Mosse are synchronous over a large area of southern Sweden is confirmed by ^{14}C dates from Bjärsjöholmssjön (Göransson 1991) and Bjäresjösjön (Gaillard et al. in press). Therefore, dating by pollen-stratigraphical correlation with the pollen diagram of Ageröds Mosse is justified for the Ystad area.

3. For the period between 6350 and 1250 BP (uncalibrated ^{14}C years), the time/depth curves are primarily based on the interpolation of the ^{14}C dates inferred from Ageröds Mosse for correlated pollen-stratigraphical key levels.

4. ^{210}Pb chronologies are available for two sites, Krageholmssjön and Bjäresjösjön, and cover the last 170 and 149 years respectively.

5. Interpolation of the ^{14}C chronologies (uncalibrated years BP) with the ^{210}Pb chronology or the present surface level (calendar years) provides a rough but satisfactory estimate of the age of levels younger than 1250 BP for pollen diagrams where the analysis resolution is not very high for that time period (Krageholmssjön, Bussjösjön, Fårarps Mosse).

6. At Bjäresjösjön, the best time-scale possible for the time 149–1300 BP (calendar years) was obtained by interpolation of calibrated ^{14}C dates with the ^{210}Pb chronology of the upper 70 cm.

7. The time-scales established for the lake-sediment-based studies of the Ystad Project can be regarded as the most accurate known today for south Scania, and cover the last 7000 calendar years. They provide a good time control for correlation of landscape-ecological changes between sites and detailed comparison of these with independent archaeological and historical records from the same area (e.g. Ch. 4 and 5.13).

8. There are still reasons for caution and discussion with regard to zonation and correlation of various records. More accurate dating of vegetation and ecological changes inferred from lake-sediment-based studies in south Scania would require additional investigations (partly in progress) including (1) AMS dating of terrestrial macroremains (Gaillard and M. Regnéll, projects in progress), (2) AMS dating of pollen concentrates (J. Regnéll, project in progress), and (3) pollen analysis of peat sequences covering the last 7000 years and ^{14}C dating of the peat at all pollen-stratigraphical key levels.

Acknowledgements. We are grateful to S. Håkansson and G. Skog who performed all ^{14}C datings at the ^{14}C Laboratory in Lund, and to F. El-Daoushy who performed the ^{210}Pb datings at the Institute of Physics in Uppsala. We are much indebted to T. Persson for his help with the computer drawings of the time/depth curves.